Aspects of the Life Record

| Major Events | Dominant Forms[b] | | Systems |
|---|---|---|---|
| | Homo | | Quaternary[c] |
| Grasses become abundant | Mammals | | Tertiary[c] |
| Horses first appear | | Flowering plants | |
| Extinction of dinosaurs | | | Cretaceous |
| Birds first appear | | | Jurassic |
| Dinosaurs first appear | Reptiles | Conifer and cycad plants | Triassic |
| | | | Permian |
| | Amphibia | | |
| Coal-forming swamps | | | Pennsylvanian |
| | | | Mississippian |
| | | Spore-bearing land plants | Devonian |
| | Fish | | Silurian |
| Vertebrates first appear (fish) | | | Ordovician |
| | Marine invertebrates | Marine plants | |
| First abundant fossil record (marine invertebrates) | | | Cambrian |
| | Primitive marine plants and invertebrates | | Precambrian |
| | One-celled organisms | | |

*Source:* L. D. Leet, S. Judson, and M. E. Kauffman, *Physical Geology*, 6th ed., copyright © 1982 by Prentice-Hall, Inc., Englewood Cliffs, N.J. Reprinted by permission of Prentice-Hall, Inc.

[c] Some geologists prefer to use the term Cenozoic for the Quaternary and the Tertiary.

[d] In most European and some American literature Pennsylvanian and Mississipian are combined in a period called the Carboniferous.

[e] Subdivisions not firmly established.

SECOND EDITION

# Geology for Engineers and Environmental Scientists

*ALAN E. KEHEW*
*Western Michigan University*

Prentice Hall, Englewood Cliffs, New Jersey 07632

*Library of Congress Cataloging-in-Publication Data*

Kehew, Alan E.
Geology for engineers and environmental scientists/Alan E. Kehew.—2nd ed.
p. cm.
Rev. ed. of: General geology for engineers. c1988.
Includes bibliographical references and index
ISBN 0-13-303538-7
1. Engineering geology. I. Kehew, Alan E. General geology for engineers. II. Title.
TA705.K38 1995 94-32126
550—dc20 CIP

Acquisitions editors: *Deirdre Cavanaugh, Robert McConnin*
Editorial/production supervision and interior design: *Kathleen M. Lafferty*
Copy editor: *Stephen C. Hopkins*
Proofreader: *Bruce D. Colegrove*
Cover design: *Tom Nery*
Manufacturing buyer: *Trudy Pisciotti*

A Simon & Schuster Company
Englewood Cliffs, New Jersey 07632

First edition published under the title
*General Geology for Engineers*



Printed in the United States of America

10 9 8 7 6 5 4 3 2 1

ISBN 0-13-303538-7

Prentice-Hall International (UK) Limited, *London*
Prentice-Hall of Australia Pty. Limited, *Sydney*
Prentice-Hall Canada Inc., *Toronto*
Prentice-Hall Hispanoamericana, S.A., *Mexico*
Prentice-Hall of India Private Limited, *New Delhi*
Prentice-Hall of Japan, Inc., *Tokyo*
Simon & Schuster Asia Pte. Ltd., *Singapore*
Editora Prentice-Hall do Brasil, Ltda., *Rio de Janeiro*

*This book is dedicated to my wife, Kay,*
*and my parents, Dick and Betty Kehew, for their love and support over the years,*
*and to my children, Melissa, Michelle, and Liz.*

# Contents

## 7 Tectonics, Structure, and the Earth's Interior 165



## 8 Earthquakes 206



## 9 Weathering and Erosion 247

# Preface

The content and technical level of a geology text for engineers is a matter of broadly differing opinion. The type of geology course offered to engineering students at universities around the United States varies from introductory physical geology to engineering geology. This situation results from differences in philosophy, engineering curricula, and geology faculty available to teach such a course. *Geology for Engineers and Environmental Scientists* is a compromise, based on my own experience, between these extremes.

The concept for the first edition evolved from a course that I taught at the University of North Dakota. The enrollment included civil and geological engineering students who were required to take the course and mechanical and miscellaneous other engineering majors who took the course as a science elective. This course was the only geology course in the civil engineering curriculum, but, for the geological engineering majors, it had to serve as a foundation for numerous future geology courses. The academic level of the students varied; the majority of them, however, completed the course in their sophomore year. As a result, most had taken little, if any, engineering mechanics, soil mechanics, or hydrology.

These conditions made structuring the course very difficult. My basic approach was to cover the essentials of physical geology and simultaneously introduce the students to the engineering and environmental applications of geology. To do this, many of the traditional topics of physical geology were omitted or only briefly discussed. On the other hand, it was apparent that students could not gain a full understanding of the applied aspects of geology without a good foundation in rocks, minerals, and structural geology. Several physical geology texts were tried, and in addition, an engineering geology text was used one year. All these proved to be unsatisfactory because they failed to cover part of the desired course content. Students requested a reference other than the lecture presentations for either the basic geology or the engineering aspects, depending on which type of text was used.

I eventually concluded that the only solution for this course was to write a text combining the two areas at an introductory level. The first edition, entitled *General Geology for Engineers*, was the result. Its objective was to provide a reasonably comprehensive introduction to physical geology and also to demonstrate the importance of geology to engineers by including introductory mechanics, hydraulics, and case studies that illustrate interactions between geology and engineering. The approach was mostly nonquantitative, although appropriate basic formulas were introduced where necessary. Enough material was presented so that individual instructors could use the book to achieve their own personal compromise between physical geology and engineering geology.

Although several changes have been made in the second edition of this textbook, its basic purpose remains the same. The book is intended to provide an introduction to geology for engineering students, with a focus on applications of geology that they are likely to use in their professional careers. The change in the title of the text, from *General Geology for Engineers* to *Geology for Engineers and Environmental Scientists*, reflects my opinion that environmental scientists could also benefit from a geology course that emphasizes the applications of the science, particularly those involving environmental problems and solutions. Curricula in environmental science commonly require at least one course in geology. It seems logical to me that the most useful type of geology course would be one in which the environmental applications would be discussed from a technical perspective. Ultimately we must address environmental problems from a technical standpoint to solve them.

With this objective in mind, a new chapter that discusses contamination and remediation of the subsurface has been added to the text. This is the most significant change to be found in the second edition, but because of other reorganization, the total number of chapters has remained the same as in the first edition.

In addition to the environmental chapter, the text has been changed in a number of subtle but significant ways, in many cases as a result of the comments of users and reviewers of the first edition. In certain chapters, examples involving quantitative concepts have been added where appropriate, and similar problems are included at the end of those chapters. The text can therefore be used in the more quantitative approach favored by some instructors. At the same time, the mathematical content has not replaced the descriptive presentations of geologic principles necessary for their understanding.

New examples and case studies have been inserted throughout the text, and some topics that were omitted in the previous edition have now been included. It is hoped that this will allow instructors more freedom to tailor the course to their own needs and preferences.

The editors and staff at Prentice Hall, including Ray Henderson and Deirdre Cavanaugh, were extremely supportive during the revision of this text. I have been especially fortunate to have a production editor as competent and congenial as Kathleen Lafferty for both editions of the book.

I feel a great debt of gratitude to individuals and organizations who provided photographs, permissions, and other materials for the book and to those who reviewed all or part of the manuscript. Among the former, numerous employees of the U.S. Geological Survey and the USDA Soil Conservation Service were exceptionally prompt, courteous, and helpful. As to the latter, the manuscript was improved considerably by suggestions and comments of those who critically reviewed successive drafts. These astute colleagues include Lawrence Herber, California State Polytechnic University; John Lewis, George Washington University; John Lemish, Iowa State University; John Williams, San Jose State University; Gary Robbins, Texas A&M University; Daniel P. Spangler, University of Florida, Gainesville; Charles Baskerville, U.S. Geological Survey; Robert B. Johnson, Colorado State University; Stan Miller, University of Idaho; Brian Stimpson, University of Manitoba; John M. Sharp, Jr., University of Texas at Austin; Jack Green, California State University, Long Beach; and David T. King, Jr., Auburn University.

*Alan E. Kehew*

## CHAPTER OPENING PHOTOGRAPHS

*Chapter 1* The Grand Canyon. (J. R. Balsley; U.S. Geological Survey).
*Chapter 2* Quartz crystals. (A. E. Kehew.)
*Chapter 3* Devil's Postpile, California. (University of Colorado.)
*Chapter 4* Bryce Canyon, Utah. (R. A. Kehew).
*Chapter 5* Migmatite. (C. C. Hawley; U.S. Geological Survey.)
*Chapter 6* Rockfall on I-70, Jefferson County, Colorado. (W. R. Hansen, U.S. Geological Survey.)
*Chapter 7* Landsat image of Ouachita Mountains, Oklahoma. (U.S. Geological Survey, EROS Data Center.)
*Chapter 8* Active faults, Hawaii Volcanoes National Park. (A. E. Kehew.)
*Chapter 9* Exfoliation in granitic rocks. (N. K. Huber, U.S. Geological Survey.)
*Chapter 10* The Leaning Tower of Pisa. (Italian Government Tourist Office, E.N.I.T.)
*Chapter 11* Comal Springs, Texas. (A. E. Kehew.)
*Chapter 12* Failure of quick clay, Nicolet, Québec. (H. Laliberté; Gouvernement du Québec.)
*Chapter 13* River terraces, Montana. (A. E. Kehew.)
*Chapter 14* Erosion from Hurricane Hugo, near Charleston, South Carolina. (R. H. Willowby and P. G. Nystrom, Jr.)
*Chapter 15* Medial moraines, Alaska. (G. Oliver.)
*Chapter 16* Landfill cell, southwestern Michigan. (A. E. Kehew.)

# 1

# Introduction

People study geology—the science of the earth—for different reasons. Even among those who become professional geologists, there is a great diversity in purpose and objective. For engineers, an understanding of geology is a tool that is utilized in the design and construction of any structure that is in contact with the earth or that interacts with its natural materials in any way. For environmental scientists, the geology of a site controls both the distribution and movement of contaminants below the ground surface and the selection of methods used for environmental cleanup.

Unfortunately, the application of geology to engineering and environmental science is not a simple task. One cannot refer to a manual for the solution to a geologic problem. Instead, the solution must come from experience and judgment, qualities that are based on a sound knowledge of the fundamental principles of geology. For this reason, engineers and environmental scientists must approach geology as they approach chemistry or physics: as a science that must be assimilated by thorough and systematic study. We will begin our study accordingly, by examining geology's origins and foundations as a science.

## GEOLOGY AS A SCIENCE

The growth of the science of geology has changed the way that we regard ourselves and our environment on the planet Earth. Modern Western Civilization evolved with a strong reliance on literal interpretations of the opening chapters of the Bible. Early scientists applied these concepts to their observations of the natural environment. Accordingly, the earth was thought to be quite young, and its formation was attributed to a series of events that were rapid and cataclysmic in intensity. This school of thought, which became known as *catastrophism*, held that the earth's surface was generally a stable, nonchanging place that had remained quiescent except for violent upheavals during which mountains were thrust upward from the plains and canyons were formed as the earth's surface was ripped apart. The related theory of *neptunism* involved the presence of a primordial ocean from which rocks at the earth's surface precipitated. A specific sequence of precipitated rocks, assumed to have formed as the ocean level dropped, was developed to account for observations of rocks that were made in western Europe.

An important early contribution to geology was made by Nicolaus Steno (1638–1687), a Danish physician working in Italy. Steno observed the rocks exposed in cliffs and in stream banks, and he also studied *sediment*, particles of rock and soil being deposited by streams and in the Mediterranean Sea. He realized that the exposed rocks had many similarities to the modern sediments he had seen, and he concluded that they had, in fact, originated as sediments at some time in the distant past. To formulate these ideas, Steno used true scientific thought because he recognized order in nature.

Steno also stated three principles that are fundamental to modern geology: First, the *Principle of Original Horizontality* states that layers of sediment are always deposited in horizontal sheets, as can be observed when they are seen to be accumulating on the bed of a stream or ocean (Figure 1-1). Thus when we see layers, or *beds*, of rock which appear to be tilted or folded (Figure 1-2), we must realize that subsequent events have deformed the originally horizontal beds. Second, Steno's *Principle of Original Continuity* states that accumulations of sediment are deposited in continuous sheets up to the point where they terminate against a solid surface (Figure 1-1). There is no better illustration

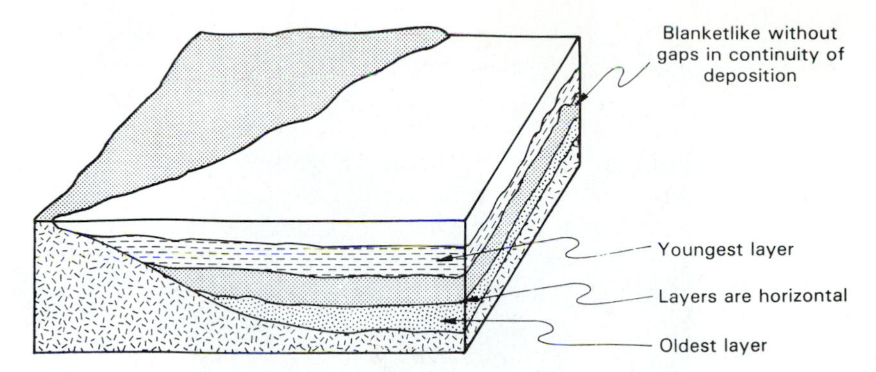


**Figure 1-1** Principles proposed by Nicolaus Steno. These principles are essential to an understanding of sedimentary rocks. (From E. A. Hay and A. L. McAlester, *Physical Geology: Principles and Perspectives*, 2d ed., copyright © 1984 by Prentice-Hall, Inc., Englewood Cliffs, N.J.

of these two principles than in the Grand Canyon (Figure 1-3), where individual horizontal rock layers can be traced for miles along the side of the canyon. Steno's third principle, known as the *Principle of Superposition*, states that in a vertical sequence of sedimentary rock layers, the oldest layer lies at the base and each successive layer above is younger than the layer it overlies. Although this principle may seem intuitively obvious to us, in Steno's era it required acceptance of the radical idea that many rocks are formed by the gradual deposition of sediment, the same processes that can be observed to be occurring in a stream, lake, or ocean.

**Figure 1-2** Beds of sedimentary rocks that have been bent into a fold from their original horizontal orientation. (A. Keith; photo courtesy of U.S. Geological Survey.)

**Figure 1-3** In the Grand Canyon, horizontal beds of sedimentary rock can be traced for great distances. (J. R. Balsley; photo courtesy of U.S. Geological Survey.)

The final great leap into the modern era of geology can be attributed to the ideas of James Hutton, an eighteenth-century Scottish farmer. Hutton challenged the ideas of catastrophism and neptunism by proposing that the earth's surface is not a stable, unchanging landscape. Instead, changes are occurring continuously, although mostly at very slow rates. Therefore, valleys are formed by the slow erosion of streams that flow within them, and mountains are uplifted gradually to their elevated positions. Hutton also suggested that the rocks exposed on the landscape had been formed by the same gradual processes operating continuously, now and in the past. In place of neptunism, Hutton advocated *plutonism*, which explains the formation of *igneous* rocks by cooling from a hot, melted state rather than by precipitation from the ocean.

These ideas are incorporated into his *Principle of Uniformitarianism*, the principle that the physical and chemical laws governing natural processes have been constant throughout the history of the earth, and that its evolution has been continuous rather than catastrophic. This principle does not mean that the rates of natural processes are always constant; we know that erosion of river valleys takes place at a much faster rate during a flood than during a normal-flow period. Thus, although the processes may vary in intensity and rate in time and space, they are still controlled by the unchanging natural laws of physics and chemistry. The Principle of Uniformitarianism is often summed up by the phrase "the present is the key to the past," because it gives geologists the ability to interpret the origin of ancient rocks in terms of natural processes that can be observed in progress at the present.

Hutton's idea of uniformitarianism required an acceptance of the great antiquity of the earth. A single rock exposure along the coast of Scotland emphatically drove this point home. The exposure consisted of horizontal beds of rocks resting upon tilted rocks of a different type (Figure 1-4). Using Steno's principles, Hutton realized that this exposure must represent a complex se-

**Figure 1-4** The exposure of rock at Siccar Point, Scotland, studied by James Hutton. The unconformity (plane marked by lens cover and arrows) represents a long period of geologic time between the formation of the vertical rock beds in the lower part of the outcrop and the nearly horizontal rock beds overlying them. The geologic events involved in the history of these rock units are shown in Figure 1-5. (Photo courtesy of John Bluemle.)

quence of events that occurred over a great length of time. First, the lower sequence of rock beds was deposited in horizontal layers of sediment (Figure 1-5). Next, these sediments hardened into rocks and became deformed and tilted into their present orientation. Deformation of this type is associated with major movements of the earth's crust, the type of movement that may result in the lateral compression and vertical uplift of rocks to form a mountain range. It is interesting to imagine the awe that Hutton must have felt as he began to realize that the next event in the history of that outcrop must have been a period of erosion representing an enormous amount of geologic time. A mountain range must have been gradually worn down by erosion to a low-lying plain on which the sediments of rock unit B were deposited, perhaps by a sea gradually spreading over the plain. The surface separating the two rock sequences is called an *unconformity*. The tremendous amount of time that these changes represent must have convinced Hutton that he was right about the great age of the earth.

By studying rock exposures in many areas, Hutton reached the conclusion that rocks in the crust follow a pattern that has been repeated over and over again throughout geologic time. Hutton's chain of events is known as the *geologic cycle* (Figure 1-6). Although the cycle has no definite beginning or end, it is convenient to think first of sediments being deposited in continuous horizontal beds according to the principles of Steno. As successive layers are deposited, the beds near the base of the sequence become compacted and begin the transformation from sediments into *sedimentary rocks*, a process known as *lithification*.

With continued deposition, the sedimentary rocks are deeply buried and subjected to higher temperatures and pressures. Under these conditions, the rocks are deformed; layers may be bent, broken, or heated to the melting point. The melting or partial melting of a rock produces *magma*, a molten liquid

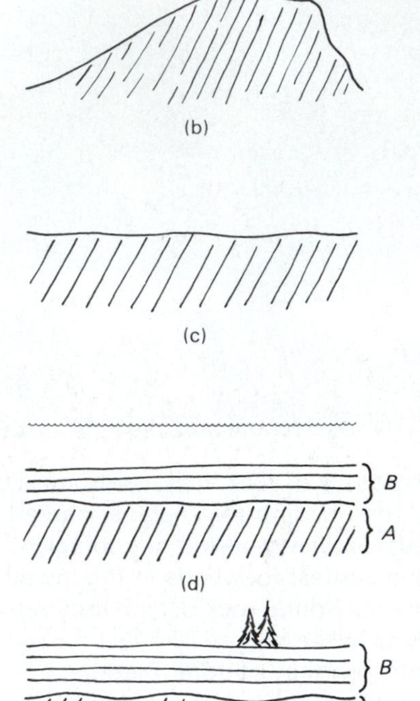


**Figure 1-5** Generalized sequence of events making up the geologic history of the rock exposure shown in Figure 1-4. (a) Deposition of the rocks of sequence *A*. (b) Deformation and uplift. (c) Erosion of the rocks of sequence *A*. (d) Deposition of the rocks of sequence *B* forming an unconformity between sequences *A* and *B*. (e) Uplift and erosion of both rock sequences as a unit.

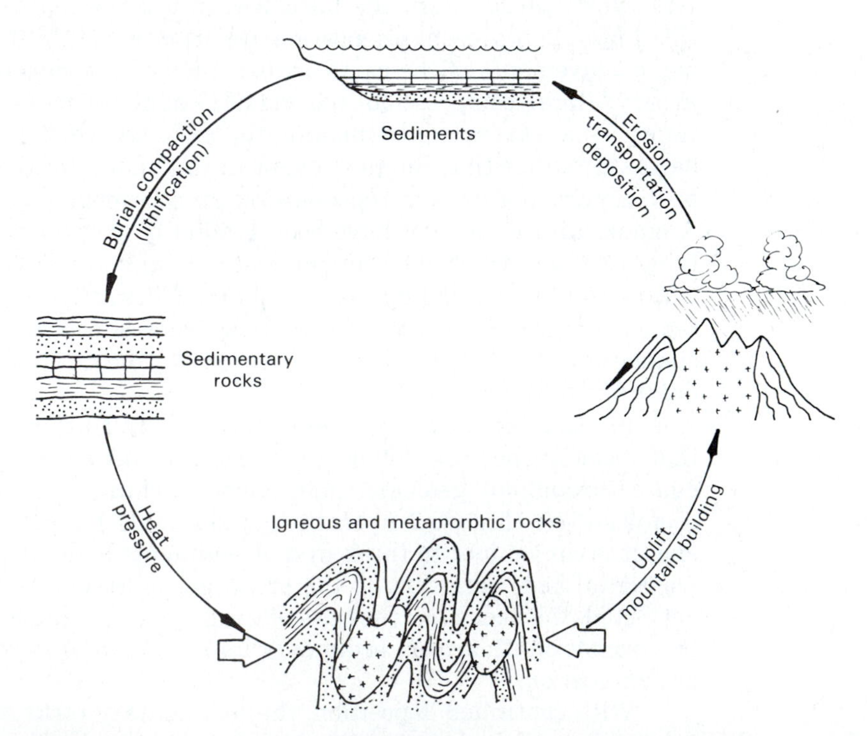


**Figure 1-6** The geologic cycle. Rocks in the earth's crust follow a cycle that includes deposition of sediments, formation of sedimentary rocks, conversion to igneous or metamorphic rocks, and erosion to form sediments.

that can later cool to form an igneous rock. Magmas can also be produced from sources other than sedimentary rocks. If any type of rock is highly altered by heat and pressure but not to the point of melting, the original rock is changed to such a great degree that it is now called a *metamorphic* rock. The close association of deformation, metamorphism, igneous activity, and mountain belts suggest that these activities are all related. It was obvious to Hutton and most other early geologists that thick sequences of sedimentary rocks are intensely deformed and altered in the process of mountain building (Figure 1-7).

The final step in the cycle occurs after the uplift of a mountain belt. Exposed to the destructive agents of gravity, wind, rain, and ice, mountains are gradually lowered by erosion. The sediments produced by erosion are transported by rivers to the ocean where the cycle can continue in a new location.

**Figure 1-7** An exposure of a metamorphic rock, produced by alteration under high temperature and pressure. (C. C. Hawley; photo courtesy of U.S. Geological Survey.)

**Figure 1-8** The Grand Tetons, a mountain range consisting of igneous and metamorphic rocks. Uplift and erosion have produced the striking, rugged topography.

Given enough time, mountain ranges can be worn down to gentle low-lying plains. The interior of Canada contains regions of igneous and metamorphic rocks that once must have been at the core of towering mountains but now are reduced to an elevation barely above sea level. In the United States, both the Appalachian and Rocky Mountains are following the path of the geologic cycle (Figure 1-8). The Appalachians are older and thus erosion has reduced them to a lower elevation than the Rockies.

The controversy set off by Hutton's ideas raged in scientific circles for many years. It was not until the mid-nineteenth century that uniformitarianism was accepted by the majority of scientists.

The work of Steno, Hutton, and many others paved the way for the recognition of geology as a modern science. The practice of geology requires the observation of rock or sediments and the interpretation of the origin and sequence of events that has led to the present condition. The only change in this approach since the time of Hutton lies in the ever-increasing sophistication and complexity of the tools and instruments that scientists have available to study the evidence found in the natural landscape. In this respect, geology differs from other sciences because the conclusions and interpretations of past geologic events cannot be verified by experimentation. Geologic study is not unlike the procedure followed by a detective who collects clues and evidence in order to solve a crime.

## GEOLOGIC TIME

Geologists are, in part, historians. Our goal is to decipher the physical, chemical, and biological evolution of our planet from its origin to the present time. Like historians who study the history of the human race, geologists must have a method of determining the time at which various events in the earth's history took place. Imagine how difficult it would be to study the history of the United States if there were no concept of time; how could we comprehend the differences between the exploration of the West and the space program of the twentieth century if we did not understand the amount of time and technological progress that separates the two periods?

Geologists in some ways face a much more formidable task in the investigation of the earth's history. There are no written records available. The only direct evidence that exists from the geologic past is the sequence of rocks and sediments that we can observe at and below the earth's surface. In the twentieth century, we have learned to use the natural radioactivity in certain types of rocks to directly determine their age; with this method we can determine the *absolute* time of an event. Geologic pioneers like Hutton, however, did not have radioactive age dating available to them. Hutton could only speculate about the amount of time represented by an unconformity in a rock exposure. Confronted with this problem, geologists began to work out the history of the rock record using *relative* time. This method is an attempt to construct a time scale based upon the relative ages of rock units by establishing their relationship to other rock units.

### *Relative Time*

Steno's Principle of Superposition is a useful tool in determining the relative age of sedimentary rocks. In an undisturbed layered sequence, the beds are successively younger from bottom to top. Another method that can be used with sedimentary rocks is *correlation*, the tracing of rock units from one area to another. In addition, the Principle of Original Continuity can be used to extend the correlation of rock units with similar characteristics across areas in which the rocks are not present or are not exposed (Figure 1-9).

More complicated rock exposures require additional techniques to establish relative time. The *Principle of Cross-Cutting Relationships*, for example, is particularly useful in some places. It states that any rock that cuts across or penetrates a second rock body is younger than the rock it penetrates. This principle is often called upon for interpretation of sequences containing igneous rocks that were forced (intruded) into existing rocks in a liquid state and then cooled to a solid state. Because the igneous rocks often cut across the layering of sedimentary or metamorphic rocks, they can be identified as younger than the rocks that enclose them (Figure 1-10). In some places a layer of igneous rocks lies between two beds of sedimentary rocks. In these instances, it is sometimes possible to determine the relative age by examining the boundaries, or *contacts*, of adjacent rock units. If the heat of the molten rock altered the rocks both above and below the igneous rock at the time it was intruded, then the igneous rock penetrated the sequence and is younger than both the overlying and underlying units. If only the upper contact of the sedimentary

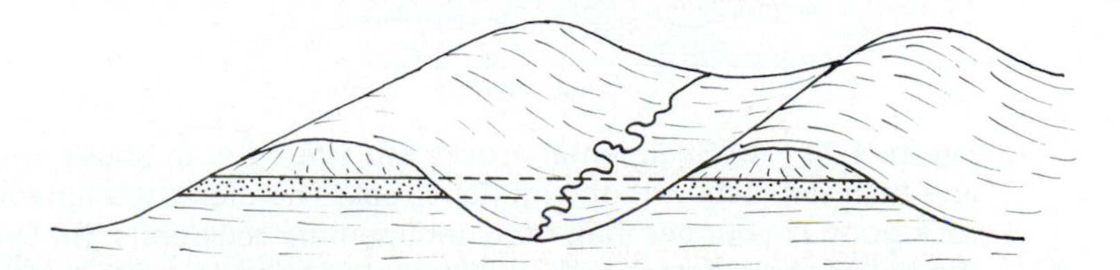

**Figure 1-9** Beds of sedimentary rock can be traced across areas in which they are not present by using correlation based on the Principles of Superposition and Original Continuity.

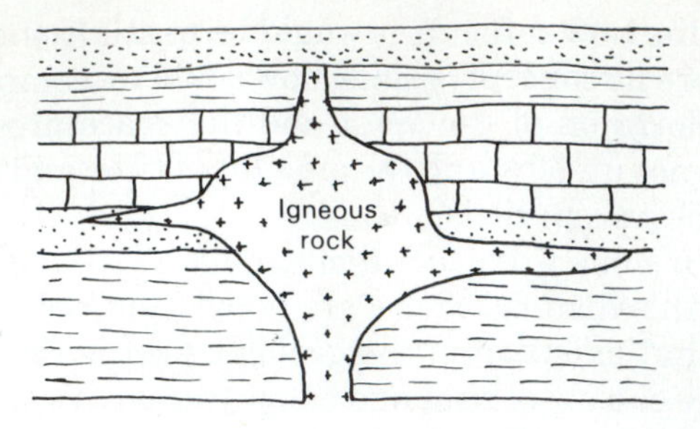


**Figure 1-10** A body of igneous rock penetrates and cuts across the bedding of a series of sedimentary rocks. It is therefore younger than the sedimentary rocks.

rock below the igneous rock was altered, then the igneous rock was probably formed as a lava flow on the ground surface and thus is younger than the rock beneath it but older than the rock above it (Figure 1-11).

It is often possible to use several of the principles described above to determine the sequence of events of a rock exposure (Figure 1-12). We therefore would be using relative time to compare the ages of rock units but still have no idea of the absolute age.

However useful the above methods may be, the foundation of relative age dating is based upon *fossils*, the visible remains of organisms preserved in rocks. William Smith (1769–1839), an English civil engineer, was one of the first people to recognize the usefulness of fossils as tools for determining relative age and correlation of sedimentary rocks. While supervising the construction of canals and other engineering works over a period of many years, Smith discovered that certain rock units always contained the same group, or *assemblage*, of fossils. The fossilized shells of marine organisms were most useful in this work. By using superposition and other principles, Smith was able to group these rock units in order of relative age on the basis of the correlation of fossil assemblages. The method was later applied throughout Europe, and a *geologic column* was constructed for all sedimentary rocks known in that region (Figure 1-13). The rocks of each *system* contained a distinct and unique assemblage of fossils. Although certain fossils were similar to those in other systems, consistent differences between systems were evident. Rocks beneath the Cambrian

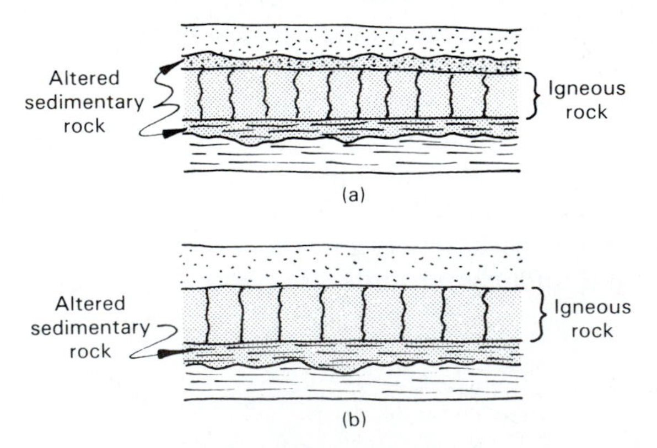


**Figure 1-11** (a) Sedimentary rocks are altered both above and below their contacts with the igneous rock. Therefore, the igneous rock body is younger than both sedimentary rock units. (b) Only the lower sedimentary rock unit is altered near its contact with the igneous rock. In this case, the igneous rock is a lava flow that is younger than the lower sedimentary rock unit but older than the upper sedimentary rock unit.

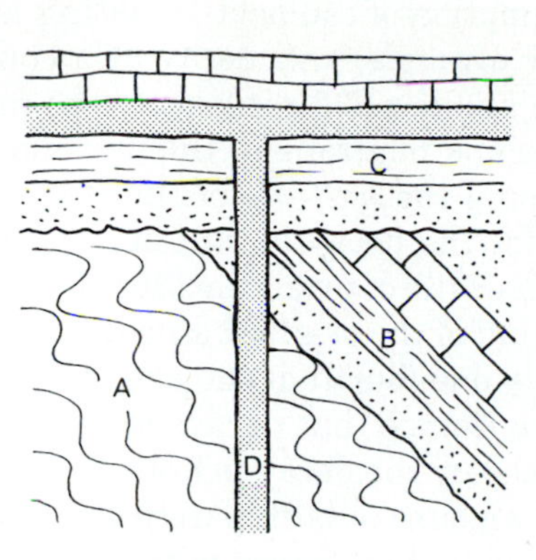


**Figure 1-12** The Principles of Original Horizontality, Original Continuity, Cross-cutting relationships, and Superposition can be used to unravel the sequence of events in this rock exposure. They include (1) deposition of rock sequence A, (2) metamorphism of A, (3) uplift and erosion of A to produce an unconformity, (4) deposition of sequence B, (5) deformation, uplift, and erosion of sequences A and B to produce an unconformity, (6) deposition of sequence C, and (7) intrusion of igneous rock body D through sequences A, B, and C.

system were generally thought to be nonfossiliferous and therefore could not easily be divided. They were given the name Precambrian.

The reasons for the changes in fossil characteristics were not known in the time of William Smith. Charles Darwin later provided an explanation for this phenomenon with his theory of evolution. The idea of the systematic change of organisms over long periods of time was consistent with the fossil record.

### *Absolute Time*

The construction of the geologic column was the culmination of relative age dating of rocks. Throughout the nineteenth century, geologists and physicists began to speculate about the age of the earth and the ages of the systems of the geologic column. The estimates became older and older. In the early twentieth century, one of the triumphs of modern geology was achieved with the application of radioactivity to the dating of rocks. Finally, a method for determining the absolute ages of certain rocks was available.

Radioactive particles (atoms) are disseminated throughout many types of rocks in trace amounts. These particles decay spontaneously by emitting atomic particles from the nucleus of the atom. Most of the radioactive particles are *isotopes* of elements; that is, they are forms of a certain element that differ slightly in atomic mass from other isotopes of the same element. (Atomic particles are more fully discussed in Chapter 2). Upon radioactive decay, isotopes are altered to different elements by the loss or gain of atomic particles by the nucleus.

| | Systems | Where First Described |
|---|---|---|
| Youngest | Quaternary | Europe |
| | Tertiary | Europe |
| | Cretaceous | England |
| | Jurassic | Jura Mts. Europe |
| | Triassic | Germany |
| | Permian | Russia |
| | Carboniferous | Great Britain |
| | Devonian | Great Britain (Devonshire) |
| | Silurian | Great Britain (Wales) |
| | Ordovician | Great Britain (Wales) |
| | Cambrian | Great Britain (Wales) |
| Oldest | Precambrian Era | Nonfossiliferous, therefore not well subdivided |

**Figure 1-13** The geologic column, showing the names of the rock systems and the regions for which they were named.

A basic assumption of radioactive dating is that each radioactive isotope (uranium 235, for example) will decay at a constant rate. In the example of uranium 235, the number 235 represents the mass of the isotope, that is, the number of protons and neutrons in the nucleus. The radioactive decay rate is often described using the *half-life* of the isotope. The half-life is the amount of time required for the decay of half of any amount of a particular isotope. For example, the half-life of uranium 235 is 0.7 billion years. After this amount of time, only one-half of the original amount of uranium 235 will remain. After two half-lives, only one-fourth of the original amount will be left.

A radioactive isotope that decays to another isotope is called the *parent*; the isotope produced by the decay is known as the *daughter*. To determine the age of a rock, the amount of both parent and daughter isotopes must be measured. This analysis is performed with a *mass spectrometer*, an instrument that can measure minute amounts of matter with great accuracy. The most useful parent-daughter isotope pairs are shown in Figure 1-14. With the exception of carbon 14, all these isotopes can be used for dating very old rocks. The date obtained, however, must be carefully analyzed. It can be considered to be accurate only if the rock has not been substantially altered since its formation. The age of an igneous rock, therefore, would represent the time that it cooled from a magma. Sedimentary and metamorphic rocks present some difficult dating problems. Sedimentary rocks consist of particles eroded from older rocks. The date obtained from an igneous grain incorporated into a sedimentary rock would thus represent the cooling of the igneous rock rather than the deposition of the sediment to form the sedimentary rock. In metamorphism, parent and daughter particles can be separated by heat and partial melting. Therefore, dates obtained from metamorphic rocks may or may not yield the correct age of the metamorphism.

| Radioactive parent nuclide | | Stable daughter nuclide | | Half-life (years) |
|---|---|---|---|---|
| Potassium 40 | K | Argon 40 | Ar | 1.3 billion |
| Rubidium 87 | Rb | Strontium 87 | Sr | 47 billion |
| Uranium 235 | U | Lead 207 | Pb | 0.7 billion |
| Uranium 238 | U | Lead 206 | Pb | 4.5 billion |
| Carbon 14 | C | Nitrogen 14 | N | 5730 |

**Figure 1-14** Radioactive isotopes used for geologic radioactive age dating.

Carbon 14 serves a different function in radioactive dating due to its short half-life of 5730 years. The value of carbon 14 lies in its utilization for dating very recent sediments, particularly those from the last part of the Ice Age. Carbon 14 is a radioactive isotope of carbon produced by the activity of cosmic rays in the earth's atmosphere. It mixes with the normal carbon (carbon 12) in the atmosphere and is taken up by plants and animals, so organic material contains carbon 14 at the same concentration as the atmosphere at the time of the death of the organism. Wood fragments and bone buried by the sediment deposited by glaciers can be used for dating the time of glacial advance. Because of the short half-life, about 75,000 years is the practical limit for dating by the carbon 14 method.

Application of radioactive dating to rocks began soon after the discovery of radioactivity. By dating rocks from all over the world and integrating these dates with the geologic column, scientists were able to construct an absolute chronology of geologic time. The geologic time scale (Figure 1-15) is the result. Geologic time is divided into two eons, Precambrian and Phanerozoic. The Phanerozoic eon is divided into eras, periods, and epochs. The periods are named after the rock systems in the geologic column; thus, finally, geologists had a means for relating evolutionary changes in organisms to absolute time.

The ages of rocks determined by radioactive dating made it necessary to greatly increase our conception of the age of the earth. Phanerozoic time alone, which includes most of the fossil record, is only a small percentage of geologic time (Figure 1-15). The oldest rocks dated extend back to about 4 billion years. If the age of Earth were reduced to a 24-hour scale, the human species would not appear until the last few seconds. Thus radioactive dating has had a major effect upon our thinking about the age of Earth and the other planets.

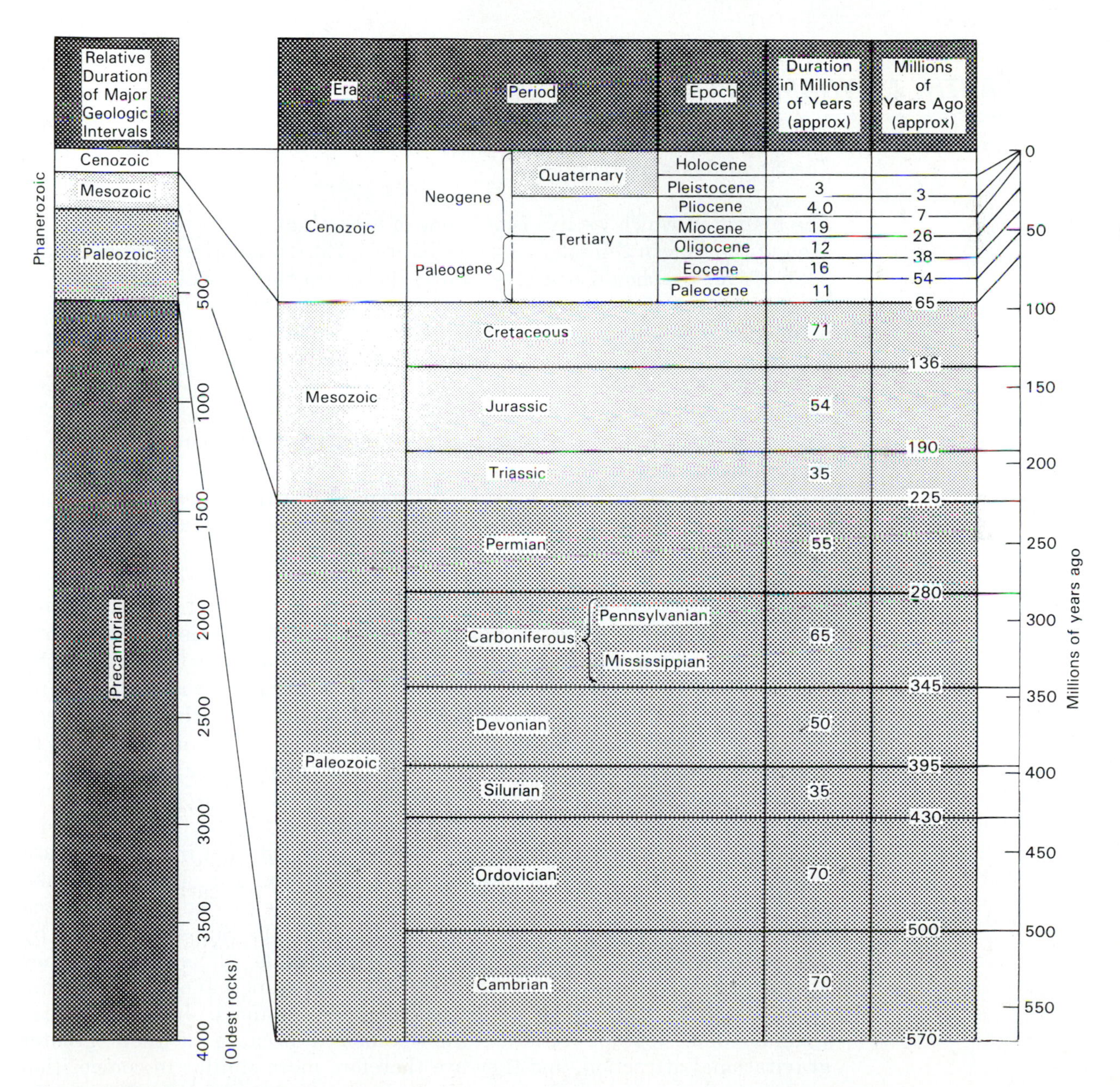


**Figure 1-15** The geologic time scale.

Although the oldest rocks on Earth yield radioactive ages of about 4 billion years, Earth is considered to be about 4.7 billion years old. Our estimate for this age comes from dating meteorites that fall to Earth. These rocky fragments consistently yield ages of 4.5 to 4.7 billion years. They are considered to be composed of material formed at the time of origin of the solar system. The oldest rocks dated from the Moon have similar ages, suggesting that the Moon was formed about the same time as the meteorites. Why are there no rocks on Earth from the period of time between 4.7 and 4 billion years ago? The answer is that Earth, as will be emphasized throughout this text, is an extremely dynamic planet. Rocks of the outermost layer of Earth, the crust, are continually transformed through the geologic cycle. The rocks of that early period of Earth's history have been melted and remelted so many times that no original rocks remain.

## *THE EARTH: A DYNAMIC PLANET*

### *The Solar System*

The planet on which we live is only one of nine planetary bodies that revolve around the Sun in the solar system (Figure 1-16). To put Earth in its proper perspective, we should briefly examine the characteristics of our planetary neighbors. One of the most basic observations is that there are two groups of planets: four smaller planets orbiting close to the Sun and four larger planets occupying the outer reaches of the solar system. The ninth planet, Pluto, is somewhat of an exception to this grouping. The inner, or *terrestrial*, planets are dense bodies mainly composed of iron and silicate (containing the element silicon) rocks. The four *giant* planets are much lighter in density because of their gaseous composition.

Although many ideas have been proposed to explain the origin of the solar system, this question is still a matter of debate. A possible hypothesis begins with a large diffuse mass of gas and dust slowly rotating in space. For some reason, this *solar nebula* began to contract, due to gravitational forces, and to increase its rotational velocity. At some point, a concentration of matter formed at the center of the nebula. Compression of matter raised its temperature to the point at which nuclear fusion was initiated. Thus the Sun, a body composed of 99% hydrogen and helium, was born. The matter rotating around the newly formed Sun gradually cooled and condensed into the nine planets. The final composition of the planets was controlled by their initial position in the nebula with respect to the Sun. Because the temperature was highest in the vicinity of the Sun and decreased with distance away from the Sun, the terrestrial planets were formed of material with relatively high boiling points. Thus Mercury, Venus, Earth, and Mars are dense bodies of iron and silicate rocks. In fact, four elements—iron, oxygen, silicon, and magnesium—make up about 90% of these planets. Volatile elements were carried away from the inner part of the nebula by matter streaming from the Sun and perhaps by the inadequate gravitational pull of the terrestrial planets before they attained their total mass. Volatile gases like water, methane, and ammonia were driven to the cooler regions of the giant planets. Jupiter, Saturn, Uranus, and Neptune retained their volatiles because of their greater gravitational attraction, and they are therefore more similar in composition to the original solar nebula.

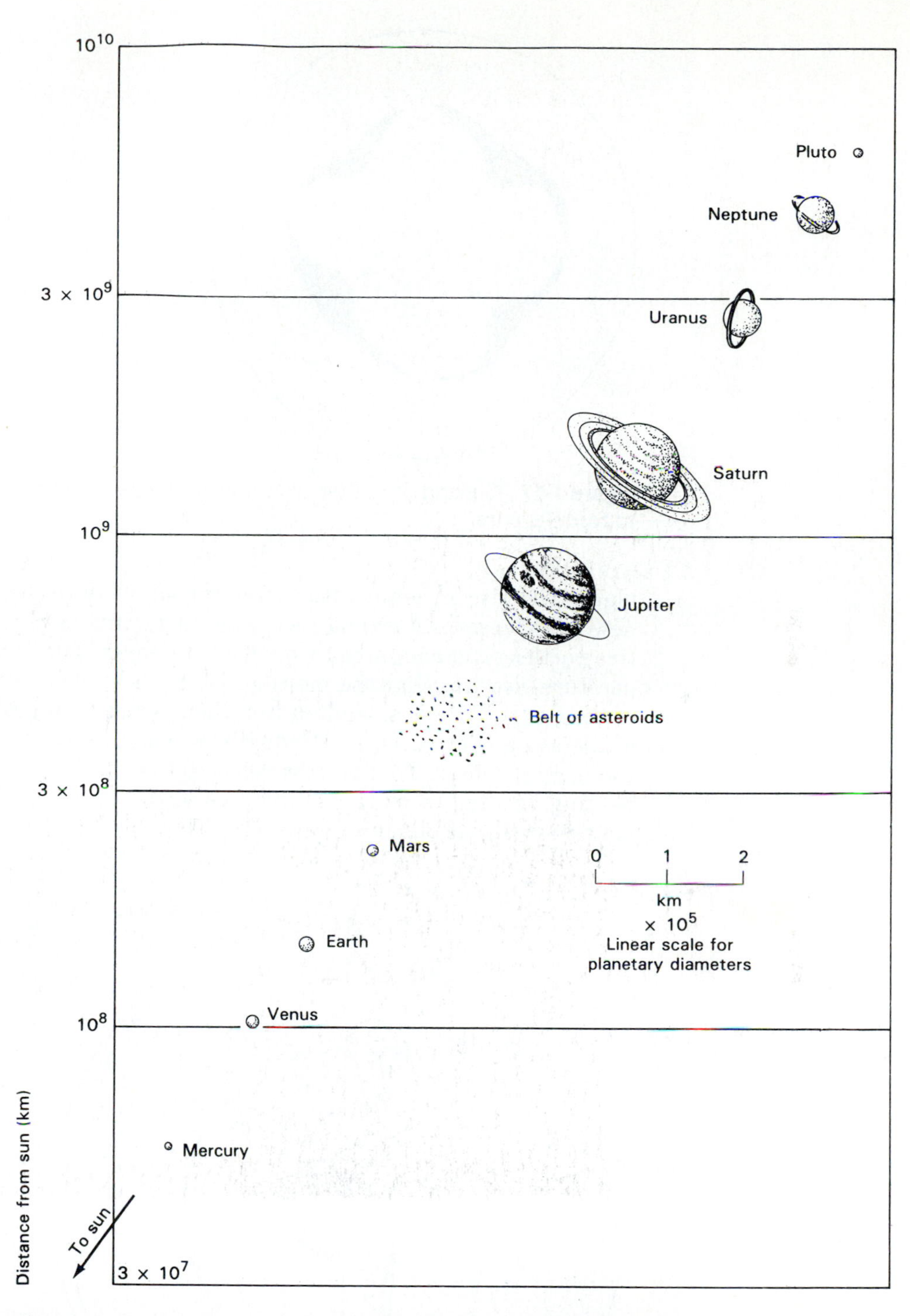


**Figure 1-16** Relative sizes and positions of the planets. (From S. Judson, M. E. Kauffman, and L. D. Leet, *Physical Geology*, 7th ed., copyright © 1987 by Prentice-Hall, Inc., Englewood Cliffs, N.J.)

### *Differentiation of the Earth*

After the initial condensation of the earth, its internal structure was very much different than at present. It probably consisted of a homogenous accumulation of rock material. At this point, a gradual heating of the earth must have taken

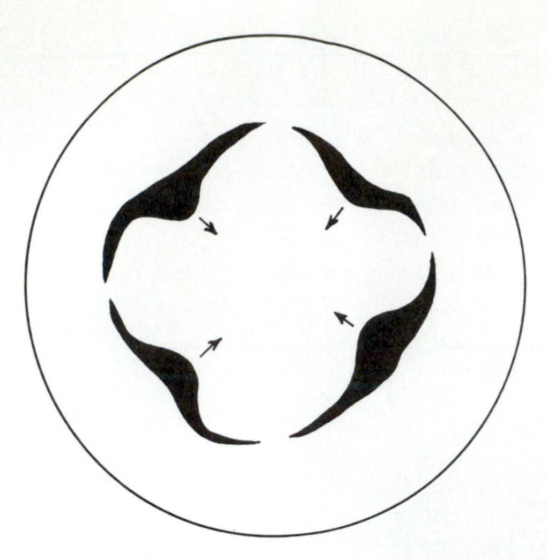

**Figure 1-17** Sinking of molten iron toward the center of the earth to form its core.

place. The rise in temperature was probably caused by compression of the newly condensed matter and also by the energy released by the decay of radioactive particles disseminated throughout the earth. The consequence of the temperature rise was that the melting point of iron was exceeded in some parts of the earth. The dense molten iron then began to migrate toward the center of the earth (Figure 1-17). Gradually a molten iron core began to develop at the earth's center. This process released gravitational energy in the form of heating, leading to more melting and partial melting. The end result was a density-stratified planet (Figure 1-18). The dense core was overlain by a mantle

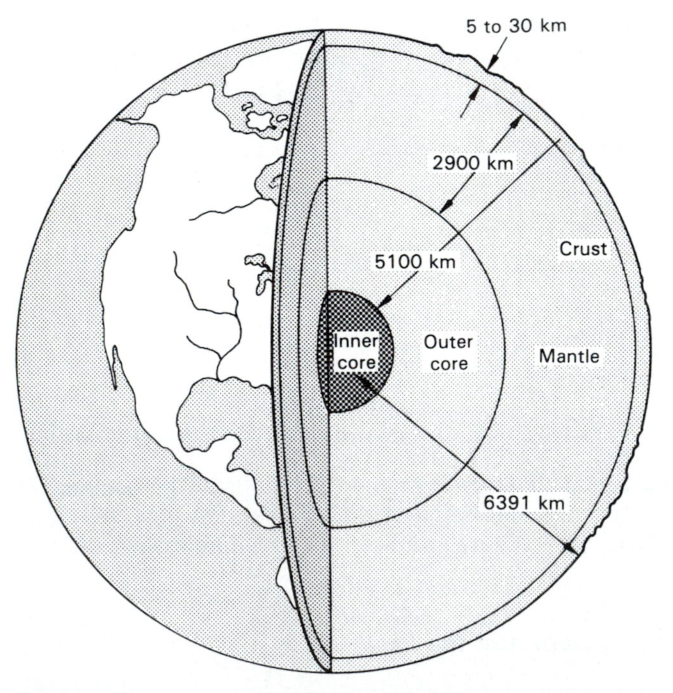


**Figure 1-18** Internal structure of the earth. (From S. Judson, M. E. Kauffman, and L. D. Leet, *Physical Geology*, 7th ed., copyright © 1987 by Prentice-Hall, Inc., Englewood Cliffs, N.J.)

composed of iron and magnesium silicate rocks. The lightest materials accumulated in a very thin layer near the surface to form the crust. Earlier we mentioned that no rocks older than 4.0 billion years remain at the earth's surface. The period of time between the formation of the earth at about 4.7 billion years and 4.0 billion years may be accounted for by this zonation or *differentiation* of the earth. The original rocks may have been melted and their components segregated by density during this interval.

After differentiation the earth consisted of a molten core, which is only partially molten now, and a cool exterior. Heat therefore flows from the core to the surface. This flow of heat is extremely important because it indirectly or directly causes many of the processes that occur in the earth's crust.

### *Plate Tectonics*

One of the greatest accomplishments in geology during the past few decades has been the formulation of a new theory called *plate tectonics*. Although this model will be discussed in greater detail later in this text, it is introduced now to reinforce the concept of Earth as a dynamic, ever-changing planet.

According to this theory, the earth is composed of an upper, rigid layer, the *lithosphere*, which includes the crust and upper mantle. The lithosphere is broken into about 12 major *plates*, which slowly move laterally over the earth's surface (Figure 1-19). Movement of these plates may be driven by the flow of material in a layer that directly underlies the lithosphere called the *asthenosphere* (Figure 1-20). Rock in the asthenosphere is soft and plastic because it is near its melting point. Heat flow and plastic flow of rock in the asthenosphere is thought to occur in cells consisting of rising segments, in areas of lateral flow beneath the lithosphere, and in descending limbs. The rising material (just below its melting point) spreads out laterally at the base of the lithosphere, carrying the rigid lithospheric plates along its path. Volcanoes erupt to form new lithosphere in the cracks formed as the plates split apart. The plates move until they reach a zone of downward flow in the asthenosphere. Here, the lithospheric plates are dragged down into the asthenosphere and may become partially melted (Figure 1-20).

The movement and interaction of lithospheric plates has probably been continuous since the differentiation of the earth. The interaction of plates can

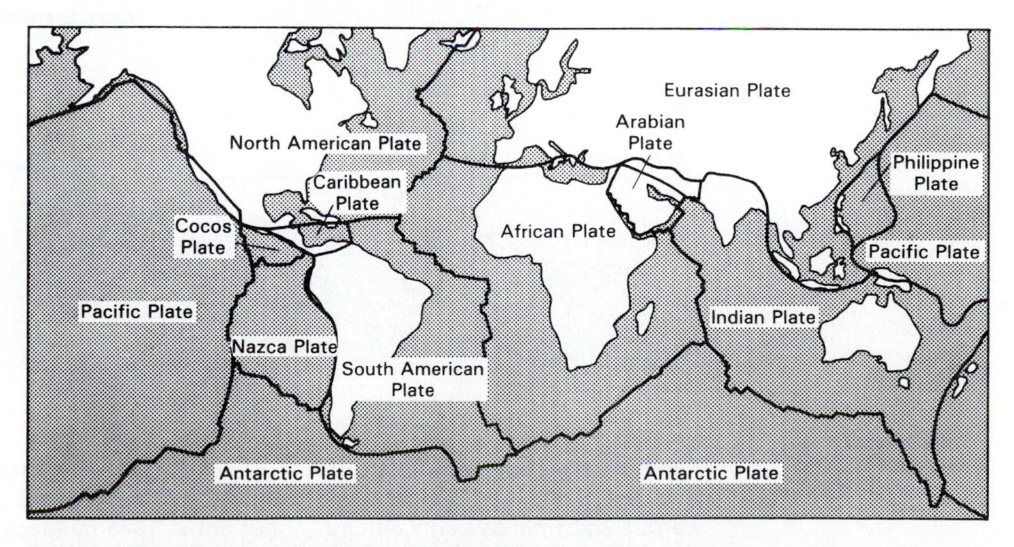


**Figure 1-19** Major plates of the lithosphere.

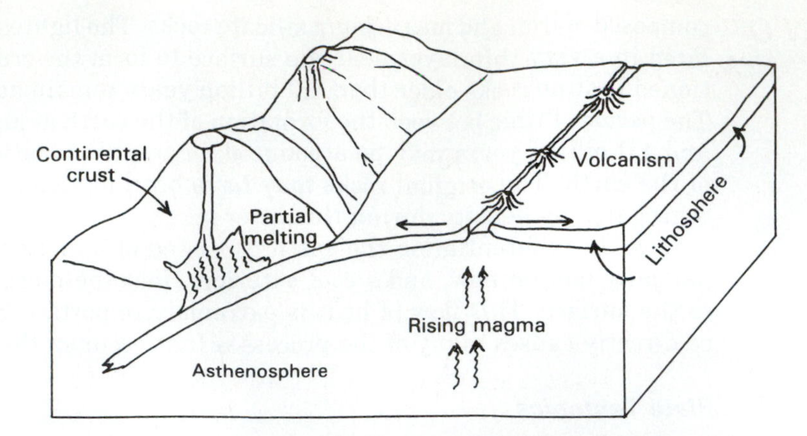


**Figure 1-20** Configuration of the lithosphere and asthenosphere.

explain the location of many geologic phenomena, including the distribution of volcanoes, earthquakes, and most mountain ranges. In fact, the movement of lithospheric plates can be considered to be a driving force in Hutton's geologic cycle.

### *Major Features of the Earth's Surface*

The earth's surface can be divided into two dominant topographic features: the continents and the ocean basins (Figure 1-21). The rocks that lie beneath the continents differ greatly from the rocks of the ocean basins. Continental rocks are low in density and include the oldest rocks on the earth. These are the lowest density materials concentrated by differentiation and subsequent plate interactions. The ocean basins are composed of denser, younger rocks. Volcanic eruptions continually produce new oceanic lithosphere at places where plates crack and split apart. In later sections of this book, the rocks of the continents and ocean basins will be described in more detail.

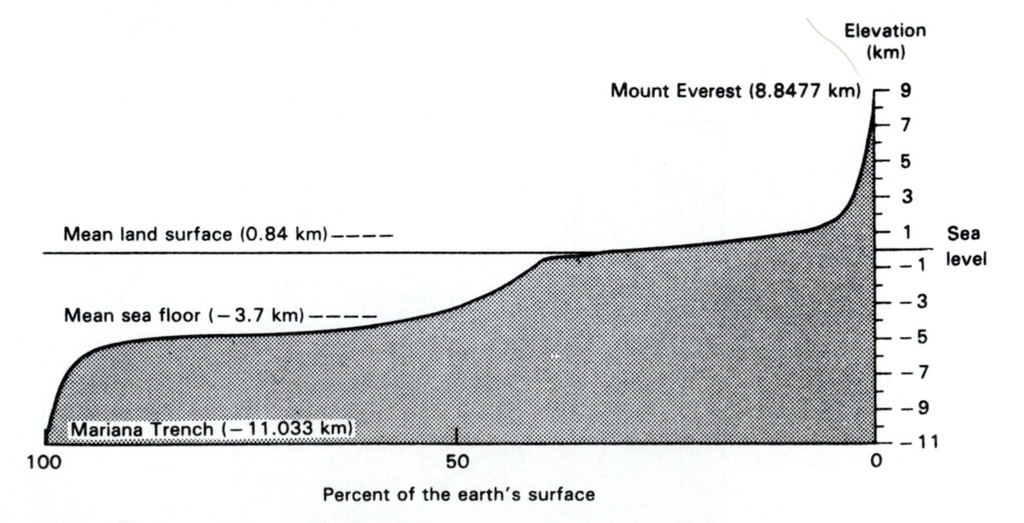


**Figure 1-21** The two predominant levels of the earth's surface: the continents and ocean basins. (From W. K. Hamblin, *The Earth's Dynamic Systems*, 4th ed., copyright © 1985 by Macmillan Publishing Co., Inc., New York.)

## APPLICATION OF GEOLOGY TO ENGINEERING AND ENVIRONMENTAL SCIENCE

Engineering is the utilization of science in the solution of practical or technological problems. Practical problems that are most likely to require the application of geology include problems found in the traditional areas of construction, mining, petroleum development, and water resources development and management. In the past several decades, however, environmental problems have increased in importance. An excellent example is the problem posed by the disposal of high-level nuclear waste. The solution to this critical problem is going to require an intensive collaboration between engineers, geologists, and other environmental scientists. The increasing population of the world is going to compel the construction of more roads and buildings; it will necessitate the location and extraction of more metallic, nonmetallic, and energy resources; it will increase the risk of loss of life and property in natural disasters; and it will intensify the pressure on our natural environment. For all these reasons, geology will continue to play a critical role in engineering and environmental science in the immediate and distant future.

### Geology and the Construction Site

In the realm of construction of civil engineering works, there are two basic ways in which geology must be considered.

First, the geologic setting of the construction site must be described and characterized. This is accomplished by mapping the distribution of geological materials and by collecting and testing samples of rock and soil from the surface as well as from below the surface of the site. The problem with these procedures is that natural materials are rarely homogeneous or continuous. Because of the high costs involved in drilling, sampling, and testing during site investigation (Figure 1-22), prediction of certain engineering properties of the geologic materials and their response to engineering activity often relies on a thorough understanding of the origin and geologic history of the rocks

**Figure 1-22** Collection of a soil sample by using a truck-mounted auger. (Photo courtesy of USDA Soil Conservation Service.)

and soils present. The amount of geologic investigation that is necessary depends upon the cost, design, purpose, and desired level of safety of the project. A dam across a canyon above a city requires a much more detailed investigation than a typical suburban housing development. Failures of large engineering projects, although very rare, often can be attributed to an inadequate understanding of the geology of the site.

The second major interaction between geology and engineering concerns the threat of hazardous geologic processes. This problem is often more relevant to the location of the structure rather than to its design or construction. The list of hazards includes spectacular geologic events such as volcanic eruptions, earthquakes, landslides, and floods (Figure 1-23). Other less sensational processes, such as land subsidence, soil swelling, frost heave, and shoreline erosion, result in billions of dollars of property damage in the United States every year.

**Figure 1-23** Damage from the Big Thompson flood of 1976 in Colorado, an event of high magnitude and low frequency. (W. R. Hansen; photo courtesy of U.S. Geological Survey.)

The critical task in dealing with geologic hazards is how to predict where and when these events will take place. Often the problem of when the hazardous process will occur is more difficult than where. Although geologists have made great strides in understanding the causes and effects of natural hazards, the ability to predict the place and, particularly, the time of a hazardous event remains an elusive goal.

An important characteristic of natural hazards is that there is usually a statistical relationship between the magnitude and frequency of the event. *Magnitude* is an indication of the size or intensity of an event; the amount of energy released by an earthquake, the size of a landslide, or the amount of flow in a flood are examples of magnitude. The *frequency,* on the other hand, is a measure of how often, on the average, an event of a particular size will occur. Frequency can be expressed as the probability of occurrence within a specific period, or the *recurrence interval* of the event. A flood with a 1% chance of taking place within a period of any one year has a recurrence interval of 100 years.

Careful recording of floods and other types of hydrologic data indicates that there is an inverse relationship between magnitude and frequency (Figure 1-24). These data can be fitted with the equations of theoretical statistical distributions with the result that the probability of events of high magnitude and low frequency can be predicted. Other geologic processes, including earthquakes and landslides, seem to follow a similar pattern. Engineers should understand this relationship and remember not to overlook the possibility of a destructive event of large magnitude. Critical structures such as nuclear power plants, schools, and hospitals must be designed to withstand large, infrequent events.

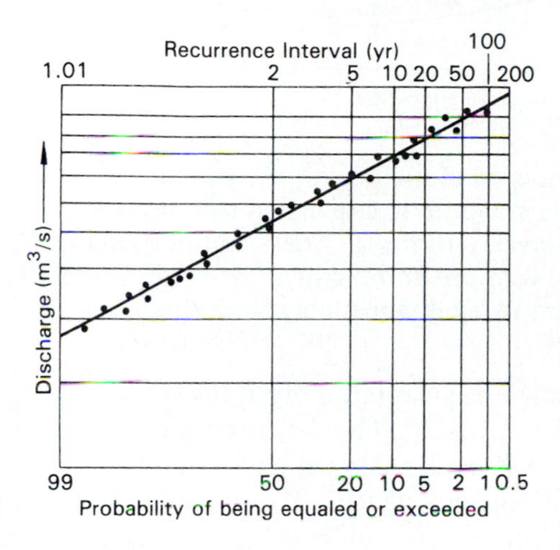


**Figure 1-24** The relationship between magnitude and recurrence interval for floods from a hypothetical river. Each point represents the magnitude and calculated recurrence interval of a particular flood.

When certain or extensive loss of life is not a prime consideration in an engineering project, the economics of the project must also be considered in terms of the magnitude-frequency relationship. For example, a bridge that is built to withstand a flood of very high magnitude and low frequency would have high initial costs but low maintenance costs because the smaller, more frequent floods would cause little damage. Alternatively, the bridge could be built to withstand only small floods. In this case, the initial construction costs would be lower, but the maintenance costs would be higher because smaller floods that might damage the bridge are more frequent than large floods. In order to minimize the long-term costs of the structure, both initial costs and maintenance must be considered (Figure 1-25). The optimal total cost would be realized by designing the bridge for a flood of moderate recurrence interval.

### *Development of Natural Resources*

The search for mineral deposits and petroleum is carried out by geologic exploration. When a resource is located, it must be delineated and assessed in order to determine if it is economically recoverable. If it is determined to be so, the extraction of the resource is conducted by mining and petroleum engineers. These engineers must have a sound knowledge of geology because of the geologic control of the location and the characteristics of the deposit.

Mining engineers supervise mine design and operation, including reclamation of the site (Figure 1-26). Environmental scientists are heavily involved in the reclamation process, both from industry and regulatory positions. The geologic and hydrologic setting of the deposit and the site influence many decisions that must be made in the process.

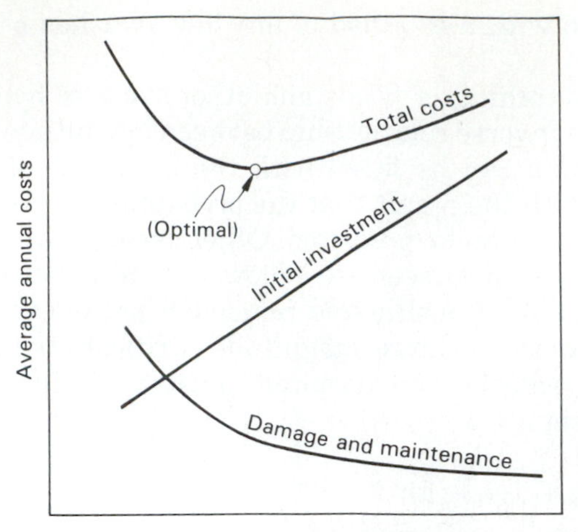


**Figure 1-25** Economics of the design of hydraulic structures. The optimal cost of the structure is associated with an event of moderate recurrence interval. (From J. E. Costa and V. R. Baker, *Surficial Geology: Building with the Earth*, copyright © 1981. Reprinted by permission of John Wiley & Sons, Inc., New York.)

The production of petroleum often involves drilling through great thicknesses of rock (Figure 1-27). The drilling process requires the petroleum engineer to be familiar with all relevant physical and chemical conditions that may be encountered at various depths. This geologic knowledge must encompass a broad range of solutions to problems associated with rocks, liquids, and gases that are present in the subsurface. If a petroleum reservoir is located, petroleum engineers must develop it in the most efficient and economic manner possible. Extensive information must be gathered concerning the size, shape, and rock characteristics of the reservoir. New technologies are constantly being tested to produce the maximum amount of oil possible from each reservoir.

**Figure 1-26** Extraction of coal from an open-pit mine in central North Dakota. About 20 m of overburden has been removed above the coal bed.

**Figure 1-27** A drilling rig exploring for oil in the Williston Basin, North Dakota.

### *Water Resources and the Environment*

The development and protection of water resources is becoming one of society's most important concerns. Future increases in water supply are going to rely more heavily upon ground water than surface water. The occurrence, movement, and quality of ground water is therefore a matter of great significance. Engineers and geologists together are developing techniques to manage the utilization of ground-water supplies to ensure their availability for the future. Overuse of ground water for irrigation or other needs can lead to depletion and exhaustion of the supply (Figure 1-28).

**Figure 1-28** Crop irrigation places a heavy demand upon water resources. (Photo courtesy of USDA Soil Conservation Service.)

Along with the need to wisely develop ground-water resources is the increasing necessity of protecting the chemical quality of ground water. Past, and perhaps present, waste disposal and land-use activities have contaminated ground-water supplies in all parts of the United States. Development of new methods to rehabilitate ground-water reservoirs will require the talents of a wide variety of environmental scientists and engineers. Geology will be critical to the effort because an understanding of the subsurface geologic environment is necessary for dealing with any aspect of ground-water occurrence or movement.

## *SUMMARY AND CONCLUSIONS*

The development of geology over the past four centuries has been a remarkable scientific journey. We have seen our conception of the earth's history change greatly. Hutton realized the role of continuing change in the evolution of the earth's surface. He recognized geologic processes in action, as well as the evidence of past geologic processes, in the rocks that he studied.

Although Hutton and others had begun to suspect that the earth was very old, its true age was not determined until the method of radioactive age dating was developed in the early twentieth century. This dating provided a method of quantifying the divisions of the geologic column. These divisions were incorporated into the geologic time scale.

The early history of the planet Earth includes a period of density stratification. Since the formation of the core, crust, and mantle, continual circulation and alteration of material has taken place. Hutton's geologic cycle is now integrated into the model of plate tectonics.

The need for the integration of geology with engineering and environmental science is increasing. Detailed geologic input is required in construction and development, in the recovery of natural resources, and in solving problems of ground-water contamination. The geologic information required by engineers and environmental scientists includes a comprehensive study of surface and subsurface geologic materials, and also a thorough analysis of hazardous geologic processes that could affect an engineering project.

### PROBLEMS

1. In what way do the early concepts of catastrophism and neptunism differ from modern scientific thought?
2. What were the contributions of Nicolaus Steno to the science of geology?
3. Why is Hutton's Principle of Uniformitarianism so important to modern geology?
4. What forms of evidence are used in relative age dating?
5. Discuss the limitations of radioactive age dating.
6. Choose a recently constructed engineering project in your area (such as a building, highway, or power plant). Describe all the ways in which this project interacts with the geologic environment both during construction and throughout the life of the structure.
7. What role should engineers play in the development and protection of water resources?
8. Summarize the earth's early geologic evolution following its initial formation.
9. What are the major ideas of the theory of plate tectonics?

10. Future changes in the earth's climate are of concern to the human population. In what ways can geology help us to understand the magnitude and timing of changes in global climate?
11. Make an inventory of the types of mineral resources produced in your region. What role do these operations play in the regional economy? What environmental problems are associated with production of the resources?

## REFERENCES AND SUGGESTIONS FOR FURTHER READING

COSTA, J. E., and V. R. BAKER. 1981. *Surficial Geology—Building with the Earth*. New York: John Wiley.

HAY, E. A., and A. L. MCALESTER. 1984. *Physical Geology: Principles and Perspectives*, 2d ed., Englewood Cliffs, N.J.: Prentice-Hall, Inc.

JUDSON, S., M. E. KAUFFMAN, and L. D. LEET. 1987. *Physical Geology*, 7th ed. Englewood Cliffs, N.J.: Prentice-Hall, Inc.

# 2

# Minerals

The minerals, rocks, and soils that occur at and beneath the earth's surface are the materials with which the engineer and environmental scientist must work. Unlike the materials used in buildings and other structures, which have uniform, homogeneous properties, these natural materials are notoriously variable and nonhomogeneous. The task for engineers designing any structure is to evaluate the distribution and properties of the natural materials present at the site and then base the design upon this assessment. It is impossible to evaluate natural materials at specific sites without a general understanding of the physical and chemical characteristics of earth materials. In addition, the engineering properties must be ascertained. In this part of the text we will concentrate upon the origin and characteristics of the minerals and rocks that make up the earth's crust. In the following sections, we will turn our attention to geologic processes operating both deep within the earth and at its surface.

An in-depth study of geology usually begins with an introduction to minerals. This particular point of departure should not come as a surprise, considering that the earth's solid surface is composed of rocks and soils that are primarily mineral aggregates. An understanding of mountains, volcanoes, earthquakes, and all other geologic phenomena could never be complete without some knowledge of the types and states of matter that make up the solid part of the earth.

Knowledge of minerals is essential for the engineer who deals with earth materials. Minerals are partially responsible for the physical and mechanical properties of rock and soil encountered in mines, tunnels, excavations, and environmental cleanups. Likewise, dams, embankments, and other structures built from earth materials function because engineers have successfully utilized the properties of rock and soil in the project design. In industry, minerals are directly incorporated into chemicals, abrasives, and fertilizers, and are processed into thousands of other useful products.

## THE NATURE OF MINERALS

*Minerals* are inorganic, naturally occurring solids. In order to define the term mineral completely, however, we must further state that minerals are crystalline substances that have characteristic internal structures and chemical compositions that are fixed or that vary within fixed limits. The distinct internal structure of each mineral results in diagnostic external properties that can be used in mineral identification. In subsequent sections, we will examine each component of the definitions of a mineral.

### Internal Structure

At the heart of the definition of a mineral is the concept of a regular internal structure. It is this characteristic that distinguishes minerals from other types of solid, inorganic matter. A mineral, therefore, is a chemical compound in which elements react and combine to form a regular arrangement of particles within the solid.

The fundamental unit of each chemical element in the periodic table is the *atom*, the smallest amount of the element that retains its characteristic properties. An atom, however, is composed of even smaller particles. The most common model of atomic structure consists of a central concentration of mass, the *nucleus*, surrounded by concentric shells inhabited by minute charged subatomic particles called *electrons* (Figure 2-1). Electrons are held in the

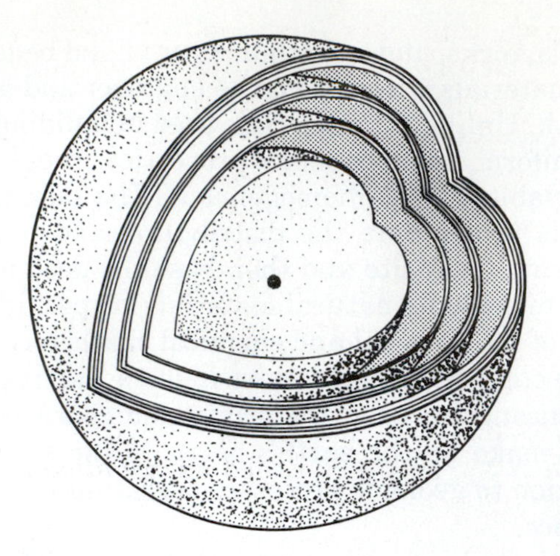

**Figure 2-1** Electron cells around the nucleus of an atom. (From L. D. Leet, S. Judson, and M. E. Kauffman, *Physical Geology*, 5th ed., copyright © 1978 by Prentice-Hall, Inc., Englewood Cliffs, N.J.)

vicinity of a nucleus because of the attraction of opposite electrical charges. By convention, electrons are considered to be negatively charged and the nucleus is considered to be positively charged.

The nucleus of the atom can also be subdivided. Positive charges are concentrated in particles called *protons*. The number of protons in an atomic nucleus defines the *atomic number* of the element. Thus an atom of sodium has 11 protons in the nucleus balanced electrically by 11 electrons orbiting around the nucleus in three shells (Figure 2-2). Particles that are neither positively nor negatively charged also reside in the nucleus. These *neutrons* have about the same mass as a proton, so the *atomic mass* of an atom is basically the mass of the protons plus the neutrons. The mass of an electron is about three orders of magnitude less than the mass of a proton or neutron and can be neglected in determining the atomic mass. Unlike the atomic number, the atomic mass of an element is not fixed.

Nuclei of an element may vary slightly in the number of neutrons that they contain. *Isotope* is the name given to atoms of the same element that differ in the number of neutrons in the nucleus. For example, hydrogen, with one proton in the nucleus, forms the isotope protium with an atomic mass of approximately one. With the addition of one neutron to the nucleus, the hydrogen isotope deuterium is formed. Another hydrogen isotope, tritium, contains two neutrons in the nucleus. Even though the properties of these three forms of hydrogen vary slightly, they are all unmistakably hydrogen. The proportions of the three isotopes of hydrogen vary in substances such as water as a function

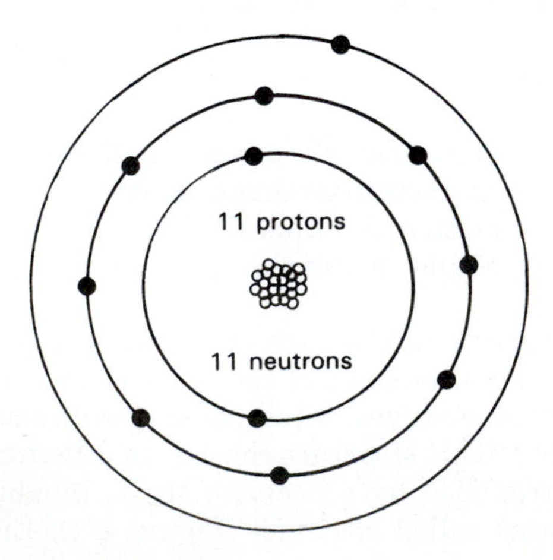


**Figure 2-2** Atomic structure of sodium.

of temperature, age, and other variables. Thus the measurement of the isotopic content of a particular body of water is an important research tool in hydrology and other branches of the earth sciences.

A final type of atomic particle with which we must be familiar is the *ion*. Ions form when an atom gains or loses electrons, so the number of electrons is then no longer equal to the number of protons. When an ion is formed, the atomic particle then becomes a positively charged *cation* or a negatively charged *anion*. The formation of ions is caused by the tendency of atoms to attain a state in which the outer electron shell is completely filled with electrons. Thus atoms like sodium with a small number of electrons in their outer shells will readily lose them and become cations. Elements that lack only one or two electrons to complete their outer shells will gain electrons and become anions. The charge of a cation or anion, called its *oxidation number*, is determined by the number of electrons that it gains or loses. The oxidation number of the sodium ion is +1 because sodium loses the single electron from its outer shell.

In a mineral, atoms or ions are arranged in an orderly fashion to form the regular internal structure. Among the reasons that particles assume a certain structure are *chemical bonds* between adjacent atoms or ions in the internal framework. The presence of interparticle bonds indicates that minerals are no different from any other type of *chemical compound*. As compounds, the elemental composition of minerals can be described by chemical formulas that specify the exact proportions of elements in the mineral. *Quartz*, for example, has the formula $SiO_2$, meaning that there are two oxygen ions for every silicon ion in the structure.

Two types of bonds are of particular importance for geologic applications (Figure 2-3). An *ionic bond* is formed when one atom donates one or more electrons to an atom of another element. The first atom becomes a cation and the second, an anion. As illustrated in Figure 2-3, when an electron is transferred from an atom of sodium to an atom of chlorine, both ions then have achieved a stable state because their outer shells are filled with electrons. The resulting compound is called sodium chloride (common table salt), has the formula NaCl, and is given the mineral name *halite*. Elements that do not have as strong a tendency to lose or gain electrons may form *covalent bonds*, in which electrons are shared between atoms to complete the outer shell.

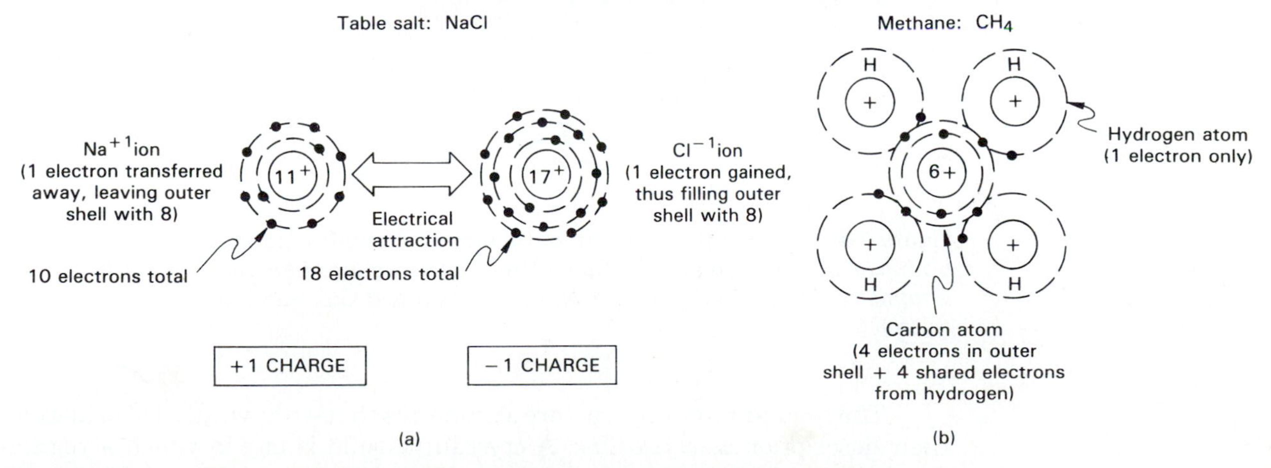


**Figure 2-3** Common types of atomic bonding in minerals. (a) Ionic bonding (electron transfer). (b) Covalent bonding (electron sharing). (From E. A. Hay and A. L. McAlester, *Physical Geology: Principles and Perspectives*, 2d ed., copyright © 1984 by Prentice-Hall, Inc., Englewood Cliffs, N.J.)

Examples of covalent bonding are methane (Figure 2-3) and water. The type and number of bonds that particles in a mineral form with adjacent particles have a strong influence on the properties of the mineral. In fact, two entirely different minerals can have the same chemical composition. *Diamond* and *graphite*, for example, are both composed only of carbon. In diamond, the carbon atoms form a three-dimensional framework with each atom bonded strongly to four adjacent atoms (Figure 2-4). The graphite structure consists of strong bonds between each carbon atom and three neighbors, all within a single plane. This results in sheets or layers of graphite with much weaker bonds *between* layers than *within* layers. Because of these differences in internal structure, diamond is the hardest mineral known and graphite is one of the softest.

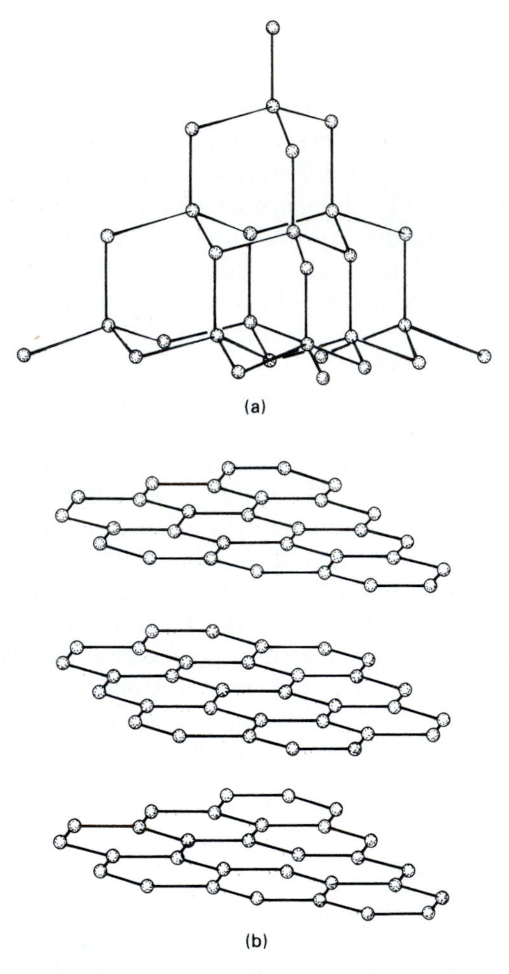


**Figure 2-4** The internal structure of (a) diamond and (b) graphite, both composed entirely of carbon. (After Linus Pauling, *General Chemistry*, copyright © 1953 by W. H. Freeman and Co., San Francisco.)

The regular internal structure of minerals that we have alluded to justifies their description as *crystalline*. A crystalline solid is one in which a regular arrangement of atoms is repeated throughout the entire substance. The smallest unit of this structure is called the *unit cell*. The relative size of the ions plays an important role in the manner in which ions are packed in a particular structure. The unit cell of halite is composed of six chloride ions surrounding

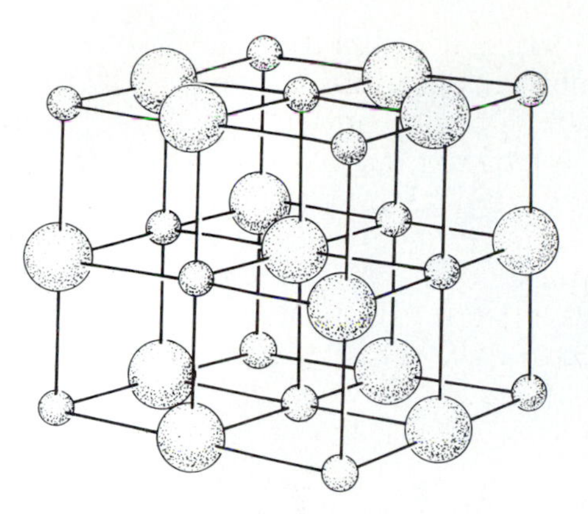

**Figure 2-5** Structure of halite, showing arrangement of chloride ions (large spheres) and sodium ions (small spheres). (From S. Judson, M. E. Kauffman, and L. D. Leet, *Physical Geology*, 7th ed., copyright © 1987 by Prentice-Hall, Inc., Englewood Cliffs, N.J.)

each sodium ion. The repetition of the unit cell in three directions produces a cubic structure as shown in Figure 2-5. The presence of a regular, unique internal structure for each mineral was suspected by mineralogists for several centuries before proof was obtained in the early part of the twentieth century. In the process called *x-ray diffraction*, x-rays are passed through mineral samples. Photographic images produced by the x-rays as they emerge from the sample show the regular pattern of atomic particles in the structure.

As we have seen in the cases of quartz and halite, the chemical composition of a mineral can be expressed by a chemical formula. For some minerals, the chemical formula is fixed. There are minerals, however, that have a variable composition. If two ions have similar size and charge, they may occupy the same types of sites in a mineral structure without altering the structure and without greatly affecting the properties of the mineral. Magnesium and iron form cations that have the same charge and the same approximate size. In the silicate mineral *olivine*, magnesium and iron can substitute freely. The varying proportions of magnesium and iron constitute a *solid-solution series*. Thus olivine can range in composition from $Mg_2SiO_4$ to $Fe_2SiO_4$. These two end members represent olivine with all magnesium or all iron. In addition, any intermediate composition is possible. A sample of olivine with 50% iron and 50% magnesium would have the formula $(Mg_{0.5}, Fe_{0.5})_2SiO_4$. Although the composition of a solid-solution–series mineral is variable, it varies within fixed limits. The limits for olivine are the magnesium and iron end members.

### Crystals

We have now reached the point in our discussion where we can consider the exterior appearance of minerals. Normally, minerals will be present as irregular grains in a rock. Occasionally, however, we will be lucky enough to find a mineral specimen bounded by smooth, plane surfaces arranged in a symmetrical fashion around the specimen (Figure 2-6). This type of sample is said to be a *crystal*. The plane, exterior surfaces, or *crystal faces*, are manifestations of the internal structure of the mineral.

The unique internal arrangement of atoms in each mineral determines the possible crystal faces that can be present. Crystals form slowly by adding unit cells of the mineral to the outer faces. Only when crystals are allowed to grow in this manner in an uncrowded environment, that is, unhindered by the growth of neighboring crystals, are well-shaped crystal faces developed. These conditions are somewhat rare, explaining the occurrence of most minerals as irregular grains. Despite the lack of crystal development, the characteristic internal structure is always present in any mineral.

**Figure 2-6** A crystal of garnet. The exterior surface of the specimen is composed of smooth, planar crystal faces arranged in a regular geometrical pattern.

Crystals have been treasured as gems for thousands of years. The scientific study of crystals took great strides when Nicolaus Steno recognized in 1669 that the angles between corresponding crystal faces of the same mineral are always the same. This holds true no matter what size crystals are measured. Steno's observation is now known as the *Law of Constancy of Interfacial Angles* (Figure 2-7).

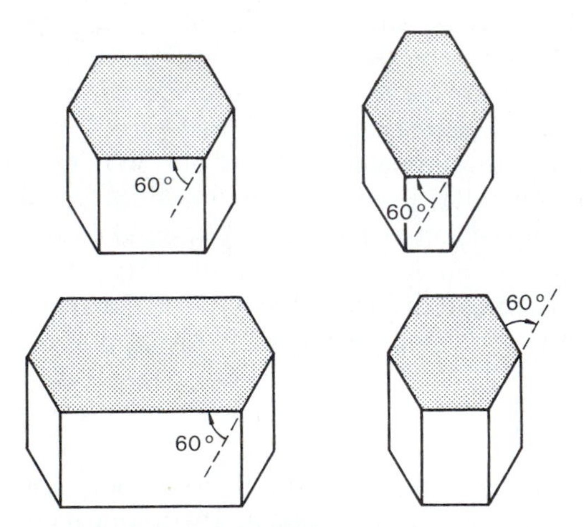


**Figure 2-7** Crystals of the same mineral, no matter what size, have the same angle between corresponding crystal faces. (From E. A. Hay and A. L. McAlester, *Physical Geology: Principles and Perspectives*, 2d ed., copyright © 1984 by Prentice-Hall, Inc., Englewood Cliffs, N.J.)

The location of crystal faces on a crystal is quantified by their orientation with respect to the *crystallographic axes*, which are imaginary lines drawn through the crystal connecting the centers of corresponding crystal faces on opposite sides of the crystal. Despite the large number of possible crystal-face orientations that have been observed in natural crystals, all crystals have been grouped into six *crystal systems* based upon their geometrical properties. The

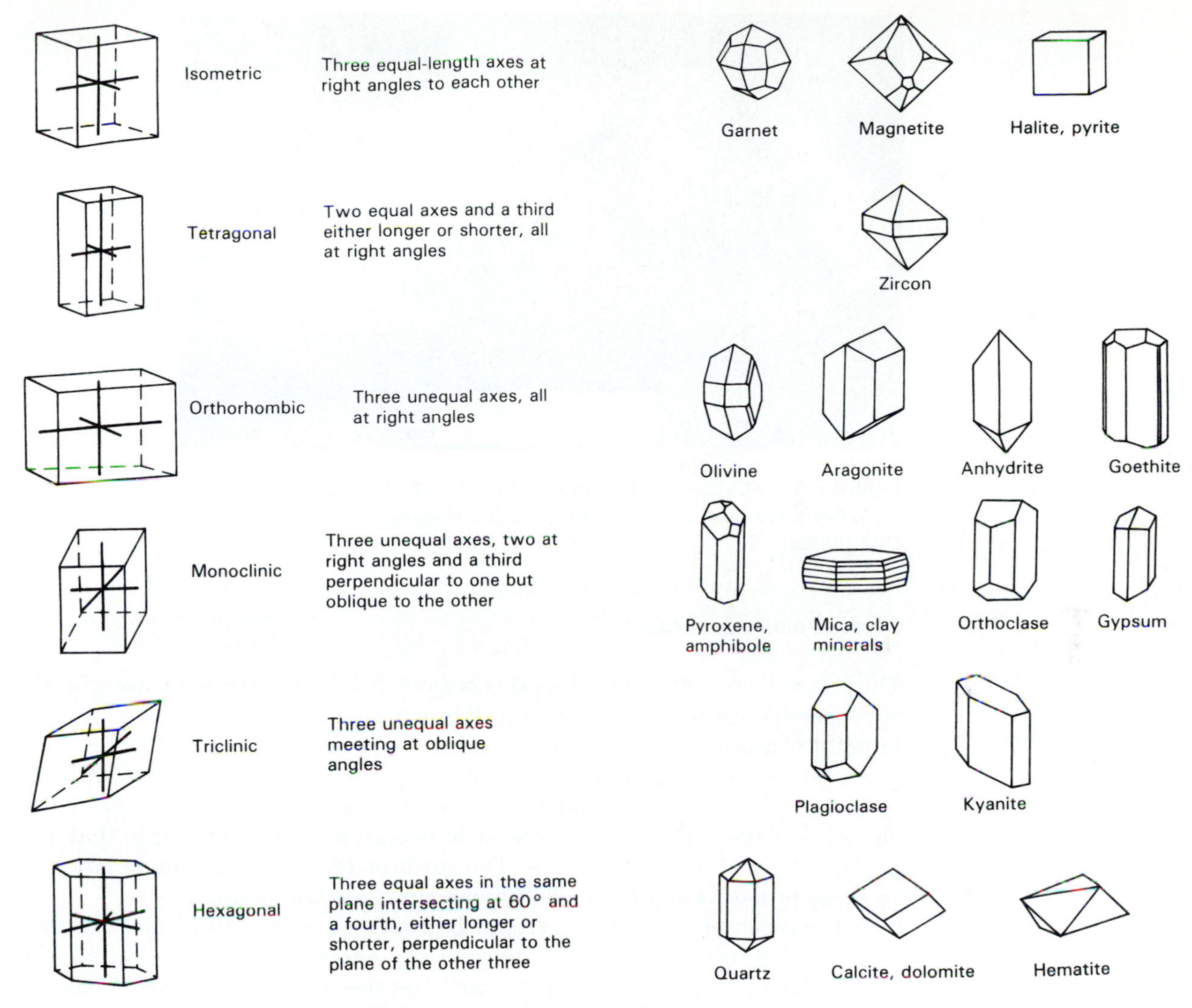


**Figure 2-8** The six crystal systems with representative minerals and crystal forms. (From E. A. Hay and A. L. McAlester, *Physical Geology: Principles and Perspectives*, 2d ed., copyright © 1984 by Prentice-Hall, Inc., Englewood Cliffs, N.J.)

definition of each system is based upon the number of crystallographic axes and the angles between the axes. Illustrations and examples of the crystal systems are presented in Figure 2-8.

## PHYSICAL PROPERTIES

Because minerals are rarely present as crystals, we must use criteria other than the shape and arrangement of crystal faces for identification. Although there are about 2000 minerals known, only a few are abundant in the most common rocks of the earth's crust. The common rock-forming minerals can be identified by their *physical properties*, which are characteristics that can be observed or determined by simple tests.

**Figure 2-9** Muscovite, showing excellent cleavage in one direction. Samples split readily parallel to the cleavage planes into very thin plates.

### *Cleavage and Fracture*

The internal structure of a mineral is responsible for an extremely useful group of properties involving its strength. For example, there two ways in which a mineral can break. First, the breakage can take place along regularly spaced planes of weakness caused by weak bonds in those directions within the internal structure. This type of breakage is called *cleavage* and the planes are called cleavage planes. The second type of breakage occurs in minerals that lack preferred directions of weakness. The surfaces of rupture are more irregular in these minerals and the type of breakage is known as *fracture.*

Cleavage planes can sometimes be confused with crystal faces because both develop in response to the internal arrangement of atoms in a mineral. Crystal faces, however, do not necessarily represent planes of weak bonds. For example, quartz has no cleavage but often exhibits well-developed crystal faces. Cleavage is classified as *perfect* if the cleavage planes are level and smooth. *Biotite* and *muscovite*, the micas, are examples of minerals with excellent cleavage in one direction (Figure 2-9). Cleavage can be developed in as many as six

**Figure 2-10** Halite, a mineral with well-developed cubic cleavage.

**Figure 2-11** Rhombohedral cleavage in the mineral calcite. The three directions of cleavage intersect at angles of 75 and 105 degrees.

directions, creating a variety of cleavage types. Several common minerals develop cleavage in three directions. When the three planes are mutually perpendicular, the cleavage form is called *cubic* (Figure 2-10). If the three directions of cleavage do not intersect at right angles, *rhombohedral cleavage* can be formed (Figure 2-11).

Most fractures are relatively rough and irregular. Fracture is the form of breakage displayed by materials composed of atoms that are evenly spaced and equally attracted to adjacent particles in all directions. Occasionally, when materials of this type break, the surface is marked with smooth, concentric depressions (Figure 2-12). This type of fracture is called *conchoidal fracture*. Quartz sometimes displays this surface feature.

**Figure 2-12** Obsidian, a type of volcanic glass that produces conchoidal fracture upon breakage.

### *Hardness*

The strength of atomic bonds, along with the size and density of the packing of atoms or ions in a mineral, determines its *hardness*. The physical test used to determine the hardness of a mineral involves scratching it with various materials. A mineral with a high hardness value cannot be easily scratched by most substances. A soft mineral, on the other hand, can be scratched by any harder material. Tests of this type have shown that minerals vary quite widely in hardness. A relative scale used to compare minerals is the *Mohs Hardness Scale* (Table 2-1). The scale is made up of 10 levels arranged in order of increasing hardness, each represented by a common mineral. It is important to remember, however, that the scale is relative and the hardness increments between the levels are not necessarily equal.

**TABLE 2-1**

**The Mohs Hardness Scale**

| *Mineral* | *Common object* |
|---|---|
| 1. Talc | |
| 2. Gypsum | Fingernail ($2\frac{1}{2}$) |
| 3. Calcite | Penny (3) |
| 4. Fluorite | |
| 5. Apatite | Knife blade ($5\frac{1}{2}$) |
| 6. Orthoclase | File (6), window glass (6) |
| 7. Quartz | Unglazed porcelain streak plate (7) |
| 8. Topaz | |
| 9. Corundum | |
| 10. Diamond | |

### *Color and Streak*

It would simplify mineral identification greatly if each mineral had a unique color that never varied from sample to sample. Unfortunately, this is not the case; many minerals occur in a wide variety of colors.

The property of color is caused by the absorption of selective wavelengths of white light. The color of a mineral is associated with the wavelengths that are not absorbed. These portions of the spectrum are reflected from the surface of the mineral. The absorption of light energy is related to the configuration of electrons around the nucleus in certain elements. In the transition elements, the outer electron shell contains one or two electrons before the adjacent inner shell is filled. Upon stimulation by light energy, the loosely held electrons in the outer shell vibrate easily and in doing so readily absorb energy. Even a minute percentage of such elements present in a mineral as an impurity can supply color to the mineral. For example, pure quartz is colorless, but only a small amount of manganese or titanium can impart a pink hue to the mineral. Quartz of this color is known as rose quartz.

The color of finely powdered mineral particles produced by scraping the specimen across a porcelain (*streak*) plate is often more diagnostic than the bulk color of the mineral. The mineral *hematite*, for example, produces a reddish brown streak, even though the sample may have a metallic gray appearance. The limitation of a streak plate is that it can only be used on minerals with a hardness less than seven.

### Luster

*Luster* is a property that results from the manner in which light is reflected from a mineral. All minerals can be classified with respect to luster as either *metallic* or *nonmetallic*. Metallic luster is caused by high surface reflectivity of light by the opaque minerals, minerals that strongly absorb light. The native metals such as gold and silver, as well as numerous other minerals (Figure 2-13), have the appearance that we normally associate with metals because of their metallic luster.

The remaining minerals have various types of nonmetallic luster. These include the brilliant, reflective luster of diamond and other gems known as *adamantine luster*. One of the most common varieties of nonmetallic luster is *vitreous luster*, which is best illustrated by common glass. Quartz is a mineral with vitreous luster. Other minerals have luster that can be described as *greasy*, *waxy*, *pearly*, or *dull (earthy)*.

### Specific Gravity

Density provides a simple way of identifying some minerals that look similar simply by lifting the specimen. When the density is expressed as the ratio of the weight of a mineral to the weight of an equal volume of water, it is called the *specific gravity*. It is most useful for minerals composed of heavy elements such as iron, nickel, and lead. The mineral *galena* (Figure 2-13), which contains lead and sulfur, has a specific gravity of 7.57. A sample of galena is much heavier than a specimen of quartz of about the same size because quartz has a specific gravity of 2.65.

**Figure 2-13** The mineral galena has metallic luster. Cubic cleavage is also evident.

### Other Properties

A few other properties may be useful for identifying specific minerals. Examples are magnetism, radioactivity, and taste, smell, or touch. *Magnetite*, an iron-bearing mineral, is the best example of a magnetic mineral. Highly magnetic varieties of magnetite called lodestone were used as the first compasses in the early days of navigation.

## MINERAL GROUPS

Minerals are classified according to chemical composition and structure. The composition of the most common rock-forming minerals is limited by the abundance of elements in the earth's crust. In fact, only eight elements constitute about 98% of the weight of the earth's crust (Table 2-2). Most of the minerals are members of a group characterized by combinations of the two most abundant elements, oxygen and silicon. This group is called the *silicate group* because all its members contain a specific structural combination of silicon and oxygen, even though most silicate minerals also contain other elements. Similarly, the other major mineral groups are composed of specific compositional units, usually anions or ion groups. We will begin our description of these groups with the silicates because of their abundance in rock-forming minerals.

**TABLE 2-2**
**Elemental Abundance in the Earth's Crust**

| *Element* | *Symbol* | *Weight percent* |
|---|---|---|
| Oxygen | 0 | 46.60 |
| Silicon | Si | 27.72 |
| Aluminum | Al | 8.13 |
| Iron | Fe | 5.00 |
| Calcium | Ca | 3.63 |
| Sodium | Na | 2.83 |
| Potassium | K | 2.59 |
| Magnesium | Mg | 2.09 |

SOURCE: From Brian Mason, *Principles of Geochemistry*, © 1958 by John Wiley & Sons, New York.

### *Silicates*

Every silicate mineral contains a basic structural unit called the *silica tetrahedron*. As shown in Figure 2-14, its structure involves a central silicon cation with an oxidation number of +4 surrounded by four oxygen anions, each with an oxidation number of −2. The four oxygen anions form the corners of a four-sided geometric form called a tetrahedron.

In its isolated state, the silica tetrahedron has a charge of −4. To form a mineral with an overall neutral charge, the negative charge must be balanced by adding cations or by forming linked tetrahedra in which the oxygen ions

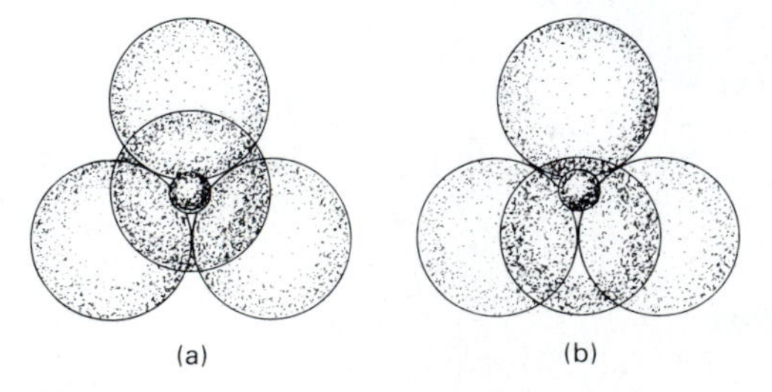


**Figure 2-14** The silica tetrahedron shown from (a) top and (b) side views. A central silicon ion is surrounded by four oxygen ions. (From S. Judson, M. E. Kauffman, and L. D. Leet, *Physical Geology*, 7th ed., copyright © 1987 by Prentice-Hall, Inc., Englewood Cliffs, N.J.)

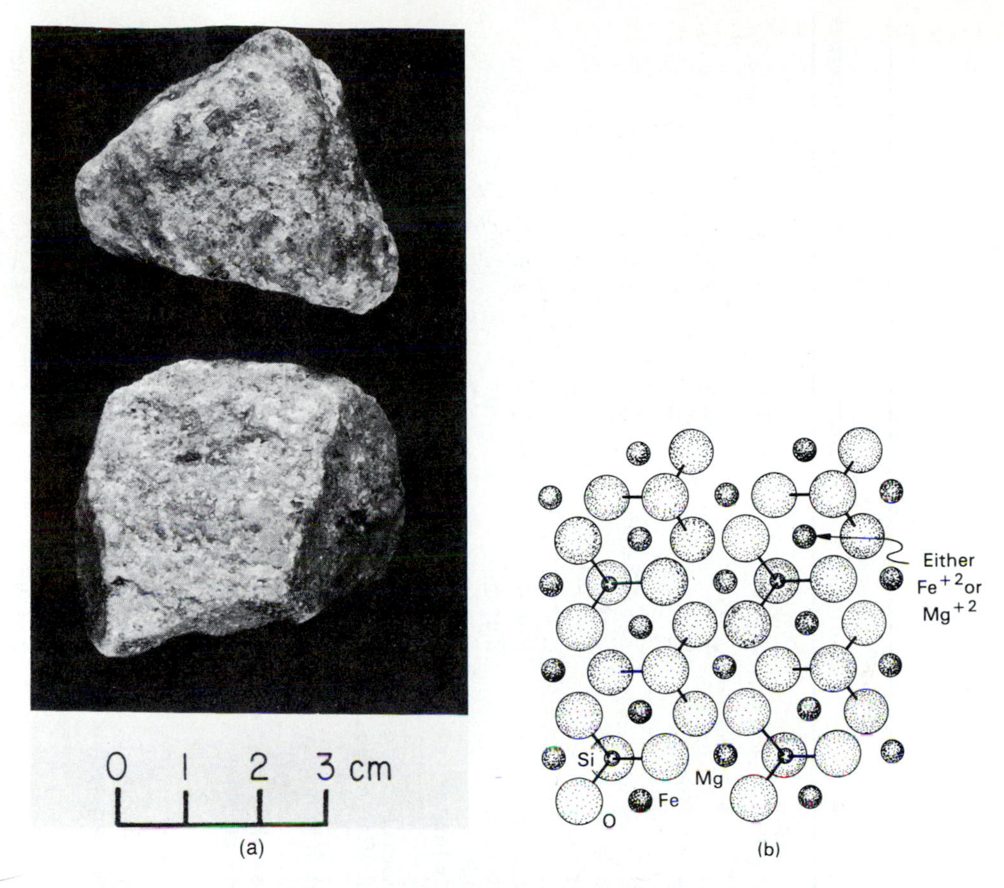


**Figure 2-15** (a) Photo of the mineral olivine. (b) Structure of olivine is characterized by isolated silica tetrahedra and regularly spaced magnesium and/or iron ions. (From E. A. Hay and A. L. McAlester, *Physical Geology: Principles and Perspectives*, 2d ed., copyright © 1984 by Prentice-Hall, Inc., Englewood Cliffs, N.J.)

are shared by adjacent tetrahedra. The vast number of silicate minerals are formed by utilizing both of these mechanisms.

The silicate group is subdivided by the way in which silica tetrahedra interact within the structure. The most basic structural form consists of isolated tetrahedra with no sharing of oxygens. Olivine (Figure 2-15) is an example of a mineral with isolated tetrahedra. Magnesium and iron are added in various proportions to balance the charge, forming the solid-solution series that we previously considered. The physical properties of olivine are listed along with other minerals in Table 2-3.

In other subdivisions of the silicate group, the linking pattern of silica tetrahedra becomes successively more complex. One common type of linkage produces chains of tetrahedra. Two types of chains are formed: the *pyroxene* group consists of silicate minerals with single chains (Figure 2-16), and the *amphibole* group contains double chains of silica tetrahedra (Figure 2-17). *Augite* and *hornblende* are common representatives of the pyroxene and amphibole groups, respectively. Hornblende has a particularly recognizable form of cleavage with two directions intersecting at angles of 56 degrees and 124 degrees.

The *sheet silicates* are composed of thin layers, or sheets, of silica tetrahedra in which three oxygens are shared with adjacent tetrahedra (Figure 2-18). These sheets are described as tetrahedral sheets because the silicon atoms are

**TABLE 2-3**
**Physical Properties of Common Rock-Forming Minerals**

| *Mineral* | *Chemical formula* | *Color* | *Cleavage directions* | *Hardness* | *Specific gravity* | *Other properties* |
|---|---|---|---|---|---|---|
| Silicates | | | | | | |
| Augite (pyroxene) | $Ca(Mg,Fe,Al)(Al,Si_2O_6)$ | Dark green to black | 2 at 90° | 5–6 | 3.2–3.6 | |
| Biotite (mica) | $K(Mg,Fe)_3AlSi_3O_{10}(OH)_2$ | Black | 1 | $2\frac{1}{2}$–3 | 2.8–3.2 | |
| Garnet | $(Ca,Mg,Fe,Mn)_3(Al,Fe,Cr)_2(SiO_4)_3$ | Dark red, brown, green | 0 | $6\frac{1}{2}$–$7\frac{1}{2}$ | 3.5–4.3 | |
| Hornblende (amphibole) | $(Na,Ca)_2(Mg,Fe,Al)_5Si_6(Si,Al)_2O_{22}(OH)_2$ | Dark green to black | 2 at 56° and 124° | 5–6 | 2.9–3.2 | |
| Muscovite (mica) | $KAl_3(AlSi_3O_{10})(OH)_2$ | Colorless to pale green | 1 | 2–$2\frac{1}{2}$ | 2.8–2.9 | |
| Olivine | $(Mg,Fe)_2SiO_4$ | Pale green to black | None | $6\frac{1}{2}$–7 | 3.3–4.4 | |
| Orthoclase (feldspar) | $KAlSi_3O_8$ | White, gray, or pink | 2 at 90° | 6 | 2.6 | |
| Plagioclase (feldspar) | $(Ca,Na)(Al_2Si)AlSi_2O_8$ | White to gray | 2 at 90° | 6 | 2.6–2.7 | Striations |
| Quartz | $SiO_2$ | Colorless to white but often tinted | None | 7 | 2.6 | |
| Oxides | | | | | | |
| Hematite | $Fe_2O_3$ | Reddish brown to black | None | $5\frac{1}{2}$–$6\frac{1}{2}$ | 5.26 | |
| Goethite (limonite) | $FeO \cdot OH$ | Yellowish brown to dark brown | None | 5–$5\frac{1}{2}$ | 4.37 | Limonite is noncrystalline |
| Magnetite | $Fe_3O_4$ | Black | None | 6 | 5.18 | Strongly magnetic |
| Halides and sulfides | | | | | | |
| Halite | $NaCl$ | Colorless or white | 3 at 90° | $2\frac{1}{2}$ | 2.16 | |
| Pyrite | $FeS_2$ | Pale brassy yellow | 3 at 90° | 6–$6\frac{1}{2}$ | 5.02 | Sometimes called "fool's gold" |
| Chalcopyrite | $CuFeS_2$ | Brassy yellow | 2 at 90° | $3\frac{1}{2}$–4 | 4.1–4.3 | Ore of copper |
| Sphalerite | $ZnS$ | Brown to yellow | 6 at 120° | $3\frac{1}{2}$–4 | 3.9 | Ore of zinc |
| Galena | $PbS$ | Lead gray | 3 at 90° | $2\frac{1}{2}$ | 7.54 | Ore of lead |
| Sulfates and carbonates | | | | | | |
| Gypsum | $CaSO_4 \cdot 2H_2O$ | Colorless to white | 1 | 2 | 2.32 | |
| Calcite | $CaCO_3$ | White to colorless | 3 at 75° | 3 | 2.72 | Forms limestone |
| Dolomite | $CaMg(CO_3)_2$ | Pink, white, or gray | 3 at 74° | 3.5–4 | 2.85 | |

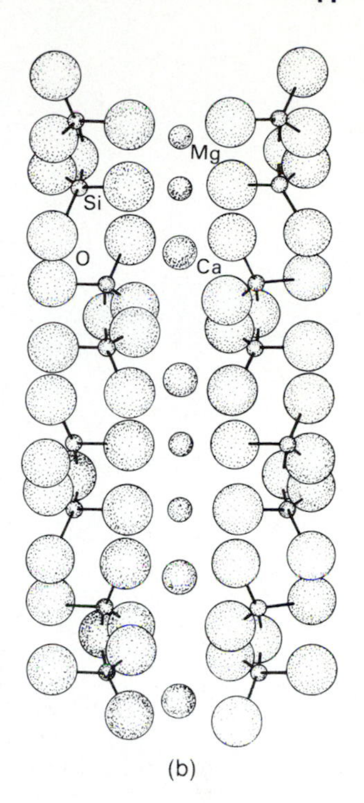


**Figure 2-16** (a) Photo of pyroxene. (b) In pyroxene, single chains of silica tetrahedra are bound together by calcium and magnesium ions. (From E. A. Hay and A. L. McAlester, *Physical Geology: Principles and Perspectives*, 2d ed., copyright © 1984 by Prentice-Hall, Inc., Englewood Cliffs, N.J.)

surrounded by four oxygens. In addition, the sheet silicates contain other structural types of sheets. For example, octahedral sheets (Figure 2-18) contain aluminum, magnesium, or other cations coordinated by six oxygen of hydroxyl atoms. The geometric figure defined by the arrangement of oxygen atoms is an octahedron.

Important minerals within the sheet silicate group include the micas and the clay minerals. Different minerals are formed by different ways of combining sheets and by variations in charge caused by ionic substitutions within the

**Figure 2-17** Hornblende, a member of the amphibole group of silicate minerals. The distinctive cleavage of hornblende consists of two planes intersecting at angles of 56 and 124 degrees.

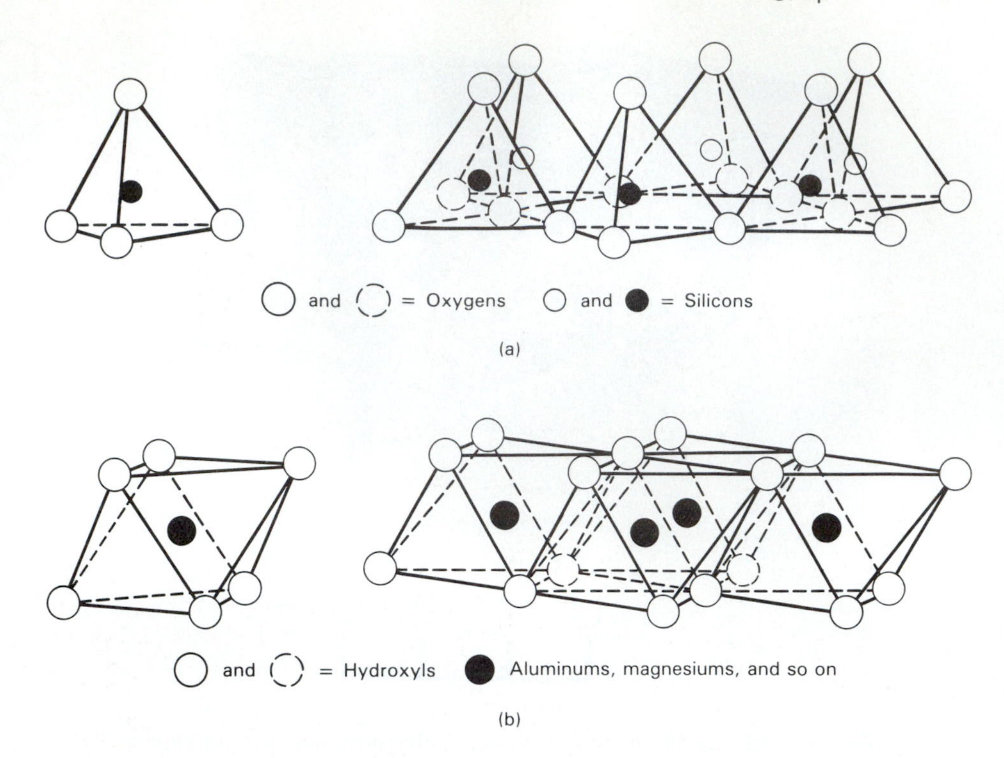


**Figure 2-18** The sheet silicates are composed of two types of sheets: (a) A tetrahedral sheet of linked silica tetrahedra (single tetrahedron shown at left) and (b) an octahedral sheet (single octahedron shown at left) in which cations are surrounded by oxygens in octahedral coordination. (From R. E. Grim, *Clay Mineralogy*, 2d ed., copyright © 1968 by McGraw-Hill, Inc., New York.)

structures. The clay minerals are especially significant to engineering because of the tendency of some clays to absorb water and swell. The effects of swelling soils on foundations can be very destructive.

One of the most basic structural configurations is the combination of one tetrahedral sheet and one octahedral sheet to form the clay mineral *kaolinite* (Figure 2-19). Aluminum ions fill the octahedral sites. Hydroxyls substitute for some of the oxygens to balance the charge of the structural unit. Adjacent double-sheet layers of kaolinite are attracted to each other by very weak forces called *van der Waals* bonds.

When two tetrahedral sheets crystallize and are separated by an octahedral sheet, several minerals can form. The *smectite* group of clay minerals, which includes the mineral *montmorillonite*, crystallizes with this structure (Figure 2-20). Substitutions of cations in the smectite structure commonly include aluminum (+3) for silicon (+4) in the tetrahedral sheets; and magnesium (+2), iron (+2), or other cations, for aluminum (+3) in the octahedral sheets. The result of these substitutions is that there is a net negative charge on the structure that must be balanced to maintain electroneutrality. Charge balance is maintained by the incorporation of cations into positions between the layers (Figure 2-20). These cations, which usually include calcium and sodium in montmorillonite, serve to balance the positive charge deficiency and, in addition, bond adjacent montmorillonite layers together. Because the charge imbalance is relatively small for montmorillonite, the interlayer cations are weakly held and can be exchanged for other cations that may be brought into contact with the clay. *Cation exchange* is a very important clay property. In

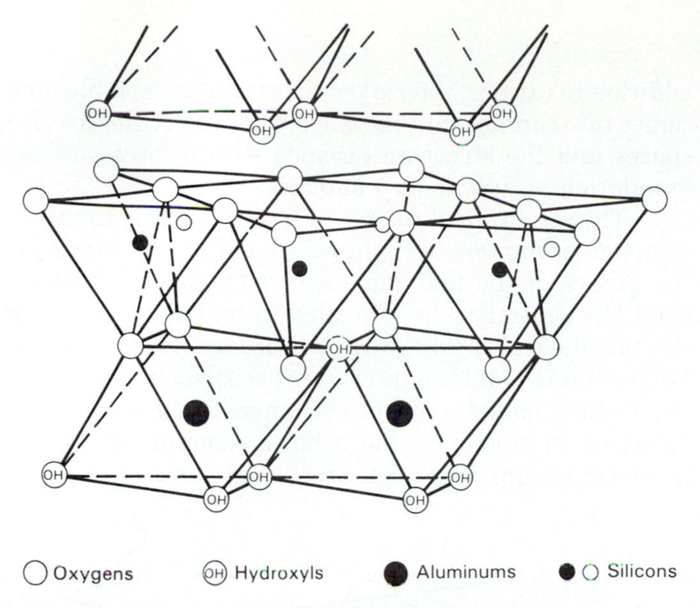


**Figure 2-19** The kaolinite structure forms layers composed of one octahedral sheet and one tetrahedral sheet. Adjacent layers are held together by van der Waals bonds. (From R. E. Grim, *Clay Mineralogy*, 2d ed., copyright © 1968 by McGraw-Hill, Inc., New York.)

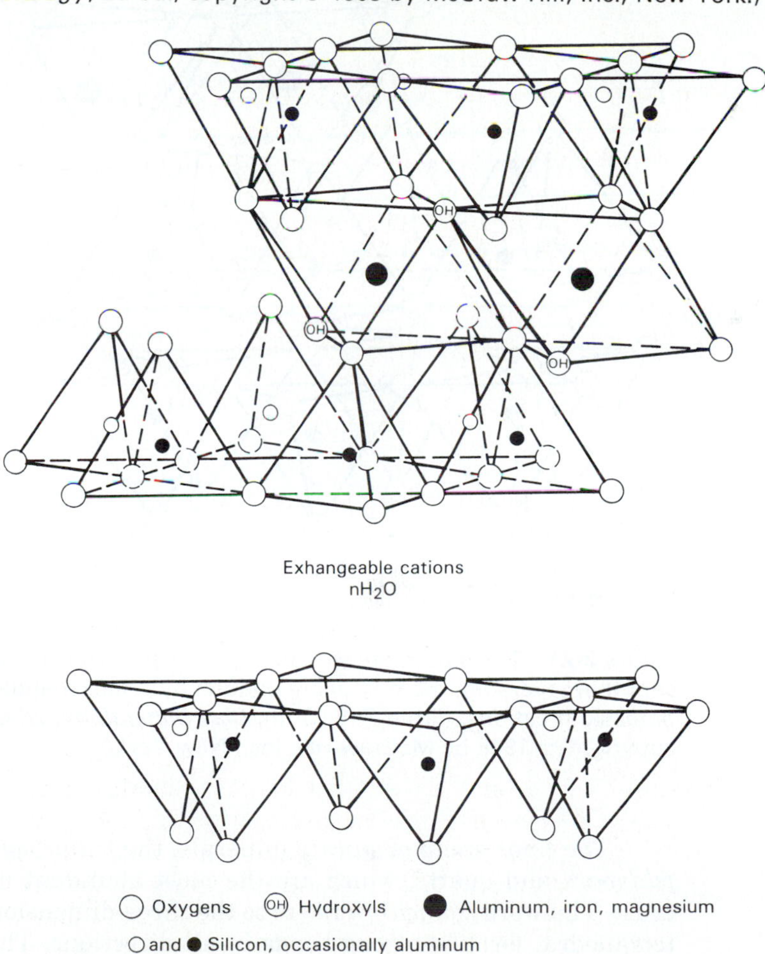


**Figure 2-20** Two tetrahedral sheets and one octahedral sheet form the smectite structure. Layers are separated by exchangeable cations and water molecules. (From R. E. Grim, *Clay Mineralogy*, 2d ed., copyright © 1968 by McGraw-Hill, Inc., New York.)

addition to cations, interlayer sites contain variable amounts of water. If water comes into contact with the clay, water molecules are drawn into the interlayer spaces and the structure expands with a force sufficient to lift or crack the foundation of fully loaded buildings.

The structure of the micas is very similar to the smectites. The excellent cleavage of muscovite (light mica) and biotite (dark mica) is developed along the planes of the individual layers (Figures 2-9 and 2-21). The micas differ from the smectites in that aluminum ions always substitute for silicon in the tetrahedral layers in the micas, resulting in a larger charge imbalance. Potassium ions fill the interlayer positions in micas and, unlike the smectites, are tightly bound and not exchangeable. *Illite* is a clay mineral similar in structure to muscovite, but it has a structure in which the exchange of interlayer potassium does occur because of variations in ion substitutions.

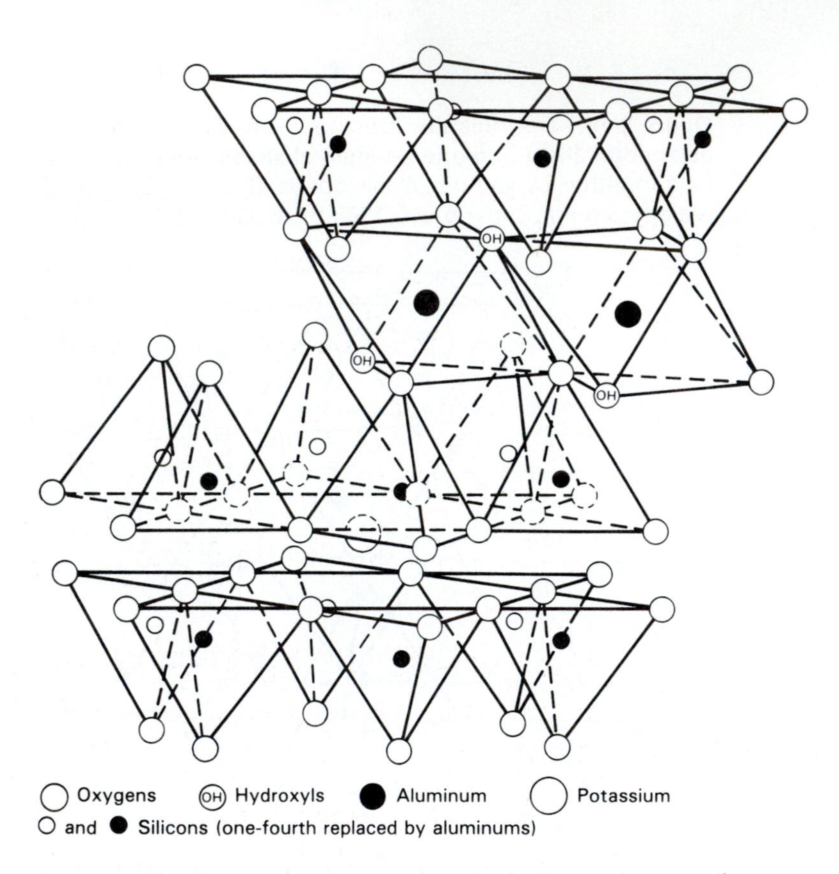


**Figure 2-21** The muscovite structure is similar to the smectites except that layers are more strongly bonded by nonexchangeable potassium cations. (From R. E. Grim, *Clay Mineralogy*, 2d ed., copyright © 1968 by McGraw-Hill, Inc., New York.)

The final group of silicate minerals, the *framework silicates,* contain the *feldspars* and quartz, which are the most abundant minerals in the earth's crust. The word *framework* refers to the three-dimensional linking of the silica tetrahedra, forming strong bonds in all directions. There are two important types of feldspar. *Plagioclase feldspar* includes a number of minerals in a solid-solution series with sodium and calcium as end members. Plagioclase can sometimes be recognized by fine, closely spaced lines, or *striations,* on the cleavage faces of the specimen. Striations are caused by defects in the crystal

**Figure 2-22** Orthoclase, one of the two major types of feldspar, is a framework silicate mineral. Note two directions of cleavage at right angles.

structure that formed as the mineral was crystallizing. *Orthoclase* is the other main type of feldspar. It differs from plagioclase in its lack of striations and in its high potassium content (Figure 2-22).

Quartz (Figure 2-23) is a framework silicate composed entirely of silica tetrahedra in which all four oxygen ions are shared. The development of strong bonds in all directions and the lack of additional elements in the structure results in the absence of cleavage. The color of quartz is extremely variable due to minute amounts of impurities within the structure.

**Figure 2-23** Quartz, one of the most common silicate minerals, occurs as well-developed crystals (lower left), but more commonly in massive form (upper right).

The silicate minerals are used for many purposes in our society. Most gem stones, including emerald, topaz, jade, and turquoise are silicate minerals. In addition, silicate minerals are mined as ores for various metals and used in a variety of industrial processes. The clay minerals are particularly valuable in industry and manufacturing, with uses in the production of brick, tile, plastics, rubber, paint, ceramics, and paper.

### Oxides

Minerals that form by combination of various cations with oxygen are called *oxides*. Among the oxide minerals are important ore minerals of iron, aluminum, chromium, and other metals. The iron oxide minerals are particularly common. Iron and oxygen can form several different minerals depending on the oxidation number of the iron ions. In hematite (Figure 2-24) and *geothite*, iron is present in the oxidized (+3) state, while in magnetite, iron is present in both the oxidized and reduced (+2) states.

**Figure 2-24** Hematite, a common iron oxide mineral.

### Halides and Sulfides

Minerals of the halide and sulfide group contain anions of fluorine, chlorine, bromine, iodine, and sulfur as the framework anion. Minerals composed of the first four elements of the preceding list constitute the halide minerals. An example of a halide mineral is halite, or common table salt (Figure 2-10). The sulfide mineral group constitutes a very important source of the metallic ores of iron, copper, zinc, and lead. *Pyrite* (Figure 2-25) is a common mineral often mistaken for gold because of its metallic luster. For this reason, it is sometimes called fool's gold.

**Figure 2-25** Pyrite, an important iron sulfide mineral, can be recognized by its brassy yellow color, cubic cleavage, and metallic luster. The specimen on the left contains thin parallel lines, or striations, on its exterior surfaces.

### *Sulfates and Carbonates*

Sulfates and carbonates consist of framework radicals similar to the silica tetrahedra in that an anion group is the basis of the structure. The sulfate anion group consists of sulfur and four oxygen ions ($SO_4^{2-}$). An important mineral in this group is *gypsum*, the main ingredient in building materials such as sheet rock.

In carbonate minerals, the basic building block is the carbonate ion ($CO_3^{2-}$). The most important carbonate minerals are *calcite* (Figure 2-11), which combines calcium with the carbonate ion, and *dolomite*, which contains calcium and magnesium in its structure. Calcite is often found in hot-spring deposits and caves. Large amounts of *travertine* (a variety of calcite) have been deposited at Mammoth Hot Springs in Yellowstone National Park (Figure 4-13) in a beautiful series of terraces and pools.

Among the many uses of calcite and dolomite are building stones and the production of lime and Portland cement. Lime is valuable for its acid neutralizing properties with applications in agriculture and industry, and Portland cement is an essential ingredient of concrete.

### *Native Elements*

Certain minerals consist of a single element. Examples are gold, silver, copper, sulfur, and carbon. Elemental gold, silver, and copper, are relatively rare in nature; they more commonly occur as components of other minerals. Sulfur is utilized in the production of sulfuric acid and other chemicals. Graphite and diamond (Figure 2-4) are two minerals composed of carbon. The difference in structure between graphite and diamond is largely due to pressure at the time of formation. Diamond forms in high-pressure geologic environments deep within the earth and therefore develops a dense structure. Its great hardness is attributed to a three-dimensional network of strong covalent bonds. Graphite is a soft, greasy mineral with perfect cleavage in one direction.

## SUMMARY AND CONCLUSIONS

The rocks of the earth's crust are aggregates of minerals, which are inorganic crystalline solids with a chemical composition that is fixed or varies within fixed limits. Within each mineral, atoms and ions are arranged in a specific structural pattern that determines the properties we can observe on the exterior surfaces of each sample.

When a mineral slowly crystallizes in unobstructed surroundings, crystals form. The smooth, planar crystal faces reflect the internal structure of the mineral and, for a specific mineral, maintain uniform angles between corresponding faces on crystals of any size. The conditions necessary for the formation of crystals are rare, so, instead, minerals usually occur as irregular masses in rocks because of competition for space with other minerals during crystallization.

Identification of most common rock-forming minerals is possible by using observation and simple tests to determine the physical properties of the specimen. The most useful physical properties include color, streak, luster, cleavage, hardness, and specific gravity.

Composition is the main criterion for dividing minerals into classes of chemically similar minerals. The basic structural unit used in classification is the major anion or anion group in the mineral. Silicates, constituting the most abundant group, are defined by the silica tetrahedron, a structural unit of four oxygen anions and one silicon cation. Subgroups of silicates are formed by different ways of arranging and linking silica tetrahedra in the structure. Oxides, halides and sulfides, sulfates and carbonates, and native elements are the classes of nonsilicate minerals.

## PROBLEMS

1. List the characteristics of each major atomic particle.
2. How do ionic bonds and covalent bonds differ?
3. Why are only crystalline solids included in the definition of a mineral?
4. How does the internal structure of a mineral relate to its external appearance?
5. What is mineral cleavage? How could cleavage affect the engineering properties of a rock composed of minerals with strong cleavage?
6. For which common minerals is hardness a very useful diagnostic physical property?
7. Why are the silicate minerals so important?
8. Briefly describe the structure of clay minerals. What is it about these minerals that makes their presence in soil so important to engineering?
9. What is cation exchange?
10. If you allow a bucket of seawater to totally evaporate, various minerals will precipitate. Into which mineral groups do you think most of these minerals will fit?

## REFERENCES AND SUGGESTIONS FOR FURTHER READING

GRIM, R. E. 1968. *Clay Mineralogy*, 2d ed. New York: McGraw-Hill.

HAY, A. E., and A. L. MCALESTER. 1984. *Physical Geology: Principles and Perspectives*, 2d ed. Englewood Cliffs, N.J.: Prentice-Hall, Inc.

JUDSON, S., M. E. KAUFFMAN, and L. D. LEET. 1987. *Physical Geology*, 7th ed. Englewood Cliffs, N.J.: Prentice-Hall, Inc.

MASON, B. 1958. *Principles of Geochemistry*. New York: John Wiley.

TENISSEN, A. C. 1983. *Nature of Earth Minerals*. Englewood Cliffs, N.J.: Prentice-Hall, Inc.

# 3

# Igneous Rocks and Processes

Let us assume that you are in charge of constructing a highway tunnel through a mountain. Obviously, you will need to know as much as possible about the properties and characteristics of the rocks along the route of the tunnel. This information will influence the design and construction of the tunnel as well as the safety of the project. The tunneling conditions would be much different in a sequence of volcanic lava flows than they would be in alternating beds of sandstone and shale. With an understanding of how rocks form, it is possible to make general predictions about their spatial distribution and their physical properties.

Typically, preliminary information available for such a tunneling project consists of rock samples taken from holes drilled at intervals along the proposed path of the tunnel. Our task is to predict the rock types and rock conditions between the test hole locations. The only way to do this is to use geological reasoning and interpretation based on sound knowledge of how different rock types are formed. Thus although it may seem academic to study the processes of rock formation, an understanding of these processes must be applied in every major rock-engineering project.

## PROPERTIES OF IGNEOUS ROCKS

When a rock sample is examined, much can be inferred about its origin by observing its color, texture, and mineral content. The term *texture* refers to the size, shape, and arrangement of the mineral grains in a rock. In this chapter we are concerned with igneous rocks, rocks that were formed by cooling and solidification from a molten liquid called *magma*. These rocks are known as *intrusive* igneous rocks if the cooling takes place below land surface. If the molten liquid flows to the surface, it is called *lava;* here cooling can also take place to form *extrusive* igneous rocks. In subsequent chapters we will examine the other two fundamental rock types, sedimentary and metamorphic.

### Texture

The texture of igneous rocks (Table 3-1) is variable but usually consists of an interlocking network of mineral crystals that grew from the magma or lava during cooling. If the crystals are too small for observation by the naked eye, the texture is called *aphanitic* (Figure 3-1). These extremely fine-grained rocks are associated with volcanic processes because of the relatively rapid cooling

**TABLE 3-1**

**The Texture of the Igneous Rocks**

| *Rock type* | *Description and interpretation* |
|---|---|
| Extrusive rocks | |
| Glassy | Noncrystalline, very fine grained; very rapid cooling |
| Aphanitic | Uniformly fine grained; rapid cooling |
| Porphyritic | Large phenocrysts within fine-grained groundmass; two-stage cooling process |
| Vesicular | Numerous small holes on surface; gas escape during cooling |
| Intrusive rocks | |
| Phaneritic | Uniformly coarse grained; slow, gradual cooling in subsurface |

**Figure 3-1** Rhyolite, an igneous rock with aphanitic texture. Individual crystals are not visible with the unaided eye.

that occurs when a magma reaches the earth's surface during a volcanic eruption and becomes lava. Crystal size is one of the major differences between extrusive, rocks and intrusive rocks. Growth of large crystals in a rock takes long periods of time and conditions of gradual, undisturbed cooling. The intrusive rocks that formed under these conditions have a texture called *phaneritic* (Figure 3-2). Phaneritic rocks are composed of an interlocking network of large crystals that are easily visible.

When a volcano erupts, the outpouring lava is quickly cooled by contact with the atmosphere, which is much cooler than the subsurface *magma chamber* that contained the magma prior to eruption. At times, this cooling can be so rapid that crystallization is prevented. The silica tetrahedra are frozen in place before they can attain an orderly arrangement necessary for mineral growth. Cooling at these rates leads to the formation of *glassy* texture, the glass in this case being a natural, noncrystalline material. *Obsidian* (Figure 2-12) is a good example of a rock composed primarily of volcanic glass.

Several other types of textures are associated with volcanic rocks. Rocks that are composed of crystals of two sizes have a *porphyritic* texture (Figure 3-3). The interpretation of this texture is that crystals began to separate from the magma in the magma chamber or within the conduit leading from the magma chamber to the surface. These earlier-formed crystals grew to a substantial size by the time they reached the surface, enclosed in the still-liquid magma. Once exposed to the air, the remainder of the magma then crystallized rapidly as much smaller crystals. The resulting rock therefore consists of large *phenocrysts*, the early crystals, surrounded by a fine-grained crystalline *groundmass*.

**Figure 3-2** Phaneritic texture is characteristic of coarse-grained rocks. This specimen of granite contains large grains of orthoclase, quartz, and biotite.

In addition to liquid, magmas contain significant amounts of dissolved gases that are kept in solution by the great pressure exerted on the magma chamber by the overlying rocks. Water vapor is the most abundant gas, although carbon dioxide ($CO_2$), sulfur dioxide ($SO_2$), and other gases are present. When a magma reaches the earth's surface, the pressure decreases and the dissolved gases subsequently migrate to the surface of the lava and escape to the atmosphere. The upper surfaces of solidified lava flows are given an appearance much like Swiss cheese by the small holes, or *vesicles*, through

**Figure 3-3** Porphyritic texture is composed of large phenocrysts embedded in a finer-grained groundmass.

**Figure 3-4** A sample of vesicular basalt. As this lava cooled, gases escaped to the air through the holes (vesicles) in the rock.

which the gases escape (Figure 3-4). These rocks are said to have a *vesicular* texture. The types of igneous rock texture are summarized in Table 3-1.

### *Color and Composition*

The chemical and mineralogical composition of igneous rocks is a reflection of the composition of the magma from which the rocks crystallized. Magmas are variable in composition, most importantly in the amount of silica ($SiO_2$) that they contain. The silica content ranges from less that 24% to more than 66%. Rocks that are rich in silica are called *silicic*, or *felsic*, rocks; and those that are low in silica are called *mafic* rocks. Fortunately, color provides a valuable clue for identification of igneous rocks because the silicic rocks are mainly composed of light-colored minerals like quartz and feldspar; whereas the mafic rocks are dark colored because of the abundance of *ferromagnesian* minerals. The dark-colored ferromagnesian minerals, which, as their name implies, are rich in iron and magnesium, include olivine, pyroxene, and hornblende. The major igneous rock types fall into categories of high, intermediate, and low silica content. Figure 3-5 shows the principal minerals contained in each of these rock types.

Texture and silica content provide a basis for classifying the igneous rocks. Aphanitic and phaneritic rocks of the same silica and mineral content have different rock names. For example, *granite* and *rhyolite* are both light-colored felsic rocks containing mostly quartz and orthoclase. They receive different names, however, because granite is a phaneritic intrusive rock and rhyolite is an aphanitic extrusive rock. In the intermediate range of silica content (52–66%), *diorite* and *andesite* are the coarse-grained and fine-grained equivalents, respectively. The common intrusive rock *granodiorite* falls between granite and diorite in silica content. As Figure 3-5 indicates, granodiorite contains more plagioclase feldspar than orthoclase.

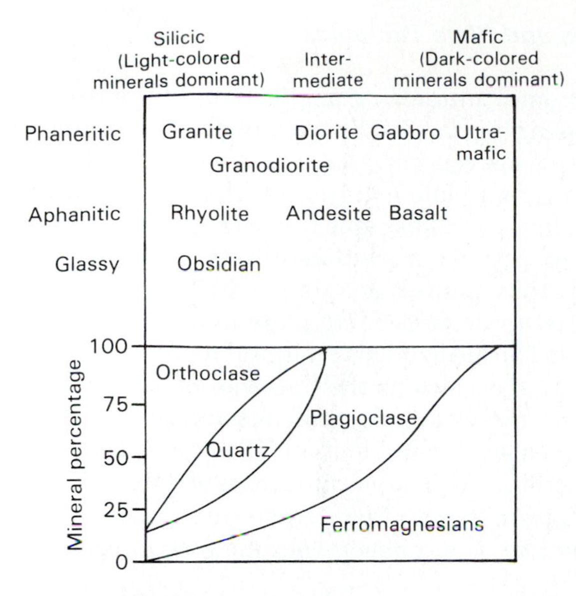


**Figure 3-5** Classification of the igneous rocks. (Modified from L. Pirsson and A. Knopf, 1926, *Rocks and Rock Minerals*, John Wiley & Sons, Inc. New York.)

On the mafic side of the chart, *gabbro* and *basalt* (Figure 3-4) are important rock types. Basalt is the most common type of volcanic rock present on the earth's surface. The equivalent intrusive rock is gabbro. Rocks that are even lower in silica than gabbro have been discovered in the crust and are inferred to be much more common in the mantle. These rocks, which are largely composed of ferromagnesian minerals, are known as *ultramafic* rocks.

The most common rock with a glassy texture is obsidian (Figure 2-12). Despite its high silica content, obsidian is usually black in color. This dark-color exception to the usual light color of silica-rich rocks is due to the presence of impurities.

Figure 3-5 is a handy guide to the igneous rocks. By observing the color, it is possible to estimate the silica content of an igneous rock and thereby to predict its major minerals. The texture of the rock indicates whether it is intrusive or extrusive. Together, the color and texture can be used to assign the proper rock name.

## VOLCANISM

Volcanism is the most obvious igneous process. Major eruptions generate worldwide interest because of their spectacular power and beauty and, as Mount Pinatubo in the Philippines and Mount St. Helens have recently proven so clearly, their danger and destructiveness. In addition to the areas of the earth that have been affected by volcanic activity in historic times, vast regions of the earth, including all the ocean basins, are underlain by volcanic rocks that were erupted earlier in geologic time.

### *Volcanoes and Plate Tectonics*

One of the most interesting applications of the theory of plate tectonics is its use in explaining the distribution of active volcanoes on the earth's surface. Most of these are concentrated in narrow bands along the edges of lithospheric plates. Divergent plate margins are characterized by voluminous outpourings of basaltic lavas at midoceanic ridges (Figure 3-6). These lavas form the foundation of new oceanic crust that fills the void left as the plates move away from each other. Convergent plate margins (Figure 3-6) also generate volcanic activity. Here, the descending plate in a subduction zone is partially melted. The magmas formed at depth migrate to the surface to form chains of volcanic mountain ranges such as the Cascades of the northwestern United States or to form *island arcs* composed of volcanic accumulations on the sea floor that rise above sea level. The Aleutian Islands of Alaska form an island arc adjacent to a subduction zone. The volcanic rocks of convergent plate margins are usually intermediate in composition. Andesite, named for the Andes Mountains of South America, is a common volcanic rock at convergent plate margins.

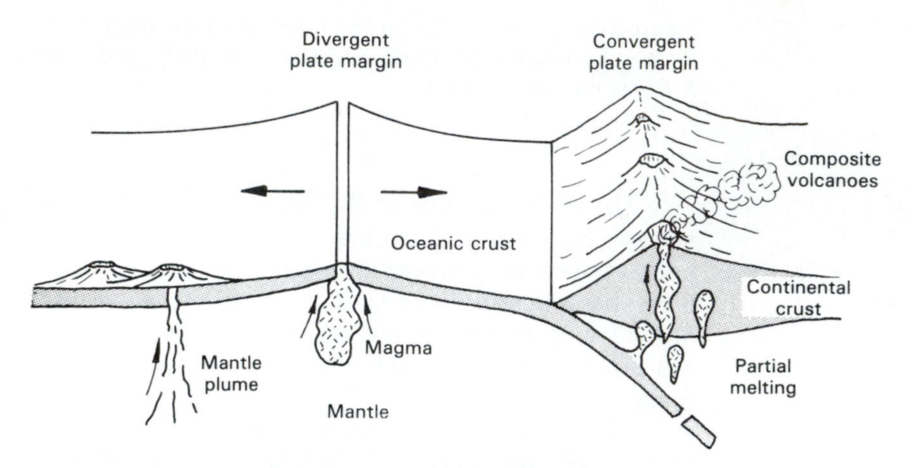


**Figure 3-6** Volcanism is common at both divergent and convergent plate margins. Fissure eruptions produce new oceanic crust at divergent margins; composite volcanoes of andesitic composition are common at convergent margins, where partial melting of the crust occurs.

Volcanism is not restricted to plate margins. A number of volcanic areas occur at some distance from plate margins. The volcanic activity is attributed to localized zones of high heat flow extending downward, deep into the mantle. These *mantle plumes* are assumed to remain in fixed positions as the plates above move along their current paths. As long as the plate movement is in a straight line, a linear chain of volcanoes is produced. The Hawaiian Ridge–Emperor Seamount chain records the passage of the Pacific plate over the Hawaiian hot spot (Figures 3-7 and 3-8). The bend in the ridge indicates a change in direction of plate movement from northward, the orientation of the Emperor Seamount chain, to northwestward, the bearing of the Hawaiian Ridge. Seamounts in the ridge and seamount chain are extinct volcanoes that have subsided below sea level. The Hawaiian Islands themselves are located near and over the hot spot. Individual islands increase in age from Hawaii, which lies over the plume and is the location of the active volcanoes Mauna Loa and Kilauea, to Kauai at the northwest end of the chain. The size of these

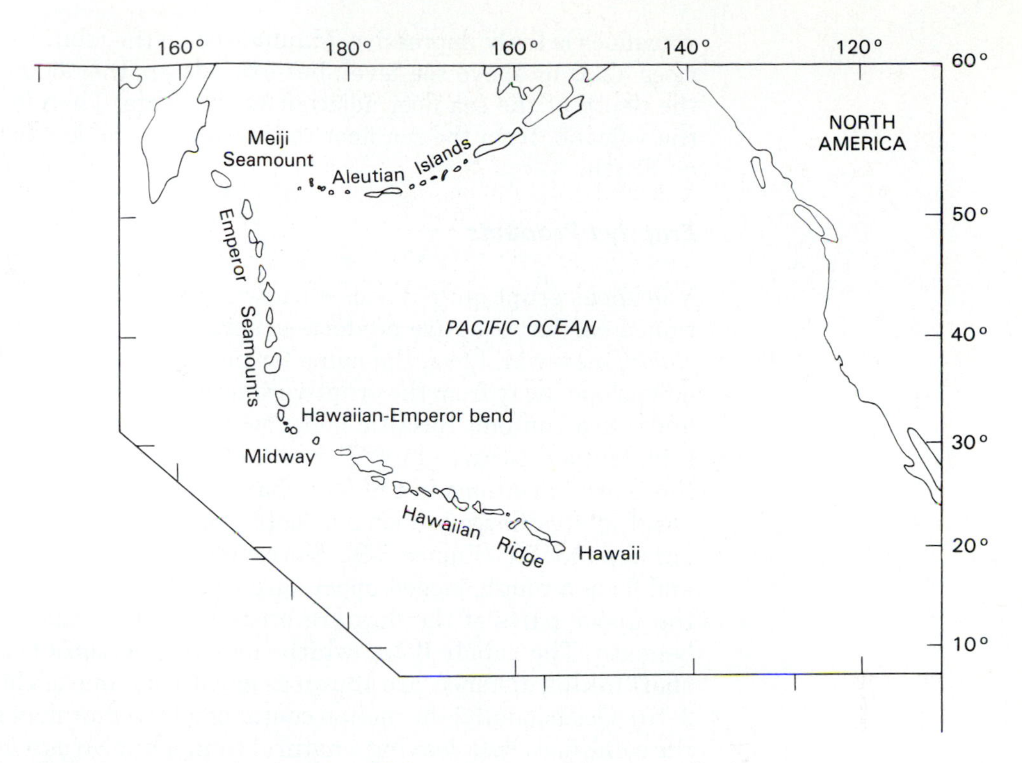


**Figure 3-7** Map of Hawaiian Ridge–Emperor Seamount Chain.

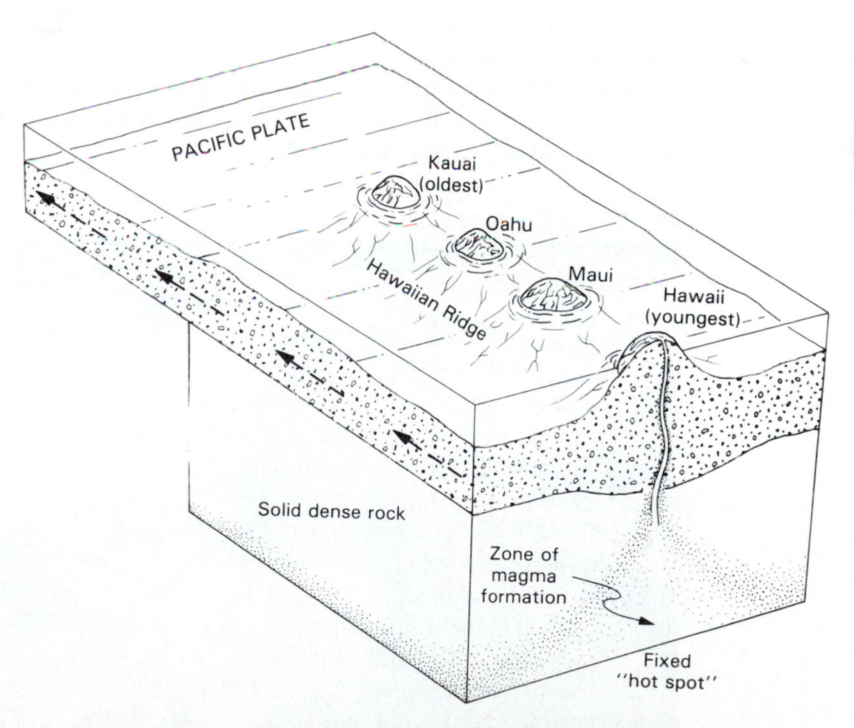


**Figure 3-8** Diagram of Pacific Plate moving over hot spot to form Hawaiian islands. (From R. I. Tilling, et al., 1987, *Eruptions of Hawaiian Volcanoes: Past, Present and Future*, U.S. Geological Survey; based on a drawing by the late Maurice Krafft.)

volcanoes is truly impressive. Mauna Kea on the island of Hawaii, for example, rises 4200 m above sea level, but extends another 6000 m below sea level to the depth of the sea floor adjacent to the ridge. Therefore, the total height of the volcano, from the sea floor to the summit, makes it the largest mountain on Earth.

### *Eruptive Products*

Volcanoes erupt quite a wide variety of materials. Aside from the gases mentioned earlier, eruptive products generally can be divided into lava and *pyroclastic* material. Lava, the name for magma after it reaches the surface, flows downslope away from the eruptive vent at a rate dependent on its temperature and silica content. Basaltic lavas, which are low in silica and are erupted at high temperatures (~1100°C) are fluid and may flow great distances. Even so, there are variations in the flow characteristics of basaltic lavas. Highly fluid basaltic lava flows develop a smooth, ropy surface that is given the Hawaiian name *pahoehoe* (Figure 3-9). More viscous lava flows move at a slower rate and form a rough, jagged upper surface as blocks of partially solidified lava in the upper parts of the flow are broken apart by the slowly advancing mass beneath. The rubbly flows, which can cut a person's boots to shreds within a short hiking distance, are known as *aa*, also a name of Hawaiian origin (Figure 3-10). Occasionally, the molten center of a lava flow may flow out from beneath the solidified crust, leaving a natural tunnel known as a *lava tube* (Figure 3-11).

In addition to temperature and the amount of silica, the volatile content of a magma exerts an important influence on lava characteristics. Siliceous magmas are generally higher in volatile content than mafic magmas, but the volatiles tend to boil off as the magma approaches the surface. The resulting lavas become more viscous after the degassing of volatiles. Rhyolite, one of the most silicic types of lava, is much more viscous than basaltic lava. Rhyolite lava can often be found in the form of plugs or domes that remain close to the vent area of the volcano because the lava is too viscous to flow away (Figure 3-12). The rhyolite flows that do occur move as thick, sluggish masses of lava with very steep fronts (Figure 3-13).

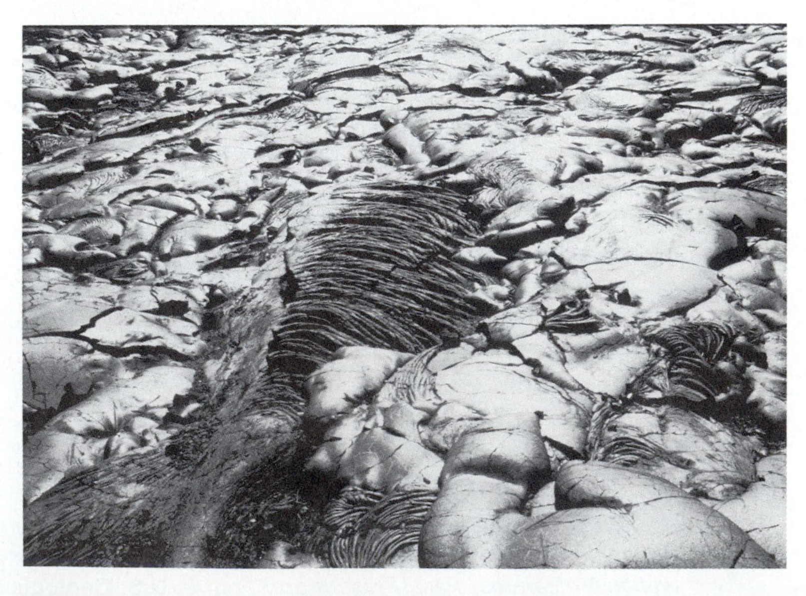

**Figure 3-9** Recently erupted pahoehoe lava; Hawaii.

**Figure 3-10** Aa lava; Hawaii.

The separation of gas from magma as it rises in a volcano often occurs with explosive force. Particles of all sizes are blasted into the air and carried away from the vent. Collectively, all materials of this type are called *pyroclastics*, or *tephra*. The individual particles may consist of rock, minerals, or glass. The smaller particles, called *volcanic ash*, may be carried upward thousands of meters into the atmosphere and then transported around the earth several

**Figure 3-11** Roof collapse of a lava tube, Idaho. The tube forms by cooling of the lava-flow surface and outflow of the molten interior.

**Figure 3-12** A lava dome that formed in the crater of Mount St. Helens after the May 18, 1980, eruption. (R. Tilling; photo courtesy of U.S. Geological Survey.)

times as clouds of dust. In sufficient quantities, volcanic dust can even affect short-term climatic conditions by reflecting solar radiation. The abnormally cool summer of 1992 in North America was attributed by some meteorologists to the eruption of Mount Pinatubo in 1991 in the Philippines.

The coarsest particles, which range in size from about 6 cm to many meters in diameter, fall close to the volcano or upon the mountain itself, where they then may bound or roll downward to the base of the slope (Figure 3-14). Rocks formed from volcanic ash are given the name *tuff*, whereas *breccias* are rocks formed from angular coarse-grained pyroclastic material.

**Figure 3-13** Margin of rhyolite flow; Iceland.

**Figure 3-14** Two large pyroclastic fragments on the flank of a volcano in the Canary Islands. (University of Colorado.)

One of the most destructive volcanic phenomena is a type of pyroclastic flow called *nuée ardente*, or fiery cloud (Figure 3-15). In these events, pyroclastic material, buoyed by gases and dust at an extremely high temperature, flows down the sides of a volcano at great speeds. In 1902, a nuée ardente from the volcano Mount Pelée on the Caribbean island of Martinique killed 28,000 people in the town of St. Pierre (Figure 3-16). Flows of pyroclastic material form distinctive rocks called *welded tuffs*. Particles of rock and volcanic glass contained in these flows are still very hot when they are deposited on the ground surface. Thus, as the material cools, the soft, high-temperature particles become "welded" together to form a hard, dense rock unlike other pyroclastic material.

**Figure 3-15** A nuée ardente flowing down the flank of a volcano. (University of Colorado.)

**Figure 3-16** The remains of St. Pierre, Martinique, West Indies, after the nuée ardente of May 8, 1902. (Howell Williams.)

Another type of flow produced by volcanic eruptions is called a *lahar*. Lahars, which are similar to mudflows, are generated when pyroclastic material becomes saturated with water from melting snow, glacial ice, or rainfall. The resulting mass of debris flows down the mountainside in existing stream valleys, burying roads, bridges, and buildings in its path. The deposition of debris from lahars in stream channels leads to increased flooding because of the resulting reduction in flow capacity of the river channels.

### *Types of Eruptions and Volcanic Landforms*

Volcanoes erupt in a variety of styles. The type of eruption is governed by the composition of the magma and the characteristics of the vent. Characteristic landforms are formed by combinations of these two factors, as well as by the number of eruptions from a single vent and relative proportions of lava and pyroclastics (Figure 3-17). The basic forms of surface ruptures are *central vents* and *fissures*. A central vent constitutes a point source for eruptive products, and the resulting landforms are symmetrical accumulations of material. If the eruption produces lava of basaltic composition, the fluid lavas will spread laterally from the vent in all directions, producing a symmetrical cone with gently sloping sides (Figure 3-18). The slope angles of these *lava cones* and *shield volcanoes* are low because the basaltic lava flows can travel great distances because of their low viscosity. Smaller counterparts of lava cones are called *spatter cones*. Basaltic lava flows frequently exhibit a distinctive joint pattern called *columnar jointing*. During cooling of the lava, contraction initiates a series of cracks that form a polygonal pattern on the flow surface. Viewed from the side of the lava flow, the jointed rock resembles a series of columns (Figure 3-19). Lava of a more silica-rich composition than basalt is more likely to form domes and plugs in a vent area (Figure 3-12) than a complete cone.

*Composite volcanoes* are formed when a volcano erupts both lava and abundant pyroclastic material. The resulting cone has a layered internal structure (Figure 3-20) consisting of alternating layers of lava and pyroclastics. The lava of composite volcanoes is more viscous than that of shield volcanoes and falls into the intermediate range of composition. Andesite is a common rock type associated with composite volcanoes. Because of their high viscosity, andesitic

| | Eruptional products | Number of eruptions | Form of rupture | |
|---|---|---|---|---|
| | | | Central vent | Fissure |
| CONSTRUCTIONAL | Basaltic lava | 1 | Lava cone | Lava fissure<br>Lava cone row |
| | | >1 | Shield volcano | Basalt plateau |
| | Andesitic-rhyolitic lava | 1 | Domes, plugs | |
| | Lava and tephra | 1 | Spatter cone | Spatter cone row |
| | | >1 | Composite volcano | Stratified ridge |
| | Tephra | 1 | Cinder cone (Tephra ring) | Cinder cone row |
| DEPRESSIONAL | Location | Size | Type | |
| | Cones | Small | Crater | |
| | | Small | Pit crater | |
| | | Large | Caldera | |
| | Lowlands | Small-Large | Maar | |
| Erosional | | Volcanic necks and diatremes | | |

**Figure 3-17** Landforms of volcanic vent areas.

lavas do not flow very far, and the side slopes of composite volcanoes are quite steep. Volcanoes in the Cascade Range, including Mount St. Helens, are composite volcanoes (Figure 3-21). The tephra associated with very silicic eruptions, called *pumice,* is very distinctive in appearance. It is nearly white in color and highly vesicular. Vesicles lower the density of some pumice fragments to the point where they will float.

The origin of andesitic lava is somewhat different than basaltic lava. Composite volcanoes occur above descending plates at convergent plate bound-

**Figure 3-18** The gentle slopes and symmetrical profile of this small volcano in Iceland are typical of shield volcanoes.

**Figure 3-19** Columnar jointing at Devil's Postpile, California. (University of Colorado.)

aries (Figure 3-6). As oceanic crust is forced downward into the mantle, where the temperature is much higher, *partial melting* may occur. In the process of partial melting, silica-rich minerals in the rock melt before other minerals, so the magma produced is more silicic in composition than the host rock. By partial melting, basaltic oceanic crust may produce magma of andesitic composition. The composition of the magma may also be influenced by partial melting of sediments on the subducted slab or of other crustal rocks.

In some volcanic eruptions, the gas content is so high that the eruption is extremely explosive. This situation produces mainly pyroclastic eruptive products. The larger clasts that fall around the eruptive vent form a *cinder cone*, or *tephra ring* (Figure 3-22). The finer particles may be carried by winds for thousands of kilometers.

In contrast to eruptions from central vents, the surface rupture may be a linear crack, or *fissure*. Fissure eruptions are common along the midoceanic

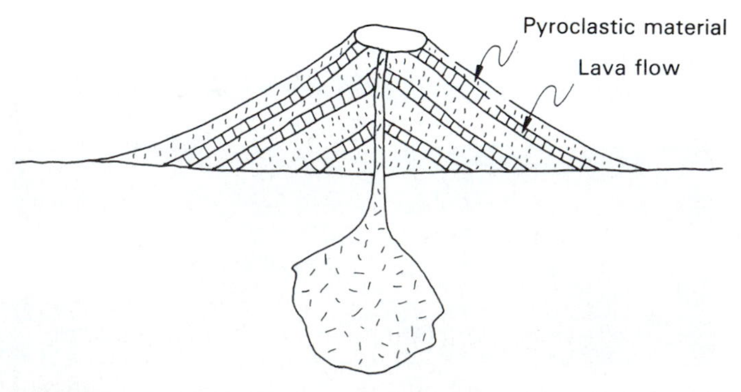


**Figure 3-20** Alternating strata of lava and pyroclastics in a composite volcano.

**Figure 3-21** Mount Hood, a composite volcano in the Cascade Range. (Photo courtesy of Glenn Oliver.)

ridges, where lithospheric plates are being pulled apart. Basaltic lava formed by partial melting of the asthenosphere moves upward to the surface to form new crustal material as the plates move away from each other. The entire oceanic crust is generated in this fashion. Iceland is a portion of a midoceanic ridge that projects above sea level (Figure 7-45). Here, fissure eruptions sometimes produce *lava cone rows* (Figure 3-23). Voluminous fissure eruptions of basaltic composition also occur away from plate boundaries. The highly fluid lava spreads out from fissures over vast areas. The Columbia River and Snake River *basalt plateaus* in the northwestern United States (Figure 3-24) are underlain by immense volumes of basalt that originated from fissure eruptions. Beneath these plateaus, basalt flows are stacked on top of each other to form a basalt sequence thousands of meters thick (Figure 3-25). A generalized cross section of a Columbia River basalt flow (Figure 3-26) shows that columnar jointing is best developed near the top and bottom of the flow. The pillow-palagonite complex at the base of the flow is formed by movement of the flow into a body of water. Pillows are formed underwater as blobs of molten basalt

**Figure 3-22** Hverfall, a large tephra ring in northern Iceland.

**Figure 3-23** The line of small cones marks the site of a fissure eruption that produced the surficial lava flows in the valley.

are squirted out through cracks in the flow front, and they then rapidly cool and settle to the bottom. The lava is cooled more rapidly in water than in air, and much of the lava and pyroclastic material chills to a basaltic glass, which is later altered to form *palagonite*.

The volcanic landforms that we have described thusfar are constructional; that is, they are associated with accumulations of volcanic debris. Topographic depressions of volcanic origin also occur both on and off of volcanic cones. The most common depression is a *crater*, which is simply the vent area at the top

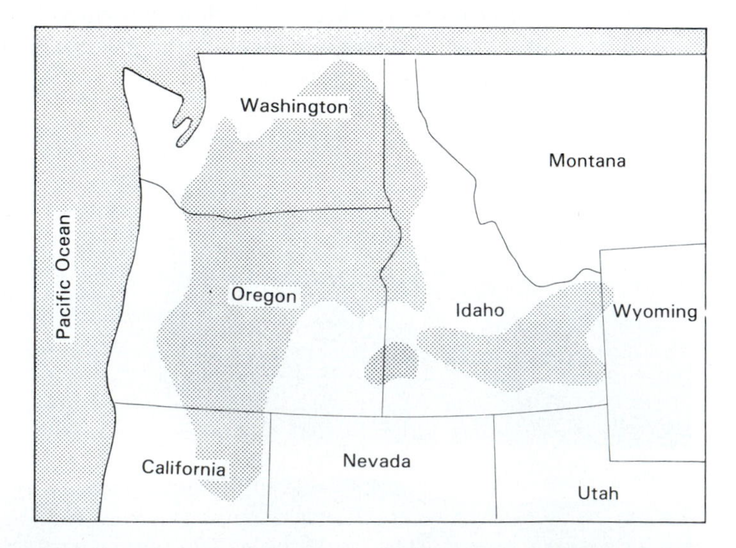


**Figure 3-24** Map showing the extent of the Columbia River and Snake River flood basalts. (From S. Judson, M. E. Kauffman, and L. D. Leet, *Physical Geology*, 7th ed., copyright © 1987 by Prentice-Hall, Inc., Englewood Cliffs, N.J.)

**Figure 3-25** The Columbia River basalts in eastern Washington consist of a sequence of nearly horizontal lava flows, stacked one upon another.

of a volcanic cone. If the vent area has been subjected to collapse or subsidence following eruption, the name *pit crater* is used. Subsidence, which is caused by removal of some of the underlying magma by eruption, is indicated by a circular pattern of fractures with near-vertical walls that formed as the central region of the crater sank downward toward the magma chamber (Figure 3-27). In some cases, a larger part of the roof of the magma chamber collapses to form a large depression that sometimes resembles a pit crater. This type of depression, called a *caldera*, is typified by Kilauea caldera in Hawaii and Crater Lake in Oregon. To be classified as a caldera, the depression must be equal to

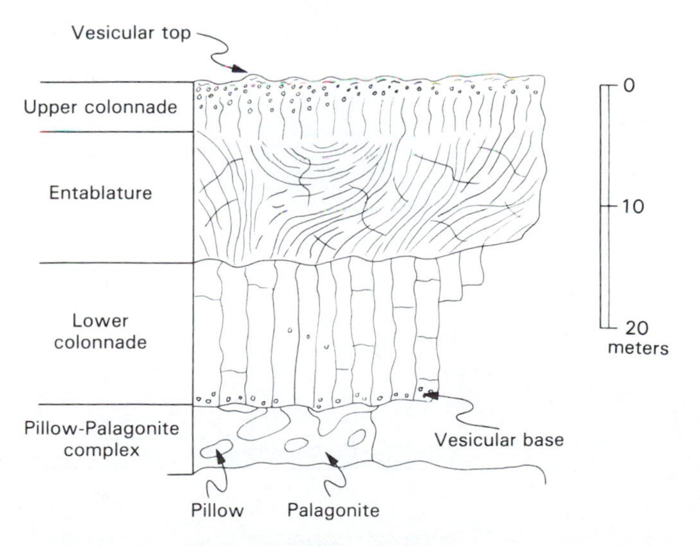


**Figure 3-26** Generalized cross section of basalt flow. Columnar jointing present in upper and lower colonnades. Irregular jointing in entablature. Pillow-palagonite complex is glassy zone formed by flow of lava into body of water. (Modified from V. R. Baker, 1987, Dry falls of the channeled scabland, Washington. *Geological Society of America Centennial Field Guide—Cordilleran Section*, p. 369.)

**Figure 3-27** Oblique aerial photo of Kilauea caldera, looking north. Halemaumau pit crater in left-center. (Photo courtesy of U.S. Geological Survey.)

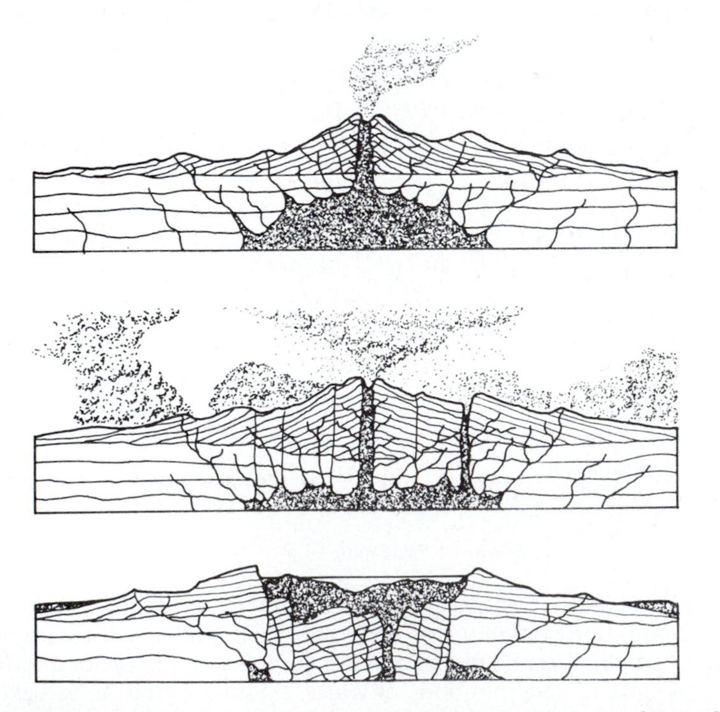

**Figure 3-28** Hypothetical sequence of events in the formation of a caldera. (After Howell Williams, 1941, Calderas and their origin, *Bull. Univ. Calif. Dept. Geol. Sci.*, vol. 25, pp. 239–346.)

**Figure 3-29** Shiprock, a diatreme in New Mexico. The thin, dark, linear feature extending outward from the diatreme is a dike. (University of Colorado.)

or greater than an arbitrary 1.6 km in diameter. Caldera formation is illustrated in Figure 3-28.

Other types of depressions are more likely to be found in lowland areas adjacent to volcanic cones. The ways in which these craterlike depressions form include eruptions caused by the rise of gases from a magma, without the accompanying lava or pyroclastics, and the explosion generated when a magma or lava comes in contact with ground water or surface water. The explosiveness in the latter case is caused by the vaporization of the water into steam. These depressions often form lakes in humid regions and are known as *maars*.

After a volcano becomes dormant, erosion begins to attack the volcanic cone. During long periods of geologic time, entire volcanoes can be eroded away, along with the rock surrounding the volcano. Erosion to this degree may expose the pipe or conduit that led from the magma chamber to the vent. These pipes are filled either with solidified magma or breccia that was formed by the explosive escape of gases from the magma chamber. In either case, the rock in the volcanic pipe may be more resistant to erosion than the surrounding rock and may therefore project above the landscape as an imposing monument to past volcanism. Shiprock is a breccia-filled pipe, or *diatreme*, that towers above the New Mexico landscape (Figure 3-29). *Volcanic necks* are similar features composed of magma that cooled and solidified prior to reaching the surface.

## CASE STUDY 3-1
## Volcanic Hazards

Four recent volcanic eruptions and a release of gas from a volcanic crater serve as awe-inspiring examples of the hazards associated with volcanic processes. The 1980s was a deadly decade for volcanic eruptions. The 28,500 people killed in this decade represents a greater loss of life than found in the preceding 78 years, since the eruption in 1902 of Mount Pelée in the Caribbean killed 28,000 people.

The type and degree of hazards vary greatly between volcanoes. The Hawaiian volcanoes represent the lowest degree of hazard because of the volatile-poor, basaltic magmas. The lava flow hazards on the island of Hawaii, the youngest island in the chain, are shown in Figure 3-30. Kilauea, which has been erupting intermittently since 1982, is studied continuously by scientists at the Hawaiian Volcano Observatory run by the U.S. Geological Survey. Because of knowledge gained from this study, loss of life is usually avoided and damage to roads or subdivisions (Figure 3-31) has been the most serious consequence of recent eruptions.

In contrast to the Hawaiian shield volcanoes, composite volcanoes generate more explosive, violent eruptions with correspondingly greater hazards. A well-known example in the United States is the eruption of Mount St. Hel-

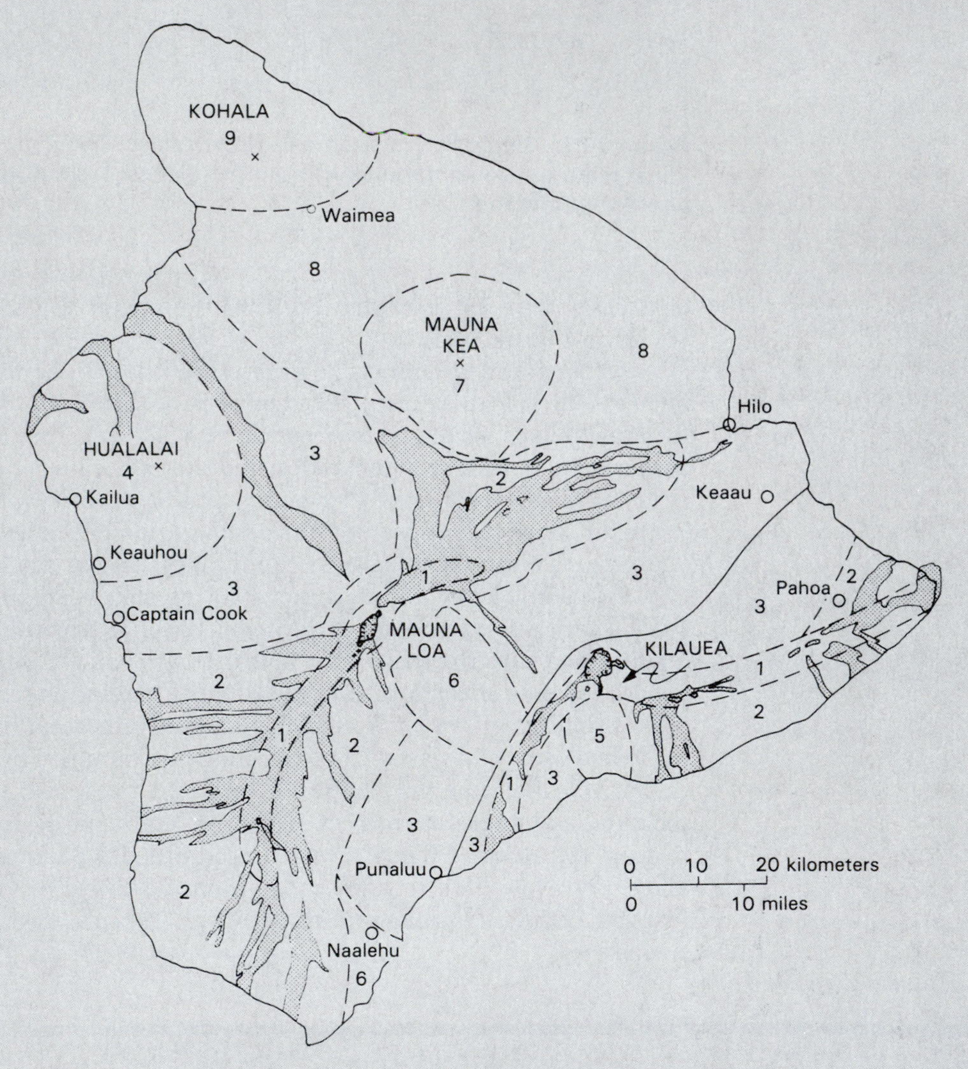


**Figure 3-30** Hazards Map of the island of Hawaii. Relative hazard from lava flows ranges from 1 (high) to 9 (low). Thin, solid line marks boundary between Mauna Loa and Kilauea Volcanoes. Stipple pattern indicates areas covered by pre-1975 historical lava flows. Nearly 30% of the land surface in hazard zone 2, south of the East Rift Zone of Kilauea, has been covered with lava flows since 1955. (From D. R. Mullineaux et al., 1987, U.S. Geological Survey Professional Paper 1350.)

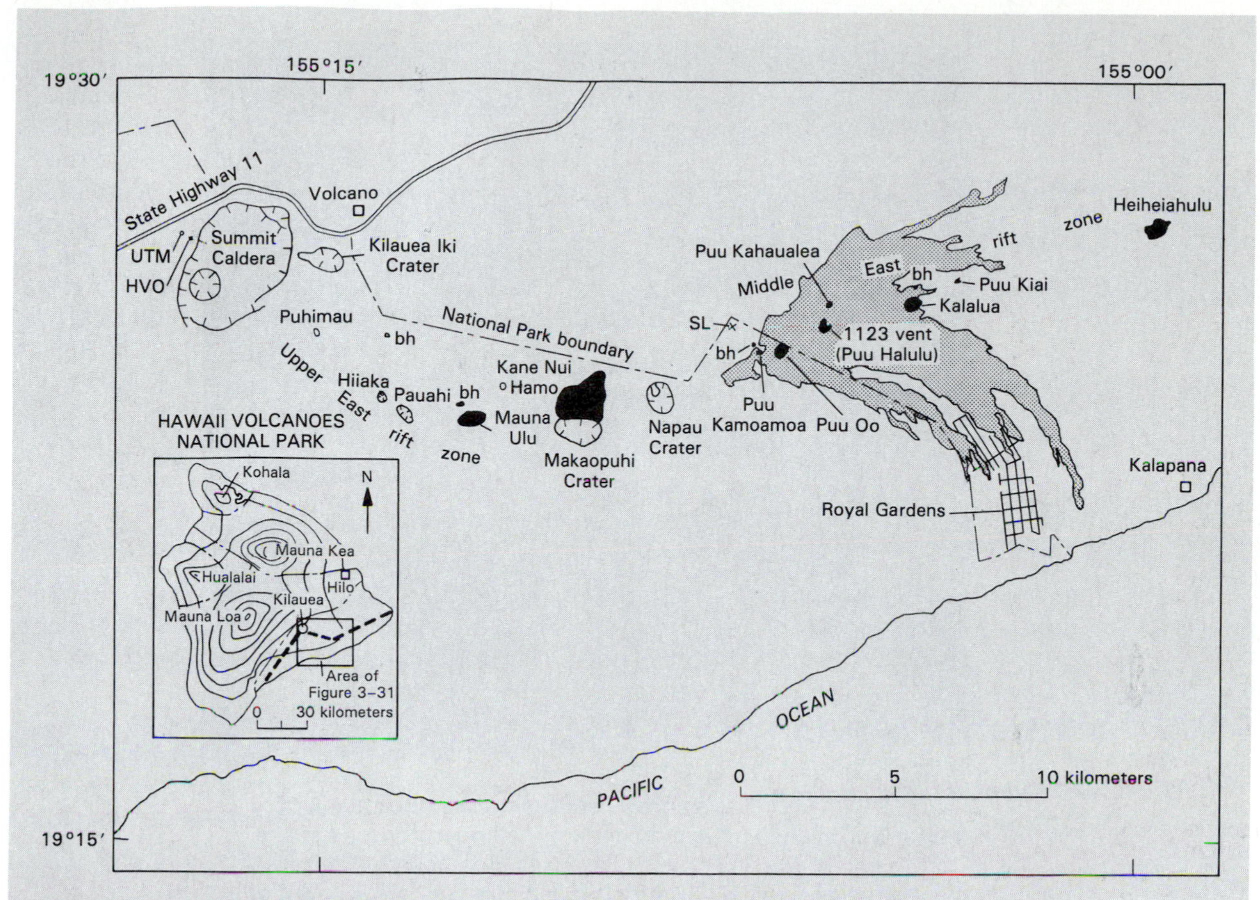


**Figure 3-31** Distribution of lava flows from the Puu Oo eruption from the East Rift Zone of Kilauea in 1983 and 1984. Flows covered parts of the Royal Gardens subdivision. Black areas are large shields or sputter cones. (From E. W. Wolfe et al., 1987, U.S. Geological Survey Professional Paper 1350.)

ens on May 18, 1980 after a 123-year period of inactivity. After several months of preliminary signals, the eruption began, but in an unpredicted fashion, the north side of the mountain gave way in a massive debris avalanche that buried the upper part of the North Fork of the Toutle River valley beneath 3 billion cubic meters of rock, ash, pumice, snow, and ice (Figure 3-32). The sudden release of pressure caused by the debris avalanche unleashed a lateral blast of hot gases and ash. This unexpected directed blast of pyroclastic material devastated the forests to the north of the mountain for a distance of 25 km. Trees within the blast zone were blown over like toothpicks (Figure 3-33). Subsequent to the lateral blast, ash was erupted vertically into the atmosphere, where it began to drift eastward across the United States as a dense cloud.

Among the most damaging long-term effects of the Mount St. Helens eruption were lahars mobilized from the debris avalanches; within hours after these huge landslides came to rest, waves of fluidized debris moved down the major rivers draining the volcano. The largest lahar swept down the North Fork of the Toutle River, into the Cowlitz River, and eventually into the Columbia River (Figure 3-34). The channels of the Toutle and Cowlitz rivers were filled by the surging mudflows, which occasionally overflowed onto the flood plain, where roads and buildings were buried by the debris (Figure 3-35).

Deposition within the channel itself was

**Figure 3-32** Debris-avalanche deposits on the north flank of Mount St. Helens and extending into the upper portions of the North Fork Toutle River valley. Deposits are about 200 m thick in the area shown. (R. M. Krimmel; photo courtesy of U.S. Geological Survey.)

**Figure 3-33** Blown-down timber in the Green River valley from the Mount St. Helens eruption. (Washington Dept. of Natural Resources.)

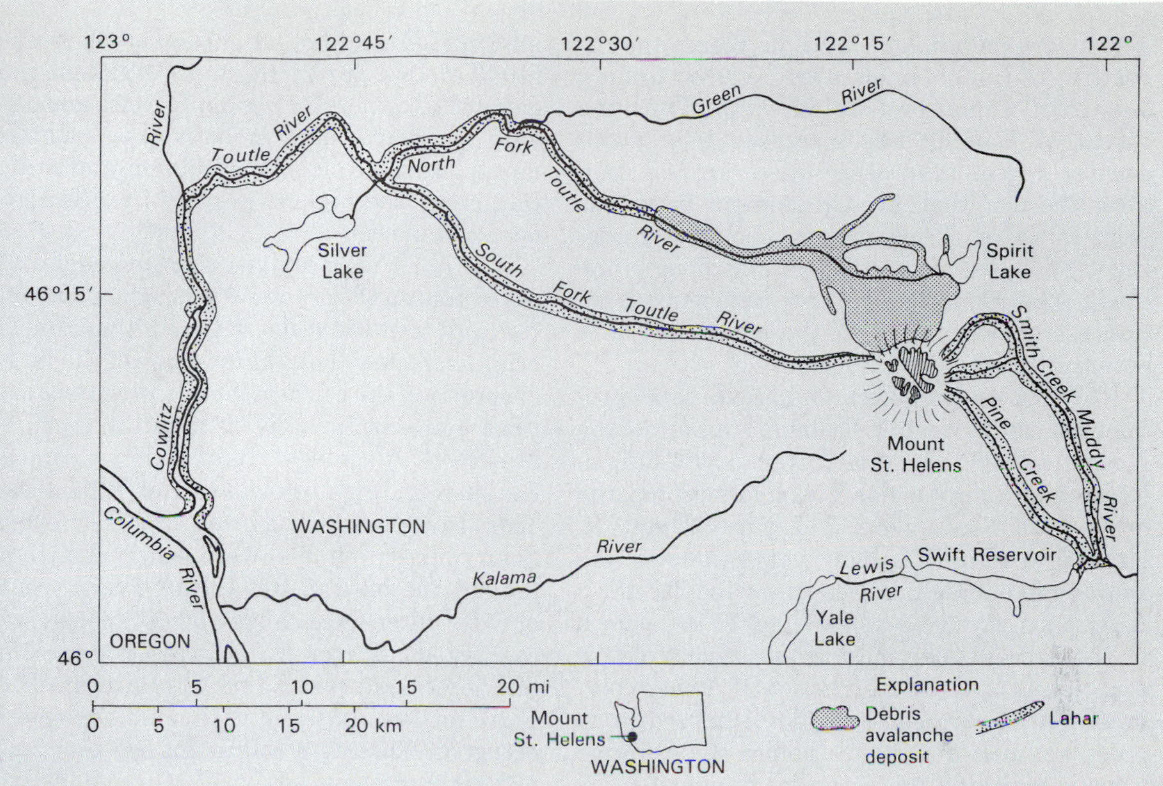


**Figure 3-34** Map of the Mount St. Helens region showing the extent of debris avalanches and lahars. (Modified from U.S. Geological Survey Circular 850-B.)

**Figure 3-35** Building partially buried by lahar deposits, North Fork Toutle River valley. (R. L. Shuster; photo courtesy of U.S. Geological Survey.)

also a serious problem. In some places the capacities of the channels were reduced to only one-tenth of their preeruption value. Thus, the threat of flooding was increased to a much greater degree than before the eruption. Soon after the eruption, flood-abatement programs were initiated. These projects included dredging and levee improvement in flood-prone areas. The Mount St. Helens experience is an impressive warning to all those who live near potentially active volcanoes.

When warnings are not heeded, the consequences can be deadly. By most standards, the November 1985 eruption of Nevado del Ruiz in Columbia was not major. The volume of magma erupted was only about 3% of the Mount St. Helens eruption. Lahars generated on the snow-covered peak, however, buried the town of Armero (Figure 3-36), killing 23,000 people in the process. The most tragic aspect of this eruption is that the destruction of Armero perhaps could have been avoided. A hazards map published just one month before the eruption clearly indicated the potential danger for the town. Even more lamentable is that there was a historical record of lahar damage in Armero. Mudflows similar to those of 1985 had inundated the town twice before, in 1595 and 1845, with hundreds dead in both cases. Unfortunately, natural hazard prediction and mitigation is not the highest priority in underdeveloped countries.

A much more active response undoubtably saved thousands of lives during the 1991 eruption of Mount Pinatubo in the Philippines. This eruption is now thought to be one of the largest eruptions of the century. Small, precursor eruptions and seismic activity began in April and May 1991. A geologic study of the volcano and the surrounding area was begun immediately. This investigation showed that pyroclastic flows (nuées ardentes) were widely distributed around the base of the volcanic cone. As the activity intensified, about 50,000 people were evacuated in stages, including 14,000 personnel and dependents from Clark Air Base, a U.S. Air Force facility. On June 12, a massive explosion occurred, blasting a column of ash to a height of 19,000 m above sea level. Three days later, on June 15 a lateral blast sent huge pyroclastic

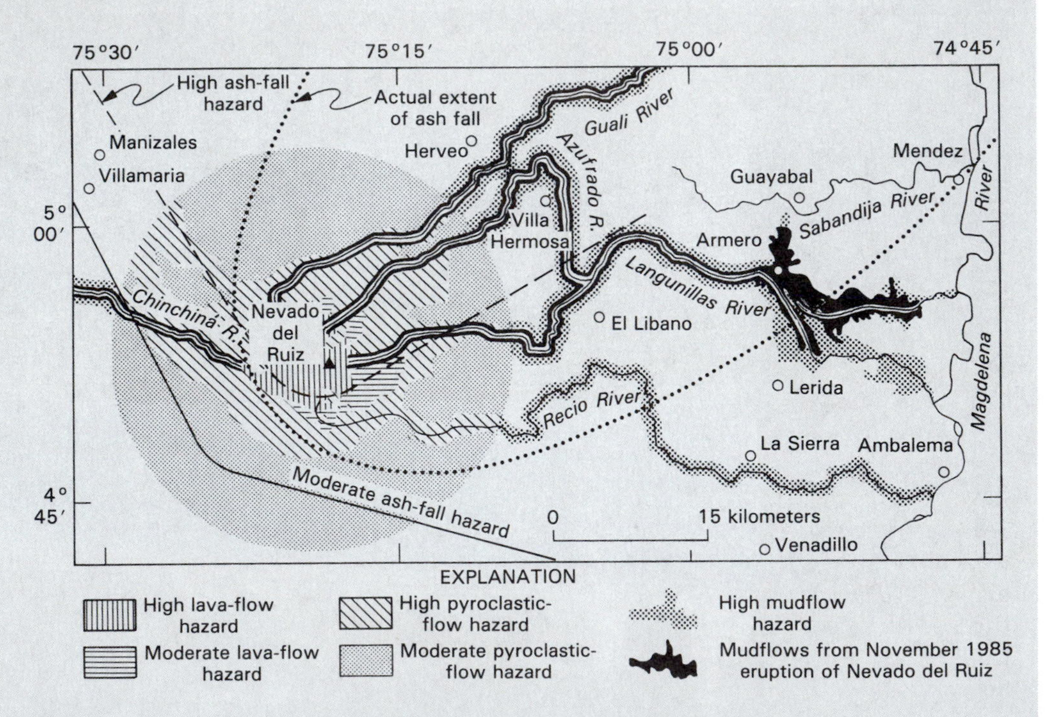


**Figure 3-36** Hazards map of Nevado del Ruiz, Columbia, published one month before the 1985 eruption. Distribution of mudflows from the eruption shown. (From T. L. Wright and T. C. Pierson, *Living with Volcanoes, The U.S. Geological Survey's Volcano Hazards Program*, U.S. Geological Survey Circular 1073.)

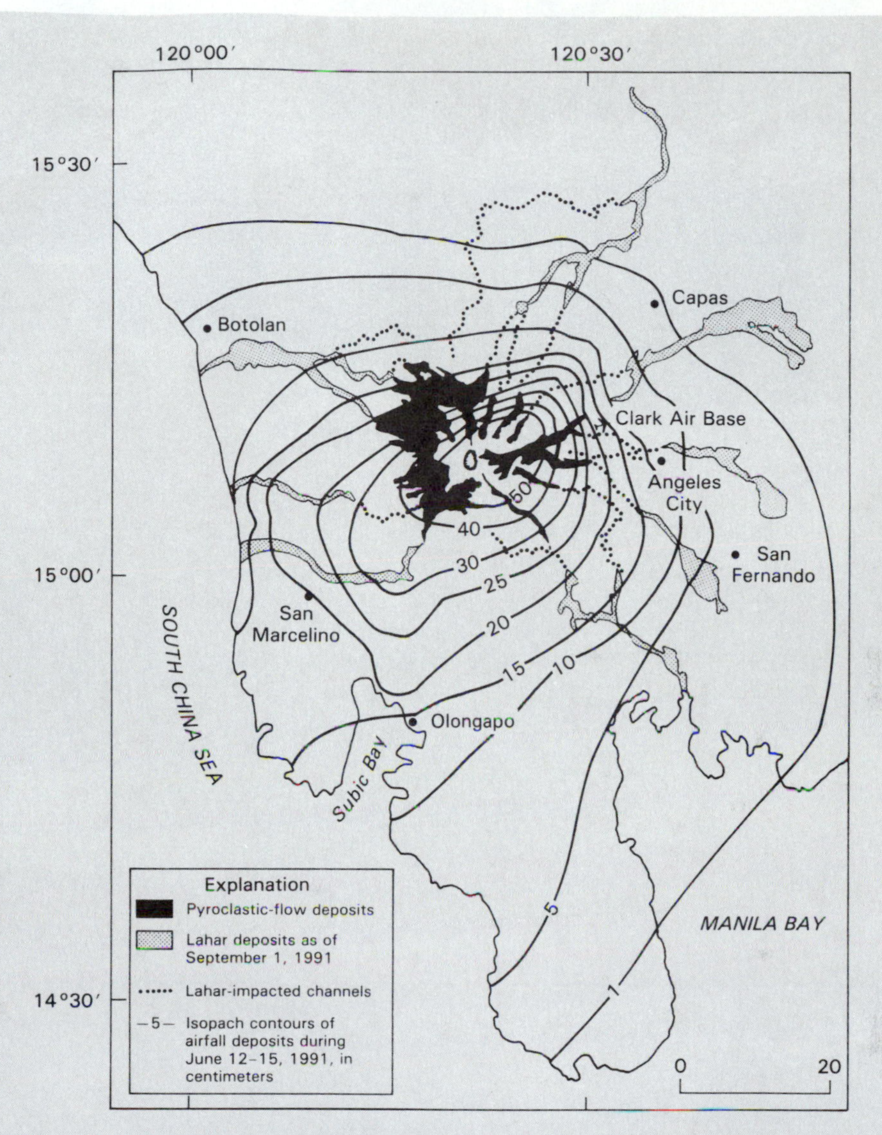


**Figure 3-37** Map showing pyroclastic and lahar deposits from the 1991 eruption of Mount Pinatubo. Contours show thickness of airfall deposits. (From E. W. Wolfe, The 1991 eruptions of Mount Pinatubo, Philippines, *Volcanoes and Earthquakes*, 23:5–37.)

flows in all directions down the volcano (Figure 3-37). Evacuations after this event pushed the total to 200,000 people, the largest number ever evacuated for a volcanic eruption. By an unfortunate coincidence, a typhoon was moving through the area at the same time as pyroclastic flows were moving across the landscape. Torrential rains soaked the ash, making it much heavier. The accumulation of this water-logged pumice on rooftops led to their collapse and to many of the deaths that resulted from the eruption (Figure 3-38). The total accumulations of tephra and the distribution of pyroclastic flow and lahar deposits are shown in Figure 3-37. The immense amount of tephra erupted during this event had many unanticipated ramifications from the near crash of jetliners flying through the plume to worldwide weather changes. The loss of life from the eruption, about 300 people were killed, was immeasurably less because of the ability and willingness to respond to the hazard effectively.

Not all hazards in volcanic areas are associated with actual eruptions. A very unusual

**Figure 3-38** Thick accumulation of pyroclastics near Mt. Pinatubo.

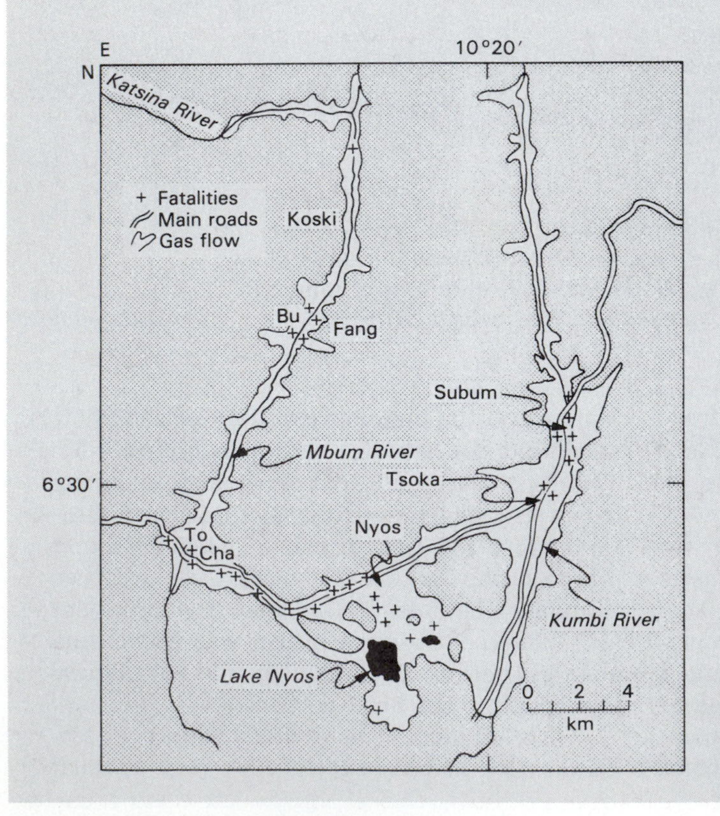


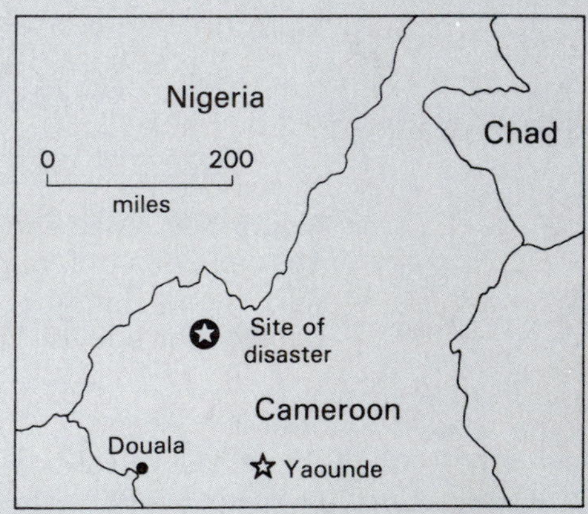


**Figure 3-39** Map of Lake Nyos area, showing areas covered by gas cloud. (From *Eos, Transactions, American Geophysical Union*, June 9, 1987.)

type of event occurred at two lakes in the west African county of Cameroon in the mid-1980s. The lakes involved are located in volcanic maars that are less than 1000 years old. Gases that rise through the volcanic vent conduits accumulate in the bottom of these deep lakes, one of which is called Lake Nyos. The gas, much of which is carbon dioxide, tends to build up at the bottom of the lakes, perhaps because lakes in the tropical climate typical of Cameroon are not subject to yearly overturn as are lakes in temperate climates. On August 21, 1986, Lake Nyos released a large quantity of carbon dioxide as a giant bubble, which caused a splash as high as 100 m on the steep walls of the crater. The mechanism causing the sudden rise of gases from the bottom of the 208 m-deep lake is still not definitely known. At any rate, the gas release formed a lethal cloud of $CO_2$ that flowed as a density current near ground surface down several river valleys (Figure 3-39). The momentum of the gas cloud was sufficient to flatten stalks of corn due to the fact that $CO_2$ is 1.5 times denser than air. When the gas encountered several villages along its path, people were overwhelmed and killed by asphyxiation. The survival rate was less than 1% in the village of Nyos. In all 1746 people and 8300 livestock were lost. Deaths were caused as far as 23 km from Lake Nyos. A similar event of this type at a nearby lake occurred in 1984, although with a much lower loss of life. This indicates that gas releases must be considered to be a continuing threat in Cameroon and in other tropical volcanic regions.

## INTRUSIVE PROCESSES

*Intrusion* refers to the movement of magma from a magma chamber to a different subsurface location. Bodies of rock formed by the intrusion of magma are called *plutons*. Rocks that make up plutons usually have phaneritic texture because the cooling time was sufficient to allow the formation of large crystals. It is only after erosion has removed the overlying rocks that we are able to observe intrusive rocks at the earth's surface.

### Types of Plutons

Plutons differ in terms of size, shape, and relationship from the rocks that were intruded by the magma, which are older rocks known as *country rocks*. A major group of plutons is classified as *tabular* because they are thin in one dimension as compared with the other two dimensions (Figure 3-40). If a tabular pluton is roughly parallel to the layering of the country rocks, it is said to be *concordant* and is called a *sill*. *Discordant* plutons cut across the layering of the country rock. *Dikes* (Figure 3-41) are vertical or near vertical tabular igneous bodies that fill planar cracks through which magma was moved to the surface in fissure eruptions. At the close of the eruption, the magma remaining in the crack connecting the magma chamber with the fissure solidifies to become a dike.

Nontabular plutons include *laccoliths*, *stocks*, and *batholiths*. The shapes of these plutons are thick as well as broad. A laccolith is a relatively shallow concordant pluton that causes doming of the overlying country rocks (Figure 3-42). The rocks above are bowed upward during the intrusion of the magma. Nontabular, discordant plutons are called stocks if their surface exposure covers an area of less than 100 km$^2$, and batholiths if they are exposed over a larger area. Batholiths are truly immense bodies of rock that form the cores of entire

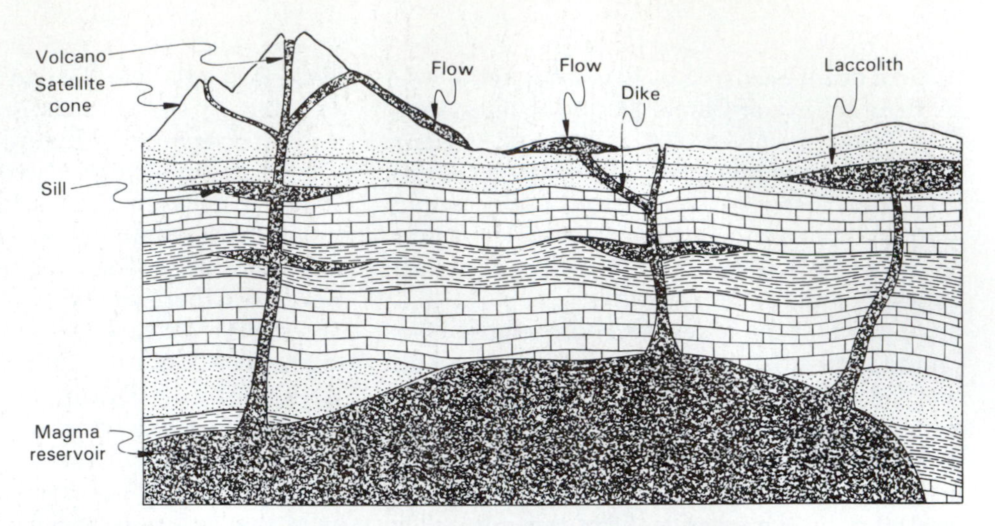


**Figure 3-40** Types of plutons. (From S. Judson, M. E. Kauffman, and L. D. Leet, *Physical Geology*, 7th ed., copyright © 1987 by Prentice-Hall, Inc., Englewood Cliffs, N.J.)

mountain ranges. The Idaho and Sawtooth batholiths in central Idaho are excellent examples (Figure 3-43).

The origin of batholiths has been a controversial subject in geology for many years. The rock type of most batholiths is granite or granodiorite; this highly silicic composition rules out a mantle source for the magma. Therefore, the magma must have originated in the continental crust, the part of the crust that approximates the composition of granite. The major problem concerning batholiths is to account for the huge volume of rock that previously occupied the space now filled by the batholith. A possible explanation for some batholiths is that existing rock was converted to granite, essentially in place. Ideas to explain the mechanism for this conversion range from actual melting to a solid-state process in which existing rocks react at high temperature with

**Figure 3-41** A basaltic dike (dark rock at center), cutting metamorphic rocks near Bar Harbor, Maine.

**Figure 3-42** Laccoliths in Texas exposed by erosion of the overlying sediments. (C. C. Albritton, Jr.; photo courtesy of U.S. Geological Survey.)

solutions and gases from great depths. This theory for the origin of granitic batholiths is known as *granitization.*

### *Crystallization of Magmas*

The great variety in the mineral and chemical composition of igneous rocks suggests that the crystallization of magma is not a simple process. Pioneering experiments by N. L. Bowen in the early 1900s (Bowen, 1922) demonstrated

**Figure 3-43** Granitic rocks of the Sawtooth batholith, central Idaho. (T. H. Kilsgard; photo courtesy of U.S. Geological Survey.)

that minerals crystallize sequentially as the temperature drops in a silicate magma, and that solid crystals can react with the liquid phase of the magma to form new minerals during the crystallization process.

To explain the crystallization process, let us assume that we have a silicate melt of basaltic composition at about 1500°C. As the temperature is slowly lowered, crystals begin to separate from the liquid. There are two crystallization sequences that are observed as the melt cools. The first sequence can be illustrated by the crystallization of plagioclase, which as indicated in Chapter 2, forms a solid-solution series between calcium-rich and sodium-rich compositions. The first plagioclase crystals to form are higher in calcium content than the calcium content of the liquid phase. As the mixture continues to cool, the crystals that form have progressively less calcium and more sodium than the original plagioclase crystals. In addition, the earlier-formed crystals react with the liquid by exchanging sodium from the liquid for calcium in the solid crystals. The internal structure of the mineral remains the same during this process. The crystallization of plagioclase follows what is called a *continuous reaction series*, in which the liquid and the crystals continuously change in composition until no liquid remains.

The ferromagnesian minerals follow a second type of crystallization sequence. In this series, olivine is the first ferromagnesian mineral to crystallize. As the temperature decreases, no change in the olivine crystals occurs until a critical temperature is reached. At this point, pyroxene rather than olivine begins to crystallize and the earlier-formed olivine crystals react with the liquid to form pyroxene. These reactions are different from the continuous reaction of plagioclase because entirely new minerals with different internal structures form at specific temperatures. For this reason the ferromagnesian crystallization sequence is called a *discontinuous reaction series*. The same type of reaction occurs between pyroxene and the liquid to form amphibole at a lower temperature.

The entire sequence of mineral crystallization is known as *Bowen's Reaction Series* (Figure 3-44). Although the series demonstrates the trend of mineral

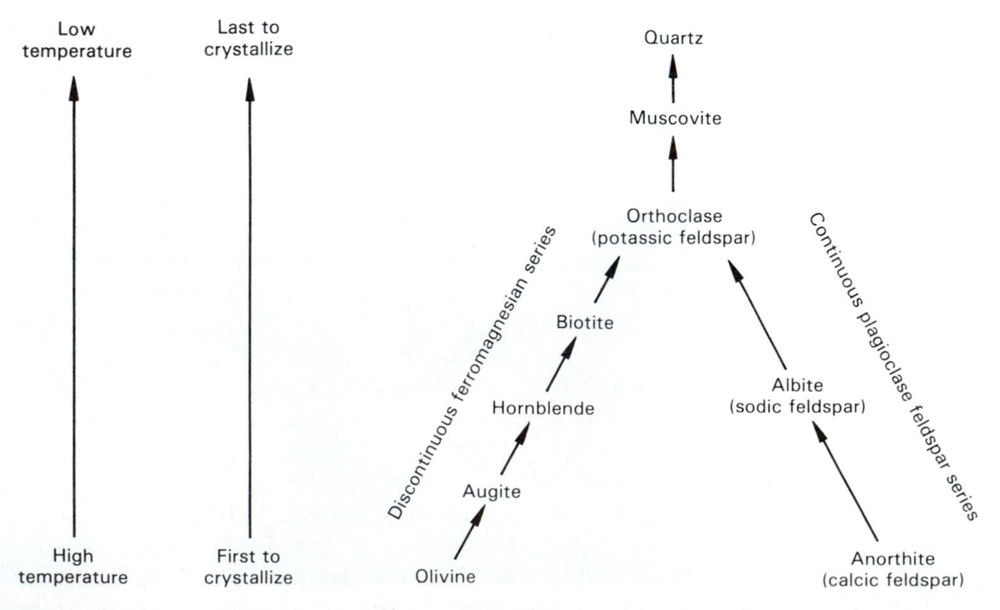


**Figure 3-44** Bowen's reaction series describes the sequence of minerals that crystallize from a basaltic melt.

changes in magmatic crystallization, most individual magmas would not progress through the entire sequence to quartz and muscovite. A magma of basaltic composition, for example, may be totally solidified before its composition could significantly change.

Bowen's work did suggest a way in which magmas could evolve from mafic to silicic composition. This mechanism, known as *fractional crystallization*, involves a separation of crystals from the remaining liquid. If crystals of olivine and pyroxene were separated from a basaltic magma, by gravitational settling for example, they would be prevented from further reactions with the liquid phase. The remaining liquid would be enriched in silica and would progress further along the reaction series. By fractional crystallization, a small amount of granite could be produced from a basaltic magma if crystals were progressively separated from the magma as they formed.

## ENGINEERING IN IGNEOUS ROCKS

Igneous rocks vary greatly in suitability for various types of engineering projects. An engineering site investigation must answer two questions: First, what rock types are present and how are they distributed; and second, how have the rocks been changed or altered since formation?

The first question deals directly with the topic of this chapter. The geologists and engineers working on the project must determine the nature of the igneous rock, the contacts it has with the adjoining rock types and their conditions, and the mineralogy of the rocks.

Unaltered intrusive igneous rocks generally are very suitable for most types of engineering projects. The interlocking network of mineral crystals gives the rock great strength. These rocks thus provide adequate support for building or dam foundations, can remain stable at high angles in excavations, and require minimal support in tunnels. Because of the dense interlocking of crystals within the rock, very little water can move through. Therefore, unaltered intrusive rocks are well suited for the construction of reservoirs because of the low potential for leakage.

The engineering properties of extrusive rocks are much less uniform. Sequences of extrusive rocks contain pyroclastic materials and lahar deposits, which are much weaker than the interlayered lava flows. These rocks may be susceptible to slope failures in excavations and also provide more variable and generally weaker foundation support. In general, the water-bearing capacity of extrusive rocks is much greater than intrusive rocks. Sequences of basaltic lava flows on the Columbia Plateau and elsewhere are known for their great supplies of ground water. This same property can render the rocks unsuitable for reservoir or tunnel construction. The presence of lava tubes in a series of basalt lava flows would prove to be an especially serious problem.

The second question concerns the geologic history of the igneous rocks since the time of their formation. Several types of changes have significant engineering implications. Any process that tends to fracture the rocks will cause weakening of a large rock mass with respect to engineering suitability. Under the forces imposed by the interaction of lithospheric plates, fracturing of crustal rocks is quite common. These processes will be discussed in Chapter 7. It is sufficient to note at this point that a network of fractures within a rock mass can greatly increase the potential for failures of natural or excavated slopes and also increase the construction problems of dams, tunnels, and other structures. Fractured welded tuffs were implicated in the failure of the Teton

Dam in 1976 in Idaho. According to a probable scenario for the failure of the dam, ground water flowing through the fractured volcanics came in contact with erodible silt in the core of the dam and washed the silt through fractures in the rock. Larger and larger channels in the silt were eroded, leading to collapse of the embankment (Chapter 11).

Although tectonic fracturing is important, there are several other mechanisms that cause fracturing. Fracturing of the extrusive rocks, as we have seen, occurs during cooling, in addition to any later processes that may take place. An additional class of fracturing mechanisms becomes possible when the rock mass is at or near the earth's surface. These processes are collectively called *weathering*. A detailed discussion of weathering is found in Chapter 9. Rocks can be fractured in the near-surface environment by freezing and thawing as well as by other means. Thus a mass of igneous rocks generally is more fractured near the surface than at depth.

Weathering produces other changes in the rock besides fracturing. Chemical reactions between the minerals within the rock and air and water gradually form new minerals. Clay minerals are a common product of these alteration processes. The result is a significant loss of strength as the feldspars and ferromagnesian minerals are converted into clay. In warm, humid climates, igneous rock bodies may be mantled with tens of meters of weathered material. The engineering properties of this material are totally different from the properties of unaltered rock.

## SUMMARY AND CONCLUSIONS

Igneous rocks can be divided into extrusive rocks that crystallized at the earth's surface and intrusive rocks that crystallized below the surface. Texture is the major criterion used to assign rocks to one of these categories. Texture and mineralogical composition form the basis for classification of igneous rocks. Based on mineralogy, composition also refers to the amount of silica in the rock, ranging from the low-silica mafic rocks to the silicic rocks. Each major compositional group (silicic, intermediate, and mafic) has both intrusive and extrusive rock types. With the exception of obsidian, color is a good indication of composition because the silicic minerals like quartz and feldspar are mostly light in color.

Volcanoes erupt many types of materials and produce a variety of landforms. Most active volcanoes are associated with tectonic-plate margins, although intraplate volcanism is known to occur; mantle plumes are thought to be the cause of some of this activity.

Volcanic products include gas, lava, and pyroclastic material. Lava flows range from high-temperature, low-viscosity flows of basaltic composition to highly viscous, silicic flows. When gas is explosively released during an eruption, pyroclastic materials are abundant. Large pyroclasts fall in the immediate vicinity of the vent, whereas volcanic ash can be transported thousands of kilometers by winds. Occasionally, destructive pyroclastic flows are generated during eruptions. These nuée ardentes flow down the mountainsides at high velocities and with great force. Lava and pyroclastic materials sometimes mix with snow, ice, or water to form lahars, flows that are similar to mudflows in terms of velocity and the mechanics of movement.

Volcanic products can be extruded from fissures or central vents. Fissure eruptions produce highly fluid lavas of basaltic composition that flow great distances. By successive eruptions, vast plateaus composed of thick vertical sequences of flood basalts are built up. Eruptions from central vents, on the other hand, tend to produce conical mountains composed of lava, pyroclastic materials, or both. Shield volcanoes are composed of lava flows only, whereas composite volcanoes contain alternating layers of lava and pyroclastic materials. Cinder cones form when the eruption is limited to pyroclastic material.

Intrusions are formed by migration of magma from a magma chamber to a subsurface location higher in the crust. The resulting plutons range from moderate-size laccoliths and stocks to giant batholiths that form the cores of major mountain ranges. Some batholiths probably formed by melting or alteration of existing rocks rather than by intrusion of magma from a separate magma chamber.

Igneous rocks crystallize in a complex fashion. The order in which minerals crystallize from a silicate melt is known as Bowen's Reaction Series. This sequence includes both a discontinuous and a continuous series, the latter represented by plagioclase, in which the earlier-formed crystals and the liquid change composition gradually while retaining the internal structure of plagioclase. The ferromagnesian minerals follow the discontinuous series, in which reactions between crystals and liquid produce entirely new minerals at specific temperatures.

Engineering conditions in igneous-rock terrains depend on the composition and type of rock as well as on the amount and type of alteration that has occurred since intrusion or extrusion. Weathering and fracturing are two types of alteration that can weaken the rock and render it less suitable for engineering projects.

## PROBLEMS

1. If an igneous rock has phaneritic texture, what origin can be presumed for the rock?
2. List all the criteria you might use to identify a volcanic rock.
3. What criteria are used to classify the igneous rocks?
4. How can the distribution of volcanoes on the earth be explained by the theory of plate tectonics?
5. If someone told you you were walking across rocks that cooled from pahoehoe lavas, what would you know about the characteristics of the lava at the time of eruption?
6. Nuées ardentes and lahars both involve flow of material from a volcano. What are the differences between the two processes?
7. What factors control the type of volcanic landform that develops from an eruption?
8. What factors determine whether a dike or a sill will form?
9. Summarize Bowen's reaction series.
10. Contrast the general conditions relative to engineering projects that would be encountered in intrusive and extrusive rock terrains.

## REFERENCES AND SUGGESTIONS FOR FURTHER READING

BOWEN, N. L. 1922. The reaction principle in petrogenesis. *Journal of Geology*, 30:177–198.

EICHELBERGER, J., ed. 1987. Lethal gas bursts from Cameroon crater lakes. *Eos, Transactions, American Geophysical Union*, June 9, 1987, p. 568.

HAY, E. A., and A. L. MCALESTER. 1984. *Physical Geology: Principles and Perspectives*, 2d ed. Englewood Cliffs, N.J.: Prentice-Hall, Inc.

JUDSON, S., M. E. KAUFFMAN, and L. D. LEET. 1987. *Physical Geology*, 7th ed. Englewood Cliffs, N.J.: Prentice-Hall, Inc.

MULLINEAUX, D. R., D. W. PETERSON, and D. R. CRANDALL. 1987. Volcanic hazards in the Hawaiian Islands, in *Volcanism in Hawaii*, R. W. Decker, T. L. Wright, and P. H. Stauffer, eds. U.S. Geological Survey Professional Paper 1350.

TILLING, R. I., C. HELIKER, and T. L. WRIGHT. 1987. *Eruptions of Hawaiian Volcanoes: Past, Present, and Future*. U.S. Geological Survey.

U.S. GEOLOGICAL SURVEY. 1980. *Hydrologic Effects of the Eruptions of Mount St. Helens, Washington, 1980*. U.S. Geological Survey Circular 850.

WOLFE, E. W. 1992. The 1991 eruptions of Mount Pinatubo, Philippines. *Earthquakes and Volcanoes*, 23:5–37, U.S. Geological Survey

WOLFE, E. W., M. O. GARCIA, D. B. JACKSON, R. Y. KOYANAGI, C. A. NEAL, and A. T. OKAMURA. 1987. The Puu Oo eruption of Kilauea Volcano, Episodes 1–20, January 3, 1983, to June 8, 1984, in *Volcanism in Hawaii*, R. W. Decker, T. L. Wright, and P. H. Stauffer, eds. U.S. Geological Survey Professional Paper 1350.

WRIGHT, T. L., and T. C. PIERSON. 1992. *Living with Volcanoes. The U.S. Geological Survey's Volcano Hazards Program*. U.S. Geological Survey Circular 1073.

# 4

# Sedimentary Rocks and Processes

Early in the development of geology as a science, James Hutton and others recognized that rocks exposed on the continents, particularly in mountainous areas, are continuously broken down into small particles under the relentless attack of the atmosphere. These finer constituents of the rock are gradually transported to lower elevations, where they are deposited as sediments. This phase of the geologic cycle leads to the formation of sedimentary rocks, the second major rock group. The paths leading from existing rocks to sediments and finally to sedimentary rocks may be long and complicated. Our understanding of these rocks is aided, however, by using the Principle of Uniformitarianism. By studying the deposition and characteristics of sediments on the earth today, sedimentologists can reconstruct the geologic history of ancient sedimentary rocks.

Although sedimentary rocks constitute less than 10% of the rocks in the crust, their importance is emphasized by the fact that they cover about 70% of the surface of the continents. As a result, sedimentary rocks and modern sediments are the most common materials encountered in construction and waste disposal projects. In addition, sedimentary rocks contain the deposits of petroleum, coal, and other energy resources upon which our society depends.

The benefits of utilizing geology in engineering practice have never been better illustrated than by the work of the English civil engineer William Smith (1769–1839). Smith compiled the first real geologic map, which showed the surficial sedimentary rock units in southern England, as a result of his observations of rocks and fossils in connection with the construction of canals and roads. Since the time of Smith, sedimentary rocks have been intensely studied for clues into the history of the earth as well as to locate the resources they contain.

## ORIGIN OF SEDIMENTARY ROCKS

The processes involved in the formation of sediment from existing rocks, in addition to those processes relating to the transportation and deposition of sediment, will be described in considerable detail in later chapters of this text. For this reason, we briefly introduce these concepts here and concentrate on the characteristics and properties of the sedimentary rocks that result from these processes.

The decomposition and disintegration of igneous, sedimentary, and metamorphic rocks near the earth's surface is the result of physical and chemical *weathering* processes. The products of weathering are detached rock fragments, or *clasts,* that can be transported by gravity, water, wind, or ice from higher elevations to lower elevations (Figures 4-1 and 4-2). The ultimate repository of sediments is the ocean basin, but by the time sediments reach the oceans, they have been reduced to extremely small particles. There are many intermediate points of deposition along the path from mountain peak to ocean basin. These *depositional environments* can include almost any part of the continent or ocean floor. On the continents, depositional environments include alluvial (stream) (Figure 4-3), lacustrine (lake), paludal (swamp), desert (Figure 4-4), and glacial types. Along the continental margins a variety of shallow marine depositional environments can be recognized (Figure 4-5). Finally, sediment can be deposited in deep marine settings away from the influence of continents.

The nature of the depositional environment determines the characteristics of the resulting sedimentary rock. Some of the factors that influence rock type are listed in Table 4-1. These factors leave their geologic signature upon the

**Figure 4-1** The weathering of rocks in mountainous regions commonly begins by the formation of large rock fragments, or talus, that accumulate on slopes. (Shenandoah Valley, Virginia; J. T. Hack; photo courtesy of U.S. Geological Survey.)

sediments and resulting rocks that form in a particular environment. One of the most frequent objectives of studies of sedimentary rocks is to reconstruct the conditions of an ancient depositional environment from the rocks that remain.

A final step is required to convert an aggregate of particles into a sedimentary rock. This step is called *lithification*, and it can be accomplished in several

**Figure 4-2** The river plain in the foreground is composed of sediments transported by the stream from uplands in the background. (Southern Iceland.)

**Figure 4-3** Meandering stream at base of valley, Mono County, California. (W. T. Lee; photo courtesy of U.S. Geological Survey.)

ways. The most basic lithification process is *compaction*. Sediment deposited in a loose condition is gradually compacted to a denser state as the weight of additional material is added from above. Glaciers and tectonic forces can impose additional stress upon the sediment. The result of compaction is a reduction in void space between particles. One measure of the relative amount of void space is *porosity* (Chapter 6). In a deep sedimentary basin, decreases in porosity with depth are significant (Figure 4-6). A process that accompanies compaction is the expulsion of water from the void spaces between particles as they are forced closer together. Clay-rich sediments undergo a much greater degree of compaction than sands. Interparticle attractive forces between clay particles also aid in the lithification process.

**Figure 4-4** Some deserts are depositional environments for sand transported by strong winds. (Sonora, Mexico; E. D. McKee; photo courtesy of U.S. Geological Survey.)

**Figure 4-5** Coral reefs constitute important shallow marine depositional environments in tropical areas. (Aga Point, Guam; J. I. Tracey, Jr.; photo courtesy of U.S. Geological Survey.)

Along with simple compaction of sediments to a denser state, a wide variety of physical and chemical processes contribute to lithification. Over long periods of time, the pore spaces remaining after compaction may be gradually filled by precipitation of solid material from ground water circulating through the sedimentary sequence. The filling of void spaces by chemical precipitation is appropriately called *cementation*. Cementation is one of the most effective lithification processes because the chemical cement bonds the particles in the rock together. Most commonly, the cements are composed of either silica ($SiO_2$) or calcium carbonate ($CaCO_3$). Iron-bearing compounds are also fairly common cements. Rocks cemented by silica are among the hardest and strongest sedimentary rocks known. As such, they make excellent foundation materials for large structures.

In addition to cementation, *crystallization* can also contribute to the lithification of a rock. Crystallization usually refers to crystal growth within the void spaces in a rock. Under high pressures and temperatures, however, some of the original materials in the rock may chemically react to form crystals of new minerals that are more stable under the prevailing conditions. If tempera-

**TABLE 4-1**

**Important Factors of Depositional Environments**

| |
|---|
| Type of transporting agent (water, wind, ice). |
| Flow characteristics of depositing fluid (velocity, variation in velocity). |
| Size, shape, depth of body of water, and circulation of water (in lacustrine and marine basins). |
| Geochemical parameters (temperature, pressure, oxygen content, pH). |
| Types and abundances of organisms present. |
| Type and composition of sediments entering environment. |

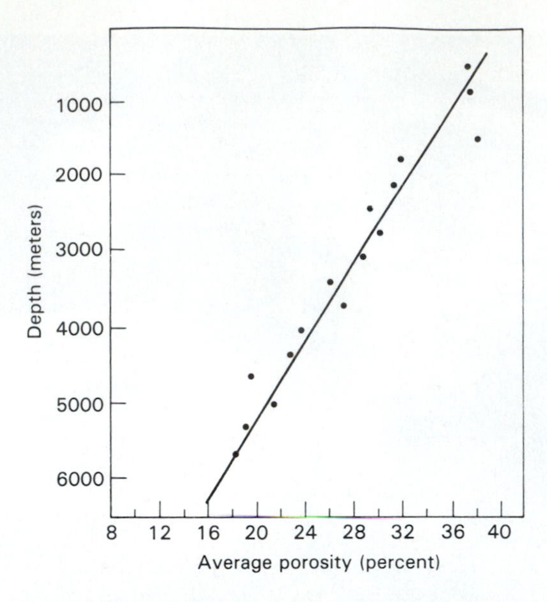


**Figure 4-6** Decrease in porosity with depth in South Louisiana Tertiary sands; data include 17,367 samples averaged for each 1000 ft interval. (Modified from H. Blatt, G . Middleton, and R. Murray, *Origin of Sedimentary Rocks*, 2d ed., copyright © 1980 by Prentice-Hall, Inc., Englewood Cliffs, N.J.; data from G. I. Atwater and E. E. Miller, 1965, unpublished manuscript.)

tures and pressures reach sufficiently high levels, the alteration of sedimentary rocks is great enough to be called metamorphism. Because the transition between the two is gradual, it is sometimes hard to draw the line between sedimentary and metamorphic rocks. The boundary between sediments and sedimentary rocks is also difficult to define. Accordingly, some poorly lithified sedimentary rocks may be classified as soils for engineering purposes because their mechanical behavior may be closer to particle-aggregate systems than to well-cemented rock.

## CHARACTERISTICS OF SEDIMENTARY ROCKS

Sedimentary rocks exhibit a variety of distinctive features. By carefully observing these characteristics, it is possible to assign the proper rock name to a specimen as well as to deduce the origin and depositional environment of the sediment that was lithified to form the sedimentary rock.

### Texture and Classification

Perhaps the most important characteristic of a sedimentary rock is its texture, which, as is true for the igneous rocks, refers to the size and arrangement of the particles or grains that make up the rock. There are only two main types of texture with which we need to be concerned: *clastic* and *nonclastic*. Clastic rocks are composed of aggregates of individual mineral or rock fragments. When these fragments have been eroded, transported, and then deposited, the origin of the rock can be described as *detrital* (Table 4-2). Individual grains can be seen with the naked eye in conglomerates, breccias, and sandstones,

**TABLE 4-2**

**Classification of Sedimentary Rocks**

| *Origin* | *Texture* | *Particle size or composition* | *Rock name* |
|---|---|---|---|
| Detrital | Clastic | Granule or larger | Conglomerate (round grains) |
| | | | Breccia (angular grains |
| | | Sand | Sandstone |
| | | Silt | Siltstone |
| | | Clay | Mudstone and shale |
| Chemical | | | |
| Inorganic | Clastic or nonclastic | Calcite, $CaCO_3$ | Limestone |
| | | Dolomite, $CaMg(CO_3)_2$ | Dolomite |
| | | Halite, NaCl | Salt |
| | | Gypsum, $CaSO_4 \cdot 2H_2O$ | Gypsum |
| Biochemical | Clastic or nonclastic | $CaCO_3$ (shells) | Limestone, chalk |
| | | | Coquina |
| | | $SiO_2$ (diatoms) | Diatomite |
| | | Plant remains | Coal |

SOURCE: From S. Judson, M. E. Kauffman, and L. D. Leet, *Physical Geology*, 7th ed., copyright © 1987 by Prentice-Hall, Inc., Englewood Cliffs, N.J.

which make up the coarse-grained clastic rocks. Hand-lens or microscopic observation is necessary for clasts in fine-grained sandstone and siltstone. Clay particles in shale and mudstone are so small that electron microscopes are used to study them.

If the texture of a rock is nonclastic, it means that the grains form an interlocking network similar to igneous rocks with crystalline texture. Nonclastic rocks are formed by chemical and biochemical precipitation from aqueous solutions. This type of texture can be observed in rocks formed by both inorganic and organic processes (Table 4-2). Organic precipitation is common to organisms that secrete shells composed of calcium carbonate or silica. However, if a rock is composed of an accumulation of shell fragments, its texture is considered to be clastic rather than nonclastic.

Once the texture of a rock is identified, it is necessary to measure the size of the grains. Geologists often use the Wentworth scale for classification of particle sizes (Table 4-3). Another commonly used measure of grain size is the phi unit (Table 4-3). This unit is defined by the formula

$$\phi = -\log_2 X, \qquad \text{(Eq. 4-1)}$$

where $X$ is the grain size in millimeters. Phi units are useful for statistical analysis because the grain-size distribution is often normally distributed when expressed in phi units.

The names of several particle sizes are used to form the rock name for the corresponding rocks. For example, sandstone is composed of particles between $\frac{1}{16}$ and 2 mm in diameter. Actually there may be a wide range in grain sizes. Finer particles that fill the void spaces between larger grains are called *matrix*. Because of its resistance to erosion, sandstone often forms cliffs or upland areas (Figure 4-7). Detrital rocks with most particles larger than sand size are called

**TABLE 4-3**

**Wentworth Scale of Particle Sizes for Clastic Sediments**

| *Size (mm)* | ∅ | *Size name* |
|---|---|---|
| | | Boulder |
| 256 | −8 | |
| | | Cobble |
| 64 | −6 | |
| | | Pebble |
| 4 | −2 | |
| | | Granule |
| 2 | −1 | |
| | | Sand |
| 1/16 (0.0625) | 4 | |
| | | Silt |
| 1/256 (0.0039) | 8 | |
| | | Clay |

SOURCE: From C. K. Wentworth, A Scale of Grade and Class Terms for Clastic Sediments. *Journal of Geology*, 30:381, copyright © 1922, University of Chicago Press.

*conglomerate* if the particles are rounded, or *breccia* if the particles are angular (Figure 4-8). Fine-grained clastic rocks are called *shale* or *mudstone* (Figure 4-9). The term shale usually refers to a rock that tends to split into thin slabs parallel to the depositional layering of the sediment. These fine-grained rocks are typically weaker than sandstone and form low-angle, covered slopes or valleys in sedimentary rock terrains.

The particle size of detrital rocks gives an important clue about the depositional environment. In the case of water-laid materials, the ability to transport particles is dependent upon the current velocity of the transporting fluid. Therefore, conglomerates represent sediment deposits of high-velocity streams, like those that rush down steep mountain slopes or form at the edge of a melting

**Figure 4-7** Resistant sandstone (Entrada Formation) exposed in Arches National Park, Utah. Natural arches are formed by weathering processes and wind erosion. (Photo courtesy of R. A. Kehew.)

(a)

(b)

**Figure 4-8** (a) Conglomerate is a detrital sedimentary rock composed of large, rounded particles (south-central Norway). (b) Breccia is similar in grain size to conglomerate but differs because the grains are angular (western Maryland).

glacier. In contrast, silt and clay accumulate in low-energy environments like lakes or sea floors.

The chemical and biochemical sedimentary rocks include a wide variety of materials. Chemical precipitation from water is responsible for deposits of limestone (Figure 4-10). Deposits of marine origin are the most common type of limestone. Most of Florida is underlain by fossiliferous limestone deposited under shallow marine conditions on a carbonate shelf.

Classification of these rocks is based upon the types of grains and the amount of matrix making up the rock. Grain types include *bioclastic debris*, *oolites*, *intraclasts*, and *pellets*. Bioclastic debris includes the skeletal remains of organisms that secrete shells of calcium carbonate. Paleontologists, geologic

(a)

(b)

**Figure 4-9** (A) An outcrop of shale showing characteristic layering. (Mancos Shale near Aspen, Colorado; B. Bryant; photo courtesy of U.S. Geological Survey.) (b) Under the microscope, the mineral composition of shale can be observed. Minerals in the light-colored bands are mostly quartz, whereas the dark-colored layers are made up of platy crystals of muscovite. (Kuskokwim region, Alaska; W. M. Cady; photo courtesy of U.S. Geological Survey.)

specialists who study fossils and their evolution through geologic time, have identified thousands of these organisms. Examples include corals that build extensive reefs composed of intergrown networks of skeletal calcite (Figure 4-11) and solitary organisms like clams and snails. Ancient reefs buried in sedimentary basins are eagerly sought by petroleum geologists because of their abundant void space, which may serve as a reservoir for oil and gas. Oolites are spherical grains formed by precipitation of concentric layers of calcium

**Figure 4-10** An exposure of limestone in a quarry in Florida showing a solution cavity filled with soil that collapsed into the void from above.

carbonate in a shallow water marine environment where waves or currents periodically roll grains across the bottom (Figure 4-12). Intraclasts are fragments of precipitated calcium carbonate that are eroded and then redeposited, and pellets are composed of the fecal matter of organisms. These four grain types may or may not be surrounded by a matrix composed of calcium carbonate mud. Rock names describe the grain type and matrix characteristics. For example, an *oomicrite* is a rock containing abundant oolites within a muddy matrix. *Micrite* is an abbreviation for microcrystalline calcite.

Areas underlain by carbonate rocks present a unique set of conditions for engineers and geologists involved in development and environmental problems. Although carbonate terrains frequently contain abundant ground-water resources, ground water may easily become contaminated due to the ability of rain to dissolve calcium carbonate and create pathways for the rapid downward

**Figure 4-11** Brain coral (*Diploria*, species) and several species of Alcynonarian corals on shallow reef in St. Croix, U.S. Virgin Islands. (Photo courtesy of W. B. Harrison, III.)

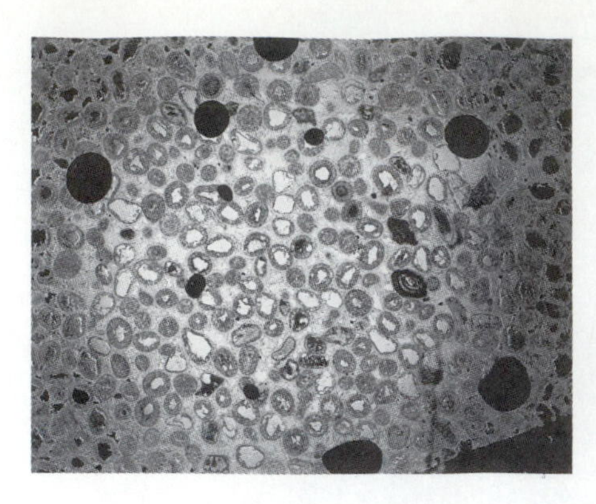

**Figure 4-12** Photomicrograph of oolitic limestone from core, × 30. Sand grains form nuclei of many oolites. (Hernando County, Florida; P. L. Applin; U.S. Geological Survey.)

movement of contaminants. Florida and other limestone terrains are also plagued by rapid subsidence of the land surface to form *sinkholes*, as subsurface rocks are slowly dissolved by infiltrating rain or snowmelt (Chapter 11). Although dissolution of the rock may take thousands of years, collapse of the surficial soil into the underground cavity may be instantaneous. A solution cavity filled with younger soil materials is shown in Figure 4-10. The actual and potential damage caused by subsidence is a major threat to development in carbonate terrains. Ground waters precipitate calcium carbonate under certain conditions as well as dissolve it; most commonly, this occurs in caves and around hot springs. This type of limestone is known as *travertine*. The spectacular travertine deposits at Mammoth Hot Springs in Yellowstone National Park (Figure 4-13) are a good example.

Minerals precipitate from water because the concentration in the liquid phase is greater than the solubility of the solid phase. The concentration at the point of equilibrium between dissolution and precipitation is a function of

**Figure 4-13** These terraces at Mammoth Hot Springs at Yellowstone National Park were formed by precipitation of travertine from the emerging spring water.

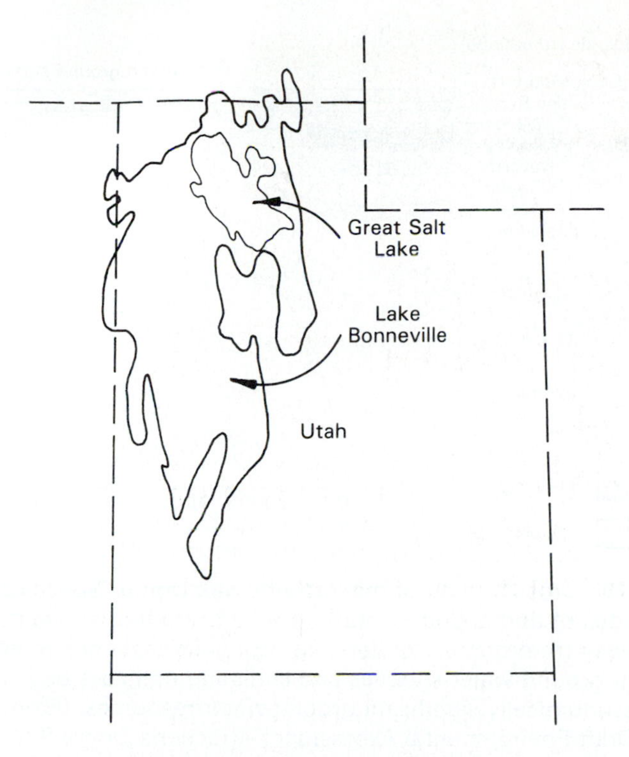


**Figure 4-14** The Great Salt Lake is a small remnant of ancient Lake Bonneville. Evaporites were deposited as evaporation caused the lake to shrink in size.

temperature, pH, and other factors. Some minerals are so soluble that they rarely precipitate. For example, halite requires a very high concentration of sodium and chloride to initiate precipitation. The necessary concentrations are sometimes achieved by evaporation of water from a lake or marine basin. The Great Salt Lake in Utah (Figure 4-14), for example, is a remnant of a much larger lake, Lake Bonneville, that existed during a cooler and wetter climatic period during the Pleistocene Epoch. Climatic changes, including warmer temperatures and less precipitation, led to a gradual shrinking of the lake by evaporation. As the lake water became concentrated, precipitation of halite and other salts formed the Bonneville salt flats, the floor of ancient Lake Bonneville. Salt deposits of this type are called *evaporites*. Evaporites can also form in marine basins that are deeper than the surrounding parts of the sea. The isolation and depth of these basins limits normal oceanic circulation, and the necessary evaporative concentration can occur. Precipitation of evaporites follows a sequence controlled by the solubility of the evaporite minerals. As water begins to evaporate from sea water, calcite is the first mineral to precipitate. Gypsum follows as evaporation continues, and finally halite and other salts drop out of solution when the remaining water is nearly gone.

When beds of halite become buried by thousands of feet of other sedimentary rocks, a curious phenomenon sometimes occurs due to the mechanical behavior of salt under high pressures. Evaporite rocks are extremely weak, and under pressure, they behave as a plastic material and can be made to deform, or flow, for great distances. From an initially uniformly thick bed subjected to small differences in overburden pressure, salt begins to gradually flow toward a point in the bed that slowly thickens. Because of the low density of salt, thickened zones bulge upward, forming a *salt dome* (Figure 4-15). Sinking occurs adjacent to the dome because of the loss of salt there, due to

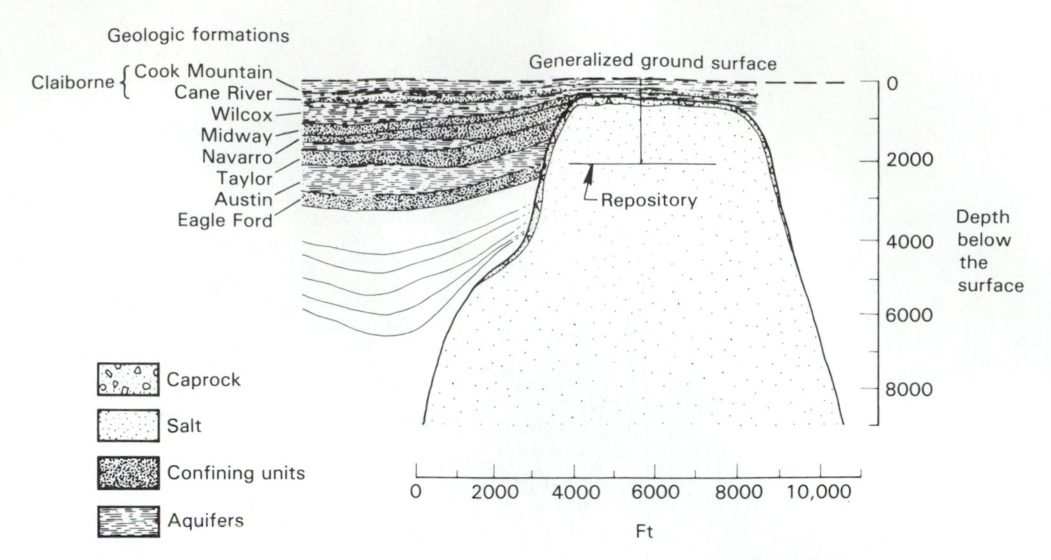


**Figure 4-15** Cross section of the Vacherie Salt Dome. Notice caprock above and on sides of dome. Sedimentary rock beds are thin above dome and thicken away from dome. Aquifers are rock units that contain economically significant ground-water supplies and confining units are beds that do not contain economically significant ground-water resources. (From U.S. Dept. of Energy, Draft Environmental Assessment—Vacherie Dome Site, Louisiana.)

the flow of salt toward the growing dome. More sediment is deposited in the low-lying areas during dome growth, so that rock formations thicken outward from the center of the dome. With continued dome expansion, salt may break through overlying rock units and rise hundreds of meters until a stable condition of equilibrium is reached and no further growth occurs. Dissolution of salt at the top and margins of the dome by ground water flowing in the adjacent rock formations leaves behind less soluble material as a caprock. The formation of a caprock isolates the soluble salt to some extent from ground water in adjoining rocks.

Salt domes sometimes form effective petroleum traps, and have been exploited for petroleum production in the U.S. Gulf Coast region for many years. Because of its low permeability, some of our emergency supplies of oil are stored in large underground cavities excavated in salt. More recently, salt beds and salts domes have been suggested for another purpose: disposal of high-level radioactive wastes. Prior to selection of the Yucca Mountain site for further study, seven of the original nine potential repository sites chosen by the U.S. Department of Energy (DOE) were sedimentary rock sequences containing evaporite beds or domes. Studies of these materials have continued, and it is likely that these sites will eventually serve as repositories for some types of hazardous wastes.

### *Sorting*

The range of particle sizes in a sedimentary rock is known as *sorting*. Rocks that contain a narrow range of sizes are described as *well sorted;* those that consist of a wide size range are termed *poorly sorted* (Figure 4-16). The value of sorting in interpretation of sedimentary rocks lies in its use as an environmental indicator. Poor sorting in a water-laid sedimentary rock suggests rapid fluctuations in current velocity and rapid deposition. Wind-deposited sedi-

(a)

(b)

**Figure 4-16** (a) Photomicrograph of well-sorted quartz sandstone, Moffat County, Colorado. (W. R. Hansen; photo courtesy of U.S. Geological Survey.) (b) A poorly sorted sandstone under the microscope. Particles range in size from silt to fine gravel, Valmy Formation, Nevada. (J. Gilluly; photo courtesy of U.S. Geological Survey.)

ments, on the other hand, tend to be well sorted because wind is capable of transporting only small particles. Thus the small particles are separated from the larger particles and accumulate in one location when the windstorm ceases. Sediment transported directly by a glacier can often be recognized by its very poor sorting. Although glaciers move very slowly, they are equally capable of picking up both huge boulders and small clay particles as they advance over an area. Very little segregation of the particles occurs during the movement of the glacier.

### *Sedimentary Structures*

Sedimentary structures set sedimentary rocks apart from other rock types. The result of sediment transport and deposition, these structures are critical for understanding the conditions under which the sediments were deposited.

Using sedimentary structures and other factors, geologists can reconstruct the environment of deposition of a sedimentary rock unit. When the depositional environment is known, variations in rock type and engineering characteristics can be predicted. If an engineering project involves subsurface data that can be very expensive to obtain, reconstructions of sedimentary environments quickly become very relevant to the investigation.

The most fundamental sedimentary structure is *bedding*. Individual layers, or beds, of sediment stacked upon each other are the most recognizable indication of sedimentary rock (Figure 4-17). A bed is a unit that differs in some way from the beds above and below it. *Laminae* are layers similar to bedding that are less than 1 cm thick. Bed boundaries may be sharp or gradational. Bedding is the result of variation in the depositional process. A slight change in current velocity is all that is necessary to change the size of the material being transported and deposited. A river may deposit sand or gravel on the bed of its channel. Later, if that reach of the channel is abandoned in favor of a new channel, beds of silt and clay representing a lower energy environment may be deposited over the coarser sediments. Thick sequences of sedimentary rocks like the Grand Canyon record much more drastic changes in depositional environments—alternating terrestrial and marine conditions, for example. Beds and other structures are the evidence of these changes.

Differences in the process of sedimentation can lead to different types of bedding. Three main types include *parallel bedding*, *cross bedding*, and *graded bedding* (Figure 4-18). Parallel bedding can be produced by transport of sediment by currents or by deposition in standing bodies of water where sediment particles fall through a column of water to their resting points on the bottom.

**Figure 4-17** Bedding stands out even at a great distance in the well-exposed sedimentary rocks of the Grand Canyon. (J. R. Balsley; photo courtesy of U.S. Geological Survey.)

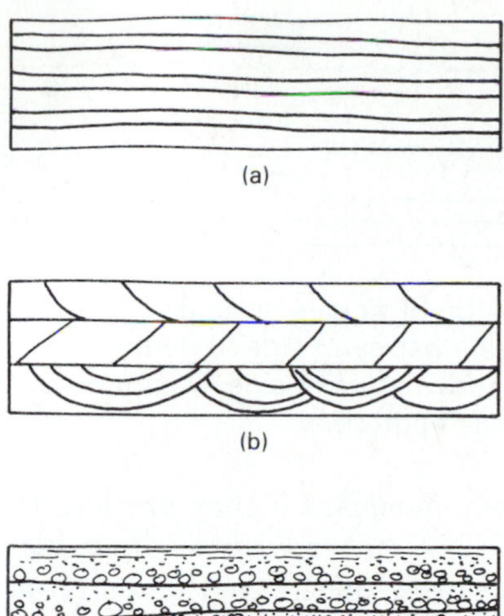


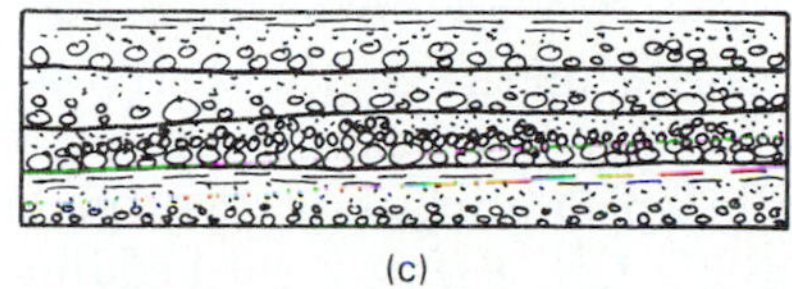


**Figure 4-18** Types of bedding: (a) parallel bedding; (b) cross bedding; and (c) graded bedding.

Cross bedding is the result of the deposition of sediment that moved in wavelike bedforms near the sediment-fluid interface. These conditions occur in streams and in wind-blown sand. Three major types of cross beds are recognized (Figure 4-19). Tabular and wedge-shaped cross beds are bounded by planar boundaries that are parallel for tabular and nonparallel for wedge-shaped cross beds. Trough cross beds have curved lower boundaries.

The formation of cross beds is illustrated in Figure (4-20). The wavelike

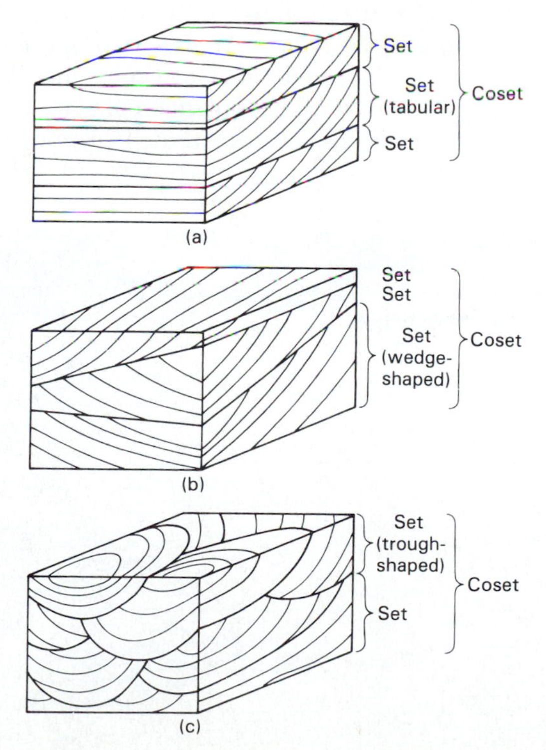


**Figure 4-19** Types of cross bedding. (a) Tabular; boundaries are parallel, planar surfaces. (b) Wedge-shaped; boundaries are nonparallel, planar surfaces. (c) Trough; boundaries are trough-shaped.

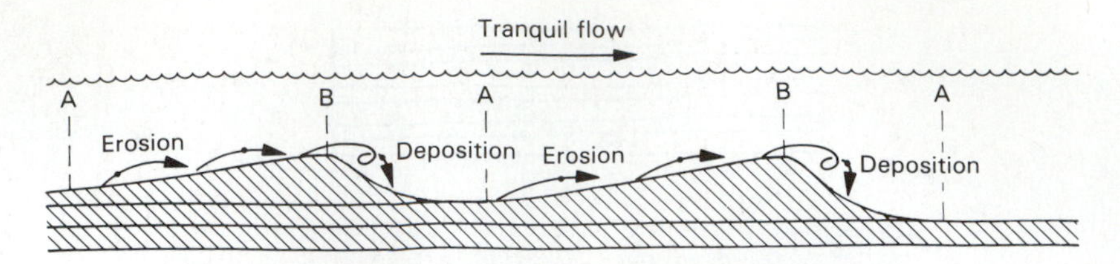


**Figure 4-20** Formation of ripples or dunes. Grains are eroded and transported up stoss (upstream) side of bedform and deposited by avalanching on lee (downstream) side. Migration of bedform truncates tops of previous bedforms.

bedforms are known as *ripples* if they are less than 0.3 m from crest to crest (Figure 4-21) and *dunes* or *megaripples* if they are longer (up to 10 m). Sand is eroded and transported up the gentle upstream slope and deposited by avalanching down the steeper lee slope. As the next bedform migrates over the bed, it truncates the upper surface of the ripple or dune to form a planar boundary (Figure 4-22). Large scale cross bedding is typical of subaerial sand dunes in desert environments. The Navajo Sandstone on the Colorado Plateau (Figure 4-23) is interpreted to be a rock formed from eolian deposits.

Graded bedding is similar to parallel bedding except that the grain size within the bed decreases systematically from the bottom toward the top. This gradual change in grain size is produced by currents in which sediment is carried partially in *suspension* rather than being transported along the bed. Such currents, called *turbidity currents* (Chapter 14), are dense, rapidly moving currents that flow down submarine slopes near the edges of continents or along lake bottoms. Graded bedding is produced by the order in which particles settle out of suspension. As the current flows out onto an area of more gentle slopes, the flow velocity gradually decreases. As this occurs, the larger particles begin to fall to the bed, so that large particles are concentrated at the bottom. The process continues, leading to a gradual decrease in grain size from bottom to top. Sequences of turbidites indicate that, in some marine environments, turbidity currents occur at regular intervals. Graded beds, therefore, alternate

**Figure 4-21** Sand ripples on a modern stream bed. The internal structure of ripples and dunes is often preserved as cross bedding.

**Figure 4-22** Cross bedding in sand. The inclined layers are remnant downstream faces of migrating dunes.

with nongraded beds, representing the intervals between turbidity currents (Figure 4-24).

Other types of sedimentary structures are also recognized in sedimentary rocks. Under some conditions, entire bedforms are preserved on a stream bed or tidal flat, for example, the ripples in Figure 4-25, rather than being truncated by subsequent erosion and deposition. The resulting sedimentary structures are called *ripple marks*. *Mud cracks* (Figure 4-26) are polygons formed at the surface of a fine-grained sediment as it gradually dries and shrinks. The presence of these structures in an ancient sedimentary rock would be an indication of periodic exposure of the sediment surface, an important clue in deducing the depositional environment of the rock.

**Figure 4-23** Checkerboard Mesa, at Zion National Park, Utah, displays large-scale cross bedding typical of wind-deposited sandstone. (Photo courtesy of R. A. Kehew.)

**Figure 4-24** Alternating beds of sandstone with graded bedding and shale in a turbidite sequence. (San Mateo County, California; D. G. Howell; photo courtesy of U.S. Geological Survey.)

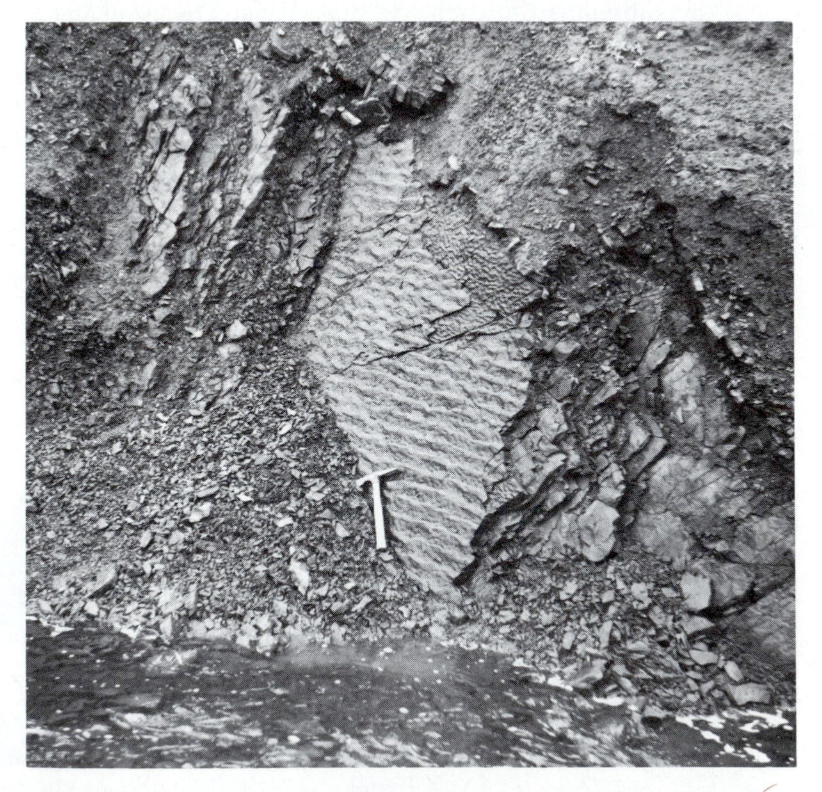

**Figure 4-25** Current ripple marks on bed of sandstone tilted upward by tectonic forces. (Northern Alaska; R. H. Campbell; photo courtesy of U.S. Geological Survey.)

**Figure 4-26** Mud cracks indicate that this bed dried in the open air soon after deposition. (Isle Royale National Park, Michigan; R. B. Wolff; photo courtesy of U.S. Geological Survey.)

### *Color*

The color of sedimentary rocks is often diagnostic of the geochemical environment at the time of formation. Shades of red or brown indicate formation of the sediment in an environment with abundant free oxygen. Under these conditions, iron exists in the ferric, or oxidized state. A small amount of ferric iron is sufficient to impart a reddish or yellowish color to the deposit. Sediments that accumulated in environments lacking oxygen are usually darker in color. The somber gray and green shades of these rocks are attributed to the presence of iron in the ferrous, or reduced, state. Marine and lacustrine environments frequently have sufficiently low dissolved-oxygen levels to ensure that iron will exist in the reduced form.

If sediment rich in organic matter is deposited in a reducing environment, the resulting sediments will be black in color. Petroleum and coal are good examples. The lack of oxygen is critical to the preservation of these materials because bacteria will decompose the organic matter if an oxygen supply exists.

### *Fossils*

The presence of organic remains in sedimentary rocks is a major concern to the science of geology. It was primarily by the study of fossils that the geologic time scale was devised.

Living organisms consist of soft and hard parts. The soft parts, including cell tissue and internal organs, decay rapidly in the presence of oxygen when the organism dies. Hard parts—bones and shell, for example—can be preserved for long periods of time under favorable conditions in sedimentary deposits (Figure 4-27). Shells are usually composed of calcium carbonate or silica, whereas bones consist of a phosphatic material.

**Figure 4-27** Coquina, a rock composed almost entirely of fossils of marine organisms. (Gila County, Arizona; C. Teichert; photo courtesy of U.S. Geological Survey.)

In anaerobic environments soft organic remains can be preserved, although they may be greatly altered from their original state. These environments account for deposits of petroleum, which are derived from the remains of microscopic marine organisms, and coal, the product of terrestrial plants.

Fossils also include material that is not necessarily organic in origin. Petrified wood, for example, is composed of silica that has been precipitated by ground water circulating through sedimentary materials. In this process, silica precipitation gradually replaces the organic material, while retaining the original cellular structure (Figure 4-28).

**Figure 4-28** Petrified stumps in western North Dakota preserved in their original growth position by the replacement of organic matter by dissolved silica carried by ground water. The stumps have been exposed by erosion of the surrounding materials.

### *Stratigraphy*

Imagine that you are assigned to a highway design project in an area of sedimentary rocks. Since the route may cross several rock types with different engineering properties, where do you get preliminary information on the distribution of these rocks? With luck, there will be a published geologic map of the area available. For this reason, engineers should be familiar with the use of these maps.

The basic map unit on geologic maps is the *formation*. In a sedimentary rock terrain, a formation is a bed or group of related beds whose contacts can be traced where they are exposed or approximated where they are covered by other units or eroded away. Other units that may be shown on geologic maps include *members*, which are significant subdivisions of formations, and *groups*, which include two or more formations that appear to have a genetic relationship. These map units are called *rock units* because they are based strictly on lithologic characteristics without any consideration of age.

The field geologist constructs the map by measuring and describing vertical sequences of beds wherever they are well exposed. These measured sections are correlated to other sections across areas where the rocks are not exposed by using color, grain size, sedimentary structures, fossils, and any other characteristic that may be distinctive (Figure 4-29). Diagrams showing vertical sections and correlations are called *cross sections*. Contacts are drawn on topographic or planimetric base maps to complete the geologic map, and the map and cross sections are often published together (Figure 4-30).

When rock units can be traced for a substantial distance, changes in lithology can often be observed. The reason for these changes is simply that the boundaries of a specific depositional environment are finite. In fact, most depositional environments can be divided into subenvironments, or *facies*, each containing lithologically distinct sediments deposited contemporaneously with those in the adjacent subenvironment. For example, facies in a meandering stream environment are shown in Figure 4-31. Stream landforms and deposits are discussed in more detail in Chapter 13. These relationships would be of great significance to the engineer building a highway and bridge across a valley containing a meandering stream because each facies would require a different foundation design.

The study of fossils in sedimentary rocks has a very practical purpose for stratigraphers. The changes in organisms over geologic time, the record of evolution, can be used to assign relative ages to rock units. For example, Ordovician fossils in North America are identical to those in the type area in Wales, whether or not specific rock units were ever physically continuous between the two regions. The use of fossils to date sedimentary rocks is known as *biostratigraphy,* and stratigraphic units defined by these methods are known as *time-rock* units. The systems that we discussed in Chapter 1 are examples of *time-rock* units.

## *SEDIMENTARY ROCKS AND RESOURCES*

Sedimentary rocks contain many types of mineral resources. Energy resources, however, are particularly important because they are confined to sedimentary rocks. Petroleum and natural gas require very complex conditions for formation and concentration where they can be economically recovered. The first step in the sequence of events involves burial of abundant marine organisms in

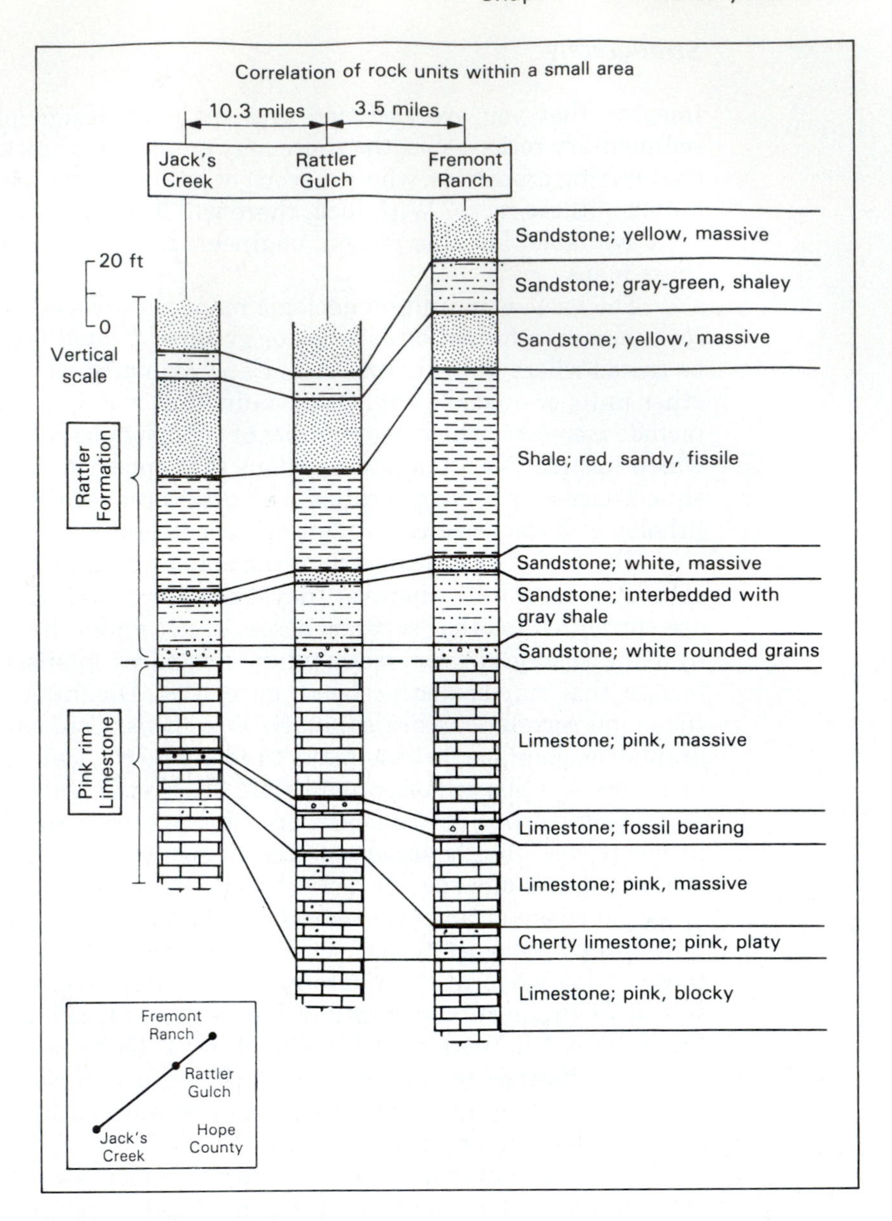


**Figure 4-29** Correlation of rock units within a small area. (From W. L. Newman, *Geologic Time*, U.S. Geological Survey Pamphlet.)

sedimentary basins that maintain low oxygen levels; in some cases this has been caused by a restricted circulation between the basin and the surrounding ocean. During the Paleozoic era, much of North America consisted of relatively stable continental crust, a *craton*, covered by shallow seas. Sediments that became *source rocks* for petroleum accumulated in intracratonic basins, which were large areas of the crust that gradually subsided relative to the surrounding craton. The Williston Basin is a good example of these sediment traps (Figure 4-32). Sedimentary basins adjacent to the continental crustal block also provide favorable conditions for petroleum source rocks. As the source rocks gradually sink to greater depths in the subsiding basins, the decaying organic matter is broken down under high temperatures and pressures to form simpler hydrocarbon compounds. These hydrocarbons then begin to migrate under hydraulic gradients until geologic circumstances produce a trap in which oil and gas may

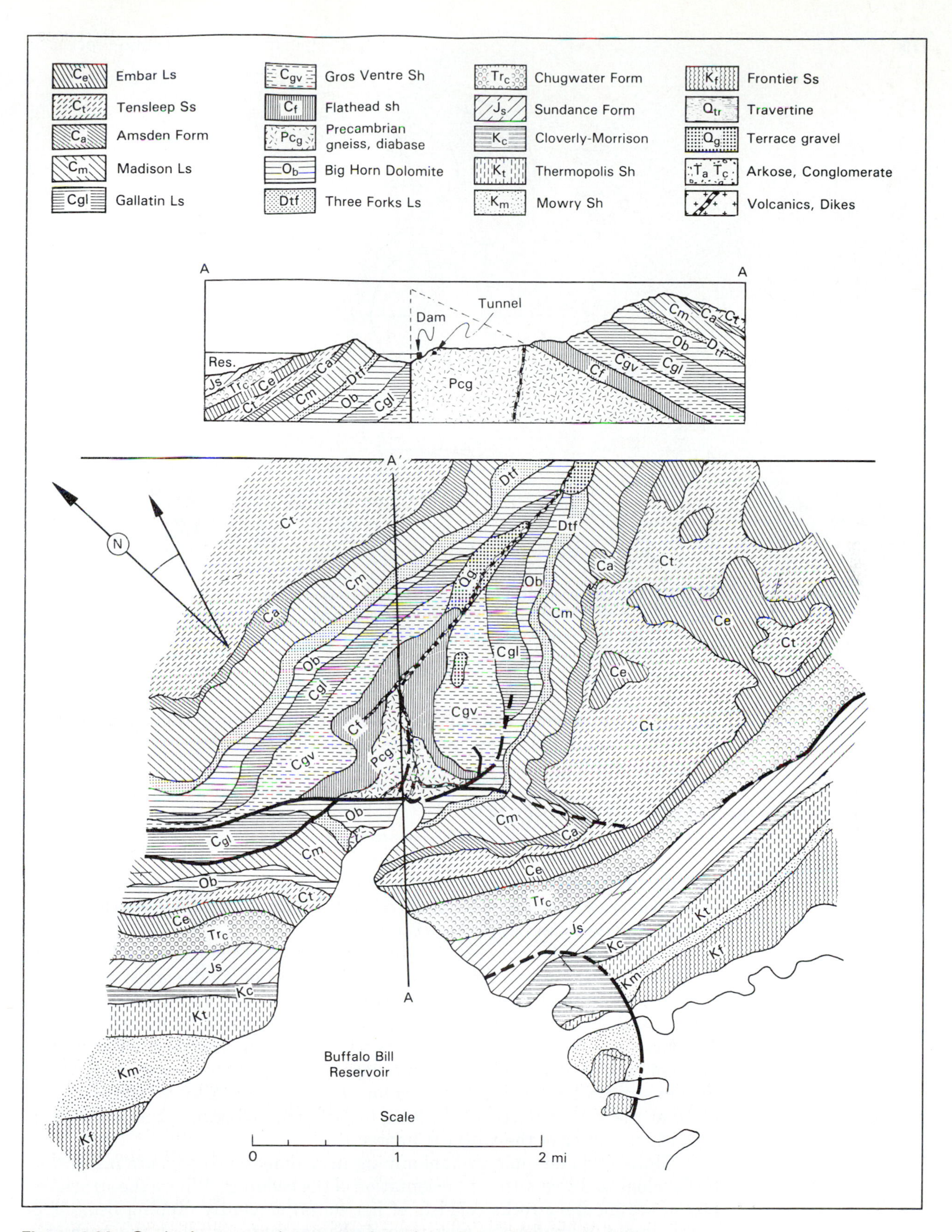


**Figure 4-30** Geologic map and cross section of the Cody Highway Tunnels area, Wyoming. (From W. F. Sherman, 1964, Engineering geology of the Cody Highway Tunnels, Park County, Wyoming, in *Engineering Geology Case Histories*, No. 4, P. D. Trask and G. A. Kiersch, eds.)

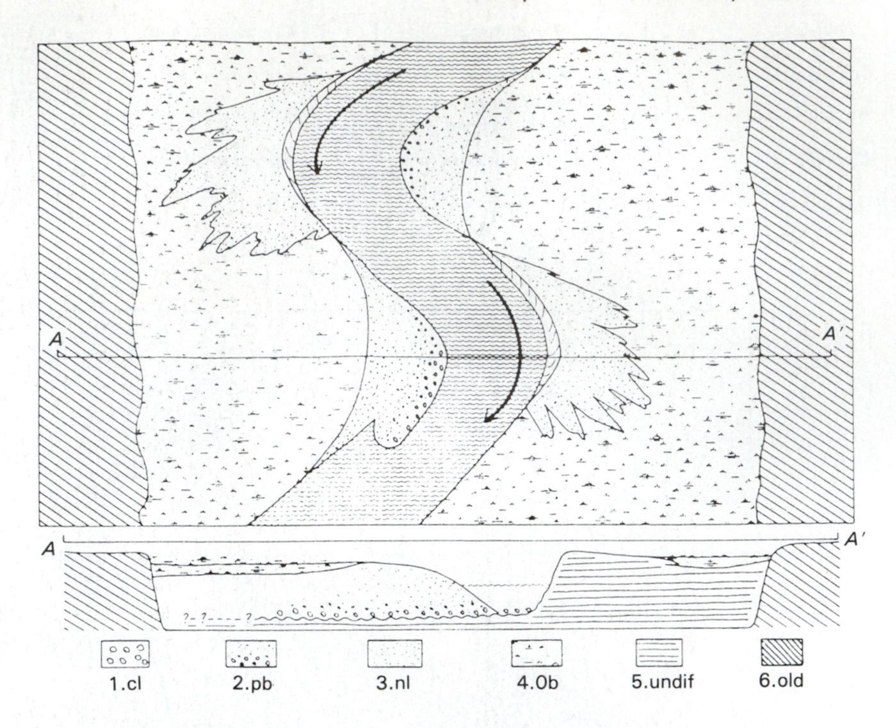


**Figure 4-31** Generalized facies in a meandering stream environment. (From R. K. Matthews, *Dynamic Stratigraphy*, 2d ed., copyright © 1984 by Prentice-Hall, Inc., Englewood Cliffs, N.J.)

accumulate in economically recoverable amounts (Figure 4-33). Figure 4-34 shows the major oil and gas provinces of the United States.

## ENGINEERING IN SEDIMENTARY ROCKS

Engineering conditions in sedimentary rocks are difficult to generalize because of the wide range in lithology, in degree of lithification, and in the orientation of bedding planes and other structures.

Stable vertical slopes can usually be excavated in well-cemented, horizontally bedded sandstones and limestones. Flatter slope angles must be cut for weaker rock types. Particularly important factors in the stability of sedimentary rock slopes are the direction and the amount of slope, or dip, of bedding. The most unfavorable situation occurs where bedding dips parallel or nearly parallel to the downslope angle of the slope. Bedding planes are zones of weakness in sedimentary rock masses and failure may occur. A huge landslide took place at the Vaiont Reservoir in Italy in 1963 partly because bedding dipped toward the center of the valley (Chapter 12).

Tunneling and underground mining in sedimentary rocks are influenced by lithology and structure (the orientation of the bedding). Where the stratigraphy of sedimentary rocks consists of horizontal or gently dipping beds, it is quite simple to predict the rock types to be encountered along the path of a tunnel. Difficulties arise in areas where the structure is more complex. Well-cemented sedimentary rocks are generally adequate for most types of building foundations. Special problems occur in limestones and evaporite deposits because these rocks are soluble under the action of flowing ground water. The

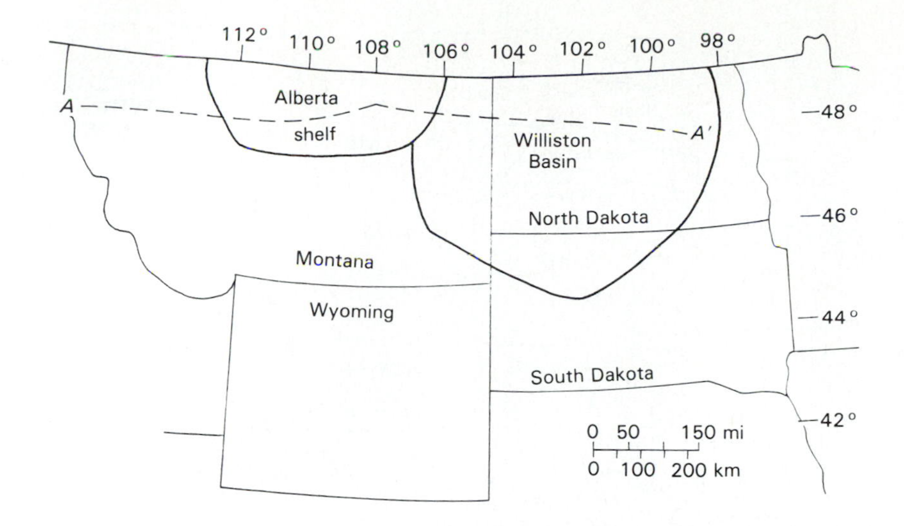


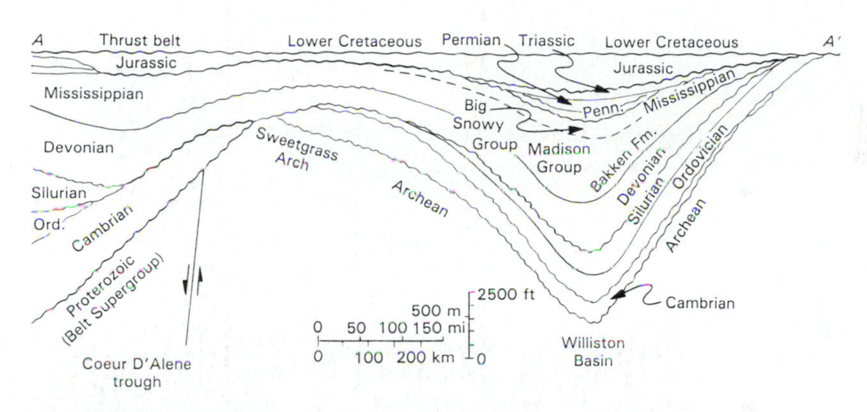


**Figure 4-32** Subsurface correlation of time-rock units across the Williston Basin. (After J. A. Peterson, 1988, Phanerozoic Stratigraphy of the northern Rocky Mountain region, in *The Geology of North America, vol. D-2: Sedimentary Cover North American Craton, U.S.*, Geological Society of America.)

soils and rocks overlying underground cavities produced by chemical dissolution may collapse into the voids, damaging or destroying buildings constructed at the surface (Chapter 11). This phenomenon is also a problem above shallow underground mines. Dams and reservoir sites are subject to similar limitations. Undesirable leakage of water may occur along bedding planes or through solution cavities in the rock.

Like igneous rocks, the engineering properties of the sedimentary rocks are influenced by geologic events that take place long after deposition of the sediments. Strength can be increased by compaction and cementation. Alternatively, sedimentary rocks can be weakened by weathering. Shales are particularly susceptible to breakdown to clayey soils upon exposure to air and water. These relatively rapid changes can lead to problems in excavations and foundations.

Discontinuities in the rock caused by tectonic events are also important in determining the engineering behavior of a sedimentary rock mass. Faults and fractures can significantly weaken a body of rock, as well as allowing the movement of water through the material.

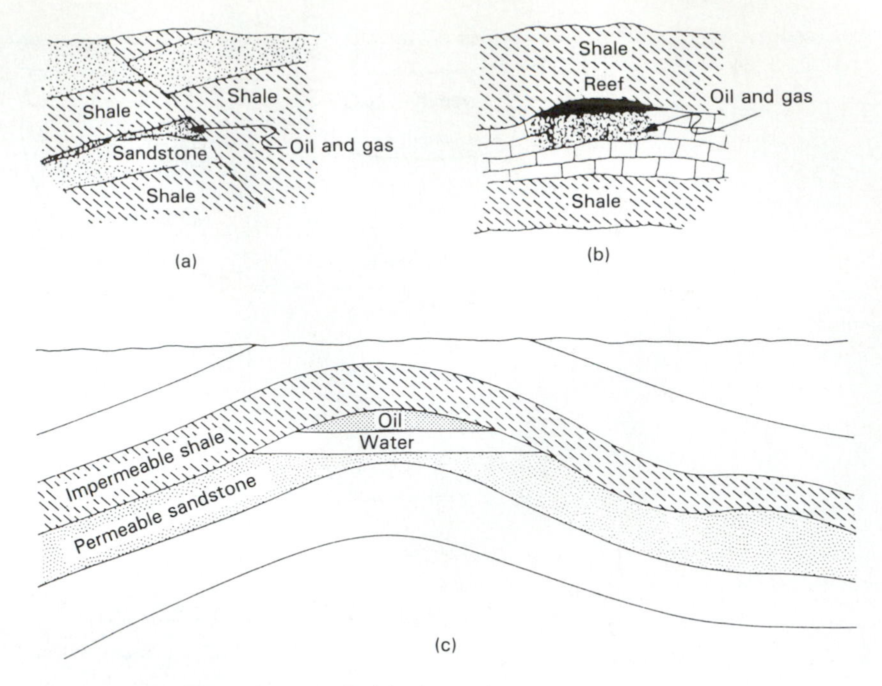


**Figure 4-33** Three types of petroleum traps.

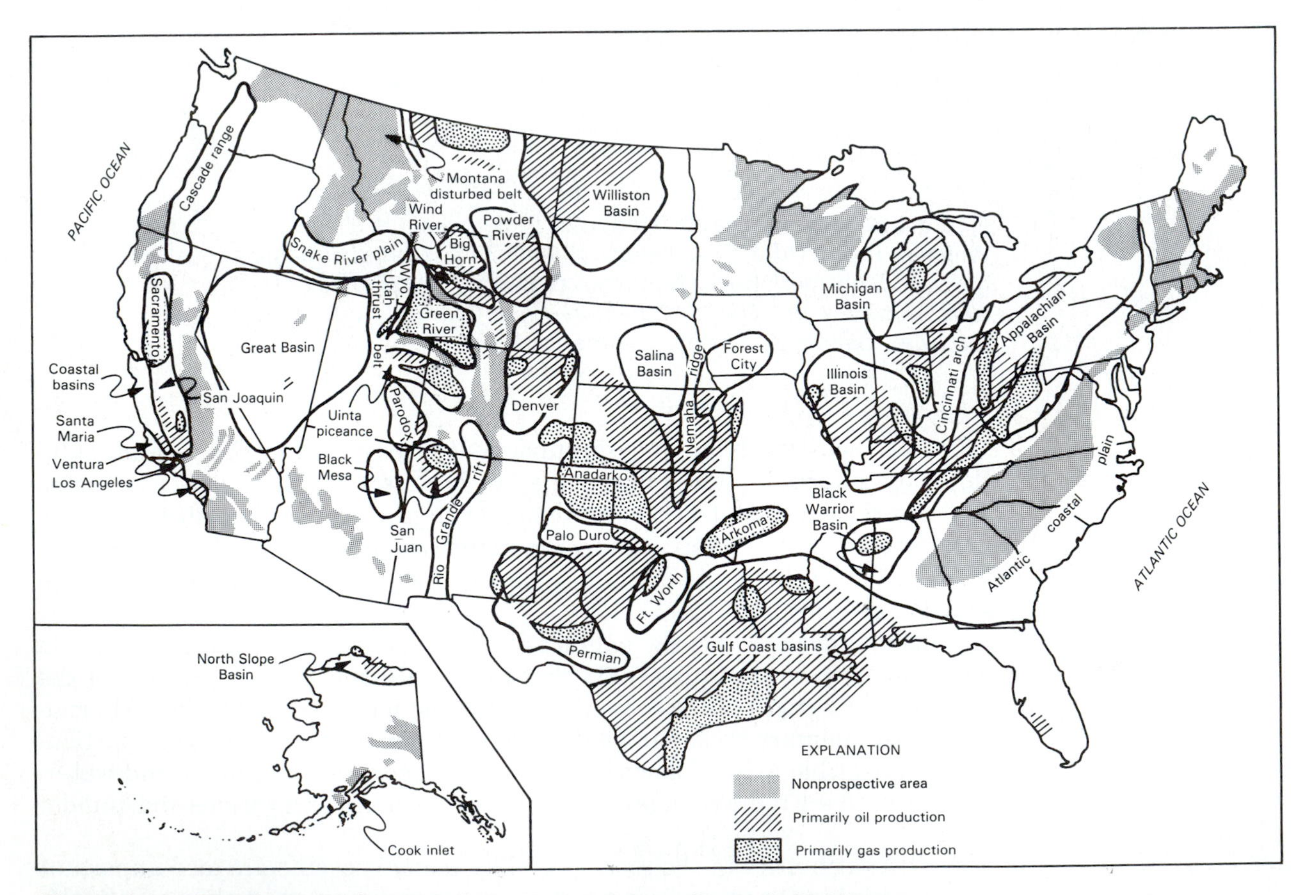


**Figure 4-34** Petroleum basin map of the United States. (From C. D. Masters and R. F. Mast, 1987, Geological Setting of U.S. Fossil Fuels, *Episodes*, vol. 10.)

## CASE STUDY 4-1
## Low-Level Radioactive Waste Disposal

One of the most controversial environmental problems facing nations that use nuclear power and possess nuclear weapons is the disposal of radioactive wastes. These wastes include high-level components, that have long half-lives and must be isolated from the environment for thousands of years, and low-level waste products that must be isolated for much shorter periods. Under current policy in the United States, a single repository is being developed for high-level commercial nuclear power plant wastes. In contrast, low-level waste, because of its lower hazard, will be disposed of in regional facilities licensed by the U.S. Department of Energy.

The history of radioactive waste disposal, like almost every other type of waste disposal, has progressed from little to no regard for health or the environment to a state approaching paranoia on the part of some individuals and groups in our society. As commercial nuclear reactors became common in the United States after World War II, low-level waste was commonly buried in shallow trenches at federally operated sites. After 1962, commercial sites were liscensed to dispose of the waste from nuclear power plants (Figure 4-35). These sites were distributed across the country from arid to humid climatic settings. Burial methods changed little from earlier operations—wastes were buried in shallow trenches and capped with nearby soil materials (Figure 4-36). Materials of low permeability (a low ability to trans-

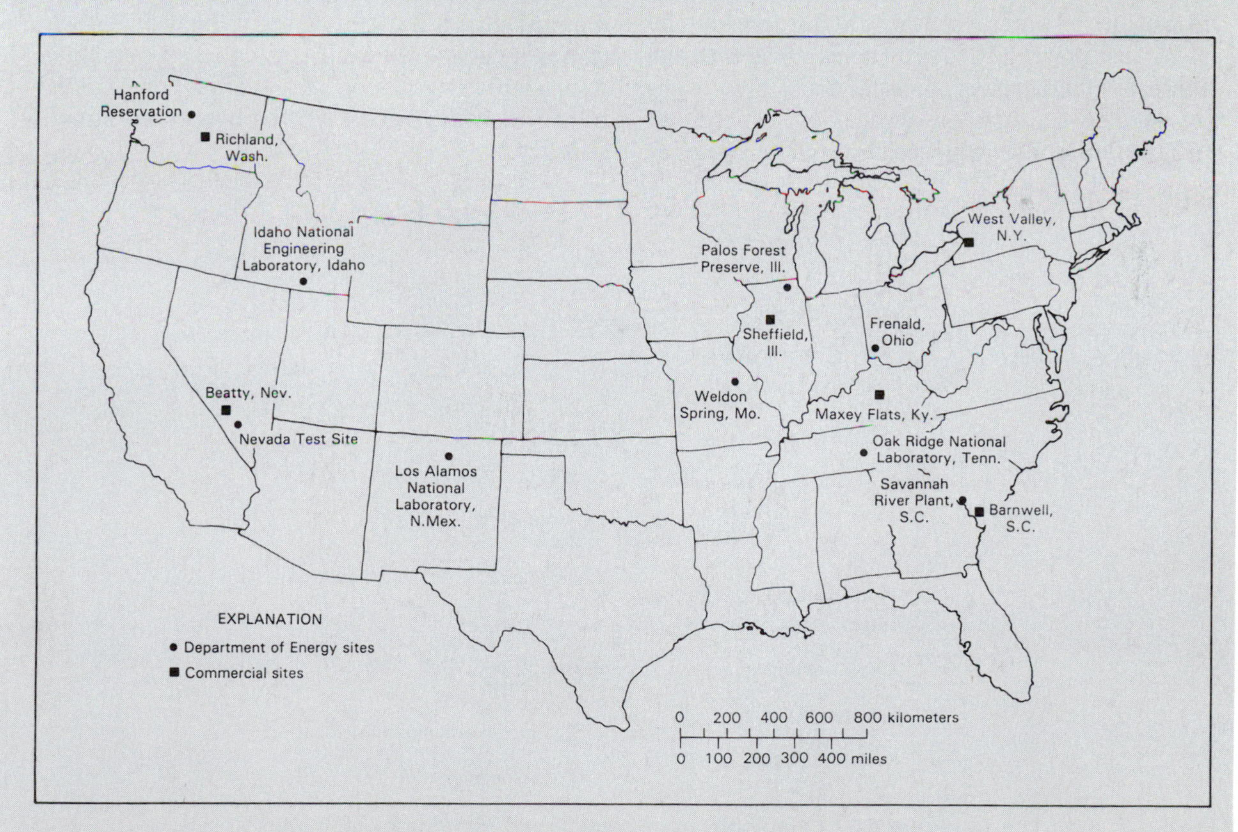


**Figure 4-35** Commercial and Department of Energy low-level radioactive-waste sites in the United States. (From M. S. Bedinger, 1989, *Geohydrologic Aspects for Siting and Design of Low-Level Radioactive-Waste Disposal*, U.S. Geological Survey Circular 1034.)

**Figure 4-36** Active trenches containing low-level radioactive waste at the Barnwell, South Carolina, site.

mit fluids) were selected if possible to minimize migration of waste that could be leached by water infiltrating through the cap and into the waste. Unfortunately, most of these sites were not able to contain the wastes as anticipated and radionuclides migrated out of the trenches for various distances. Currently, only two commercial sites remain open until the new generation of engineered facilities can be developed. A brief history of two of the sites constructed in sedimentary rocks will be presented subsequently.

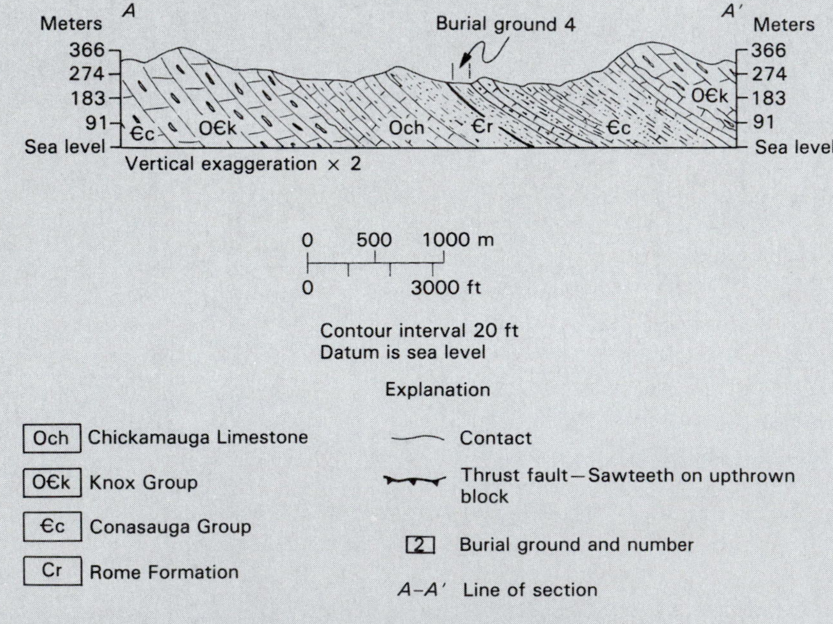


**Figure 4-37** Geologic cross section of sedimentary rock units at the Oak Ridge National Laboratory site in Tennessee. (From M. S. Bedinger, 1989, *Geohydrologic Aspects for Siting and Design of Low-Level Radioactive-Waste Disposal*, U.S. Geological Survey Circular 1034.)

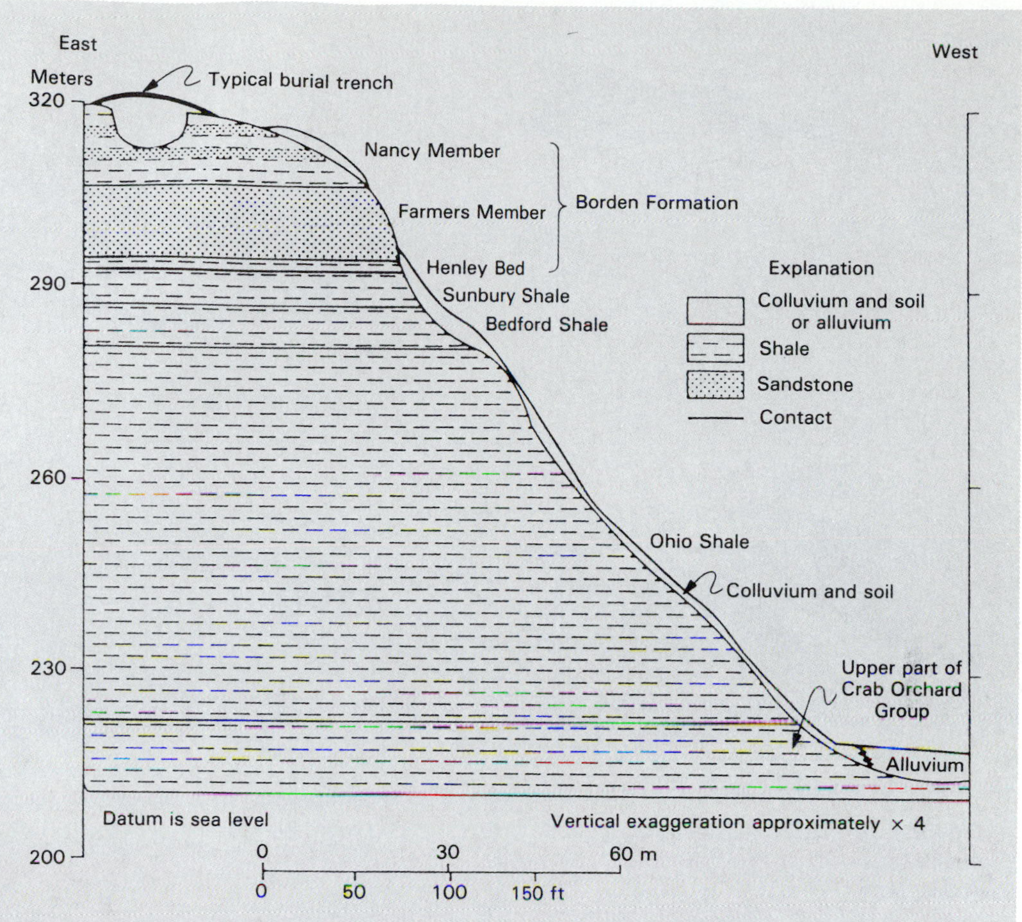


**Figure 4-38** Geologic cross section of sedimentary rock units at the Maxey Flats, Kentucky, disposal site. (From M. S. Bedinger, 1989, *Geohydrologic Aspects for Siting and Design of Low-Level Radioactive-Waste Disposal*, U.S. Geologic Survey Circular 1034.)

The Oak Ridge National Laboratory has been used for burial of low-level wastes since 1944. Trenches were excavated in weathered material overlying Paleozoic sedimentary rocks that were mainly composed of shale and limestone (Figure 4-37). Water frequently accumulated in the trenches in this humid climate and was not initially considered to be a problem. Problems developed, however, and radionuclide migration occurred in several ways. In the carbonate rocks, leached radionuclides were transported through solution cavities and fractures. In the shale sites a different process occurred. Because of the low permeability of these materials, and the higher permeability of the trench caps, the filled trenches behaved according to the "bathtub" effect: Water infiltrated from the surface more rapidly than it drained through the trench walls and bottoms, and gradually waste liquids overflowed in seeps at the contact point between the cap material and the natural soils.

Maxey Flats, Kentucky, is located about 300 km north of the Oak Ridge site. Here, a commercial site was constructed in flat-lying, late Paleozoic shale and sandstone (Figure 4-38). Trenches were excavated in the soil developed upon the Nancy Member of the Borden Formation, a unit composed of interbedded sandstone and shale. Despite grading the trenches to a sump containing pipes to drain off accumulated water, radionuclides migrated out of the trenches. The primary pathway was found to be a fractured sandstone unit that occurred near the bottom of the trenches. Certain radionuclides were estimated to be migrating

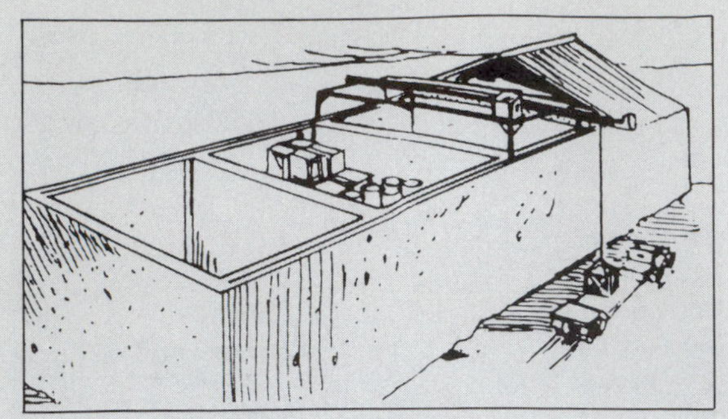

Above-ground vault

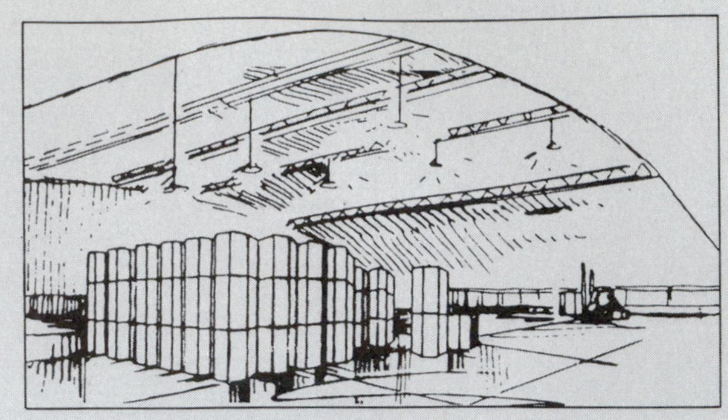

Above-ground modular concrete canister

Below-ground vault

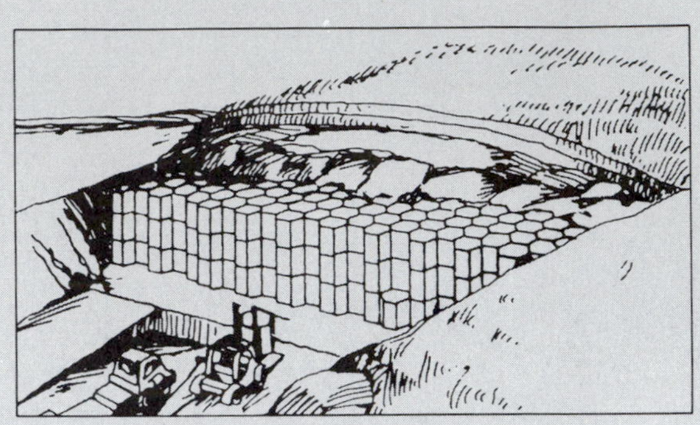

Below-ground modular concrete canister

**Figure 4-39** Potential design concepts for engineered low-level radioactive waste repository. (From Michigan Low-Level Radioactive Waste Authority, 1989, *Siting Process Overview.*)

at a rate of 17 m per year, and seeps along hillsides near the trenches showed contamination by radioactive isotopes. In an attempt to reduce infiltration into the trenches, compacted clay caps were used to cover the trenches. When the clay caps failed to solve the problem, PVC (polyvinylchoride) sheets were used to cover the trenches. Although these covers have been effective, it is obvious that the site was poorly suited for the method of waste disposal chosen.

Under the Low-Level Radioactive Waste Policy Act (1980) and Amendments Act (1985), states must take responsibility for their own waste. New facilities must meet very strict performance requirements. Siting will be based upon favorable geological and hydrological conditions as indicated by extensive geological investigations at prospective sites. In addition, the disposal facilities will be engineered structures, designed to provide multiple barriers to waste migration. Design concepts include both above- and below-ground concrete vaults, with multiple permeability barriers and drainage systems to prevent any leachate from migrating into the surface or subsurface environment (Figure 4-39). Continuous monitoring of all components of the site will be required.

As with other waste-disposal issues in our society, public opposition has been the biggest obstacle to site selection. Under the site-selection procedure specified by the federal government, states have banded together in compacts for the purpose of developing regional sites. Although progress is being made in some states, in others the process has become bogged down in a quagmire of lawsuits, delays, and public opposition whenever a specific site or area is mentioned.

## SUMMARY AND CONCLUSIONS

Sedimentary rocks originate from sedimentary processes that have been operating throughout geologic time. The first stage in the evolution of these rocks is the weathering and erosion of existing rocks. The products derived from this activity are then transported and deposited in a variety of terrestrial and marine environments. After deposition has occurred, then compaction, cementation, and crystallization cause lithification of the initially soft sediment.

Sedimentary rocks exhibit either clastic or nonclastic textures. Clastic textures are characteristic of the detrital rocks, those that have been transported to their point of deposition. The classification of the detrital rocks depends upon the size of the particles that make up the rock.

Although clastic textures may also be present in the chemical and biochemical sedimentary rocks, many of these rocks are composed of grains that have chemically precipitated at the site of deposition. Nonclastic textures, characterized by interlocking grain networks, result from this process.

The nature of each depositional environment imparts numerous characteristic features to the resulting rocks. Current strength, mode of transport, and water depth influence the sorting and bedding of the rock. Transport of sediment in ripples or dunes causes cross bedding to develop. Parallel bedding forms both by current action and by deposition in standing water. Turbidity currents transport high concentrations of suspended sediments that are deposited in graded beds.

The color and fossil content of sedimentary rocks are partially determined by geochemical factors. The presence or absence of oxygen is an extremely important variable in this regard.

Study of the stratigraphy of sedimentary rocks, using all the characteristics that have been described, has produced important information about the geologic past as well as practical data of great value to engineering projects.

## PROBLEMS

1. How do sedimentary rocks fit into the geologic cycle?
2. What characteristics of the depositional environment influence the properties of sedimentary rocks?
3. How do postdepositional changes convert sediments to sedimentary rocks?
4. What chemical conditions influence the precipitation of nonclastic sedimentary rocks?
5. How do cross beds form?
6. What does color indicate about the geochemical conditions of the depositional environment?
7. What methods are available to gather data necessary to construct a stratigraphic cross section in an area? Why are cross sections useful in highway construction and other engineering projects?
8. What properties influence the strength or possible failure of slopes underlain by sedimentary rocks?
9. What geologic conditions are required for the accumulation of petroleum?
10. What types of sedimentary rocks are considered to be potentially suitable for high-level nuclear waste disposal? Why?

11. Describe the design features of low-level radioactive waste-disposal facilities? What problems have been encountered with respect to these sites in sedimentary rocks?

## REFERENCES AND SUGGESTIONS FOR FURTHER READING


BEDINGER, M. S. 1989. *Geohydrologic Aspects for Siting and Design of Low-Level Radioactive-Waste Disposal.* U.S. Geological Survey Circular 1034.

BLATT, H., G. MIDDLETON, and R. MURRAY. 1980. *Origin of Sedimentary Rocks*, 2d ed. Englewood Cliffs, N.J.: Prentice-Hall, Inc.

GEOLOGICAL SOCIETY OF AMERICA. 1964. *Engineering Case Histories: Numbers 1–5*, P. D. Trask and G. A. Kiersch, eds. Boulder, Colo.

JUDSON, S., M. E. KAUFFMAN, and L. D. LEET. 1987. *Physical Geology*, 7th ed. Englewood Cliffs, N.J.: Prentice-Hall, Inc.

MASTERS, C. D., and R. F. MAST. 1987. Geological setting of U.S. fossil fuels. *Episodes*, 10 (no. 4):308–313.

MICHIGAN LOW-LEVEL RADIOACTIVE WASTE AUTHORITY. 1989, *Siting Process Overview.*

PETERSON, J. A. 1988. *Phanerozoic Stratigraphy of the Northern Rocky Mountain Region: The Geology of North America, v. D- 2, Sedimentary Cover North American Craton.* Geological Society of America.

U.S. DEPARTMENT OF ENERGY. 1984. Draft environmental assessment—Vacherie Dome site, Louisiana. (DOE/RW-0016.)

# 5

# Metamorphic Rocks and Processes

The igneous and sedimentary rocks that form at or beneath the earth's surface are not always the final result of the dynamic activity of the earth's crust. When exposed to the atmosphere, the minerals in these rocks may chemically alter to become new minerals that are more stable under the temperatures and pressures at the earth's surface. Similarly, when igneous and sedimentary rocks become buried deep within the crust because of the movements of lithospheric plates, they are subjected to elevated temperatures and pressures. Just as the rocks are unstable at low temperatures and pressures on the earth's surface, they are also unstable at the high temperatures and pressures existing at depths well below the surface. Under these conditions, minerals in the rock recrystallize to become new minerals, and new rock is formed. The processes that cause these changes are grouped under the term *metamorphism*. Study of these rocks yields valuable information about the effects of these conditions on rock and about the geologic history of a region.

## METAMORPHIC PROCESSES

The normal increases in temperature and pressure with increasing depth in the earth are sufficient to initiate metamorphic activity. In addition, tectonic stresses caused by the collision of two lithospheric plates, for example, can generate intense pressures oriented in a particular direction (Figure 5-1). The heat and pressure in deep crustal environments are two of the main components of metamorphic processes. A third component is the effect of liquid or gaseous solutions that move through the rocks and promote the chemical reactions that form new minerals.

Heat is perhaps the most important metamorphic agent. In Figure 5-2 the rate of increase of temperature with depth, or the *geothermal gradient*, suggests that the temperature in the crust at a depth of 15 km is approximately 300°C. This temperature is sufficient for the recrystallization of some minerals to begin. Because the geothermal gradient is highly variable from place to place, the depth associated with a particular temperature is not constant.

The effect of pressure varies at different depths in the crust. At shallow depths, rocks are relatively cold and brittle, so they can be fractured and crushed when subjected to high pressures. At greater depths, rocks are much softer because of the high temperatures. Under the action of pressure, they tend to deform by plastic flow, like modeling clay squeezed between the fingers. In the region of plastic deformation, pressure influences the types of new

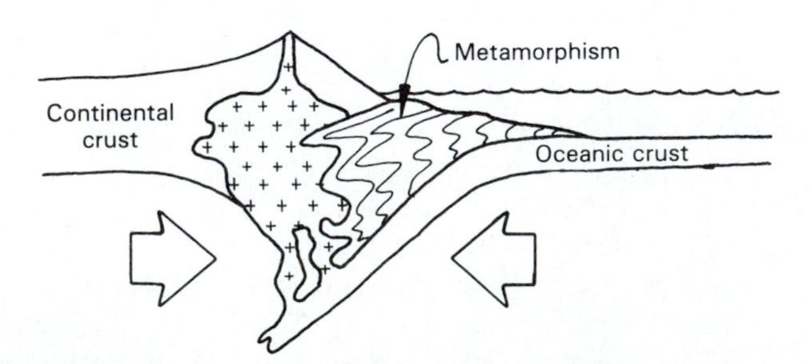


**Figure 5-1** Metamorphism at a convergent plate margin. Rocks metamorphosed may include sediments or other continental or oceanic crustal material.

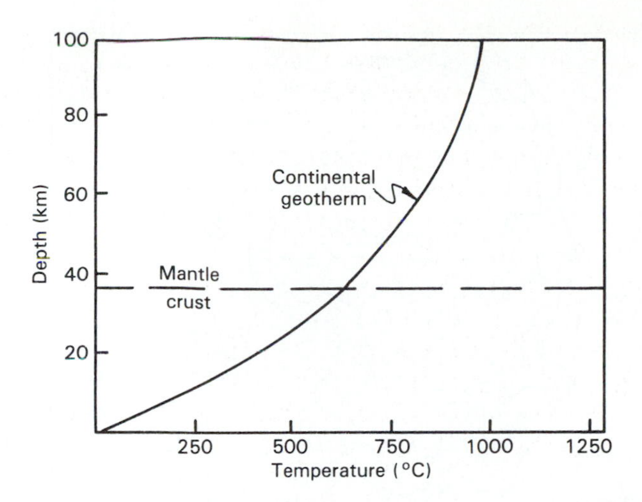


**Figure 5-2** Representative continental geotherm. (From G. C. Brown and A. E. Mussett, *The Inaccessible Earth*, copyright © 1981 by George Allen & Unwin, Ltd., London.)

minerals formed. Typically, the atoms within the mineral structure are more closely packed together when the mineral crystallizes under high pressure.

The recrystallization of minerals during a metamorphic event is largely a solid-phase process. The rock does not actually melt and then recrystallize. The solid-phase reactions between minerals are greatly facilitated by the movement of small amounts of liquid or gaseous solutions through the rock. These solutions, which travel through the pores and cracks of the rock, add and remove various ions and molecules as the reactions occur. In this way new chemical constituents can be brought into contact with mineral grains so that they may diffuse through the mineral structures during recrystallization.

### *Types of Metamorphism*

Metamorphism is associated with several types of geologic events. The intrusion of a pluton, for example, brings magma into contact with existing crustal rocks (Figure 5-3). *Contact metamorphism* is the name given to the alteration of the surrounding *country rock* by the intruding magma. Heat is the most significant influence in contact metamorphism. The effect of pressure is much less important. The extent of contact metamorphism, the zone of altered rock called the *metamorphic aureole*, rarely extends more than several hundred meters outward from the magmatic body. Thus contact metamorphism is limited in areal extent to a thin shell around the pluton. More extensive effects may occur if solutions and vapors given off by the magma penetrate into the country rock along fractures. These *hydrothermal solutions* carry volatile components that separate from the cooling magma. Important vein-type ore deposits can be formed by hydrothermal solutions that migrate during contact metamorphic events.

A second type of metamorphism is characterized by much more extensive zones of rock alteration than found in the contact metamorphism zones. Its name, *regional metamorphism*, suggests a process that may transform rocks through huge portions of the crust. The conditions that produce regional metamorphism involve the effects of both temperature and pressure. Thus regional metamorphism must occur deep within the crust, at depths of at least 10 km.

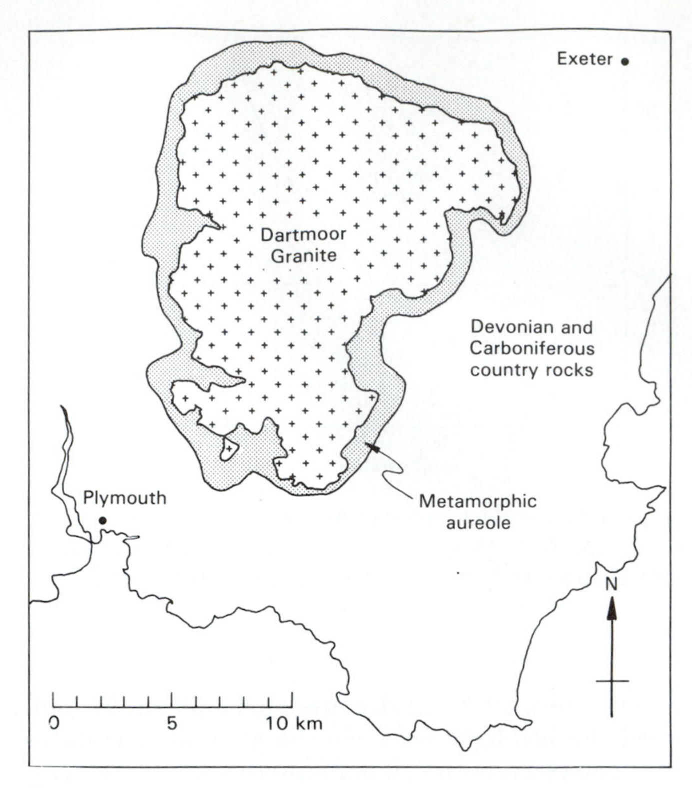


**Figure 5-3** The contact metamorphic aureole surrounding the Dartmoor Granite, Devon, England. (From R. Mason, *Petrology of the Metamorphic Rocks*, copyright © 1978 by George Allen & Unwin, Ltd., London.)

Geologic studies of metamorphic rocks indicate that the mineral content of rocks in regionally metamorphosed areas varies systematically. The specific group of minerals present in rock can be used to define a certain metamorphic *grade* in a particular area. High grade, for example, means that the rock was subjected to very high temperatures and pressures. Different metamorphic minerals can be produced from the same original rock under various metamorphic grades. Figure 5-4 shows the conditions of temperature and pressure under which the minerals kyanite, sillimanite, and andalusite form. All three metamorphic minerals have the same chemical composition ($Al_2SiO_5$) but different internal structures. Regional metamorphism produces areas in which metamorphic grade is very high. Grade decreases in all directions from the most intensely altered zones (Figure 5-5).

Regional metamorphism is often associated with the central cores of mountain ranges, whose rocks are exposed after the removal by erosion of thousands of meters of overlying rocks. This relationship suggests that regional metamorphism may be a part of the mountain-building process. A possible scenario for regional metamorphism involves the intense lateral pressure generated at convergent plate margins. Thick wedges of sediment deposited adjacent to continents or island arcs are downwarped as the plates collide. Granite batholiths are often found at the centers of these metamorphosed zones, surrounded by metamorphic rocks that decrease in grade radially outward (Figure 5-5). Rocks called *migmatites* (Figure 5-6) exhibit characteristics of both igneous and metamorphic rocks. Migmatites can often be observed to merge laterally with granite of igneous appearance in one direction and with unmistakable

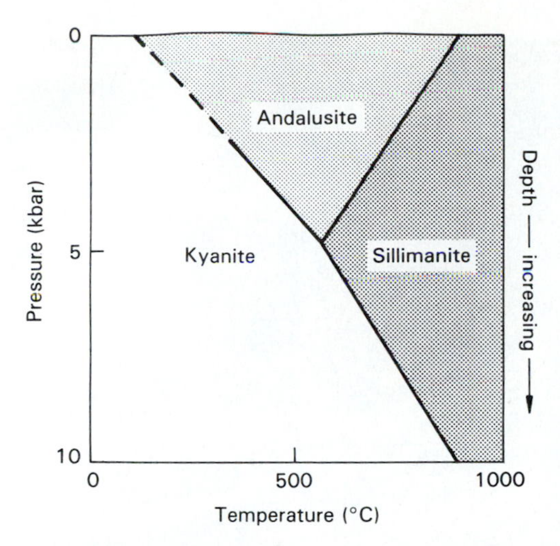


**Figure 5-4** Metamorphic minerals formed from $Al_2SiO_5$ under varying conditions of temperature and pressure. (From S. Judson, M. E. Kauffman, and L. D. Leet, *Physical Geology*, 7th ed., copyright © 1987 by Prentice-Hall, Inc., Englewood Cliffs, N.J.)

metamorphic rocks in the opposite direction. All these relationships suggest that metamorphism of downwarped crustal rocks may be so intense that remelting occurs in the zone of highest temperature and pressure. Thus granitic batholiths surrounded by metamorphic rocks may represent an episode of plate collision in the geologic past.

A third type of metamorphism is recognized when pressure is the dominant metamorphic agent. In the upper part of the crust, the intense pressure

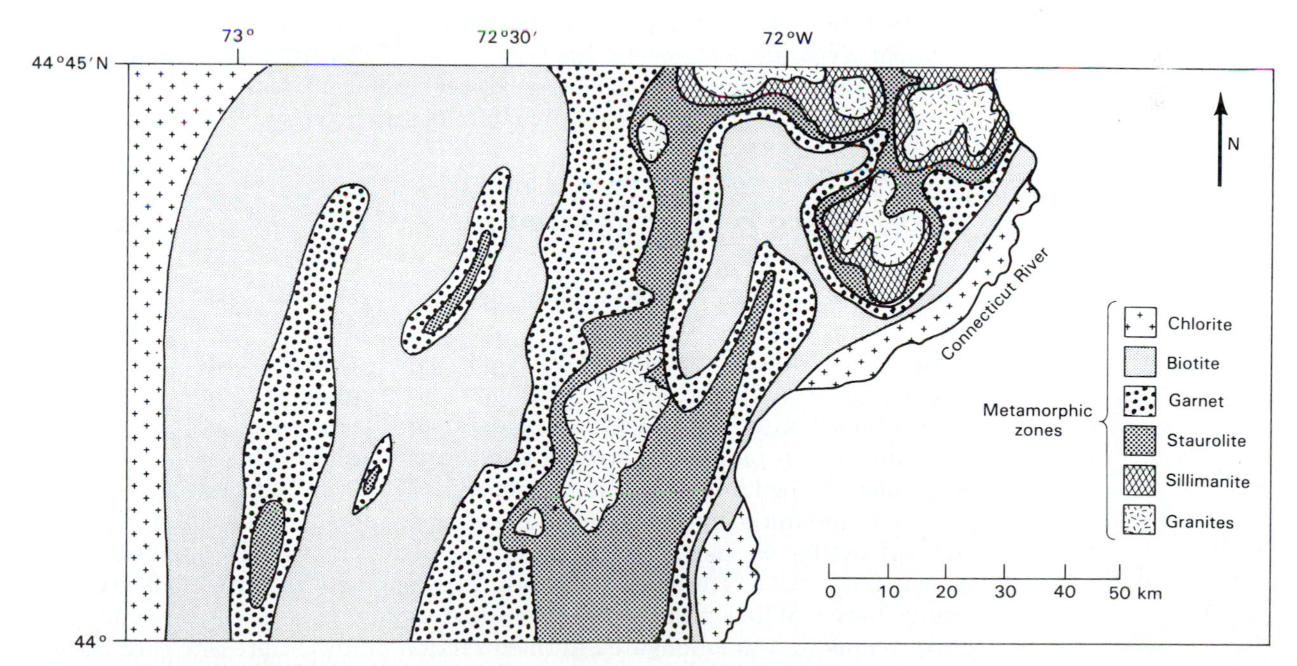


**Figure 5-5** Regional metamorphism in northern Vermont. Metamorphic grade decreases outward from granitic plutons. (From F. J. Turner, *Metamorphic Petrology*, 2d ed., copyright © 1981 by Hemisphere Publishing Corp. New York.)

**Figure 5-6** Migmatite, a metamorphic rock containing thin layers of light-colored granitic igneous rock. (Park County, Colorado; C. C. Hawley; photo courtesy of U.S. Geological Survey.)

associated with folding and faulting is sufficient to crush and pulverize the minerals along a fault plane. The alteration of rocks under these conditions is called *dynamic metamorphism*. The zones of crushing are limited to the close proximity of planes of shearing between adjacent masses of brittle rock. Temperature is a minor factor in dynamic metamorphism.

## CHARACTERISTICS AND TYPES OF METAMORPHIC ROCKS

A major division of metamorphic rocks can be made according to texture. The aspect of texture that is most important is called *foliation*, the parallel orientation of mineral grains within a metamorphic rock. Foliation gives metamorphic rocks a banded or layered appearance, and for this reason foliation resembles the bedding of sedimentary rocks (Figure 5-7). In reality, however, foliation is produced by an entirely different mechanism from bedding. Tectonic stresses acting on rocks during metamorphism are usually applied in one principal direction. The stress, or pressure, acting in other directions is considerably lower. When minerals recrystallize under these pressure conditions, platy grains that are elongated in the direction of lowest pressure are favored (Figure 5-8). The long axes of the mineral grains are therefore perpendicular to the direction of greatest pressure, and the parallel orientation of these grains, as seen in the micas for example, gives the rock foliation, which is perpendicular to the applied stress.

**Figure 5-7** Foliation, the orientation of platy minerals that gives metamorphic rocks a layered appearance. (Southern Maine.)

Metamorphic rocks can be classified as foliated or nonfoliated. Foliated rocks are associated with regional metamorphism where tectonic stress is a major factor. Nonfoliated rocks are produced during contact metamorphism or where platy minerals cannot crystallize because of a lack of necessary chemical components.

The presence of foliation also imparts to a rock a tendency to split or break along the foliation planes rather than across the planes or at some other orientation. This inclination to break along planes of weakness is called cleavage, although it must be kept in mind that the planes of weakness are not of the same origin as cleavage in a mineral.

Several common foliated metamorphic rocks are named because of the type of cleavage that they possess. If the cleavage planes are very thin and the rock is fine grained, the cleavage is known as *slaty cleavage*. The resulting rock is *slate* (Figure 5-9), a rock that has been used as roofing material for centuries because of its smooth and regular cleavage surfaces. Slate is usually produced by low-grade metamorphism of shale, primarily consisting of directed pressure and fairly low temperatures. With increasing metamorphic grade,

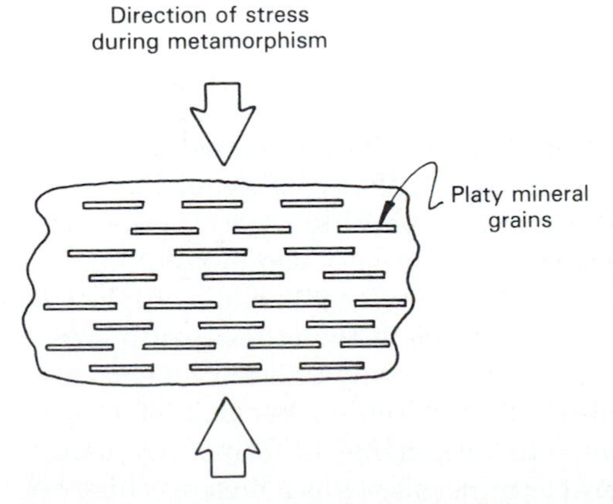


**Figure 5-8** Orientation of platy mineral grains at right angles to direction of highest pressure during metamorphism.

**Figure 5-9** Well-developed slaty cleavage in slate outcrop near Kodiak, Alaska. (G. K. Gilbert; photo courtesy of U.S. Geological Survey.)

slate can be converted to *phyllite*, so named because of its phyllitic cleavage. Phyllite differs from slate in that it is somewhat coarser grained. In addition, the flaky mineral grains of phyllite possess a distinctive silky luster.

*Schist* (Figure 5-10) is a very common foliated rock in which the grains are coarser than in slate or phyllite. In addition, the surfaces of cleavage planes are relatively rough because of the coarse grain size. Schist represents a still higher metamorphic grade. Many types of igneous and sedimentary rocks can be metamorphosed to form schist. Because of the variety of mineralogy in schists, the complete rock name is formed by using the name of the most abundant mineral or minerals. Thus a biotite schist would be mainly composed of biotite with lesser amounts of other minerals.

**Figure 5-10** An outcrop of schist near Riggins, Idaho. The well-developed foliation is evident. (W. B. Hamilton; photo courtesy of U.S. Geological Survey.)

The foliated rock that develops under high-grade metamorphic conditions is known as *gneiss*. Gneiss is a coarse-grained, coarsely banded rock (Figure 5-11). The foliation consists of alternate bands of light- and dark-colored minerals. The light-colored layers are mainly composed of quartz and feldspar; whereas the dark layers contain biotite, hornblende, augite, and other miner-

**Figure 5-11** Gneiss, a coarse-grained metamorphic rock, with broad light- and dark-colored bands of foliation. (Ferry County, Washington; R. L. Parker; photo courtesy of U.S. Geological Survey.)

**Figure 5-12** Quartzite, a very hard metamorphic rock in which recrystallization forms a dense network of quartz grains. (East of Knoxville, Tennessee; J. B. Hadley; photo courtesy of U.S. Geological Survey.)

als. Gneisses are formed from silicic igneous rocks as well as from various types of sedimentary rocks.

Of the nonfoliated metamorphic rocks, *quartzite* and *marble* are the most common. Quartzite is the name given to metamorphosed quartz sandstone (Figure 5-12). Recrystallization of quartz in the original sandstone fills voids between the existing grains and increases the strength and density of the rock. Quartzites are among the strongest and hardest rock types. Because quartzite consists mainly of nonplaty quartz grains, foliation is lacking, although some quartzite may appear foliated because of relict bedding structure in the rock.

Marble (Figure 5-13) is recrystallized limestone or dolomite. The recrystallization process produces large interlocking grains of calcite or dolomite. Impurities in the rock give marble a number of possible colors. Many of these varieties are highly sought after for use as decorative building stone.

Rocks formed by dynamic metamorphism may be foliated or nonfoliated. If rocks adjacent to a fault zone are crushed by pressure and shear displacement, a structureless *fault breccia* (Figure 5-14) may be produced. If the zone of shear is subject to an intense degree of crushing, a fine-grained rock called *mylonite* may be formed. Under these conditions, recrystallization and foliation are produced. In some areas mylonite has been linked to high levels of radon gas

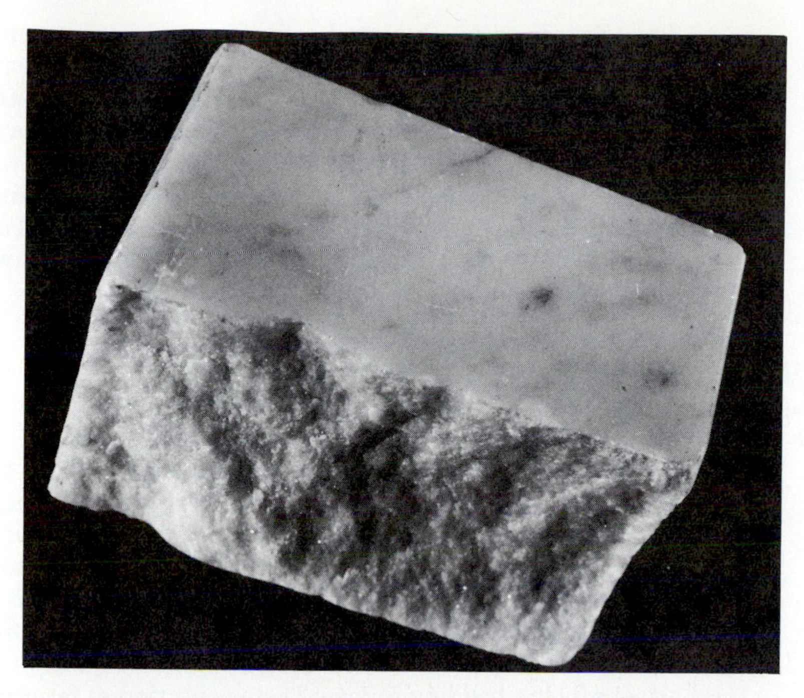

**Figure 5-13** Metamorphism of limestone or dolomite produces marble, a rock commonly used as building stone. The coarse, sugary appearance on the broken face of the specimen is characteristic.

in buildings that overlie these rocks. Radon is a radioactive gas that is produced by the decay of uranium. Fault zones, which may contain mylonite, provide conduits for the upward migration of radon from its source rocks. Migration has to be rapid for the gas to reach the surface because its half-life is only 3.8

**Figure 5-14** Fault breccia composed of dolomite fragments. (Clark County, Nevada; C. Deiss; photo courtesy of U.S. Geological Survey.)

days. The danger to human health from radon occurs when it accumulates in basements and people breathe it on a long-term basis. Once in the lungs, radon decays to polonium, another radioactive element, and the rapid decay of polonium can cause damage to lung tissues that may lead to cancer.

Although high radon levels are common in areas underlain by metamorphic rocks, particularly those with fault zones containing mylonites, radon is not limited to metamorphic rocks. A variety of igneous and sedimentary rocks also contain enough uranium to create a radon hazard when the conditions in soils and buildings in the area are also favorable.

## *ENGINEERING IN METAMORPHIC ROCK TERRAINS*

The engineering characteristics of metamorphic rocks can be generalized into two basic types. Nonfoliated rocks possess similar engineering properties to intrusive igneous rocks. In an unaltered and unfractured condition, they can be considered to be strong materials, with few limitations for foundations, tunnels, and dams. Vertical excavation slopes will remain stable. Foliated metamorphic rocks, however, are more similar to sedimentary rocks because of their tendency to fail along specific planes. Foliation planes in this instance are similar to bedding planes. The orientation of foliation planes with respect to a natural slope or excavation, therefore, becomes critical to the stability of the material.

In a way similar to the igneous and sedimentary rocks, the ultimate behavior of a metamorphic rock mass depends upon the degree and orientation of fractures and the weathering characteristics. These properties must be ascertained prior to construction of each individual engineering project.

### CASE STUDY 5-1

### Failure of the St. Francis Dam

Construction in metamorphic rock terrains requires careful mapping of rock types and foliation directions. Foliation imparts a directional weakness to the rock that can be critical to the design and operation of engineering projects. On March 12, 1928, only months after construction was completed, the St. Francis Dam in California (Figure 5-15) collapsed, releasing a wall of water 30 to 50 m high, which caused more than 400 deaths and great destruction in the valley below.

The geologic setting of the dam site is shown in Figure 5-16. On the east side of the valley, schists with foliation planes inclined toward the center of the valley formed the foundation material for the dam. The inherent weakness of the micaceous shists due to weak bonds along the foliation planes between adjacent mica grains was undoubtably a factor in the failure of the dam. Beneath the center of the dam, the schists were in fault contact with sedimentary rocks. The fault zone itself contained brecciated rocks produced during fault movement.

Although the precise cause of the dam collapse has never been proven, it is highly likely that water from the impounded reservoir seeped into the rocks beneath the dam, thus weakening the rock materials. On the east side, penetration of water into the schists may have weakened the resistance to failure along the inclined foliation planes in this abutment (Outland, 1977). At a critical point, failure of the rock abutment may have initiated the collapse of the dam.

(a)

(b)

**Figure 5-15** St. Francis Dam (a) before and (b) after failure. (Photos courtesy of Los Angeles Department of Water and Power.)

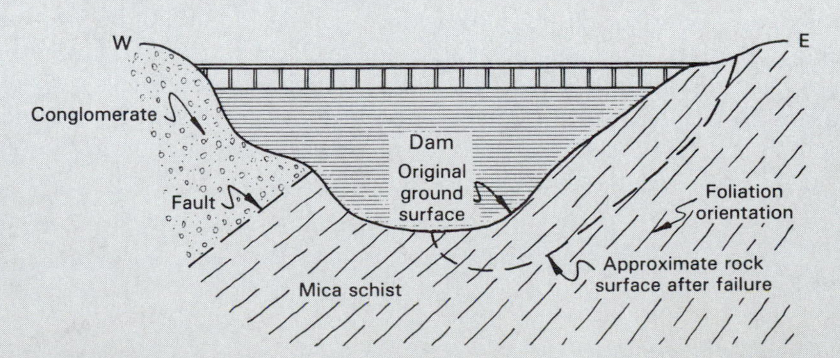


**Figure 5-16** Geologic cross section of the St. Francis Dam. Failure probably occurred in the schists beneath the east side of the dam. (Modified from C. F. Outland, *Man-made Disaster: The Story of St. Francis Dam*, copyright © 1977 by the Arthur H. Clark Co., Glendale, Calif.)

## CASE STUDY 5-2

## Tunneling Problems in Metamorphic Rocks

Tunneling often presents some of the most complex and dangerous problems in civil engineering. These problems usually result from unknown or unexpected adverse geologic conditions encountered along the path of the tunnel.

The Harold D. Roberts Tunnel was constructed to divert water from the west slope of the Continental Divide, through the Front Range of the Rocky Mountains, for municipal use in the city of Denver, Colorado. Problems experienced during the construction of this 37.5-km concrete-lined tunnel were directly related to the physical properties of the rocks encountered along the center line of the tunnel (Warner and Robinson, 1981). The geologic setting of this part of the Front Range consists of a highly complex series of igneous, metamorphic, and sedimentary rocks (Figure 5-17). The Precambrian metamorphic rocks, including schist, gneiss, quartzite, and migmatite, provided some of the most troublesome tunneling conditions along the route. Foliated and nonfoliated metamorphic rocks are present in various sections of the route, with biotite responsible for most of the foliation.

One of the major problems was the inflow of ground water at high pressure from fractured rocks at certain locations. Flows as high as 32 L/s (500 gal/min) were measured. In order to test for zones of water inflow, which could not be predicted in advance, feeler holes were drilled ahead of the advancing tunnel face. When a flow was detected, grout (a cement slurry), was injected into the rock at high pressure to seal the fractures. Relatively strong gneiss frequently yielded high inflows of water through occasional fractures. Apparently, this rock behaved as a brittle material that could maintain open fractures, unlike the weaker foliated schist, in which fractures were most likely to seal themselves under the immense weight of the mountain range.

The cost of a tunnel is directly related to the number of steel supports that are needed to prevent collapse of the rock walls into the opening. These supports were installed as needed during construction at 0.6 to 0.18 m centers along the tunnel. Where the tunnel passed through strong, competent rock, steel supports were unnecessary. The percentage of a particular segment of tunnel that required steel supports ranged from 7.3 to 100, depending upon the rock type.

When steel supports are needed at a section, the contractor must blast and excavate a larger cross-sectional area of rock than the estimated cross-sectional area, which is the area inside the *payline* (Figure 5-18). The extra area excavated is called *overbreak*. Overbreak increases the cost of a tunnel to the contractor because more concrete is needed to construct the finished tunnel lining. In Figure 5-19, mean overbreak, roughness factor, and eccentricity are plotted against percent support for a section of the Harold D. Roberts Tunnel. Roughness factor and eccentricity are parameters related, respectively, to irregularities and to differences in direction of the amount of overbreak. The graph clearly shows that, of the rock types compared, the foliated gneiss and schist had the highest values of the overbreak parameters and required the highest percentage of support. In this case, foliated metamorphic rocks were more costly for tunnel construction than other rock types.

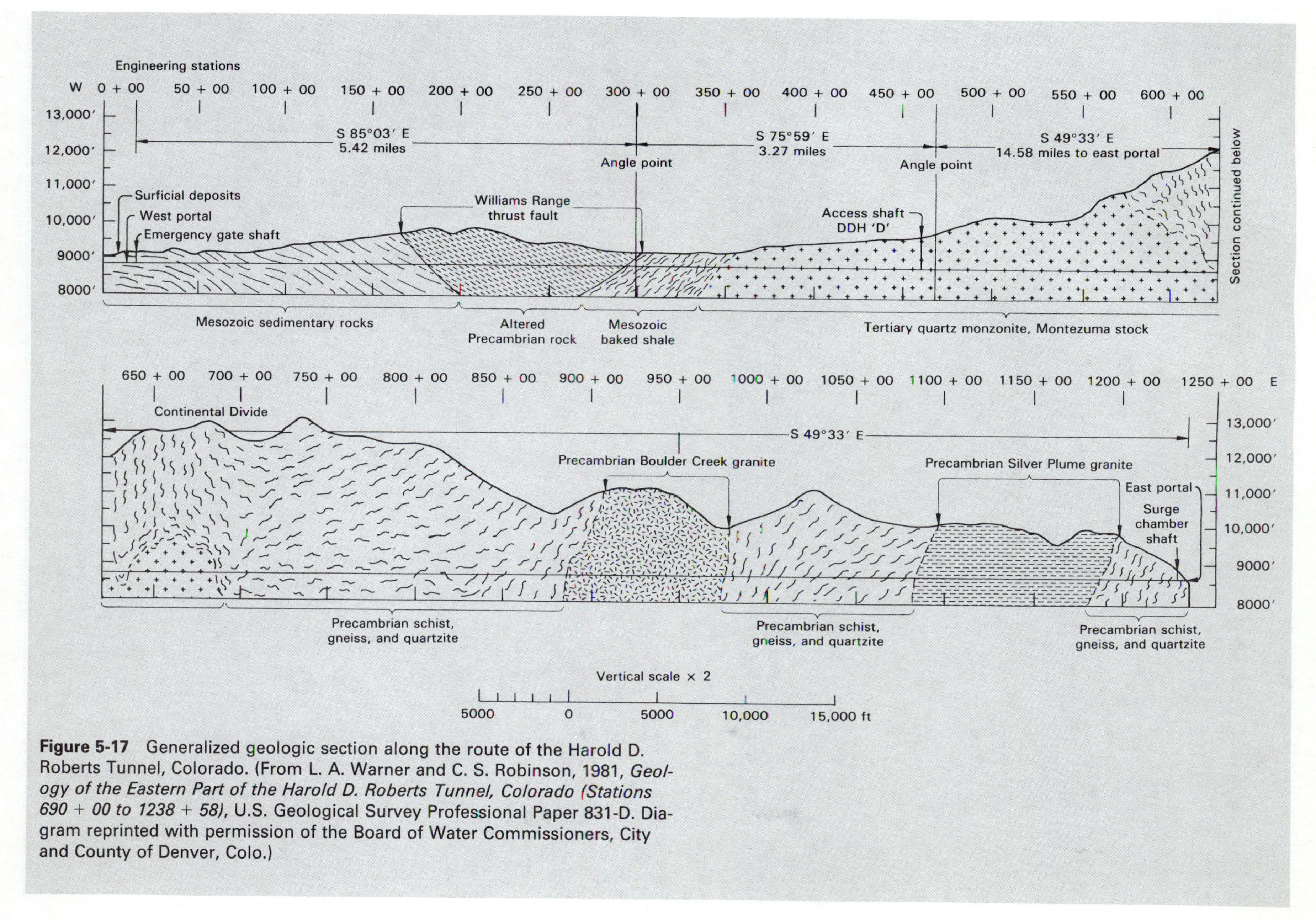


**Figure 5-17** Generalized geologic section along the route of the Harold D. Roberts Tunnel, Colorado. (From L. A. Warner and C. S. Robinson, 1981, *Geology of the Eastern Part of the Harold D. Roberts Tunnel, Colorado (Stations 690 + 00 to 1238 + 58)*, U.S. Geological Survey Professional Paper 831-D. Diagram reprinted with permission of the Board of Water Commissioners, City and County of Denver, Colo.)

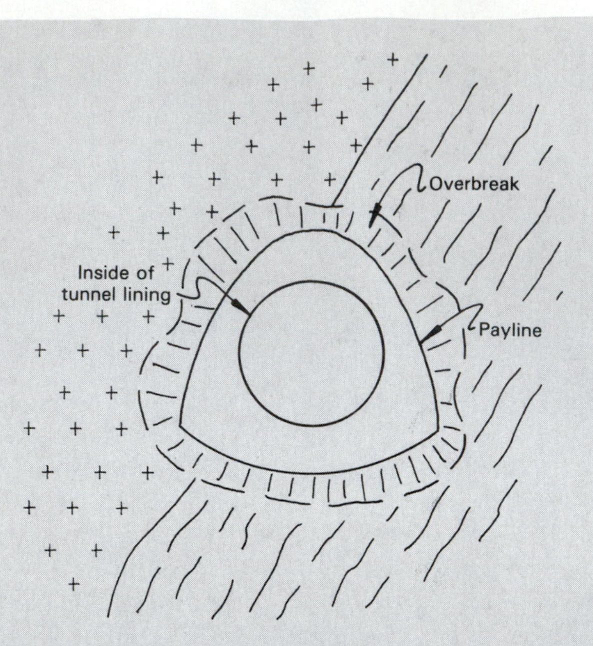


**Figure 5-18** Cross section of a tunnel showing the overbreak, or area excavated beyond the *payline* to accommodate steel supports. The *payline* is the outer boundary of the area estimated for excavation prior to construction.

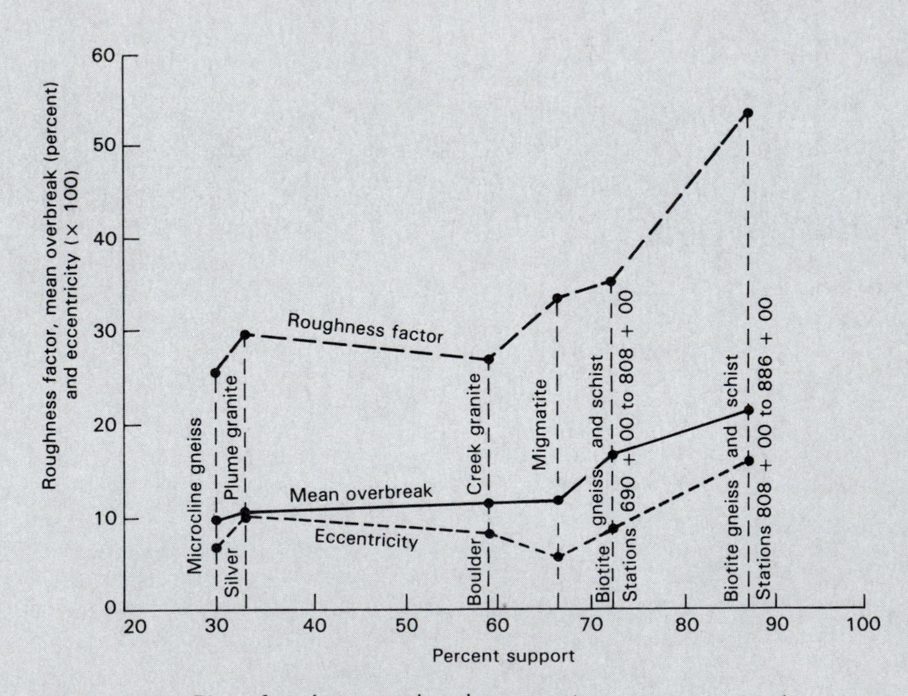


**Figure 5-19** Plot of various overbreak parameters versus percent support, showing that overbreak and therefore cost of tunneling increases in areas of foliated metamorphic rocks. (From L. A. Warner and C. S. Robinson, 1981, *Geology of the Eastern Part of the Harold D. Roberts Tunnel, Colorado (Stations 690 + 00 to 1238 + 58)*, U.S. Geological Survey Professional Paper 831-D.)

## SUMMARY AND CONCLUSIONS

Metamorphic processes produce new rocks under the influences of heat, pressure, and hydrothermal solutions. Contact metamorphism is produced by the emplacement of magma bodies within the crust. Under the action of heat from the cooling magma, rocks are altered within a thin zone around the pluton. Nonfoliated metamorphic rocks are most commonly produced by contact metamorphism because of the minor effect of pressure in the process.

Pressure plays a much more important role in the formation of foliated rocks during regional metamorphism. Tectonic stress, the result of lithospheric plate interactions, combined with heat from deep burial, produces metamorphism on a regional scale. Metamorphic grade often decreases outward from a core of intensely altered rocks, where remelting may occur. This process may explain the huge granitic batholiths of the crust associated with migmatites.

Dynamic metamorphic rocks develop by recrystallization of long, platy minerals oriented normal (perpendicular) to the direction of greatest pressure. Foliation is associated with various types of rock cleavage. With increasing metamorphic grade, slate, phyllite, schist, and gneiss can be formed.

Nonfoliated rocks, such as quartzite and marble, involve recrystallization without the development of mineral-grain alignment.

The orientation of foliation is an important factor in construction in areas of metamorphosed rocks. The strength of a rock mass is much lower in the direction of the foliation than in other directions.

### PROBLEMS

1. Where and why does metamorphism occur in the earth?
2. What are the differences between regional and contact metamorphism?
3. Explain how metamorphism can alter a rock chemically as opposed to mineralogically.
4. How is foliation produced in a metamorphic rock?
5. How do the characteristics of the original rock influence the properties of the metamorphic rock?
6. What are some of the engineering problems associated with metamorphic rocks?

### REFERENCES AND SUGGESTIONS FOR FURTHER READING

MASON, ROGER. 1978. *Petrology of the Metamorphic Rocks*. London: Allen & Unwin.

OUTLAND, C. F. 1977. *Man-made Disaster: The Story of St. Francis Dam*. Glendale, Calif.: Arthur H. Clark.

TURNER, F. J., 1981. *Metamorphic Petrology*, 2d ed. New York: McGraw-Hill.

WARNER, L. A., and C. S. ROBINSON. 1981. *Geology of the Eastern Part of the Harold D. Roberts Tunnel, Colorado (Stations 690 + 00 to 1238 + 58)*, U.S. Geological Survey Professional Paper 831-D.

# 6

# Mechanics of Rock Materials

It is not enough for engineers to have a basic understanding of the physical, chemical, and mineralogical characteristics of rocks, for in dealing with rock as an engineering material, engineers are involved with the *mechanics* of rock and other earth materials. Mechanics refers to the response of materials to applied loads.

## GENERAL TYPES OF EARTH MATERIALS

### Rocks, Soils, and Fluids

From an engineering perspective, earth materials can be subdivided into three categories: rocks, soils, and fluids. Rocks are solid, dense aggregates of mineral grains. Igneous and metamorphic rocks are composed of interlocking mineral grains that were formed by crystallization from a magma or by recrystallization of an existing rock. The degree of interlocking of the mineral grains is one of several factors that determine how strong a rock will be. Chemically precipitated sedimentary rocks can also be interlocking aggregates of mineral grains. Clastic sedimentary rocks, on the other hand, are composed of particles that were derived from a preexisting rock. These particles were then transported by wind, water, or ice, deposited at a particular location, and bound together by various types and amounts of cementing agents.

Soils, from the engineering standpoint, are similar to transported sedimentary rocks in that they consist of rock particles and minerals derived from preexisting rocks. Soils, however lack strong cementing material between grains. For this reason, they usually can be excavated easily without heavy machinery or blasting. We will turn our attention to these materials in a later chapter.

The third major category of earth materials is composed of fluids. The earth's fluids include such liquids as water, magma, and petroleum, but they also include gases, for example, natural gas beneath the land surface and the gases of the atmosphere above the surface.

### Phase Relationships

All rocks contain some void space between grains. This void space may be occupied by liquids and gases. For example, a sandstone rock unit several thousand meters below the surface may contain petroleum and natural gas. Near the surface, however, void spaces are usually filled or partially filled with water. Above the water table (Chapter 11), gases like nitrogen, oxygen, and carbon dioxide fill the remainder. Rocks must be considered as three-phase systems in many instances.

The volume of void space in an earth material can range from a very small percentage of the solid volume to several times the volume of the solid material. The relative amount of void space, along with the type and amount of fluid occupying the space, have an important influence on the mechanical behavior of the material. The relative amount of void space is quantified by the use of the parameters *porosity* and *void ratio* (Figure 6-1). The porosity, $n$, is defined as

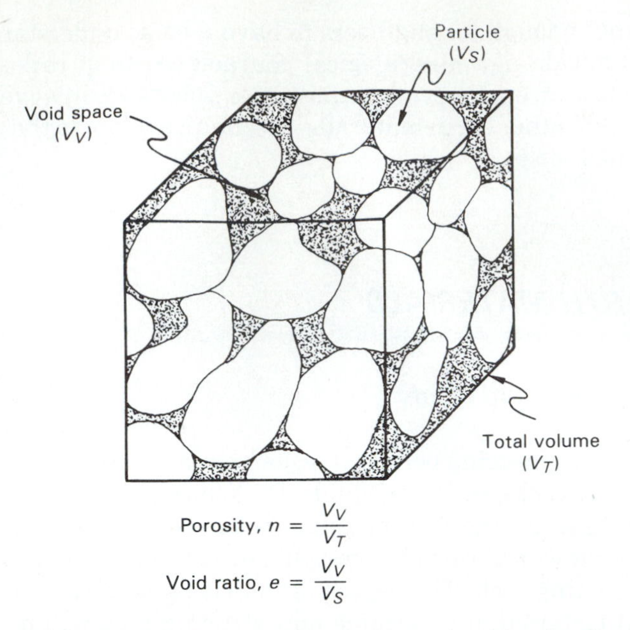


**Figure 6-1** Definition of porosity and void ratio of an earth material. Void space ($V_v$) includes liquid and gas phases.

$$n = \frac{V_v}{V_T} \tag{Eq. 6-1}$$

where $V_v$ is the volume of voids and $V_T$ is the total volume of a representative volume of rock. Void ratio, $e$, is principally used by soils engineers. It can be defined as

$$e = \frac{V_v}{V_s} \tag{Eq. 6-2}$$

where $V_s$ is the volume of soil solids. Porosity is usually expressed as a percentage, whereas void ratio is expressed as a decimal.

Although both porosity and void ratio describe the amount of void space in a soil or rock, they do not give any indication of the rate at which fluids will move through a saturated material. This property, called *permeability* (Figure 6-2), is determined by the size and degree of interconnection of the voids, as well as by the properties of the fluid, including its temperature, density, and viscosity. *Intrinsic permeability* measured in darcys or $cm^2$ (Table 6-1) refers to the material properties that influence fluid movement. The characteristics that determine intrinsic permeability include the size, shape, and packing of the grains, as well as the degree of cementation and fracturing. Intrinsic permeability is of great importance to petroleum geologists and engineers. A rock may be saturated with oil, but if the permeability of the rock is low, oil would not move rapidly enough for an oil well to be considered economically feasible.

A composite parameter, *hydraulic conductivity* (m/s), is used by groundwater hydrologists and soils engineers to measure the ability of a rock or soil to transmit water (Table 6-1). Hydraulic conductivity, $K$, is defined as

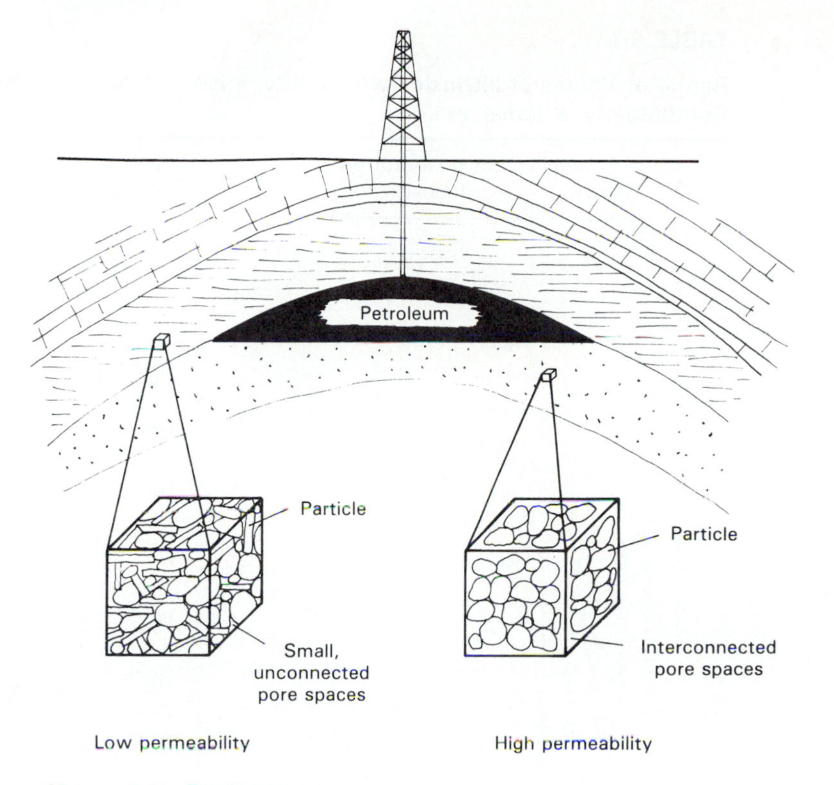


**Figure 6-2** Rocks with large, interconnected pore spaces allow the rapid migration of fluids like petroleum or water, but rocks with low permeability transmit fluids very slowly.

$$K = k\frac{\rho g}{\mu} \qquad \text{(Eq. 6-3)}$$

where $k$ is the intrinsic permeability, $\rho$ and $\mu$ are the density and viscosity of water respectively, and $g$ is the acceleration of gravity. The properties of the material and the fluid are combined in hydraulic conductivity because water at relatively constant values of density and viscosity is usually the only liquid encountered when dealing with upper part of the earth's crust. Contamination with petroleum fuels is an exception to this statement.

## STRESS, STRAIN, AND DEFORMATIONAL CHARACTERISTICS

### Stress

When a load is applied to a solid, it is transmitted throughout the material (Figure 6-3). The load subjects the material to pressure, which equals the amount of load divided by the surface area of the external face of the object over which it is applied. *Pressure*, therefore, is force per unit area. Within the material, the pressure transmitted from the external face to an internal location

**TABLE 6-1**

**Range of Values of Intrinsic Permeability, *k* ($cm^2$, darcys) and Hydraulic Conductivity, *K* (cm/s, m/s)**

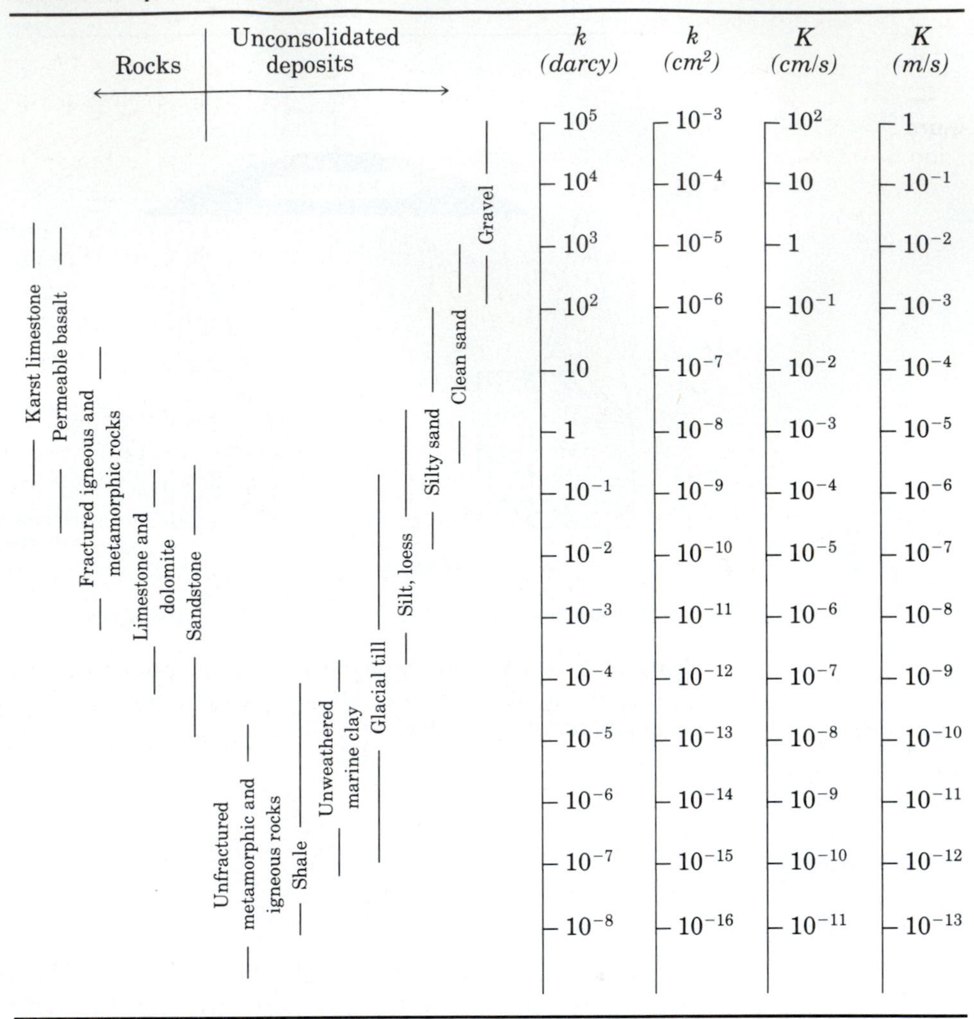


SOURCE: From R. A. Freeze and J. A. Cherry, *Groundwater*, copyright © 1979 by Prentice-Hall, Inc., Englewood Cliffs, N.J.

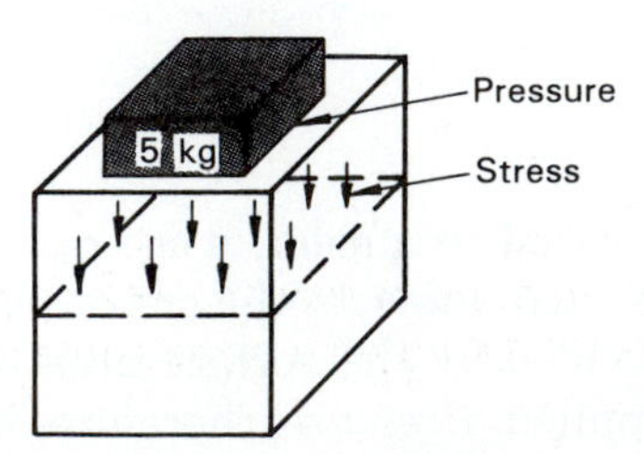


**Figure 6-3** A weight resting on a block causes pressure on the external surface of the block and stress on internal planes in the body.

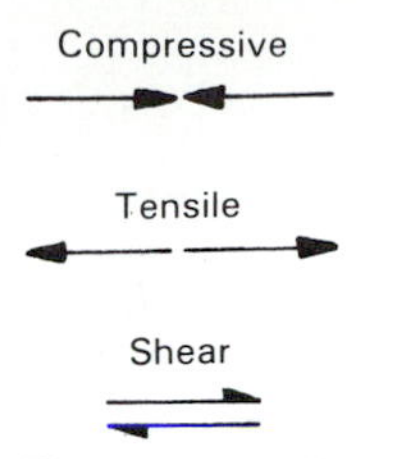


**Figure 6-4** Classification of stress.

is called *stress*. Stress, at any point within the material, can also be defined as force per unit area.

Stresses can be classified according to their orientation within a body (Figure 6-4). Stresses of equal magnitude that act toward a point from opposite directions are called *compressive stresses*. When the stresses are directed away from each other, *tensile stress* is present. The third type of stress, *shear stress*, includes stresses that are offset from each other and act in opposite directions, as in a couple.

On any plane passed through a solid body, there are stresses acting normal to the plane, either compressional or tensional, as well as shear stresses acting parallel to the plane (Figure 6-5). In mechanics, it is possible to resolve the stresses acting at any point within an object into three mutually perpendicular *principal stresses* that are called the maximum, intermediate, and minimum principal stresses (Figure 6-5). The planes perpendicular to the directions of the three principal stresses, which are known as *principal planes*, have no shear stress acting upon them.

At shallow depths beneath the earth's surface, the vertical and lateral (horizontal) stresses present are due to the weight of the overlying rocks, soil, and air. Thus in Figure 6-6, the vertical stress, $\sigma_v$, acting on a horizontal plane at a depth $h$ can be calculated as

$$\sigma_v = \gamma h + P_a \qquad \text{(Eq. 6-4)}$$

where $\gamma$ is the unit weight of rock material and $P_a$ is atmospheric pressure acting on the surface above the rock. Atmospheric pressure is usually neglected in calculating subsurface stresses because it is assumed to be a uniform compo-

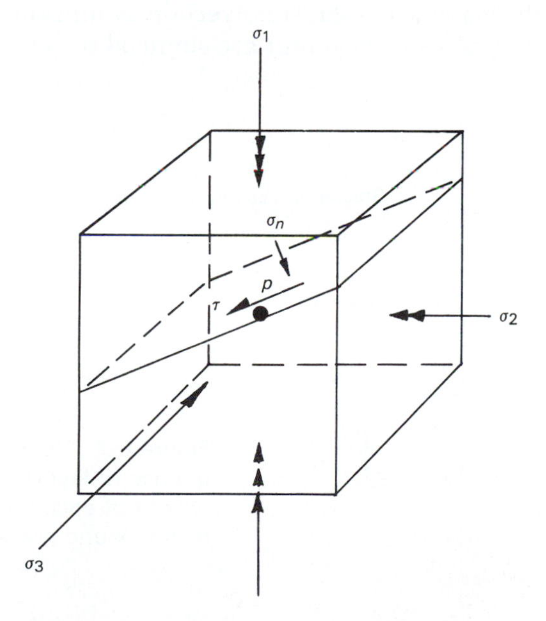


**Figure 6-5** Maximum ($\sigma_1$), intermediate ($\sigma_2$), and minimum ($\sigma_3$) principal stresses acting at point $P$ within a body produce shear ($\tau$) and normal ($\sigma_n$) stresses acting on a plane containing point $P$ passed through the body.

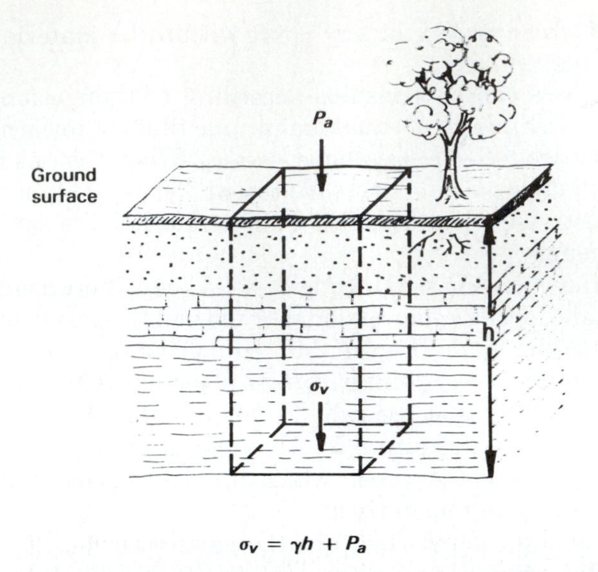


$$\sigma_v = \gamma h + P_a$$

**Figure 6-6** The vertical stress acting on shallow horizontal planes in the earth is the sum of the unit weight $\gamma$ of the material times the depth ($h$) and the atmospheric pressure ($P_a$).

nent of all subsurface stresses. The vertical stress is calculated like hydrostatic stress in a liquid, although the lateral stress in the earth may or may not be equal to the vertical stress. In most subsurface applications, the vertical stress is considered to be the maximum principal stress. The minimum and intermediate principal stresses are assumed to be horizontal.

When the horizon of interest is overlain by units of different unit weights, the stress components of each unit are summed to get the total vertical stress (Figure 6-7).

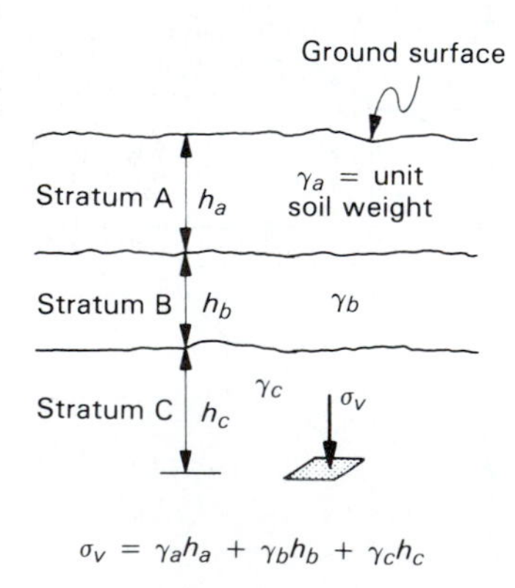


**Figure 6-7** Vertical stress beneath a sequence of layers is the sum of the unit weight of each layer times its thickness. Atmospheric pressure is neglected.

## EXAMPLE 6-1

Calculate the vertical stress at a depth of 8 m at a location where a 5 m bed of sandstone with a unit weight of 25.1 kN/m$^3$ overlies a thick shale unit with a unit weight of 27.5 kN/m$^3$.

*Solution.*

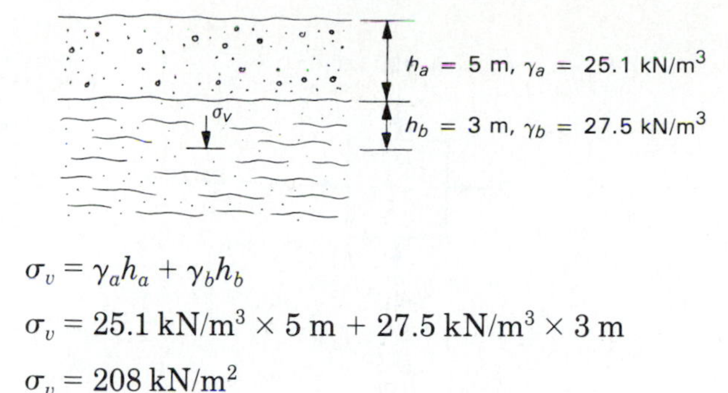


$$\sigma_v = \gamma_a h_a + \gamma_b h_b$$

$$\sigma_v = 25.1 \text{ kN/m}^3 \times 5 \text{ m} + 27.5 \text{ kN/m}^3 \times 3 \text{ m}$$

$$\sigma_v = 208 \text{ kN/m}^2$$

### *Mohr's Circle*

Example 6-1 deals with the case in which we want to measure stresses upon the major principal plane. Suppose, however, that we now wish to measure stresses on a plane inclined at some angle, $\theta$, measured counterclockwise from the major principal plane (Figure 6-8). It is possible to work in two dimensions with the assumption that, below ground surface, the intermediate principal stress, $\sigma_2$, is equal to the minimum principal stress, $\sigma_3$. The new plane is not a principal plane and therefore will be acted upon by both normal and shear stresses. The determination of stresses on planes inclined to the principal planes has applications in lab testing of rock strength, in investigating the stability of slopes, and in interpreting the geologic history of deformed rocks.

A graphical method for representing shear and normal stresses on inclined planes was devised by Otto Mohr (1835–1918). The plot, which is known as Mohr's circle, is shown in Figure 6-9. To construct the circle, values of $\sigma_1$ and $\sigma_3$ are plotted on the horizontal axis, which represents normal stress. By convention, compressive stresses are positive on the diagram. These points establish a circle of radius $(\sigma_1 - \sigma_3)/2$. The center is then plotted on the axis and the circle is drawn. Points on the circle represent the normal and shear stress on any plane oriented at angle $\theta$ from the principal plane, as measured from $\sigma_3$ (Figure 6-9). Because of the properties of a circle, the angle $\theta$ is equal to angle $2\theta$, measured from the center of the circle (Figure 6-9c).

We now have a diagram representing the stress on any plane inclined relative to the principal plane. Normal and shear stress values can be determined graphically using the circle or by using the equations shown in Figure 6-10.

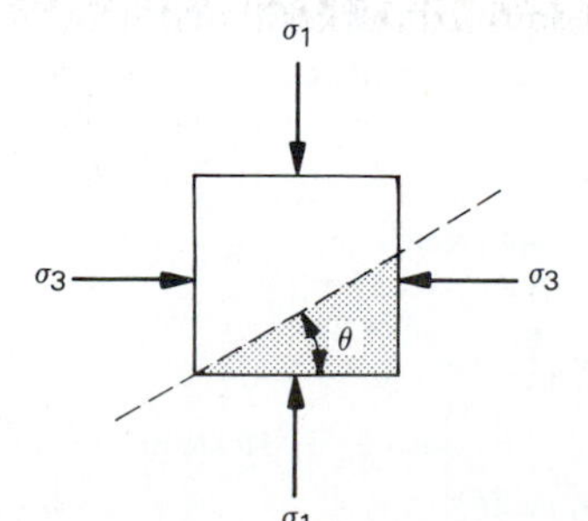


**Figure 6-8** Position of a random plane oriented at angle $\theta$ measured counterclockwise from the major principal plane.

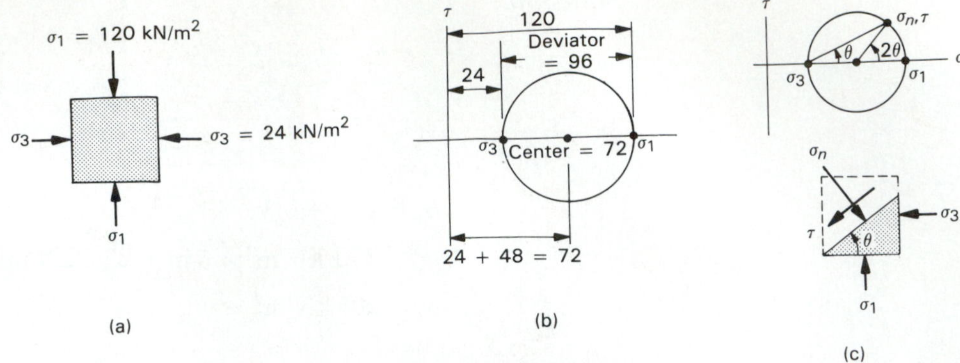


**Figure 6-9** (a) Stresses acting on an incremental element surrounding a point in the subsurface. (b) Mohr's circle plot for the stresses shown in (a). Deviator stress is the difference between $\sigma_1$ and $\sigma_3$. (c) Relationship of planes on the incremental element to points on Mohr's circle.

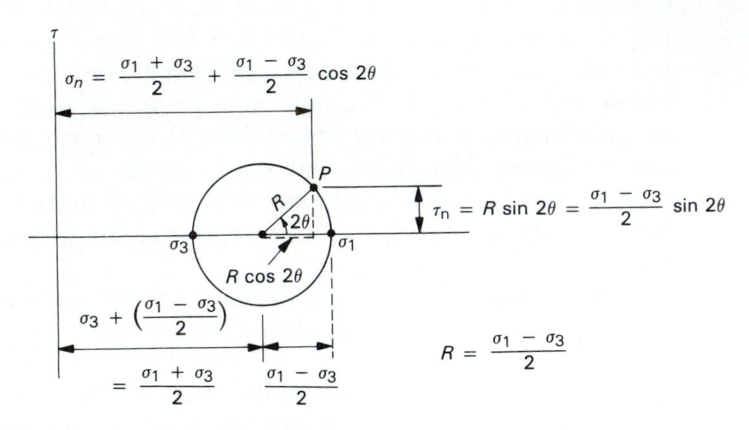


**Figure 6-10** General equations for shear and normal stress derived from Mohr's circle.

## EXAMPLE 6-2

Vertical and horizontal principal stresses are 144 kN/m² and 36 kN/m², respectively. Determine the normal and shear stresses on a plane inclined at 45 degrees to the principal plane. As shown below, $\theta$ is measured counterclockwise from the major principal plane and angle $2\theta$, or 90 degrees, is measured counterclockwise from the center of Mohr's circle. In this case, $\sigma_n$ is equal to $\sigma_3$ plus the radius, and $\tau$, the shear stress, is equal to the radius. It is evident from the diagram that a plane inclined at 45 degrees from the major principal plane will have the highest value of shear stress of any plane.

***Solution.***

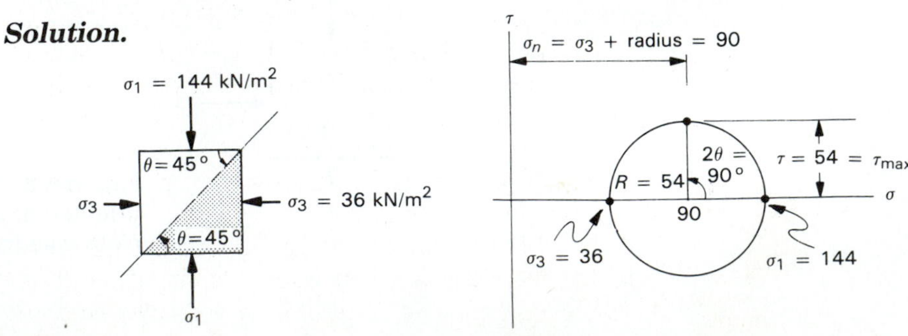

$$\tau = \frac{\sigma_1 - \sigma_3}{2} = \frac{144 - 36}{2}$$
$$= 54 \text{ kN/m}^2$$
$$\sigma_n = \sigma_3 + \text{radius} = 36 + 54$$
$$= 90 \text{ kN/m}^2$$

### *Deformation-Response to Stress*

The application of stress to a body of rock or soil causes the material to yield or deform. The amount of deformation is called *strain*. The type and amount of strain that a particular material experiences depends on the type of stresses applied, as well as the depth and temperature. An understanding of deformation is important to geologists concerned with the behavior of earth materials during geologic processes, and to engineers, who are more interested in deformation of material under the loads applied by engineering structures. Before considering natural materials, we will first review the deformation of so-called ideal materials.

Ideal Materials. Fundamental types of deformation can be demonstrated by the use of simple mechanical systems. As shown in Figure 6-11, the three basic types of material behavior are *elastic, viscous,* and *plastic*. Perfect elastic behavior, the type demonstrated by a spring, results in a linear plot of stress versus strain. The slope of the line relating stress and strain is an important material property called the *modulus of elasticity*. It can be stated as

$$E = \frac{\sigma}{\varepsilon} \qquad \text{(Eq. 6-5)}$$

where $E$ is the modulus of elasticity, $\sigma$ is the applied stress, and $\varepsilon$ is the strain. The modulus of elasticity specifies how much strain will occur under a given stress. In this case, strain ($\varepsilon$) is measured as

$$\varepsilon = \frac{\Delta L}{L} \qquad \text{(Eq. 6-6)}$$

where $\Delta L$ is the change in length of the spring and $L$ is the original length. Another characteristic of elastic materials is that they tend to return to their original condition when the stress is removed. Thus elastic strain is recoverable. Some rocks approach ideal elastic behavior during deformations of small magnitudes. If the modulus of elasticity is known, it is therefore possible to predict the amount of deformation that will occur under an applied load.

Certain fluids display the type of behavior known as *viscous* (Figure 6-11). Stress is directly proportional to *strain rate* for these fluids. The constant of proportionality is the *viscosity* of the liquid, which is the slope of the stress versus strain rate plot. Even solid earth materials may exhibit viscous behavior under certain circumstances. The model for viscous behavior is a dashpot, in which a piston is moved through a cylinder containing a viscous liquid. The liquid flows through a space between the piston and the wall of the cylinder as the piston is pulled.

Plastic behavior, as demonstrated in Figure 6-11 by a block pulled along a surface, involves continuous deformation after some critical value of stress has been achieved. There is no deformation until the critical stress has been

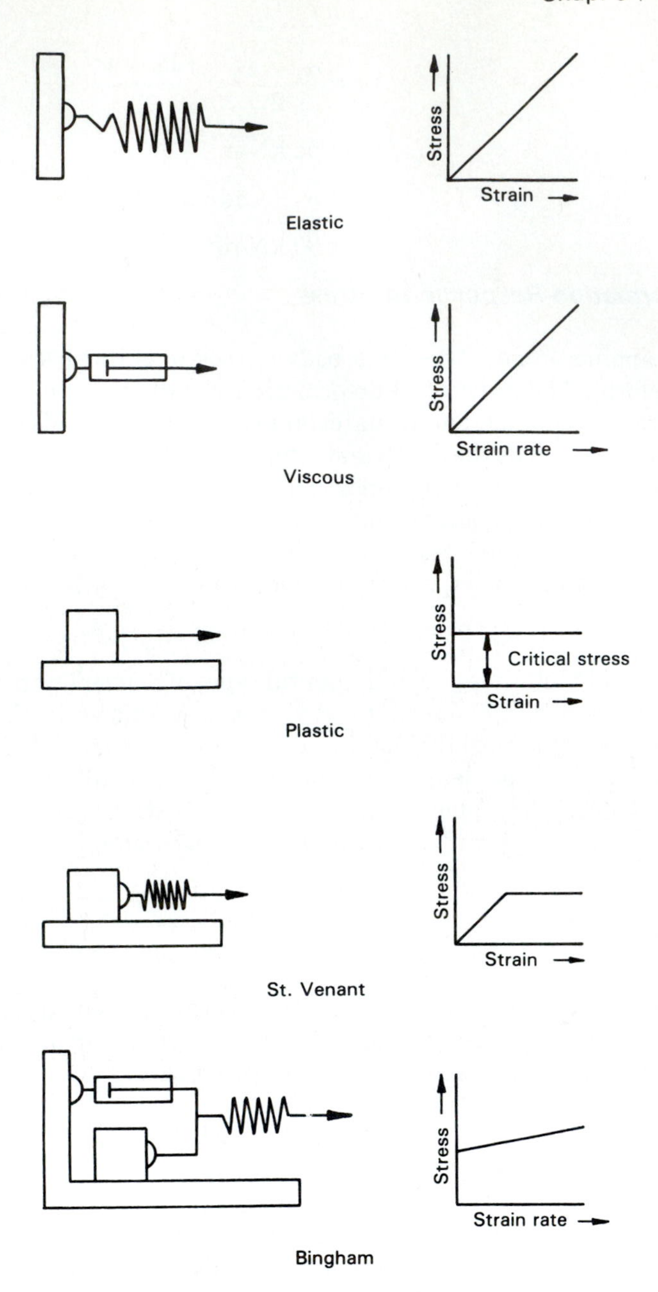


**Figure 6-11** Types of idealized material deformation as illustrated by simple mechanical systems.

reached. Many rocks display plastic deformation under stress, but they differ from ideal plastic behavior in the other types of deformation that occur before plastic behavior begins.

Models that more closely approximate the behavior of real earth materials are provided by combining basic models of ideal behavior. The St. Venant model, for example, which combines elastic and plastic behavior, approximates the deformation of many rocks under stress. The Bingham model combines all three types of ideal behavior. This response has been used as a model for mud, lava, and similar flows. These processes will be described in Chapter 12.

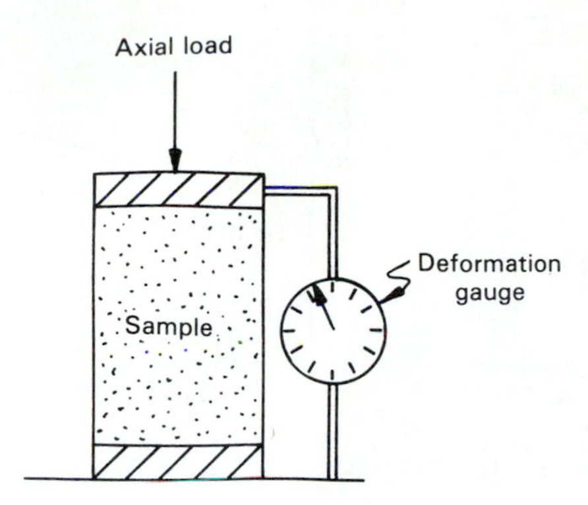


**Figure 6-12** Schematic diagram of the unconfined compression test. An axial load is applied to a rock or soil sample and the resulting strain is measured.

Stress-Strain Behavior of Rocks. Although rocks can be compared with the idealized models just discussed, their actual behavior is more complex. One common method of testing rock behavior is the *unconfined compression test*, in which a cylindrical sample of rock is subjected to an axial load applied to the ends of the sample (Figure 6-12). As the axial stress is increased during the test, the changes in length of the sample can be measured. From this value, the strain at any instant can be determined using Equation 6-6.

In Figure 6-13, a generalized stress-strain curve for rocks is shown. Unlike the ideal models, the curve is nonlinear and has three distinct segments. The first segment relates to the closing of microscopic pores and void spaces in the rocks as stress is applied at low levels. In the middle section of the curve, the nearly linear response approximates elastic behavior. The slope of this segment can be used to calculate a modulus of elasticity for the rock, even though the rock is not perfectly elastic (Figure 6-11). The final segment of the curve is most similar to plastic behavior. At the point of *failure*, where the sample crumbles and loses all resistance to stress, the curve terminates.

Different types of rocks vary considerably in their stress-strain behavior. In Figure 6-14, two types of response are shown. Rocks that display mostly

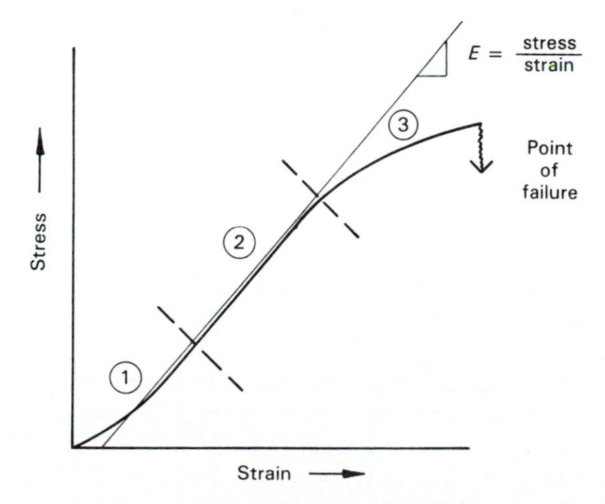


**Figure 6-13** A generalized stress-strain curve for rocks. The modulus of elasticity is measured by the slope of the tangent to segment 2 of the curve.

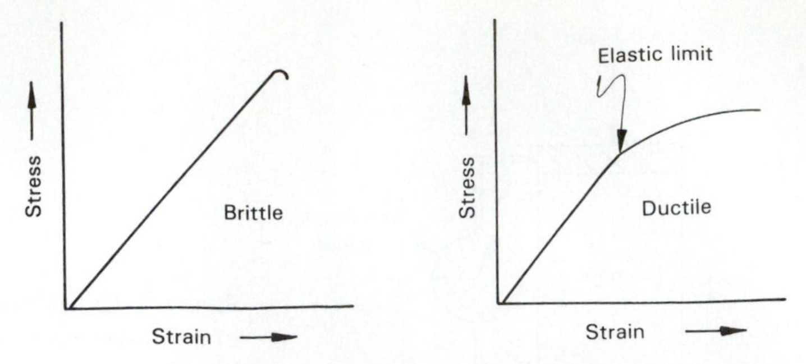


**Figure 6-14** Stress-strain curves for brittle and ductile rocks.

elastic behavior until rupture are called *brittle*. Those that exhibit a significant amount of plastic response before failure are termed *ductile*. Ductile rocks usually deform elastically to a point called the *elastic limit*, where strain becomes more plastic with increasing stress.

Strength. We have described the failure of a brittle rock as the point when the rock loses all resistance to stress and crumbles. In a plastic material, a specific point of failure is harder to identify because deformation continues indefinitely at a constant level of stress. Failure, in this case has to be defined as a certain amount, or percentage, of strain. For whatever criteria established for failure, *strength* is defined as the level of stress at failure.

When the stress at a certain point in the ground or in a laboratory test sample is resolved into its principal stresses, a *stress differential* is apparent between the three principal stresses. In an unconfined compression test, for example, the axial load becomes the greatest principal stress and the other two principal stresses are zero. When a stress differential exists, shear stresses develop on planes at all angles to the principal stress directions within the body (Figure 6-15). It is possible to resolve the principal stresses acting on the sample into shear and normal stresses acting on any plane within the sample (Figure 6-16). Mohr's circle is a handy way to calculate the value of these stresses. For an unconfined compression test, the $\sigma_3$ value would be plotted at the origin. For any plane, the shear stress, $\tau$, tends to cause failure, and the normal stress, $\sigma_n$, tends to resist failure. On two critical planes within the rock, the combination of shear and normal stresses will produce the greatest

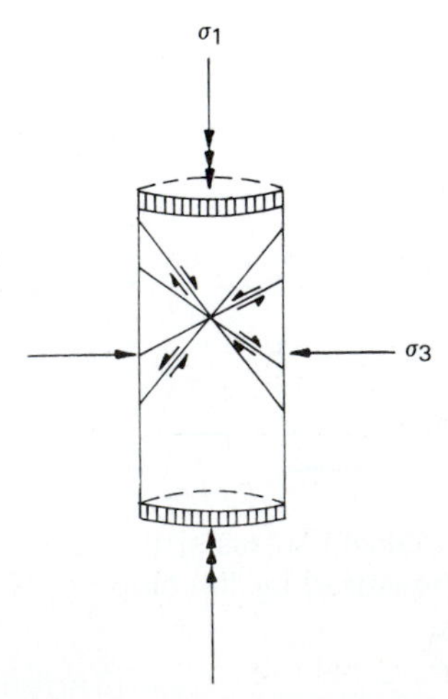


**Figure 6-15** Shear stresses on planes inclined to the principal planes produced by a stress differential.

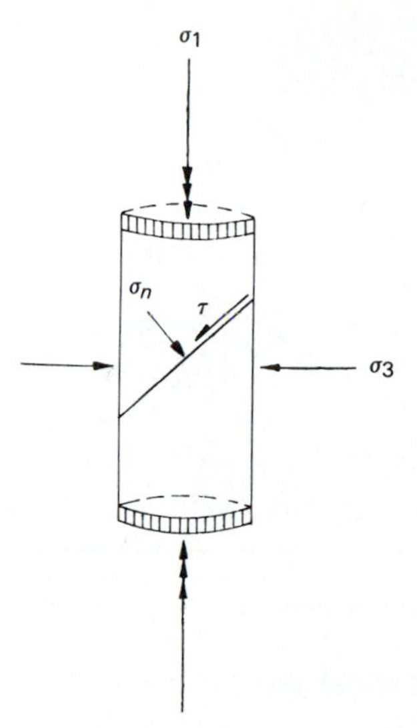


**Figure 6-16** Shear and normal stresses acting on a plane within a test sample.

tendency for failure. If the shear stress on these planes exceeds the *shear strength* of the rock in those directions, failure of the rock will occur. Therefore, the unconfined compression test indirectly determines the shear strength of the rock.

The relationship between shear and normal stresses during a strength test, and at failure, is critical to understanding the deformation behavior of the material. A test that can be used to determine this relationship directly is called, appropriately, the *direct shear* test (Figure 6-17). In the direct shear test, the failure plane is specified by the construction of the test cell so that variable shear and normal stresses can be applied to test for shear strength along the plane where the cell separates. The test is usually conducted by applying a constant value of normal stress and then increasing the shear stress until failure occurs. Successive tests are then repeated at a higher normal stress. If the cell is filled with dry sand for a strength test, the results will be as shown in Figure 6-18. The line on the graph shows the relationship between shear stress and normal stress at failure. The constant slope indicates that shear strength is directly proportional to the normal stress applied to the failure plane. This result is obtained because the strength of the sand is controlled by frictional contact between sand grains, and the shear stress necessary to cause failure is directly dependent upon the normal stress applied to the failure

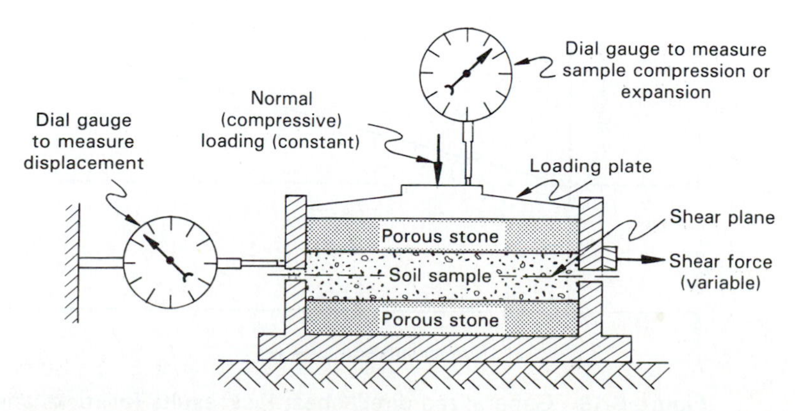


**Figure 6-17** The direct shear test.

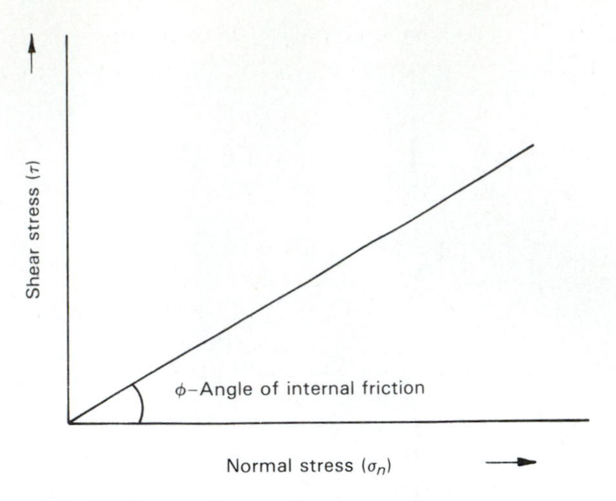


**Figure 6-18** Direct shear test results for dry sand.

plane. The angle that the plotted line makes with the horizontal axis ($\phi$) is a basic property of the specific sand tested, and it is called the *angle of internal friction*. It depends upon the grain size, distribution, shape, and packing of the sand grains in the sample tested. The equation of the line relating shear strength, $S$, and normal stress, $\sigma_n$, is

$$S = \sigma_n \tan \phi \qquad \text{(Eq. 6-7)}$$

The shear stress versus normal stress curve for dry sand passes through the origin because frictional contact between grains is the only component of strength in this material. For rocks, as well as for some soils, there are additional components of strength arising from the interlocking nature of grains in the rock, the cement in the pores of the rock, or attractive forces between grains or particles. These properties give the rock or soil a certain inherent strength that is independent of normal stress. In soils, this strength is called *cohesion*. These materials also have a strength component proportional to normal stress on the failure plane. The strength-test results, then, for rocks and some cohesive soils, are of the type shown in Figure 6-19. The intercept

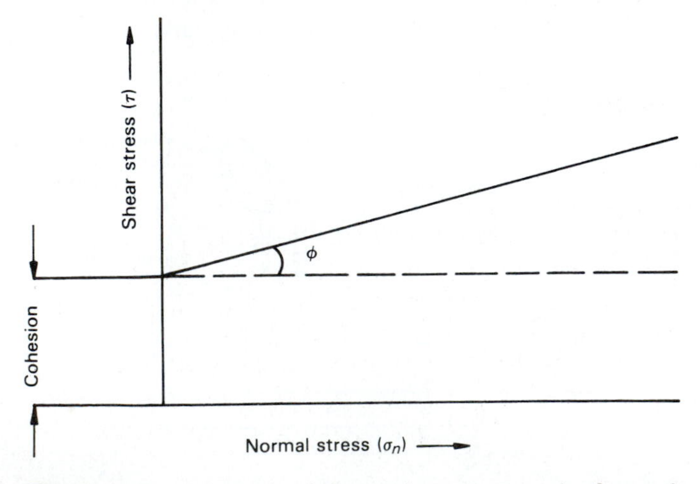


**Figure 6-19** Generalized direct shear test results for rocks and some cohesive soils.

on the vertical axis is a measure of the inherent strength or cohesion of the material. The formula for shear strength in this case can be written as

$$S = C + \sigma_n \tan \phi \qquad \text{(Eq. 6-8)}$$

where $C$ is cohesion. This equation is called the Mohr-Coulomb equation.

### EXAMPLE 6-3

A direct shear cell 25.4 cm by 25.4 cm in plan is filled with dry sand. A normal load of 6 kN is applied to the sample, and the shear force at failure is 3.5 kN. What is the angle of internal friction for the sand, based upon this one test?

***Solution.*** The normal stress on the failure plane is 6 kN/0.065 $m^2$ = 92.3 kN/$m^2$, and the shear stress at failure (shear strength) is 53.8 kN/$m^2$. When plotted on a graph of shear stress versus normal stress, $\phi$ can be determined as shown below.

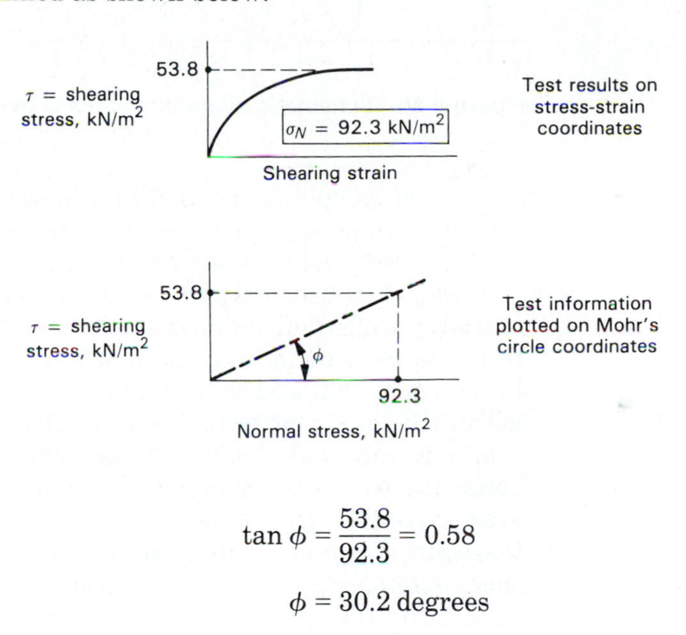


$$\tan \phi = \frac{53.8}{92.3} = 0.58$$

$$\phi = 30.2 \text{ degrees}$$

Effects of Confining Pressure. The unconfined compression test is applicable to rock engineering where rock masses are exposed at the surface. However, in the design of a tunnel, an underground mine, or an underground waste repository, we would need to take into consideration the effects of deep burial of the rock upon its strength characteristics. Similarly, geologists must understand the deformation of rocks far below the earth's surface to interpret tectonic structures (Chapter 7) that were formed by deformation of the crust in the geologic past. Under these conditions, $\sigma_3$, the minimum principal stress, would no longer be equal to zero as in the unconfined compression test. The weight of the overlying column of rock translates into a pressure applied from all directions to any given element of rock at depth. This all around stress is known as *confining pressure*. It is similar to hydrostatic pressure in that it increases with depth; however, this *lithostatic pressure* is not always equal in all directions. The major principal stress, $\sigma_1$, may be applied from a vertical or lateral direction.

A laboratory test called the *triaxial test* (Figure 6-20) has been devised to more closely simulate the behavior of rocks or soil at depth. In this test, the

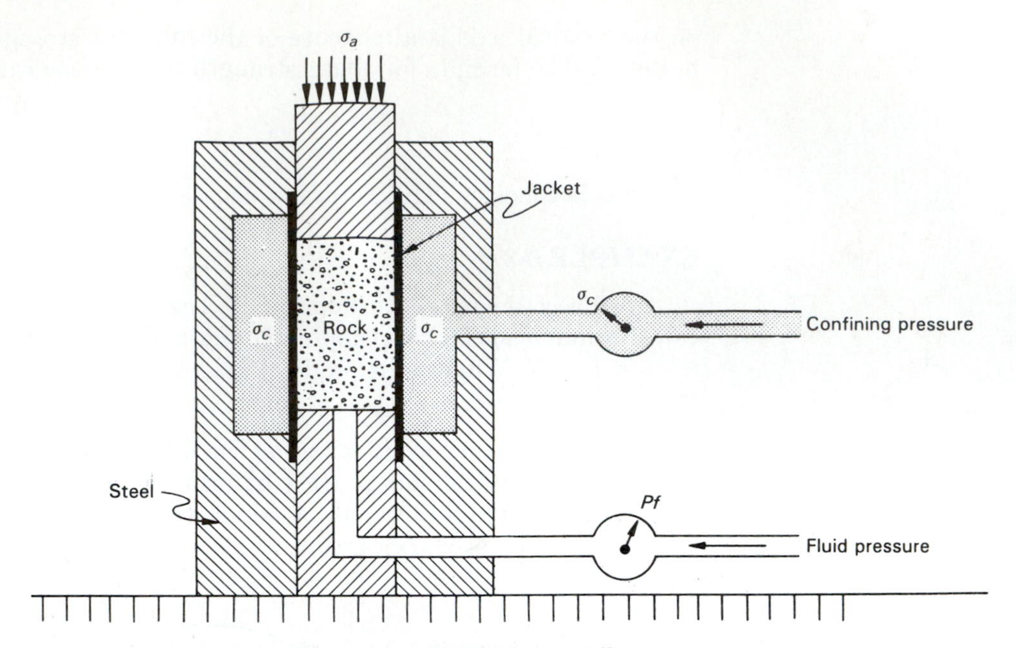


**Figure 6-20** Schematic diagram of triaxial test cell.

cylindrical sample is enclosed in a jacket through which a radial confining pressure can be applied. Gases or liquids can be used to apply this all-around pressure. An axial load can then be applied, similar to an unconfined compression test, until failure occurs. With this apparatus, tests can be run repeatedly to study the effect of confining pressure upon deformation. The results of several tests can be used to establish a *failure envelope* (Figure 6-21) by drawing a line tangent to the Mohr's circle representing each test. The point of contact with each circle represents the shear strength for the corresponding values of confining and axial pressure. A complete failure envelope would include a line below the $\sigma$ axis to define an area which would be the mirror image of the area above the axis. Notice that because the failure envelope is a sloping line, the point of contact with each circle does not occur at the highest value of shear stress, which would be at the top of the circle. Thus failure does not necessarily take place on the plane where shear stress is the maximum, but on the plane where the combination of shear and normal stresses produce optimum failure conditions. When used for soils, the triaxial test data yield the angle of internal friction (the slope of the failure envelope) and the cohesion (the intercept of the shear stress axis).

Under increasing levels of confining pressure, as we might expect with increasing depth, rocks no longer deform as they did at shallow depth. Triaxial test results plotted on stress-strain diagrams illustrate several types of changes

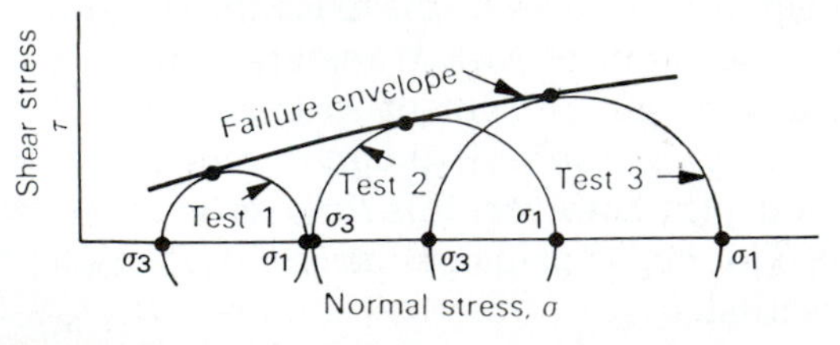


**Figure 6-21** Construction of failure envelope using Mohr's circle results from multiple tests.

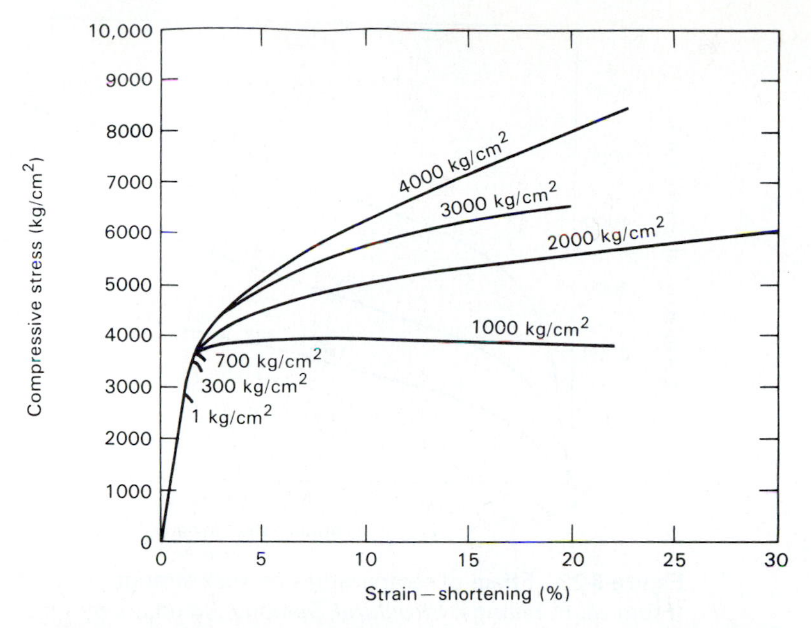


**Figure 6-22** Effect of confining pressure on a rock sample (Solenhofen limestone) in a triaxial test. Each curve represents a test conducted at a different confining pressure. (From M. P. Billings, *Structural Geology*, 3d ed., copyright © 1972 by Prentice-Hall, Inc., Englewood Cliffs, N.J.)

(Figure 6-22). First, as the confining pressure is increased, the rock passes through a transition from brittle to ductile behavior. Ductile response becomes dominant at confining pressures above 700 kg/cm$^2$. This result supports our expectation that rocks whose behavior may be brittle at the earth's surface become ductile under high confining pressures. Second, the tests run at higher confining pressures indicated that the strength of the rock increases with increasing confining pressure.

Confining pressure is not the only change in rock deformation with depth. Because the temperature increases steadily with depth, it is likely that rocks will behave differently at higher temperatures. Triaxial test cells can be constructed to investigate the effects of temperature on rock behavior under stress. Some results are shown in Figure 6-23. A particular rock type was tested at several temperatures under a constant confining pressure. Strength decreased at higher temperatures. In addition, the samples became ductile (reached their elastic limit) at a lower stress. Thus plastic deformation will become more prevalent with increasing depth because of greater temperatures and pressures.

Another factor that affects rock deformation is time. Pressures are applied to rocks over millions of years in the earth. Although we obviously cannot duplicate such conditions in a lab test, it is possible to vary the rate of strain at which a rock sample approaches failure. In this way the effect of time can be investigated, although extrapolation to geologic time intervals is still difficult. When the strain rate of a triaxial test is decreased, two observations can be made. First, rock strength decreases with decreasing stain rate (increasing the time to failure); second, the rocks become more ductile at lower strain rates. These results hold true even at stresses below the elastic limit. The stresses are applied over long periods of time. In fact, the relationship between stress and strain rate can even be compared with that of a viscous material. Rocks, therefore, that are strong and brittle at the earth's surface may deform like a fluid deep within the earth over long periods of time.

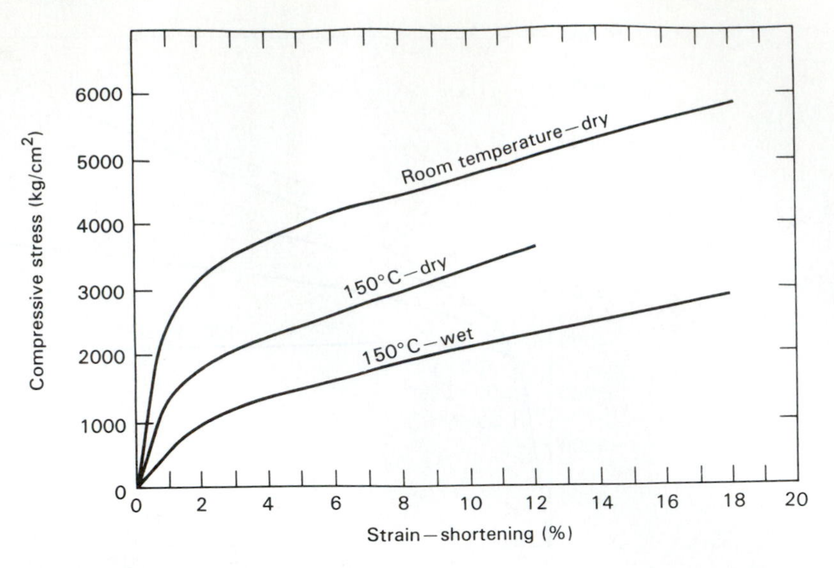


**Figure 6-23** Effect of temperature on rock strength (Yule marble). (From M. P. Billings, *Structural Geology*, 3d ed., copyright © 1972 by Prentice-Hall, Inc., Englewood Cliffs, N.J.)

## ENGINEERING CLASSIFICATION OF INTACT ROCK

Stress-strain behavior is utilized in the engineering classification of rock developed by D. U. Deere and R. F. Miller (Deere, 1968). One component of the classification is the unconfined compressive strength (Table 6-2). Five categories are established for an intact sample of rock. The modulus of elasticity forms the other aspect of the classification. This classification applies only to rock that is internally continuous, or intact. Most masses of rock contain discontinuities that strongly influence engineering behavior. Rock-mass properties will be discussed in a subsequent section.

In Figure 6-24, the definition of the particular modulus value used is illustrated. $E_{t_{50}}$ is the modulus obtained by taking the slope of a line tangent to the stress-strain curve at 50% of the unconfined compressive strength. The classes of modulus of elasticity are listed in Table 6-3.

Together, the modulus of elasticity ($E_{t_{50}}$) and the unconfined compressive strength ($\sigma_a$) can be used to calculate a modulus ratio, which is defined as

**TABLE 6-2**

**Strength Classification of Intact Rock**

| *Class* | *Description* | *Unconfined compressive strength (kg/cm²)* |
|---|---|---|
| A | Very high strength | >2250 |
| B | High strength | 1125–2250 |
| C | Medium strength | 562–1125 |
| D | Low strength | 281–562 |
| E | Very low strength | <281 |

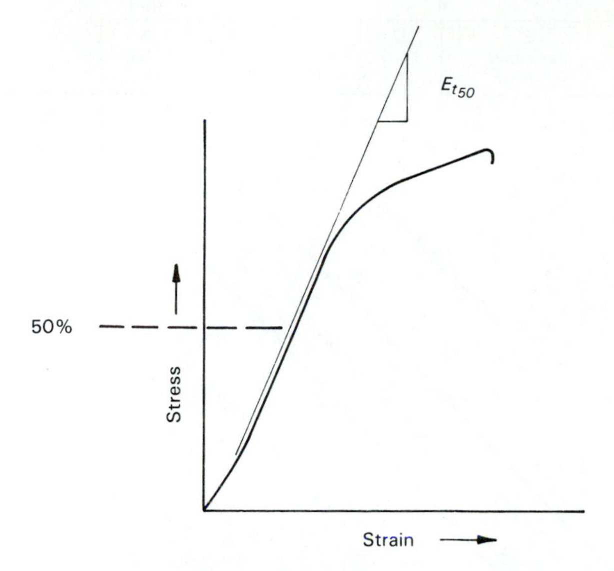


**Figure 6-24** Definition of the tangent modulus ($E_{t_{50}}$) used in the Deere and Miller classification. Tangent is drawn at 50% of unconfined compressive strength.

$$M_R = \frac{E_{t_{50}}}{\sigma_a} \qquad \text{(Eq. 6-9)}$$

Modulus-ratio values are plotted on a diagram (Figure 6-25) divided into fields of high modulus ratio (>500 : 1), medium modulus ratio (500 : 1–200 : 1), and low modulus ratio (<200 : 1). The position of plotted points on the modulus ratio diagram gives a visual comparison of rocks and indicates the strength and modulus values, the rock properties that control engineering behavior.

### *Igneous Rocks*

The strength of igneous rocks is high when the rock is composed of a dense network of interlocking crystals. This condition is usually present in the intrusive rocks, which had sufficient time for crystallization to develop a three-dimensional network. As shown in Figure 6-26, intrusive rocks generally have a high modulus of elasticity and a medium modulus ratio.

The extrusive igneous rocks have a much wider range in strength and modulus. Lava flows may develop strength and modulus properties that are

**TABLE 6-3**

**Modulus of Elasticity Classification of Rocks**

| *Description* | $E_{t_{50}}$ *(kg/cm² × 10⁵)* |
|---|---|
| Very stiff | 8–16 |
| Stiff | 4–8 |
| Medium stiffness | 2–4 |
| Low stiffness | 1–2 |
| Yielding | 0.5–1 |
| Highly yielding | 0.25–0.5 |

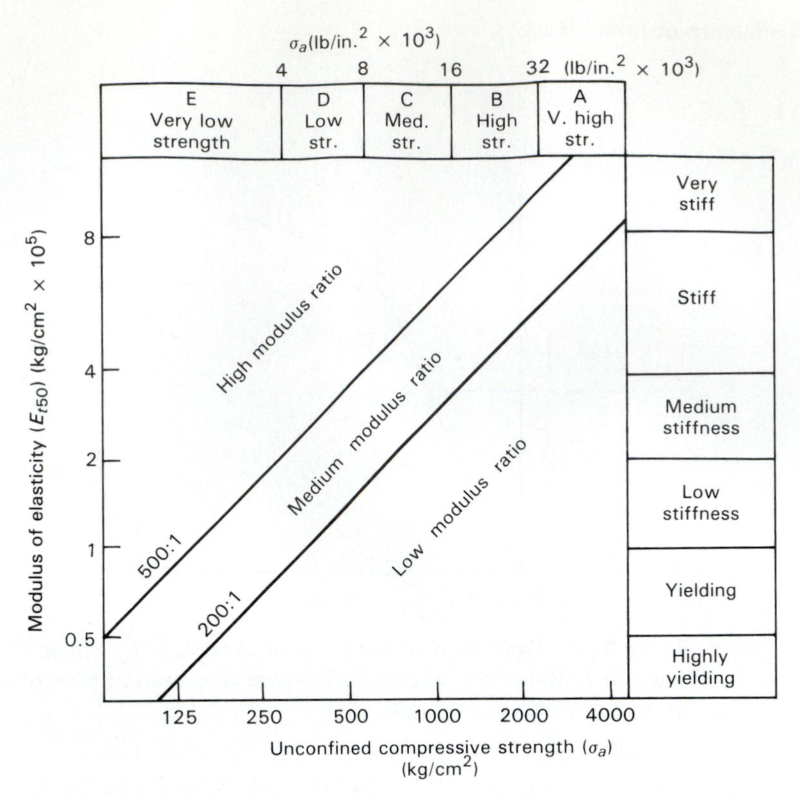


**Figure 6-25** Classification diagram for the Deere and Miller classification of intact rock. (Modified from D. U. Deere, *Rock Mechanics in Engineering Practice*, K. G. Stagg and O. C. Zienkiewicz, eds., copyright © 1968 by John Wiley & Sons, Ltd., London.)

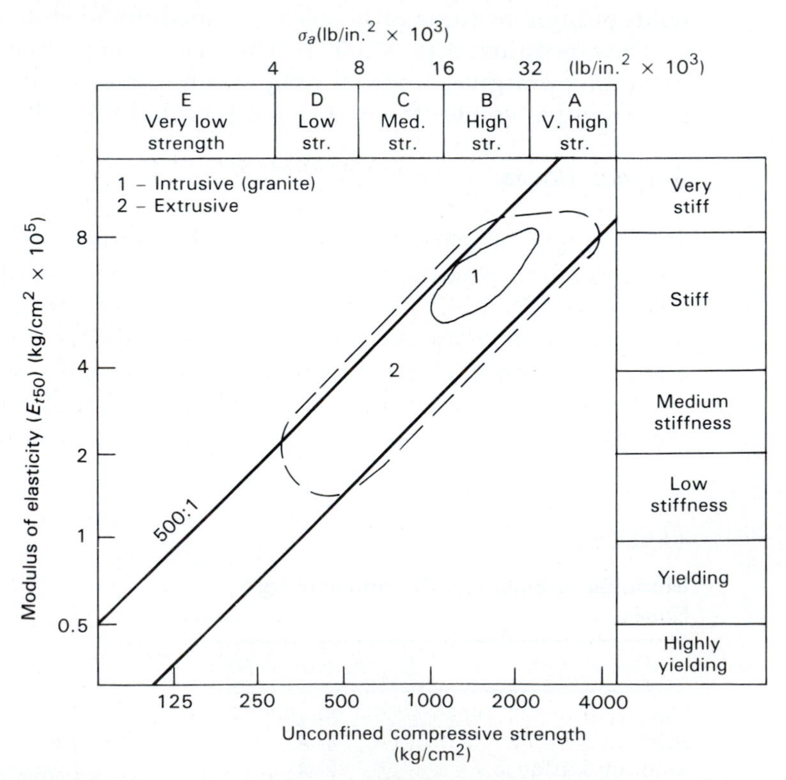


**Figure 6-26** Engineering classification of intact igneous rock. (Modified from D. U. Deere, *Rock Mechanics in Engineering Practice*, K. G. Stagg and O. C. Zienkiewicz, eds., copyright © 1968 by John Wiley & Sons, Ltd., London.)

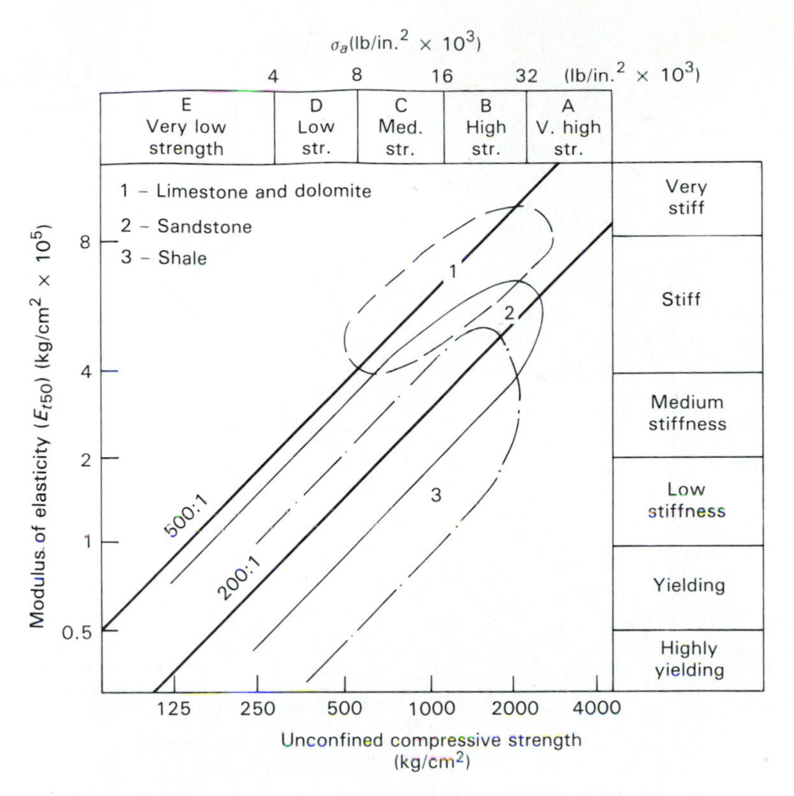


**Figure 6-27** Engineering classification of intact sedimentary rock. (Modified from D. U. Deere, *Rock Mechanics in Engineering Practice*, K. G. Stagg and O. C. Zienkiewicz, eds., copyright © 1968 by John Wiley & Sons, Ltd., London.)

similar to the intrusive rocks, but vesicular extrusive rocks are usually weaker. The pyroclastic rocks extend the field of the extrusive rocks into the area of low strength and modulus because of their low density and high porosity caused by their formation from airfall or flow processes. Most extrusive rocks, however, fall into the medium modulus ratio category.

### *Sedimentary Rocks*

Sedimentary rocks have extremely variable strength and modulus properties (Figure 6-27). For the clastic rocks, these values depend on the rock characteristics acquired in the depositional environment, as well as all changes that have affected the rock in the lithification process. Important depositional factors include grain-size distribution, sorting, rounding, and mineral composition. Lithification processes, such as compaction and cementation, usually increase the strength of the rock. Clastic rocks of medium and coarse grain size usually fall into the moderate modulus ratio class over a wide range of strength and modulus values. Shale exhibits a strong tendency for plastic deformation. As a result, the modulus of elasticity is low and the modulus ratio overlaps into the low zone.

Nonclastic rocks differ in engineering properties according to the composition of the rocks. Limestone and dolomite generally have medium to high strength and modulus ratios. Evaporite rocks exhibit weak, plastic behavior. The tendency for plastic deformation at low strength in the evaporite rocks explains why these rocks are considered to be a possible option for high-level nuclear waste repositories. Because of their stress-strain behavior, any cavities

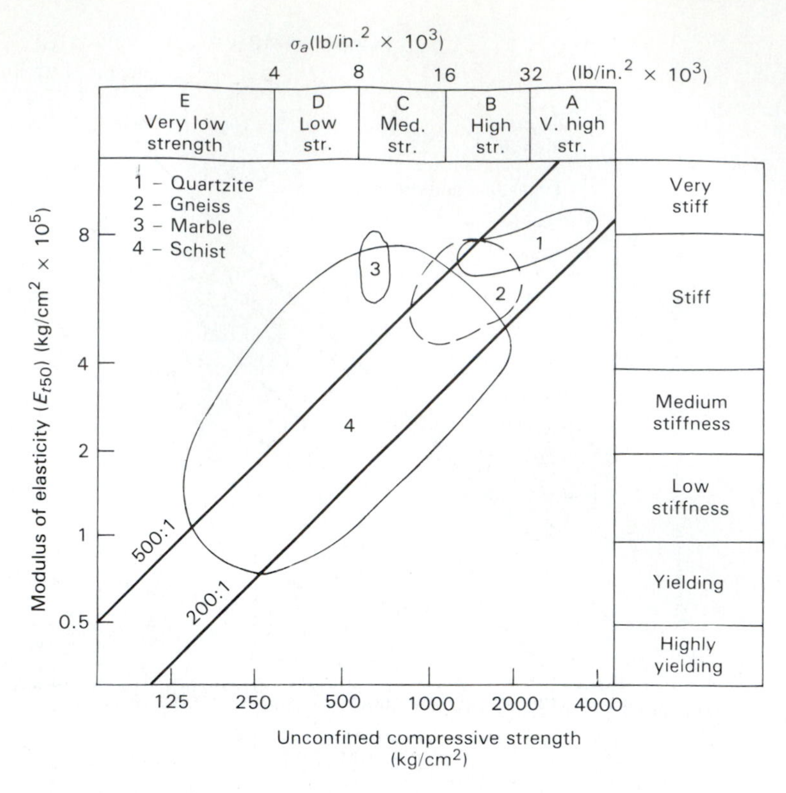


**Figure 6-28** Engineering classification of intact metamorphic rock. (Modified from D. U. Deere, *Rock Mechanics in Engineering Practice*, K. G. Stagg and O. C. Zienkiewicz, eds., copyright © 1968 by John Wiley & Sons, Ltd., London.)

or cracks in the rocks are sealed by plastic flow. This property is important in preventing migration of waste products from the disposal site.

### *Metamorphic Rocks*

Metamorphism increases the strength of some sedimentary rocks by compaction and recrystallization. Thus (Figure 6-28) the modulus ratio of quartzite is limited to the high-strength side of the diagram relative to the corresponding sedimentary rock sandstone. Marble is an exception to the trend of increased strength following metamorphism. When limestone or dolomite recrystallize, the crystals in the resulting marble are large. The strength of a rock generally is proportional to the surface area of contact between grains, and a fine-grained rock has a higher contact area than a coarse-grained rock. Therefore, limestone and dolomite lose strength when they are metamorphosed. Because only the grain size changes during metamorphism, and the mineral composition remains the same, the modulus ratio field of marble falls within the high modulus ratio section of the chart.

Schists have a wide variation in strength and modulus because of their strongly oriented foliation, which produces planes of weakness parallel to the foliation. The strength of a schist depends upon the direction from which stresses are applied. If the stress configuration tends to produce failure in the direction of the foliation, the strength is minimized. A much greater strength develops if the stresses tend to cause failure perpendicular to the foliation planes. Gneisses are limited to the high-strength and modulus regions of the chart because the coarse bands of alternating light- and dark-colored minerals in gneiss do not result in highly developed planes of weakness in the rock.

## *ROCK-MASS PROPERTIES*

The strength and deformational properties of rocks that we have discussed usually are determined by laboratory tests on rock samples. These samples are necessarily small, intact specimens taken from large bodies of rock at a field location. Although the test results obtained from intact samples are useful for comparison of properties between various rock types, the strength values cannot be directly applied to the overall rock mass in the field situation. The reason for this apparent discrepancy is that the behavior of a rock mass under load in the field is partially controlled by the strength developed along discontinuities in the rock and by the weathering characteristics, rather than by the strength of the intact portions of the rock itself. Discontinuities are present in almost every type of rock and they act to lower the strength of the rock mass. Under stress, the rock will fail along existing planes of weakness rather than develop new fracture planes within intact, solid rock (Figure 6-29). Therefore, it is very important to determine the properties of the rock mass as well as the properties of intact rock within the rock mass.

Some of the types of rock discontinuities are shown in Table 6-4. Depositional discontinuities are acquired by sedimentary rocks as their constituent particles are deposited as sediment. These discontinuities include bedding planes between beds and laminations of different lithology (Chapter 4). Sedimentary structures, including mud cracks and ripple marks, also represent discontinuities in the rock. In addition to knowing the types of discontinuities, engineers must also determine their spacing, orientation, and roughness. The orientation is particularly important in considering the ability of a rock slope to support itself under the load of the rocks and soils that make up the slope. Failure in this situation occurs when part of the rock mass breaks away along a discontinuity and moves downslope (Figure 6-30). Slope movements, which will be discussed in more detail later, constitute a very significant category of geologic hazards.

**Figure 6-29** Rock bolts used to prevent failure of sandstone slabs bounded by discontinuities in Colorado River canyon downstream from the Glen Canyon Dam.

**TABLE 6-4**
**Rock Discontinuities**

| |
|---|
| Sedimentary |
| Bedding planes |
| Sedimentary structures (mud cracks, ripple marks, cross beds, and so on) |
| Unconformities |
| Structural |
| Faults |
| Joints |
| Fissures |
| Metamorphic |
| Foliation |
| Cleavage |
| Igneous |
| Cooling joints |
| Flow contacts |
| Intrusive contacts |
| Dikes |
| Sills |
| Veins |

Quantification of rock-mass properties is very difficult because of the number of variables involved. One index that is used is called the *rock quality designation* (RQD). During site investigation for an engineering project, test holes are drilled to determine subsurface rock formations. One sampling technique is to obtain a *core*, which is a small cylindrical sample of the entire

**Figure 6-30** Rockfall on eastbound lane of Interstate 70, Jefferson County, Colorado. Failure occurred along joints and foliation planes in Precambrian gneiss. (W. R. Hansen; photo courtesy of U.S. Geological Survey.)

**TABLE 6-5**

**Rock Quality Designation (RQD)**

| *RQD (%)* | *Description of rock quality* |
|---|---|
| 0–25 | Very poor |
| 25–50 | Poor |
| 50–75 | Fair |
| 75–90 | Good |
| 90–100 | Excellent |

vertical interval drilled. The core can then be used for laboratory tests of the rocks penetrated. The length of an individual piece of core is dependent upon the degree and orientation of fractures in the rock. Cores of highly fractured rock will consist of long, unbroken segments. RQD is defined as the percentage of core recovered in pieces 10 cm or longer in length. The index is expressed as a percentage of the total depth drilled (Table 6-5). Thus in 10 m of drilling, recovery of 9.2 m in pieces 10 cm or longer in length would have an RQD of 92% and be described as excellent quality.

Several rock-mass classifications have been devised for specific applications. Few of these, if any, are suitable for all purposes. One rock-mass classification is shown in Figure 6-31. The variables usually considered in a classification include the strength of intact rock within the rock mass, the RQD, the state of weathering of the rock mass, the fracture spacing, the continuity and in-filling, the orientation of fractures, and the ground-water conditions.

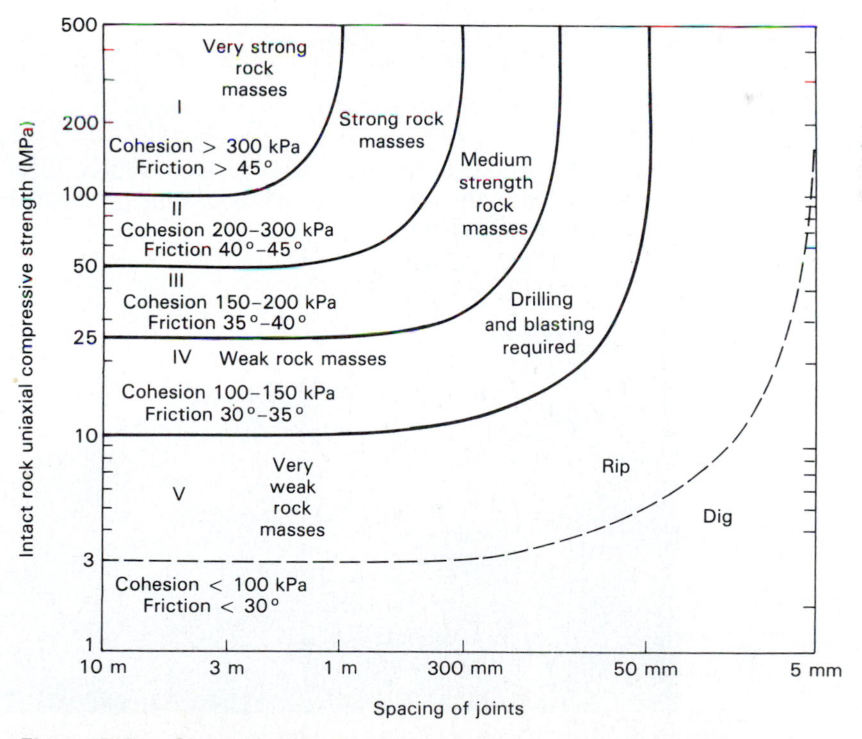


**Figure 6-31** Strength diagram for jointed rock masses showing types of excavation needed. (From Z. T. Bieniawski, 1974, Geomechanics classification of rock masses and its application to tunnelling, *Proc. 3rd Cong. Int. Soc. Rock Mech.*, 1:27–32. Reprinted with permission of Butterworth-Heinemann Limited.)

## CASE STUDY 6-1

## The Gros Ventre Slide; Role of Discontinuities in Rock-Mass Stability

A good example of the effect of discontinuity orientation upon slope movements is provided by the Gros Ventre slide, the largest historic rock slide in the United States (Figures 6-32 and 6-33). Sedimentary rock formations are oriented as shown in the Gros Ventre Valley near Jackson, Wyoming (Alden, 1928). On the south slope of the valley, bedding planes are tilted, or dip, in the same direction as the slope. The strength of the rock mass, therefore, is controlled by the strength developed along the sloping bedding planes rather than by the strength of the rock itself. After a period of heavy rains in 1925, the slope failed because the stress along one or more of the bedding planes was greater than the shear strength. As a result, about 40 million $m^3$ of rock slid rapidly into the valley along bedding-plane discontinuities.

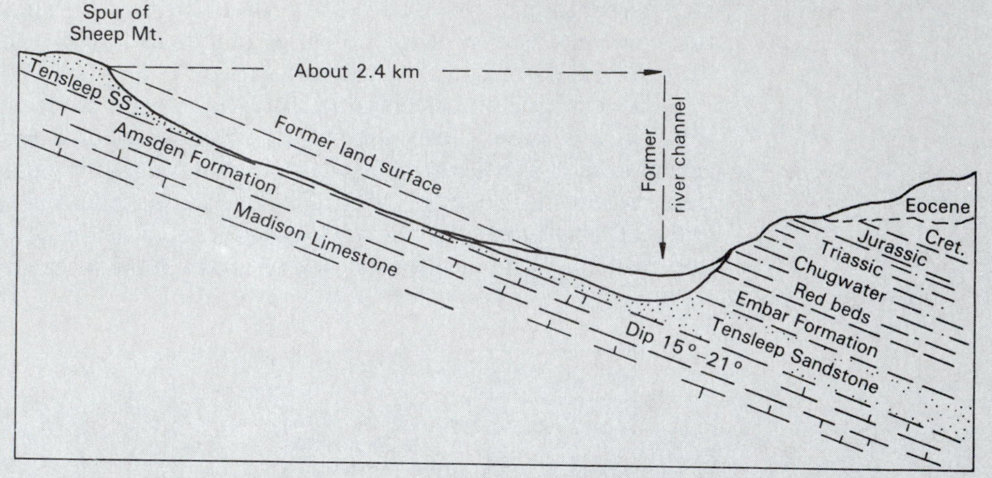


**Figure 6-32** The Gros Ventre rockslide. (From Alden, 1928; used by permission of the American Institute of Mining, Metallurgical, and Petroleum Engineers.)

**Figure 6-33** View across the Gros Ventre Valley of the scar and lower part of the rockslide. Photo taken in 1925, 3 months after slide. (W. C. Alden; photo courtesy of U.S. Geological Survey.)

Numerical values can be assigned for each variable and summed to obtain an overall rock-mass classification. Problems arise in assigning numerical values to largely descriptive variables such as weathering characteristics. Rock is divided into five classes in this classification based on the uniaxial compressive strength of intact rock and the joint spacing in the rock mass. The resulting rock-mass categories can be related to methods of excavation, types of tunnel support systems, and other applications.

## SUMMARY AND CONCLUSIONS

Earth materials are three-phase substances, with possible solid, liquid, and gas components. The volume of the void spaces in the rock or soil, along with the size and interconnectedness of pores, determines the intrinsic permeability of the material. A related parameter, hydraulic conductivity, includes the properties of the fluid in its definition.

Stress is present beneath the surface of the earth as a result of the weight of overlying rock and soil, as well as any load applied at land surface. The stress at any point can be resolved into maximum, minimum, and intermediate components, acting in mutually perpendicular directions at the point.

Strain results from the application of stress to a material. Rocks exhibit complex deformational response to stress. Models can be used to approximate the stress-strain relationships of different types of materials. Simple models that demonstrate elastic, plastic, and viscous behavior can be used to simulate deformation of rocks, although real deformation may incorporate combinations of more than one type of simple response.

The stresses applied to a rock sample may be high enough to cause failure. Failure is controlled by the shear strength of the material because the rock fails under the action of shear stresses that develop on planes inclined at various angles to the principal stresses. Shear and normal stresses at failure are plotted on diagrams to show the variations of shear strength under various stress conditions. Mohr's circle is a convenient way of graphically showing the relationships between principal stresses and stresses on planes inclined to the principal plane.

The relationship between shear and normal stresses at failure varies for different types of material. Rocks and some soils derive strength from cementation between grains, cohesion between particles, and the interlocking grain structure of minerals that crystallized from a magma or solution. These materials follow the Mohr-Coulomb strength criterion, which includes the intrinsic strength of the material as well as the frictional resistance to shear that is proportional to normal stress along the failure plane.

The engineering behavior of intact rock can be classified using the modulus of elasticity and the uniaxial compressive strength, but the behavior of rock masses is dependent on the type and spacing of discontinuities as well as the properties of the intact rock.

### PROBLEMS

1. What are the differences among porosity, void ratio, and permeability?
2. Give as many examples as you can of materials that exhibit elastic, plastic, and viscous deformation.

3. How does the response of rocks to stress differ from models of ideal behavior?
4. How are stress and strength related?
5. How is shear strength measured?
6. What are the two major components of shear strength?
7. Define modulus ratio and explain what it is used for.
8. Discuss the limitations and problems of extrapolating mechanical properties measured on small rock samples to field situations.
9. Determine the vertical stress, $\sigma_v$, at a depth of 15 m, when the overlying rocks include the following (listed in descending order from ground surface): Unit A: 9 m thick, $\gamma_a = 28$ kN/m$^3$; Unit B: 4 m thick, $\gamma_b = 22$ kN/m$^3$; Unit C: $\gamma_c = 25$ kN/m$^3$.
10. A surficial rock unit has a unit weight of 170 pounds per cubic ft (pcf) and a thickness of 20 ft. It is underlain by rocks with a unit weight of 145 pcf. What is the vertical stress at a depth of 30 ft below the original land surface after 4 ft of the upper unit have been removed in an excavation?
11. The major principal stress is 200 kN/m$^2$ compression at a point below the surface. The minor principal stress is 80 kN/m$^2$ compression.
    (a) Draw Mohr's circle for these conditions.
    (b) Find the maximum shear stress and the normal stress that acts on the plane of maximum shear stress.
    (c) Find the normal and shear stresses on a plane inclined 30 degrees from the major principal plane.
12. A rock unit is composed of rocks with a unit weight of 200 pcf. Draw Mohr's circle for a depth of 10 ft below land surface, assuming that the vertical stress at that point is the major principal stress and the minor principal stress is one-third of that value. What is the maximum shear stress acting at the point?
13. The difference between the major and minor principal stresses at a point is 70 kN/m$^2$. If the major principal stress is 140 kN/m$^2$, what are the shear and normal stresses on a plane inclined at 30 degrees to the major principal plane?
14. A material has an angle of internal friction of 33 degrees. What is the shear strength of this material at a normal stress of 2000 psf?
15. A direct shear cell has cross-sectional dimensions of 15 cm $\times$ 15 cm. In a test of dry sand, the normal load at failure is 5 kN and the shear force is 2.7 kN. What is the angle of internal friction of this material?
16. A rock unit has a unit cohesion of 400 psf and an angle of internal friction of 35 degrees. When the normal stress is 1600 psf, what is the shear strength of the rock?

## REFERENCES AND SUGGESTIONS FOR FURTHER READING

ALDEN, W. C. 1928. Landslide and flood at Gros Ventre, Wyoming. *Transactions, AIME*, 76:347–362.

BELL, F. G. 1980. *Engineering Geology and Geotechnics*. Boston: Newnes-Butterworths.

BIENIAWSKI, Z. T., 1974. Geomechanics classification of rock masses and its application to tunnelling. *Proc. 3rd Cong. Int. Soc. Rock Mech.*, 1:27–32.

COSTA, J. C., and V. R. BAKER. 1981. *Surficial Geology: Building with the Earth*. New York: John Wiley.

DEERE, D. U. 1958. Geological considerations, in *Rock Mechanics and Engineering Practice*, K. G. Stagg and O. C. Zienkiewics, eds. London: John Wiley.

MCCARTHY, D. F. 1993. *Essentials of Soil Mechanics and Foundations*, 4th ed. Englewood Cliffs, N.J., Regents/Prentice-Hall, Inc.

RAHN, R. H., 1986. *Engineering Geology: An Environmental Approach*. New York: Elsevier.

# 7

# Tectonics, Structure, and the Earth's Interior

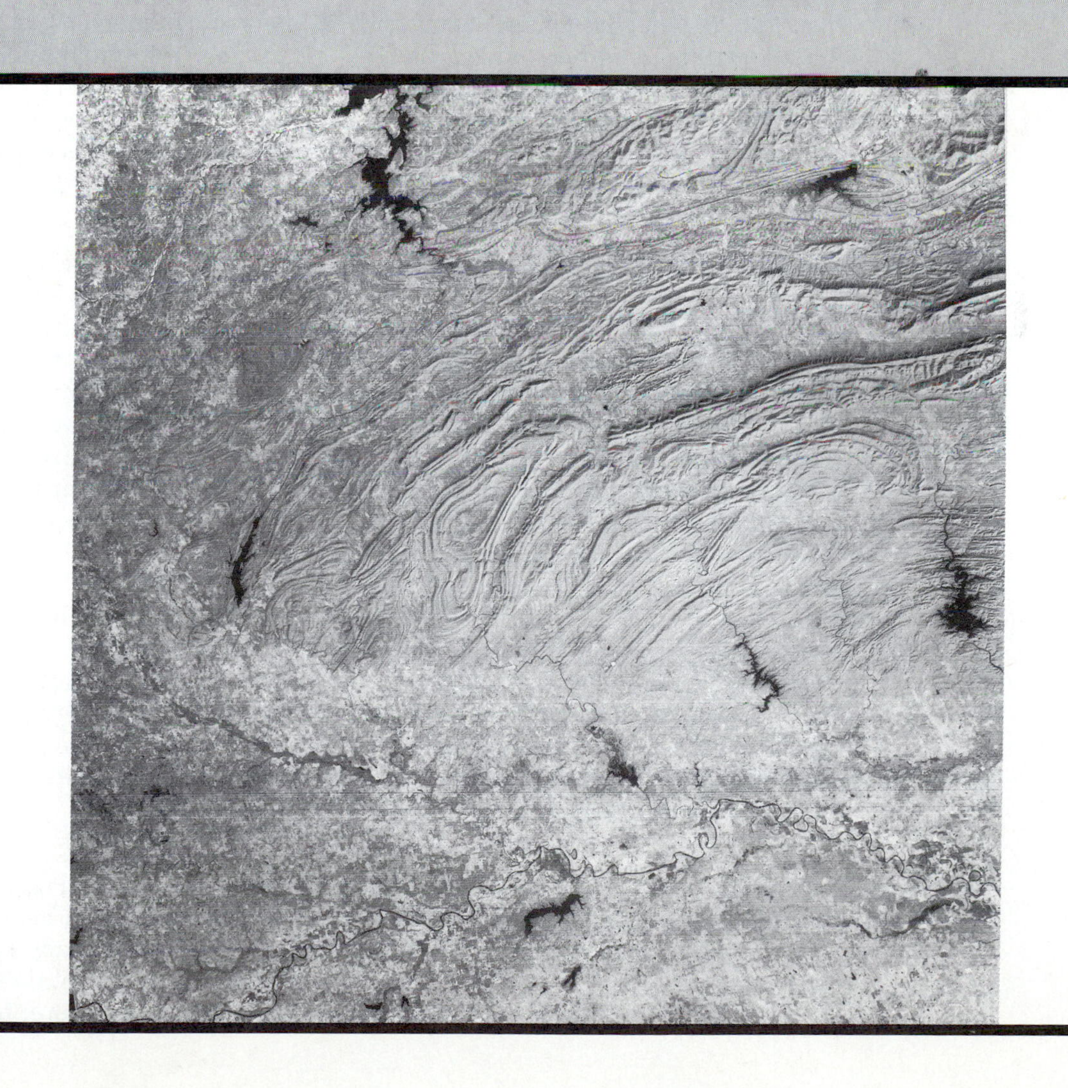

Even before satellite images, such as the chapter opening figure, it was obvious to even casual observers that many of the rocks that make up the earth's crust have been subjected to intense deformation. That these processes continue to the present is no surprise to anyone who has experienced an earthquake. In this chapter, we will examine the causes and effects of deformation of the earth's crust, which is known as *tectonic* deformation. Brittle deformation, the type that generates earthquakes, is common in the upper crust. At greater depths, plastic or ductile deformation is the norm. In the geologist's view, the rock structures produced by tectonic deformation are evidence that can be used to unravel the earth's history. From a practical standpoint, studies of geologic structure are used in exploration for petroleum and mineral deposits. Engineers must also understand rock structure because the discontinuities imparted to rocks during deformation frequently govern the engineering behavior of rock masses at the surface or in the shallow subsurface regions of the earth.

## THE EARTH'S INTERIOR

An introduction to geology, even if its purpose is the practical application of the science, is incomplete without a general understanding of the earth's interior. This region of the planet is inaccessible to direct human observation, yet the prevailing conditions and processes there exert a great influence upon the surface. To study the earth's interior, we must use indirect methods of investigation. Among these techniques are studies of heat flow, gravity, and magnetism. These topics, along with seismology, form the basis of the field of study known as *geophysics*.

### The Earth's Internal Heat

According to most recent hypotheses of the earth's origin, planetary accretion took place at a moderate temperature. During the early part of its history, the earth remained a solid, homogeneous body. Internal temperatures increased, however, as a result of the decay of radioactive elements. A crucial point was reached when the temperatures exceeded the melting point of iron in certain zones within the new planet. As iron melted, it began to separate and sink toward the center of the earth. The movement of molten iron toward the incipient core released gravitational energy in the form of heat, leading to the segregation, or *differentiation*, of the earth into its present layers (Figure 7-1). The assumed present relationship of the temperature profile and melting point curve (Figure 7-1) requires that the earth is now largely solid with the exception of the outer core. Another consequence of differentiation was that the radioactive elements became concentrated in the crust because of their chemical affinity for other elements that accumulated there. The present thermal regime of the earth therefore is dictated by a partially molten core and a more abundant radioactive heat source in the crust.

Heat Flow. The unequal distribution of volcanoes, hot springs, geysers, and other indications of high heat flow suggests that the flow of heat from the interior of the earth to the surface is not uniform. We have already mentioned the two major sources of internal heat—the molten core and crustal radioactivity. Heat flows whenever a temperature gradient exists; therefore, heat flows

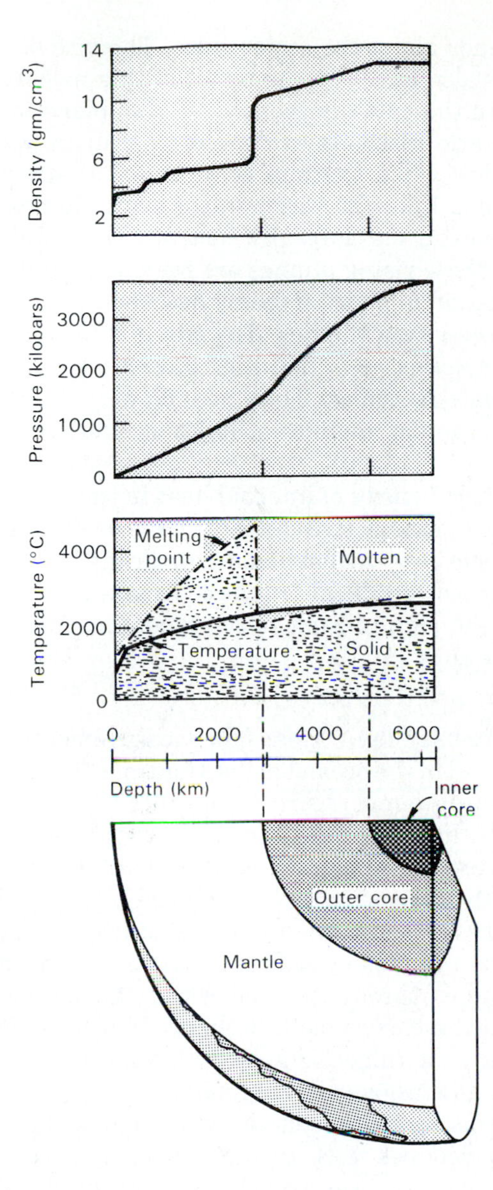


**Figure 7-1** Physical conditions in the interior of the earth. The inferred temperature profile of the earth passes above the melting point curve in the region of the earth's outer core. (From E. A. Hay and A. L. McAlester, *Physical Geology: Principles and Perspectives*, 2d ed., copyright © 1984 by Prentice-Hall, Inc., Englewood Cliffs, N.J.)

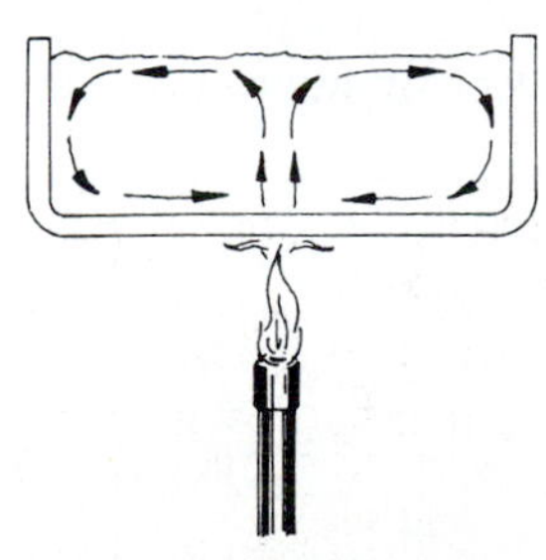

**Figure 7-2** Convective heating of water in a pan may be similar to convection cells of mantle material that rise to the base of the lithosphere, spread laterally, and then descend.

from the earth's core to its cooler surface. Three physical processes account for heat transport; these include conduction, convection, and radiation. Conduction is the atomic transfer of energy by vibrating atoms to adjacent atoms vibrating less rapidly. The rate of heat conduction is a function of the temperature gradient and the *thermal conductivity* of the rocks through which the heat is flowing. Rocks are poor heat conductors; fluctuations of temperature due to solar effects at the surface, for example, are felt only in the upper few meters of soil. By itself, conduction is inadequate to explain the distribution of heat flow in the earth. In fact, heat that began to move outward from the center of the earth at the time of its origin would not yet have reached the surface. Although conduction of heat is important in the earth, another process must be occurring. Radiation of heat is unlikely through rocks; therefore, convection must play an important role in heat transport.

Convective heat transfer may be more efficient than conduction because the material being heated moves from areas of higher to lower temperature. A good example of convection is the movement of water in a pan that is heated from below (Figure 7-2). The hottest water in the center of the pan rises to the surface and descends along the cooler sides. The rise of hot water in geysers and hot springs is also partially the result of this mechanism.

Convection in the earth is thought to take place in the mantle. Despite the rigidity of mantle rocks, as evidenced by transmission of seismic waves, these rocks can flow when thermal or mechanical stresses are applied over long periods of time. Like the water in the pan, plumes of hot mantle rocks slowly rise and spread out laterally beneath the rigid lithosphere (Figure 7-3). Areas above these rising plumes are regions of high-heat flow and active volcanism. Circulation is maintained by descending currents and lateral flow at depth toward the rising plumes. The whole mantle may be broken into a series of such *convection cells*. Mantle convection may provide the driving mechanism for plate tectonics. Thus heat flow in the mantle is ultimately responsible for volcanism, mountain building, and earthquakes in the lithosphere.

The second major source of internal heat in the earth is radioactive decay. Since differentiation, radioactive elements have been concentrated in the rocks near the earth's surface—in particular, in granitic rocks. Thus radioactive heat production is most significant in the continental crust. Heat flow measured at the surface is the sum of radioactive production from the crust and deep heat flow from the core.

Variations in Heat Flow. Heat-flow measurements have been made both on the continents and in the ocean basins. Continental heat-flow measurements are made by inserting temperature probes into holes drilled into the walls of mines and tunnels. In the oceans, temperature probes are driven into bottom sediments to measure the temperature. In both areas, measurements of temperature changes with increasing depth are used to determine the temperature gradient. When the temperature gradient is multiplied by the measured thermal conductivity of the rock or sediment, the heat flow is obtained.

Heat-flow values throughout the world can be related to the geologic setting of the point of measurement. In the ocean basins, high heat flow is found beneath the midoceanic ridges (Figure 7-3). Midoceanic ridges are thought to be the surface manifestations of rising mantle convection currents. Volcanism in these areas forms new lithosphere, which moves laterally away from the ridges. Heat flow declines with distance from midoceanic ridge crests and reaches minimum values over trenches. Trenches develop over subduction zones, where cold lithospheric slabs are forced downward into the mantle. Subduction zones constitute the descending plumes of convection current cells; thus heat flow is low in these areas.

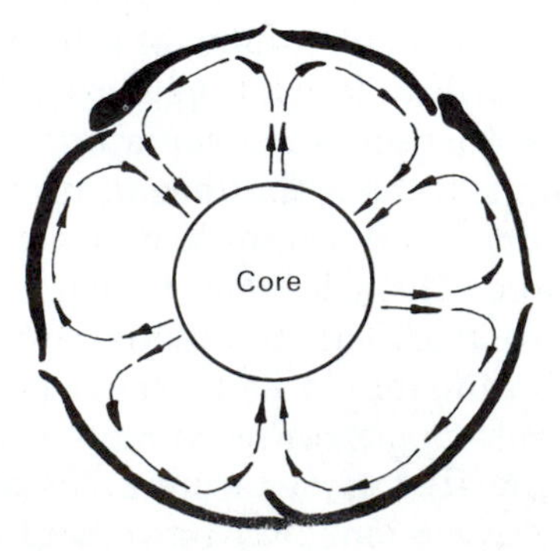


**Figure 7-3** Hypothetical configuration of convection cells in the mantle.

On the continents, heat flow also varies, even though the radioactive heat component of total heat is relatively constant. Areas of high heat flow have had recent tectonic and volcanic activity. The Basin-and-Range Province of western North America is one such area (Figure 7-4). Volcanism and recent faulting have been common in the Basin-and-Range Province, so it is possible that a rising convection current or plume is present beneath this province. The high heat flow in this region is under study for possible production of *geothermal energy*. The Geysers geothermal plant in California (Figure 7-5) already produces 600 MW of electrical power from geothermal energy. Low continental heat flow occurs in geologically old regions, like the Canadian Shield, that have been tectonically inactive for millions of years.

### Gravity

The force of gravity can be expressed as

$$F = \frac{mM}{r^2} G \qquad \text{(Eq. 7-1)}$$

where $F$ is the gravitational attraction between two bodies, $m$ and $M$ are the masses, $r$ is the distance between the bodies, and $G$ is the universal gravitational constant. For a body at the surface of the earth, $r$ and $M$ become the radius of the earth and its mass, respectively. If the earth were perfectly spherical, nonrotating, and uniform in density, the gravitation acceleration, or the acceleration of a freely falling body, would be constant anywhere upon the earth's surface. Gravitational acceleration varies from place to place, however, and it is these minor variations that make gravity measurements a very useful tool in geology.

Measurement. The unit of measurement used in gravity studies is the Gal, which equals an acceleration of 1 cm/sec$^2$. The average value of $g$ (gravitational acceleration) at the earth's surface is about 980 Gal, although the measured values vary from place to place. These variations are attributed to (1) rotation of the earth, (2) topography, and (3) variations in density of near-surface rocks. Rotation affects gravity by causing a bulge at the earth's equator that is due to centrifugal force. The radius of the earth is thus greater at the equator than at the poles. The radius also varies from highlands to lowlands or to ocean basins. These departures from the ideal ellipsoidal shape of the earth cause minor changes in gravitational attraction across the surface (Figure 7-6). Finally, lateral changes in density of near-surface rocks can affect the value of $g$. The value of $g$ would be greater over a region of dense near-surface rocks such as basalt than over lighter granitic rocks at the same elevation.

The magnitude of these variations is on the order of milliGals (0.001 Gal) or fractions of a milliGal. These minute variations are measured by instruments called *gravity meters*. A gravity meter consists of a weight suspended on a spring that expands or contracts according to the local gravitational field. Despite this simple principle, gravity meters are extremely sophisticated, expensive instruments because of the precision they must attain.

After a field measurement is taken, it must be corrected before it will yield geologically useful information. One major correction resolves the effects

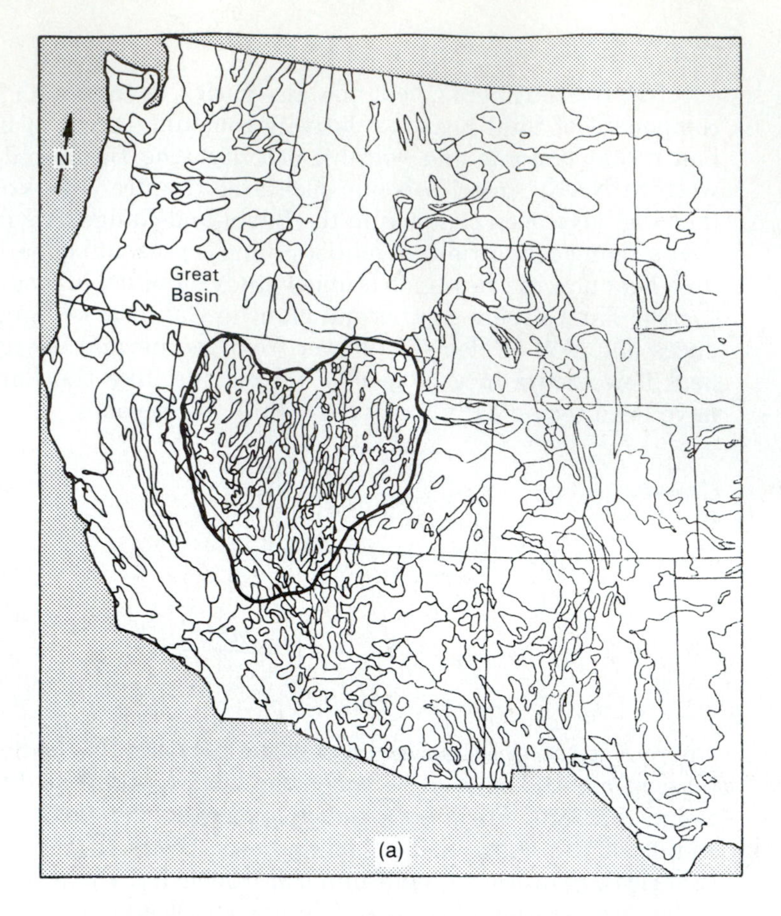


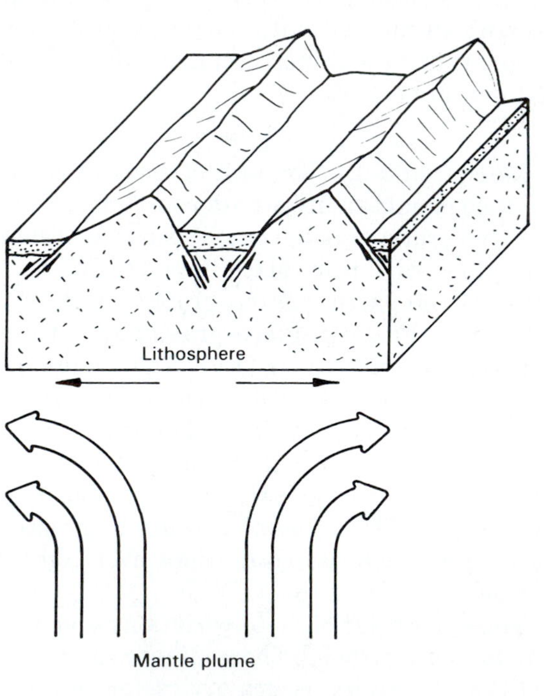


**Figure 7-4** (a) Map of the western United States showing location of Great Basin area of Basin-and-Range Province. (From E. A. Hay and A. L. McAlester, *Physical Geology: Principles and Perspectives*, 2d ed., copyright © 1984 by Prentice-Hall, Inc., Englewood Cliffs, N.J.) (b) Diagram of possible basin-and-range structure caused by rising mantle plume beneath.

**Figure 7-5** Steam wells at The Geysers geothermal area, California. (Photo courtesy of U.S. Geological Survey.)

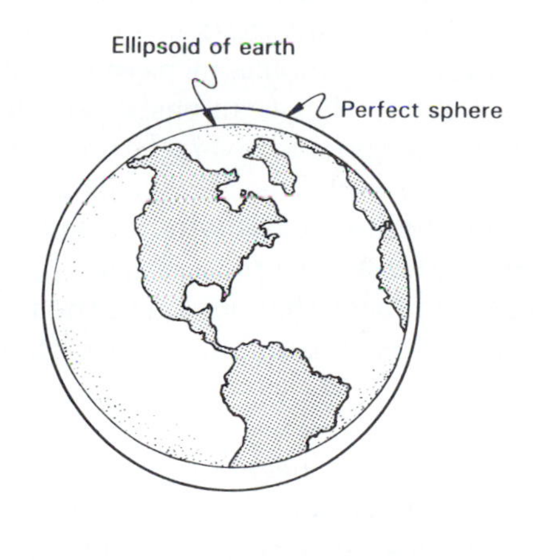


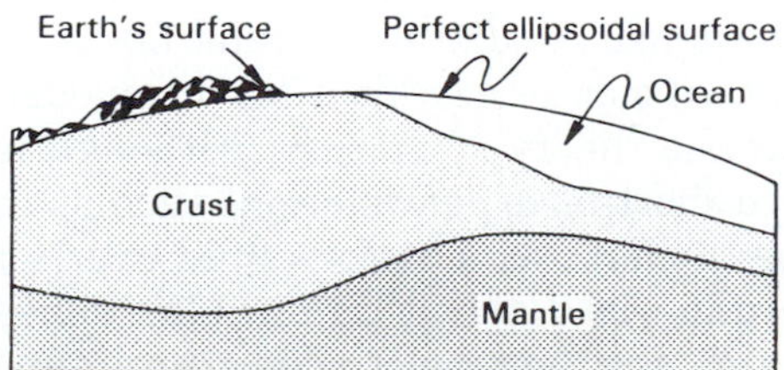


**Figure 7-6** Variations in gravitational attraction are caused by the earth's ellipsoidal shape, elevation differences, and differences in density of crustal materials. (From E. A. Hay and A. L. McAlester, *Physical Geology: Principles and Perspectives*, 2d ed., copyright © 1984 by Prentice-Hall, Inc., Englewood Cliffs, N.J.)

**Figure 7-7** Horizontal deflection of a plumb bob by the mass of a prominent mountain range.

of local topographic variations. For example, the presence of a nearby mountain range will deflect a pendulum from its vertical position because of the gravitational attraction caused by the mass of the mountains (Figure 7-7). The *Bouguer correction* adjusts the reading to represent a slab of uniform thickness above a hypothetical plane at sea level called the *geoid* (Figure 7-8). Also, a correction is applied to compensate for the material between the geoid and the point of measurement. This is known as the *free-air* correction. The final values are compared to an expected value, which is calculated for an ideal earth, the spheroid, with a homogeneous crust. Any differences between the corrected measurement and the expected value are due to local variations in density caused by the composition of near-surface rocks. These differences are called *gravity anomalies*.

The study of gravity anomalies has led to important theories about the earth's crust. For example, young mountain ranges typically yield large negative anomalies. This implies that rocks of low density extend to some depth below the mountain range. Ocean basins have positive anomalies, which indicates that the rocks beneath the ocean basins are denser than average. Thus the density of near-surface rocks is inversely proportional to their topographic position; regions of dense rocks usually lie below sea level, and mountain ranges are composed of lighter rocks. Gravity measurements also have many practical applications. Masses of igneous rock that contain ore bodies often differ in density from the rocks that surround them. In addition, sedimentary rock structures that trap petroleum may cause characteristic gravity anomalies.

Isostasy. The discovery of negative gravity anomalies beneath mountain ranges has important implications concerning the earth's crust and mantle. Negative gravity anomalies beneath young mountain ranges means that mountains must have low-density roots that extend far below the surface (Figure 7-9). In effect, mountain masses are "floating" in the denser mantle rock. This concept is known as the theory of *isostasy*. Although mantle rocks are rigid solids with respect to short-term deformations such as the propagation of seismic waves, they may behave as a viscous fluid over long periods of geologic time. Like an iceberg in water, the mountain's low density root lies well below the level of the plains surrounding it. The mountain mass is supported by buoyant forces in the mantle similar to the buoyancy that supports any floating object.

Implied in the theory of isostasy is the idea that crustal masses establish a state of dynamic equilibrium with the mantle. Thus the dense basalts of the ocean basins lie at a low elevation in comparison with lighter granitic mountain

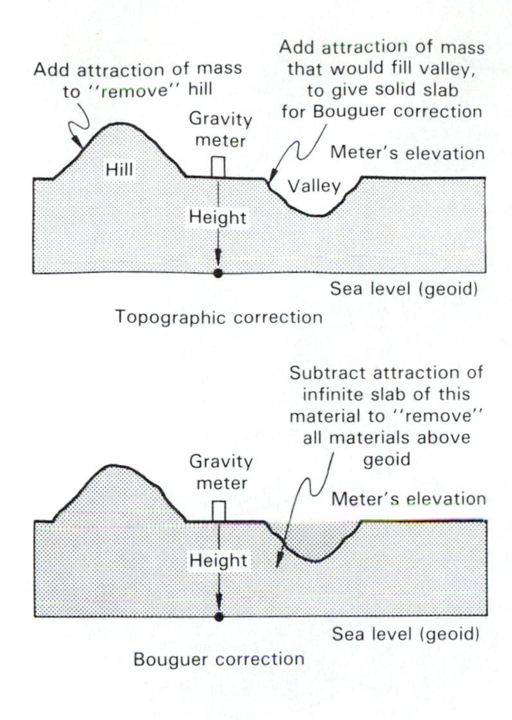


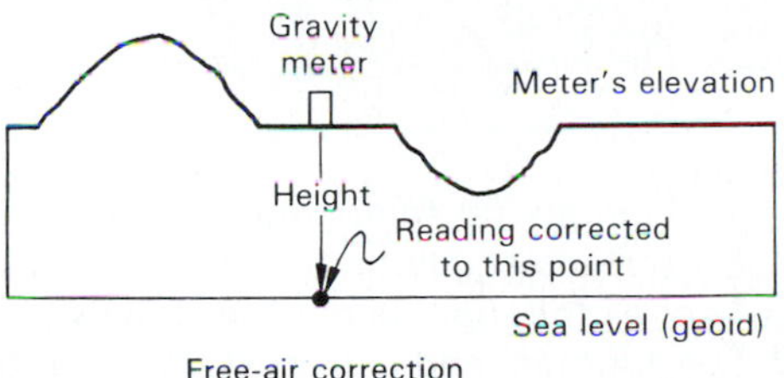


**Figure 7-8** Corrections to measured gravity readings are necessary to discover subsurface geological features. The topographic correction removes the effect of local variations in elevation. The Bouguer correction compensates for a uniform slab of material between the geoid and the point of measurement. The free-air correction results in a reading that would be obtained if the meter were at sea level. (From S. Judson, M. E. Kauffman, and L. D. Leet, *Physical Geology*, 7th ed., copyright © 1987 by Prentice-Hall, Inc., Englewood Cliffs, N.J.)

ranges, which float higher in the mantle. Any change in the mass of a crustal body will be reflected in changes in this *isostatic equilibrium*. For example, as mountains are slowly eroded, mass is removed and the low-density root is uplifted and is decreased in size because less support is required. Thus the Rocky Mountains have deep roots in comparison with the old, eroded Appalachian Mountains, whose roots are much thinner than they were during the period following their formation.

Isostatic equilibrium is also illustrated by the advances and retreats of glaciers. The heavy load of a glacier imposed upon the crust causes the crust to sink deeper into the mantle. When the ice melts and retreats, however, the load is removed much faster than the crust and mantle can reestablish isostatic equilibrium. The slow uplift of the crust after glacial retreat is called *isostatic rebound*. Slow uplift of the crust is still going on, 10,000 years after the last glaciation (Figure 7-10).

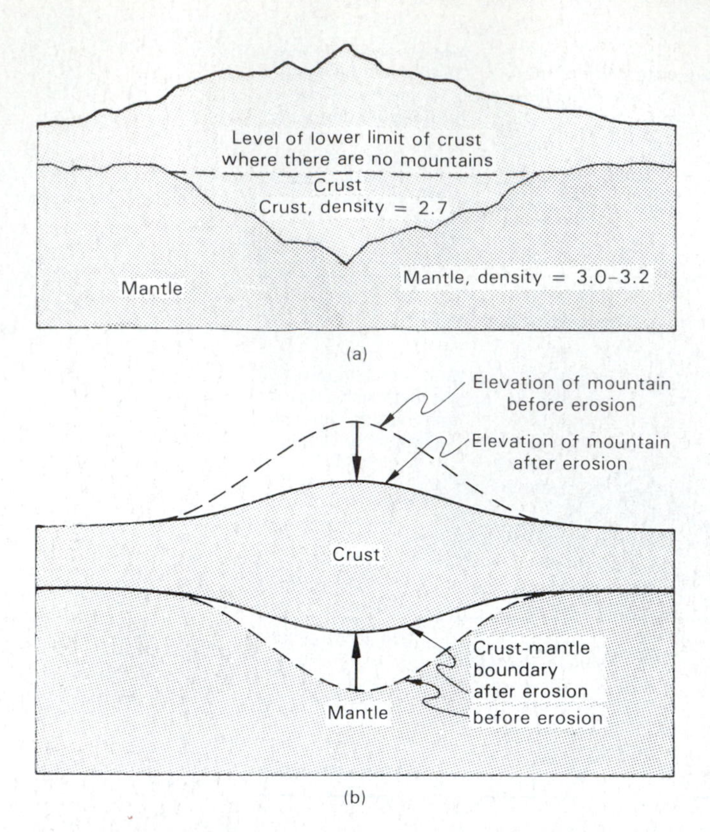


**Figure 7-9** Isostatic equilibrium. (a) A mountain range with a low-density crustal root. (b) Isostatic equilibrium is maintained by reduction in size of the root as the mountain mass is slowly eroded. (From S. Judson, M. E. Kauffman, and L. D. Leet, *Physical Geology*, 7th ed., copyright © 1987 by Prentice-Hall, Inc., Englewood Cliffs, N.J.)

### *Magnetism*

The Earth's Magnetic Field. The earth as a whole has long been recognized as a magnetic body that behaves as a simple bar magnet, or dipole, aligned at a small angle from its axis of rotation. The magnetic poles are located in the vicinity of the geographic poles. Magnetic *lines of force* associated with the earth's magnetic field cause magnetized bodies on the earth to align themselves parallel to the lines of force (Figure 7-11). The discovery of this principle made possible navigation with the magnetic compass.

The vertical angle between a line of force and the earth's surface is known as magnetic *inclination.* This angle is zero, or horizontal, at the magnetic equator and 90 degrees, or vertical, at the magnetic pole. Because the magnetic and geographic poles do not exactly coincide, there is a small angle between the direction of a compass needle pointing to the magnetic north pole and the direction of a line pointing to the geographic north pole. This angle is known as *magnetic declination* (Figure 7-12).

The cause of the earth's magnetism has been and continues to be the object of much speculation. It is now believed that this field is caused by currents moving within the liquid iron core of the earth. These currents would generate electrical currents in the highly conductive iron, which would in turn induce the magnetic field. The core therefore functions as a self-exciting dynamo.

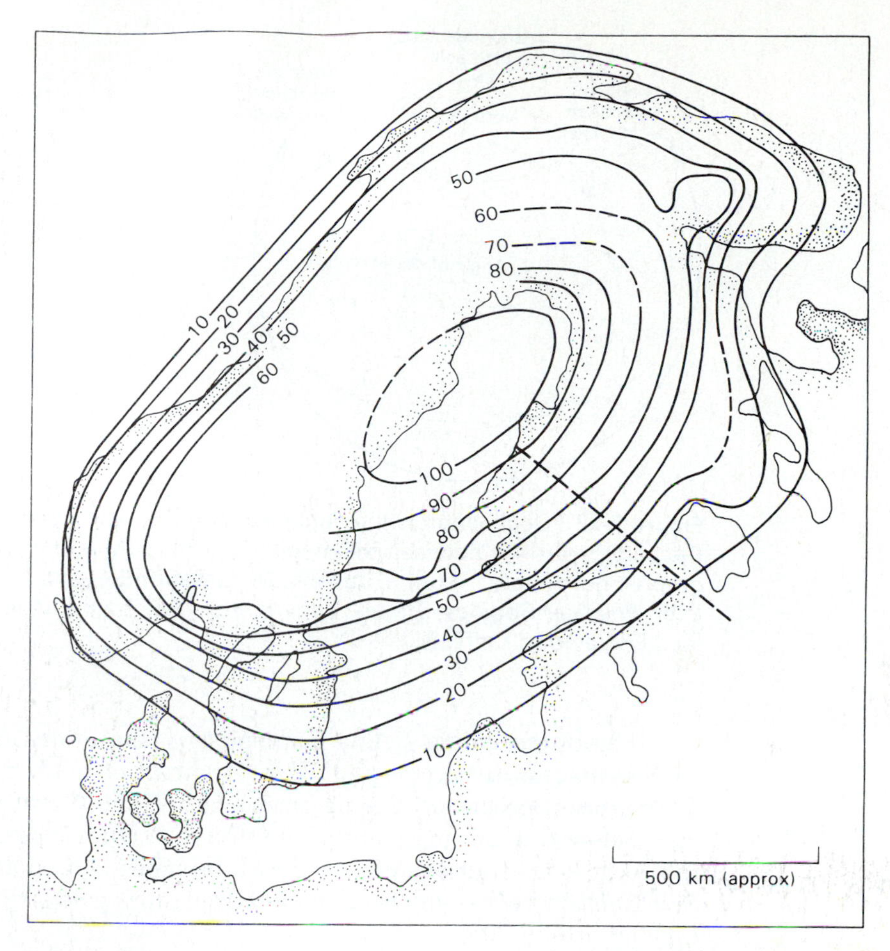


**Figure 7-10** Uplift in Scandinavia during the last 5000 years caused by glacial retreat. The contours are the elevations in meters of a shoreline that was at sea level 5000 years ago. (After R. K. McConnell, Jr., 1968, *Journal of Geophysical Research*, 73:7090.)

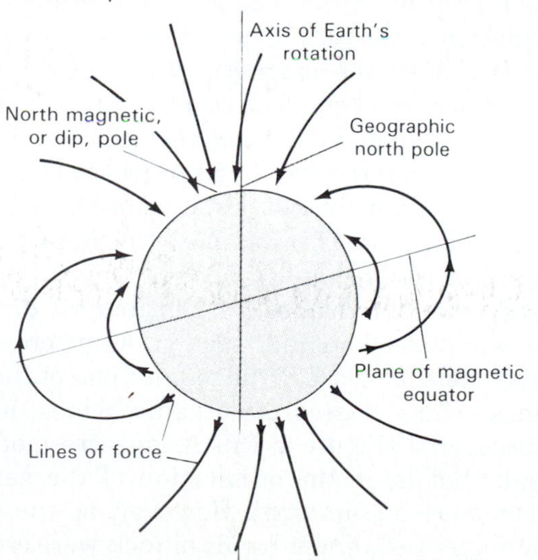


**Figure 7-11** Lines of force caused by the earth's magnetic field. (From S. Judson, M. E. Kauffman, and L. D. Leet, *Physical Geology*, 7th ed., copyright © 1987 by Prentice-Hall, Inc., Englewood Cliffs, N.J.)

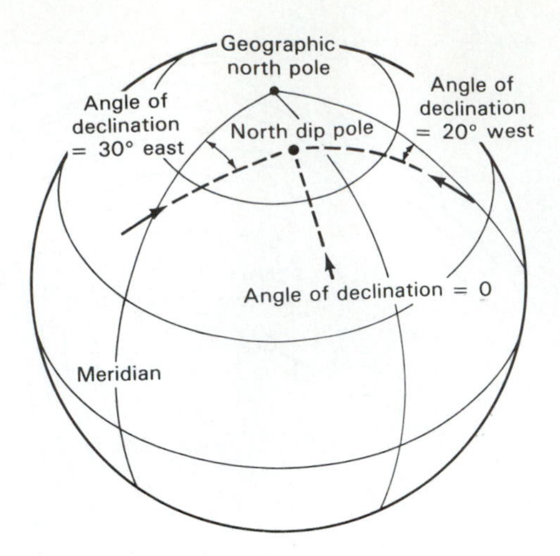


**Figure 7-12** Declination is the angle measured at any point on the earth's surface between the magnetic north pole and the geographic north pole. (From S. Judson, M. E. Kauffman, and L. D. Leet, *Physical Geology*, 7th ed., copyright © 1987 by Prentice-Hall, Inc., Englewood Cliffs, N.J.)

Paleomagnetism. As a magma cools, the atoms in magnetic minerals such as magnetite become aligned in the direction of the earth's magnetic field. This process occurs at a temperature called the *Curie point*, a temperature well below the melting point of the mineral. Once magnetism is acquired by the rock, it is retained unless the rock becomes heated above the Curie point. Sedimentary rocks can also be magnetized during deposition in some sedimentary environments.

Measurement of a rock's magnetic declination and inclination can be made; such measurements are said to indicate the *paleomagnetism* of the rock. Paleomagnetic studies have yielded dramatic results concerning the history of the earth. For example, from declination and inclination values it is possible to determine the position of the magnetic (and therefore geographic) pole at the time of formation of the rock. One of the most fascinating results of paleomagnetic studies is that the magnetic poles have apparently moved great distances through geologic time. For several reasons, movements of the magnetic poles over great distances is not thought to be possible. Therefore, to explain the paleomagnetic results, it was proposed that the continents containing the rocks, rather than the magnetic poles, have moved over the earth's surface during geologic time. This was one of the main lines of evidence supporting the theory called *continental drift.*

The second form of evidence for continental drift was gathered when *magnetometers* were towed behind research ships crossing the oceans. These instruments were able to measure the magnetism of the rocks beneath the sea floor. When these ships crossed midoceanic ridges, a remarkable magnetic pattern was discovered (Figure 7-13). At the crest of the midoceanic ridge, rocks of normal (that is, in the orientation of the earth's present magnetic field) magnetism were encountered. However, as the ship moved away from the midoceanic ridge crest, linear bands of rock were crossed that were magnetized in alternately reversed and normal polarities. Even more exciting was the realization that these magnetic bands or stripes were symmetrical on opposite sides of the ridge crests. This was strong evidence that new oceanic crust was formed by volcanism at the ridge crest in the prevailing polarity of

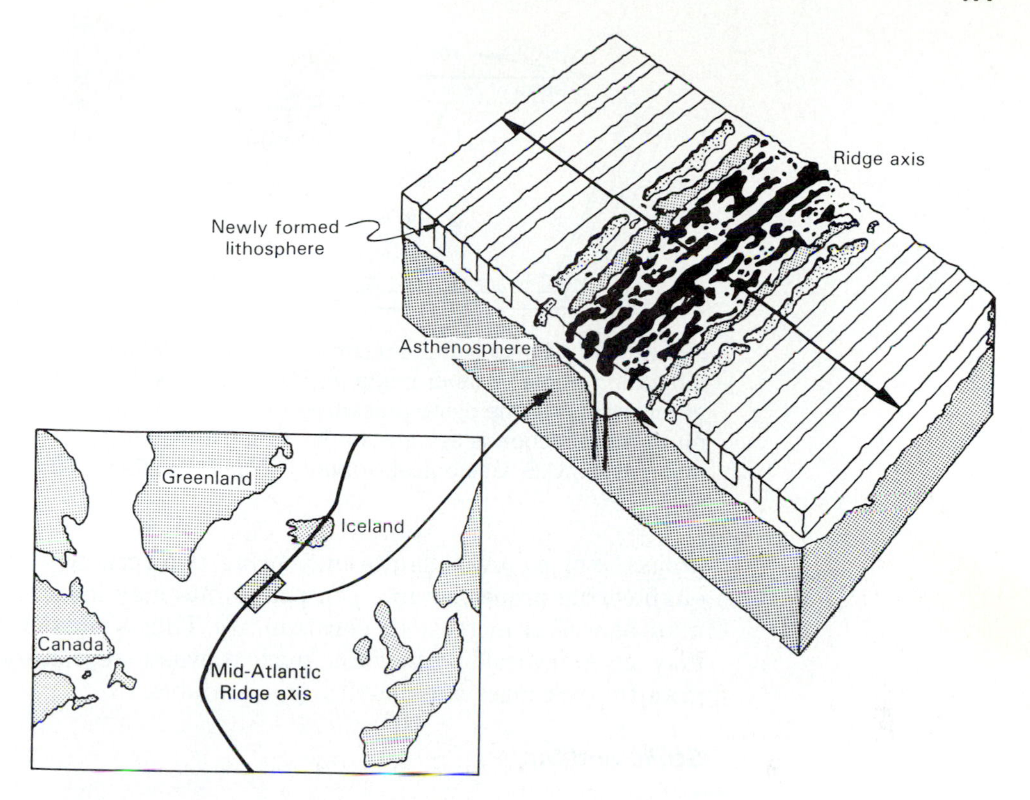


**Figure 7-13** Symmetrical magnetic patterns across the Mid-Atlantic Ridge, indicating that bands of newly formed rocks were magnetized at the ridge axis and that the bands were split apart during lateral movement in opposite directions. The black and stippled magnetic bands represent normal polarity, and the white bands beyond represent reversed polarity. (From E. A. Hay and A. L. McAlester, *Physical Geology: Principles and Perspectives*, 2d ed., copyright © 1984 by Prentice-Hall, Inc., Englewood Cliffs, N.J.)

the earth's magnetic field. The new oceanic crust was then split apart, moving in opposite directions away from the ridge. When a magnetic reversal occurred, the rocks forming at the ridge crests would be of that polarity. Thus the magnetic bands were symmetrical about the ridge. This discovery became nearly irrefutable evidence for the formation of new lithosphere at midoceanic plate margins. This theory became known as *sea-floor spreading*. It is considered to be one of the foundations of plate tectonics.

## GEOLOGIC STRUCTURES

Under the stresses that are imposed upon rocks in the earth, deformation takes place. The type of deformation is dependent upon the material properties of the rock as well as the confining pressure, temperature, strain rate, and other factors. The mechanics of deformation involve various combinations of elastic, plastic, and viscous behavior. When deformation is complete, the rock may be totally changed from its initial state. The record of deformation is retained in the altered rock in the form of geologic *structures* such as folds, faults, and joints. Engineers are more concerned, however, with the properties that these structures impart to a rock mass, especially those that impact an engineering

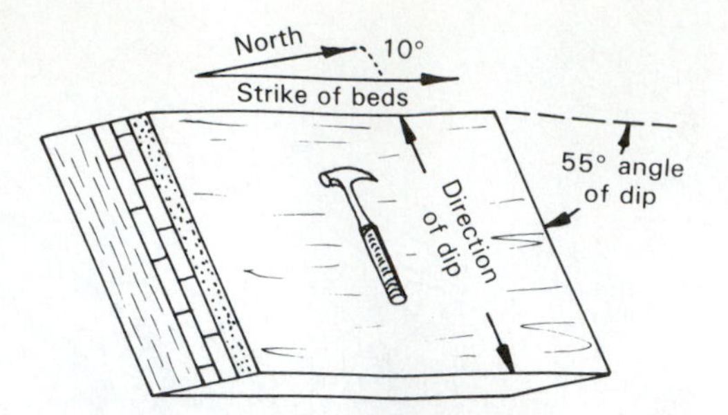


**Figure 7-14** Strike is the direction of a horizontal line intersecting a bedding plane or other plane in a body of rock. Dip is the angle of inclination of the plane measured from the horizontal. The maximum dip direction is at right angles to the strike. (After G. D. Robinson et al., U.S. Geological Survey Professional Paper 505.)

project. For example, faults may bring two rock types with vastly different engineering properties into contact. Joints may lower the strength of a rock mass as well as increase its permeability. Thus a mass of largely intact granite may be a favorable site for a nuclear waste repository, whereas a jointed granitic rock mass may be totally unsuitable.

### *Strike and Dip*

Most geologic structures are studied by measuring the orientation of planar elements within the rock. The orientation of a plane in space can be specified by measuring its *strike and dip*. (Figure 7-14). Strike is the direction of a line formed by the intersection of the plane and the horizontal. Dip is the amount of slope of the plane. It is determined by measuring the acute angle between the horizontal and the sloping plane. Dip is always measured in a vertical plane perpendicular to strike in the direction of maximum inclination of the rock plane.

In the field, strike and dip measurements are often made with a Brunton compass (Figure 7-15). The types of planes that are measured in the field vary with the types of rocks and deformational mechanisms. In folded sedimentary rocks, the orientations of bedding planes are measured throughout the area in order to reconstruct the pattern of folding. Fault and joint planes are measured directly so that their distribution throughout a rock mass can be predicted.

### *Folds*

Folds are produced by lateral compression of the crust. Under this type of stress, the crustal rocks are deformed into a series of wavelike forms oriented transversally to the direction of maximum stress (Figure 7-16). Folds are an indication that crustal shortening has occurred.

Folds are most easily visible in sedimentary rock sequences because of the bedding of the rock strata. Bedding planes were initially horizontal prior to deformation (see the Principle of Original Horizontality, Chapter 1). These planes facilitate the recognition and interpretation of folds (Figure 7-17).

Two general fold types are recognized. *Anticlines* are formed by the upward bending or buckling of strata, so the fold has the shape of an arch (Figure 7-18). The sides, or limbs, of an anticline dip downward and outward from the fold crest. The opposite of an anticline is a *syncline*, in which the central portion

**Figure 7-15** Measurement of dip with a Brunton compass at the base of a sandstone bed. (J. R. Stacy; photo courtesy of U.S. Geological Survey.)

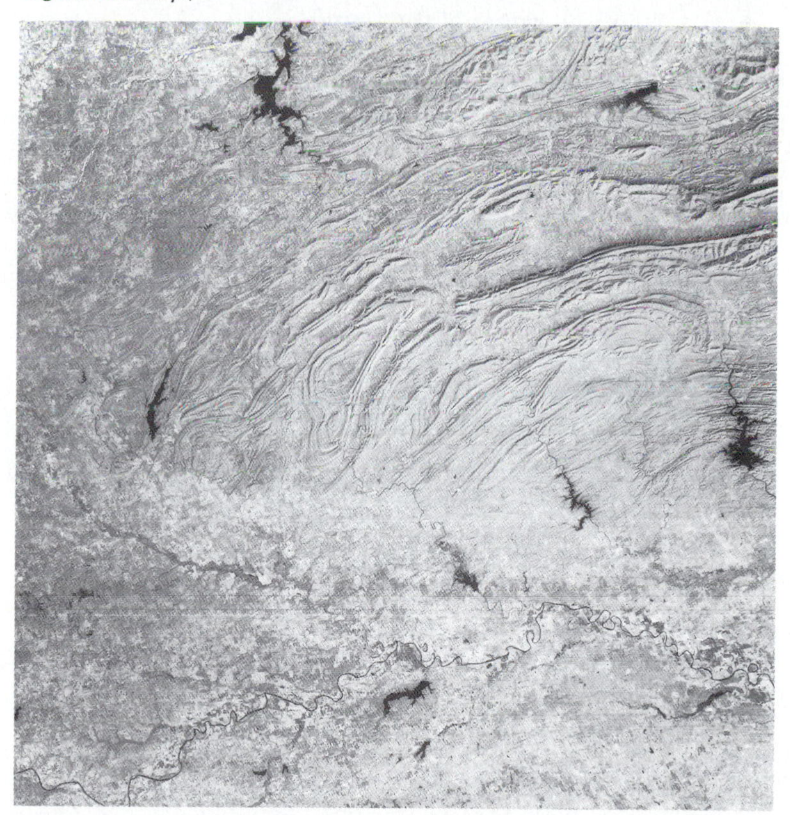

**Figure 7-16** Landsat image of Ouachita Mountains in eastern Oklahoma. Fold patterns are defined by outcrop pattern of resistant sedimentary rock units. Mountains were formed by folding and faulting of Paleozoic sedimentary rocks followed by uplift and erosion by streams. Width of scene is approximately 185 km. (U.S. Geological Survey; EROS Data Center; Landsat image 1146-16300-7; Dec. 16, 1972.)

**Figure 7-17** Folded sedimentary rocks in Hudspeth County, Texas. (C. C. Albritton, Jr.; photo courtesy of U.S. Geological Survey.)

is bent downward to form a trough (Figure 7-18). The limbs of a syncline dip toward the center of the trough.

Parts of a Fold. The terminology used to describe folds is shown in Figure 7-19. The line connecting the points of maximum curvature is called the *axis*. The *axial plane* is a plane containing the axis that divides the fold into two equal sections.

The axis of a fold or group of folds may be approximately horizontal or tilted, in which case the structure is classified as a plunging fold (Figure 7-20). The outcrop pattern of plunging folds differs from nonplunging folds as

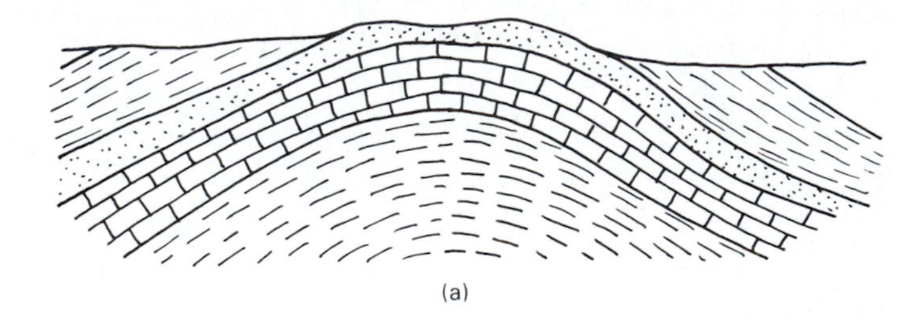

(a)

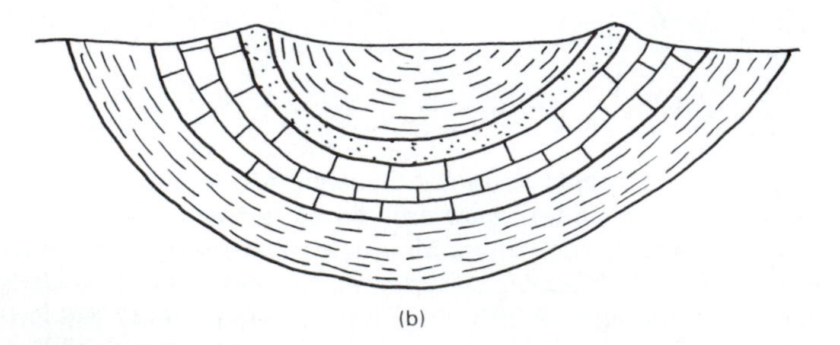

(b)

**Figure 7-18** Cross sections of (a) an anticline and (b) a syncline.

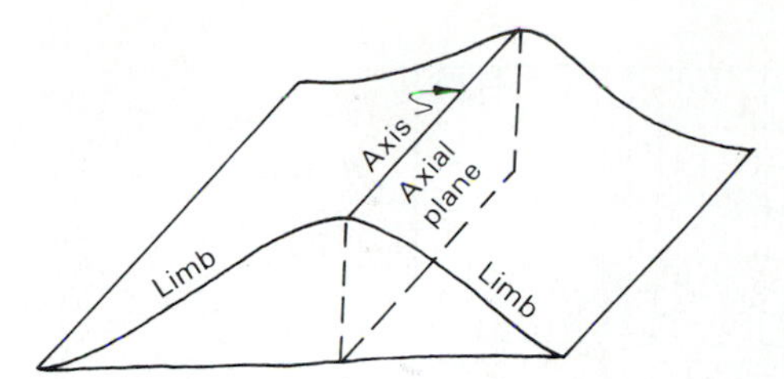


**Figure 7-19** Parts of a fold. The axial plane divides the fold into two equal portions, and the axis is a line connecting the points of maximum curvature.

illustrated in Figure 7-21. In map view, the beds on opposite limbs of the nonplunging fold are parallel, but form a sinuous pattern in the plunging fold. Erosion of the landscape accentuates these differences. The sharp bends in beds identify the points of intersection between eroded anticlines and the ground surface. Therefore, in a series of plunging folds the direction of plunge is toward the bend, or nose, of an anticline (Figure 7-22). For a syncline, the direction of plunge is toward the opposite direction from the sharp bend in the outcrop pattern of the beds.

The relative age of rock units in a folded sequence can be determined after identification of the fold as an anticline or syncline. In an eroded anticline, the oldest rocks are exposed at the center of the structure (Figure 7-23). The relationship is reversed in a syncline, where the youngest rocks are exposed at the center of the structure.

Types of Folds. Folds can vary in size from structures visible on satellite images to examples that must be observed through a microscope. In addition to this size variation, there are also many types of folds. Several of the basic forms are shown in Figure 7-24. Fold types are differentiated by the geometry of the limbs and the orientation of the axial planes. In *symmetrical* folds, the limbs dip in opposite directions in about the same amount. Dips of opposing limbs are not equivalent in *asymmetrical* folds. If both limbs dip in the same direction, the folds are said to be *overturned.* This condition can be caused by an intense stress applied in one direction. In regions of more severe deformation, axial planes may be nearly horizontal. These types of folds are termed *recumbent.*

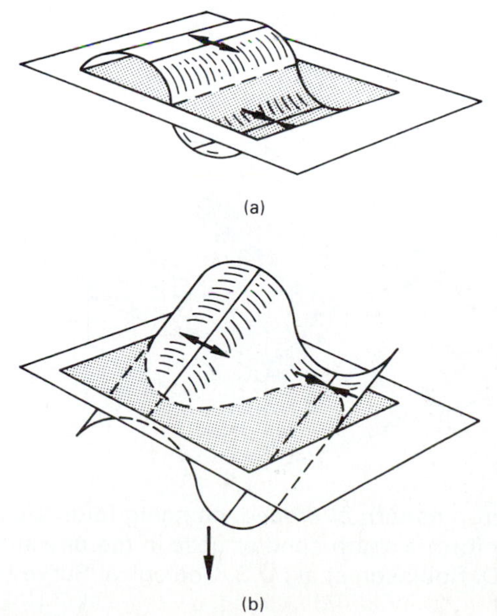


**Figure 7-20** Comparison of (a) nonplunging and (b) plunging folds. (From E. A. Hay and A. L. McAlester, *Physical Geology: Principles and Perspectives*, 2nd ed., copyright © 1984 by Prentice-Hall, Inc., Englewood Cliffs, N.J.)

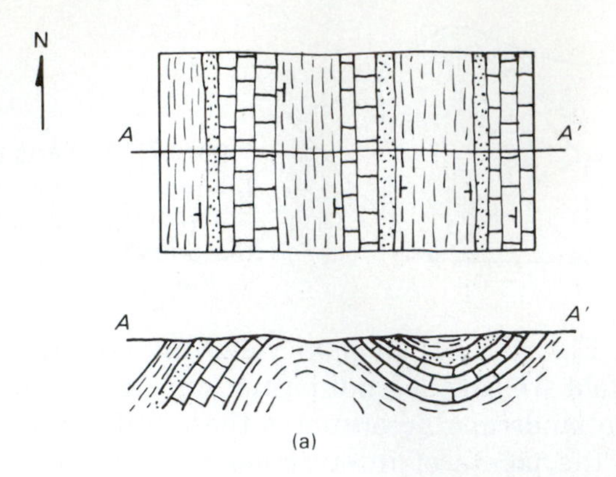


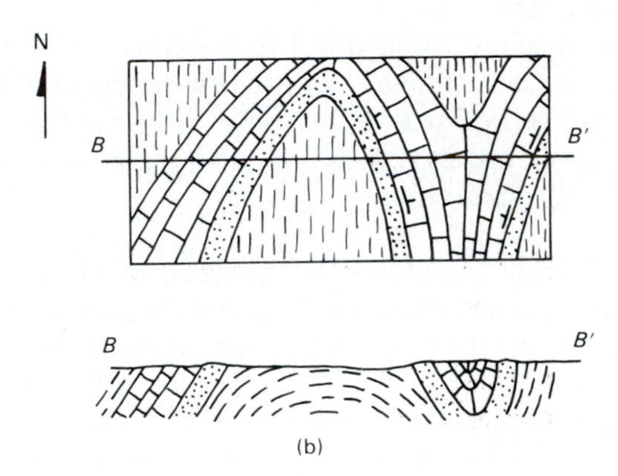


**Figure 7-21** (a) Map and cross section of nonplunging folds. (b) Map and cross section of folds plunging north.

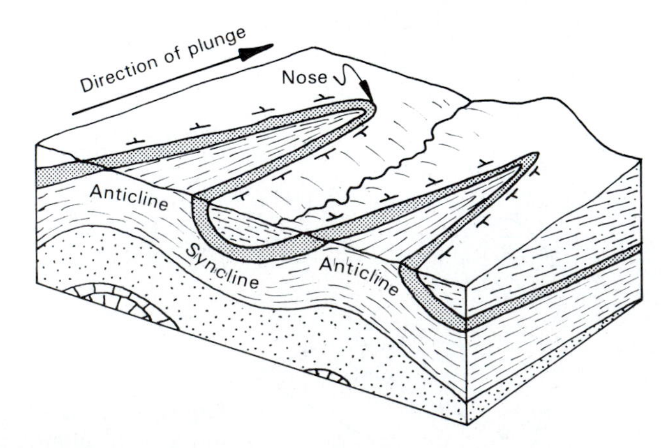


**Figure 7-22** Outcrop pattern of eroded, plunging folds. Beds in a plunging anticline form a sharp bend or nose in the direction of plunge. (After G. D. Robinson et al., U.S. Geological Survey Professional Paper 505.)

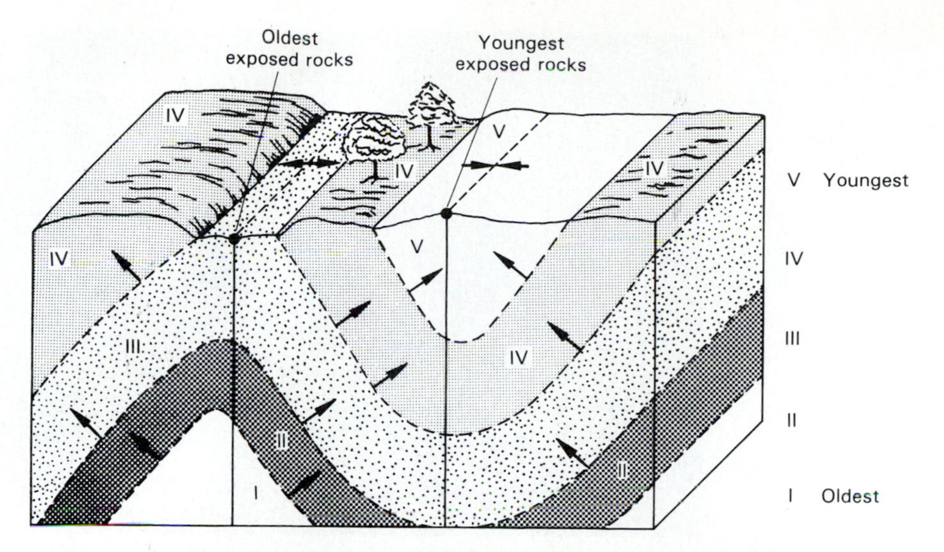


**Figure 7-23** The relative age of folded sedimentary rocks exposed at the surface depends upon the structure. The rocks exposed at the center of the anticline, Unit III, are the oldest exposed rocks in the sequence; whereas those exposed at the center of the syncline are the youngest. The arrows show the direction toward which the tops of the beds are facing: away from the axial plane of the anticline and toward the axial plane of the syncline. (From E. A. Hay and A. L. McAlester, *Physical Geology: Principles and Perspectives*, 2d ed., copyright © 1984 by Prentice-Hall, Inc., Englewood Cliffs, N.J.)

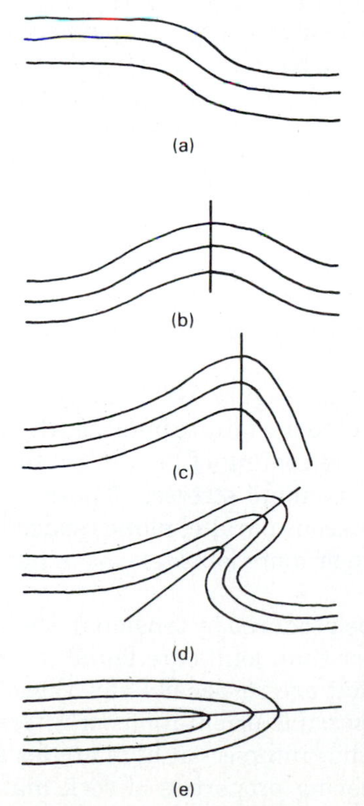


**Figure 7-24** Cross sections of fold types as defined by geometry and position of axial plane: (a) monocline, (b) symmetrical anticline, (c) asymmetrical anticline, (d) overturned anticline, and (e) recumbent anticline.

**Figure 7-25** Inclined beds in the Lizard Canyon monocline, Mesa County, Colorado. Dips flatten out to horizontal beyond the right borders of photo. (S. W. Lohman; photo courtesy of U.S. Geological Survey.)

Not all folds consist of two limbs. A steplike bend in strata without two well defined limbs is known as a *monocline* (Figures 7-24 and 7-25). Finally, large-scale warps in relatively stable continental crustal regions resemble folds. These structures are commonly circular or oval in plan view. Uplifts of this type are called *domes* (Figure 7-26), and downwarps are known as *basins* (Figure 7-27). Eroded domes are similar to anticlines in that the oldest rocks are exposed at the center; whereas, in basins, like synclines, the youngest beds are exposed at the center. Basins are particularly important for their petroleum resource potential (Chapter 4). During the slow subsidence of the crust to form basins, thick sections of sedimentary rocks are deposited. Under the proper conditions, petroleum is formed and trapped in these sequences.

### Fractures

The brittle failure of rock produces fractures. These curved or planar discontinuities can be subdivided into *joints* and *faults*. Both types are extremely important to identify and evaluate at the site of an engineering project.

Joints. Joints (Figure 7-28) are fractures in which there has been no movement parallel to the failure plane. Although joints are produced by unloading (Chapter 9) and cooling of lava (Chapter 3), our main concern here is for joints caused by tectonic stresses. These fractures commonly occur in sets consisting of numerous parallel planes spaced throughout a rock mass at regular intervals. One or more joint sets, each having a distinctive orientation, are often present.

Joints can be produced by tensional, shearing, and compressional stresses. If structures other than joints are found in an area, the joints may be related to the stresses that are responsible for those structures. For example, Figure 7-29 shows two possible orientations of intersecting joint sets associated with folds. Joint sets that intersect at high angles are known as *conjugate joint sets*.

The engineering properties of rock masses are established by the type,

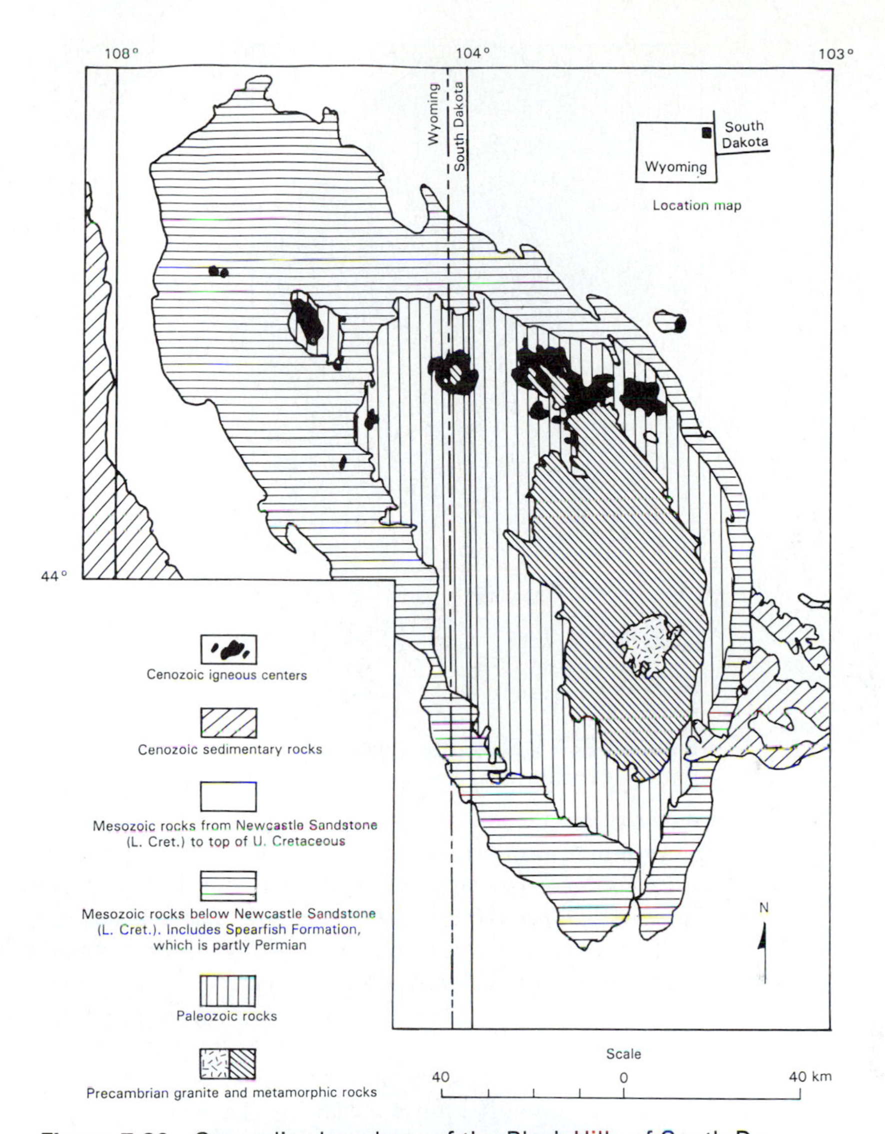


**Figure 7-26** Generalized geology of the Black Hills of South Dakota and Wyoming. Erosion has truncated the original surface of the dome and now the oldest rocks (Precambrian metamorphics) are exposed at center of structure surrounded by successively younger rock units. (From F. R. Karner and D. L. Halvorson, 1987.)

origin, and spacing of joints and other discontinuities (Chapter 6). Strength is controlled by these fractures because failure usually occurs along planes of weakness rather than within intact rock. The orientation of joints may impart a degree of *anisotropy* to the rock mass; in other words, strength may vary with the direction from which the stress is applied because of the spatial distribution of joints.

Rock-slope stability is also controlled by the spacing and orientation of joints. These factors determine the size and direction of failure of blocks that may be unstable, thus affecting the engineering of highway cuts, open-pit mine slopes, and other excavations.

Another rock-mass property that is influenced by jointing is permeability. In many rock types, most of the movement of ground water, petroleum, and also contaminants is through fractures rather than through pores in the intact rock. In rocks of low permeability, exploration for ground water involves identi-

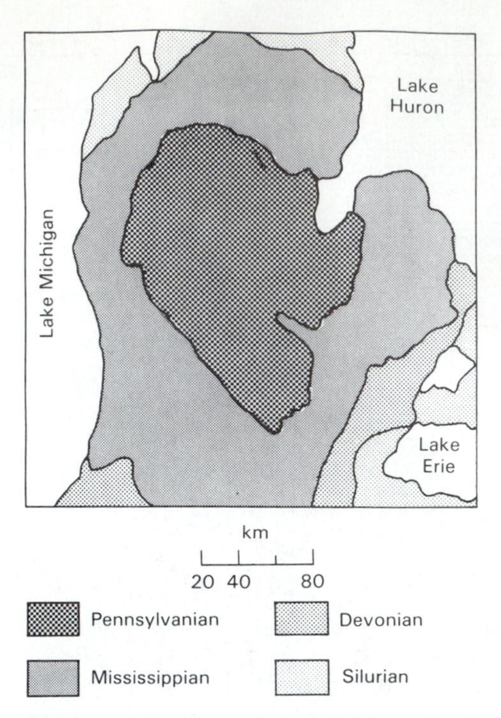


**Figure 7-27** Map of the Michigan Basin showing progressively younger rock units toward the center of the structure.

fying major joint traces on the ground surface. The point of intersection of two or more joints is a prime location for high well yields in such areas. Planning for waste-disposal facilities must also consider fracture patterns. Contaminated ground water can migrate rapidly along joints and faults; these fractures must therefore be identified and monitored.

**Figure 7-28** Closely spaced joints in an igneous rock body. (Whatcom County, Washington; M. H. Staatz; photo courtesy of U.S. Geological Survey.)

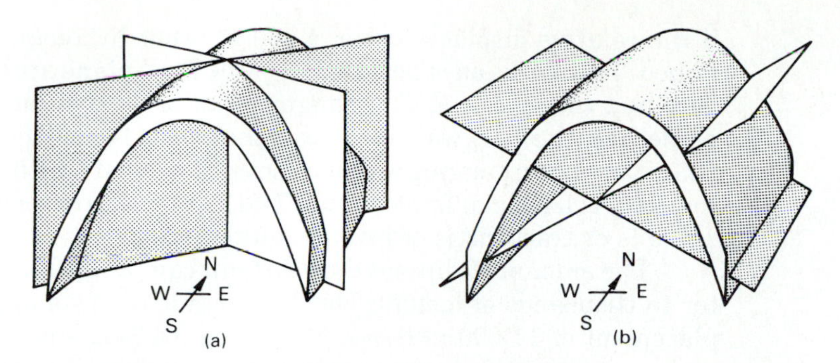


**Figure 7-29** Conjugate joint sets caused by stress that produced folding: (a) fold with vertical diagonal joints, (b) fold with strike joints dipping about 30°. (From M. P. Billings, *Structural Geology*, 3d ed., copyright © 1972 by Prentice-Hall, Inc., Englewood Cliffs, N.J.)

Faults. *Faults* are fractures along which movement of rock masses has occurred parallel to the fault plane. Faults are classified according to the type of movement, or slip, that has taken place. The two categories established on this basis include dip-slip faults and strike-slip faults.

Dip-slip faults, in which the relative movement of rock masses is in the direction of the dip of the fault plane, include *normal* faults, *reverse* faults, and *thrust* faults (Figure 7-30). Normal and reverse faults can be differentiated

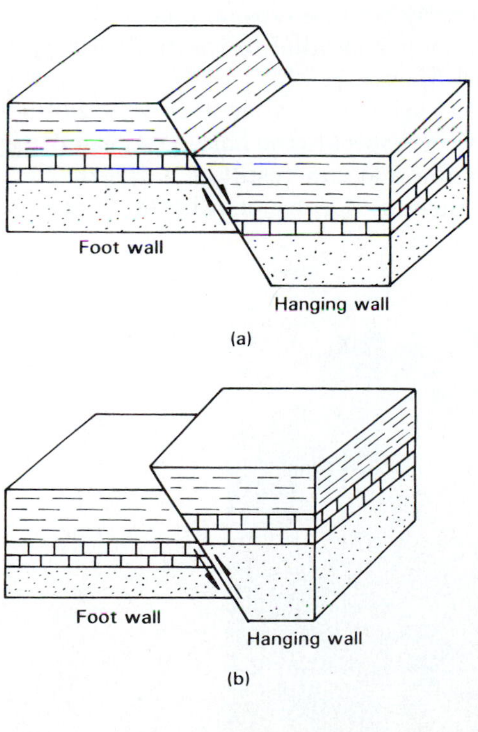


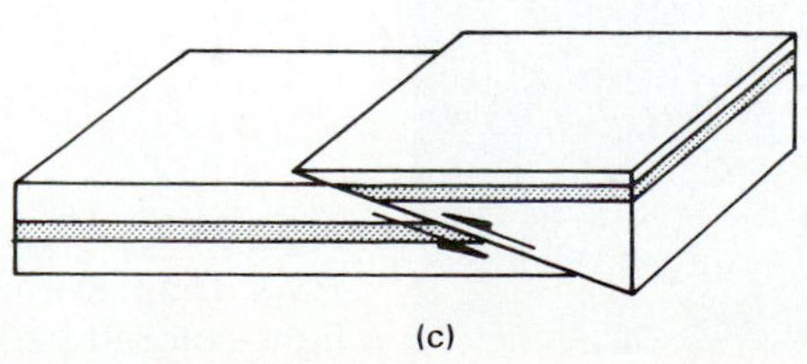


**Figure 7-30** Types of dip-slip faults: (a) normal fault, hanging wall is displaced downward with respect to foot wall; (b) reverse fault, hanging wall is displaced upward with respect to foot wall; (c) thrust fault, similar to reverse fault except that fault plane is inclined at lower angle.

if the relative displacement of a bed or other lithologic feature can be determined. The blocks on opposite sides of the fault plane are defined as the *hanging wall* and the *foot wall.* The hanging wall is always the block above the fault plane and the foot wall is always the block below the plane. Therefore, by definition, if the hanging wall has moved downward with respect to the footwall, the fault is a normal fault (Figure 7-31), and relative movement in the opposite sense is characteristic of reverse faults.

The amount of slip on a dip-slip fault can range from fractions of a centimeter to thousands of meters. The Teton fault in Wyoming has a combined displacement of 11,000 m (Love, 1987). Figure 7-32 shows a block diagram and cross section of this impressive structure. Paleozoic through Quaternary sedimentary rocks are preserved in the down-dropped hanging wall, but these formations have been eroded away on the foot-wall block. Of the total displacement on the fault, only about 1500 to 2000 m are evident at land surface in the rugged Teton mountain front (Figure 1-8), which is the steepest and youngest mountain front in the Rocky Mountains. Most of the displacement has occurred in the last 9 million years. The exposed and eroded plane of a fault, in this case the Teton mountain front, is called a *fault scarp*. The Teton fault scarp owes its justifiably famous topography to the youth of the fault and the resistance to erosion of the Precambrian metamorphic and igneous rocks in the foot wall.

It is not always possible to identify the type of fault by the slope of the topographic scarp. Although the topography of the Colorado Front Ranges is similar to the Tetons, the faults that produced these mountain fronts were reverse faults. Notice in Figure 7-33 that the topographic scarp formed by the mountain front is almost at right angles to the reverse faults at the base of the slope. This occurs because erosion has produced a more stable topographic scarp that is no longer parallel to the fault scarp. The scarp in this case is known as a *fault-line* scarp.

Thrust faults are very similar to reverse faults except that the angle of the fault plane with respect to the horizontal is quite low. They involve intense crustal compression and can result in large displacements. The McConnell thrust in the Canadian Rockies of Alberta, for example (Figure 7-34), has a

**Figure 7-31** Normal fault, northern Arizona. Displacement defined by offset of light-colored beds.

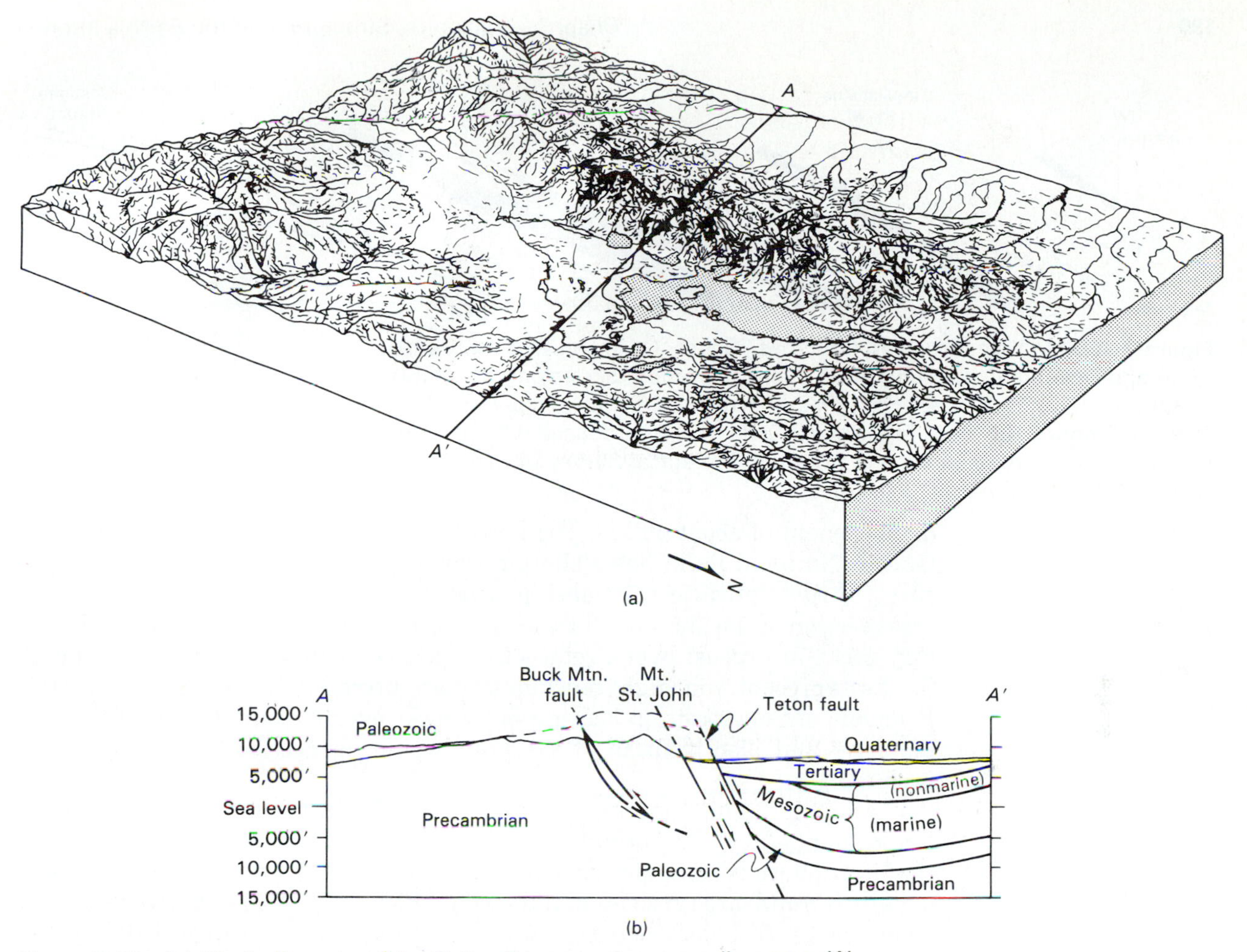


**Figure 7-32** (a) Block diagram of the Teton Range and surrounding area, Wyoming. The Teton mountain front is a fault scarp produced by erosion of resistant Precambrian rocks on the upthrown block of the Teton fault. (a) Dimensions of block shown are approximately 85 km by 70 km. (b) Cross section showing displacement across the Teton fault and associated rock units. (From Love, 1987.)

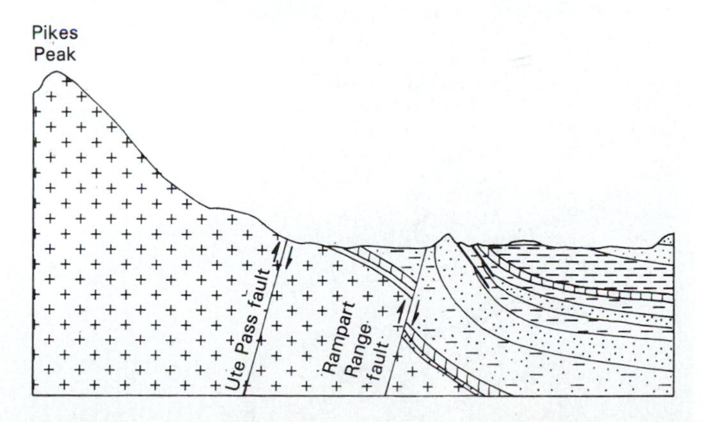


**Figure 7-33** Schematic cross section through the eastern flank of the Colorado Front Range. The primary structure, the Rampart Range fault, is a high-angle reverse fault. Precambrain granite in the hanging wall was faulted upward relative to Paleozoic sedimentary rocks to the east. Width of section approximately 25 km. (From Noblett et al., 1987.)

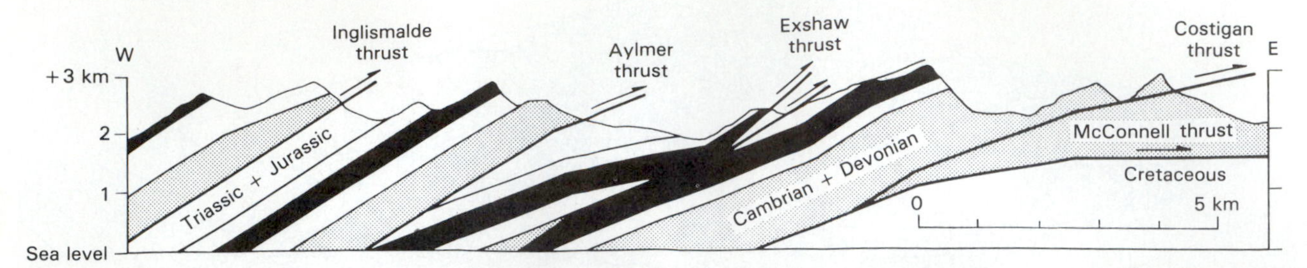


**Figure 7-34** Multiple thrusts in the Front Ranges, Canadian Rockies. Cambrian and Devonian rocks are thrust over Cretaceous rocks along the McConnell thrust. (From J. Suppe, *Principles of Structural Geology*, copyright © 1985 by Prentice-Hall, Inc., Englewood Cliffs, N.J. Reproduced with permission from Prentice-Hall, Inc., and from the Geological Survey of Canada, GSC Map 1272A.)

displacement of about 40 km. The Lewis overthrust (Figure 7-35), exposed in Glacier National Park, has a similar scale of displacement. In isolated exposures, thrusts produce relationships that can be confusing. Notice from the cross section of the McConnell thrust that Cambrian and Ordovician sedimentary rocks are thrust over Cretaceous rocks. Without knowing that a thrust fault was present, you might conclude that an outcrop of this section violates the Principle of Superposition because older rocks overlie younger rocks! Careful mapping and identification of the fault would be necessary to resolve the problem.

Dip-slip faults sometimes occur in pairs (Figure 7-36). A down-dropped block bounded by two normal faults is called a *graben*. Rift zones, in which the lithosphere is being pulled apart by plates moving away from each other, usually contain large grabens at the surface (Figure 7-37). The structure formed when the central block between two dip-slip faults is up-thrown is known as a *horst* (Figure 7-36).

*Strike-slip* faults are characterized by the lateral movement of fault blocks along the strike of the fault plane (Figure 7-38). These faults are either *right-*

**Figure 7-35** Photo of Lewis overthrust (arrows mark exposure of fault plane), Glacier National Park. Precambrian metasedimentary rocks are thrust over Cretaceous rocks.

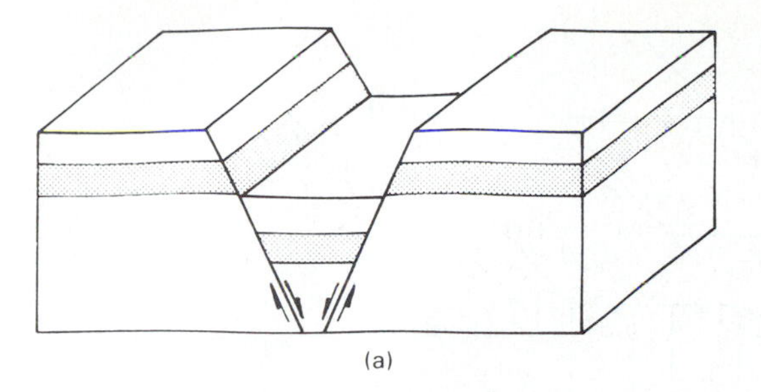

(a)

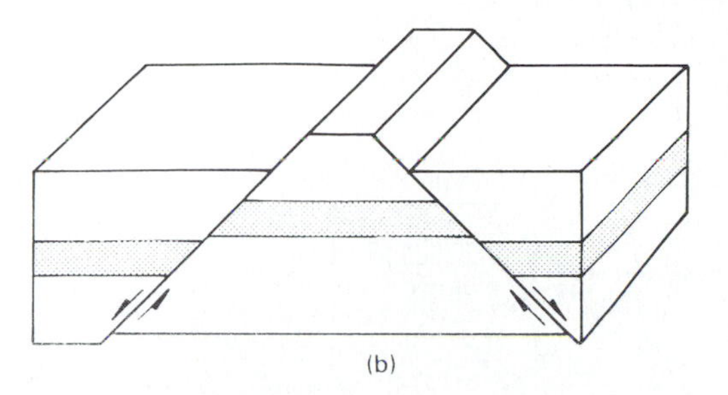

(b)

**Figure 7-36** Pairs of normal faults forming (a) a graben and (b) a horst.

*lateral* or *left-lateral* types, depending on the relative movement of fault blocks. The two varieties can be distinguished by facing the fault from either side and noting the direction, either left or right, toward the continuation across the fault of a linear feature such as a stream, fence, or road that passed straight across the fault prior to displacement. The San Andreas fault is an excellent example of a right-lateral, strike-slip fault (Figure 7-39).

Dip-slip and strike-slip faults are merely end members of fault types that may display any combination of dip-slip and strike-slip displacement. Faults that contain components of both types of movements are called *oblique-slip* faults. The actual type of movement that takes place along any individual fault is the result of the type and orientation of the principal stresses imposed

**Figure 7-37** Flank of a graben associated with one branch of the Mid-Atlantic rift in Iceland. Tilted block to the right of river is part of downdropped central block of graben.

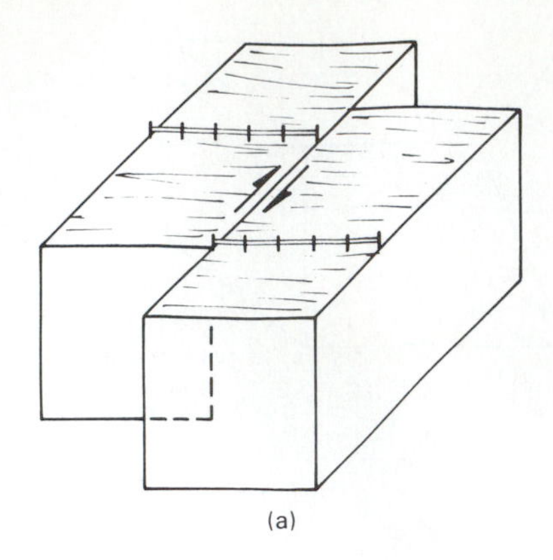


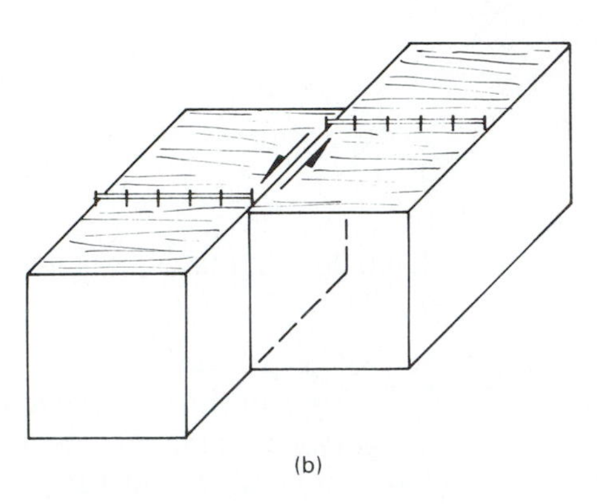


**Figure 7-38** (a) Right-lateral and (b) left-lateral strike-slip faults. Fence on surface can be used to determine amount of displacement.

**Figure 7-39** The San Andreas fault in the Carrizo Plains of California. (F. E. Wallace; photo courtesy of U.S. Geological Survey.)

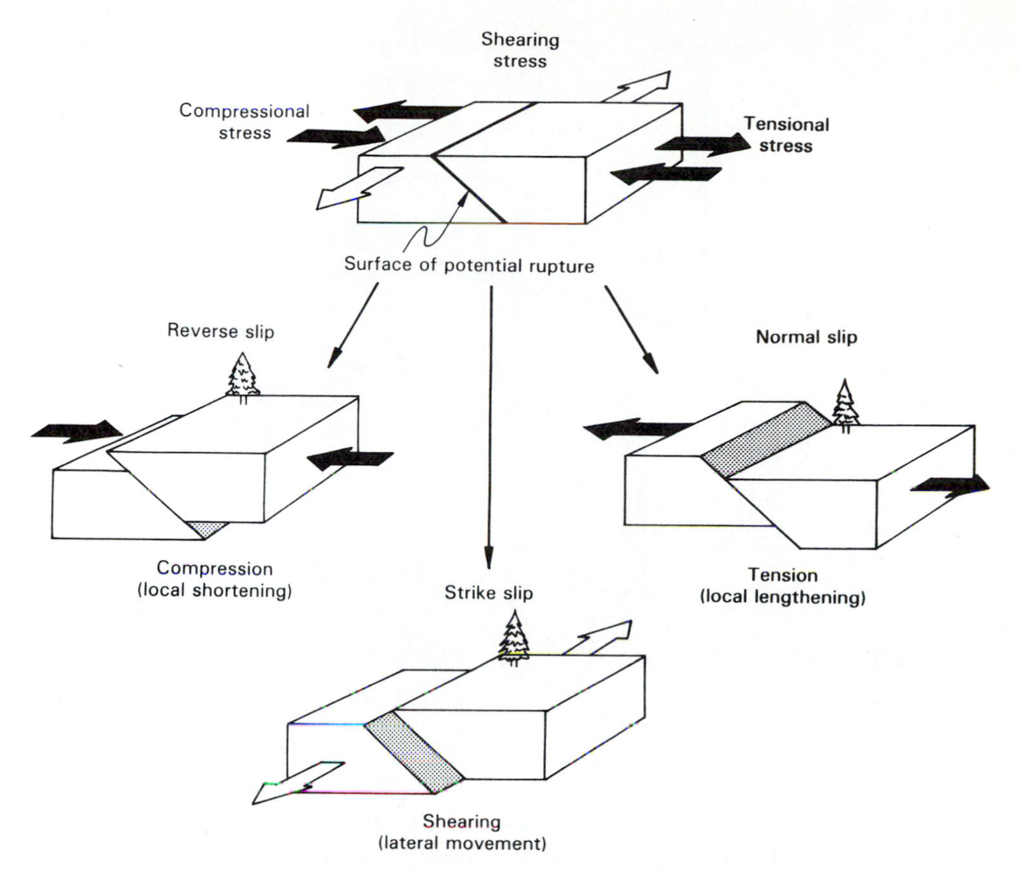


**Figure 7-40** Relationship of stress orientation to the style of faulting. (From E. A. Hay and A. L. McAlester, *Physical Geology: Principles and Perspectives*, 2d ed., copyright © 1984 by Prentice-Hall, Inc., Englewood Cliffs, N.J.)

upon a crustal block. As shown in Figure 7-40, dip-slip faults are associated with horizontal stresses. When these stresses are compressional, reverse and thrust faults will occur. When the horizontal stresses are tensional, normal faults develop. Strike-slip faults are caused by shearing stresses acting as a couple on a crustal block.

When fault movement takes place, the type and amount of displacement is readily observable. Fault scarps can be used to measure the amount of displacement on a dip-slip fault. These ruptures displace any surface features present. An active strike-slip fault can often be recognized by a characteristic set of topographic features (Figure 7-41). These include bends in stream courses or offsets of other linear topographic features, as well as depressions or *sag ponds* along the trend of the fault.

Fault scarps and other types of topography associated with active faults are rapidly modified by erosion. Because of this, there may be no scarps marking the presence of inactive faults, and as a result these structures can only be located by careful geologic mapping. Clues may be provided by linear stream courses or other topographic features. Stream courses are readily established along fault zones because of the low resistance to erosion caused by the crushing and grinding of rock along the fault plane.

The identification of active or potentially active faults is critical to the siting of such projects as nuclear power plants, schools, hospitals, and other buildings. Fault zones should be avoided for these types of structures. Inactive

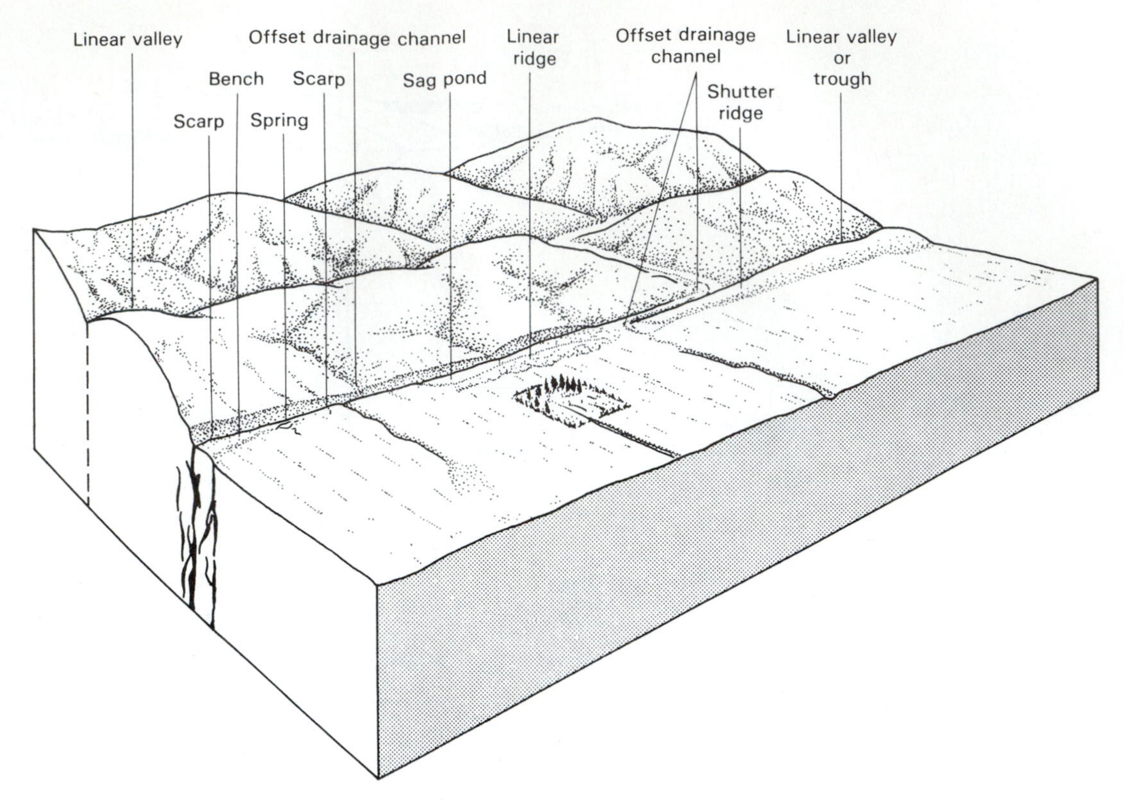


**Figure 7-41** Topography developed along an active strike-slip fault. (From R. D. Borcherdt, ed., 1975, *Studies for Seismic Zonation in the San Francisco Bay Region*, U.S. Geological Survey Professional Paper 941-A.)

faults are also relevant to engineering design because they may bring rocks of different engineering properties into contact, or they may control the permeability and ground-water movement through a rock mass.

## GEOLOGIC MAPS AND CROSS SECTIONS

Geologic maps are among the best sources of information for preliminary site location and design. For this reason, engineers need to become familiar with the construction and use of these maps. Geologic maps are drawn after extensive field work, during which the geologist observes as many outcrops as possible in the area being mapped. Lithology, fossils, contacts, and any other information are systematically depicted. Rock units are divided into formations, which are traced through the map area by plotting their contacts with other formations on a base map, which is commonly a topographic map. When the area contains folded sedimentary rocks, strike and dip are measured frequently and plotted on the base map at the location of the outcrop. These measurements form the basis for interpretation of these structures. Vertical aerial photographs are commonly used before and during field work.

The geologic map is the final result of this effort (Figure 7-42). Rock units are shown on the map by colors or patterns. Because contacts are rarely exposed continuously across an area, they are sometimes drawn by using indirect evi-

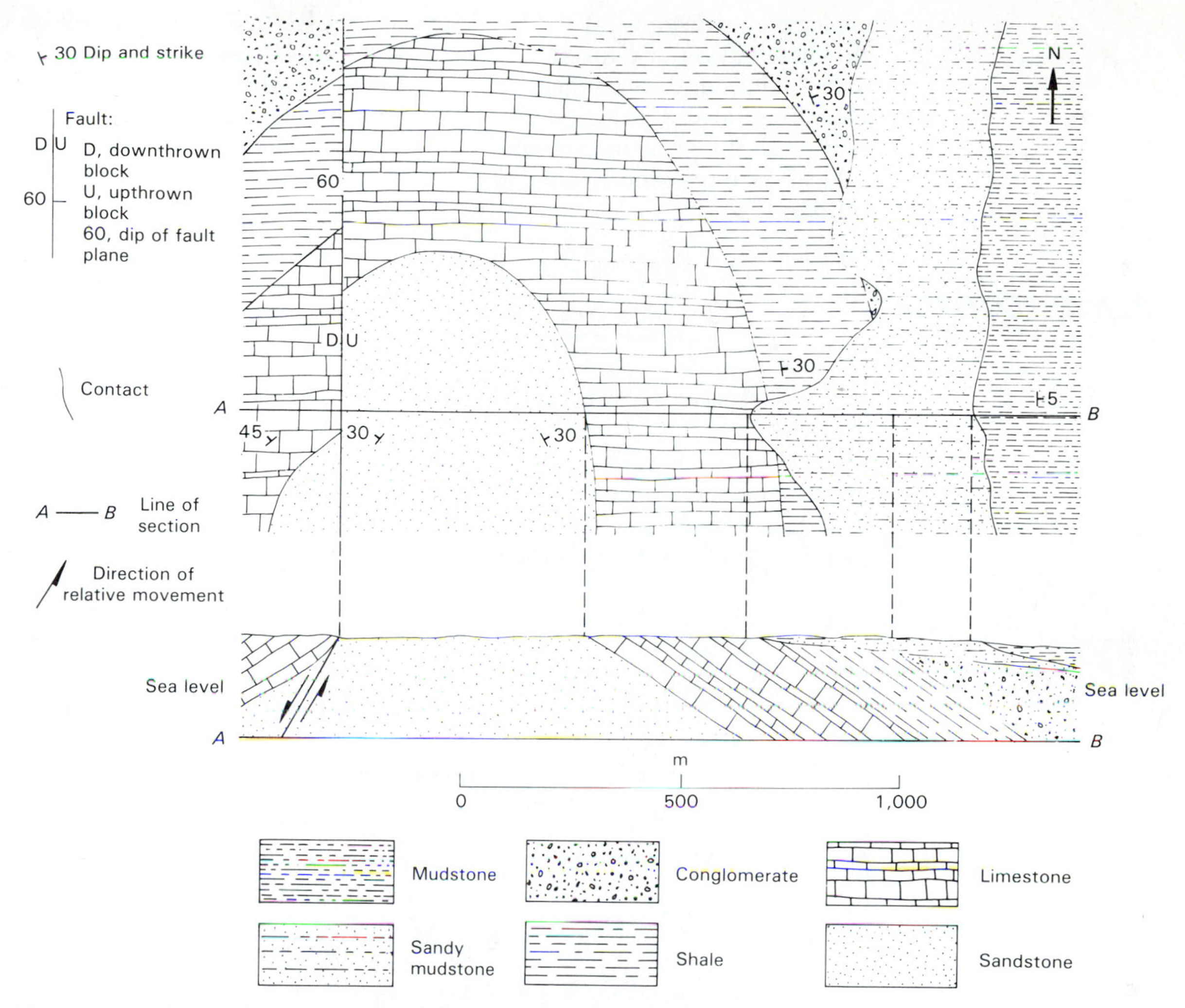


**Figure 7-42** Geologic map and cross section. (From S. Judson, M. E. Kauffman, and L. D. Leet, *Physical Geology*, 7th ed., copyright © 1987 by Prentice-Hall, Inc., Englewood Cliffs, N.J.)

dence. The geologic map, therefore, is actually the geologist's interpretation of the geology. A certain amount of uncertainty and error in placing contacts and representing structural relationships is unavoidable. For this reason, contacts should not be transferred from a geologic map drawn at a scale of 1:24,000 or smaller to an engineering site map at a much larger scale and be expected to be precisely accurate. More detailed mapping is usually required when the exact position of a contact is required.

Several types of geologic maps are made. When bedrock formations are exposed in an area, these may be the only units shown on the map, even though there may be surficial materials, like stream or wind-blown sediments, overlying the bedrock units in some parts of the map area. Even though these materials are not mapped, they may be very important for foundation design or other engineering reasons. Some geologic maps show the surficial geologic materials and some show both surficial and bedrock units. It is important to examine the map and read the report to determine what types of units were mapped by the geologist.

Cross sections are drawn by projecting contacts and other information from the map onto a vertical plane oriented along a line which is shown on the geologic map (Figure 7-42). These diagrams are the geologist's interpretation of the distribution of map units in the subsurface. Cross sections are based on outcrop data or direct subsurface data from boreholes or well records, if available. In an area of complex geology, there is almost never as much subsurface control as the geologist would like to have. For this reason, cross sections are subject to more uncertainty than geologic maps, and should be used with caution.

## PLATE TECTONICS

In several sections of this book the theory of plate tectonics has been mentioned. This theory is important because it provides a unifying model for changes that have affected the earths's crust through geologic time. In addition, it provides an explanation for the distribution of volcanoes, earthquakes, and many other geologic phenomena. In this section, we will briefly consider some of the main points of evidence for this theory.

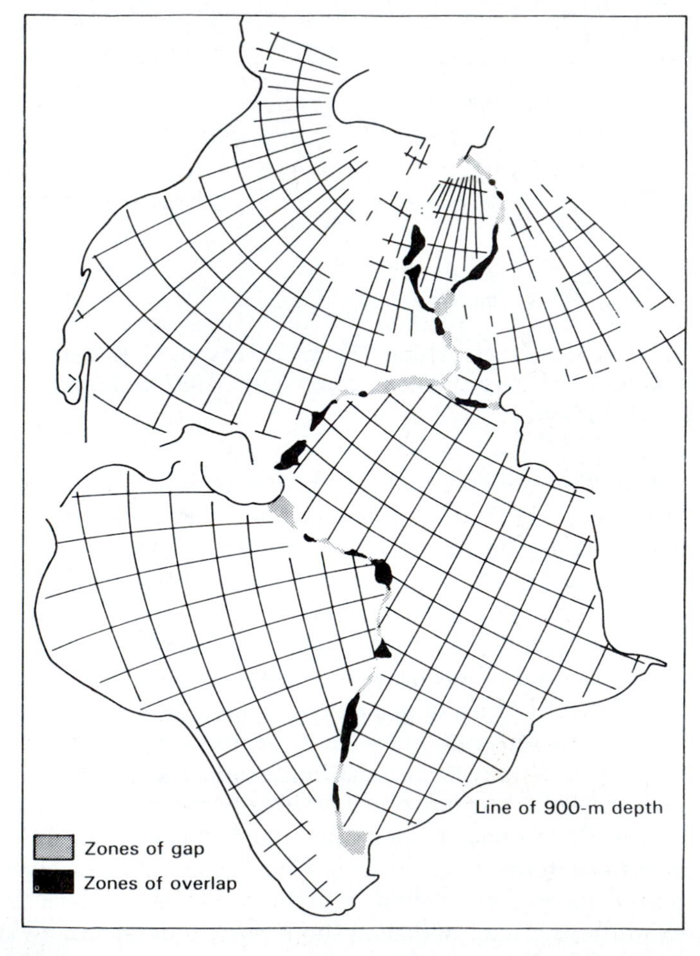


**Figure 7-43** Assembly of the continents at the 900-m depth of the ocean bottom to form Pangaea. (From E. Bullard et al., 1965, *Phil. Trans. Royal Soc.*, 1088:41–51.)

### *Continental Drift*

In 1912 Alfred Wegener, a German meteorologist, proposed that the continents had moved great distances laterally across the earth's surface through geologic time. The shapes of the coastlines of Africa and South America, which would fit together like the pieces in a jigsaw puzzle if the continents were closer, was one of the main lines of evidence for this idea. This theory, which became known as continental drift, proposed that all the continents were once assembled into a huge continental mass called *Pangaea* (Figure 7-43), and that they then drifted apart to their present positions.

In addition to noting the fit of continents, Wegener summarized geologic evidence suggesting that the continents were once joined together (Figure 7-44). This evidence included several types. First, mountain belts, structural trends, and rock types found on different continents would be continuous if the continents were assembled as in Pangaea. Second, the distribution of numerous fossil species found on several continents could be explained if the continents had been originally joined. These species included land plants and animals

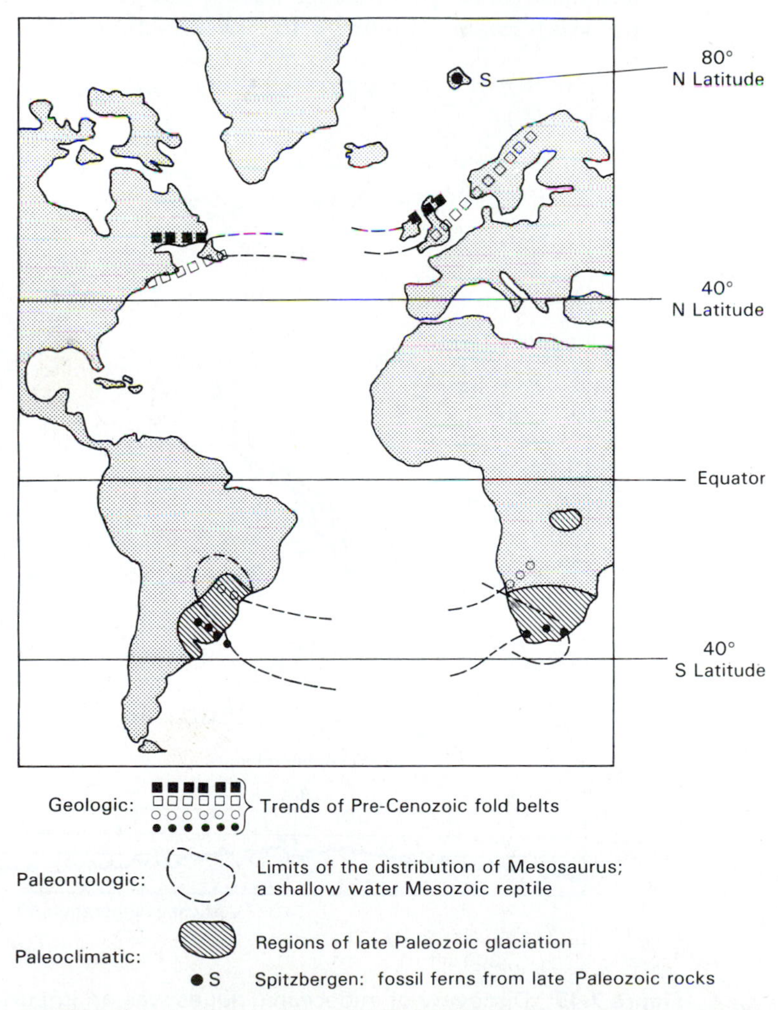


**Figure 7-44** Evidence for continental drift as proposed by Alfred Wegener. (From E. A. Hay and A. L. McAlester, *Physical Geology: Principles and Perspectives*, 2d ed., copyright © 1984 by Prentice-Hall, Inc., Englewood Cliffs, N.J.)

that could not have crossed an ocean to colonize widely separated continents. Third, rock types that form under specific climatic conditions were encountered in regions whose present climate is very different. For example, rocks containing coal and tropical plants were discovered in polar regions, and rocks formed by glaciation were mapped near the present equator. The locations of these rocks would be difficult to explain unless the land mass had moved into different climatic zones.

Wegener's hypothesis was largely discounted by scientists of the day because of the implausible mechanism to which he attributed the drift of continents. Wegener suggested that the granitic rock of the continents had moved laterally through the stronger, denser mantle rock. Other scientists quickly pointed out the mechanical impossibility of this process, and Wegener's ideas were ignored by most geologists for several decades.

### *The Theory of Sea-Floor Spreading*

In the first decades after World War II, tremendous advances in exploration of the oceans were realized. Topographic surveys of the ocean floor using sounding devices discovered the immense midoceanic ridges positioned near the center of the ocean basins (Figure 7-45). Rock samples from the ridge crests were

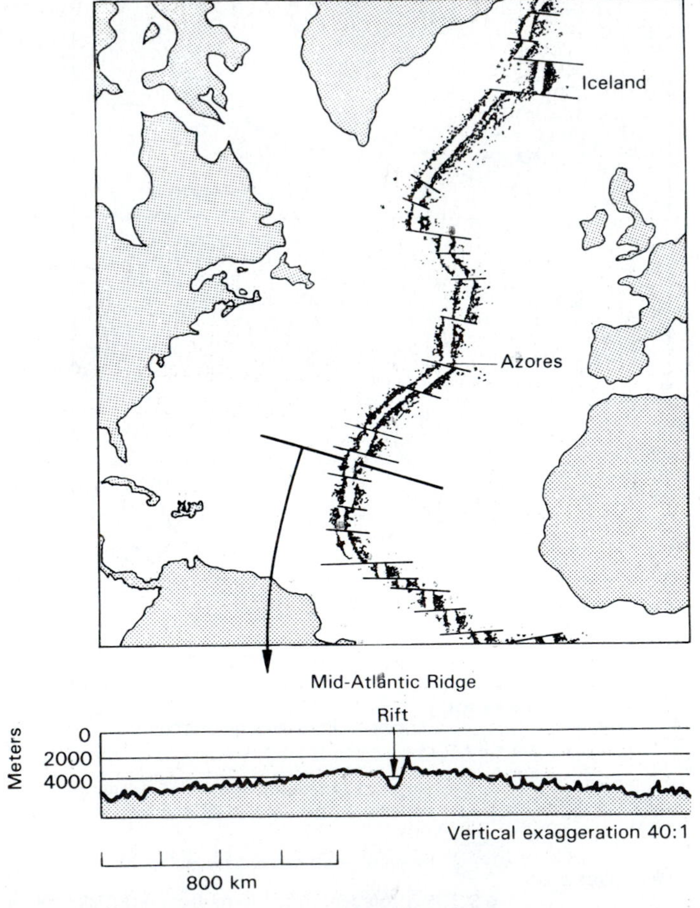


**Figure 7-45** Discovery of midoceanic ridges was an important step in the development of the sea-floor spreading hypothesis. Dots show positions of earthquakes along the crest of the Mid-Atlantic Ridge. (From B. C. Heezen, The Rift in the Ocean Floor, *Scientific American*, Oct. 1960.)

composed of fresh basaltic lava covered by very little sediment. The thickness of bottom sediment above bedrock increases laterally away from the ridge crests, implying an increase in the age of the oceanic crust with distance from the midoceanic ridges.

The most dramatic evidence pertaining to the origin of the ocean basins was obtained from paleomagnetic studies of the oceanic crust. The symmetrical bands of alternately magnetized volcanic crustal rock provided important evidence for the *sea-floor spreading* theory, which proposed that new crustal material was formed by volcanic eruptions at the crests of the midoceanic ridges and that slow lateral movement of the crust away from the ridges was occurring. Other geophysical evidence suggested that the layer in motion consisted of the crust and upper mantle, a layer named the lithosphere. This rigid slab could move laterally above a hot plastic layer called the asthenosphere. Thus Wegener's basic premise was vindicated: Continents do move laterally but as part of a thicker, rigid slab that slides along above the weak asthenosphere.

### Plates and Plate Margins

The sea-floor spreading theory quickly became integrated into a more comprehensive model called plate tectonics, which considers the rigid lithosphere to be divided into 12 major slabs, or plates (Figure 7-46). Most of the dynamic

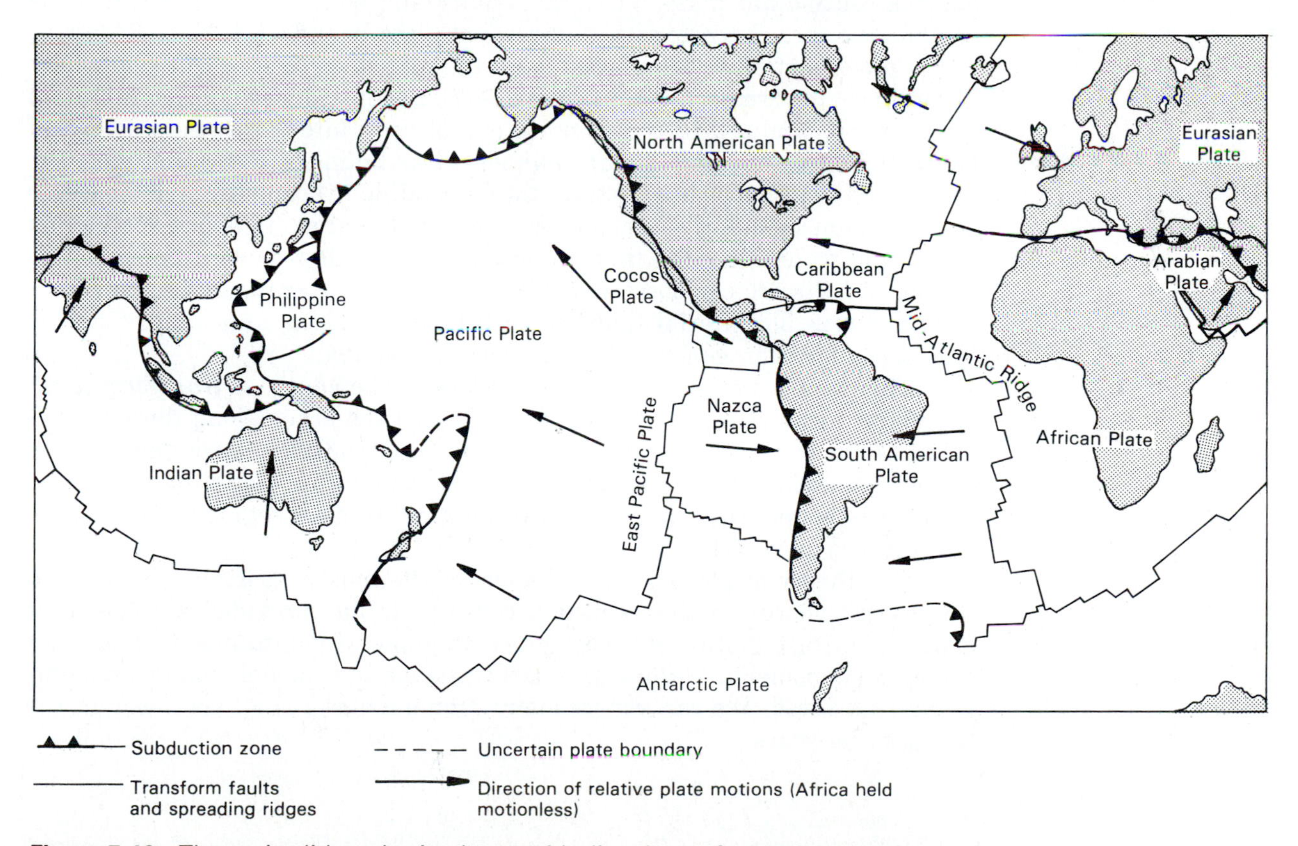


**Figure 7-46** The major lithospheric plates, with directions of movement and types of boundaries shown. (From E. A. Hay and A. L. McAlester, *Physical Geology: Principles and Perspectives*, 2d ed., copyright © 1984 by Prentice-Hall, Inc., Englewood Cliffs, N.J.)

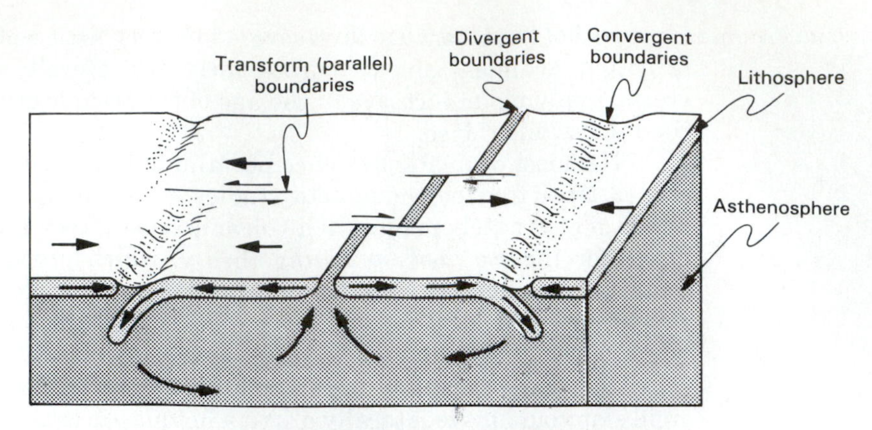


**Figure 7-47** Three types of plate boundary: convergent, divergent, and transform. (From Isacks et al., 1968. Copyright by the American Geophysical Union.)

crustal processes such as volcanic eruptions and earthquakes take place at the boundaries of the plates. Three major types of plate boundaries are recognized (Figure 7-47). At *divergent* plate boundaries, new lithosphere is produced by volcanic eruptions. Midoceanic ridges fall into this category. Divergent boundaries also develop beneath continents. The Red Sea, separating Africa and the Arabian Peninsula, is considered to be a spreading center beginning to split apart a continental mass. Volcanic activity and shallow- to moderate-focus earthquakes are characteristic of divergent plate boundaries.

Convergent plate boundaries mark the zone of contact or collision where plates move toward each other (Figure 7-47). Because of the enormous crustal compression localized in these zones, one of the plates is thrust downward below the other one in a *subduction zone*. Earthquakes associated with the subduction zone can be used to trace the descending lithospheric slab. Features of convergent boundaries include oceanic trenches that lie above subduction zones, and mountain belts that are formed at the edges of continents. These mountain belts are composed of andesitic composite volcanoes (Chapter 3) or deformed crustal material that has been folded, faulted, and uplifted by the lateral compression generated by the colliding plates.

The third major type of plate boundary is the *shear,* or *transform fault,* boundary. Lithospheric plates are sliding past each other along these boundaries. Major strike-slip faults, such as the San Andreas, mark the location of shear boundaries. Earthquakes are very common as along divergent and convergent boundaries. Numerous transform faults offset segments of midoceanic ridges (Figure 7-47).

The theory of plate tectonics is one of the most important advances in the earth sciences in the twentieth century. It has provided a model that explains the distribution of volcanoes, earthquakes, mountain belts, and other geologic phenomena. Despite the usefulness of this model, many problems remain unsolved. We can expect many refinements of the hypothesis in the forthcoming years.

## CASE STUDY 7-1

## Effects of Rock Structure on Dam Construction

Careful geologic mapping was necessary prior to construction of the Arbuckle Dam near Ardmore, Oklahoma, which was completed in 1966 by the U.S. Bureau of Reclamation. The complex geologic setting of this dam consists of a series of tightly folded and faulted Paleozoic sedimentary rocks and beautifully illustrates the role of geology in a major engineering project (Jackson, 1969).

The Arbuckle Dam is located in a region of rolling hills known as the Arbuckle Mountains. Rocks of the Arbuckle Mountains consist primarily of a sequence of shales and limestones deposited in a Paleozoic sea (Table 7-1). In late Paleozoic time, the rocks were deformed by compression from the southwest into a series of broad anticlines and synclines. Some of the folds were overturned and some rocks in the anticlines were thrust over younger rocks in the adjacent synclines. A later sequence of events included uplift, high-angle normal faulting, and erosion. Sediments eroded from the rising mountain mass were deposited around the flank of the uplift. During Mesozoic time, marine rocks were deposited over the Paleozoic sequence and then removed by erosion. Continued erosion during the Cenozoic era has produced the present landscape. Major structural elements of the region are shown in Figure 7-48.

Geologic conditions at the dam site were

**TABLE 7-1**

**Generalized Stratigraphic Section of Arbuckle Dam and Reservoir**

| | *Formation* | *Thickness (feet)* |
|---|---|---|
| Quaternary | ALLUVIUM | 0–30 |
| Pennsylvanian | VANOSS FM.—Poorly to moderately cemented conglomerate, shale, and sandstone of outwash debris on north flank of mountains. | 650–900 |
| | DEESE FM.—Varicolored shale; well-cemented conglomerate; thin beds of hard limestone and sandstone | up to 2000 |
| | SPRINGER FM., GODDARD MBR.—Black shale, noncemented but very compact, bituminous, appears graphitic. | 1000 to 3000 |
| Mississippian | CANEY FM.—Dark-colored shale, moderately hard, limy, fissile. | 150 |
| | SYCAMORE FM.—Blue to weathered limestone, hard, tough. | 0–50 |
| | WOODFORD FM.—Dark-colored shale, siliceous and cherty, platy and brittle | 400–500 |
| Devonian-Silurian | HUNTON GR.—Thick to medium bedded limestone, marlstone | 250 |
| Cambrian-Ordovician | SYLVAN FM.—Gray-green fissile shale. | 200 |
| | VIOLA FM.—Bituminous limestone. | 450 |
| | ARBUCKLE GR.—Limestone, dolomite, shale, sandstone. | 5000 to 8000 |

SOURCE: From J. L. Jackson, 1969, Geologic studies at Arbuckle Dam, Murray County, Oklahoma, *Bulletin of the Association of Engineering Geologists*, 5:79–100.

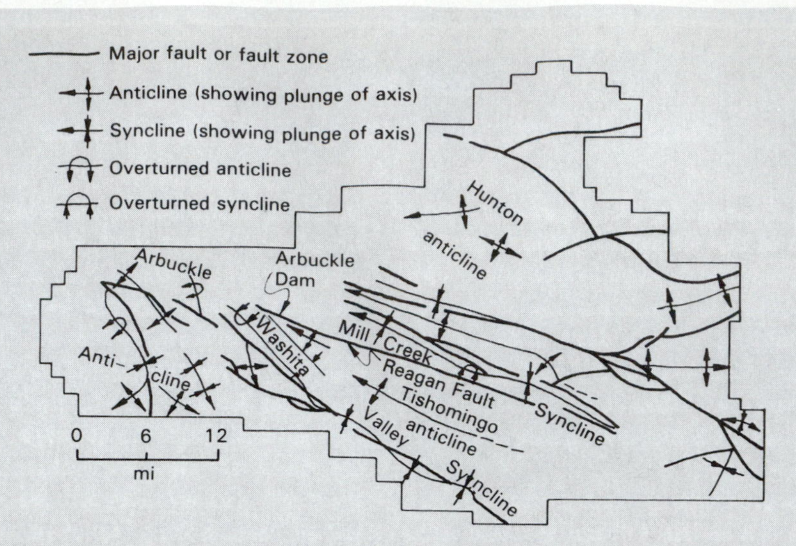


**Figure 7-48** Geologic setting of the Arbuckle Mountains, showing major faults and folds. (From J. L. Jackson, *Bulletin of the Association of Engineering Geologists*, 5:79–100, copyright © 1969 by the Association of Engineering Geologists.)

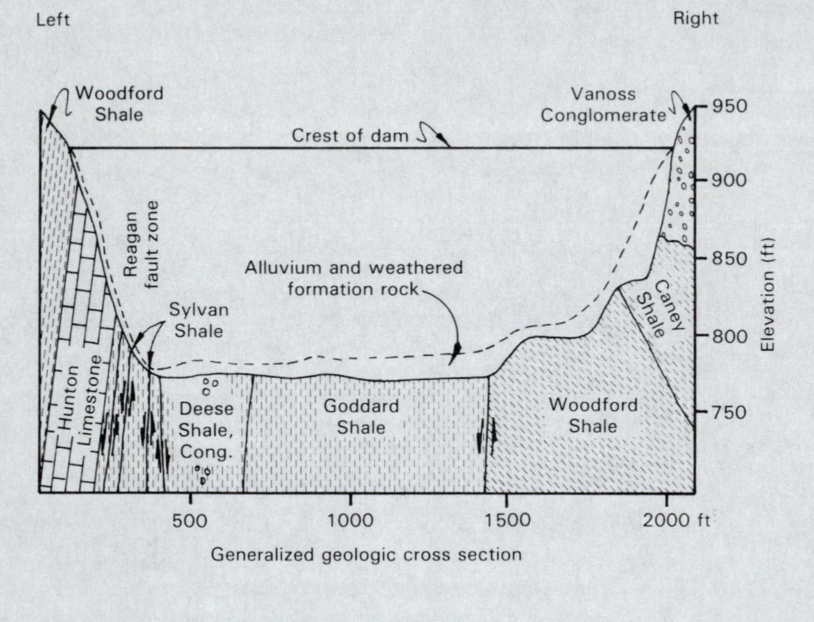


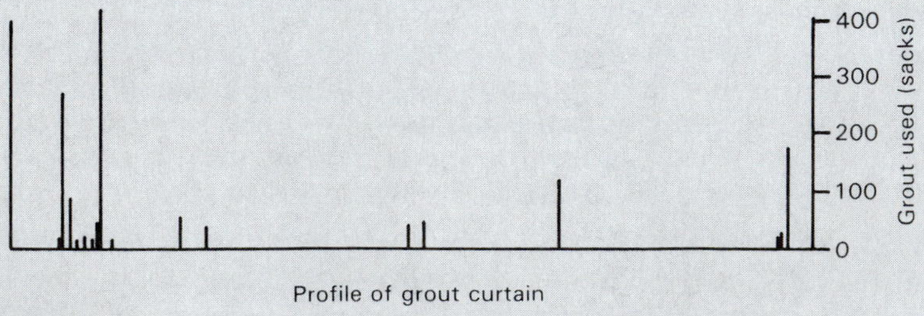


**Figure 7-49** Cross section of the Arbuckle Dam site and profile of the grout curtain emplaced beneath the dam. (From J. L. Jackson, *Bulletin of the Association of Engineering Geologists*, 5:79–100, copyright © 1969 by the Association of Engineering Geologists.)

defined by careful geologic mapping and test drilling (Figure 7-49). The stream valley is established in a sequence of steeply dipping, faulted shales and limestones. The center of the valley is a graben formed by several normal faults, including the highly deformed Reagan fault zone at the base of the left abutment (the section of the dam that meets the valley wall). Rocks are fractured, crushed, and altered along the fault and shear zones.

One of the major concerns in dam construction is to prevent excessive leakage from the reservoir through the foundation rocks beneath the base of the dam. Leakage can result in the loss of reservoir storage as well as in the possibility of failure and collapse of the dam. A common procedure to prevent such leakage involves sealing dam foundations with a grout curtain. The grout curtain is produced by injecting grout (cement slurry) under high pressure into holes drilled in a line across the valley floor. Grout is forced into pores, cavities, and fractures within the rock and then hardens to form a low permeability seal for the dam. At the Arbuckle Dam, 135 grout holes were used, ranging in spacing from 25 to 110 ft (7.6 m to 33.5 m). The profile of the grout curtain shown in Figure 7-49 illustrates the amount of grout needed to seal the foundation. The most permeable area was the Reagan fault zone at the base of the left abutment. Relatively small amounts of grout were needed across the valley, except at a fault and unconformity in the vicinity of the right abutment.

Construction of the Arbuckle Dam required a thorough understanding of the geology of the site. At this project the distribution of folds, faults, and other structures was critical to the design of the dam.

## SUMMARY AND CONCLUSIONS

Deformation of the earth's crust is intimately linked with processes that occur in the earth's interior, a region that is explored using geophysical investigations of heat flow, gravity, and magnetism. The earth achieved its current internal structure through differentiation; this process separated the iron core, the dense silicate mantle, and the low-density crust. Heat flow from the core generates convection currents in the mantle. Where rising plumes impact the base of the lithosphere, the lithosphere is broken apart and dragged laterally. The point of upwelling frequently coincides with midoceanic ridges, zones of high heat flow where new lithosphere is formed.

Heat flow beneath continents is variable according to the age and tectonic activity of individual geologic provinces. Continental heat flow is the sum of mantle heat and radioactive crustal heat. Tectonically active provinces are centers of high heat flow, whereas stable shield regions yield low heat-flow values.

The study of gravity is mainly used to detect minor variations in the density of near-surface rocks. These variations remain after latitudinal and topographical effects have been corrected from measured gravity values. Interpretation of many gravity readings led to the theory of isostasy. This important concept compares the mantle to a viscous fluid, when considered in terms of geologic time. Crustal masses reach a condition of equilibrium by "floating" in the mantle to a degree determined by their size and density. High mountain ranges are composed of low-density rocks that extend as roots well below the earth's surface. Any rapid change in load causes crustal blocks to rise or sink into the mantle; thus one example of isostatic rebound is the crustal rise of glaciated areas after the last glacial retreat.

The earth's magnetic field can be approximated as a dipole bar magnet oriented near the axis of rotation. It is probable that electrical currents in the outer core induce this magnetic field.

Studies of paleomagnetism have shown that the earth's magnetic field periodically reverses its polarity. This discovery has been most useful in providing evidence for sea-floor spreading. Symmetrical zones of alternating normal- and reversed-polarity rocks adjacent to midoceanic ridges strongly suggest that oceanic lithosphere is produced at midoceanic ridges and then moves laterally away from these spreading centers.

In order to understand rock structures, we must take into account the high confining pressures and temperatures at the depths where deformation occurs, as well as the long periods of time available for the application of stress. The interpretation of aerial photographs and the construction of geologic maps is the first step in characterizing the structure and geologic history of an area. These maps are made by plotting the distribution of rock types and the spatial orientation of structural elements on base maps. The strikes and dips of planes associated with bedding, joints, and faults are measured in order to reconstruct the deformation of the area prior to erosion. Cross sections utilizing these data can be derived from geologic maps. Geologic maps and cross sections are essential tools for engineering site investigations.

Folding encompasses a range of mechanical processes in which horizontal sequences of rocks are deformed into undulating, wavelike forms. Anticlines and synclines are the main types of folds, although folds can be described as symmetric, asymmetric, overturned, or recumbent, depending upon the orientations of the axial plane and the limbs. Domes and basins are large-scale structures whose bilateral cross sections are anticlines and synclines, respectively. Plunging folds develop complex outcrop patterns after erosion.

Fractures can be classified as either joints or faults. Both types control the strength and permeability of rock masses. Fault terminology is based on the type of slip displayed and the relative movement of opposing blocks. The types of faults that occur are governed by the type and orientation of stresses in the deforming rock mass. Thrust and reverse faults develop under compression, whereas tensional stresses often are responsible for normal faults. Strike-slip faults are found where shear stresses dominate between crustal blocks.

During the past several decades geologists have formulated a unifying model to explain many of the geologic processes observed in the earth's crust. Predecessors of this model were the theories of continental drift and sea-floor spreading. Continental drift was discounted because of the lack of a mechanism to explain the lateral movement of the continents. Sea-floor spreading provided an explanation for the lateral movement of the rigid lithosphere above the plastic asthenosphere, which rafts the granitic continental masses along in the process.

The currently accepted plate tectonics model includes about a dozen major lithospheric plates that move on the asthenosphere. Tectonic and volcanic processes are concentrated at plate boundaries, which may be of three types: divergent, convergent, or transform.

## PROBLEMS

1. How can the deformation of rocks deep within the earth's crust be simulated in laboratory tests of rock samples?
2. What evidence leads us to believe that rocks exposed at the surface were deformed at great depths?

3. What field measurements are made in the study of rock structures? What do they indicate about the structures?
4. In what type of geologic setting or province would you be likely to find highly folded rocks?
5. What is the difference between faults and joints?
6. Describe the relationship between fault type and stress orientation.
7. Summarize the ideas that led up to the theory of plate tectonics.
8. List as many ways as you can that folds and faults would influence the construction of tunnels and dams.
9. Describe the major structural features of your state, province, or region.

## REFERENCES AND SUGGESTIONS FOR FURTHER READING

BILLINGS, M. P. 1972. *Structural Geology*, 3d ed. Englewood Cliffs, N.J.: Prentice-Hall, Inc.

BULLARD, E. J., E. EVERETT, and A. G. SMITH. 1965. The fit of the continents around the Atlantic. *Philosophical Transactions of the Royal Society of London*, 1088:41–51.

HAY, E. A., and A. L. MCALESTER. 1984. *Physical Geology: Principles and Perspectives*, 2d ed. Englewood Cliffs, N.J.: Prentice-Hall, Inc.

ISACKS, B., J. OLIVER, and L. R. SYKES. 1968. Seismology and the New Global Tectonics. *Journal of Geophysical Research*, 73:5855–5899.

JACKSON, J. L. 1969. Geologic studies at Arbuckle Dam, Murray County, Oklahoma. *Bulletin of the Association of Engineering Geologists*, 5:79–100.

JUDSON, S., M. E. KAUFFMAN, and L. D. LEET. 1987. *Physical Geology*, 7th ed. Englewood Cliffs, N.J.: Prentice-Hall, Inc.

KARNER, F. R., and D. L. HALVORSON. 1987. The Devils Tower, Bear Lodge Mountains, Cenozoic igneous complex, northeastern Wyoming, in *Geological Society of America Centennial Field Guide—Rocky Mountain Section*. Boulder, Colo.: Geological Society of America, pp. 161–164.

LOVE, J. D. 1987. Teton mountain front, Wyoming, in *Geological Society of America Centennial Field Guide—Rocky Mountain Section*. Boulder, Colo.: Geological Society of America, pp. 173–176.

NOBLETT, J. B., A. S. COHEN, E. M. LEONARD, B. M. LOEFFLER, and D. A. GEVIRTZMAN. 1987. The Garden of the Gods and basal Phanerozoic nonconformity in and near Colorado Springs, Colorado, in *Geological Society of America Centennial Field Guide—Rocky Mountain Section*. Boulder, Colo.: Geological Society of America, pp. 335–342.

PRESS, F., and R. SIEVER. 1982. *Earth*, 3d ed. San Francisco: W. H. Freeman.

ROBINSON, G. D., A. A. WANEK, S. H. HAYS, and M. E. MCCALLUM. 1964. *Philmont Country: The Rocks and Landscape of a Famous New Mexico Ranch*, U.S. Geological Survey Professional Paper 505.

SUPPE, J. 1985. *Principles of Structural Geology*. Englewood Cliffs N.J.: Prentice-Hall, Inc.

WESSON, R. L., E. J. HELLEY, K. R. LAJOIE, and C. M. WENTWORTH. 1975. Faults and future earthquakes, in *Studies for Seismic Zonation in the San Francisco Bay Region*, R. D. Borcherdt, ed., U.S. Geological Survey Professional Paper 941-A, pp. A5–A30.

# 8

# Earthquakes

At 5:12 A.M. on April 18, 1906, a great earthquake struck San Francisco. During the next 60 seconds, violent ground shaking devastated portions of the city and caused panic and terror. The ground was actually observed to be moving in waves in some places; train tracks were bent like wire, and whole buildings or chimneys collapsed to the ground (Figure 8-1). The aftermath of the earthquake was even more destructive because fires broke out throughout the city and firefighters were helpless because the city water main had been ruptured by the shaking. In all, at least 700 people were killed and 250,000 were left homeless.

The 1906 earthquake was the result of a sudden rupture along a 400-km-long segment of the San Andreas fault, the boundary between two lithospheric plates. Since that time, the United States has experienced only one other great earthquake, the Alaska earthquake of 1964. Fortunately, it occurred in a sparsely populated area. In recent years building codes have been continually modified to minimize damage from earthquakes. In nearly every metropolitan area, however, many buildings predate modern building practices.

Eighty-three years after the 1906 earthquake, many Americans were settling in on the evening of October 17, 1989, to watch the third game of the World Series to be played in San Francisco, California. What they saw instead, was the gradual unfolding of the effects of another severe earthquake in the San Francisco Bay region, this one centered under Loma Prieta mountain near the town of Santa Cruz. This earthquake, while not as strong as the 1906 temblor, still wrecked havoc in the Bay area (Figure 8-2). Sixty-two people were killed, 3757 injured, 12,053 displaced, and property damage amounted to a staggering $6 billion. Perhaps even more frightening because an anxious nation watched on television as the drama of rescue and human suffering unfolded, the Loma Prieta event, along with an even more destructive earthquake near Los Angeles in January 1994, were ominous warnings of things to come. While geologists work to develop the ability to predict earthquakes, the role of engineers to design and build structures to resist strong ground motion is crucial.

**Figure 8-1** Damage to the San Francisco City Hall by the 1906 earthquake. (W. C. Mendenhall; U.S. Geological Survey.)

**Figure 8-2** Rubble from collapsed building on Jefferson Street in the Marina District of San Francisco resulting from the Loma Prieta earthquake. (D. Perkins; U.S. Geological Survey.)

## OCCURRENCE

Earthquakes are caused primarily by the rupture of rocks in the earth's crust along faults. Most earthquakes take place along faults in the vicinity of plate boundaries (Figure 8-3). Earthquakes also occur far from plate boundaries. Often the faults associated with these intraplate earthquakes are deeply buried and not visible at land surface.

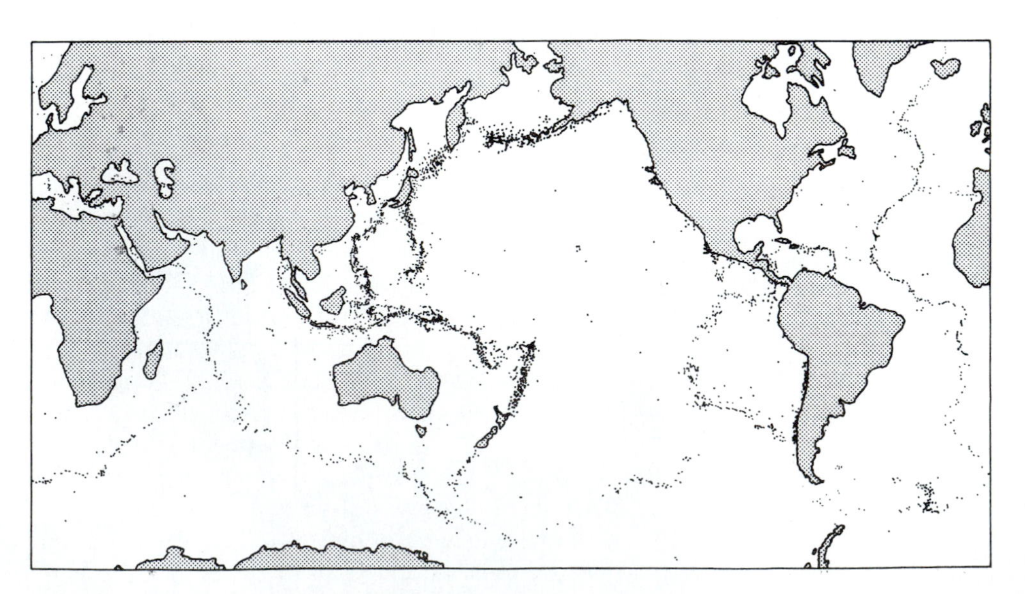

**Figure 8-3** Locations of shallow earthquakes during a 7-year period. The epicenters are mainly concentrated on tectonic plate margins. (Data from ESSA, U.S. Coast and Geodetic Survey.)

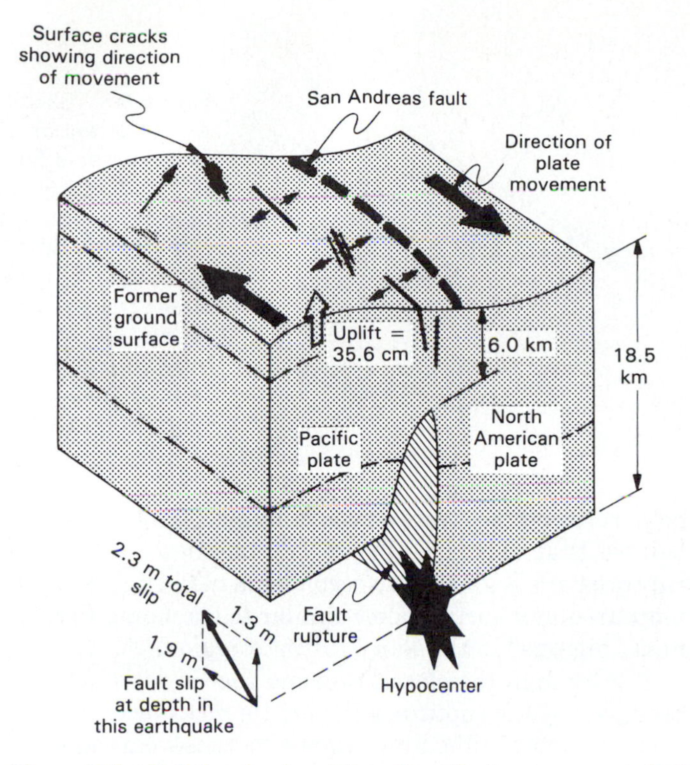


**Figure 8-4** Relative horizontal and vertical movement of plates during Loma Prieta earthquake. (From P. L. Ward and R. A. Page, *The Loma Prieta Earthquake of October 17, 1989*, U.S. Geological Survey Pamphlet.)

### *Elastic Rebound Theory*

During earthquakes, bodies of rock on opposite sides of faults move relative to each other. Both dip-slip and strike-slip displacements are recognized, as well as combinations of the two. During the Loma Prieta earthquake, for example, the relative strike-slip displacement between the Pacific and North American plates was 1.9 m. In addition, there was 1.3 m of uplift of the Pacific plate relative to the North American plate (Figure 8-4), despite the common perception of the San Andreas fault as a strike-slip fault. Vertical displacement may be due to a bend in the trend of this section of the fault.

A longstanding theory in geology, formulated by the analysis of surface rupture after the 1906 earthquake, holds that rocks deform in an elastic manner until brittle failure takes place (Figure 8-5). Elastic strain that had been accumulating on either side of the fault is suddenly released causing the rocks on either side of the fault to abruptly slide past each other. Elastic strain energy is released in the form of *seismic waves*, which radiate outward through the rocks in all directions from the *hypocenter*, or point of rupture. The energy that can be obtained from elastic strain can be painfully experienced by holding a rubber band against someone's arm, stretching it outward, and then releasing it suddenly from the opposite end. The propagation of seismic waves during fault movement is a consequence of elastic rebound, and the hypothesis is known as the *elastic rebound theory*. It is significant that seven damaging earthquakes occurred in the San Francisco Bay region during the 83 years prior to the great earthquake of 1906, but only two have occurred in the 83 years since. Release of strain energy during the 1906 earthquake may thus

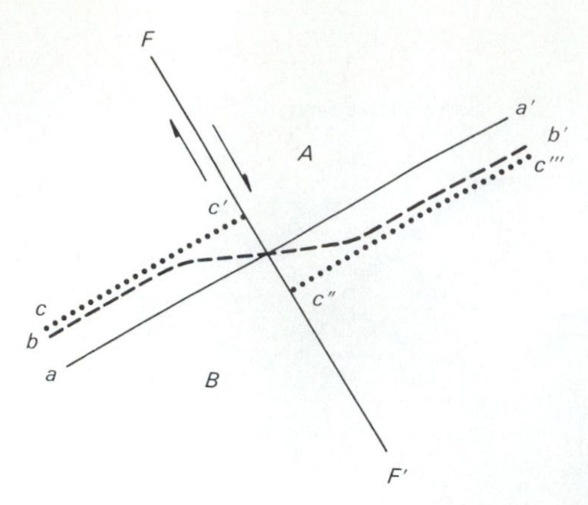


**Figure 8-5** Elastic rebound theory. Line *a–a'* is a real or imaginary linear feature that crosses a fault (*F–F'*) such as a road or fence. Elastic strain produces distortion of the line to the form shown as line *b–b'*. At the time of rupture, block *A* slides rapidly past block *B* and the elastic strain energy is now represented by segments *c–c'* and *c''–c'''*. (From M. P. Billings, *Structural Geology*, 3d ed., copyright © 1972 by Prentice-Hall, Inc., Englewood Cliffs, N.J.)

have relieved the stress on the fault for a period of decades. Some seismologists believe that elastic strain energy has now accumulated to the point where ruptures are beginning to occur. This may suggest that in the coming decades more frequent earthquakes similar to the Loma Prieta event may precede the next "big one" in the San Francisco region.

Although the elastic rebound theory satisfactorily explains earthquakes produced by the rupture of the brittle rocks near the earth's surface, the theory becomes more difficult to apply to those earthquakes generated at depths greater than a few tens of kilometers beneath the surface. At these depths the pressure exerted by the overlying rocks makes elastic behavior less likely. A more comprehensive theory for the origin of earthquakes in regions of plastic deformation is thus needed. Intensive research underway at present may lead to an improved explanation for deep earthquakes.

### Detection and Recording

During rupture at the hypocenter of an earthquake, the elastic strain energy stored in the rocks is suddenly released in the form of *seismic waves*. These waves are propagated outward from the hypocenter through the rocks in all directions. This seismic energy causes rapid vibrations in the rocks in the subsurface as well as on the land surface itself. The instrument designed to measure and record the characteristics of seismic waves is called a *seismograph* (Figure 8-6). It consists of a weight suspended on a string or wire that acts as a pendulum. The inertia of the mass resists its motion relative to the rigid frame to which it is attached. When the pendulum is connected either mechanically or electrically to a pen in contact with a rotating drum and chart, a record of the wave vibrations, called a *seismogram*, is produced.

The pendulum of a seismograph is constrained to move in only one direction. Therefore, a complete seismological station must contain three seismographs. One measures the vertical component of motion, and the other two measure horizontal motion in directions at right angles to each other.

Ground motion from an earthquake usually is most severe near the *epicenter*, the point on the earth's surface directly above the hypocenter. Instruments called *accelerographs* are used to measure the strong motion of an earthquake in the vicinity of the epicenter. Accelerographs differ from seismographs in that they begin recording only during strong earth movements. Seismographs record seismic waves continuously.

Accelerographs measure ground acceleration relative to $g$, the acceleration of gravity, which has a value of 9.8 m/s$^2$. The highest acceleration recorded

**Figure 8-6** A seismograph is the instrument used to measure seismic waves. (Photo courtesy of U.S. Geological Survey.)

was 1.25 *g* during the 1971 San Fernando, California, earthquake. Accelerations are measured in three directions at right angles to each other (Figure 8-7). Commonly, accelerographs are installed at different levels in a building or in different geologic settings in order to evaluate the response to seismic waves of different types of structures, soils, or rocks.

Seismograph stations are now integrated into regional and worldwide networks. These networks use similar equipment and process data uniformly in order to standardize data gathering and analysis.

### *Magnitude and Intensity*

The strengths and effects of earthquakes are determined in two ways. *Magnitude* is a quantitative measure of the energy released by an earthquake, as obtained from seismograph records. A method for determining magnitude developed by seismologist Charles F. Richter defines magnitude as the logarithm of the largest seismic wave amplitude produced by the trace of a standard seismograph located at a distance of 100 km from the epicenter. Magnitudes obtained by this method are commonly referred to as the magnitude on the *Richter scale*. The highest magnitude values measured have been approximately 8.9. The logarithmic aspect of magnitude means that an earthquake

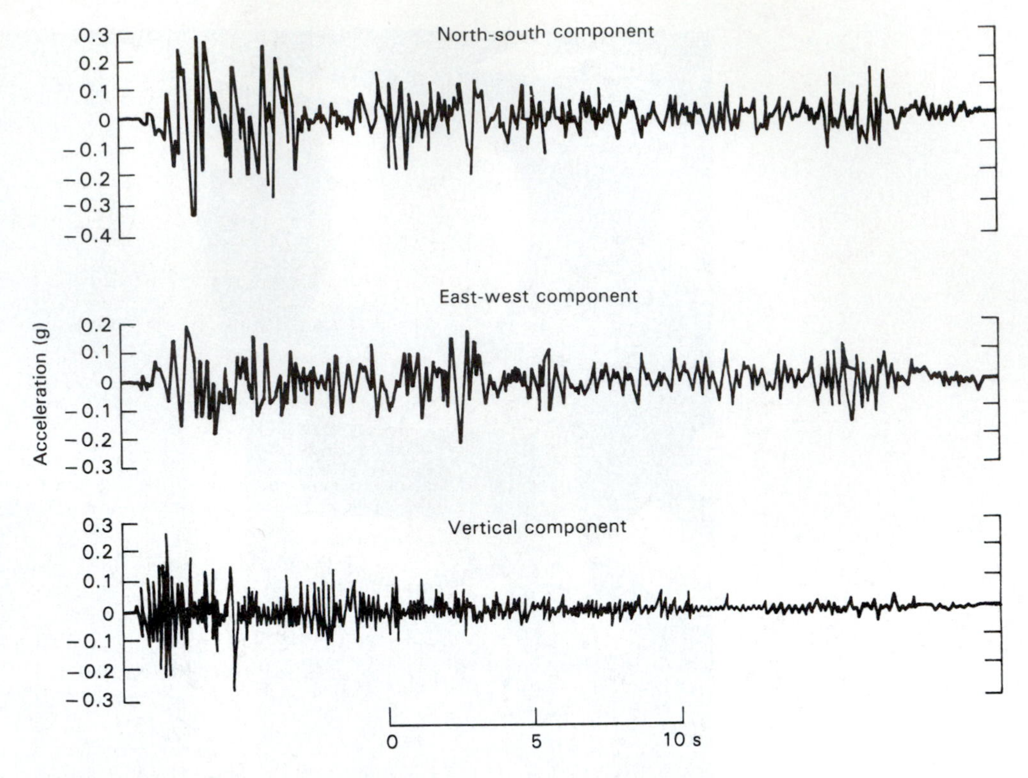


**Figure 8-7** Accelerogram of the 1940 Imperial Valley, California, earthquake recorded at El Centro, California. (From W. W. Hays, 1980, *Procedures for Estimating Earthquake Ground Motions*, U.S. Geological Survey Professional Paper 1114.)

of magnitude 7, for example, produces 10 times the wave amplitude of an earthquake of magnitude 6. Great earthquakes are classified as those with a magnitude of 8 or more.

The amount of seismic energy released in an earthquake is a function of the volume of rock releasing strain energy during the event. Thus, the amount of energy liberated is maximized in great earthquakes. In fact, the small number of high-magnitude earthquakes that occur every year release more energy than the combined energy of the hundreds of thousands of smaller earthquakes that occur annually.

Unlike earthquake magnitude, *intensity* is a subjective measure that is determined by observing the effects of the earthquake on structures and by interviewing people who experienced the event. The most commonly used intensity scale is the *Modified Mercalli* scale, which has 12 divisions (Figure 8-8). Each division describes a particular level of human response and building performance. Intensity generally decreases with distance from the epicenter (Figure 8-9); local geologic conditions, however, also influence the amount of ground shaking. We will discuss these factors in more detail later.

Earthquakes cannot be predicted with accuracy. On a global basis, however, there is a consistent relationship between magnitude and *frequency* (Figure 8-10). The inverse relationship illustrated in Figure 8-10 shows that whereas on an average there are about 10,000 earthquakes a year of magnitude 4, there are fewer than 10 earthquakes every year above magnitude 7.

MODIFIED MERCALLI INTENSITY SCALE OF 1931

(Abridged)

I. Not felt except by a very few under especially favorable circumstances.

II. Felt only by a few persons at rest, especially on upper floors of buildings. Delicately suspended objects may swing.

III. Felt quite noticeably indoors, especially on upper floors of buildings, but many people do not recognize it as an earthquake. Standing motor cars may rock slightly. Vibration like passing of truck. Duration estimated.

IV. During the day felt indoors by many, outdoors by few. At night some awakened. Dishes, windows, doors disturbed; walls made cracking sound. Sensation like heavy truck striking building. Standing motor cars rocked noticeably.

V. Felt by nearly everyone; many awakened. Some dishes, windows, etc., broken; a few instances of cracked plaster; unstable objects overturned. Disturbance of trees, poles and other tall objects sometimes noticed. Pendulum clocks may stop.

VI. Felt by all; many frightened and run outdoors. Some heavy furniture moved; a few instances of fallen plaster or damaged chimneys. Damage slight.

VII. Everybody runs outdoors. Damage negligible in buildings of good design and construction; slight to moderate in well-built ordinary structures; considerable in poorly built or badly designed structures; some chimneys broken. Noticed by persons driving motor cars.

VIII. Damage slight in specially designed structures; considerable in ordinary substantial buildings with partial collapse; great in poorly built structures. Panel walls thrown out of frame structures. Fall of chimneys, factory stacks, columns, monuments, walls. Heavy furniture overturned. Sand and mud ejected in small amounts. Changes in well water. Disturbed persons driving motor cars.

IX. Damage considerable in specially designed structures; well designed frame structures thrown out of plumb; great in substantial buildings, with partial collapse. Buildings shifted off foundations. Ground cracked conspicuously. Underground pipes broken.

X. Some well-built wooden structures destroyed; most masonry and frame structures destroyed with foundations; ground badly cracked. Rails bent. Landslides considerable from river banks and steep slopes. Shifted sand and mud. Water splashed (slopped) over banks.

XI. Few, if any (masonry), structures remain standing. Bridges destroyed. Broad fissures in ground. Underground pipe lines completely out of service. Earth slumps and land slips in soft ground. Rails bent greatly.

XII. Damage total. Waves seen on ground surfaces. Lines of sight and level distorted. Objects thrown upward into the air.

**Figure 8-8** The Modified Mercalli scale (abridged) for determining earthquake intensity.

## SEISMIC WAVES

### Wave Types

Seismic waves generated during an earthquake fall into two basic categories: *body waves* that travel through the earth's interior from the earthquake's hypocenter, and *surface waves* that move along the earth's surface from the epicenter. Body waves can be subdivided further into *P* (*primary*) and *S* (*second-

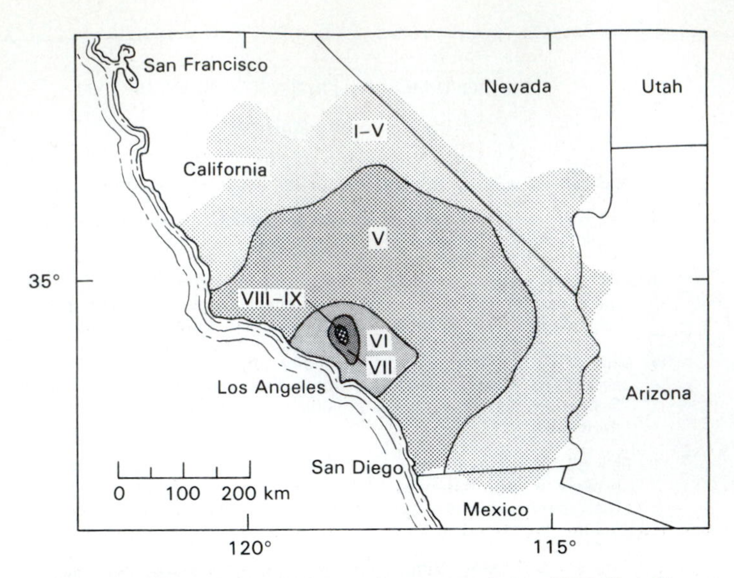


**Figure 8-9** Seismic intensity map of the 1971 San Fernando, California, earthquake. (From W. W. Hays, 1980, *Procedures for Estimating Earthquake Ground Motions*, U.S. Geological Survey Professional Paper 1114.)

*ary*) waves. *P* waves are also known as compressional waves because they travel through rocks as alternate compressions and expansions of the material (Figure 8-11). The movement of individual rock particles is a back-and-forth motion in the direction of propagation. *P* waves are the fastest moving seismic waves. Their velocity ($\alpha$), in m/s is expressed as

$$\alpha = \sqrt{(\kappa + \tfrac{4}{3}\mu)\rho} \qquad \text{(Eq. 8-1)}$$

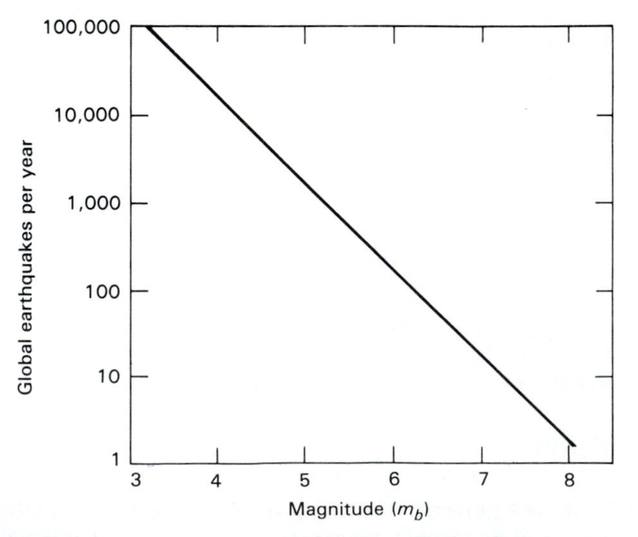


**Figure 8-10** Average relationship between magnitude and frequency on a global basis. (From B. A. Bolt, *Nuclear Explosions and Earthquakes: The Parted Veil*, copyright © 1976 by W. H. Freeman and Co., San Francisco.)

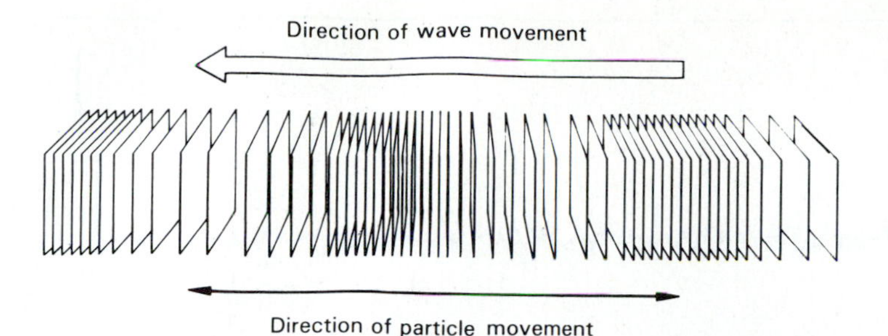


**Figure 8-11** The motion of *P* waves. The wave moves through rock as a sequence of compressions and expansions parallel to the direction of wave movement. Particles vibrate in back-and-forth fashion in the direction of wave movement. (After E. A. Hay and A. L. McAlester, *Physical Geology: Principles and Perspectives*, 2d ed. copyright © 1984 by Prentice-Hall, Inc., Englewood Cliffs, N.J.)

where ($\kappa$) is the bulk modulus of the rock (resistance to volume change), $\mu$ is the modulus of rigidity (resistance to change in shape), and $\rho$ is the density ($g/cm^3$).

*S* waves are also known as shear waves because of their mode of travel (Figure 8-12). The travel of *S* waves is analogous to the movement of a rope anchored to a wall at one end and shaken up and down with your hand at the other end. Although the wave motion moves from your hand toward the wall, the movement of the rope is up and down as the wave passes. This shear deformation, which is transverse to the direction of wave propagation, explains why *S* waves cannot be transmitted through a liquid: A liquid has no resistance to shear stresses. The discovery that *S* waves do not travel through the earth's outer core is the main evidence for the hypothesis that this part of the core is liquid-like.

The velocity of *S* waves ($\beta$), which is lower than that of *P* waves, can be expressed as

$$\beta = \sqrt{\frac{\mu}{\rho}} \qquad \text{(Eq. 8-2)}$$

The modulus of rigidity, $\mu$, is equal to zero in liquids; therefore, the *S*-wave velocity is zero.

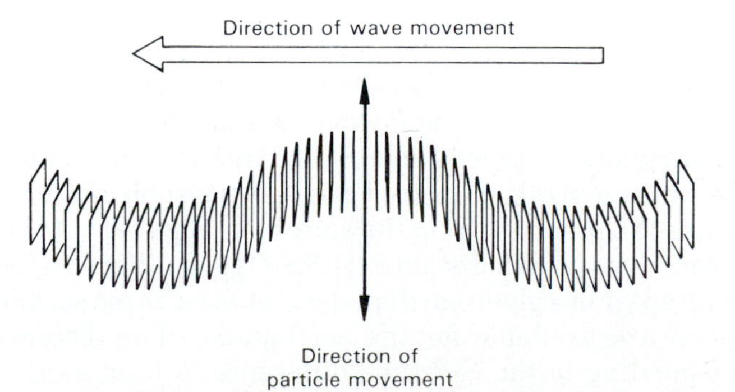


**Figure 8-12** Transmission of *S* waves. Particle movement is perpendicular to direction of wave movement through rock. (After E. A. Hay and A. L. McAlester, *Physical Geology: Principles and Perspectives*, 2d ed., copyright © 1984 by Prentice-Hall, Inc., Englewood Cliffs, N.J.)

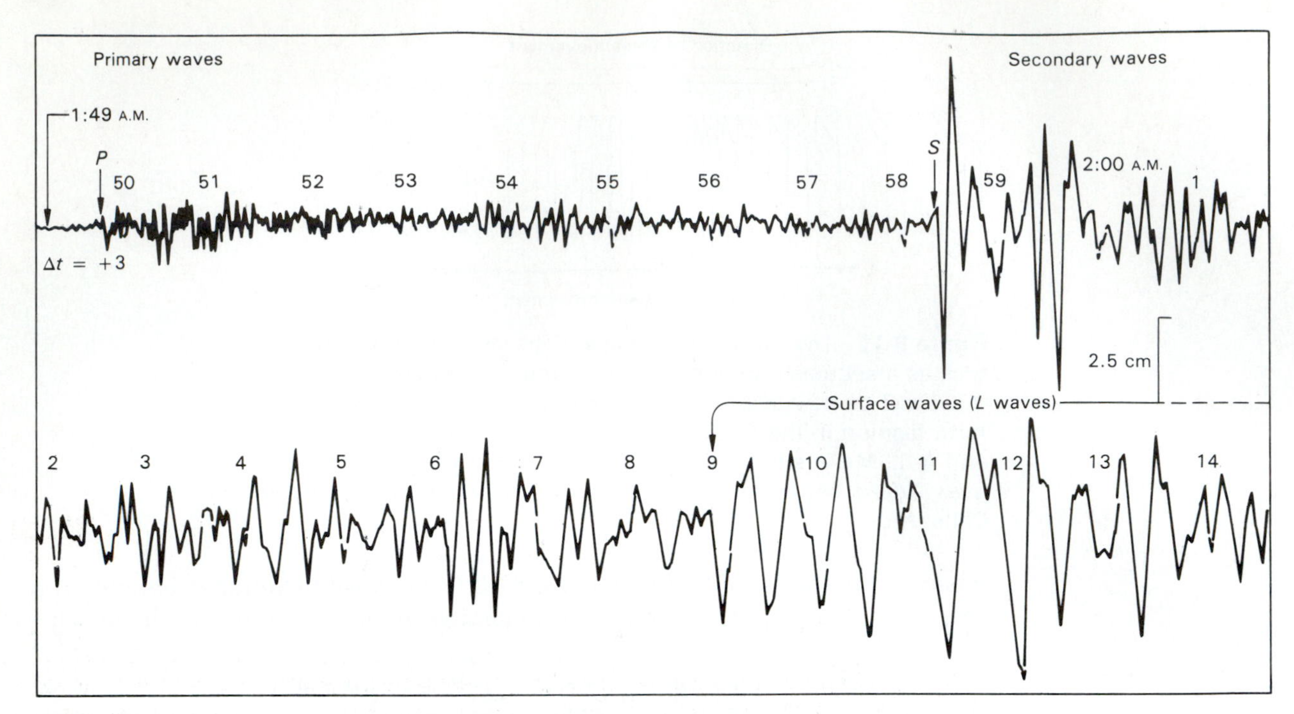


**Figure 8-13** Portion of a seismogram showing the arrival of *P*, *S*, and surface waves. (From S. Judson, M. E. Kauffman, and L. D. Leet, *Physical Geology*, 7th ed., copyright © 1987 by Prentice-Hall, Inc., Englewood Cliffs, N.J.)

Surface waves, which arrive at seismograph stations after both *P* and *S* waves, include *Love* and *Rayleigh* waves. Surface waves are characterized by long periods (10 to 20 s) and wavelengths of 20 to 80 km. The period measures the amount of time that elapses between the arrival of two successive wave crests, and wavelength measures the distance separating identical points on successive waves. Love waves originate from *S* waves that reach the surface at the epicenter. As they move outward along the surface of the earth, they produce only horizontal ground motion. Rayleigh waves resemble the motion of ocean waves, with predominantly vertical displacement. A seismogram showing the arrival of *P* waves, *S* waves, and surface waves is shown in Figure 8-13.

### *Time-Distance Relations*

The difference in velocity between *P* and *S* waves means that the time interval between the first arrival of *P* waves and *S* waves increases with distance from the earthquake source (Table 8-1). Thus the time interval between *P*-wave and *S*-wave arrivals measured at a seismograph station is an indication of the distance of the station from the epicenter. Figure 8-14 illustrates how *S-P* time differences vary with distance. The *S-P* time difference can be used for locating the epicenter of a given earthquake if at least three seismograms from different stations are available for the earthquake. The distance from the epicenter corresponding to the *S-P* time difference is used as the radius of a circle to plot a circular arc in the vicinity of the epicenter (Figure 8-15). The point where three circular arcs intersect is the epicenter.

The *S-P* time interval, along with maximum trace amplitude, is also useful in determining the magnitude (Figure 8-16). If the plotted points representing the trace amplitude and *S-P* time interval are connected with a straight

**TABLE 8-1**

**Travel Time and Time Interval for Arrival of Seismic Waves**

| *Distance from source, km* | *Travel time* | | | | *Interval between P and S (S−P)* | |
|---|---|---|---|---|---|---|
| | *P* | | *S* | | | |
| | *min* | *s* | *min* | *s* | *min* | *s* |
| 2,000 | 4 | 06 | 7 | 25 | 3 | 19 |
| 4,000 | 6 | 58 | 12 | 36 | 5 | 38 |
| 6,000 | 9 | 21 | 16 | 56 | 7 | 35 |
| 8,000 | 11 | 23 | 20 | 45 | 9 | 22 |
| 10,000 | 12 | 57 | 23 | 56 | 10 | 59 |
| 11,000 | 13 | 39 | 25 | 18 | 11 | 39 |

Distance
Time (*S-P*)
km 500
50 sec
40
300
30
200
20
100
10
75
8
50
6
4
25
20
10
2
5
0
0

SOURCE: From S. Judson, M. E. Kauffman, and L. D. Leet, *Physical Geology*, 7th ed., copyright © 1987 by Prentice-Hall, Inc., Englewood Cliffs, N.J.

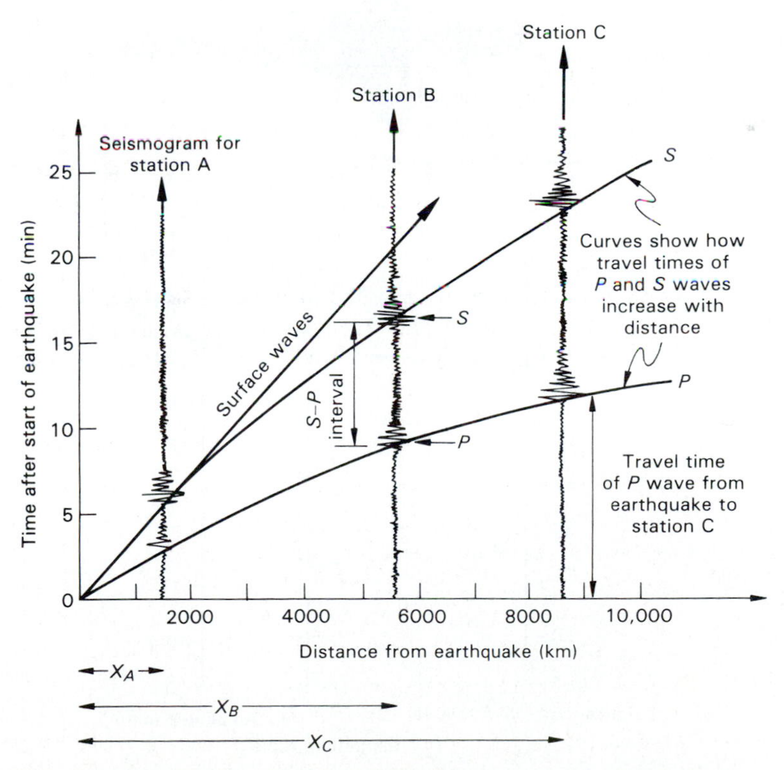


**Figure 8-14** Variation of the *S-P* time interval with distance from the epicenter. As the distance from the epicenter increases from seismograph station $X_A$ to $X_C$, the *S-P* time interval increases. (From F. Press and R. Siever, *Earth*, 3d ed., copyright © 1982 by W. H. Freeman and Co., San Francisco.)

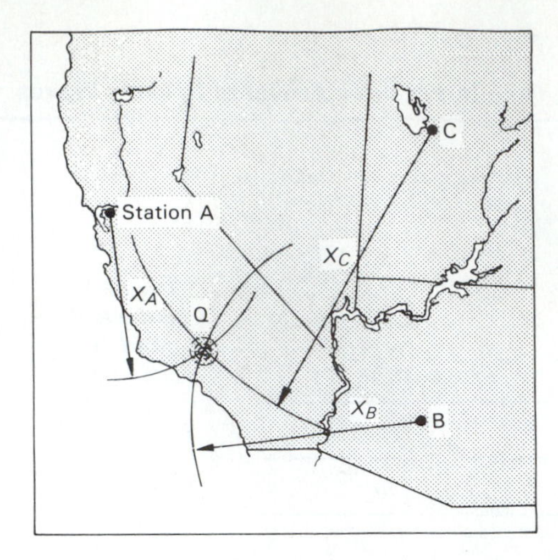


**Figure 8-15** The epicenter of an earthquake can be located by plotting three circular arcs with radii determined from the *S-P* time interval as shown in Figure 8-14. The epicenter is located at the intersection of the three arcs. (From F. Press and R. Siever, *Earth*, 3d ed., copyright © 1982 by W. H. Freeman and Co., San Francisco.)

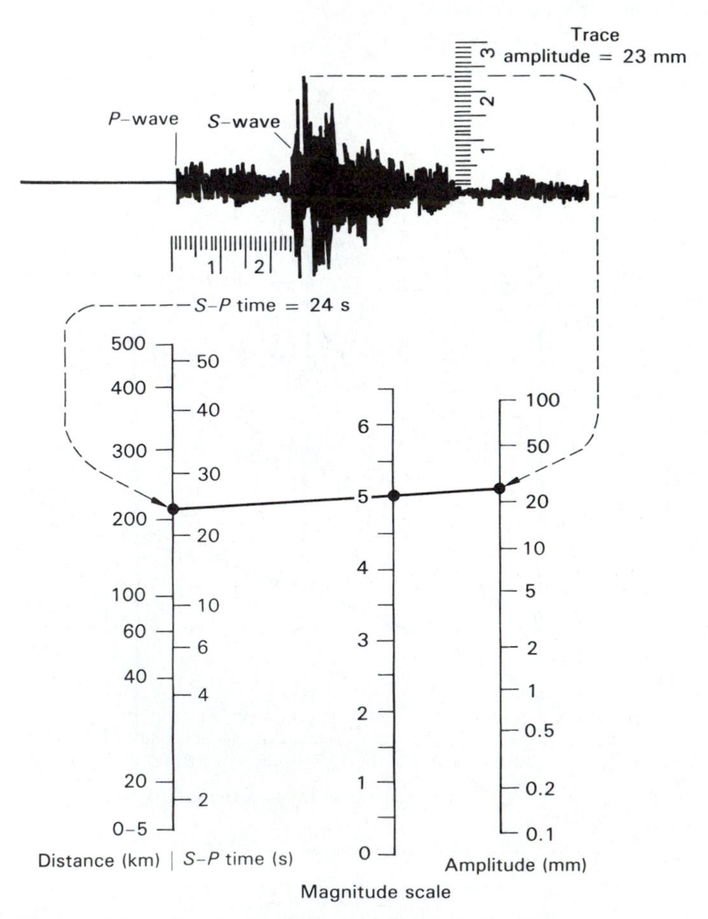


**Figure 8-16** Determination of the Richter magnitude from the *S-P* time interval and the maximum trace amplitude measured from a seismogram. (From W. W. Hays, 1980, *Procedures for Estimating Earthquake Ground Motions*, U.S. Geological Survey Professional Paper 1114.)

line on a nomogram, the magnitude of the earthquake is indicated where the line crosses the center scale.

## EARTHQUAKE HAZARDS

Most earthquakes accompany major movements of crustal blocks along faults. Crustal blocks can also move without generating seismic waves. Such movement, called *aseismic creep*, is desirable from the standpoint that it relieves accumulating strain gradually, thus preventing a buildup of strain leading to a major earthquake. Aseismic creep also has detrimental effects because of the associated displacement that takes place along faults. This displacement can rupture utility lines and wells and cause damage to streets, sidewalks, and other structures. A particularly dramatic example of aseismic creep is the 1963 failure of the Baldwin Hills Reservoir in southern California, an impoundment that stored water for the City of Los Angeles (Figure 8-17). Aseismic displacement along faults located beneath the reservoir caused fracturing of the reservoir's underdrain system and impermeable asphalt membrane. Water leakage through the reservoir floor and dam eventually led to the complete failure of the dam and the draining of the impoundment.

Hazards associated with earthquakes include surface faulting, ground shaking, ground failure, and destructive ocean waves called *tsunamis*. Rupture of the ground surface by fault displacement includes both vertical and lateral

**Figure 8-17** Aerial view of the Baldwin Hills reservoir after failure. Aseismic creep along faults beneath the impoundment ruptured the impermeable seal of the bottom. Water then migrated through the dam at an increasing rate until a section of the dam collapsed and the reservoir drained rapidly. (Photo courtesy of Los Angeles Dept. of Water and Power.)

**Figure 8-18** Fault scarp formed during the 1959 Hebgen Lake earthquake in Montana. (I. J. Witkind; photo courtesy of U.S. Geological Survey.)

motion. Fault scarps are the result of surface rupture formed by dip-slip displacement during an earthquake (Figure 8-18). Obviously, any structure built directly across a fault that sustains surface rupture during an earthquake is likely to be damaged or destroyed (Figure 8-19). By mapping faults known to be active (those in which movement has occurred within recent geologic time), engineers can avoid these hazardous areas for some or all types of structures. Unfortunately, development has occurred over active faults in some areas (Figure 8-20). In addition, surface displacement frequently has occurred along faults that were thought to be inactive.

Ground shaking is usually the greatest threat to buildings and human life associated with earthquakes. The peak acceleration attained by a structure multiplied by its mass defines the dynamic force applied to the structure during an earthquake. Ground acceleration depends on the distance from the epicenter, the depth of the hypocenter, and the characteristics of the soils and rock beneath the foundation. The response of the structure depends upon its size and design. We will examine these factors more closely in the following sections.

Ground failure includes a variety of processes. During intense shaking, soil structure and behavior can be drastically altered. Seismic stresses often cause soils to densify and compact. Certain saturated soils lose strength and liquefy almost instantaneously during seismic shaking. In this state the soil can no longer support buildings, which then may sink and rotate into the soil as complete units (Chapter 10). This process, called *liquefaction*, was responsible for extensive damage to buildings in both the Loma Prieta earthquake and the June 21, 1990, earthquake in northern Iran (Figure 8-21). This 7.7 magnitude earthquake left 40,000 dead, 60,000 injured, and 500,000 homeless. A major factor in the loss of life was the poor performance of buildings during

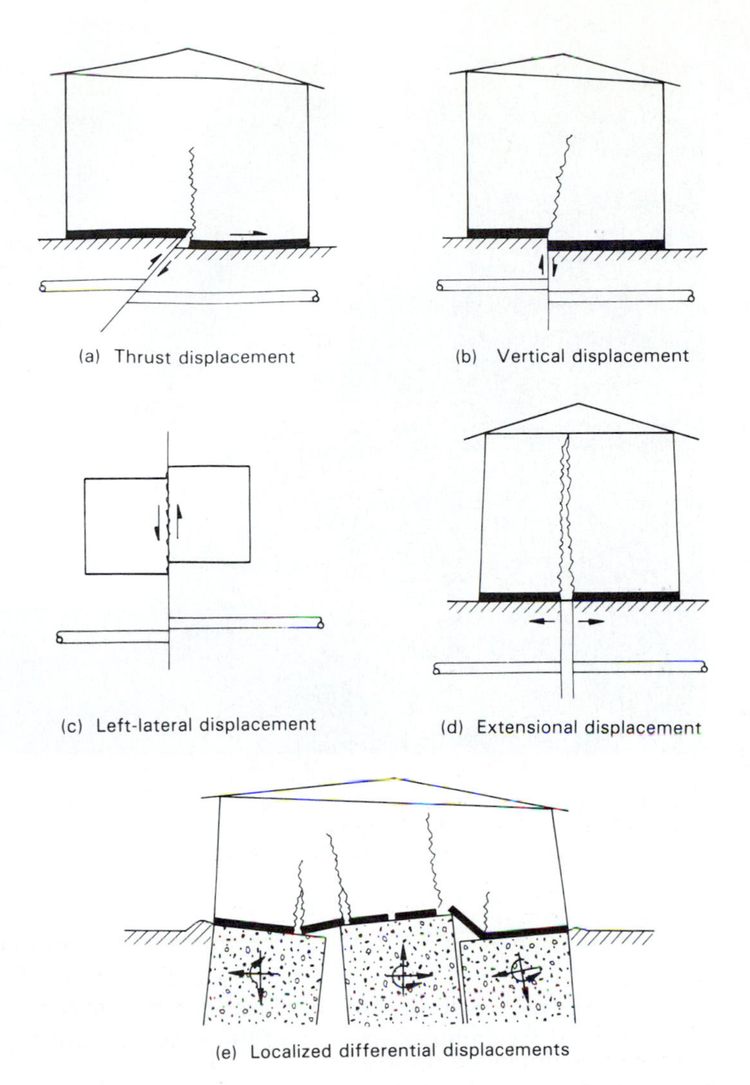


**Figure 8-19** Types of structural damage caused by ground rupture during the San Fernando earthquake. (From U.S. Geological Survey, 1971, *The San Fernando, California, Earthquake of February 9, 1971*, U.S. Geological Survey Professional Paper 733.)

ground shaking. Downslope mass movements, including slumps, slides, flows, and avalanches can be triggered by earthquakes. Often, these movements involve huge masses of material and are responsible for great damage and loss of life. The Lower Van Norman Dam (Figure 8-22) nearly failed during the 1971 San Fernando, California, earthquake because of a large slump that developed on the upstream side. If the reservoir level had not been maintained at a lower elevation than normal design limits called for (because of concerns for the safety of the dam), it most likely would have totally failed during the earthquake. The imminent failure during the earthquake necessitated the evacuation of 80,000 people downstream. Other examples of mass movements caused by earthquakes are discussed in Chapter 12.

An earthquake hazard that is greatly feared in coastal areas is the tsunami, or seismic sea wave. These are high-energy waves generated by earthquakes whose epicenters lie beneath the sea bed. Tsunamis travel rapidly across the ocean basins and cause great damage when they impact coastlines. Tsunamis are described more fully in Chapter 14.

**Figure 8-20** Urban development along the San Andreas fault, San Mateo County, California. Arrows show location of fault. (R. W. Wallace; photo courtesy of U.S. Geological Survey.)

In addition to the hazards just mentioned, other problems are encountered during earthquakes. These include fires, disruption of water supplies, and disease. As the population of seismically active areas continues to increase, the potential for damage and loss of life grows accordingly.

## SEISMIC RISKS AND LAND-USE PLANNING

Assessment of seismic hazard for an area is a complex procedure that must take into account the history of seismic events in the area, the location of active faults, the regional and local responses to seismic events, and the seismic

**Figure 8-21** Differential settlement due to liquefaction during the 1990 northern Iran earthquake. The building to the left sank relative to the small attached structure at center, which has a separate foundation. (M. Mehrain.)

**Figure 8-22** Lower Van Norman Dam near San Fernando, California, after damage by the 1971 earthquake. Slumping of parts of the dam into the reservoir is evident. (Photo courtesy of U.S. Geological Survey.)

characteristics of the surficial geologic materials. If the evaluation of all these factors indicates that the area may be subjected to a potentially damaging earthquake, measures can be taken to reduce the risks associated with the event. These measures can include land-use planning and zoning as well as the adoption of stringent building codes.

### *Location of Earthquakes*

Locations of earthquake epicenters recorded during a particular period of time provide a good indication of seismically active areas. Such maps (Figure 8-3) show a high concentration of earthquakes along lithospheric plate boundaries. Earthquakes are not limited to plate boundaries, however. Figure 8-23 indicates that damaging earthquakes have struck many areas in the United States during the past 200 years. Table 8-2 provides information on the magnitude and effects of these events.

In addition to determining the location of earthquake epicenters, it is also important to map and identify active faults. A number of specific criteria are used for recognition of an active fault. For example, the U.S. Nuclear Regulatory Commission (NRC) defines a *capable* fault as one which has experienced displacement at least once within the past 35,000 years. A map of major faults in California is shown in Figure 8-24. If a particular fault segment has not undergone rupture within a certain period of time, this does not necessarily mean that the segment is inactive. Instead, it may indicate that that section

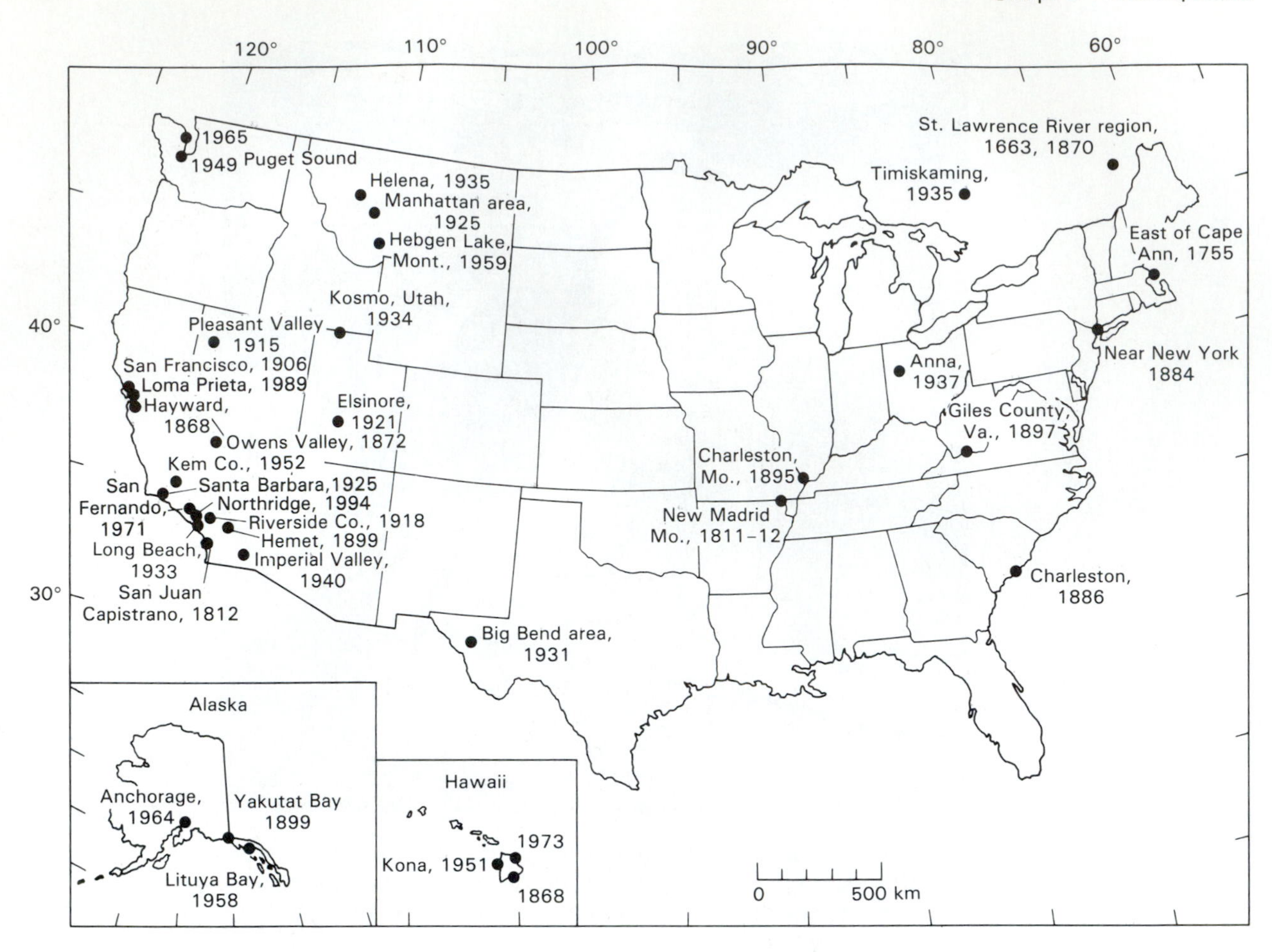


**Figure 8-23** Location of past destructive earthquakes in the United States. (From W. W. Hays, 1980, *Procedures for Estimating Earthquake Ground Motions*, U.S. Geological Survey Professional Paper 1114.)

of the fault is *locked*, a condition in which large amounts of elastic strain accumulate prior to infrequent, but large-magnitude earthquakes. Locked sections of faults are called *seismic gaps* because plots of earthquake hypocenters show a lack of events in these areas. A cross section of the San Andreas fault from northern to central California (Figure 8-25) indicates three seismic gaps: the San Francisco Peninsula gap, the southern Santa Cruz Mountains gap, and the Parkfield gap. The lower cross section shows that the southern Santa Cruz Mountains gap was filled by the Loma Prieta earthquake and its aftershocks. Seismic gaps are closely monitored because the longer a gap exists, the stronger the eventual earthquake may be in that area.

### Ground Motion

The response of the earth's crust to an earthquake varies both on a regional and local scale. The decrease in amplitude of seismic waves as they move outward from the epicenter is known as *seismic attenuation*. The seismic attenuation characteristics of the San Francisco earthquake and the New Madrid, Missouri, earthquake are shown in Figure 8-26. The contours define zones of equal seismic intensity as measured by the Modified Mercalli scale. Even though the San Francisco earthquake was much larger in magnitude than the

**TABLE 8-2**

**Property Damage and Lives Lost in Notable U.S. Earthquakes**

| *Year* | *Locality* | *Magnitude* | *Damage (million dollars)* | *Lives lost* |
|---|---|---|---|---|
| 1811–12 | New Madrid, Mo. | 7.5 (est.) | — | — |
| 1865 | San Francisco, Calif. | 8.3 (est.) | 0.4 | — |
| 1868 | Hayward, Calif. | — | 0.4 | 30 |
| 1872 | Owens Valley, Calif. | 8.3 (est.) | 0.3 | 27 |
| 1886 | Charleston, S.C. | — | 23.0 | 60 |
| 1892 | Vacaville, Calif. | — | 0.2 | — |
| 1898 | Mare Island, Calif. | — | 1.4 | — |
| 1906 | San Francisco, Calif. | 8.3 (est.) | 500.0 | 700 |
| 1915 | Imperial Valley, Calif. | — | 6.0 | 6 |
| 1925 | Santa Barbara, Calif. | — | 8.0 | 13 |
| 1933 | Long Beach, Calif. | 6.3 | 40.0 | 115 |
| 1935 | Helena, Mont. | 6.0 | 4.0 | 4 |
| 1940 | Imperial Valley, Calif. | 7.0 | 6.0 | 9 |
| 1946 | Hawaii (tsunami) | — | 25.0 | 173 |
| 1949 | Puget Sound, Wash. | 7.1 | 25.0 | 8 |
| 1952 | Kern County, Calif. | 7.7 | 60.0 | 8 |
| 1954 | Eureka, Calif. | — | 2.1 | 1 |
| 1954 | Wilkes-Barre, Pa. | — | 1.0 | — |
| 1955 | Oakland, Calif. | — | 1.0 | 1 |
| 1957 | Hawaii (tsunami) | — | 3.0 | — |
| 1957 | San Francisco, Calif. | 5.3 | 1.0 | — |
| 1958 | Khantaak Island and Lituya Bay, Alaska | — | — | 5 |
| 1959 | Hebgen Lake, Mont. | — | 11.0 | 28 |
| 1960 | Hilo, Hawaii (tsunami) | — | 25.0 | 61 |
| 1964 | Prince William Sound, Alaska | 8.4 | 500.0 | 131 |
| 1965 | Puget Sound, Wash. | — | 12.5 | 7 |
| 1971 | San Fernando, Calif. | 6.5 | 553.0 | 65 |

SOURCE: From Office of Emergency Preparedness, 1972, *Disaster Preparedness*, Report to Congress of the United States, Vol. 3.

New Madrid event (Table 8-2), it was felt over a much smaller area. The regional crustal properties of the eastern and western parts of the United States are therefore inferred to be very different.

On a local scale the intensity of ground motion is largely a function of the surficial geology. Soft soils tend to vibrate much more strongly than stiff soils or rock. This relationship is clearly illustrated by the response to earthquakes in San Francisco. The geologic setting of the city consists of deposits of bay mud and alluvium (river-deposited soils) in low-lying areas (Figure 8-27) and areas of bedrock at shallow depths throughout the rest of the city. Figure 8-28 shows the distribution of intensity during the 1906 earthquake. The areas of high intensity correspond closely to the areas underlain by bay mud and alluvium. Damage from the Loma Prieta earthquake closely followed the pattern of damage from the 1906 temblor. The Marina district (Figure 8-28) was the most heavily damaged sector of the city, with structural damage to 150 buildings. This area was a lagoon underlain by bay mud at the time of the 1906 earthquake. It was later filled by rubble from destroyed buildings to create a fairground for the 1915 Panama-Pacific International Exposition. In 1989, land use in the Marina district was dominated by expensive residential

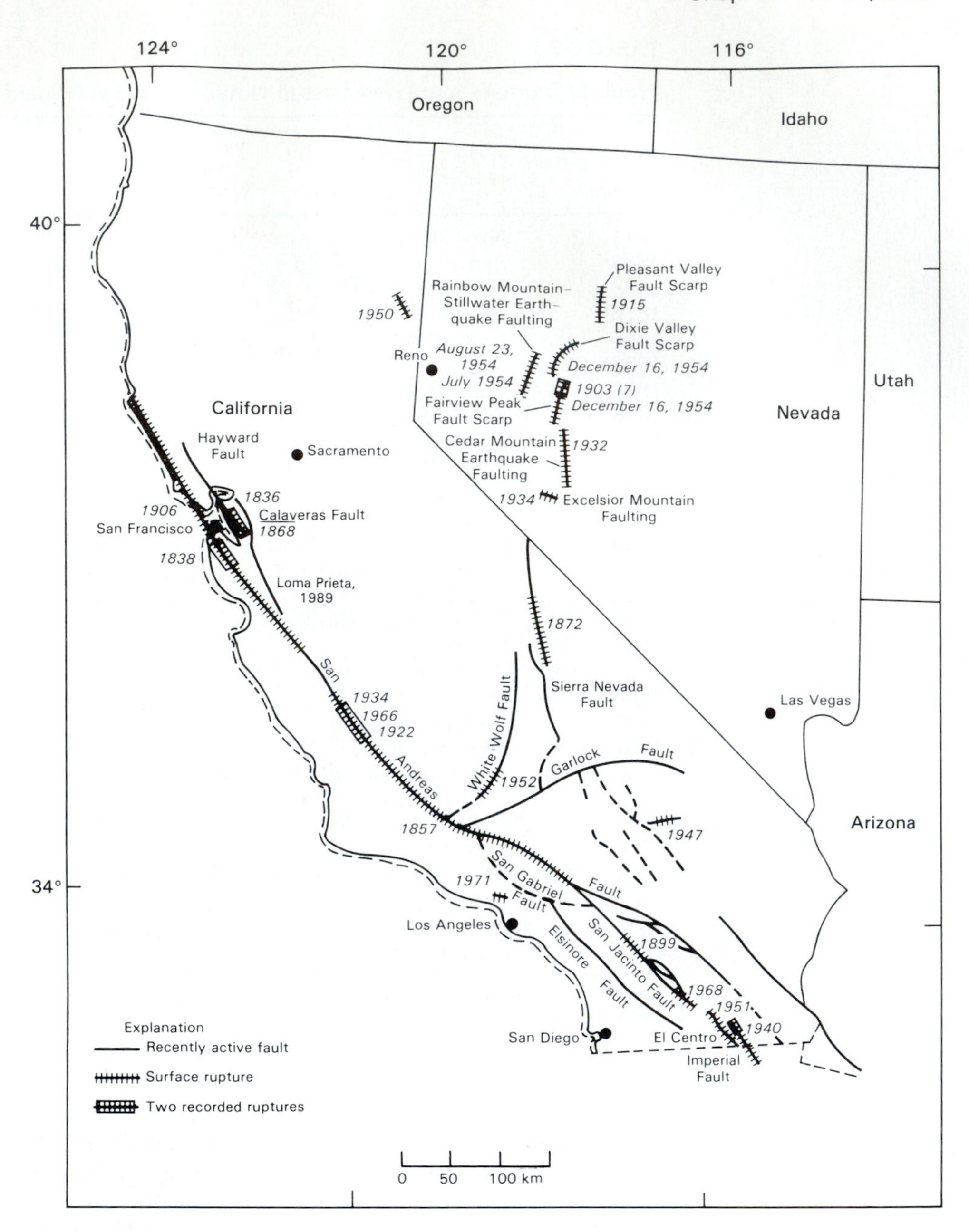


**Figure 8-24** Major faults and locations of past ruptures in California and Nevada. (From W. W. Hays, 1980, *Procedures for Estimating Earthquake Ground Motions*, U.S. Geological Survey Professional Paper 1114.)

dwellings. During the earthquake, ground motion caused liquefaction of the bay muds and damage to many buildings (Figure 8-29).

The relationship between surficial material and ground motion is further illustrated by the recordings of accelerographs in San Francisco of a distant underground nuclear explosion (Figure 8-30). Horizontal ground motion was much greater for bay mud and alluvium than for bedrock when subjected to the same seismic waves. Similar results were obtained by accelerographs that recorded some of the larger aftershocks of the Loma Prieta earthquake. Soft soils actually amplify incoming seismic waves, a condition that may have

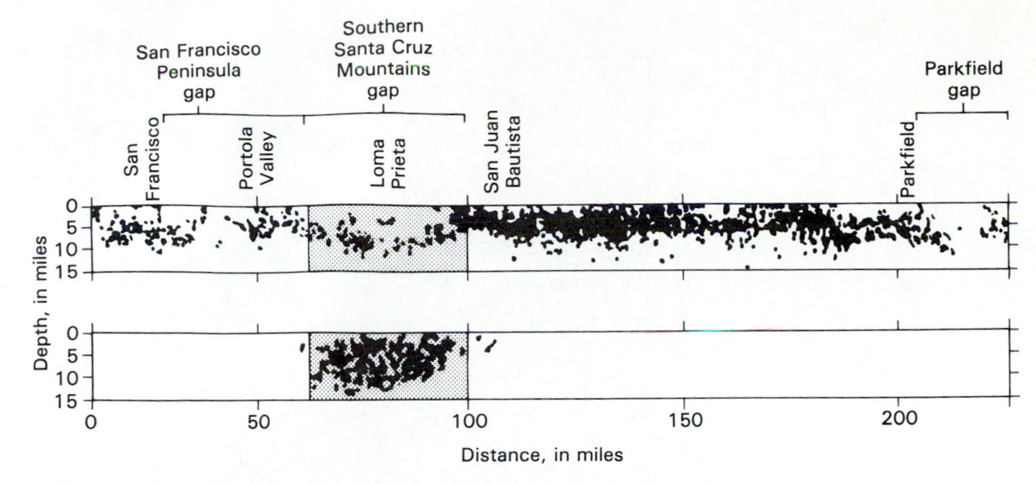


**Figure 8-25** Cross sections along the San Andreas fault from north of San Francisco to Parkfield. Upper section shows locations of earthquakes from January 1969 through July 1989, and lower section shows Loma Prieta earthquake and its aftershocks, which filled one of three gaps in the cross section. (From P. L. Ward and R. A. Page, *The Loma Prieta Earthquake of October 17, 1989*, U.S. Geological Survey Pamphlet.)

contributed to the collapse of part of the Nimitz Freeway (Interstate 880) in Oakland, California. The section that failed was a two-tiered roadway known as the Cypress structure (Figure 8-31). The mode of failure involved collapse of the upper roadway onto the lower section. Vehicles on the lower section were crushed and 41 people were killed. Although the entire Cypress structure was damaged, only that portion constructed on bay mud collapsed (Figure 8-31).

One of the most effective ways to reduce future losses in areas of high seismic risk is land-use planning. Zoning ordinances of various types can pre-

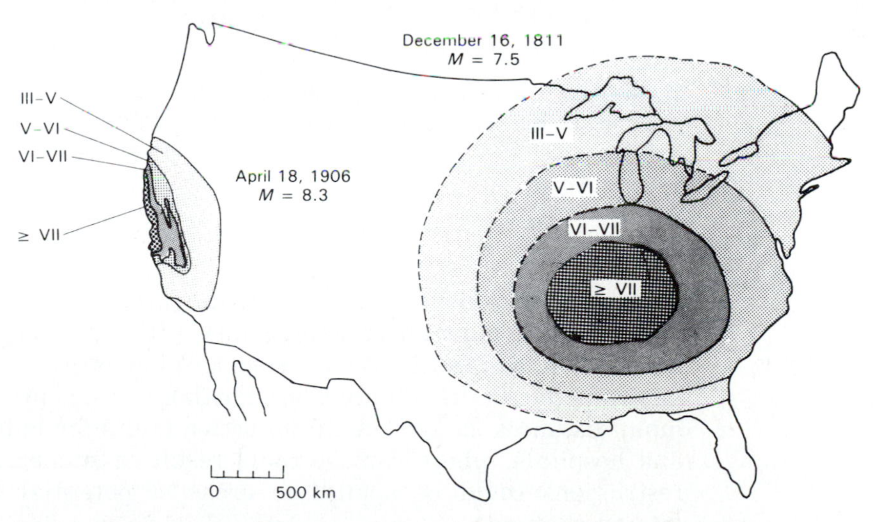


**Figure 8-26** Seismic intensity plotted for the 1906 San Francisco earthquake and the 1811 New Madrid, Missouri, event. Even though the San Francisco earthquake had a larger magnitude, it affected a smaller area because of differences in regional seismic attenuation. (From W. W. Hays, 1980, *Procedures for Estimating Earthquake Ground Motions*, U.S. Geological Survey Professional Paper 1114.)

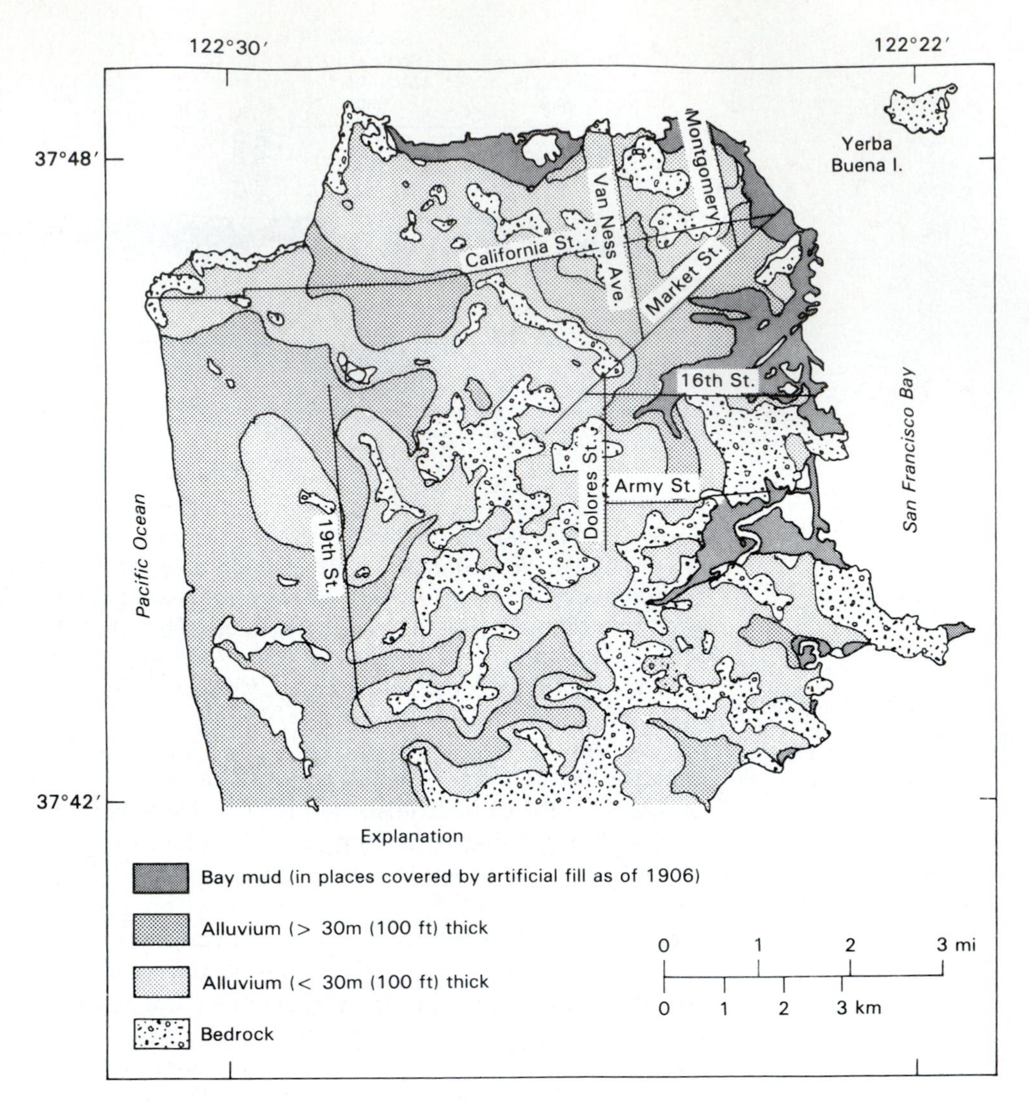


**Figure 8-27** Generalized geology of the San Francisco area. (From R. D. Borcherdt, ed., *Studies for Seismic Zonation of the San Francisco Bay Region*, U.S. Geological Survey Professional Paper 941-A.)

vent or restrict development in critical areas. An obvious restriction is to prevent development over active faults. Although such a restriction seems to be only common sense, past development has largely ignored the presence of known active faults (Figure 8-20). Further zoning options can establish minimum distances, or *setbacks*, from active faults for buildings such as schools and hospitals, where damage could result in numerous casualties. Similar restrictions could be applied to areas of potential landsliding or severe ground shaking.

The most stringent standards for site location and construction should be established for engineering projects such as dams and nuclear power plants, whose satisfactory performance in an earthquake is necessary to prevent massive losses in human life and property. In most parts of the United States, these types of structures now require the most detailed site investigations and seismic risk assessment procedures possible.

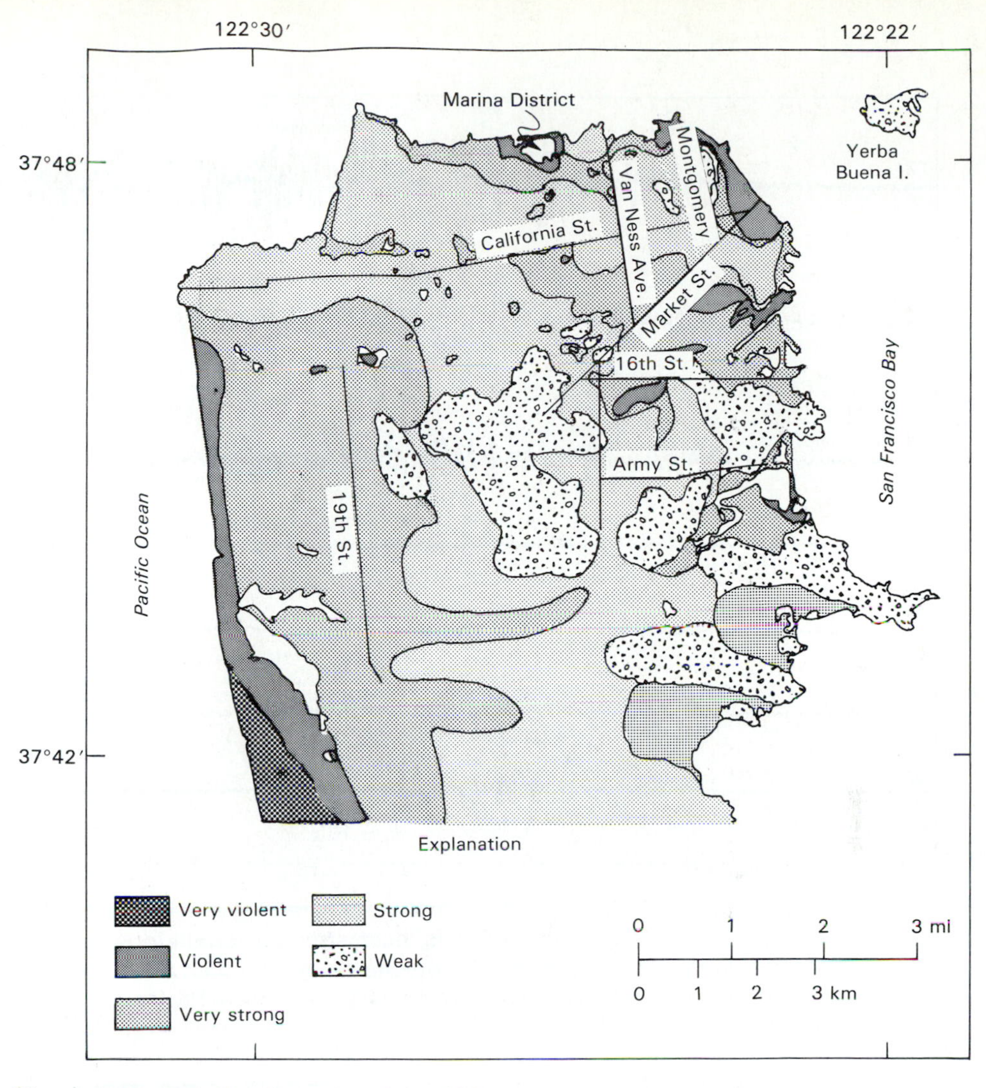


**Figure 8-28** Relative intensity of the 1906 earthquake in San Francisco. The areas of highest intensity correspond to areas of bay mud and alluvium, as shown on Figure 8-27. (From R. D. Borcherdt, ed., *Studies for Seismic Zonation of the San Francisco Bay Region*, U.S. Geological Survey Professional Paper 941-A.)

**Figure 8-29** The third story of a three-story building in the Marina district of San Francisco. The first and second stories collapsed due to ground shaking and liquefaction. (D. Perkins; U.S. Geological Survey.)

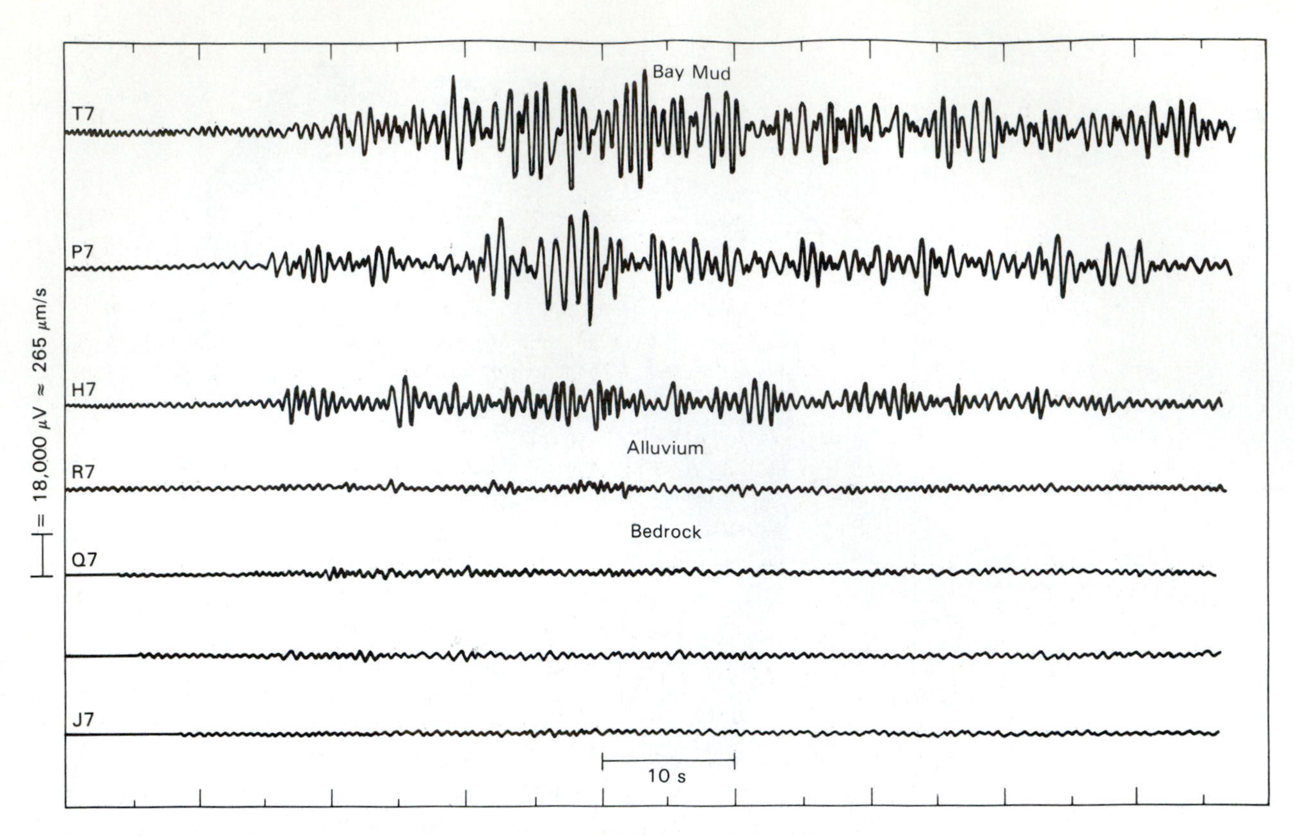


**Figure 8-30** Recordings of horizontal ground motion generated by an underground nuclear explosion. The ground motion is much stronger for stations located on bay mud and alluvium. (From R. D. Borcherdt, ed., *Studies for Seismic Zonation of the San Francisco Bay Region*, U.S. Geological Professional Paper 941-A.)

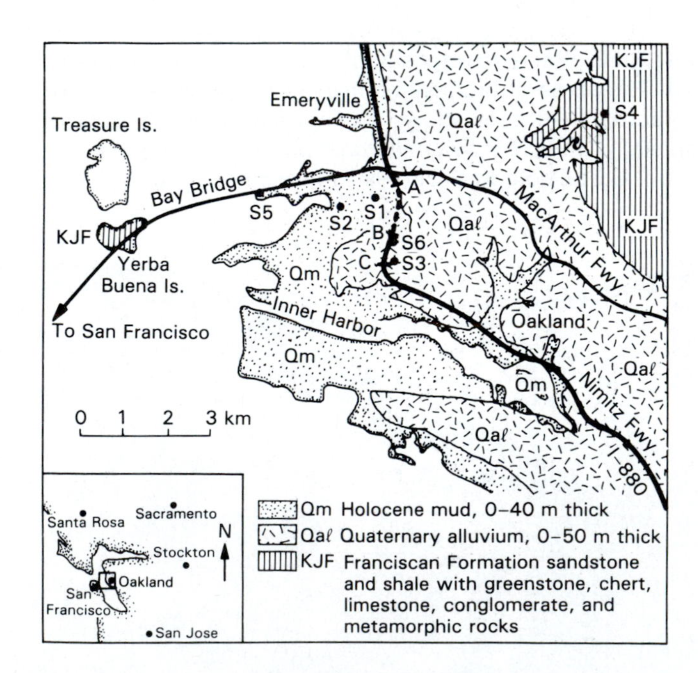


**Figure 8-31** Surficial geology of the Oakland, California, area. Segment AB of the Nimitz freeway, constructed on bay mud, collapsed. Segment BC was damaged but did not collapse. (From S. E. Hough, P. A. Fregerg, R. Busby, E. F. Field, and K. H. Jacob, *Eos, Transactions of the American Geophysical Union*, Nov. 21, 1989.)

## EARTHQUAKE ENGINEERING

The effect of an earthquake upon a building is determined by the relationship and interaction between the characteristics of the seismic waves that reach the site, the response of the materials beneath the building, and the design and construction of the building itself.

Horizontal ground motions often present the most severe test of a building's ability to withstand an earthquake without failure. The resulting lateral accelerations impose inertial forces on a structure. In the absence of strict building codes, structures rarely are designed to resist lateral forces of the magnitude imposed by strong ground motion. Figure 8-32 illustrates the nature of these inertial forces. For low, rigid structures (Figure 8-32, left) the peak lateral force can be calculated as the product of the mass of the building and the peak lateral acceleration. The lateral force for flexible buildings subjected to short-duration seismic shaking can be less than the product of mass times acceleration. This force decrease is achieved because part of the energy is absorbed by the bending action of the frame (Figure 8-32, center). In tall, very flexible buildings, the lateral force can even exceed $M \times A$ because of the increase in inertial force caused by the oscillation of the structure by repeated vibrations (Figure 8-32, right).

One of the most important factors in earthquake damage is the relationship between the vibrational period of a structure and the period of the material upon which it is built. When set into motion, geologic materials, as well as buildings, tend to vibrate at a certain rate. This characteristic rate is defined by the *fundamental period*. The period of a wave or vibration is the time interval between the passage of two corresponding points on the waveform, the time between two successive crests, for example. Fundamental periods of geologic materials range from well below 1 s for bedrock and stiff soils to

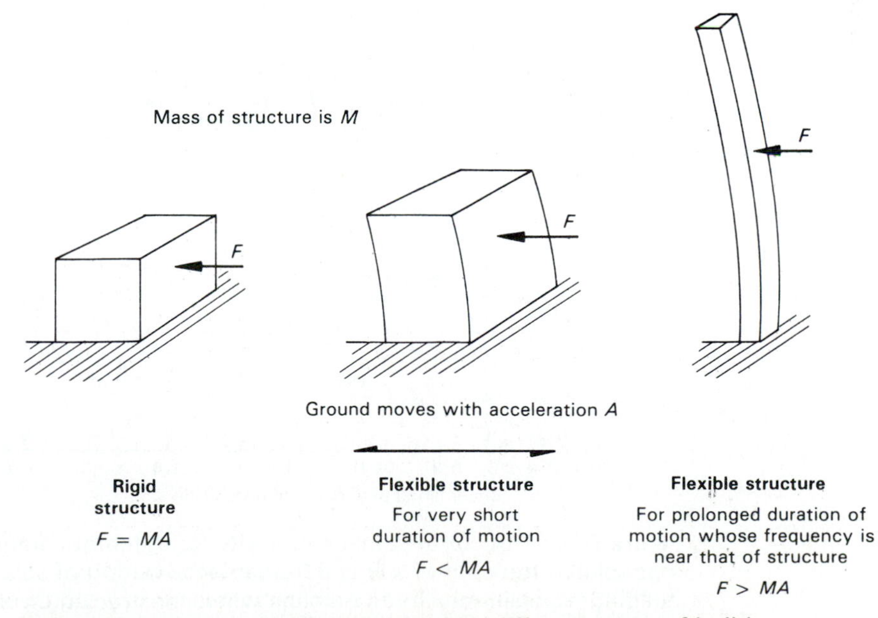


**Figure 8-32** Inertial forces developed by different types of buildings when subjected to earthquake shaking. (From H. J. Degenkolb, 1977, *Earthquake Forces on Tall Structures*, Booklet 2717A, Bethlehem Steel Corp., Bethlehem, Pa.)

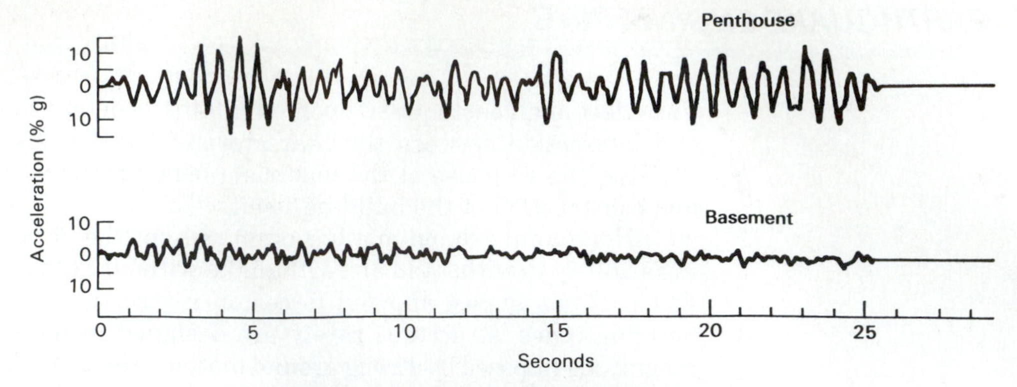


**Figure 8-33** Differences in acceleration between the basement and penthouse of a 10-story reinforced concrete building. (From F. W. Housner, Design spectrum, in R. L. Wiegel, *Earthquake Engineering*, copyright © 1970 by Prentice-Hall, Inc., Englewood Cliffs, N.J.)

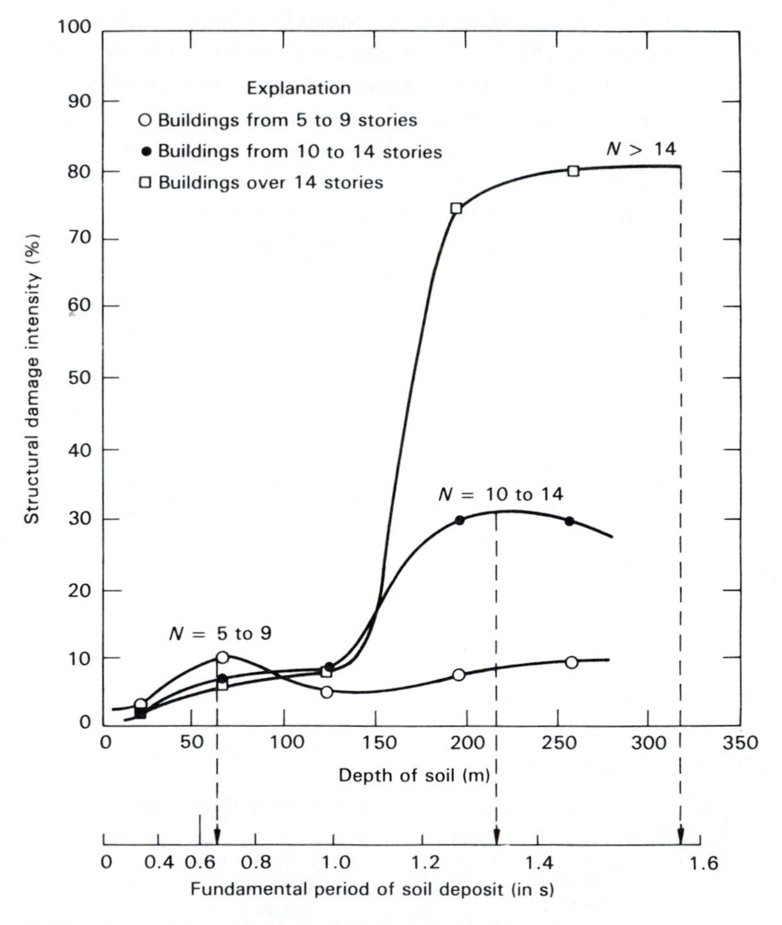


**Figure 8-34** Structural damage intensity for different height buildings related to depth of soil and fundamental period of soil. Tall buildings sustain maximum damage when constructed on long-period soil deposits. (From H. B. Seed et al., 1972, Soil Conditions and Building Damage in 1967 Caracas Earthquake, *Journal of the Soil Mechanics and Foundations Division*, American Society of Civil Engineers, SM-8:787–806.)

several seconds or more for deep, soft soils. The fundamental period of a building varies with the height of the structure. Tall buildings have long periods (several seconds), and low buildings have short periods. The effect of building height upon acceleration is illustrated in Figure 8-33. Accelerographs were installed in the basement and penthouse of a 10 story warehouse that was shaken by the 1952 Tehachapi, California, earthquake. Accelerations were much greater in the penthouse than in the basement. In addition, the penthouse accelerations clearly show the effect of the fundamental period of the building, with maximum accelerations occurring every few seconds. No such pattern is evident in the basement accelerogram.

When the period of a soil and a building are similar, the most hazardous situation occurs. Resonance between the building and the soil leads to more violent shaking and greater damage is usually the result. Tall buildings sustain greater damage when constructed on deep, soft soils because of their similarity in vibrational period. Similarly, small, rigid buildings may perform more poorly when sited upon short-period materials such as bedrock. Figure 8-34 illustrates the effect of building height upon relative seismic damage. The damage intensity for tall structures rises dramatically when the fundamental period of the foundation soil is longer than 1 s.

Construction materials and design often play a critical role in the performance of a building during an earthquake (Table 8-3). Small, wood frame structures are generally the safest type as long as they are anchored securely

**TABLE 8-3**

**Earthquake Ratings for Common Building Types**

| *Simplified description of structural types* | *Relative damageability (in order of increasing susceptibility to damage)* |
|---|---|
| Small wood-frame structures, i.e., dwellings not over 3000 sq. ft, and not over 3 stories | 1 |
| Single or multistory steel-frame buildings with concrete exterior walls, concrete floors, and concrete roof. Moderate wall openings | 1.5 |
| Single or multistory reinforced-concrete buildings with concrete exterior walls, concrete floors, and concrete roof. Moderate wall openings | 2 |
| Large area wood-frame buildings and other wood-frame buildings | 3 to 4 |
| Single or multistory steel-frame buildings with unreinforced masonry exterior wall panels; concrete floors and concrete roof | 4 |
| Single or multistory reinforced-concrete frame buildings with unreinforced masonry exterior wall panels, concrete floors and concrete roof | 5 |
| Reinforced concrete bearing walls with supported floors and roof of any materials (usually wood) | 5 |
| Buildings with unreinforced brick masonry having sandlime mortar; and with supported floors and roof of any materials (usually wood) | 7 up |
| Bearing walls of unreinforced adobe, unreinforced hollow concrete block, or unreinforced hollow clay tile | Collapse hazards in moderate shocks |

SOURCE: From D. Armstrong, 1973, *The Seismic Safety Study for the General Plan*, Sacramento, Calif.: California Council on Intergovernmental Relations.

NOTE: This table is not complete. Additional considerations would include parapets, building interiors, utilities, building orientation, and frequency response.

**Figure 8-35** Devastation of adobe houses in the town of Tecpan, Guatemala, by the 1976 earthquake. (Photo courtesy of U.S. Geological Survey.)

to their foundations. Improperly anchored houses have been observed to shear apart from their foundations during lateral ground accelerations. Steel-frame or reinforced-concrete construction methods are least hazardous for multistory buildings. Buildings that provide the least degree of protection include those of nonreinforced masonry, brick and mortar, and adobe construction. Structures of this type are literally deathtraps because of their heavy weight and brittle behavior. The Guatemalan earthquake of 1976 provides a good example of the high correlation between adobe construction and nearly total destruction of villages in areas of moderate- to high-intensity earthquakes. (Figure 8-35). Likewise, in the 1990 northern Iran earthquake, destruction was overwhelmingly linked to structures with unreinforced masonry walls and roofs.

Although building performance is certainly a major concern, the effects of earthquakes upon *lifeline* systems are also of vital importance. Lifeline systems include electricity and fuel supplies, water and wastewater facilities, transportation lines, and communication networks. Loss of these facilities in a large urban area can lead to greater problems than the earthquake itself. Emergency rescue and relief efforts are severely hampered by damage to energy, transportation, and communication systems. The breakdown of water and sanitary networks can lead to disease and the inability to combat fires that may break out. Eighty-three years after uncontrollable fires destroyed the city of San Francisco after the earthquake of 1906, fires in the Marina district could not be fought with city water due to the rupture of water mains during the 1989 earthquake.

## EARTHQUAKE PREDICTION

Prediction of earthquakes is a scientific goal that, if achieved, could prevent widespread loss of life in a major earthquake. Unfortunately, accurate prediction is not currently possible, although intensive research is proceeding in many areas.

Two types of earthquake prediction are theoretically possible. The first type is long-term forecasting, in which the probability of an earthquake along a particular fault segment within a certain time interval is calculated by

**TABLE 8-4**

**Probabilities of Major Earthquakes on California's Primary Fault Segments**

| *Fault segment* | *Date of most recent event* | *Expected magnitude* | *Estimated recurrence interval (yr)* | *Probability of occurrence between 1988 and 2018 (percent)* | *Level of reliability (A, most reliable)* |
|---|---|---|---|---|---|
| | | San Andreas fault | | | |
| North Coast | 1906 | 8 | 303 | <10 | B |
| San Francisco Peninsula | 1906 | 7 | 169 | 20 | C |
| Southern Santa Cruz Mountains | 1906 | 6.5 | 136 | 30 | E |
| Central creeping | — | — | — | <10 | A |
| Parkfield | 1966 | 6 | 21 | >90 | A |
| Cholame | 1857 | 7 | 159 | 30 | E |
| Carrizo | 1857 | 8 | 296 | 10 | B |
| Mojave | 1957 | 7.5 | 162 | 30 | B |
| San Bernardino Mountains | 1812(?) | 7.5 | 198 | 20 | E |
| Coachella Valley | 1680±20 | 7.5 | 256 | 40 | C |
| | | Hayward fault | | | |
| Northern East Bay | 1836(?) | 7 | 209 | 20 | D |
| Southern East Bay | 1868 | 7 | 209 | 20 | C |
| | | San Jacinto fault | | | |
| San Bernardino Valley | 1890(?) | 7 | 203 | 20 | E |
| San Jacinto Valley | 1918 | 7 | 184 | 10 | C |
| Anza | 1892(?) | 7 | 142 | 30 | D |
| Borrego Mountain | 1968 | 6.5 | 189 | <10 | B |
| | | Imperial fault | | | |
| Imperial | 1979 | 6.5 | 44 | 50 | C |

SOURCE: From P. L. Ward and R. A. Page, *The Loma Prieta Earthquake of October 17, 1989*, U.S. Geological Survey Pamphlet.

studying seismic gaps and historical records of earthquakes that have occurred along that fault segment. Table 8-4 presents this type of analysis for major faults in California, and Figure 8-36 graphically illustrates the probabilities. Obviously, residents of the Parkfield area must become well prepared for a fairly large earthquake. That the Parkfield segment of the San Andreas fault is locked is quite clearly illustrated by the delineation of the Parkfield gap on Figure 8-25.

By plotting the numbers of earthquakes within specific time intervals against their magnitudes, diagrams similar to Figure 8-10 can be constructed for a local area. From this plot it is possible to determine the *recurrence interval*, or the average time interval between earthquakes of a specific magnitude. Predictions can then be made that an earthquake of that magnitude has a high probability of occurrence within a specified time interval, if the date of the last earthquake is known.

Research leading to short-term predictive ability, involving a time interval small enough for evacuation of an area, for example, has focused on *precursors* that have been observed prior to previous earthquakes. Precursors are physical or chemical phenomena that occur in a typical pattern before an earthquake. These phenomena include changes in the velocity of seismic waves, the electrical resistivity of rocks, the frequency of preliminary earthquakes

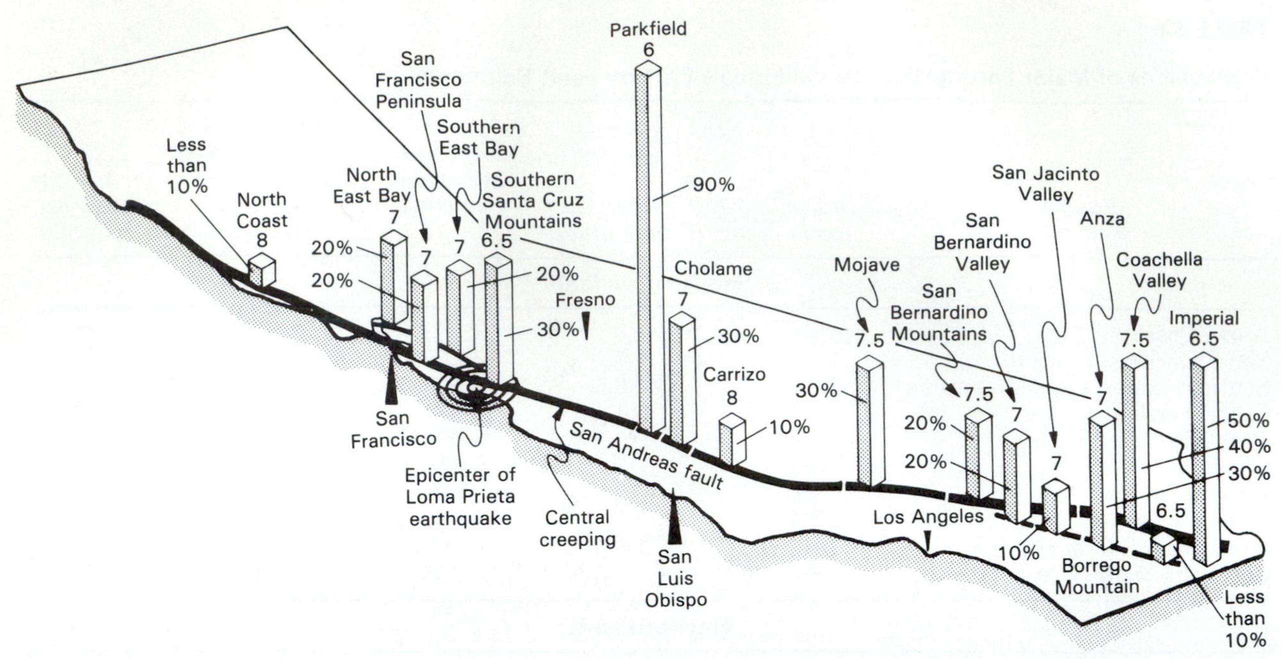


**Figure 8-36** Probabilities of earthquakes between 1988 and 2018 for segments of major faults in California. Box size is proportional to probability and expected magnitude is shown above each box. Probabilities were determined one year before Loma Prieta earthquake. (From P. L. Ward and R. A. Page, 1990, *The Loma Prieta Earthquake of October 17, 1989*, U.S. Geological Survey Pamphlet.)

(foreshocks), the deformation of the land surface, and the water level or water chemistry of wells in the area.

Many of these precursors can be explained by a theory called the *dilatancy model*. Under this hypothesis, rocks in the process of strain along a fault show significant dilation or swelling before rupture. This volume increase is caused by the opening of *microcracks*, which are minute failure zones in weaker mineral grains in the rock and along grain boundaries. As the porosity increases during the formation of microcracks, ground water flows into the highly stressed areas.

These changes in density and water content affect the ability of the rock to transmit seismic waves and conduct electricity. Therefore, seismic-wave velocity and electrical resistivity progressively change as the overall rupture along the fault draws near. Localized changes in land-surface elevation are also related to volume changes at depth. An area of recent uplift along the San Andreas fault near Los Angeles, which has been named the Palmdale Bulge (Figure 8-37), is being monitored in great detail as a possible indicator of a future earthquake.

Volume changes and ground-water movement may be reflected by changes in water levels in wells as well as by changes in the chemical composition of ground water. Radon gas has been observed to increase in wells prior to earthquakes. These increases are perhaps related to the release of radon gas from rocks during the formation of microcracks. The pattern of seismic activity in the vicinity of an area of imminent fault rupture is also significant. This pattern consists of an initial rise in the number of small events, followed by a decline in foreshocks just prior to the major earthquake. The decline may

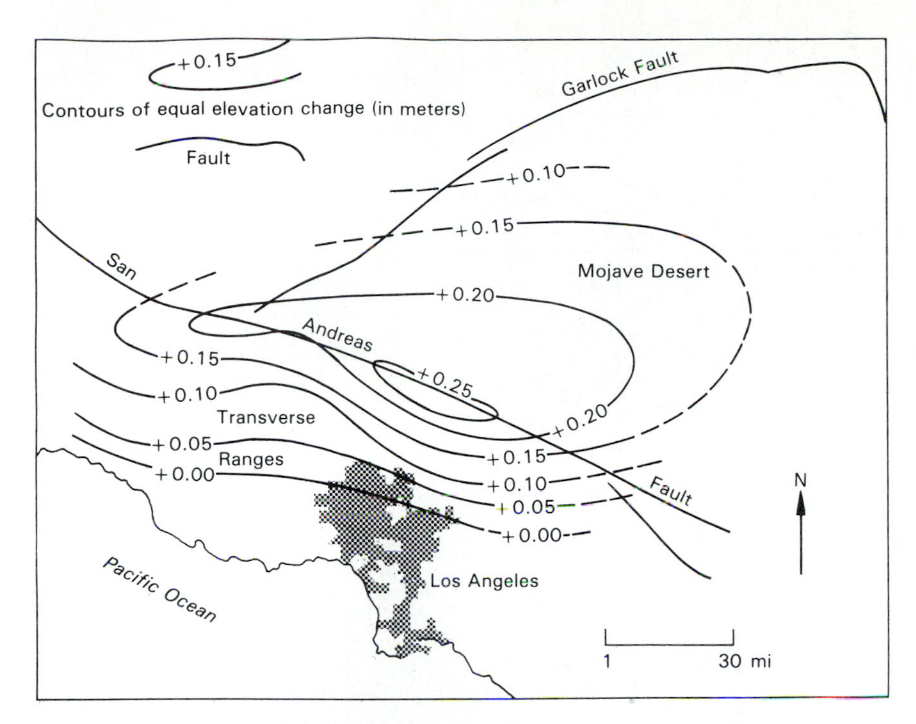


**Figure 8-37** Uplift along the San Andreas fault in the vicinity of Palmdale, California. This activity may be a precursor to a future earthquake. (From F. Press and R. Siever, *Earth*, 3d ed., copyright © 1982 by W. H. Freeman and Co., San Francisco.)

represent a temporary increase in rock strength before the newly formed microcracks are filled with water.

The precursor phenomena can be grouped into stages according to the dilatancy model (Figure 8-38). Stage I consists of a gradual stress buildup along the fault. Stages II and III are correlated with dilatancy and water influx. Stage IV is the major earthquake, and Stage V is the aftermath of the event.

If every earthquake followed the sequence shown in Figure 8-38, with uniform stage duration, earthquake prediction would be a simple matter. Instead, each earthquake is unique in terms of specific precursor behavior patterns and length of precursor stages. Two magnitude-5 earthquakes preceded the Loma Prieta event by 15 and 2 months. In each case a public advisory was issued stating that the earthquakes could be foreshocks to a stronger earthquake within 5 days. However, the fault did not cooperate and prediction was not successful. Continued research and study of future earthquakes will certainly lead to refinement of the dilatancy model, or to a replacement model with more accurate predictive capabilities.

The strange behavior of animals before an earthquake has been recognized for centuries and is now being studied scientifically for use in earthquake prediction. The range of unnatural behavior patterns is extremely diverse. Many animals seem restless and frightened. Snakes have come out of hibernation during the winter and frozen to death. Normal eating and sleeping patterns are abandoned. Although the exact means by which animals sense an impending earthquake are unknown, suggestions have included changes in subsurface water content, acoustic emissions, and electrostatic effects.

As a result of past earthquake disasters, the People's Republic of China

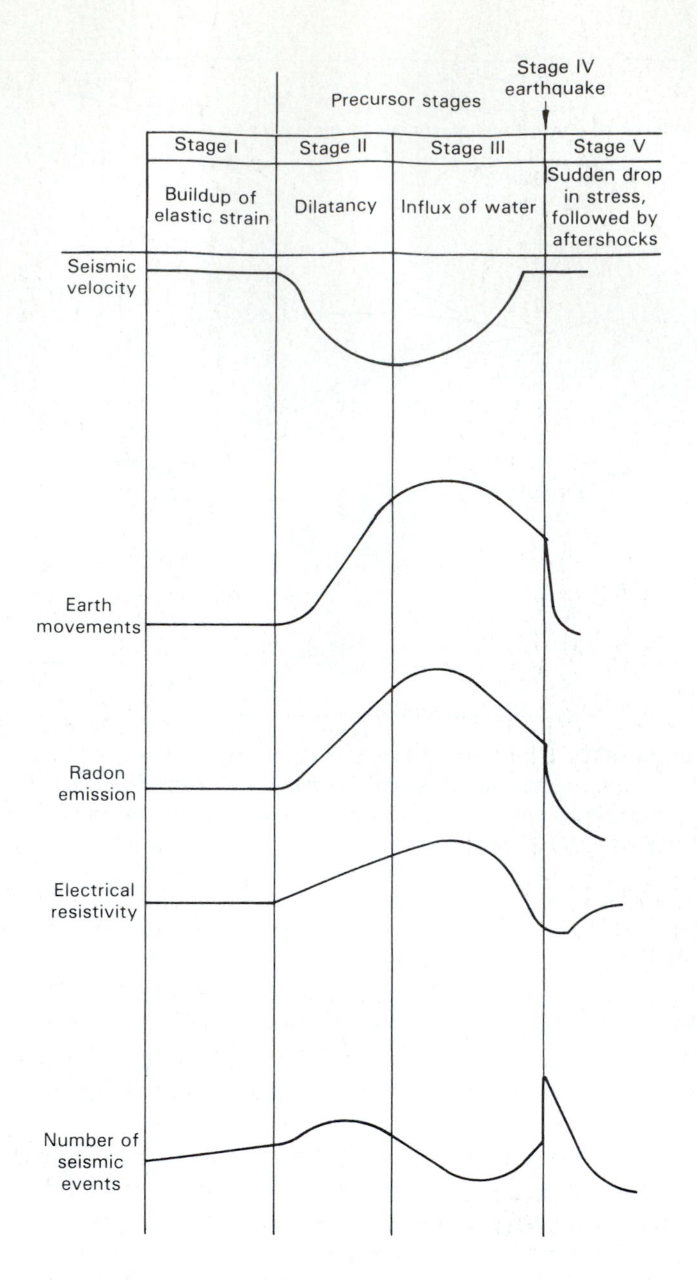


**Figure 8-38** Trends in precursor activity that have been related to major earthquakes by use of the dilatancy model. (From F. Press, Earthquake Prediction, *Scientific American*, May 1975, with permission.)

has developed the world's most advanced earthquake prediction program. Since the 1970s, a number of successful predictions have been made. In 1975, prediction of the Haicheng earthquake (magnitude 7.3) was responsible for saving tens of thousands of lives by evacuation from hazardous buildings. A notable failure, however, was the inability to predict the 1976 Tangshan event (magnitude 7.6), in which perhaps as many as three-quarters of a million people were killed.

**CASE STUDY 8-1**

## The San Fernando and Northridge Earthquakes: (Seismic) History Repeats Itself

Among recent earthquakes in the United States, the February 9, 1971, earthquake in San Fernando, California, and the nearby January 17, 1994, Northridge earthquake stand out for several reasons. First, both events struck the same outlying vicinity of a major metropolitan area, Los Angeles. Thus despite the relatively moderate magnitude of these events (Table 8-5), great damage was caused to the densely populated area. Second, although the earthquakes occurred in a region known to be seismically active, displacement took place along faults that either were not known or considered to be active prior to the earthquakes. Third, the network of seismograph stations that recorded the 1971 event provided the most detailed and abundant data ever gathered for an earthquake. These data provide many interesting facts and interpretations, including the highest ground accelerations ever recorded. These accelerations commonly ranged from 0.5 g to 0.75 g, with peak recorded values of more than 1 g. Accelerations of this magnitude are particularly significant because they exceed by several times the design values used for earthquake-resistant structures. Finally, the earthquakes provided a performance test for many modern buildings constructed under recent building codes. Unfortunately, many of the structures did not perform well (Table 8-6). Even worse, some critical highway bridges that were destroyed and rebuilt after the 1971 earthquake collapsed again in 1994.

The San Fernando earthquake was generated by movement along a number of minor faults near a bend in the San Andreas fault. The locations of the main shock and major aftershocks are shown in Figure 8-39. Fortunately, the earthquake took place about 6:00 A.M.; at this hour schools were unoccupied and streets and highways were nearly deserted. Had the earthquake struck just 1 h or 2 h later, the loss of life would have been much greater. Ground shaking from the main shock lasted about 1 min; the duration of strong motion was only about 10 s.

For a moderate event, the destruction inflicted by the San Fernando earthquake was quite striking. Sixty-four people died, and property damage estimates ran into the hundreds

**TABLE 8-5**

**Comparison of the San Fernando Valley and Northridge Earthquakes**

| | *1971 San Fernando Valley* | *1994 Northridge* |
|---|---|---|
| Deaths | 64 (most occurred in one building) | 51 (as of January 20, 1994) |
| Damage | $511 million | Estimates up to $30 billion |
| Magnitude | 6.5 | 6.6 |
| Hypocenter | 12–13 km | 14 km |
| Fault | Reverse fault. Rupture began at depth but ruptured the surface for 15 km. North dipping. | Previously unmapped blind reverse fault. Rupture began at depth and did not break surface. South dipping. |
| Ground movement | About 2 m of offset, primarily vertical, with some left lateral slip. | Approximately 1 m of vertical uplift. |

SOURCE: *Eos, Transactions of the American Geophysical Union,* v. 75, no. 4, January 25, 1994.

**TABLE 8-6**

**Estimate of Damage from the San Fernando Earthquake**

| *Structure* | *Number damaged* | *Amount* |
|---|---|---|
| Schools | 180 | $ 22,500,000 |
| Hospitals | 4 | 50,000,000 |
| Residential: | | |
| Homes | 21,761 | |
| Apartment houses | 102 | |
| Mobile homes | 1,707 | 179,500,000 |
| Commercial buildings | 542 | |
| Miscellaneous structures | 250 | |
| Highways and roads | | 27,500,000 |
| Dams | | 36,500,000 |
| Other public structures | | 145,000,000 |
| Utilities | | 42,000,000 |
| Personal property | | 50,000,000 |
| Total | | $553,000,000 |

SOURCE: From R. Kachadoorian, 1971, An estimate of damage in the San Fernando, California, earthquake of Feb. 9, 1971, in *the San Fernando, California, Earthquake of February 9, 1971*, U.S. Geological Survey and National Oceanic and Atmospheric Administration, 1971, U.S. Geological Survey Professional Paper 733.

NOTE: Data supplied by the city of Los Angeles, city of San Fernando, county of Los Angeles, Los Angeles Unified School District, U.S. Army Corps of Engineers, Los Angeles Food Control System, State of California, and region 7 of the Office of Emergency Preparedness.

of millions of dollars (Table 8-5). These damages would have seemed insignificant had the Lower Van Norman Dam failed (Figure 8-22), an event that was narrowly avoided. By comparison, the Northridge earthquake, an event of approximately the same magnitude only 16 km from the epicenter of the 1971 event, caused property damage that may amount to 60 times greater than the 1971 earthquake (Table 8-5). Aside from inflation, the dramatic increase in losses can be attributed to an explosion of population growth and development in the area between 1971 and 1994.

Surface ruptures associated with the earthquake were mapped in a zone approximately 15 km in length. The specific types of surface displacements exhibited a complex and variable pattern (Figure 8-19). Damage to buildings, roads, and utilities was intense in these areas (Figure 8-40).

Ground shaking provided the other main cause of damage. High-rise buildings in the area consisted mainly of medical facilities. Of four hospitals damaged, two were total losses. Parts of Veterans's Hospital, a pre-1930 nonreinforced concrete structure, collapsed, killing 45 people (Figure 8-41). The newly constructed $25 million Treatment and Care Facility at the Olive View Hospital also sustained severe damage and had to be demolished. The building was a five-story reinforced concrete structure. The first- and second-story columns were particularly hard hit (Figure 8-42). Two other relatively new hospitals were also severely affected.

One hundred eighty school buildings were damaged in the earthquake. Seismic standards were established for schools in California with the enactment of the Field Act in 1933. Schools built before the Field Act performed much less satisfactorily than post–Field Act buildings. Seven pre–Field Act schools in the Los Angeles Unified School District suffered sufficient damage to warrant demolition after the earthquake. Luckily, no schools were occupied at the time of the earthquake.

Other buildings erected prior to 1933 fared poorly during the shaking. Most of these, including Veteran's Hospital, were constructed of nonreinforced concrete.

Wood-frame buildings were damaged throughout the area of ground shaking, with more severe damage occurring in older houses that lacked lateral bracing and that may have been weakened by dry rot or termite damage. Two-story houses sustained more damage than

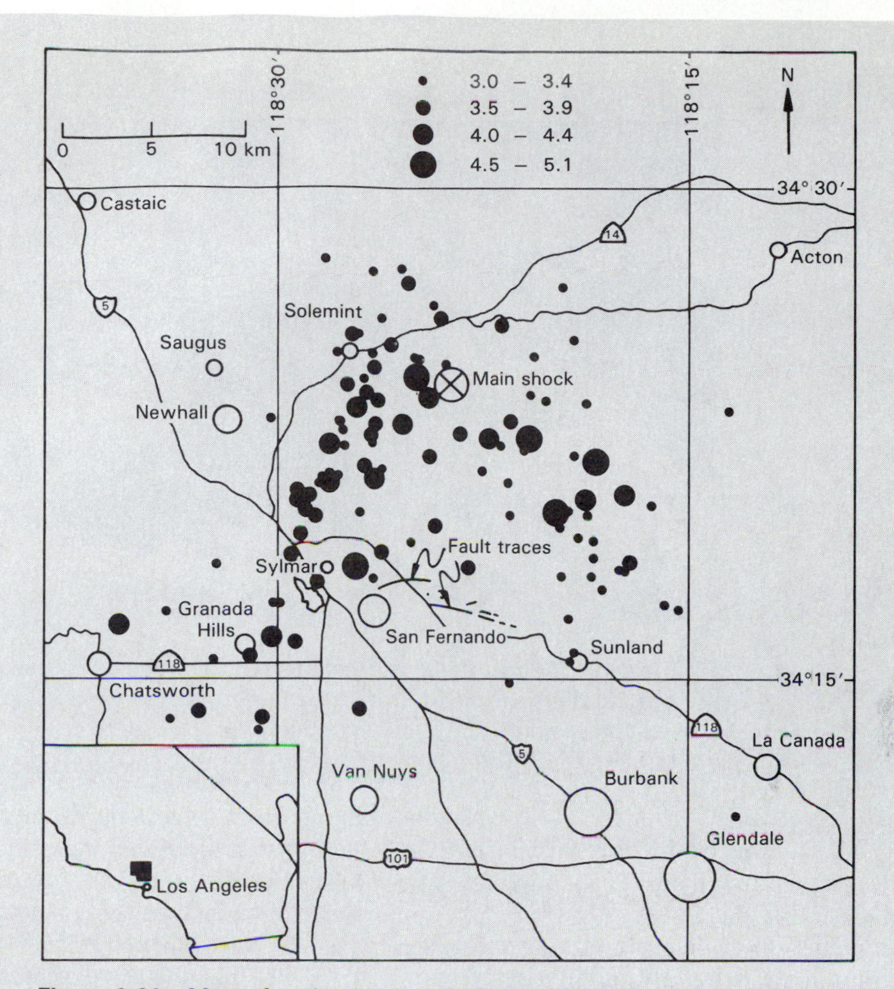


**Figure 8-39** Map of main shock and aftershocks greater than magnitude 3.0 for a 3-week period following the San Fernando earthquake. (From U.S. Geological Survey, 1971, *The San Fernando, California, Earthquake of February 9, 1971*, U.S. Geological Survey Professional Paper 733.)

**Figure 8-40** Damage to highway by surface rupture from the 1971 San Fernando, California, earthquake. (Photo courtesy of U.S. Geological Survey.)

**Figure 8-41** Cleanup following collapse of Veterans's Hospital during the San Fernando earthquake. Forty-five people were killed. (Photo courtesy of U.S. Geological Survey.)

one-story structures (Figure 8-43). Mobile homes were shaken from their foundations in many developments because of inadequate anchoring.

Damage to lifeline systems was extensive in the areas of strongest ground motion. Water-supply systems were disrupted in several cities, including the failure of dams, and breaks in water lines, tunnels, and aqueducts. The water distribution system for the city of San Fernando (population 17,000) was completely destroyed. Also affected were gas systems, sewage lines, and highways. Several major highway interchanges were blocked by overpass collapse (Figure 8-44). The failure of some of the same overpasses in 1994 was a sobering reminder that earthquake-proof design is not yet a reality in Southern California.

**Figure 8-42** Damaged column of reinforced concrete ambulance port at Olive View Hospital. (National Oceanic and Atmospheric Administration.)

**Figure 8-43** Damage to two-story house near San Fernando. (Los Angeles Dept. of Building and Safety.)

**Figure 8-44** Collapse of an overpass caused by the San Fernando earthquake. (Photo courtesy of U.S. Geological Survey.)

Overall, the San Fernando earthquake produced widespread damage that could have been inestimably worse if it had struck during midday or if the Lower Van Norman Dam had totally failed and released its reservoir upon the large population below. Similarly, the Northridge earthquake struck very early in the morning when schools and public buildings were unoccupied. Although the loss of life was severe, the vast majority of dwellings performed well enough to limit casualties to fewer than were sustained in 1971 with a lower population. We can only speculate what the cost in human lives would have been without the advances in building codes, land-use planning, and seismic monitoring that have taken place over the past few decades. We can only hope that the combination of improved building codes, strict land-use planning, and advances in earthquake prediction will minimize loss of life and damage from future earthquakes.

## SUMMARY AND CONCLUSIONS

When crustal blocks suddenly slip past one another, elastic strain energy is released in the form of seismic waves. According to the elastic rebound theory, elastic strain energy is gradually accumulated until brittle rupture occurs. Seismic waves are recorded by seismographs or, near the epicenter, as ground accelerations by accelerographs.

Earthquake magnitude is a quantitative measure of the amount of energy released, whereas intensity is a descriptive value assigned by the assessment of earthquake damage at a given location. Magnitude and frequency are inversely proportional; only a few great earthquakes strike each year over the entire earth.

Much information about earthquake processes and the earth's interior is gained by the study of seismic waves. The compressional (*P*) waves travel at the highest velocity; shear (*S*) waves travel more slowly and terminate abruptly at the earth's liquidlike outer core. Love and Rayleigh waves travel outward from the epicenter near the earth's surface. The *S-P* time interval is useful for determining the magnitude of the earthquake as well as the location of the epicenter.

A variety of hazardous processes are initiated by earthquakes. Surface rupture is common along the fault that generated the earthquake. Ground shaking, which affects a much larger area, is responsible for much of the damage to structures. Shaking also can cause various types of ground failure. When earthquakes take place offshore, tsunamis radiate outward to cause great damage in coastal areas. Finally, earthquakes are responsible for human catastrophes due to disease and fire when urban and suburban lifeline systems are severed.

Massive governmental and private efforts have been directed toward reducing earthquake damage. These programs include the identification of hazardous areas such as fault zones and the prediction of localized ground motion. Soil and rock types govern the response of a particular site to ground shaking. Specifically, buildings constructed on thick, soft soils sustain maximum damage when set into vibration. Risks can be reduced by zoning ordinances that prevent construction in high-hazard areas or specify construction techniques for critical structures.

Earthquake engineering is the analysis of the effect of vibrations upon structures and the design of earthquake-resistant buildings. Each building has a fundamental period of vibration. When this period corresponds closely to the period of the foundation materials, maximum damage is likely. Modern steel, wood-frame, and reinforced-concrete designs perform well during vibrations; heavy, rigid buildings including those of nonreinforced concrete, brick, and adobe fare poorly. Increasingly strict building codes eventually will prove worthwhile in savings of life and property.

Earthquake prediction holds promise for future damage reduction. It is hoped that the increasing sophistication of seismic detection instruments will allow the recognition of predictable patterns of precursor phenomena. These data, along with the study of animal behavior, someday may yield a method for preventing the great disasters to which some regions are now accustomed.

## PROBLEMS

1. Why doesn't the elastic rebound theory satisfactorily explain all earthquakes?
2. Describe the instruments used to measure earthquakes.
3. Contrast earthquake magnitude and intensity.
4. Why can't *S* waves travel through the earth's core?
5. A city near the epicenter of an earthquake sustains major damage to poorly constructed buildings, although hospitals and schools designed to resist damage from ground motion sustain only minor damage. Tall, thin structures such as chimneys and towers topple throughout the city, and newly deposited mounds of sand are found on the ground surface near the river. What was the intensity of the earthquake?
6. Based on Figure 8-9, what were the probable effects of the San Fernando earthquake in San Diego, California?
7. Using Table 8-1, estimate the distance between the earthquake epicenter and the point at which the seismogram shown in Figure 8-13 was recorded.
8. If the *S-P* time interval measured from a seismogram is 32 s and the maximum amplitude is 50 mm, what was the Richter magnitude of the earthquake?
9. How is a tsunami generated?
10. How do the geologic materials beneath a site influence its response to an earthquake?
11. A 12-story building is constructed in an area in which the soils are 600 ft deep. What percentage of structural damage can be expected?
12. Approximately how many earthquakes of magnitude 7 occur worldwide every year?
13. Suggest two likely locations for future earthquakes along the San Andreas fault south of San Francisco. How are these areas identified as possible earthquake epicenters?
14. What are the major areas of research concerned with earthquake predictions?
15. How might the effects of the San Fernando earthquake have differed if the magnitude were 8.0 instead of 6.5?

## REFERENCES AND SUGGESTIONS FOR FURTHER READING

BAROSH, P. J. 1969. *Use of Seismic Intensity Data to Predict the Effects of Earthquakes and Underground Nuclear Explosions in Various Geologic Settings.* U.S. Geological Survey Bulletin 1279.

BERLIN, G. L. 1980. *Earthquakes and the Urban Environment*, vols. 1, 2, 3. Boca Raton, Fla.: CRC Press.

BILLINGS, M. P. 1972. *Structural Geology*, 3d ed. Englewood Cliffs, N.J.: Prentice-Hall, Inc.

BLAIR, M. L., and W. E. SPANGLE. 1979. *Seismic Safety and Land-use Planning—Selected Examples for California*. U.S. Geological Survey Professional Paper 941-B.

BOLT, B. A. 1976. *Nuclear Explosions and Earthquakes: The Parted Veil*. San Francisco: W. H. Freeman.

BORCHERDT, R. D., ed. 1975. *Studies for Seismic Zonation of the San Francisco Bay Region*. U.S. Geological Survey Professional Paper 941-A.

COSTA, J. E., and V. R. BAKER. 1981. *Surficial Geology: Building with the Earth*. New York: John Wiley.

DEGENKOLB, J. J. 1977. *Earthquake Forces on Tall Structures*, Booklet 2717A. Bethlehem, Pa.: Bethlehem Steel Corp.

ECKEL, E. B. 1970. *The Alaska Earthquake March 27, 1964: Lessons and Conclusions*. U.S. Geological Survey Professional Paper 546.

ESPINOSA, A. F., ed. 1976. *The Guatemalan Earthquake of February 4, 1976: A Preliminary Report*. U.S. Geological Survey Professional Paper 1002.

HAYS, W. W., ed. 1980. *Procedures for Estimating Earthquake Ground Motions*. U.S. Geological Survey Professional Paper 1114.

HOUGH, S. E., P. A. FREGERG, R. BUSBY, E. F. FIELD, and K. H. JACOB. 1989. Did mud cause freeway collapse? *Eos, Transactions of the American Geophysical Union*, 70, no. 47:1497.

JUDSON, S., M. E. KAUFFMAN, and L. D. LEET. 1987. *Physical Geology*, 7th ed. Englewood Cliffs, N.J.: Prentice-Hall, Inc.

NICHOLS, D. R., and J. M. BUCHANAN-BANKS. 1974. *Seismic Hazards and Land-use Planning*. U.S. Geological Survey Circular 690.

OFFICE OF EMERGENCY PREPAREDNESS. 1972. *Disaster Preparedness*. Report to Congress of the United States, vol. 3.

PRESS, F. 1975. Earthquake prediction. *Scientific American*, 232, no. 5:14–23.

PRESS, F., and R. SIEVER. 1982. *Earth*, 3d ed. San Francisco: W. H. Freeman.

SEED, H. B., R. U. WHITMAN, H., DEZFULIAN, R. DOBRY, and I. M. IDRISS. 1972. Soil conditions and building damage in the 1967 Caracas earthquake. *Journal of the Soil Mechanics and Foundations Division*. American Society of Civil Engineers, SM-8: 787–806.

U.S. GEOLOGICAL SURVEY. 1976. *Earthquake Prediction: Opportunity to Avert Disaster*. U.S. Geological Survey Circular 729.

U.S. GEOLOGICAL SURVEY. 1982. *The Imperial Valley, California, Earthquake of October 15, 1979*. U.S. Geological Survey Professional Paper 1253.

U.S. GEOLOGICAL SURVEY, and NATIONAL OCEANIC AND ATMOSPHERIC ADMINISTRATION. 1971. *The San Fernando, California, Earthquake of February 9, 1971*. U.S. Geological Survey Professional Paper 733.

WALLACE, R. E. 1974. *Goals, Strategy, and Tasks of the Earthquake Hazard Reduction Program*. U.S. Geological Survey Circular 701.

WALLACE, R. E. 1984. *Faulting Related to the 1915 Earthquake in Pleasant Valley, Nevada*. U.S. Geological Survey Professional Paper 1274-A.

WALLACE, R. E., ed. 1990. *The San Andreas Fault System, California*. U.S. Geological Survey Professional Paper 1515.

WARD, P. L., and R. A. PAGE. 1989. *The Loma Prieta Earthquake of October 17, 1989*. U.S. Geological Survey Pamphlet.

WIEGEL, R. L., ed. 1970. *Earthquake Engineering*. Englewood Cliffs, N.J.: Prentice-Hall, Inc.

YOUD, T. L., and S. N. HOOSE. 1978. *Historic Ground Failures in Northern California Triggered by Earthquakes*. U.S. Geological Survey Professional Paper 993.

# 9

# Weathering and Erosion

We have followed the various igneous, metamorphic, and sedimentary processes leading to the formation of rocks on the surface and below the surface of the earth's crust. We have seen how tectonic forces driven by the earth's internal energy elevate deeply formed rocks to the earth's surface. Progression through the geologic cycle now brings us to the processes acting upon, or just below, the surface of the earth that are associated with the action of water, ice, wind, and gravity. These processes, fueled by external energy derived from the Sun, are components of the hydrologic cycle. Together, they lead to the disintegration and decomposition of continental rocks, followed by the transportation of debris and chemical elements to the ocean basins, where sedimentation precedes the ultimate formation of new continental rocks.

*Weathering* is the mechanical and chemical breakdown of rocks exposed at the earth's surface into smaller particles that may differ in composition from the original substance. Although the mechanical and chemical aspects of weathering will be discussed separately, they occur simultaneously and are related in many ways.

Weathering processes are of interest and importance to a broad range of scientists and engineers. Soil scientists study weathering processes in relation to the genesis of agriculturally productive soils from rocks and unconsolidated deposits. Hydrogeologists investigate the chemical constituents that are released by weathering reactions in the upper few meters of the soil zone and are then carried downward to influence the chemical composition of ground water. Geologists and engineers have demonstrated the role of weathering in landslides and other types of slope failure. In the construction industry, the effect of weathering on building stone is of great concern, an effect obvious to anyone who has observed the faded and indistinct inscriptions on old tombstones. The deterioration of building stone, road aggregate, and other construction materials is a serious and expensive consequence of weathering phenomena. Perhaps the most important influence of weathering is its relevance to the design and performance of structures built upon weathered rock. Weathering gradually weakens rock; in some climates the end result is a deep mantle of residual soil with vastly different engineering properties from the parent rock. The effects of weathering must be assessed at every foundation site where design requires the presence of predominantly unaltered rock.

The types and intensity of weathering processes that occur in a particular area are primarily the result of climate. Temperature controls the mechanical processes caused by the freezing and thawing of water, as well as influencing the rates of chemical weathering reactions. The other main climatic variable is precipitation, because water plays a part in both physical and chemical weathering processes. The dominant type of weathering that can be expected in a particular climatic region is indicated in Figure 9-1, which utilizes mean annual rainfall and mean annual temperature to characterize climate.

## MECHANICAL WEATHERING

Mechanical weathering includes processes that fragment rocks into smaller particles by exerting forces greater than the strength of the rock. Usually, these forces act within rock masses to overcome the tensile strength of the rock, which is much lower than its compressive strength. The principal mechanical and chemical weathering types are shown in Table 9-1.

*Frost action* is one of the most effective mechanical weathering processes. When water freezes within a rock mass, its volume expansion exerts pressure

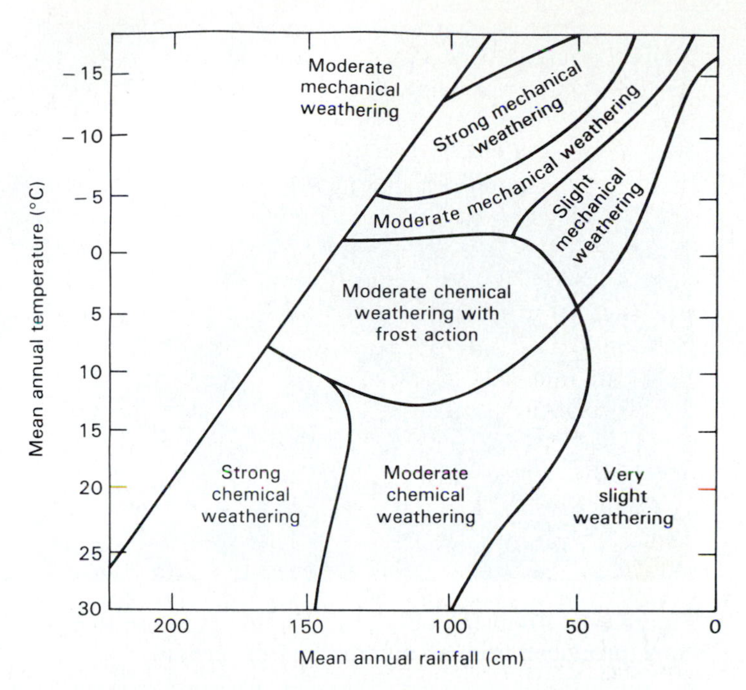


**Figure 9-1** Climatic influences on types of weathering processes. (From Peltier, 1950; reproduced by permission from the *Annals of the Association of American Geographers*, 40:219, Fig. 3.)

**TABLE 9-1**

**Mechanical and Chemical Weathering Processes**

Mechanical
- Frost action
- Salt weathering
- Temperature changes
- Moisture changes
- Unloading
- Biogenic processes

Chemical

Solution:

$$CaCO_3 + H_2CO_3 \rightarrow Ca^{2+} + 2HCO_3^-$$

(Calcite) (Carbonic acid) (Calcium ion) (Bicarbonate ion)

Hydrolysis:

$$2KAlSi_3O_8 + 2H^+ + 9H_2O \rightarrow Al_2Si_2O_5(OH)_4 + 4H_4SiO_4 + 2K^+$$

(Orthoclase) (Kaolinite) (Dissolved silica) (Dissolved potassium)

Hydration:

$$CaSO_4 + 2H_2O \rightarrow CaSO_4 \cdot 2H_2O$$

(Anhydrite) (Gypsum)

Oxidation:

$$4FeSiO_3 + O_2 + H_2O \rightarrow 4FeO(OH) + 4SiO_2$$

(Pyroxene) (Limonite) (Dissolved silica)

**Figure 9-2** Angular blocks of rock in the Madison Range, Montana, produced from an outcrop by frost action.

greater than the tensile strength of most rocks. Often, water penetrates into existing cracks and joints in the rock. When it freezes, the rock walls on opposite sides of the joint are forced apart. The size and shape of the angular rock fragments that are produced depends upon the joint spacing and orientation in the rock (Figure 9-2). On slopes, rock fragments produced in this manner may fall, slide, or roll to lower elevations where they form a continuous sloping mantle of loose rock called a *talus* slope (Figure 9-3). Water within individual

**Figure 9-3** Accumulation of loose rock on a talus slope, southeastern Iceland.

pores of a rock may also fracture the rock during freezing, but this is a complex and poorly understood process.

Crystallization of salt within a rock is somewhat similar to the effects of freezing water. The expansion and weakening of the rock caused by salt crystallization is called *salt weathering,* and it presents one of the most serious threats to the durability and appearance of stone buildings. The salt that crystallizes from solution in the rock includes not only sodium chloride but also a number of chloride, sulfate, and carbonate salts that can enter the rock in several ways. Sedimentary rocks may retain salts from their original deposition. Other dissolved constituents may be contributed by chemical weathering processes. The salt may also be introduced after the rock is quarried. Polluted urban air and rainfall, for example, are excellent sources of salt.

Salt can crystallize from an aqueous solution as a crust on the rock surface or within rock pores and voids. Often, a concentration of salt forms in a thin layer just below the rock surface. In this case, the surficial layer gradually deteriorates until it splits, or spalls, off from the rock face, at which time the process begins again. Inside the rock, salt crystals can weaken the rock in at least three ways. The first way is the pressure and expansion caused by the initial growth of the salt crystal. Second, upon heating, the salt crystals expand more than the surrounding minerals of the host rock, so that pressure is created. And third, salt crystals tend to absorb water into their structure to form hydrated crystals. This entails a volume expansion and an increase in pressure upon the adjacent grains, which can reach extremely high values (Figure 9-4). These hydration pressures can at times exceed the tensional strength of the rock, causing spalling of thin, exterior layers. Unfortunate demonstrations of salt weathering were initiated by the removal of carved stone monuments from Egypt and other arid regions and their relocation to more humid climatic regimes in Europe and the United States. Inscriptions that were clearly readable after thousands of years in their original climate are destroyed in a hundred years of weathering in cool, humid climates.

*Temperature changes* have long been suspect as a weathering mechanism because of differential thermal expansion and contraction of the minerals within a rock. Although lab experiments tended to dispute the effectiveness of heating and cooling in rock weathering, more recent work suggests that the experiments did not fully duplicate natural conditions and that the mechanism may be significant.

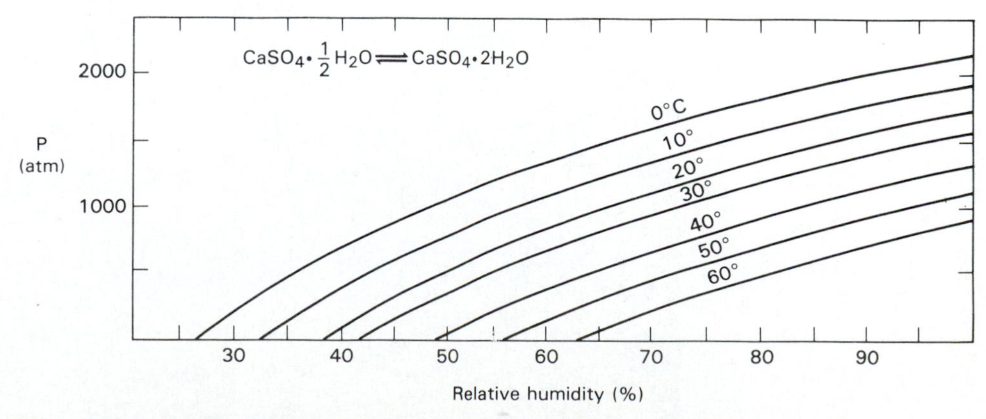


**Figure 9-4** Hydration pressures produced by hydration of $CaSO_4 \cdot \frac{1}{2}H_2O$. (From E. M. Winkler and E. J. Wilheim, 1970, *Geological Society of America Bulletin*, 81:567–572.)

*Moisture changes*, in the form of alternate wetting and drying, have experimentally been shown to cause expansion and contraction that may lead to weakening of the overlying rock. This mechanism may be even more effective when combined with temperature changes.

A peculiar type of weathering develops on natural rock faces composed of rock that is susceptible to spalling. In a process that is not as yet totally understood, rounded cavities or hollows form and gradually enlarge. The hollows may be found individually, or in a dense network of closely spaced holes (Figure 9-5). A variety of names has been applied to these features, but *honeycomb weathering* is a useful descriptive term for examples such as seen in Figure 9-5. The origin of the pits and hollows probably involves the breakdown of a protective mineral coating composed of iron and manganese oxides and the spalling off of thin layers of rock. Moisture and wind may then be involved in removal of detached grains. The action of lichens and other organisms has also been invoked in some theories. Although honeycomb weathering is not restricted to dry climates, the fact that it is common in arid regions suggests that mechanical weathering plays a significant role in its formation.

Rock, like any other material, deforms under the application of stress. Deeply buried rock experiences both elastic and plastic compression under the weight of the overlying rocks. Upon removal of the overburden, which could occur during stream erosion of a valley or by deep excavation for a construction project, the elastic component of the deformation is recovered and the rock expands. The expansion caused by *unloading* is often sufficient to fracture the rock along planes parallel to the surface on which stress has been released. This type of weathering, also called *exfoliation*, produces thin slabs of rock bounded by joints oriented parallel to the ground surface as the overlying material is removed and stress is released (Figure 9-6). The jointed rock slabs generated by exfoliation are susceptible to further weathering and erosional

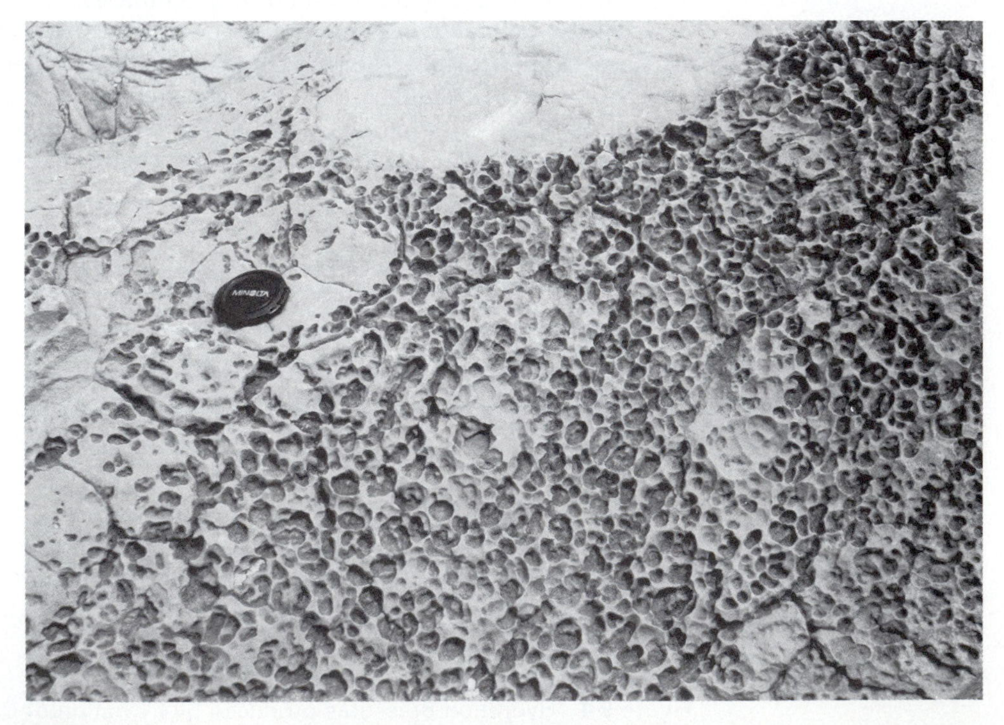

**Figure 9-5** Honeycomb weathering, Meteor Crater, Arizona. (Camera lens cap for scale.)

**Figure 9-6** Exfoliation results in the separation of thin slabs of rock from a rock mass along planes parallel to the surface. Sierra Nevada Mountains, California. (N. K. Huber; photo courtesy of U.S. Geological Survey.)

processes. As the rock slabs are gradually removed from the slope, the unloading action continues and new joints and slabs form. A process that produces a somewhat similar result is called *spheroidal weathering*, in which thin concentric layers form and gradually split off the outside of weathering boulders. Unlike exfoliation, the forces that cause the curved slabs to deteriorate and

**Figure 9-7** Rounded boulders (upper surface) produced by spheroidal weathering. Road cut below exposes rounded corestones surrounded by weathered residual soil. Near Lake Tahoe, California.

break away from the central core are due to chemical weathering reactions. Exposures of deeply weathered granitic rock commonly contain spheroidal cores of unweathered rock surrounded by weathered rock (Figure 9-7). Over time, the cores are gradually decreased in size by spheroidal weathering.

Organisms participate in both mechanical and chemical weathering processes. The mechanical effects are dominated by plant roots that exploit thin joints or cracks in rock as they grow. The pressures exerted by the growing roots can wedge apart blocks of rock, leading to acceleration of other weathering processes in the larger openings.

## CHEMICAL WEATHERING

Most rocks originally formed under conditions very different from those that exist at the earth's surface. In particular, the igneous and metamorphic rocks crystallized at very high temperatures and pressures. When these rocks are exposed to the lower temperatures and pressures present at the surface, they are unstable and tend to chemically react with components of the atmosphere to form new minerals that are more stable under those conditions. The most important atmospheric reactants are oxygen, carbon dioxide, and water. In polluted air, however, other reactants are available.

Figure 9-8 illustrates conditions in the zone of plant growth—the upper meter or so of the soil zone—as rainfall or snowmelt percolates downward from the surface. Decaying organic matter in this zone, under aerobic conditions, generates carbon dioxide. For example, the reaction

$$O_2 + CH_2O = CO_2 + H_2O \qquad \text{(Eq. 9-1)}$$

indicates that a carbohydrate ($CH_2O$) will react with oxygen to produce carbon dioxide. Carbon dioxide in the soil zone can reach levels of several orders of

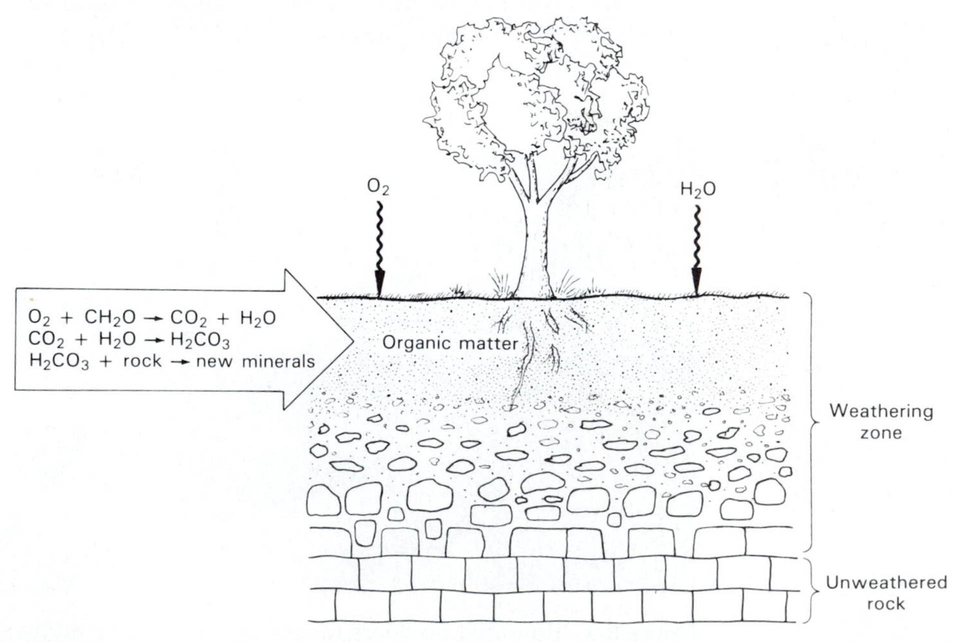


**Figure 9-8** Production of acidic weathering solutions by formation of $CO_2$ in the soil zone.

magnitude greater than its concentration in the atmosphere, and then react with infiltrating rainfall or snowmelt by the reaction

$$H_2O + CO_2 \rightleftharpoons H_2CO_3 \qquad \text{(Eq. 9-2)}$$

to produce carbonic acid. Production of carbonic acid lowers the pH by partial dissociation according to the reaction

$$H_2CO_3 \rightleftharpoons H^+ + HCO_3^- \qquad \text{(Eq. 9-3)}$$

The resulting increase in hydrogen-ion concentration (lower pH) causes a more intense attack on minerals by the solution. Decomposition of organic matter can also produce *humic* acids which, like carbonic acid, yield hydrogen ions to lower the pH.

The specific types of weathering reactions that have been recognized are listed in Table 9-1 along with example chemical reactions. When a mineral completely dissolves during weathering, the reaction is known as *solution.* The tendency of a mineral to dissolve in weathering solutions is described as its *solubility.* Evaporite minerals dissolve readily in water, whereas carbonate minerals are somewhat less soluble. Even so, solution of limestone and dolomite is extremely important. When limestone is exposed at the surface, pitting and etching of the rock is obvious (Figure 9-9). Even more significant is the subsurface solution of carbonate rocks to form caves and a type of landscape known as *karst* topography. Karst differs from other landscapes in that surface streams are rare and drainage takes place underground. The characteristics of karst areas will be described in Chapter 11. Some silicate minerals weather by solution, but their solubilities are very low. Quartz, for example, usually dissolves appreciably only under the intense weathering conditions found in

**Figure 9-9** Solution of exposed limestone outcrop to produce etched, pitted surface. Lee's Ferry, Arizona.

the tropics. Quartz differs from the carbonate minerals in that its solubility increases as the pH increases.

*Hydrolysis* is the reaction between acidic weathering solutions and many of the silicate minerals, including the feldspars. The reaction illustrated in Table 9-1 indicates that in hydrolysis a feldspar mineral reacts with hydrogen ions to form several dissolved products as well as a solid product, the clay mineral kaolinite. Other clay minerals are produced by similar weathering reactions. This reaction is an example of the breakdown of a mineral that is stable at high temperature and pressure to form a new mineral that is stable under conditions near the earth's surface. Notice that hydrolysis produces dissolved silica and potassium in the orthoclase example. These constituents are carried downward out of the weathering zone in the water in which they are dissolved toward the water table, where they become part of the groundwater flow system. Hydrolysis reactions are responsible for producing deposits of clay that are mined for use in many industrial processes.

The adsorption of water into the lattice structure of minerals is called *hydration*. The formation of gypsum from anhydrite by hydration is illustrated in Table 9-1. Clay minerals are also susceptible to hydration. The volume expansion that accompanies hydration is an important contributor to the physical weakening and breakdown of a rock. The role of hydration in salt weathering was previously described.

The reaction of free oxygen with metallic elements is familiar to everyone as rust. This process, an example of *oxidation*, affects rocks containing iron and other elements. In an oxidation reaction, iron atoms contained in minerals lose one or more electrons each and then precipitate as different minerals or amorphous substances. For example, in the weathering of pyroxene by oxidation, as illustrated in Table 9-1, iron is oxidized and hydrated to form the mineral limonite. The presence of limonite in rocks or soils is indicated by brownish or reddish staining.

## STABILITY

The stability of minerals under the action of chemical weathering agents at the earth's surface depends upon the difference between the conditions at the surface and the conditions under which the mineral originally crystallized. Minerals occurring in igneous rocks can be grouped into a sequence of relative weathering stability that is the exact opposite of their order of crystallization from the magma or lava (Figure 9-10). This relationship is known as *Goldich's stability series*. For example, olivine is the first mineral to crystallize from a silicate melt and is therefore the most unstable mineral in a weathering environment. Olivine will weather rapidly when exposed to the atmosphere. Quartz, the last mineral to crystallize from a magma, is very stable in most weathering environments. It dissolves so slowly that, for all practical purposes, it is considered to be insoluble except under humid, tropical conditions. The stability of quartz explains why it is so common in sedimentary rocks such as sandstone. Feldspars and ferromagnesian minerals from the original rocks weather relatively rapidly to clay minerals, leaving quartz to be transported and deposited in depositional environments. The clay minerals that form by the weathering of feldspars and other minerals are stable weathering products that are deposited as muds in low-energy depositional environments and that later are consolidated to form shales.

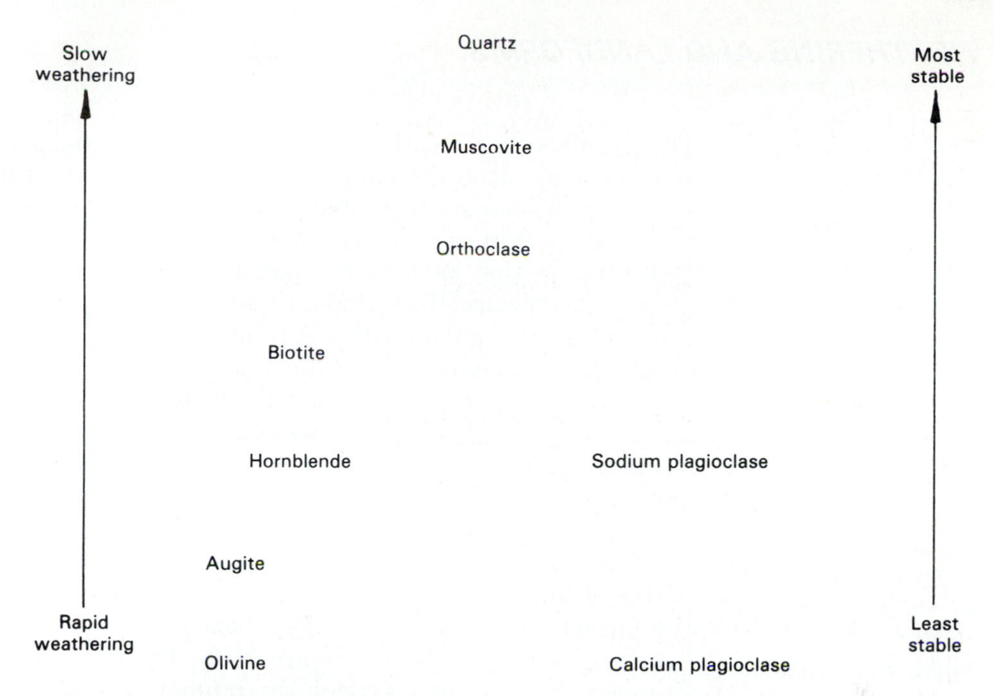


**Figure 9-10** Goldich's stability series, which shows the relative stability under weathering conditions of primary minerals in igneous rocks.

The weathering products of rock-forming minerals vary as a function of climate and other factors. Table 9-2 shows the secondary mineral composition of residual soils weathered from rocks of initially similar composition in three different types of climate in Australia. The intense leaching of the humid tropical climate left a weathering residue dominated by iron and aluminum-hydroxide secondary minerals (goethite and gibbsite, respectively). Most of the quartz was removed from the original rocks. Weathering products in the hot arid and humid temperate climates contain greater percentages of clay minerals and quartz. Determination of the mineralogical weathering products is important because these minerals influence the strength and other engineering properties of residual soils.

**TABLE 9-2**

**Mineral Composition of Weathered Rocks as a Function of Climate**

| *Climate* | *Hot arid* | *Humid tropical* | *Humid temperature* |
|---|---|---|---|
| Mineral | | Percentages | |
| Quartz | 28 | 8 | 50 |
| Illite | 32 | — | 24 |
| Chlorite | 15 | — | — |
| Montmorillonite | 12 | — | 7 |
| Kaolinite | — | 12 | 14 |
| Gibbsite | — | 68 | — |
| Geothite | 5 | 10 | — |
| Gypsum | 5 | — | — |
| Calcite | 2 | — | — |
| Others | 1 | 2 | 5 |

SOURCE: From F. C. Beavis, *Engineering Geology*, copyright © 1985 by Blackwell Scientific Publications, Inc., Melbourne.

## WEATHERING AND LANDFORMS

The significance of weathering, as can be surmised from the preceding discussion, depends upon the climate and lithology of the rocks being weathered. When conditions are favorable, and sufficient time is available, weathering can produce distinctive landscapes. To illustrate, we will focus upon igneous rock terrains that have been exposed to active chemical weathering for long periods of geological time. Under these circumstances, the effects of weathering can penetrate to great depths. Alteration of the granitic rocks decreases with depth until a boundary called the *weathering front* is reached. This boundary, which can be sharp or transitional, separates weathered from unweathered rock (Figure 9-11). When the spacing of joints is irregular, the depth of the weathering front can be highly variable, penetrating to great depths where joints are closely spaced, and extending to much shallower depths where joints are far apart. If such an area is subjected to increased erosion, perhaps due to uplift or climate change, the weathered residual materials are removed, leaving plains of low relief broken only by isolated hills of granite, which represent the former zones of shallow subsurface weathering. These striking residual hills are called *inselbergs* (Figure 9-12). Inselbergs can be composed of any lithology, but they are particularly common in arid, granitic terrains. Figure 9-13 illustrates the presumed steps in the formation of such a landscape.

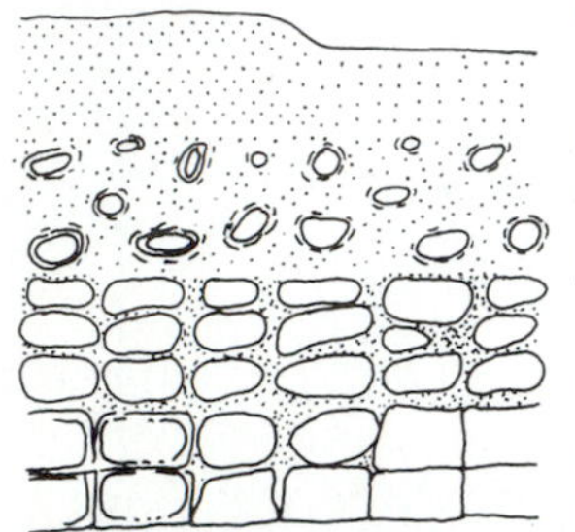

**Figure 9-11** Weathering front developed on granite. Material changes from residual sandy clay at top of section to increasing proportion of unaltered rock in corestones with increasing depth. Jointed rock at base of section is only slightly weathered. Depth to weathering front may be tens of meters. (After Gerrard, 1988.)

**Figure 9-12** Inselberg composed of granitic rocks projecting above plain. Formation illustrated in Figure 9-13. Joshua Tree National Monument, California.

## ROCK WEATHERING AND ENGINEERING

The degree and pattern of weathering are among the most important factors to be determined in an on-site engineering investigation. The effects of weathering on the engineering properties of rocks include a decrease in strength, a loss of elasticity, a decrease in density, and increases in moisture content and porosity. These detrimental changes can be critical to the suitability of a site for structures such as arch dams, which require maximum strength and elasticity.

The subsurface distribution of joints and faults is particularly important to determine in on-site engineering investigation because of the control these discontinuities exert upon the depth of the weathering front. Rock masses may be deeply weathered along fractures and unaltered in intact rock at shallow depth between the fractures. Weathering conditions at an underground hydroelectric power plant in Australia are shown in Figure 9-14. Weathering zones are classified as fresh (F), slightly weathered (SW), highly weathered (HW), and completely weathered (CW). Tunnel and underground excavations for the project were sited in order to avoid faults and associated deep-weathering zones.

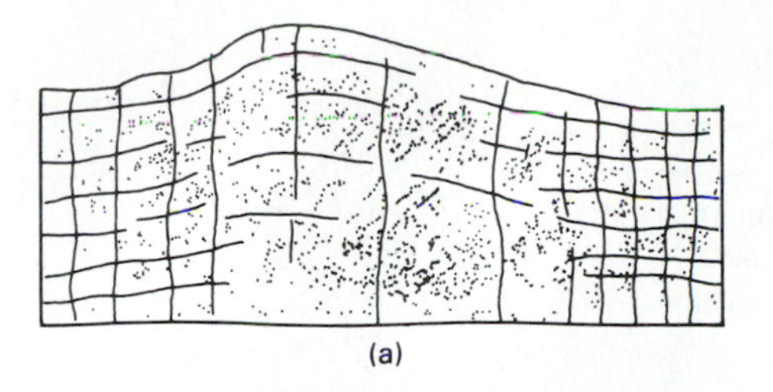

(a)

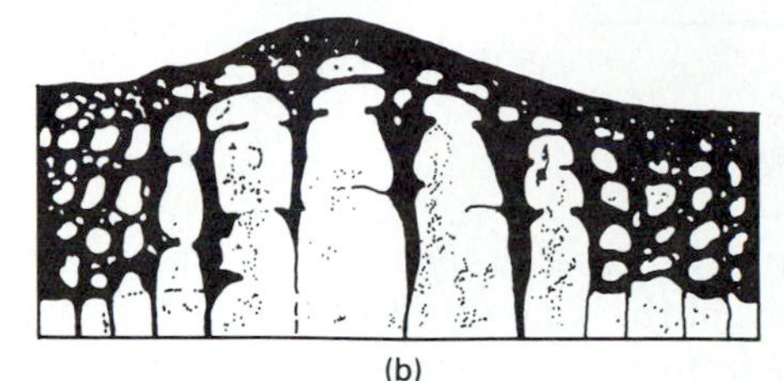

(b)

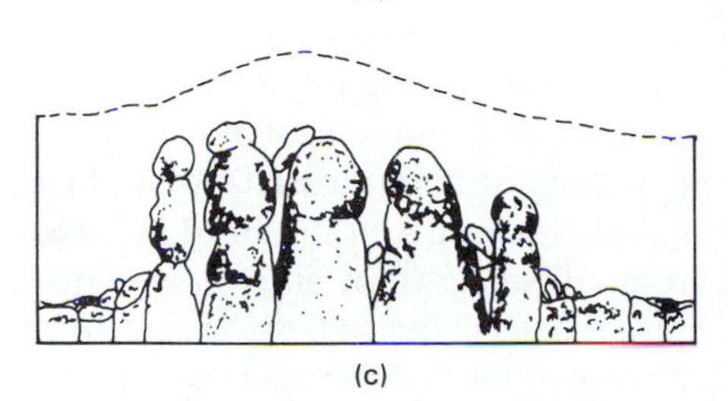

(c)

**Figure 9-13** Formation of inselbergs by two-stage process: (a) vertical section through granitic rocks with varied spacing of joints; (b) deep weathering under humid or subhumid climate; (c) erosion of residual soil under arid conditions leaving isolated remnants of unweathered rock. (Modified from D. D. Trent, 1984, Geology of the Joshua Tree National Monument, *California Geology*, 37:75–86.)

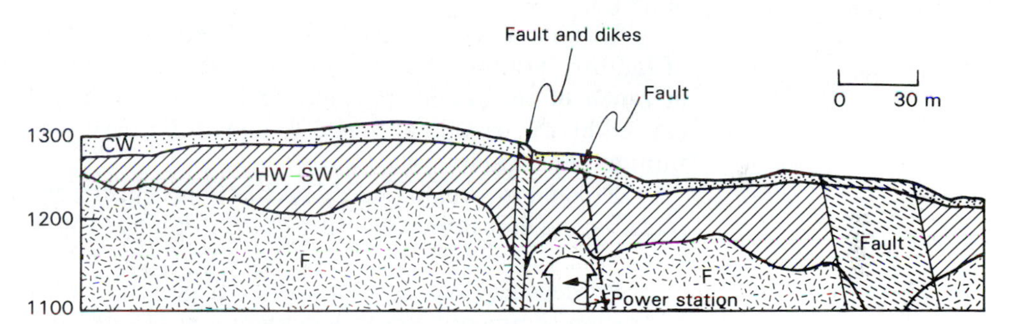


**Figure 9-14** Degree of rock weathering at the site of an underground power station in Australia. Weathering zones include fresh rock (F), slightly weathered rock (SW), highly weathered rock (HW), and completely weathered rock (CW). (From F. C. Beavis, *Engineering Geology*, copyright © 1985 by Blackwell Scientific Publication, Inc., Melbourne.)

## EROSION

The chemical and physical breakdown of rocks by weathering processes leads to the formation of materials that can be easily transported by processes operating at the earth's surface. *Erosion* is the detachment and transportation of surface particles under the action of one or more of the forces and agents listed in Table 9-3. The materials that are susceptible to erosion include not only weathered residual products developed from rocks but any unconsolidated surficial deposit. We will refer to these materials as soil in the engineering sense, that is, any unconsolidated material regardless of thickness or origin.

**TABLE 9-3**

| Erosional Agents |
|---|
| Gravity (mass wasting) |
| Water (fluvial erosion) |
| Wind (eolian erosion) |
| Glacial ice |
| Waves |

### *Erosion by Water*

Throughout the world, erosion by running water is the most important type of erosion in terms of the amount of sediment removed from the land surface. Sediment eroded from slopes is subsequently transported by streams to the oceans. The amount of slope erosion is a complex function of rock type, climate, and topography. A dramatic example of erosion is shown in Figure 9-15. The slope in the background is composed of a bed of resistant conglomerate at the top of the slope underlain by nonresistant shale. In the arid climate of the area, erosion of the shale is rapid and the slope is continually retreating. Blocks of conglomerate break off and roll or slide downslope as the underlying shale is eroded. When these blocks come to rest, they protect the shale beneath from erosion. As the shale continues to be eroded outside the protected area, *pedestals* are formed which consist of conglomerate blocks supported by thin columns of shale.

Erosion of land altered by human activity is much more rapid than erosion of land in its natural state. Unfortunately, agricultural lands are the source of much of the sediment removed by erosion. The loss of soil productivity caused by the erosion of topsoil is one of the most serious problems facing the human population.

Erosion in stream and river channels is also a significant problem. Bridge piers, levees, and other structures can be damaged or destroyed by river erosion.

Processes. Soil erosion by water occurs in a variety of ways. Initiation of erosion commonly includes detachment of soil particles from the surface by

**Figure 9-15** Erosional retreat of slope composed of shale with resistant conglomerate at top. Conglomerate blocks protect shale from weathering to form pedestals. Lee's Ferry, Arizona.

**Figure 9-16** Severe rill erosion on a steeply sloping crop field. Notice sediment deposition at base of slope. (Photo courtesy of USDA Soil Conservation Service.)

*raindrop erosion*. The energy transferred to the surface by the impact of raindrops is sufficient to dislodge individual soil particles from particle aggregates at the soil surface. Downslope movement of particles also begins by the splash of the raindrop.

In the early stages of a rainstorm, most of the precipitation infiltrates into the ground. If the duration and intensity of the precipitation are sufficient, a saturated zone develops just below the soil surface and water begins to pond on the surface. We will discuss these relationships in more detail in Chapter 11. When ponded water reaches a depth greater than the height of surface irregularities, water flows downslope in a thin, wide sheet as *overland flow*. You may have noticed this type of flow on gently sloping, paved parking lots. On soil surfaces, this shallow flow of water can transport detached soil particles and is given the name *sheet erosion*.

Sheet erosion is a rather rare and inefficient erosional process and is soon replaced by *rill erosion* (Figure 9-16). The channelized flow of water in rills has a higher velocity and a greater ability to transport particles downslope. With time, rills enlarge into gullies (Figure 9-17), which gradually extend themselves *headward* (upstream) and may render the land unfit for agriculture.

Not all erosion takes place at the surface. When sediments rich in swelling clays (Chapter 10) are exposed in valley slopes, vigorous erosion can take place below the surface, a process called *piping*. Although piping has been observed in many climatic settings, it is particularly prevalent on the steep, dry slopes of arid regions. The combination of clay-rich soils and active erosion both above and below land surface leads to a distinctive type of topography called *badlands*. The term refers to a deeply eroded landscape riddled with steep, bare slopes that is exceptionally unpleasant to cross on foot or horseback if you have to go anywhere in a hurry. The exposed clay surfaces expand when wet and crack as they shrink and dry. Over time, cracks are enlarged into a network of

**Figure 9-17** Progressive loss of agricultural or grazing land is caused by development and enlargement of gullies. (Photo courtesy of USDA Soil Conservation Service.)

tunnels just under the surface that convey water and sediment to exit holes near the base of the slopes (Figure 9-18). Piping can lead to costly damage when it develops under roads and other structures. Enlargement of the pipes eventually leads to roof collapse and surface subsidence, affecting any structures which may lie above the pipes (Figure 9-19).

**Figure 9-18** Piping in clayey slopes in an arid climate. Man points to exit hole near base of slope. Sediment transported through pipe visible at base of slope. Petrified Forest National Park, Arizona.

Erosional Rates. Attempts have been made to predict the amount of erosion that will occur within a given amount of time for various types of land settings. These methods yield only rough estimates because of the large number of variables in the soil erosion process. Some of the main variables are listed in Table 9-4. Some of these factors have been combined into the *Universal Soil Loss Equation*, which is stated

$$A = RKLSCP \qquad \text{(Eq. 9-4)}$$

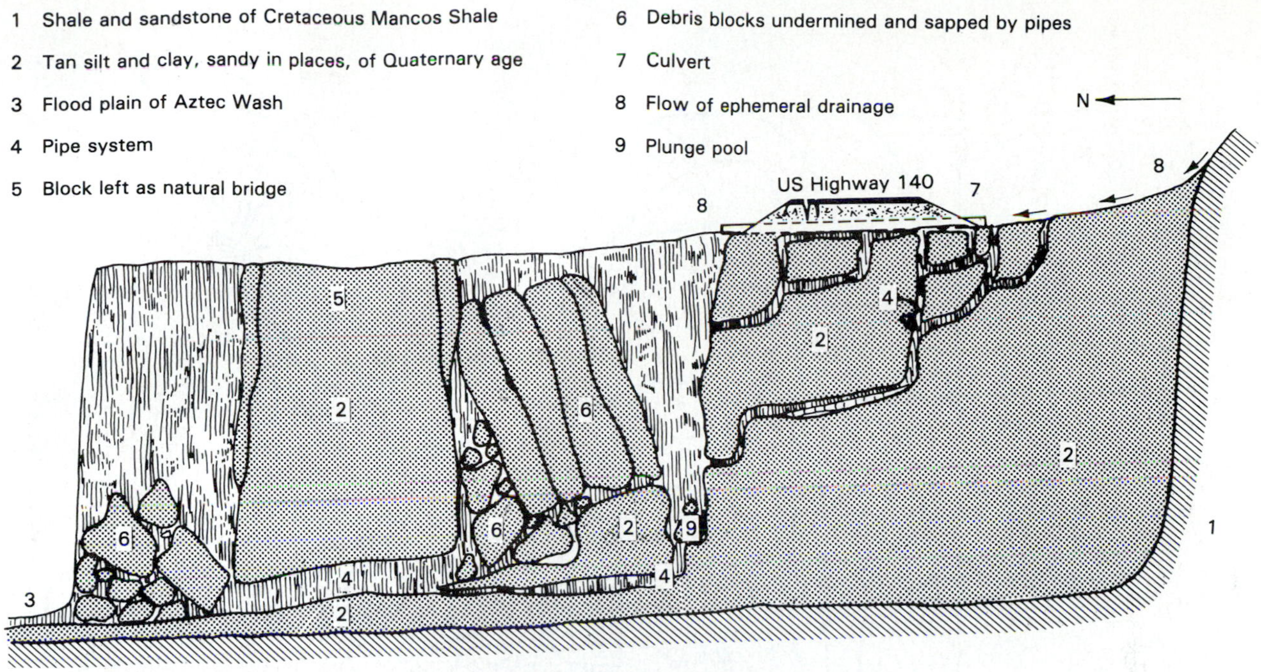

1 Shale and sandstone of Cretaceous Mancos Shale
2 Tan silt and clay, sandy in places, of Quaternary age
3 Flood plain of Aztec Wash
4 Pipe system
5 Block left as natural bridge
6 Debris blocks undermined and sapped by pipes
7 Culvert
8 Flow of ephemeral drainage
9 Plunge pool

US Highway 140

**Figure 9-19** Schematic diagram of piping beneath U.S. Route 140, southwestern Colorado. (From G. G. Parker and E. A. Jenne, 1967, 46th Annual Meeting, Highway Research Board, Washington, D.C., U.S. Geological Survey, Water Resources Division, p. 27.)

where

$A$ = average annual soil loss (tons/acre)
$R$ = rainfall factor
$K$ = soil-erodibility factor
$LS$ = slope length—steepness factor
$C$ = cropping factor
$P$ = conservation factor

**TABLE 9-4**

**Variables Affecting Soil Erosion**

Rainfall
  Intensity
  Duration
Soil Characteristics
  Porosity
  Permeability
  Moisture content
  Grain size and shape
Topography
  Orientation of slope
  Slope angle
  Length of slope
Vegetation
  Type and distribution on slope

**Figure 9-20** Average values of the rainfall-erosion index for the eastern United States (From D. D. Smith and W. H. Wischmeier, 1962, *Advances in Agronomy*, 14:109–148.)

The rainfall factor in this equation is obtained from the product of the kinetic energy of raindrops and the maximum 30-min intensity of the storm. Values for this index in the eastern United States are shown in Figure 9-20. It is apparent from this map that the most intense rainfall events occur in the Gulf Coast region. The factor $K$, relating to the characteristics of the soil, is a function of the grain-size distribution. Soils composed of silt and fine sand are the most erodible types. The topography of the area is evaluated by the product of $L$ (slope length) and $S$ (slope angle expressed as a percent). Each factor is calculated as the ratio of the actual value to standardized plots of land 72.6 feet in length and 6 feet in width, with 9% slope. For example, a slope 50 feet in length would have an $L$ value of $50/72.6 = 0.69$. The cropping factor ranges from high values (severe erosion) for barren land and row crops, to low values for natural grasses and woodlands. Vegetated land that is cleared for construction will have a substantial increase in this factor. Crop management techniques are incorporated into the conservation factor. Erosion-control measures such as contour plowing and terracing reduce the value assigned to this factor from its maximum, in the case where no erosion control methods are implemented. Tables and nomographs are available for obtaining values for the factors in the Universal Soil Loss Equation.

The Universal Soil Loss Equation was developed mainly for agricultural lands. Other methods of estimating erosion rates include measuring the amount of sediment carried by rivers downstream from the eroded area and measuring

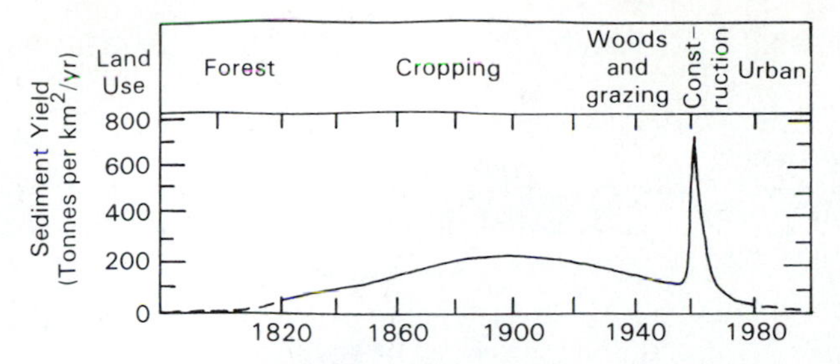


**Figure 9-21** Sediment yield from lands undergoing various land-use changes. (From M. G. Wolman, 1967, *Geografiska Annaler*, 49-A:385–395. Used by permission of the Swedish Society for Anthropology and Geography.)

the amount of sediment trapped in reservoirs constructed along river courses. These methods indicate that erosion is greatly accelerated by certain changes in land use. An excellent example of these changes is illustrated in Figure 9-21. When Europeans arrived in the United States, they initiated a series of land-use changes. One of the first changes to occur in arable areas was the conversion of land to crop production. This process involved clearing forested land or breaking sod in prairie regions. Both activities caused a decrease in infiltration, an increase in runoff, and an increase in soil erosion. Marginal agricultural areas often reverted back to forest or grazing as more productive farm land was developed. The erosion was correspondingly decreased in those areas.

The most drastic increases in soil erosion and sediment yield in streams occur during urbanization of drainage basins. Erosion is most severe during and immediately after construction (Figure 9-22). After the construction phase, erosion decreases because much of the land surface is covered by buildings and paved areas. In addition, runoff is collected and conveyed to stream channels through an artificial network of drains and storm sewers.

**Figure 9-22** Soil erosion from cleared, unprotected slopes during housing construction. (Photo courtesy of USDA Soil Conservation Service.)

**Figure 9-23** Newly constructed street (notice curb and partially buried drain in foreground) filled with sediment eroded from adjacent housing lots. (Photo of USDA Soil Conservation Service.)

Erosion Control. The use of erosion-control methods must be increased throughout the world in an attempt to combat excessive soil erosion. In agricultural lands, these methods include *contour farming*, *strip cropping*, and *terracing*. Their use becomes more important as the slope of the land increases. Contour farming is the orientation of crop rows parallel to the elevation contours of the land. This practice disrupts overland flow of water and thereby inhibits sheet and rill erosion. Strip cropping is the alternation of strips of erosion-inhibiting crops, such as grasses, with more erodible row crops. Grasses trap sediment eroded from crop rows and also increase moisture infiltration. The optimum width of the strips is determined by the amount of land slope. Terracing is a more intensive method of erosion control requiring construction of alternating embankments and terraces on the slope.

Although the area converted to urban use is small in comparison with the amount of farmland, the sediment yield from urbanizing areas can be greater by a factor of 100 or more than the sediment yield of natural or agricultural areas nearby. The problem of controlling erosion in land undergoing urbanization is directly related to the problem of excessive sedimentation, particularly in undesirable locations. Excessive sedimentation can clog highways, ditches, and drains, causing expensive cleanup operations (Figure 9-23). Increased sediment yield to stream systems kills fish, increases flooding, and decreases the capacity of reservoirs. Chemical pollutants, including pesticides and herbicides, are often adsorbed on soil particles and carried by erosion to streams and lakes.

In urban areas, new construction is likely to pose the most serious erosion and sedimentation problems. Sediment yields from areas that annually produce from 70 to 280 tonnes/km$^2$ can increase to as much as 28,000 tonnes/km$^2$ during certain types of construction. Guy (1976) has documented many examples of effective and ineffective erosion and sediment-control measures and has also suggested methods to reduce erosion by applying sound hydrologic and geomor-

**Figure 9-24** A sediment detention basin used to trap sediment in an urban area. (Photo courtesy of USDA Soil Conservation Service.)

phic principles to the design of these systems. Soil erosion can often be drastically reduced by planning construction during periods of low expected rainfall, dividing the construction site so that only small areas are cleared of vegetation at any one particular time, and utilizing temporary vegetation cover on soil stockpiles and other exposed areas whenever possible.

When additional erosion control measures are necessary, *diversions*, *bench terraces*, and *detention basins* can be constructed. Diversions consist of channels and ridges designed to collect overland flow and carry it away from areas that must be protected from erosion or sedimentation. Bench terraces reduce the length of the slope, and, as implied in Equation 9-4, reduce erosion. Diversions are often routed to detention basins (Figure 9-24), which are designed to trap sediment carried by runoff from the construction site. Temporary sediment-control measures are often installed until vegetation or paving reduces the area of unprotected ground at a site. Straw or hay bales, for example, form effective sediment traps around storm-drain inlets or culverts, when flow velocities are not extremely high. These temporary measures, fail, however, when the amount and velocity of overland flow overwhelm the straw-bale structure (Figure 9-25).

### *Erosion by Wind*

The impact of wind erosion has never been more clearly demonstrated than in the Great Plains of the United States during the 1930s (Figure 9-26). Dry conditions coupled with steadily increasing destruction of the natural grass cover devastated the agricultural economy of the region.

The erosion of soil by wind is similar to erosion by water. Wind, however, is not as effective an erosional agent as water because of its lower density. Variables in the wind-erosion system can be divided into characteristics of the wind and characteristics of the soil and land surface. Wind variables include wind velocity and duration, and the length of the open area without obstacles over which the wind blows. Important properties of the soil and land surface include particle size and size distribution, moisture content, and vegetation. Particles of typical density within the size range of 0.1 to 0.15 mm in diameter

**Figure 9-25** Failure of a straw-bale filter designed to trap sediment in an unvegetated ditch along a newly constructed highway. Flow in the ditch overtopped the straw-bale dam and eroded a new channel adjacent to the straw bales.

(very fine to fine sand) are most susceptible to wind erosion. The distribution of grains upon the surface is also very important. Soils containing large particles gradually become protected as wind erosion proceeds. As the fine particles are removed, coarser grains become concentrated at the surface, preventing further erosion of the soil.

Very fine particles in the soil, including silt and clay, promote the formation of soil aggregates. These clumps composed of smaller particles are held together by the cohesion of the silt and clay. In order for soil aggregates to be eroded by wind, they must first be broken down by raindrop impact, abrasion by wind-transported sediment, and other processes.

Moisture is one of the most important soil properties. Moist soil is cohesive

**Figure 9-26** A dust storm in Colorado during the 1930s. Notice the drifts of sand around fences and buildings. (Photo courtesy of USDA Soil Conservation Service.)

**Figure 9-27** A shelterbelt planted to minimize erosion of soil by wind. (Photo courtesy of USDA Soil Conservation Service.)

and resistant to wind erosion. The cohesive forces provided by the soil water are lost when the soil dries. This explains why droughts and wind erosion go hand in hand.

Vegetation is the final soil and land-surface variable. Vegetation provides physical protection for the soil; it also holds moisture, increases the roughness of the surface, and adds organic binding agents to the soil. The removal of vegetation during plowing and other activities is one of the primary factors in increased wind erosion.

Wind erosion can be greatly reduced by proper farming and land-use practices. *Shelterbelts* are one of the most common attempts to control wind erosion (Figure 9-27). These linear bands of trees or other plants decrease wind velocity and the unobstructed distance over which the wind blows. Farming methods for wind-erosion control include strip cropping and the planting of temporary cover crops, rather than allowing the soil to remain bare between periods of crop production.

## CASE STUDY 9-1
## Erosion, Sedimentation, and Reservoir Capacity

The interrelationship between erosion and deposition is well illustrated by the accompanying photos of Lake Ballinger Dam in Texas. Figure 9-28a shows the dam and reservoir prior to its abandonment in 1952. Lake Ballinger Dam was built in 1920 to provide a municipal water supply. Sedimentation gradually reduced the maximum depth of the reservoir from about 11 m to slightly more than 1 m by 1952. The excessive sedimentation was caused by runoff from cultivated fields in the small drainage basin of the reservoir. Figure 9-28b, taken after the final drainage of the reservoir, shows the greatly decreased capacity caused by excessive sedimentation. Better soil conservation practices in the drainage basin could have alleviated the problem. Thus the problem of erosion in one area is compounded by undesirable sedimentation in another area.

(a)

(b)

**Figure 9-28** (a) Lake Ballinger Dam, Texas, before abandonment. (b) The dam after it was abandoned in 1952 because of excessive sedimentation. (Photos courtesy of USDA Soil Conservation Service.)

## *SUMMARY AND CONCLUSIONS*

Rocks in contact with the atmosphere at the earth's surface are subjected to a wide variety of physical and chemical weathering processes that lead to their gradual decay and disintegration. Expansive pressures within rocks are caused by frost action, salt weathering, moisture-content changes, temperature changes, and the removal of overburden load. These physical processes are complemented by chemical reactions between atmospheric constituents and unstable minerals to form new compounds that are more stable under the surface conditions of low temperature and pressure. Chemical weathering reactions include complete dissolution of the original mineral (solution), formation of new minerals (hydrolysis), addition of water to the mineral structure (hydration), and reaction with oxygen to form new substances (oxidation). Minerals derived from igneous rocks weather in reverse order from their order of crystallization. The resistance of quartz to weathering explains its abundance in sedimentary rocks produced by the weathering and erosion of previously existing rocks.

Erosion by water and wind is most severe when land is altered from its natural state. Prediction of rates of fluvial erosion by such methods as the Universal Soil Loss Equation is difficult because of the large number of variables involved in the process. Erosion damage is often initiated by raindrop impact and sheet erosion. Sheet erosion progresses to rill and gully erosion during the gradual destruction of arable land by fluvial action. Land cleared for construction presents the most vulnerable condition for soil erosion. Various farming practices can be adopted to minimize erosion. Similarly, erosion of lands under development can be greatly limited by construction planning and utilization of erosion-control structures designed with an understanding of the hydrologic and geomorphic processes involved.

## PROBLEMS

1. Describe the specific climatic conditions in which the following types of weathering are dominant:
   (a) Strong chemical weathering.
   (b) Moderate chemical weathering with frost action.
   (c) Strong mechanical weathering.
2. How does salt weathering cause the deterioration of building stone?
3. What is the relationship between temperature and hydration pressure of gypsum?
4. Write a chemical equation for the solution of dolomite by carbonic acid.
5. Write a chemical reaction for the hydrolysis of albite ($NaAlSi_3O_8$).
6. Why do minerals fit into their respective positions in Goldich's stability series?
7. Summarize the formation and function of carbon dioxide in weathering processes.
8. What engineering properties of rock are affected by weathering, and what changes take place?
9. During construction of a highway, a wooded slope is cleared and excavated to increase the slope from 6% to 14%. The $C$ factor increases from 0.1 (mature woodland) to 1.0 (bare ground) prior to stabilization of the cut with vegetation. Assuming that $R$, $L$, and $P$ do not change, how much of an increase in erosion can be expected? (Hint: Calculate the erosion before and after road construction, and from these values determine the factor by which erosion has increased.)
10. What is the range in the rainfall erosion index along the Mississippi Valley of the eastern United States?
11. What methods are used to minimize erosion during construction?
12. What are the consequences of excessive soil erosion? Which one(s) would you expect to be most severe in your area?

## REFERENCES AND SUGGESTIONS FOR FURTHER READING

BEAVIS, F. C. 1985. *Engineering Geology*. Melbourne: Blackwell Scientific.

COOKE, R. V., and J. C. DOORNKAMP. 1974. *Geomorphology in Environmental Management*. Oxford: Clarendon Press.

COSTA, J. E., and V. R. BAKER. 1981. *Surficial Geology: Building with the Earth.* New York: John Wiley.

GERRARD, A. J. 1988. *Rocks and Landforms.* London: Unwin Hyman.

GUY, H. P. 1976. Sediment-control methods in urban development: some examples and implications, in *Urban Geomorphology*, D. R. Coates, ed. Geological Society of America Special Paper 174, pp. 21–35.

PARKER, G. G., and E. A. JENNE. 1967. Structural failure of western U.S. highways caused by piping. 46th Annual Meeting, Highway Research Board, Washington, D.C., U.S. Geological Survey, Water Resources Division, p. 27.

PELTIER, L. 1950. The geographical cycle in periglacial regions as it is related to climatic geomorphology. *Annals of the Association of American Geographers*, 40:214–236.

SMITH, D. D., and W. H. WISCHMEIER. 1962. Rainfall erosion. *Advances in Agronomy*, 14:109–148.

TRENT, D. D. 1984. Geology of the Joshua Tree National Monument, *California Geology*, 37:75–86.

WINKLER, E. M., and E. J. WILHELM. 1970. Salt burst by hydration pressures in architectural stone in urban atmosphere. *Geological Society of America Bulletin*, 81:567–572.

WOLMAN, M. G. 1967. A cycle of sedimentation and erosion in urban river channels. *Geografiska Annaler*, 49-A:385–395.

# 10

# Soils, Soil Hazards, and Land Subsidence

Perhaps no term causes more confusion in communication between various specialized groups of earth scientists and engineers than the word *soil.* The problem arises in the reasons for which different groups of professionals study soils. Soil scientists, or *pedologists*, constitute a group interested in soils as a medium for plant growth. For this reason, pedologists focus most of their attention on the organic-rich, weathered zone that supports plant growth—the upper meter or so beneath the land surface and refer to the rocks or sediments below the weathering zone as *parent material*. Soil scientists have developed a complex system of classification for soils that is based on the physical, chemical, and biological properties that can be observed and measured in the soil.

Soils engineers, the corresponding group of engineering soil specialists, take an entirely different approach to the study of soil. To soils engineers, the word soil connotes any material that can be excavated with a shovel. This definition places no limitations on depth, origin, or ability to support plant growth. The engineering classification of soil is based on the particle size, the particle-size distribution, and the plasticity of the material. These characteristics relate closely to the behavior of soil under the application of load.

Most geologists fall somewhere between pedologists and soils engineers in their approach to soils. Geologists are interested in soils and weathering processes as indicators of past climatic conditions and in relation to the geologic formation of useful materials ranging from clay deposits to metallic ores. The pedological soil classification is too complicated for nonspecialists and, therefore, more basic classifications are still used by geologists. Geologists usually refer to any loose material below the plant growth zone as sediment or *unconsolidated material*. The term *unconsolidated* may be confusing to engineers because *consolidation* specifically refers to the compression of saturated soils in soils engineering, in contrast to its more general usage in geology.

An important aspect of soil with respect to engineering applications is whether the material can be classified as *transported* or *residual*. Transported soils are deposits of rivers, glaciers, and other surficial geologic processes. Residual soils are those that developed in place by the weathering processes discussed in Chapter 9. This distinction is important because the origin of the material is relevant to its thickness, continuity or discontinuity over a landscape, and the nature of the lower contact with unweathered bedrock.

## THE SOIL PROFILE

In the first part of this chapter we shall investigate soil from the viewpoint of soil scientists. *Pedogenic*, or soil-forming, processes include the chemical weathering processes that we discussed in Chapter 9, as well as certain additional physical, chemical, and biological processes. The result of this activity is a sequence of recognizable layers, or *horizons*, that constitute the *soil profile*.

A number of possible horizons can be described in the field. Natural exposures can be used or pits can be dug to examine the soil profile (Figure 10-1). Profile descriptions are based on recognition of *master horizons* and subdivisions of master horizons. The criteria for master horizons are given in Table 10-1. A surficial horizon, the O, is used to designate fresh or partly decomposed organic matter. The A horizon, recognized by its dark color, is a zone of mixed mineral material and partly decomposed organic matter called *humus*. The E horizon is a zone of *eluviation*, which describes the leaching of metal oxides and clays. As a result, it is often light colored in comparison to

**Figure 10-1** A soil profile exposed in a pit dug for soil classification. (Photo courtesy of USDA Soil Conservation Service.)

**TABLE 10-1**

**General Description of Soil Horizons**

| Soil Horizon Nomenclature | |
|---|---|
| | MASTER HORIZONS |
| O horizon | Surface accumulation of mainly organic matter overlying mineral soil. |
| A horizon | Accumulation of humified organic matter mixed with mineral fraction; the latter is dominant. Occurs at the surface or below an O horizon; Ap is used for those horizons disturbed by cultivation. |
| E horizon | Usually underlies an O or A horizon, characterized by less organic matter and/or fewer sesquioxides (compounds of iron and aluminum) and/or less clay than the underlying horizon. Horizon is light colored due mainly to the color of the primary mineral grains because secondary coatings on the grains are absent. |
| B horizon | Underlies an O, A, or E horizon, shows little or no evidence of the original sediment or rock structure. Several kinds of B horizons are recognized, some based on the kinds of materials illuviated into them, others on residual concentrations of materials. Selected subdivisions are: |
| | *Bh horizon* Illuvial accumulation of organic matter-sesquioxide complexes that either coat grains, form pellets, or form sufficient coatings and pore fillings to cement the horizon. |
| | *Bk horizon* Illuvial accumulation of alkaline earth carbonates, mainly calcium carbonate; the properties do not meet those for the K horizon. |
| | *Bo horizon* Residual concentration of sesquioxides, the more soluble materials having been removed. |
| | *Bq horizon* Accumulation of secondary silica. |
| | *Bt horizon* Accumulation of silicate clay that has either formed *in situ* or is illuvial; hence it will have more clay than the assumed parent material and/or the overlying horizon. Illuvial clay can be recognized as grain coatings; bridges between grains; coatings on ped surfaces or in pores; or thin, single or multiple near-horizontal discrete accumulation layers of pedogenic origin (clay bands or lamellae). In places, subsequent pedogenesis can destroy evidence of illuviation. |
| | *By horizon* Accumulation of gypsum. |
| | *Bz horizon* Accumulation of salts more soluble than gypsum. |
| K horizon | A subsurface horizon so impregnated with carbonate that its morphology is determined by the carbonate. Carbonate coats or engulfs all |

**TABLE 10-1** *(Continued)*

| Soil Horizon Nomenclature | |
|---|---|
| MASTER HORIZONS | |
| | primary grains in a continuous medium to make up 50 percent or more by volume of the horizon. The uppermost part of a strongly developed horizon commonly is laminated. The cemented horizon corresponds to some caliches and calcretes. |
| C horizon | A subsurface horizon, excluding R, like or unlike material from which the soil formed or is presumed to have formed. Lacks properties of A and B horizons, but includes materials in various stages of weathering. |
| | *Cox and Cu horizons* In many unconsolidated Quaternary deposits, the C horizon consists of oxidized C overlying seemingly unweathered C. It is suggested the Cox be used for oxidized C horizons and Cu for unweathered C horizons. |
| | *Cr horizon* In soils formed on bedrock, there commonly will be a zone of weathered rock between the soil and the underlying rock. If it can be shown that the weathered rock has formed in place, and has not been transported, it is designated Cr. |
| R horizon | Consolidated bedrock underlying soil. It is not unusual for this and the Cr horizon to have illuvial clay in cracks; the latter would be designated Crt. |
| SELECTED SUBORDINATE DEPARTURES | |
| | Lower-case letters follow the master horizon designation. Those that are mainly specific to a particular master horizon are given above. Some can be found in a variety of horizons; they are listed below: |
| b | Buried soil horizon. May be deeply buried and not affected by subsequent pedogenesis; if shallow, they can be part of a younger soil profile. |
| c | Concretions or nodules cemented by iron, aluminum, manganese, or titanium. |
| f | Horizon cemented by permanent ice. Seasonally frozen horizons are not included, nor is dry permafrost material, that is, material that lacks ice but is colder than 0°C. |
| g | Horizon in which gleying is a dominant process, that is, either iron has been removed during soil formation or saturation with stagnant water has preserved a reduced state. Common to these soils are neutral colors, with or without mottling. Bg is used for horizon with pedogenic features in addition to gleying; however, if gleying is the only pedogenic feature, it is designated Cg. |
| k | Accumulation of alkaline earth carbonates, commonly $CaCO_3$. |
| m | Horizon that is more than 90 percent cemented. Denote the cementing material (km, carbonate; qm, silica; kqm, carbonate and silica; etc.). |
| n | Accumulation of exchangeable sodium. |
| v | Horizon characterized by iron-rich, humus-poor, reddish material that hardens irreversibly when dried. |
| x | Subsurface horizon characterized by a bulk density greater than that of the overlying soil, hard to very hard consistence, brittleness, and seemingly cemented when dry (fragipan character). |
| y | Accumulation of gypsum. |
| z | Accumulation of salts more soluble than gypsum (for example, NaCl). |

SOURCE: Modified from P. W. Birkeland, *Soils and Geomorphology*, copyright © 1984, Oxford University Press, Inc. Reprinted by permission.

the A and B horizons. Substances leached from the E horizon accumulate in the B horizon, which is known as an *illuvial* zone. A horizon that is strongly cemented by calcium carbonate is labeled a K horizon. Below the B or K horizons, there are several possible horizons that may be somewhat weathered but do not have the properties of the previously described horizons. The C horizon generally consists of the parent material from which the soil was formed. A horizon composed of consolidated bedrock is described as R.

Subdivisions of master horizons can be identified by adding a lower case letter to the master horizon designation (Table 10-1). Further subdivisions can be made for zones that receive the same letter or letters, but differ slightly in color or some other observable property. For example, Bt1 and Bt2 would be used to differentiate two parts of a clay-rich B horizon.

Many pedogenic processes are involved in the formation of soil horizons. For example, in humid climates soluble salts such as calcium carbonate are leached out of the profile and carried downward to the water table. In arid climates, however, calcium carbonate and other salts are leached from the upper horizons and redeposited in the C or K horizons. When this accumulation forms a thick cemented zone, it is known as *caliche*. This material may be so hard that blasting is required for excavation. Other pedogenic processes are involved in the transfer of material from one soil horizon to another. The movement of clay and metal oxide from the A to the B horizon includes several pedogenic processes. Dissolved organic decomposition products are very important in the removal of iron and aluminum from the A horizon. Organic molecules bond to normally insoluble metal ions in a process called *chelation*. In the B horizon, the organic molecules are destroyed, and iron and aluminum oxides precipitate from solution.

In the tropics, warm, moist conditions combine to produce thick, deeply weathered residual soils. Intense leaching removes even quartz, which is practically insoluble in arid and temperate climates. Tropical soils, therefore, contain high percentages of residual accumulations of iron and aluminum oxides (Table 9-2). The concentrations of these metals are so high in some places that the soil is mined. In climates with alternating wet and dry seasons, iron is leached and becomes mobile during the wet season, but is reprecipitated during the dry season. The iron oxide precipitants gradually build up to form soils known as *laterites*. Where the climate is wet year round, iron is removed from the profile, leaving an abundance of aluminum oxide. These soils are known as *bauxites*.

The number and type of soil horizons present are not constant from one area to another. In the next section we will examine the major factors that determine which pedogenic processes will be dominant and which soil horizons will develop.

## SOIL-FORMING FACTORS

Soil forms in response to conditions existing in its physical, chemical, and biological environment. The factors of greatest importance in soil formation include climate, parent material, organisms, relief, and time (Figure 10-2). Of these factors, climate is the only one that can be identified as being more important than all the others.

The climatic elements that influence soil formation are temperature and precipitation, the same controls that govern weathering processes. In the early days of soil science, pedologists believed that parent material—rock or sediment from which the soil developed—was most important in determining the characteristics of the resulting soil. Many studies have shown, however, that similar soils occur upon varied parent materials under constant climatic conditions. The influence of parent material is greatest in young soils that developed in mild climates. For example, soils that formed from glacial sediment in the American Midwest tend to be clayey and slightly alkaline because of shale

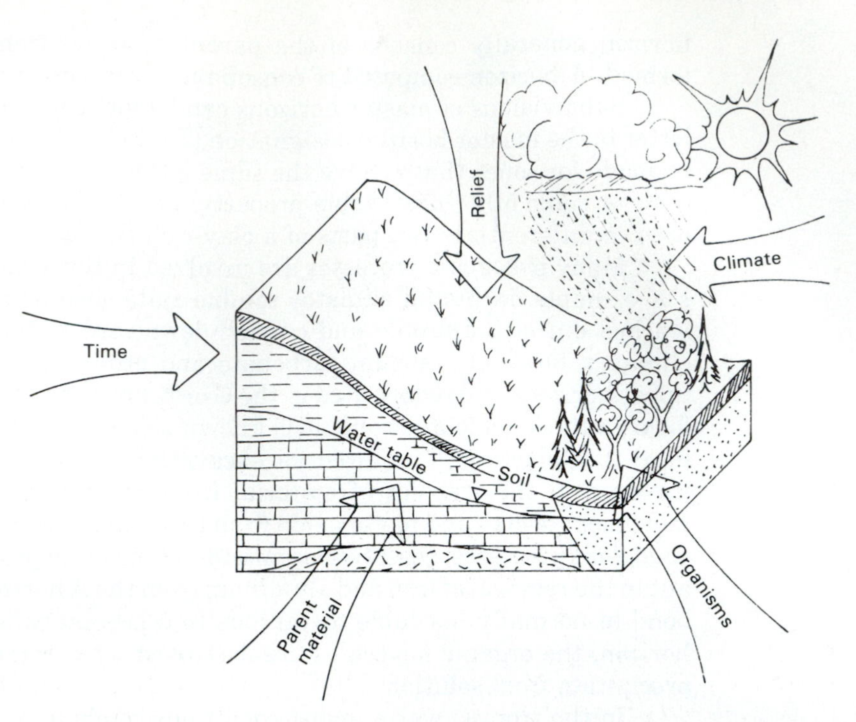


**Figure 10-2** Schematic illustration of the five soil-forming factors.

and limestone fragments in the parent material. Soils occurring on glacial sediments in New England, on the other hand, are sandier and acidic, reflecting the composition of the predominantly siliceous igneous rocks from which the glacial sediment was derived.

Organisms include both plants and animals that inhabit a particular soil. If all other soil-forming factors were held constant, soils developed under different vegetative communities would exhibit numerous differences. A good example is the difference in soil properties observed across the transition from prairie to forest vegetation. As shown in Figure 10-3, prairie soils have thicker A horizons containing more organic matter. Forest soils have a lower pH than prairie soils, have highly leached zones within the A or E horizons, and display greater accumulation of clay and iron oxide in the B horizon.

Relief, the factor that relates to the position of the soil in a particular landscape, can cause drastic changes in soil characteristics within a short distance. The aspects of relief that influence soil development include slope and topographic position. Slope is important because of erosional and depositional processes. Soil on a steep slope is likely to be thin because of more intense erosion (Equation 9-4) in comparison with soils at the base of the slope, where soil eroded from the hillslope accumulates. Topographic position refers to the location of the soil within the local landscape. The water table, for instance, may be much closer to the surface in a valley than at the top of a ridge. If the water table lies within several meters of the land surface in an arid or semiarid area, soluble salts will accumulate in the soil as moisture is evaporated from the soil.

Time, the final soil-forming factor, determines the length of the period during which climatic and vegetative processes are acting upon the parent material to produce soil. The pedogenic clock starts at the instant a rock or sediment is exposed to or deposited at the earth's surface. This point in time could include many possible events: the retreat of a glacier that has eroded or deposited material on a landscape, the deposition of sediment by a mudflow in

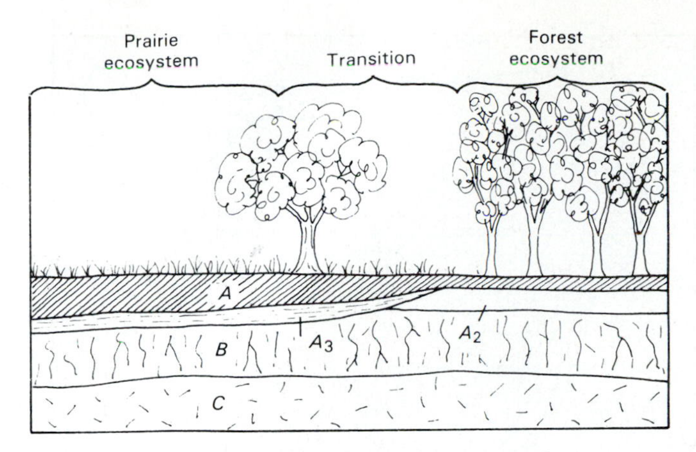


**Figure 10-3** Changes in soil from a forest to prairie area. (Reprinted with permission from S. W. Buol, F. D. Hole, and R. J. McCracken, *Soil Genesis and Classification*, 2d ed., copyright © 1980 by the Iowa State University Press, Ames, Iowa.)

the tropics, or the extrusion of a lava flow. The rate of soil development is mainly determined by climate. Thousands of years may be required for the production of a mature soil. The time factor in soil formation is the reason that prevention of erosion, particularly of agricultural lands, is such an important goal.

## PEDOGENIC SOIL CLASSIFICATION

Classification of soils is necessary to develop a better understanding of the relationships among the large number of soil types that can be produced by variation in the soil-forming factors. Soil classification in the United States has an interesting history because, rather than gradually modifying an older classification, soil scientists developed a new and radically different classification in the 1950s and 1960s. The older system divided so-called normal (zonal) soils that formed in response to the typical climatic conditions of an area from azonal soils developed under the influence of factors other than climate. Thus the same soil could be considered to be zonal in one area and azonal in another. This led to a certain amount of confusion in the use of the older classification. In addition, the highest divisions of the older system were based on factors presumed to be important in soil genesis rather than soil characteristics that could be precisely measured and compared from soil to soil. Despite these limitations, some aspects of the older classification are still used in many parts of the world. For example, the division of zonal soils into *pedalfers* and *pedocals* is a useful way of comparing soils over large areas. Pedalfers, with the abbreviations for aluminum (Al) and iron (Fe) incorporated into the name, are soils that accumulate iron and aluminum oxides within the profile. These soils are most common in moist, forested regions. Pedocals, which contain calcium carbonate (the *cal* in pedocal) within the profile, occur primarily in arid and semiarid areas. The distribution of pedalfers and pedocals is shown in Figure 10-4.

The new classification system, called Soil Taxonomy, was developed by the U.S. Soil Conservation Service to rectify the problems and confusion caused by the older system. It is based on measurable properties of soil profiles. The divisions of the classification are shown in Table 10-2. Ten soil orders constitute

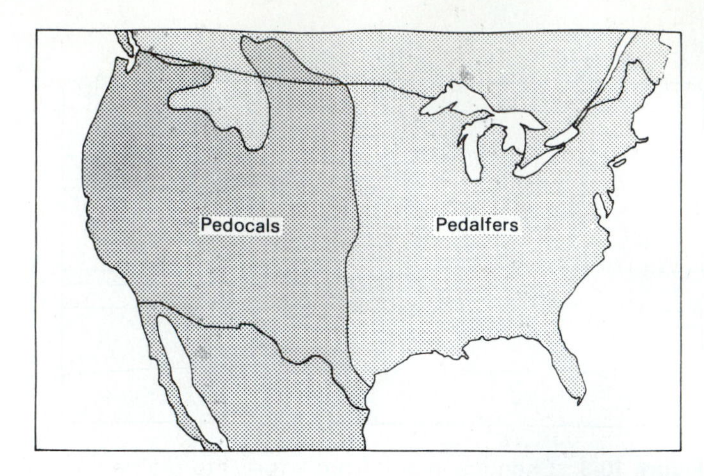


**Figure 10-4** Generalized distribution of pedalfers and pedocals in the United States. (From S. Judson, M. E. Kauffman, and L. D. Leet, *Physical Geology*, 7th ed., copyright © 1987 by Prentice-Hall, Inc., Englewood Cliffs, N.J.)

**TABLE 10-2**

**Subdivision of Soil Taxonomy Classification System**

| *Category* | *Number of taxa* |
|---|---|
| Order | 10 |
| Suborder | 47 |
| Great group | Approximately 206 |
| Subgroup | |
| Family | |
| Series | Approximately 10,000 in United States |

**TABLE 10-3**

**Simplified Definitions of Soil Orders in United States Comprehensive Soil Classification System**

| | |
|---|---|
| Alfisol | Soil with gray to brown surface horizon, medium to high base supply, and a subsurface horizon of clay accumulation. |
| Aridisol | Soil with pedogenic horizons, low in organic matter, usually dry. |
| Entisol | Soil without pedogenic horizons. |
| Histosol | Organic (peat and muck) soil. |
| Inceptisol | Soil with weakly differentiated horizons showing alteration of parent materials. |
| Mollisol | Soil with a nearly black, organic-rich surface horizon and high base supply. |
| Oxisol | Soil that is a mixture principally of kaolin, hydrated oxides, and quartz. |
| Spodosol | Soil that has an accumulation of amorphous materials in the subsurface horizons. |
| Ultisol | Soil with a horizon of clay accumulation and low base supply. |
| Vertisol | Cracking clay soil. |

SOURCE: From A. L. Bloom, *Geomorphology*, 2d ed., copyright © 1991 by Prentice-Hall, Inc. Englewood Cliffs, N.J.

NOTE: Base supply refers to amount of exchangeable cations such as calcium, magnesium, sodium, and potassium that remain in the soil.

**TABLE 10-4**

**Formative Elements in Names of Soil Orders of Soil Taxonomy**

| *No. of order*[1] | *Name of order* | *Formative element in name of order* | *Derivation of formative element* | *Mnemonicon and pronunciation of formative elements* |
|---|---|---|---|---|
| 1 | Entisol ......... | ent | Nonsense syllable. | recent. |
| 2 | Vertisol ....... | ert | L. *verto,* turn. | invert. |
| 3 | Inceptisol .... | ept | L. *inceptum,* beginning. | inception. |
| 4 | Aridisol ....... | id | L. *aridus,* dry. | arid. |
| 5 | Mollisol ....... | oll | L. *mollis,* soft. | mollify. |
| 6 | Spodosol ...... | od | Gk. *spodos,* wood ash | Podzol; odd. |
| 7 | Alfisol .......... | alf | Nonsense syllable. | Pedalfer. |
| 8 | Ultisol ......... | ult | L. *ultimus,* last. | Ultimate. |
| 9 | Oxisol .......... | ox | F. *oxide,* oxide. | oxide. |
| 10 | Histosol ....... | ist | Gk. *histos,* tissue. | histology. |

SOURCE: From Soil Survey Staff, 1960.

[1] Numbers of the orders are listed here for the convenience of those who become familiar with them during development of the system of classification.

the first category of the classification. The characteristics of the orders are illustrated in Table 10-3. The names of these orders include syllables (formative elements) that help to describe the order (Table 10-4). Names of great groups are created by combining new formative elements with the formative elements of the soil order. For example, Alfisols that develop in dry climates belong to the Suborder Ustalf. "Ust" is derived from the Latin *Ustus*, meaning burnt, and "Alf" is the formative element for the Order Alfisol. When further modifications to the name are made for the Great Group level, more than 200 such unfamiliar names are generated. At the lowest level of the classification, about 10,000 soil series have been identified in the United States. It is easy to see why the classification is not used by professionals other than soil scientists. Despite this challenge, however, it is worth the effort to become somewhat familiar with soil nomenclature. Soils in the United States are mapped at the county level and reports called *Soil Surveys* are published for each county. These reports are valuable because they include the engineering properties and land-use limitations for each soil type. If surficial geologic maps are not available, Soil Surveys may provide the best information on subsurface and hydrological conditions of an area.

## ENGINEERING PROPERTIES OF SOIL

The engineering approach to the study of soil focuses on the characteristics of soils as construction materials and the suitability of soils to withstand the load applied by structures of various types. For these purposes, the physical properties that describe soils and their components must be quantified and understood. In addition, classifications more relevant to the engineering properties of soils must be used.

### Weight-Volume Relationships

As we mentioned earlier, earth materials are three-phase systems. In most applications, the phases include solid particles, water, and air. Water and air occupy *voids* between the solid particles. For soils in particular, the physical

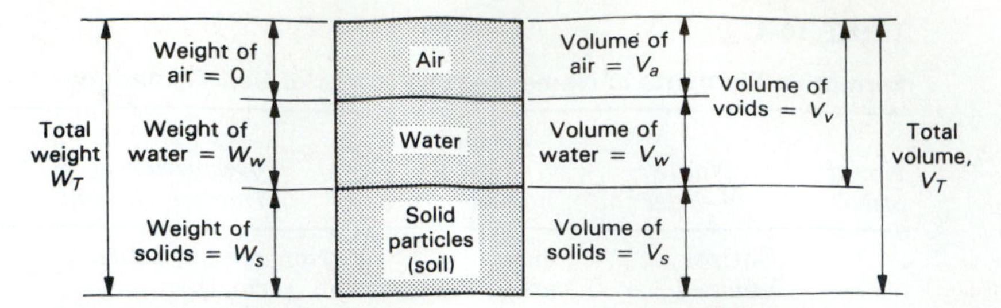


**Figure 10-5** Block diagram illustrating volumes and weights of a soil mass.

relationships between these phases must be examined. A mass of soil can be conveniently represented as a block diagram, with each phase shown as a separate block (Figure 10-5). The volume of the soil mass is the sum of the volumes of the three components, or

$$V_T = V_a + V_w + V_s \qquad \text{(Eq. 10-1)}$$

The volume of voids is the sum of $V_a$ and $V_w$. The weight of the solids is determined on a slightly different basis. Because any weighing of air in the soil voids would be done within the earth's atmosphere as with other weighings, we consider the weight of air in a soil mass to be zero. Thus the total weight is expressed as the sum of the weights of the soil solids and the water:

$$W_T = W_s + W_w \qquad \text{(Eq. 10-2)}$$

The relationship between weight and volume can be expressed as

$$W_m = V_m G_m \gamma_w \qquad \text{(Eq. 10-3)}$$

where

$W_m$ = weight of the material (solid, liquid, or gas)

$V_m$ = volume of the material

$G_m$ = specific gravity of the material (dimensionless)

$\gamma_w$ = unit weight of water

Unit weights are commonly used in the United States as a measure of density, which in this case becomes weight density rather than mass density. The unit weight of water is usually considered to be 62.4 pounds per cubic foot (pcf). The mass density of water is 1.0 g/cm$^3$. From the relationship shown in Equation 10-3, the weight of solids is

$$W_s = V_s G_s \gamma_w \qquad \text{(Eq. 10-4)}$$

and the weight of water is

$$W_w = V_w G_w \gamma_w = V_w \gamma_w \qquad \text{(Eq. 10-5)}$$

since $G_w = 1$. When dealing with a soil sample from the field that contains both water and soil particles, the unit weight can be expressed both with and without the water contained in the soil. The unit wet weight is given by

$$\gamma_{wet} = \frac{I_T}{V_T} \qquad \text{(Eq. 10-6)}$$

This quantity is determined by weighing a known volume of soil without allowing any drainage or evaporation of water from the voids. Alternatively, the unit dry weight is expressed as

$$\gamma_{dry} = \frac{W_s}{V_T} \qquad \text{(Eq. 10-7)}$$

Unit dry weight is determined by oven-drying a known volume of soil. The resulting weight will be the weight of solids ($W_s$). Both unit wet weight and unit dry weight are expressed in pcf or g/cm$^3$.

The weight of water in a soil sample that was oven dried is the difference between the weight before drying and the weight of solids measured after drying. This relationship should be evident from Figure 10-5. The *water content* of a soil, which is expressed as a decimal or percent, is defined as

$$w = \frac{W_w}{W_s} \times 100\% \qquad \text{(Eq. 10-8)}$$

Relationships between volumes of soil and voids are described by the *void ratio*, *e*, and *porosity*, *n*, which were first defined in Chapter 6. The void ratio is the ratio of the void volume to the volume of solids:

$$e = \frac{V_v}{V_s} \qquad \text{(Eq. 10-9)}$$

whereas the porosity is the ratio of void volume to total volume:

$$n = \frac{V_v}{V_T} \times 100\% \qquad \text{(Eq. 10-10)}$$

These terms are related and it is possible to show that

$$e = \frac{n}{1 - n} \qquad \text{(Eq. 10-11)}$$

One additional relationship can be developed from the block diagram in Figure 10-5. This term, known as the *degree of saturation*, *S*, relates the volume of water in the void space to the total void volume:

$$S = \frac{V_w}{V_v} \times 100\% \qquad \text{(Eq. 10-12)}$$

With the relationships described above, a variety of useful problems can be solved. It is helpful to draw a block diagram similar to Figure 10-5 and to label the diagram with the known quantities.

**EXAMPLE 10-1**

A 110 cm$^3$ sample of wet soil has a mass of 212 g and a degree of saturation of 100%. When oven-dried, the mass is 162 g. Determine the unit dry weight (dry density in this case because of the units used), water content, void ratio, and specific gravity of the soil particles.

***Solution.***

Water
Soil
$V_T$ = 110 cc
$W_W$
$W_T$ = 212 g
$W_S$ = 162 g

$$\gamma_{dry} = \frac{W_s}{V_T} = \frac{162 \text{ g}}{110 \text{ cm}^3} = 1.47 \text{ g/cm}^3$$

$$w = \frac{W_w}{W_s} = \frac{212 \text{ g} - 162 \text{ g}}{162 \text{ g}} = 0.309 = 30.9\%$$

$$V_w = \frac{W_w}{G_w \gamma_w} = \frac{212 \text{ g} - 162 \text{ g}}{(1.0)(1.0 \text{ g/cm}^3)} = 50 \text{ cm}^3$$

Since $S = 100\%$,

$$V_w = V_v$$

$$V_s = V_T - V_w = 110 \text{ cm}^3 - 50 \text{ cm}^3 = 60 \text{ cm}^3$$

$$e = \frac{V_v}{V_s} = \frac{50 \text{ cm}^3}{60 \text{ cm}^3} = 0.83$$

$$G_s = \frac{W_s}{V_s \gamma_w} = \frac{162 \text{ g}}{(60 \text{ cm}^3)(1.0 \text{ g/cm}^3)} = 2.70$$

### *Index Properties and Classification*

In the previous section, no mention was made of the characteristics of the particles within the soil. Factors such as the size and type of particles in the soil, as well as density and other characteristics, relate to shear strength, compressibility, and other aspects of soil behavior. These *index properties* are used to form engineering classifications of soil. They can be measured by simple lab or field tests called *classification tests*. Table 10-5 lists index properties and their respective classification tests. An important division of soils for engineering purposes is the separation of coarse-grained, or *cohesionless* soils, from

**TABLE 10-5**

**Index Properties of Soils**

| *Soil type* | *Index property* |
|---|---|
| Coarse-grained (cohesionless) | Particle-size distribution<br>Shape of particles<br>Clay content<br>In-place density<br>Relative density |
| Fine-grained (cohesive) | Consistency<br>Water content<br>Atterberg limits<br>Type and amount of clay<br>Sensitivity |

**Figure 10-6** Sieves used for determining grain-size distribution of soil. (Photo courtesy of Soiltest Inc.)

fine-grained, or *cohesive*, soils. Cohesive soils, which contain silt and clay, behave much differently from cohesionless materials. The term cohesion refers to the attractive forces between individual clay particles in a soil. The index properties that apply to cohesionless soils refer to the size and distribution of particles in the soil. These characteristics are evaluated by *mechanical analysis*, a laboratory procedure that consists of passing the soil through a set of sieves with successively smaller openings (Figure 10-6). The size of sieve opening determines the size of the particles that may pass through. After the test, the particles retained on each sieve are converted to a weight percentage of the total and then plotted against particle diameter as determined by the known sieve opening size. The result is a *grain-size distribution curve* (Figure 10-7),

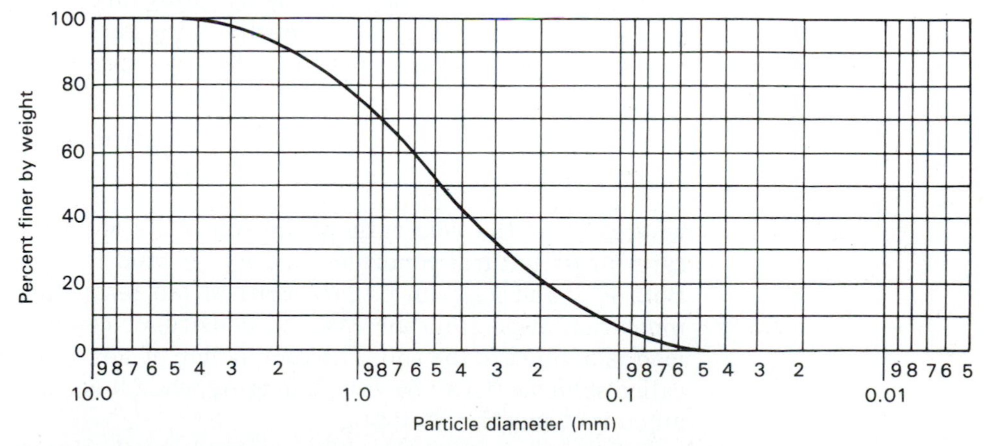


**Figure 10-7** A cumulative grain-size distribution curve drawn from data obtained in a mechanical analysis of a soil sample. (From D. F. McCarthy, *Essentials of Soil Mechanics and Foundations*, 4th ed., copyright © 1993 by Prentice-Hall, Inc., Englewood Cliffs, N.J.)

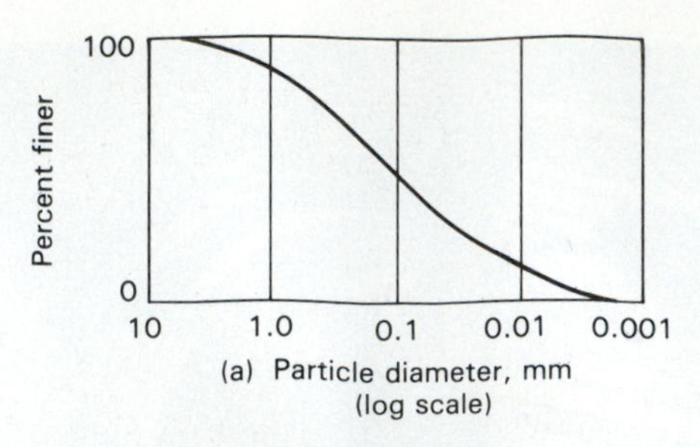


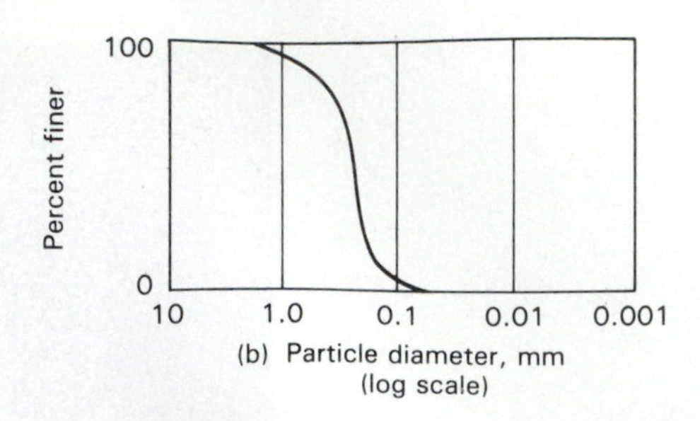


**Figure 10-8** Grain-size distribution curves of a well-graded soil (left) and a poorly graded soil (right). (From D. F. McCarthy, *Essentials of Soil Mechanics and Foundations*, 4th ed., copyright © 1993, by Prentice-Hall, Inc., Englewood Cliffs, N.J.)

plotted as the cumulative weight percent versus particle size. The cumulative weight percent is expressed as the percentage finer than a corresponding particle size. For example, if 20% of the soil by weight were retained on all the sieves with openings of 1 mm or larger in diameter, a point at 80% finer by weight would be plotted at the 1-mm size.

The shape of the grain-size distribution curve is a very important soil characteristic. A curve that covers several log cycles of the graph, as shown on the left in Figure 10-8, contains a variety of sizes. This type of soil would be called *well graded.* The opposite type of soil, composed of a very narrow range of particle sizes, would be classified as *poorly graded* (Figure 10-8). These terms can be easily confused with the term *sorting,* the geologic designation for grain-size distribution. Geologists are concerned with various depositional processes, such as river flow, that tend to separate particle sizes during transportation and deposit particles in beds composed of particles of similar sizes. For this reason, geologists would refer to a *poorly graded* soil with a narrow range of sizes as *well sorted.* The opposite, a *well-graded* soil, would be considered *poorly sorted* by geologists.

The other index properties describing cohesionless soils include *particle shape*, *in-place density*, and *relative density*. These properties are related because particle shape influences how closely particles can be packed together. The in-place density refers to the actual density of the soil at its particular depth in the field. It is measured by weighing an oven-dried sample taken from a known volume. The relative density is the ratio of the actual density to the maximum possible density of the soil. It is expressed in terms of void ratio:

$$D_R = \frac{e_{max} - e_o}{e_{max} - e_{min}} \times 100\% \qquad \text{(Eq. 10-13)}$$

where $e_{max}$ is the void ratio of the soil in its loosest condition, $e_o$ is the void ratio in its natural condition, and $e_{min}$ is void ratio in its densest condition. Both $e_{max}$ and $e_{min}$ can be measured in lab tests. Relative density is a good indication of possible increases in density, or compaction, that may occur if load is applied to the soil. The compaction of soil under the load of structures, called *settlement*, can be very damaging when it occurs in excessive or variable amounts beneath a building.

The index properties of fine-grained, or cohesive, soils are somewhat more complicated than the index properties of cohesionless soils because of the influence of clay minerals. The type and amount of clay minerals are, therefore, very significant. With respect to engineering behavior, *consistency* is the most

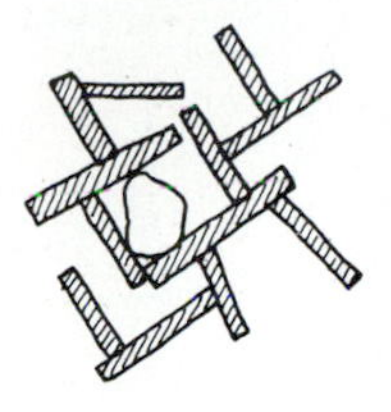

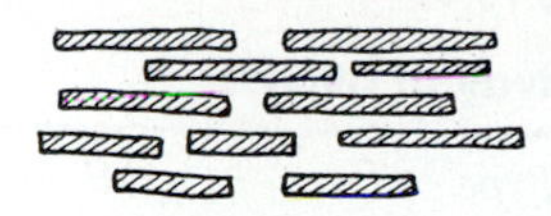



**Figure 10-9** Clay particles displaying flocculated and dispersed fabrics.

important characteristic of cohesive soils. It refers to the strength and resistance to penetration of the soil in its in-place condition. Consistency is determined by the arrangement of soil particles, particularly clay particles, in the soil, which is called the soil's *fabric*. Although many types of fabric are possible, soils in which edge-to-face contact of clay particles exists (Figure 10-9), or soils which have *flocculated* fabrics, are much stronger than soils which have the parallel arrangement of particles found in *dispersed* fabrics. Flocculated fabrics change to dispersed fabrics during *remolding*, which involves disturbance and alteration of the soil by natural processes or during various lab tests. The consistency of a soil can be determined by field tests in which the soil is evaluated in place, or by lab tests on samples that have been carefully handled to avoid remolding. The unconfined compression test is often used as an indication of consistency. In practice, the relative terms *soft*, *medium*, *stiff*, *very stiff*, and *hard* are applied to describe consistency. The relationship between these terms, unconfined compressive strength values, and the results of simple tests made with the hand and other objects are shown in Table 10-6.

The ratio of unconfined compressive strength in the undisturbed state to strength in the remolded state defines the index property called *sensitivity*.

**TABLE 10-6**

**Consistency and Strength for Cohesive Soils**

| *Consistency* | *Shear strength ($kg/cm^2$)* | *Unconfined compressive strength ($kg/cm^2$)* | *Feel or touch* |
|---|---|---|---|
| Soft | <0.25 | <0.5 | Blund end of a pencil-size item makes deep penetration easily |
| Medium (medium stiff or medium firm) | 0.25–0.50 | 0.50–1.0 | Blunt end of a pencil-size object makes 1.25-cm penetration with moderate effort |
| Stiff (firm) | 0.50–1.0 | 1.0–2.0 | Blunt end of pencil-size object can make moderate penetration (about 0.6 cm) |
| Very stiff (very firm) | 1.0–2.0 | 2.0–4.0 | Blunt end of pencil-size object makes slight indentation; fingernail easily penetrates |
| Hard | >2.0 | >4.0 | Blunt end of pencil-size object makes no indentation; fingernail barely penetrates |

SOURCE: From D. F. McCarthy, *Essentials of Soil Mechanics and Foundations*, 4th, ed. copyright © 1993 by Prentice-Hall, Inc., Englewood Cliffs, N.J.

**TABLE 10-7**

**Sensitivity of Clays**

| *Type* | *Sensitivity value* |
|---|---|
| Nonsensitive | 2–4 |
| Sensitive | 4–8 |
| Highly sensitive | 8–16 |
| Quick | >16 |

The sensitivity, $S_t$, can be expressed as

$$S_t = \frac{\text{strength in undisturbed condition}}{\text{strength in remolded condition}} \quad \text{(Eq. 10-14)}$$

Soils with high sensitivity (Table 10-7) are highly unstable and can be converted to a remolded or dispersed state very rapidly, with an accompanying drastic loss of strength. The disastrous slope movements that follow these transformations are discussed in Chapter 12.

The water content, defined in the previous section, is an important influence upon the bulk properties and the behavior of a soil. In the remolded state, the consistency of the soil is defined by the water content. Four consistency states are separated by the water content at which the soil passes from one state to another (Figure 10-10). These water content values are known as the *Atterberg limits*. For example, the *liquid limit* is the water content at which the soil-water mixture changes from a liquid to a plastic state. As the water content decreases, the soil passes into a semisolid state at the *plastic limit*,

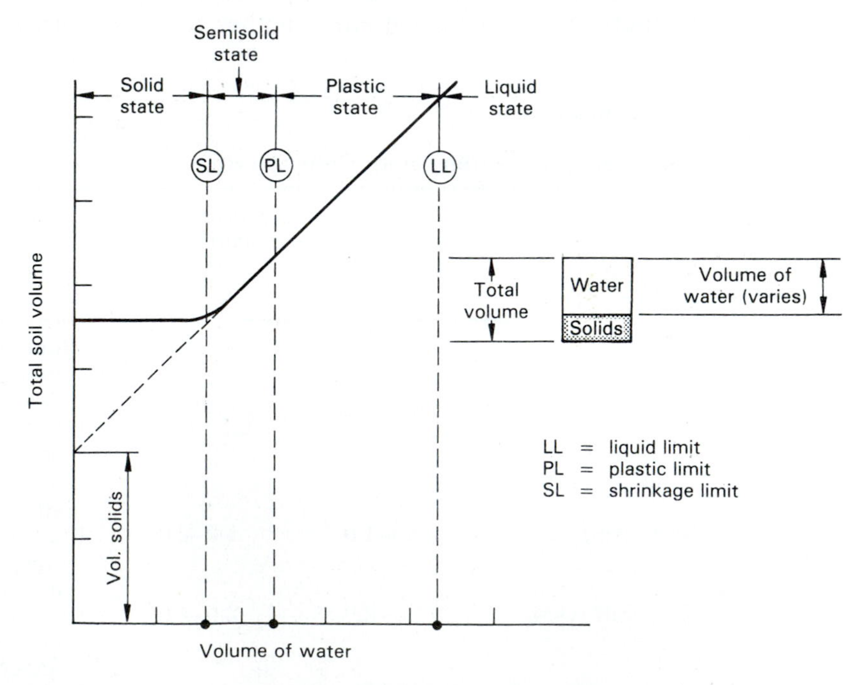


**Figure 10-10** Atterberg limits. As the volume of water (water content) in a volume of soil decreases, the soil passes through four states, separated by the Atterberg limits. (From D. F. McCarthy, *Essentials of Soil Mechanics and Foundations*, 4th ed., copyright © 1993 by Prentice-Hall, Inc., Englewood Cliffs, N.J.)

Unified Soil Classification System
(ASTM Designation D-2487)

| Major division | | | Group symbols | Typical names |
|---|---|---|---|---|
| Coarse-grained soils More than 50% retained on No. 200 sieve | Gravels 50% or more of coarse fraction retained on No. 4 sieve | Clean gravels | GW | Well-graded gravels and gravel-sand mixtures, little or no fines |
| | | | GP | Poorly graded gravels and gravel-sand mixtures, little or no fines |
| | | Gravels with fines | GM | Silty gravels, gravel-sand-silt mixture |
| | | | GC | Clayey gravels, gravel-sand-clay mixtures |
| | Sands More than 50% of coarse fraction passes No. 4 sieve | Clean sands | SW | Well-graded sands and gravelly sands, little or no fines |
| | | | SP | Poorly graded gravels and gravel-sand mixtures, little or no fines |
| | | Sands with fines | SM | Silty sands, sand-silt mixtures |
| | | | SC | Clayey sands, sand-clay mixtures |
| Fine-grained soils 50% or more passes No. 200 sieve | Silts and clays Liquid limit 50% or less | | ML | Inorganic silts, very fine sands, rock flour, silty or clayey fine sands |
| | | | CL | Inorganic clays of low to medium plasticity, gravelly clays, sandy clays, silty clays, lean clays |
| | | | OL | Organic silts and organic silty clays of low plasticity |
| | Silts and clays Liquid limit greater than 50% | | MH | Inorganic silts, micaceous, or diatomaceous fine sand or silts, elastic silts |
| | | | CH | Inorganic clays of high plasticity, fat clays |
| | | | OH | Organic clays of medium to high plasticity |
| Highly organic soils | | | Pt | Peat, muck, and other highly organic soils |

**Figure 10-11** Engineering classification of soils by the Unified Soil Classification System. (Modified from D. F. McCarthy, *Essentials of Soil Mechanics and Foundations*, 4th ed., copyright © 1993 by Prentice-Hall, Inc., Englewood Cliffs, N.J.)

and a solid state at the *shrinkage limit*. The shrinkage limit defines the point at which the volume of the soil becomes nearly constant with further decreases in water content. The Atterberg limits can be determined with simple laboratory tests. The use of the Atterberg limits in predicting the behavior of natural, in-place soil is limited by the fact that they are conducted on remolded soils. The relationship between moisture content and consistency defined by the Atterberg limits may not be the same in soils in the undisturbed state. Therefore, the Atterberg limits are mainly used for classification rather than for the prediction of soil behavior under field conditions.

The most useful engineering classification of soils is the *Unified Soil Classification System* (Figure 10-11). This classification gives each soil type a

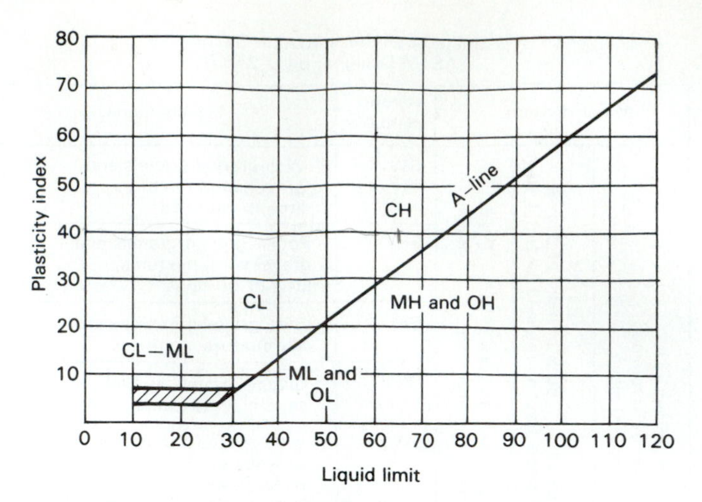


**Figure 10-12** Chart for determining classification of fine-grained soils by the Unified Soil Classification System. (From D. F. McCarthy, *Essentials of Soil Mechanics and Foundations*, 4th ed., copyright © 1993 by Prentice-Hall, Inc., Englewood Cliffs, N.J.)

two-letter designation. For coarse-grained soils, the first letter, either *G* for gravel or *S* for sand, refers to the dominant particle size in the soil. The second letter is either *W*, for well graded, *P*, for poorly graded, or *M* or *C*, for coarse-grained soils that contain more than 12% of silt or clay, respectively. The first letter of the designation for fine-grained soils is *M* or *C*, for silt or clay, respectively. The second letter, either *H* (high) or *L* (low), refers to the plasticity of the soil as defined in Figure 10-12. The chart is a plot of *plasticity index* (*PI*) against liquid limit (*LL*). The plasticity index is defined as

$$PI = LL - PL \qquad \text{(Eq. 10-15)}$$

The difference between the liquid and plastic limits (*PI*) is a measure of the range in water contents over which the soil remains in a plastic state. The plotted position of a soil with respect to the *A-line* on the plasticity chart determines whether the soil receives the letter *H* for high plasticity or the letter *L* for low plasticity. Highly organic soils, which form a final category in the classification, are also subdivided into high- and low-plasticity types. Once a soil has been classified by the Unified system, predictions can be made of the soil's permeability, strength, compressibility, and other properties.

### *Shear Strength*

The shear strength of a soil determines its ability to support the load of a structure or remain stable upon a hillslope. Engineers must therefore incorporate soil strength into the design of embankments, road cuts, buildings, and other projects. The division of soils into cohesionless and cohesive types for classification is also useful for discussing strength.

In the Mohr-Coulomb theory of failure, shear strength has two components—one for inherent strength due to bonds or attractive forces between particles, and the other produced by frictional resistance to shearing movement (Equation 6-7). The shear strength of cohesionless soils is limited to the frictional component. When the direct shear test is used to investigate a cohesionless soil, successive tests with increasing normal stress will establish a straight

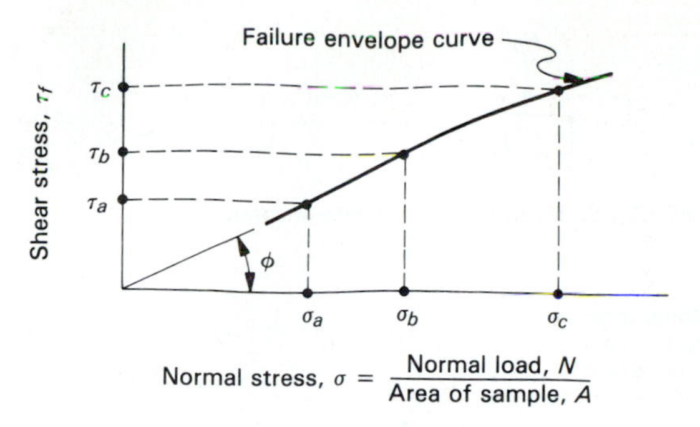


**Figure 10-13** Determination of shear strength from direct shear data for cohesionless soil.

line that passes through the origin (Figure 10-13). The angle of inclination of the line with respect to the horizontal axis is the angle of internal friction. Example 6-3 illustrates the determination of angle of internal friction from the results of direct shear tests. Values of the angle of internal friction are given in Table 10-8. If the soil is dense when tested, initially higher values for the angle of internal friction will be measured, but with increasing amounts of strain, the angle will decline to the approximate ranges seen in Table 10-8.

The shear strength of a cohesive soil is more complicated than a cohesionless material. The differences are due to the role of pore water in a cohesive soil. Most cohesive soils in field conditions are at or near saturation because of their tendency to hold moisture and their low permeability. When load is applied to a soil of this type, the load is supported by an increase in the pore-water pressure until pore water can drain into regions of lower pressure. At that point, soil particles are forced closer together and the strength increases, just like a cohesionless soil. Time is an important factor, however, because it takes longer for water to move out of a low-permeability material.

Strength tests for cohesive soils are usually made in triaxial cells in which the drainage of the sample can be controlled. Test conditions can allow (1) no drainage of the soil during loading; (2) drainage during an initial phase of loading, followed by failure in an undrained condition; and (3) complete drainage during very slow loading to failure. The response of the sample is different in each case. For the most basic case, in which the sample is undrained throughout the test, the results will yield a straight line on a plot of normal stress versus shear stress (Figure 10-14). The reason for this consequence is that the soil particles cannot be forced closer together without drainage of pore water, and thus cannot develop greater resistance to shear failure. The shear-strength

**TABLE 10-8**

**Values of $\phi$ for Cohesionless Soils**

| *Soil type* | *Angle $\phi$, degrees* |
|---|---|
| Sand and gravel mixture | 33–36 |
| Well-graded sand | 32–35 |
| Fine to medium sand | 29–32 |
| Silty sand | 27–32 |
| Silt (nonplastic) | 26–30 |

SOURCE: From D. F. McCarthy, *Essentials of Soil Mechanics and Foundations*, 4th ed., copyright © 1993 by Prentice-Hall, Inc., Englewood Cliffs, N.J.

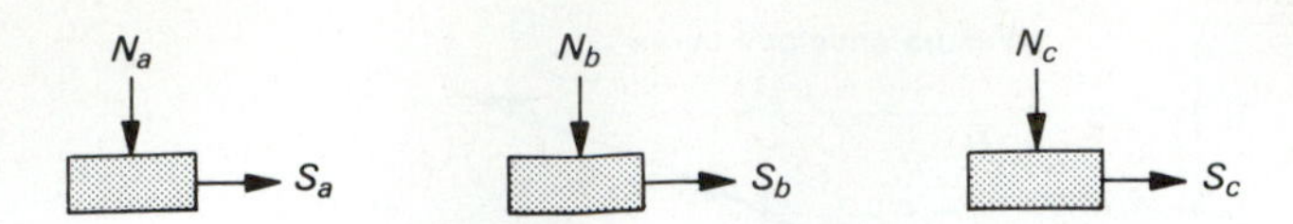


$N_c > N_b > N_a$ ($S_c$, $S_b$, $S_a$ represent ultimate shearing forces)

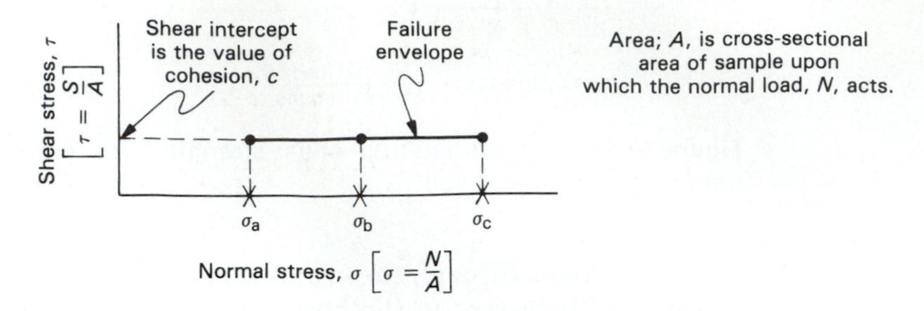


**Figure 10-14** Determination of cohesion from undrained triaxial test results for cohesive soil.

value determined in this type of test is termed cohesion (Chapter 6). The property of soils in the field that controls cohesion is consistency. Stiff or hard soils are compacted in their natural state and therefore have more shear strength than soft soils. The values of shear strength given in Table 10-6 are comparable to cohesion values obtained from an undrained triaxial test. Strength tests of soils that are mixtures of cohesive and cohesionless material yield failure envelopes similar to Figure 6-19.

## EXAMPLE 10-2

A saturated, cohesive soil is tested in a triaxial cell. During the test, no drainage is allowed from the sample. At failure, the major and minor principal stresses are 2600 and 1000 psf, respectively. What is the cohesion of the sample?

***Solution.***
Because the sample is undrained, increases in the normal stress will have no effect on the strength, and the failure envelope will be a straight line. $\sigma_1$ and $\sigma_3$ are plotted on the horizontal axis to form a diameter of a Mohr's circle plot. The failure envelope will be tangent to the top of the circle at a distance from the origin on the vertical axis equal to the radius of the circle. Thus

$$c = r = \frac{\sigma_1 - \sigma_3}{2}$$

$$c = \frac{2600 \text{ psf} - 1000 \text{ psf}}{2} = 800 \text{ psf}$$

$\sigma_1 = 2600$ psf

$\sigma_3 = 1000$ psf

$\tau$

$c = 800$ psf

$r$

$90°$

$r = \frac{\sigma_1 - \sigma_3}{2}$

$\sigma$

$\sigma_3$ (1000 psf)

$\sigma_1$ (2600 psf)

It is evident from Table 10-6 that unconfined compressive strength is equal to twice the cohesion (shear strength) of cohesive soil. This can be shown on a plot similar to Example 10-2. The unconfined compressive test is one in which $\sigma_3$ is zero and $\sigma_1$ at failure is the unconfined compressive strength. When the Mohr's circle is plotted with $\sigma_3$ at the origin, it is apparent that $\sigma_1$ is the diameter of the circle and the cohesion is the radius.

When a landslide occurs, it is an indication that the stress within the soil mass exceeded the shear strength of the soil at the time of failure. The load imposed by buildings and other structures on most soils is rarely great enough to overcome the ultimate shear strength or *bearing capacity* of the soil. Case Study 10-1 describes an example of a building that actually did cause failure of the soil upon which it was constructed.

### *Settlement and Consolidation*

Despite the lack of large-scale bearing-capacity failures caused by the load upon the soil exerted by structures, the soil does deform under the load applied. The usual response is a volume decrease in the soil beneath the foundation. *Settlement* is the vertical subsidence of the building as the soil is compressed. Excessive settlement, particularly when it is unevenly distributed beneath the foundation, can result in serious damage to the structure (Figure 10-15).

The tendency of a soil to decrease in volume under load is called *compressibility*. This property is evaluated in the *consolidation test*, in which a soil sample is subjected to an increasing load. The change in thickness is measured after the application of each load increment (Figure 10-16). During the consolidation test, soil particles are forced closer together as in strength tests. Thus compression of the soil causes a decrease in void ratio. Data from the consolidation test are plotted as void ratio against the log of vertical pressure applied to the sample (Figure 10-17). The slope of the resulting straight line is called the *compression index*, $C_c$. The compression index is an important property of a soil because it can be used to predict the amount of settlement that will occur from the load of a building or other structure. Clay-rich soils and soils high in organic content have the highest compressibility.

The compression of a saturated clay soil is a slightly different process from the compression of an unsaturated material. As we mentioned earlier, when the pores of a soil are filled with water, compression is impossible because the void volume is totally occupied. The increase in load upon the saturated clay during a consolidation test or actual field situation is initially balanced by an increase in fluid pressure in the pores of the soil (Figure 10-18). Higher fluid pressure in the clay causes the pore water to flow out of the clay toward any direction where fluid pressure is lower. Only after pore fluid is removed, can the soil compress to a lower void ratio. This process is called *consolidation*. It is an important phenomenon to consider when constructing a building on a saturated clay because consolidation is very slow. The movement of pore water out of the soil is a function of the permeability, and clay has the lowest permeability of any soil. Therefore, it may take years for the soil to reach equilibrium under the load imposed. The settlement occurring during this period can be dangerous to the building. A famous example of consolidation is the Leaning Tower of Pisa (Figure 10-19). Tilting of the tower is the result of nonuniform consolidation in a clay layer beneath the structure. This ongoing process may eventually lead to failure of the tower.

**Figure 10-15** Damage to a building by settlement. The house has been raised to its original elevation. The gap between the wall and the foundation shows the amount of settlement that has occurred. (Photo courtesy of USDA Soil Conservation Service.)

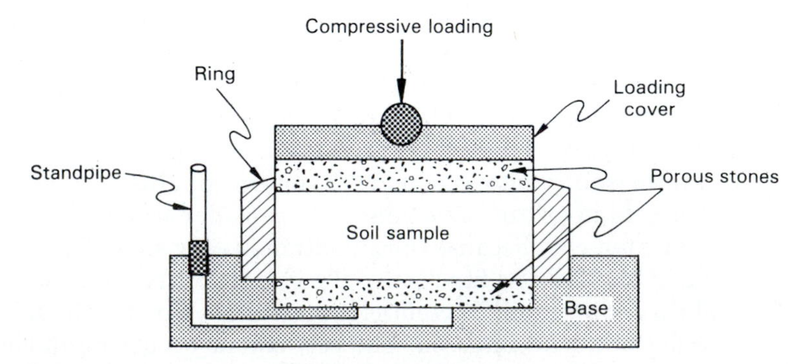


**Figure 10-16** Schematic diagram of a consolidation test to determine soil compressibility. (From D. F. McCarthy, *Essentials of Soil Mechanics and Foundations*, 4th ed., copyright © 1993 by Prentice-Hall, Inc., Englewood Cliffs, N.J.)

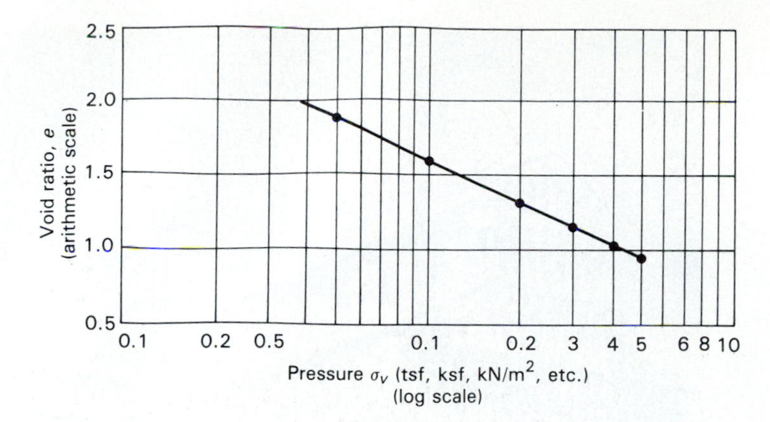


**Figure 10-17** Consolidation test data plotted as decrease in void ratio versus log of vertical pressure. (From D. F. McCarthy, *Essentials of Soil Mechanics and Foundations*, 4th ed., copyright © 1993 by Prentice-Hall, Inc., Englewood Cliffs, N.J.)

### *Clay Minerals*

Clay minerals are so important to soils engineering that we must take a closer look at their properties. Clays are good examples of *colloids*, particles so small that their surface energy controls their behavior. In particular, the electrical charge on the particle surface influences the interaction between adjacent clay particles. In Figure 10-20 the effect of the clay-surface charge upon the pore-water solution surrounding it is shown. Surface charges on clay particles are negative under most conditions. In order to balance these charges, cations from the pore-water solution and water molecules are attracted to the particle surface. The attraction of water is possible because of the polar nature of the water molecule. The positive side of the water molecule is, therefore, attracted to the negative clay surface. The negative charges on the surface of the clay

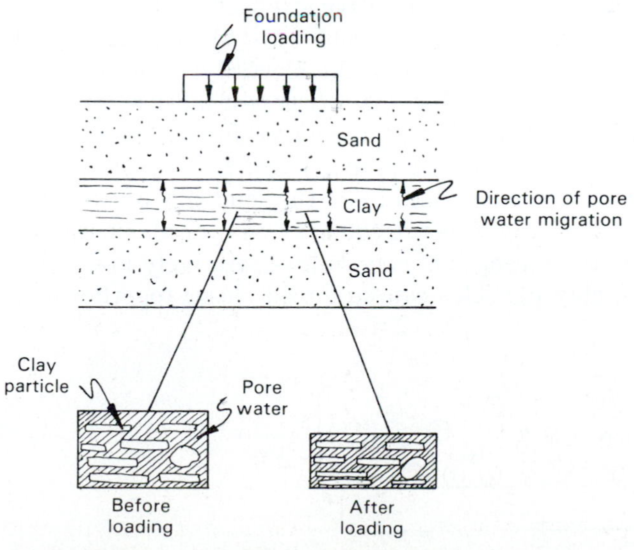


**Figure 10-18** Consolidation of saturated clay under the load exerted by a foundation. Clay particles are forced closer together after pore water is forced out of the unit.

**Figure 10-19** The Leaning Tower of Pisa (tower at right), whose tilt is the result of nonuniform consolidation beneath the structure. (Photo courtesy of the Italian Government Travel Office, E.N.I.T.)

particle in combination with the attracted cations and water molecules are called the *diffuse double layer*. Water molecules in diffuse double layers behave somewhat differently from water that is beyond the double layer in pore spaces. In addition, the positively charged sides of the water molecules cause repulsive forces between the double layers of adjacent clay particles.

When clays are initially deposited by settling to the bottom in a lake or the ocean, for example, both attractive and repulsive forces between clay particles are present. The relative strength of these forces determines which of the fabrics shown in Figure 10-9 will develop during sedimentation. Repulsive forces are stronger when the electrolyte content (ionic concentration) of the solution is weak. Therefore, *dispersed* fabrics are more common in clays deposited in fresh water, and *flocculated* fabrics characterize seawater deposition.

The electrostatic properties of clays explain many aspects of their engineering behavior. Because of the tendency of clays to absorb water, the water content of clay soils is frequently near or above the plastic limit. In the plastic state, the strength of clay soils is relatively low. When load is applied to clay soils, clay particles can be forced closer together or forced to reorient them-

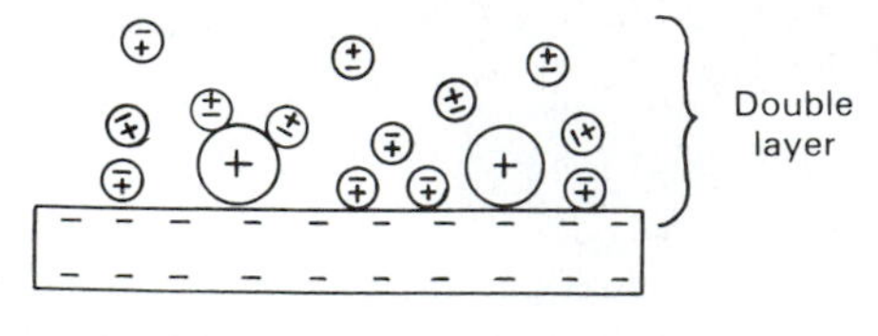


**Figure 10-20** Attraction of cations and water molecules to the charged surface of a clay particle.

selves, causing a new fabric to develop. The large decreases in void ratio that accompany these adjustments to load are responsible for the high compressibility of clay. As we have seen, when pore water is expelled during consolidation, the decrease in void ratio may be a very slow process. Because of these properties, clay soils can create troublesome engineering problems. Unstable slopes and settlement of foundations are examples. Site investigations for heavy structures must not neglect the presence of clay soils at depth; even though surficial soils may have low compressibility, a clay bed in the subsurface may lead to settlement of the structure (Figure 10-18).

## CASE STUDY 10-1
## Bearing Capacity Failure in Weak Soil

A rare example of a true bearing-capacity failure of a soil beneath a foundation is provided by the Transcona, Manitoba, grain elevator (White, 1953). This structure (Figure 10-21) was built in 1913 by the Canadian Pacific Railway. During the initial filling of the elevator, the underlying soils failed along a nearly circular surface, causing a rotation of the elevator to an angle of 27° from the vertical (Figure 10-22). Despite the sinking and rotational movement, the elevator was largely undamaged. By excavating beneath the side that was rotated upward and by underpinning the structure with piles, engineers were able to restore the elevator to an upright position, where it now rests at a depth of 7 m lower than the depth of the original foundation. Years later, when testing and analysis of the soils were done, it was determined that the pressure of the fully loaded elevator exceeded the shear strength of the soil. The soils in the area are clay-rich sediments deposited in a Pleistocene glacial lake. These materials have a very low strength and high compressibility. Failures similar to the Transcona grain elevator are less likely today because of great progress in the field of soil mechanics with respect to testing and analysis of soils.

**Figure 10-21** The Transcona grain elevator, located near Winnipeg, Manitoba, as it appears today.

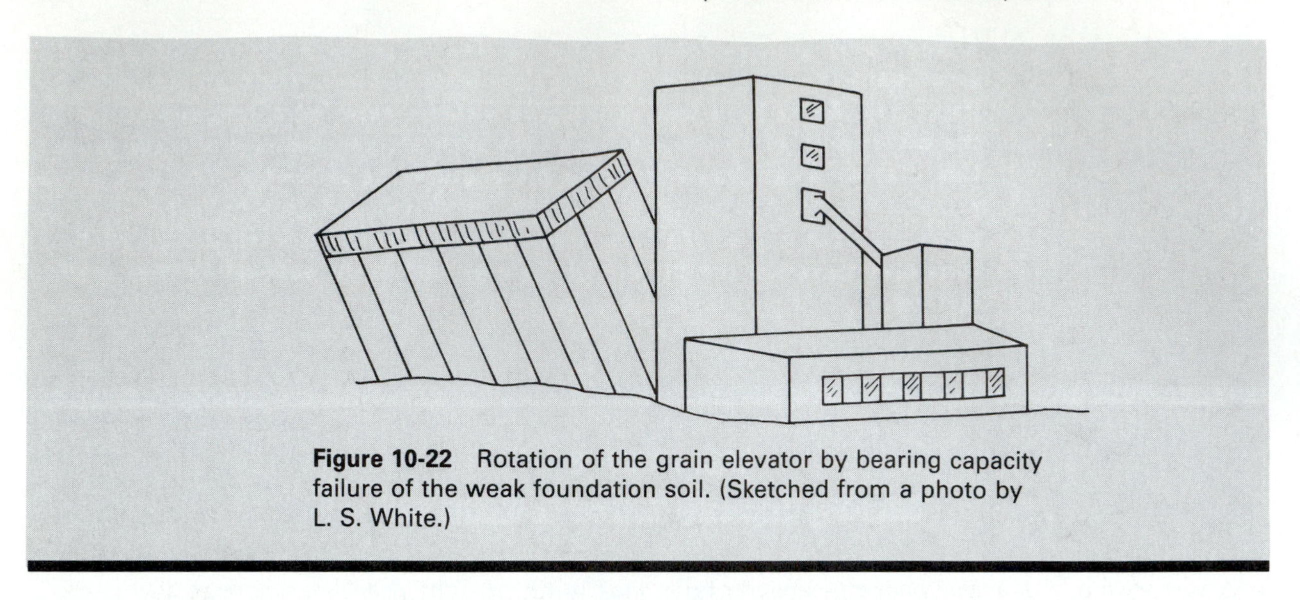

**Figure 10-22** Rotation of the grain elevator by bearing capacity failure of the weak foundation soil. (Sketched from a photo by L. S. White.)

## SOIL HAZARDS

The basic engineering principles of soil behavior provide the theory for understanding the interactions between engineering structures and soils. In practice, however, the performance of soil under imposed engineering load can be quite complex. In order to identify and predict potentially hazardous soil processes, the engineer must determine the relationships and interactions between the soil and the rock units, the ground-water flow system, the climate, and the vegetation of an area. All these factors can influence soil processes that can damage existing structures or require special designs for new projects.

### *Expansive Soil*

Surprisingly, damage to structures caused by soils that expand by absorbing water is the most costly natural hazard in the United States on an average annual basis. The annual bill for repair and replacement of highways and buildings affected by expansive (swelling) soils runs into billions of dollars. The portions of the United States underlain by expansive soils are shown in Figure 10-23.

The tendency for soils to swell can be explained by the characteristics of clay particles that we have already considered. The unbalanced electrostatic charges on clay-particle surfaces draw water molecules into the area between silicate sheets, thus forcing them apart. The cations attracted to the clay surfaces provide another factor in swelling behavior. Because of the attraction of the negatively charged clay-particle surfaces for cations, small spaces within or between clay particles may contain a higher concentration of cations than larger pores within the soil. These conditions (Figure 10-24) create an *osmotic potential* between the pore fluids and the clay-mineral surfaces. Normally, cations diffuse from a higher concentration to a lower concentration in order to evenly distribute the ions throughout the solution. In expansive soils, because ions are held by the clay particles, water moves from areas of low ionic concentration (high concentration of water) to areas of high ionic concentration (low concentration of water) within clay particles or aggregates. This influx of water exerts pressure, which causes the clay to swell.

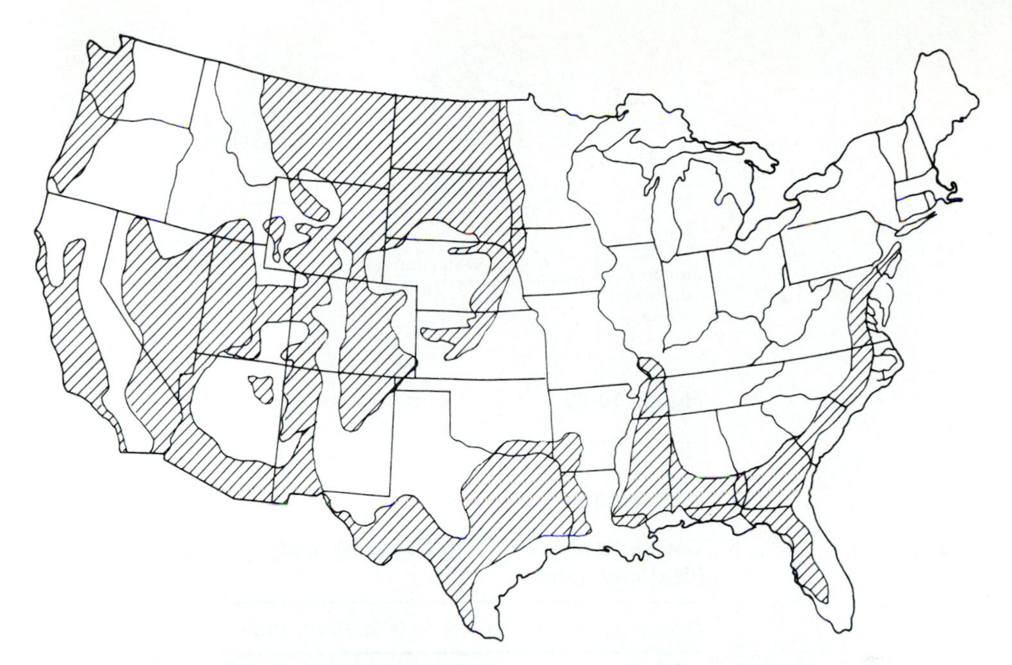

**Figure 10-23** Distribution of expansive soils in the United States. (From H. A. Tourtelot, 1974, *Bulletin of the Association of Engineering Geologists*, 11. Used by permission of the Association of Engineering Geologists.)

If a clay soil is subjected to drying conditions, for example, when evaporation is removing water from the soil near land surface, a suction effect is exerted on the soil that causes water molecules that are not held tightly to clay particles to be drawn out into the large pores of the soil and to move upward to replace the evaporated water. This loss of water from the clay leads to shrinkage, the reversal of the swelling process.

Swelling and shrinkage of a soil can occur only with changes in water content. Therefore, the potential for damage from expansive soil is limited to the upper zone of the soil in which seasonal changes in moisture content take place. This zone is called the *soil-moisture active zone* (Figure 10-25). Below this zone, even though the soil may have the potential to shrink and swell,

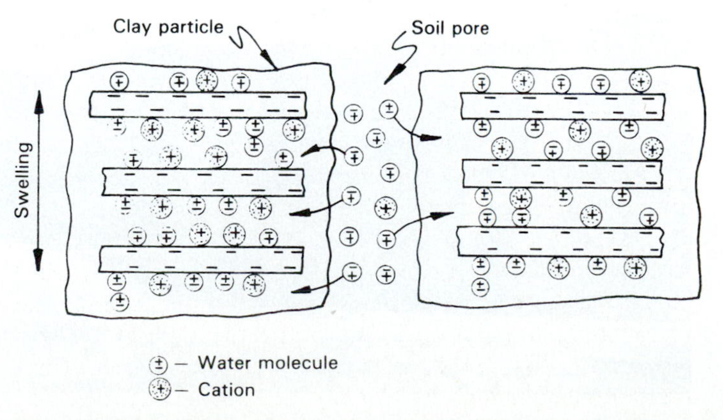


**Figure 10-24** Swelling of clay-rich soils. Water molecules diffuse into clay particles in order to equalize the ionic concentration between the particle and the soil pore.

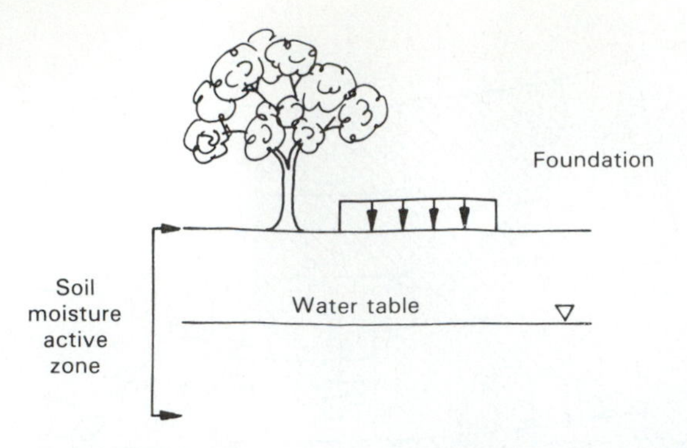


**Figure 10-25** The soil-moisture active zone.

**TABLE 10-9**

**Correlation of Swelling Behavior with Plasticity Index**

| *Degree of expansion* | *Plasticity index* |
|---|---|
| Very high | >35 |
| High | 25–41 |
| Medium | 15–28 |
| Low | <18 |

SOURCE: From R. D. Holtz and W. D. Kovacs, *An Introduction to Geotechnical Engineering*, copyright © 1981 by Prentice-Hall, Inc., Englewood Cliffs, N.J.

**Figure 10-26** Structural damage to a group of apartment buildings in San Antonio, Texas, caused by shrink-swell behavior of expansive soils. (Photo courtesy of USDA Soil Conservation Service.)

**Figure 10-27** Crack in wall of office building caused by swelling clays, San Antonio, Texas. Office building and adjacent shopping plaza were demolished because of structural damage by swelling soils.

volume changes will not take place because the water content of the soil is constant.

Prediction of swelling behavior can be accomplished by lab tests designed to measure the pressure generated by soil being wetted. A quick estimate of swelling behavior can be made using the Atterberg limits of the soil. Table 10-9 shows a classification of swelling potential based on the plasticity index.

Damage to a structure is possible when as little as 3% volume expansion takes place (Figures 10-26 and 10-27). Failure results when the volume changes are unevenly distributed in the soil beneath the foundation. For example, water-content changes in the soil around the edge of a building can cause swelling pressure beneath the perimeter of the building, while the water content of the soil beneath the center remains constant (Mathewson et al., 1975). The type of failure resulting from this situation is called *end lift* (Figure 10-28). The factors that could cause end lift include excessive lawn or shrub watering beside the building and insufficiently long drain spouts that do not carry drainage from the roof far enough away from the structure. If a house lot is not graded properly, the land may slope toward the house. This would concentrate drainage at the edge of the building and high infiltration could occur. The opposite of edge lift is *center lift* (Figure 10-28), where swelling is focused beneath the center of the structure or where shrinkage takes place under the edges. Either process causes the foundation to bulge beneath the center. A possible cause of soil shrinkage near the edges of a structure is the close proximity of trees. Certain trees remove large quantities of water from the soil by the process of *transpiration*; soil shrinkage in the vicinity of the roots is the result.

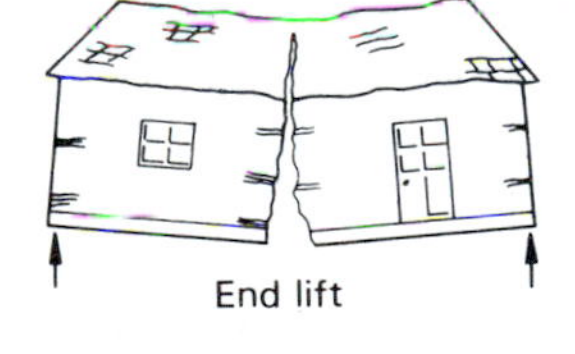


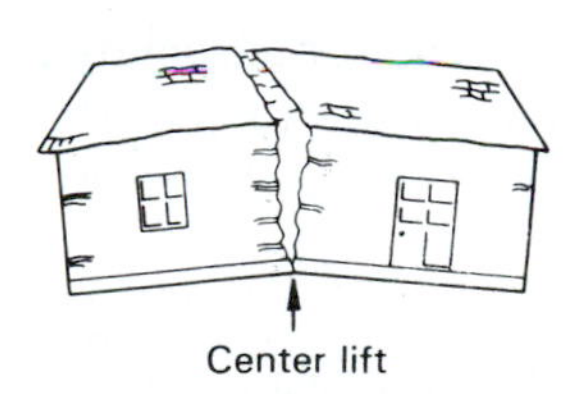


**Figure 10-28** Two types of structural failure caused by swelling soils. (From C. C. Mathewson et al., 1975, *Bulletin of the Association of Engineering Geologists*, 12:275–302. Used by permission of the Association of Engineering Geologists.)

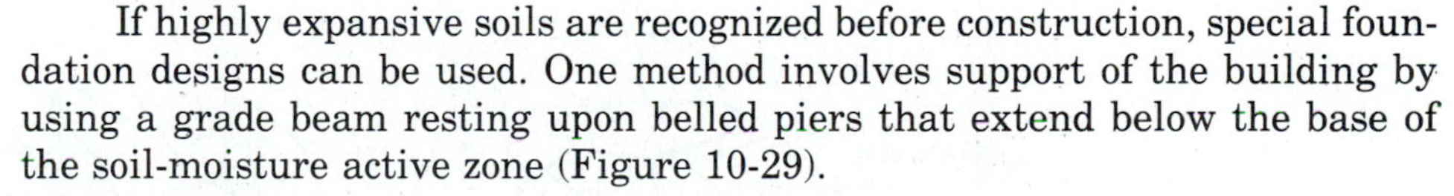

If highly expansive soils are recognized before construction, special foundation designs can be used. One method involves support of the building by using a grade beam resting upon belled piers that extend below the base of the soil-moisture active zone (Figure 10-29).

### *Hydrocompaction*

A different type of soil problem is presented by certain soils in arid regions. When water and load are applied to these soils, their structure collapses to a more dense state. This type of behavior, called *hydrocompaction*, is the result of the origin and history of the soils. Sediments deposited by wind or water usually develop a very loose, open structure. Soils susceptible to hydrocompaction, which are predominantly composed of silt and sand grains rather than

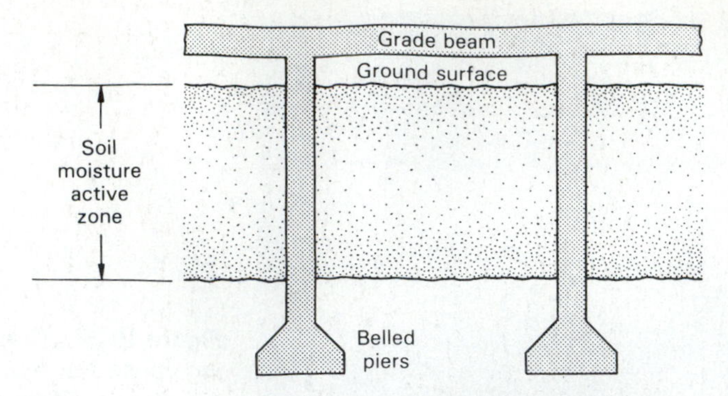


**Figure 10-29** Belled pier and grade beam foundation utilized to prevent damage from expansive soils. (From B. M. Das, *Principles of Foundation Engineering*, copyright © 1984 by Wadsworth, Inc., Belmont, Calif.)

clay particles, exist in an unsaturated state. The low-density structure is maintained by weak bonds formed by water, clay, or soluble precipitants that bridge pores between particles (Figure 10-30). When the pores are only partially filled with water, the water exerts tensional forces that tend to hold adjacent particles together. In addition, small amounts of clay or soluble precipitants (such as gypsum) between particles act as a temporary cementing agent. When water saturates the soil, both the water and clay bonds are destroyed. Pore water loses its tensional effect when the pores become saturated, and clay bonds are suspended in the solution upon wetting. The soil will then collapse to a denser structure under its own weight or the weight of a structure built upon the soil. Saturation is often caused by lawn watering around a foundation and leakage of water lines, storm sewers, and canals. When the collapse of the soil occurs, settlement of foundations or rupture of utility lines is usually the result.

### *Liquefaction*

The term liquefaction is commonly applied to the conversion of saturated sand or silt to a liquid state under rapid or cyclic stresses of the type that might be caused by earthquakes, vibrations, or explosions. Normally, a sand would respond to increased stress or loading by the expulsion of pore water associated with a decrease in void ratio. Water can be driven out readily because of the high hydraulic conductivity of sand. Under rapid or cyclic loading, however,

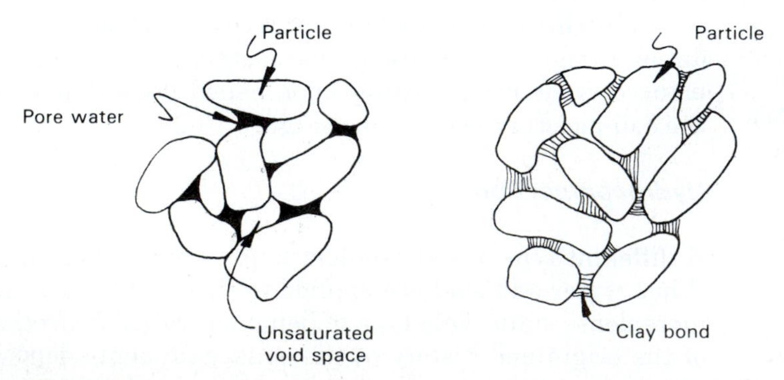


**Figure 10-30** Loose, unsaturated soil structure maintained by pore water and clay bonds.

**Figure 10-31** Rotation of buildings due to liquefaction of saturated sand during the 1964 Niigata, Japan, earthquake. (Photo courtesy of the U.S. Geological Survey.)

the increased stress is transferred to the pore water as the particles are forced closer together. If the pore-water pressure is great enough to suspend the particles within the pore fluid, total loss of shear strength occurs and fluid behavior is the result. An excellent example of liquefaction occurred in the Niigata, Japan, earthquake of 1964 (Figure 10-31). Liquefaction of a saturated sand beneath the foundations of the apartment buildings was responsible for a temporary loss of shear strength in the soils supporting the buildings. When this happened, the buildings rotated or sank into the fluidized soil. Little structural damage resulted from these movements, and the buildings were later jacked upright and new foundations were constructed.

## LAND SUBSIDENCE

*Subsidence* describes several related localized or areally widespread phenomena associated with sinking of the land surface. The vertical movements involved range from sudden collapse to slow, gradual declines in surface elevation. In a general way, subsidence can be subdivided into types caused by (1) removal of subsurface fluids, (2) drainage or oxidation of organic soil, and (3) surface collapse into natural or excavated subsurface cavities.

Subsidence attributed to withdrawal of fluid is a widely distributed occurrence in the United States and other countries. Although it is most commonly heavy pumping of ground water that leads to this type of subsidence, depletion of oil and gas reservoirs has been responsible for subsidence in some areas. A prominent example is the Wilmington oil field in Long Beach, California, in which maximum elevation decreases of about 9 m were attributed to the

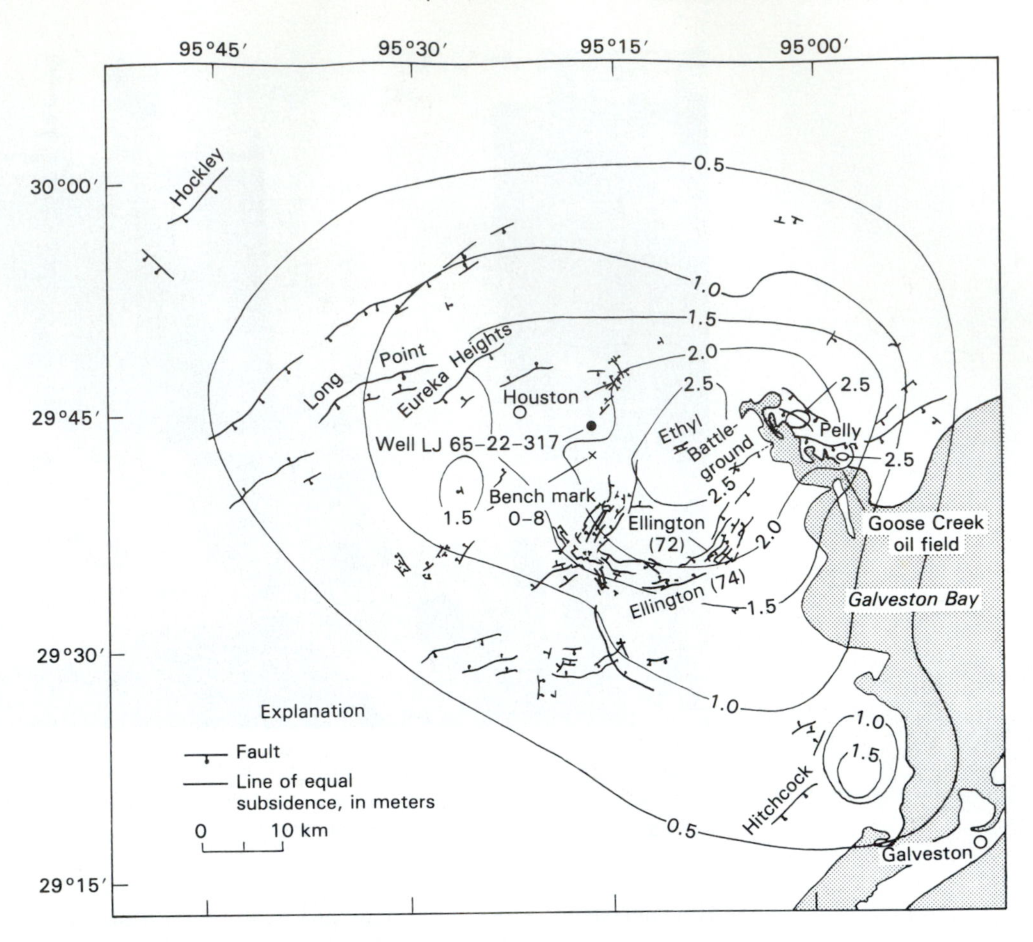


**Figure 10-32** Land subsidence and surface faults in the Houston-Galveston, Texas, area, 1906–1978. (From T. L. Holzer, 1984, Ground failure induced by ground-water withdrawal from unconsolidated sediment, in *Man-Induced Land Subsidence*, T. L. Holzer, ed., Geological Society of America, Reviews in *Engineering Geology*, v. 6.)

withdrawal of petroleum. Direct damage to structures and indirect damage because of increased flooding proved very costly to the city.

Ground-water withdrawals have accounted for subsidence over thousands of square kilometers of land in the western United States and other areas. The problem occurs in agricultural areas like the San Joaquin Valley in California, where the water is used for irrigation, as well as in urban areas worldwide, including Las Vegas, Nevada; Houston, Texas (Figure 10-32); Mexico City, Mexico; and Venice, Italy, where the ground water is withdrawn for municipal water supplies. The subsidence bowl (area undergoing subsidence) in Houston is one of the largest in the United States. The subsidence is correlated with accelerated ground-water pumping since the 1930s, although petroleum withdrawals may also be significant, at least on a local basis. Subsidence is most prevalent in geologically young, poorly lithified sequences of saturated sands and clays. Ground water is pumped from the sandy units, but these materials have low compressibilities. The lowered hydraulic pressure in the sands, however, causes the fine-grained beds to slowly consolidate as pore water is drained into the more permeable beds above and below. The process is closely analogous to a consolidation test. Because fine-grained beds are more compressible, most of the ultimate subsidence can be accounted for by consolidation of silt and

**Figure 10-33** Fissures formed during land subsidence in southern Arizona. (Photo courtesy of Thomas Holzer.)

clay beds. This process is irreversible and slow because of the low permeability of the materials.

Accompanying subsidence in many locations are ruptures of the land surface. These ground failures take the form of fissures as long as several kilometers, in which the surface cracks or separates laterally (Figure 10-33), or faults, in which vertical displacement across the rupture occurs. More than 160 faults in the Houston-Galveston area have been mapped (Figure 10-32). Levelling surveys have shown that recent movement across the faults corresponds to the long-term trends of ground-water pumpage. In some instances, it has been possible to demonstrate that the faults existed prior to subsidence and were simply reactivated by ground-water withdrawal. It is difficult to quantify the component of total subsidence caused by pumpage because some natural subsidence is also occurring due to processes such as salt-dome growth and active tectonic warping.

Subsidence by a different mechanism occurs when highly organic soils are drained during land-use conversion to agriculture or construction. Soils of this type, frequently termed *peat* or *muck*, are formed in marshes, bogs, and other poorly drained settings with water tables near the land surface. The high water table inhibits oxidation of organic matter, so this material accumulates in the soil. Characteristics of the soils include dark color, low density, and high compressibility. When drained, organic soils are exposed to oxidizing conditions as the water table drops. The resulting decomposition of the organic content, in combination with some physical compaction due to the removal of pore fluids, leads to the gradual subsidence of the land surface. Damage to buildings can be caused by settlement of the structure if it bears upon the organic soils. If the building is constructed on foundations that bear upon stronger materials below the organic soils, subsidence of the soil around the structure can leave a gap between the floor slab and the ground surface (Figure 10-34 and Case Study 10-2).

**Figure 10-34** Subsidence of peat and muck soils around a house built on pilings. Land surface around house has subsided about 0.6 m in 2 years. Porch and steps have fallen off. (Photo courtesy of USDA Soil Conservation Service.)

The third major type of subsidence is associated with the collapse of overlying materials into large underground cavities. Although the processes that cause this type of subsidence generally occur in bedrock below the soil zone, collapse of the surface soils into the voids below is the usual manifestation. The cavities may be excavated or natural. Subsidence over underground coal mines constitutes the most common example of subsidence over excavated cavities. This process has affected about 8000 $km^2$ in the United States, mostly in the eastern part of the country. The effects of subsidence depend upon the method of mining and the thickness of the overburden. When the coal is entirely removed from a seam during mining, surface subsidence is usually contemporaneous with mining. A basin-shaped depression commonly develops over the mined-out area if the coal seam is deeper than about 30 meters. Alternatively, columns or *pillars* of coal are left in place to provide support for the mine roof. Subsidence above mines of this type may take place many years later after the mine is abandoned. Unfortunately, buildings, or even towns, have been built unknowingly over abandoned underground mines. The modes of subsidence are shown in Figure 10-35. Two of the most common surficial impacts of subsidence over abandoned mines are *sinkholes* or *troughs*. Sinkholes (Figure 10-36) are circular pits formed by collapse of the surface soil into the mine voids when the coal seam is fairly shallow. Troughs are larger subsidence features that develop over deeper mines.

Natural cavities form in rock primarily by dissolution of the rock by circulating ground water. Limestone caves are the most common example of solution cavities, although other soluble rock types behave similarly. Surface collapse, leading to the development of sinkholes, often occurs in dry periods when water tables are declining, or when water tables are lowered artificially by pumping. Ground water below the water table provides a measure of support for the sediment above a limestone cavern by the hydrostatic pressure exerted by the fluid. When the water table drops, this support is lost and surface collapse is initiated (Figure 10-37). The Winter Park, Florida, sinkhole (Figure 10-38), which formed in 1981, provides a spectacular example of urban disruption by subsidence over a cavity in soluble rock. The collapse occurred over a period of several days with no prior warning.

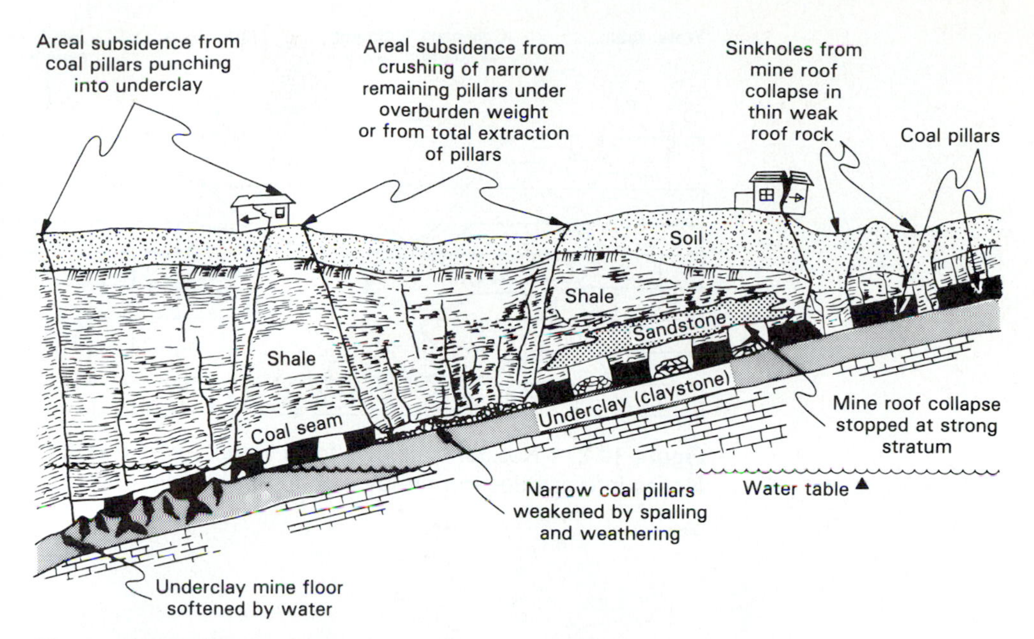


**Figure 10-35** Mechanisms of subsidence associated with subsidence over abandoned underground coal mines. (From R. E. Gray and R. W. Bruhn, 1984, Coal mine subsidence—eastern United States, in *Man-Induced Land Subsidence*, T. L. Holzer, ed., Geological Society of America, Reviews in *Engineering Geology*, v. 6.)

**Figure 10-36** Sinkholes formed by subsidence over abandoned coal mines, western North Dakota. (Photo courtesy of North Dakota Public Services Commission.)

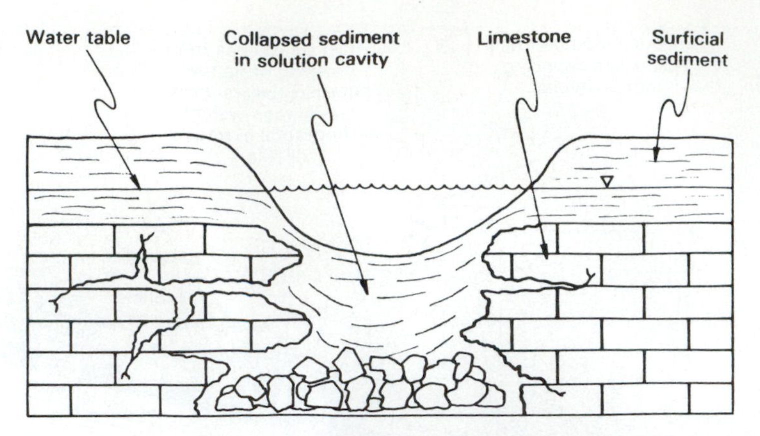


**Figure 10-37** Natural sinkholes form when surficial sediment collapses into cavities formed by solution in soluble rocks.

**Figure 10-38** The Winter Park, Florida, sinkhole as it appeared just after formation in 1981. (Photo courtesy of Jammal & Associates.)

## CASE STUDY 10-2

## Land Subsidence in New Orleans

With respect to construction and development, the geologic setting of New Orleans is one of the most challenging in the world. Its major problems stem from its location on the low-lying Mississippi Delta. Despite being 75 km from the mouth of the river, the average elevation of the City is only 0.4 m above the elevation of the Gulf of Mexico. It is of more than slight concern that periodic floods of the river rise to as much as 6.5 m above the Gulf of Mexico through the heavily diked urban area. Hurricanes, which are common to the Gulf coast, provide another potentially devastating hazard for which the city must prepare.

As if those problems were not enough, the foundation conditions of the city are extremely poor, particularly throughout the extensive areas of swamp and marshland occupied by urban development. The general subsurface conditions are shown in Figure 10-39. The Holocene deposits are composed for the most part of fine-grained deltaic and fluvial sediments. In cypress swamps and other low-lying areas, these materials are rich in organics and highly

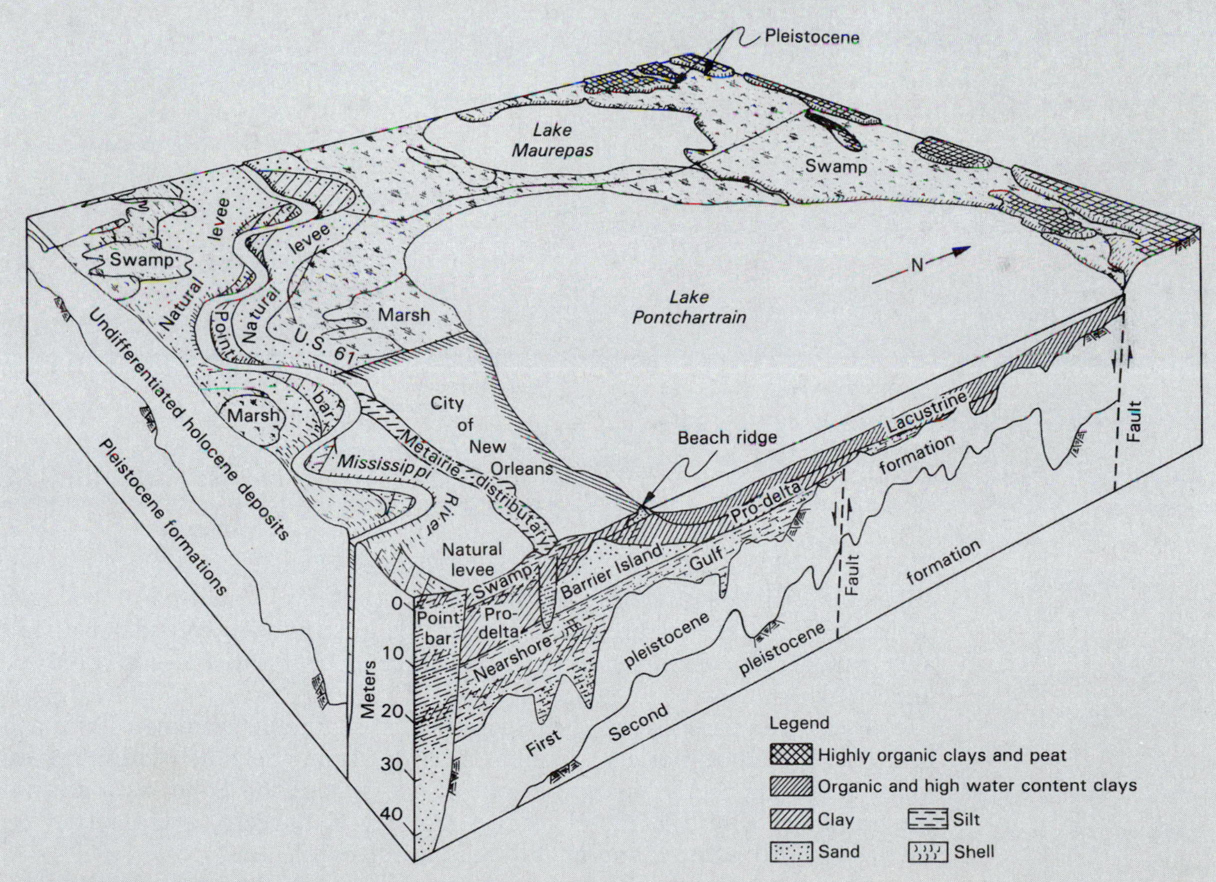


**Figure 10-39** Block diagram showing surface and subsurface geological conditions of New Orleans, La. (From C. R. Kolb and R. T. Saucier, 1982, Engineering geology of New Orleans, in *Geology Under Cities*, R. F. Legget, ed., Geological Society of America, Reviews in *Engineering Geology*, v. 5.)

**Figure 10-40** Subsidence of organic soils in New Orleans after drainage. Photo on left shows house built on pilings just after construction in 1941. Photo on right shows same house in 1992. Front-porch steps were added over the years as ground surface subsided. (Photo courtesy of J. O. Snowden.)

compressible. Large, heavy structures are built on piles that are driven into the more competent Pleistocene sediments that usually lie within 25 m of the surface. Smaller structures built in former swamp areas rely on shorter timber piles limited to the Holocene sediments.

To proceed with expansion of the city into marsh areas, drainage of the surface soils and lowering of the water table was carried out. This led to rapid subsidence because the organic rich soils occupy one-half or less the volume in a dry condition compared with their natural state. Lowering of the water table also initiates oxidation and decomposition of the soils, with further, slow subsidence. For single-family homes and other lightly loaded structures, the consequence is a sinking land surface relative to the house, which rests on piles. Left untouched, it is possible to see under the entire foundation slab. The efforts to deal with the problem at one home are shown in Figure 10-40. As the years went by, the homeowners simply added steps in increments to reach the gradually sinking front yard. Total subsidence at the house is 0.81 m. Subsidence up to 2 m has been documented in other areas.

## SUMMARY AND CONCLUSIONS

The term soil has different meanings to various groups of earth scientists and engineers. Soil scientists are concerned with material produced by pedogenic processes that can sustain plant growth. Soils engineers consider soil to be any transported sediment or residual material that can be excavated with hand tools.

Pedogenic processes acting upon rock or sediment produce recognizable horizons. The main horizons include the A horizon, the zone of high organic-matter content; the E horizon, a zone of eluviation; the B horizon, a zone of illuviation of clay and metal oxides; and the C horizon, a transition zone between highly altered and nonaltered parent material.

The development of horizons in the soil profile is controlled by the five soil-forming factors: *Parent material* and *topographic position* (relief) set the stage for the action of *climate* and *organisms* over a period of *time*. Of these factors, climate is dominant. The possible variations and interactions between the soil-forming factors can produce a vast range of soil types. As a result, pedogenic classification of soils has been difficult. A modern classification system based entirely on measurable soil properties has replaced an older, inadequate system. Because of the complexity of the new system, some terms and concepts from the older system are still in use.

The objective of engineering soils classification is to predict soil behavior in an engineering project. The index properties, which are used to classify soils, can be correlated with the strength and compressibility of the soil. For cohesionless soils, the relevant index properties are grain-size distribution, grading, in-place density, and relative density. Cohesive soils are more difficult to characterize because of the interaction between clay minerals and water. The index properties used include clay content, consistency, water content, sensitivity, and the Atterberg limits. The index properties are used to classify the soil in the Unified Soil Classification System.

The performance of a soil as a foundation support material is dependent upon its shear strength and compressibility. The bearing capacity of the soil is rarely exceeded by a structure, as it was beneath the Transcona grain elevator. Even though a soil may not fail, damage to structures will be experienced if settlement takes place. Settlement is the subsidence of a foundation because of compression of the underlying soil. Consolidation is the time-dependent compression of clay soil as pore water is slowly expelled from the loaded soil.

Many of the engineering problems associated with clay can be explained by its colloidal properties. The surface charge of clay particles attracts water and cations from the pore solution. The attraction or repulsion of adjacent particles during deposition determines the initial fabric. During consolidation, reorientation of particles often occurs. The low strength of clay is indicative of high Atterberg limits and high water contents.

Engineering projects must often be built in areas where hazardous soil processes are active. The tendency of clay soils to absorb water leads to volume expansion that can be very damaging to structures. Within the soil-moisture active zone, changes in water content of the soil will lead to swelling or shrinking. Land-use changes are often responsible for this behavior. Construction and landscaping practices that promote swelling include improper lot grading and excessive lawn watering. If highly expansive soils are present, special foundation designs may be necessary.

Hydrocompaction is a type of soil collapse that takes place when arid-region soils are saturated and loaded. Saturation destroys soil moisture or clay bonds that impart temporary strength to the soils. When these bonds are removed, the soil collapses to a more dense state.

Subsidence causes numerous engineering problems due to the collapse or sinking of the land surface. These phenomena originate by the withdrawal of fluids from the soil, oxidation, or drainage of organic soils, and the subsequent collapse of surface material into subsurface cavities.

## PROBLEMS

1. Why is there confusion or disagreement in the use of the term *soil*?
2. Summarize the processes that lead to the formation of soil horizons.
3. Under what conditions do soils become mineral deposits?
4. Explain the process of soil swelling.
5. Under what conditions are cohesionless soils susceptible to hydrocompaction?
6. Using the volumetric relationships in soils presented in Figure 10-5, show that $e = \frac{n}{1-n}$.

7. A sample of undisturbed soil has a dry weight of 110 lb and a volume of 1 $ft^3$. If the specific gravity of the soil solids is 2.65, determine the void ratio and the porosity.
8. A container with a volume 0.0084 $m^3$ is filled with 13.62 kg of dry sand. The container is then filled carefully with water so that the condition of the sand is not changed. When filled the total weight (mass) of the soil and water is 16.98 kg. What is the void ratio of the sand, and what is the specific gravity of the soil particles?
9. An undisturbed sample of clay has a volume 0.52 $ft^3$. The wet weight is 64 lb and the dry weight is 50 lb. If the specific gravity of the clay particles is 2.66, find the water content, void ratio, and degree of saturation.
10. A sample of saturated clay taken from an excavation has a volume of 500 $cm^3$. If the sample has a wet weight of 810 g and a dry weight of 500 g, what is the specific gravity of soil solids?
11. How does the engineering classification of soils differ from the engineering classification of rocks?
12. What are the Atterberg limits, and what is their usefulness?
13. Give the USCS symbol for the following soils.
    - **(a)** 40% of sample passes the No. 200 sieve; 30% of coarse fraction passes the No. 4 sieve; contains significant silt fraction.
    - **(b)** 4% passes the No. 200 sieve; 60% of coarse fraction passes the No. 4 sieve; little or no fines; poorly graded.
    - **(c)** 96% passes the No. 200 sieve; PI-40%; LL-65%.
14. Give some examples of the significance of the properties of clay minerals in soils engineering.
15. What soil characteristics influence compressibility?
16. Why is consolidation a time dependent process?
17. A saturated cohesive soil is tested in an undrained strength test in a triaxial cell. The major and minor principal stresses at failure are 1.5 $kg/cm^2$ and 0.44 $kg/cm^2$, respectively. Determine the cohesion and the unconfined compressive strength of the material. What is the consistency classification of the soil?
18. What geological conditions are likely to lead to land subsidence upon groundwater withdrawals?

## REFERENCES AND SUGGESTIONS FOR FURTHER READING

BIRKELAND, P. W. 1984. *Soils and Geomorphology*. New York: Oxford University Press.

BLOOM, A. L. 1991. *Geomorphology: A Systematic Analysis of Late Cenozoic Landforms*, 2d ed. Englewood Cliffs, N.J.: Prentice-Hall, Inc.

BUOL, S. W., F. E. HOLE, and R. J. MCCRACKEN. 1973. *Soil Genesis and Classification*. Ames, Iowa: Iowa State University Press.

DAS, B. M. 1984. *Principles of Foundation Engineering*. Monterey, Calif: Brooks/Cole Engineering Division.

GRAY, R. E., and R. W. BRUHN. 1984. Coal mine subsidence—eastern United States, in *Man-Induced Land Subsidence*, T. L. Holzer, ed. Geological Society of America, Reviews in *Engineering Geology*, 6:123–149.

HOLTZ, R. D., and W. D. KOVACS. 1981. *Introduction to Geotechnical Engineering*. Englewood Cliffs, N.J.: Prentice-Hall, Inc.

HOLZER, T. L. 1984. Ground failure by ground-water withdrawal from unconsolidated sediment, in *Man-Induced Land Subsidence*, T. L. Holzer, ed. Geological Society of America, Reviews in *Engineering Geology*, 6:67–101.

JUDSON, S., M. E. KAUFFMAN, and L. D. LEET. 1987. *Physical Geology*, 7th ed. Englewood Cliffs, N.J.: Prentice-Hall, Inc.

KOLB, C. R., and SAUCIER, R. T. 1982. Engineering geology of New Orleans, in *Geology Under Cities*, R. F. Legget, ed. Geological Society of America, Reviews in *Engineering Geology*, 5:75–93.

MATHEWSON, C. C., J. P. CASTLEBERRY, and R. T. LYTTON. 1975. Analysis and modeling of the performance of home foundations of expansive soils in central Texas. *Bulletin of the Association of Engineering Geologists*, 12:275–302.

MCCARTHY, D. F. 1993. *Essentials of Soil Mechanics and Foundations*, 4th ed. Englewood Cliffs, N.J.: Prentice-Hall, Inc.

PECK, R. B., and F. G. BRYANT. 1953. The bearing capacity failure of the Transcona elevator. *Geotechnique*, 3:201–208.

SOIL SURVEY STAFF. 1960. *Soil Classification, A Comprehensive System—7th Approximation.* U.S. Department of Agriculture, Soil Conservation Service.

TOURTELOT, H. A. 1974. Geologic origin and distribution of swelling clays. *Bulletin of the Association of Engineering Geologists*, 11:259–275.

WHITE, L. S. 1953. Transcona elevator failure: eyewitness account. *Geotechnique*, 3:209–214.

# 11

# Ground Water

The importance of ground water to our society has never been greater than it is today. There is no more fundamental natural resource than fresh water; we depend on it daily for drinking, sanitation, agriculture, industry, and recreation. In the past, most of our needs could be met by such easily accessible surface-water sources as rivers, lakes, and reservoirs. Those days have passed however; the potential for expansion of surface-water sources is limited. Ground water is our only alternative for obtaining large amounts of fresh water at a reasonable cost. Currently, nearly one-half the population of the United States uses ground water as a drinking water source; soon, one-half the total water usage will come from ground water.

The steadily increasing use of ground water has raised important concerns about this resource. First, we must learn to evaluate and manage ground-water reservoirs so that they are not mined to exhaustion like ore deposits or other nonrenewable resources. Second, we have realized that ground water is not a pristine substance that exists in total isolation from activities taking place on the land's surface. Instead it is now known that past waste disposal practices as well as the storage and handling of hazardous materials have contaminated ground-water supplies in many locations. Extensive governmental and private efforts are now underway to clean up the most severe contaminant sites and to protect the vast majority of unaffected ground-water reservoirs. Contamination and remediation in the subsurface environment is the subject of Chapter 16.

The occurrence and movement of ground water were once thought to be mysterious and unpredictable. Even today, misconceptions about ground water are held by the general public, the media, and even some geologists and engineers. A thorough understanding of ground water is critical for engineers because subsurface fluids are of major importance in civil, environmental, mining, and petroleum engineering.

## GROUND-WATER FLOW

### Darcy's Law

A nineteenth-century French engineer named Henri Darcy laid the groundwork for the modern study of ground water with experiments involving the flow of water through a column filled with sand. These experiments established that the volumetric flow rate of water through saturated sand was proportional to the energy gradient, or the loss of energy per unit length of flow path. A schematic of Darcy's apparatus is illustrated in Figure 11-1. To express *Darcy's Law* in a commonly used manner, we will define the *specific discharge*, $v$, as the volumetric flow rate, $Q$, measured in cubic meters per second or similar units, divided by the cross-sectional area of the flow tube $A$. Thus

$$v = \frac{Q}{A} \qquad \text{(Eq. 11-1)}$$

The form of energy involved in the flow process is the mechanical energy possessed by the fluid at each point in its flow path. This energy is the sum of three components: elevation, or position (potential energy); movement (kinetic energy); and pressure. When dealing with ground water, we can neglect the kinetic energy contribution because of the very slow velocities of ground-water flow. The remaining components, elevation and pressure, need not be evaluated individually but, instead, can be measured together by the level to which water

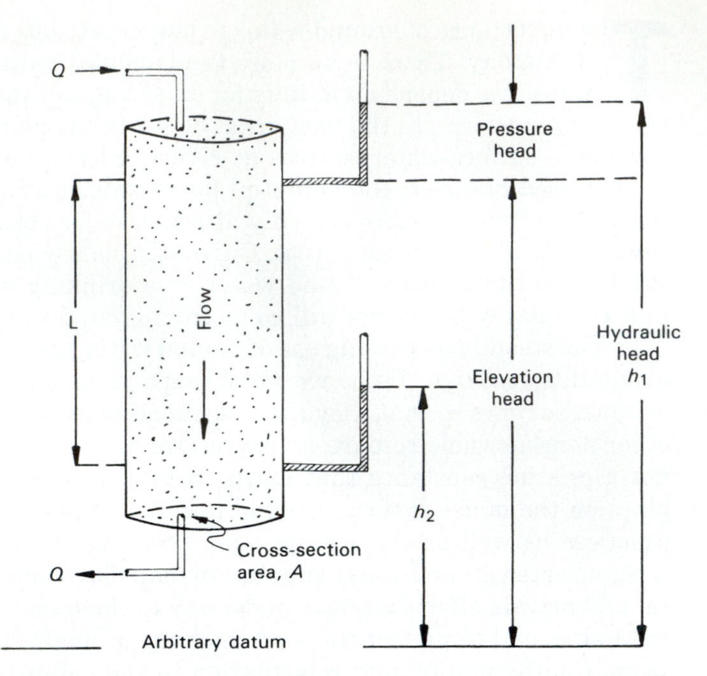


**Figure 11-1** Schematic diagram of a lab apparatus that illustrates the parameters involved in Darcy's Law.

at any point in the flow system will rise above an arbitrary datum. This parameter, as indicated in Figure 11-1, is called *hydraulic head.* The thin tubes through which water rises to measure the head are called *manometers.*

Also shown in the diagram are the two components of hydraulic head. Elevation head is the distance above the datum to the level of the manometer intake, and pressure head is the height of rise of water in the tube above the intake. Notice that the elevation above the datum, or head, in the manometers decreases in the direction of flow. This is a consequence of the loss of mechanical energy along the flow path as it is converted to heat through friction between the fluid and the sand grains and also because of friction between water molecules in the fluid. Darcy's Law requires evaluation of the *hydraulic* (energy) *gradient* and this is expressed as

$$I = \frac{h_1 - h_2}{L} \qquad \text{(Eq. 11-2)}$$

where $I$ is the hydraulic gradient, $h_1$ and $h_2$ are head values at the points where the manometers are inserted into the flow system, and $L$ is the distance between the manometers measured in the flow direction.

A final parameter is required for a complete expression of Darcy's Law, and that is the constant of proportionality between the specific discharge and the hydraulic gradient. With this term, $K$, Darcy's Law can be stated as

$$\frac{Q}{A} = v = -K\frac{h_2 - h_1}{L} \qquad \text{(Eq. 11-3)}$$

or, in differential form,

$$v = -K\frac{dh}{dl} \qquad \text{(Eq. 11-4)}$$

The minus sign simply indicates that ground-water flow is a mechanical process with an irreversible loss of mechanical energy, or head, in the direction of flow.

The constant $K$, known as the *hydraulic conductivity*, is a very important parameter in ground-water hydrology. It is related to intrinsic permeability, the ability of a porous medium, like the sand in the flow tube, to transmit a fluid under a given hydraulic gradient (Equation 6-3). A distinction is often made between the terms hydraulic conductivity and permeability because *hydraulic conductivity* is defined to include the properties of the fluid as well as the properties of the medium, whereas *permeability* is restricted to the properties of the medium. It is easy to visualize that more water would flow through a tube containing gravel than one filled with silt or clay, under the same hydraulic gradient. It is also true, however, that more water than molasses would flow through sand, again with a constant hydraulic gradient. Therefore, it is important to remember that the proportionality constant in Darcy's Law, $K$, encompasses both the characteristics of the fluid and the properties of the medium. The combination of the two into one constant is convenient in ground-water work because the density and viscosity of water in most near-surface ground-water reservoirs do not vary greatly. Hydraulic conductivity can then be assumed to represent the permeability of the units through which the ground water flows.

### *Darcy's Law under Field Conditions*

One of the reasons Darcy's Law is so useful is that it can be directly applied to field situations. The device used to measure head in the field is called a *piezometer* (Figure 11-2), and, as in a manometer, the head measurement depends on the level to which water in a tube rises above a datum. In this case, the datum used is sea level and the tube is a pipe inserted into a hole drilled into the ground, which is much like a normal water well. A piezometer differs from a well in several respects, however. A well usually contains a water intake section, or *screen*, that is as long as possible so that the yield of the well can be maximized. A piezometer, on the other hand, has a very short

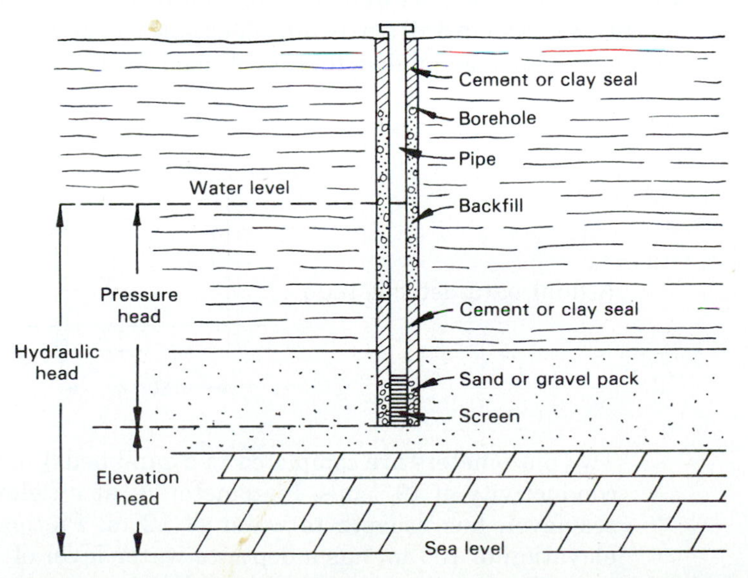


**Figure 11-2** Design of a piezometer, the instrument used for measuring hydraulic head in the field.

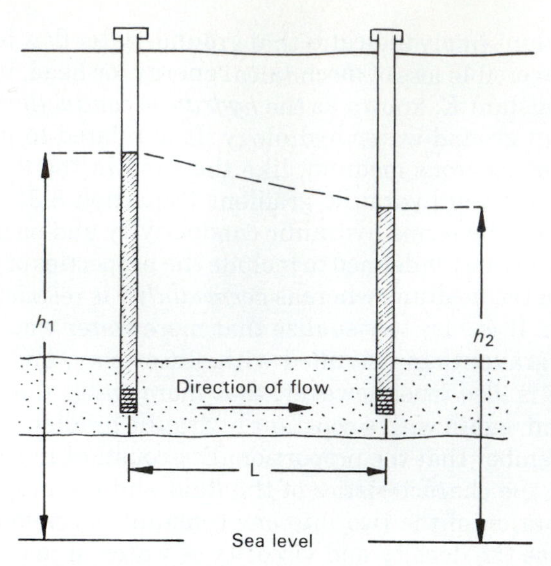


**Figure 11-3** Ground-water flow through a bed of rock or sediment. The flow direction is indicated by a decrease in head from left to right. Hydraulic gradient can be determined in the same way as in the Darcy experiment.

screened interval in order to obtain a head measurement at one specific point in a flow system. Also, piezometers are usually smaller in diameter than wells and should be carefully sealed with cement or clay above the screen to isolate the point of measurement from other parts of the ground-water flow system.

With head measurements obtained from several piezometers, it is possible to calculate the hydraulic gradient. Head measurements in a piezometer are made by determining the depth to water from the top of the pipe with a tape and then subtracting the depth from the elevation of the top of the pipe. Figure 11-3 illustrates the similarity between ground-water flow through a subsurface rock unit and flow through the Darcy apparatus (Figure 11-1). The hydraulic gradient indicates the direction of ground-water flow. In many places, ground water has a vertical as well as a horizontal component of flow. In these areas, it is necessary to install a network of piezometers at various depths to determine the three-dimensional distribution of head in the flow system.

**EXAMPLE 11-1**

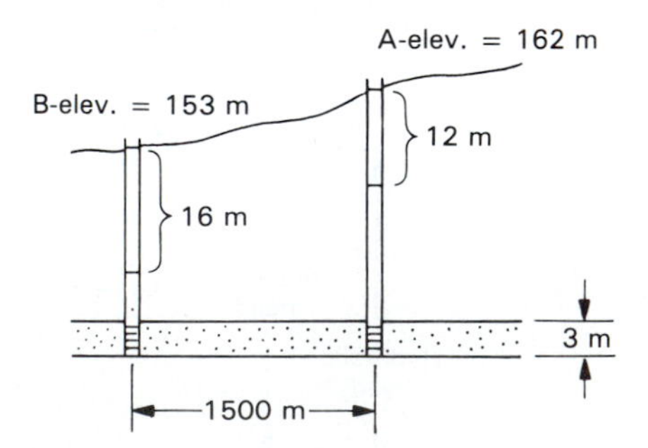


Two piezometers are completed in a sand bed 3 m thick with a hydraulic conductivity of $10^{-4}$ m/s. Piezometer A, at an elevation of 162 m above sea level, has a depth to water of 12 m. Piezometer B, located at an elevation of 153 m, has a depth to water level of 16 m. The direction of ground-water flow is in the same direction as a straight line drawn between A and B, which are separated by a distance of 1500 m. What is the rate of flow (in $m^3/s$) through a cross-section of the bed perpendicular to flow equal to its thickness and 1 m wide?

***Solution.***

In piezometer A, the hydraulic head, $h_A$, = 162 m − 12 m = 150 m. In piezometer B, the head, $h_B$, = 153 m − 16 m = 137 m. Since the two piezometers are installed along the direction of flow, we can apply Darcy's Law directly:

$$v = -K\frac{h_A - h_B}{L} = \frac{-(10^{-4}\ \text{m/s})\ 150\ \text{m} - 137\ \text{m}}{1500\ \text{m}}$$

$$= 8.7 \times 10^{-7}\ \text{m/s}$$

We can neglect the negative sign because it merely tells us that flow is in the direction from higher head to lower head (from A to B). Finally, because we need to determine the total flow through a column of sand 1 m wide and equal to the thickness of the bed (3 m), and because $v = Q/A$,

$$Q = v \times A = 8.7 \times 10^{-7}\ \text{m/s} \times 3\ \text{m}^2 = 2.6 \times 10^{-6}\ \text{m}^3/\text{s}$$

### The Water Table

During drilling for installation of a well or piezometer, water is encountered under different physical conditions. The uppermost zone contains water in the pores of rock and soil along with a gas phase. The incomplete filling of void spaces by water gives rise to the name *unsaturated zone* for this part of the subsurface (Figure 11-4a). The unsaturated zone is also referred to as the *vadose zone*. Below the unsaturated zone, at a depth determined by many factors (including climate, topography, and geologic setting), void spaces in the material are filled with water; this region is called the *saturated zone*. The saturated zone can be subdivided into the *phreatic zone*, below a surface known as the *water table*, and the *capillary fringe*, which lies above the water table. The capillary fringe and phreatic zones can be separated by the value of fluid

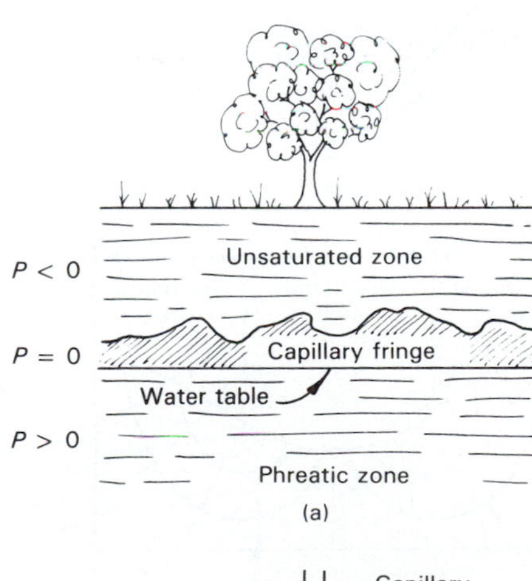


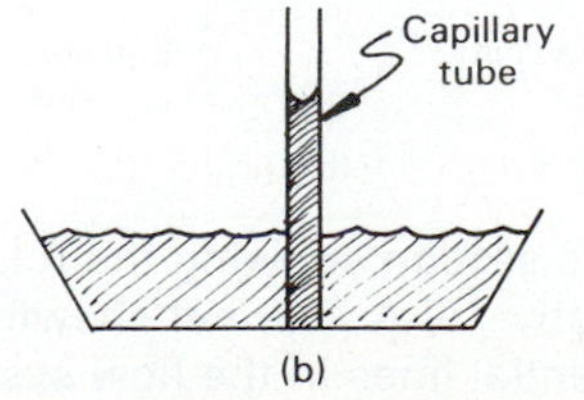


**Figure 11-4** (a) Zones of subsurface water. The relative value of fluid pressure is shown to the left of each zone. Zero represents atmospheric pressure. (b) Rise of water in a capillary tube—the same phenomenon that produces the capillary fringe.

pressure in the pore water at a particular point. Just as fluid pressure in a lake increases from zero at the surface (that is, normal atmospheric pressure) to greater values with depth, the fluid pressure in the saturated zone increases with depth from the surface where it is equal to atmospheric pressure, which is the horizon that is defined as the *water table*. The water table can be located by measuring the water-level elevation in a shallow well that only extends several meters into the saturated zone. The water level in such a well corresponds to the water table at that point.

The saturated material above the water table contains fluid in which the fluid pressure is less than atmospheric. This zone, the *capillary fringe*, is analogous to the rise of water above the free-water surface in a capillary tube (Figure 11-4b). The capillary rise is caused by adhesion between the water and the capillary tube and *surface tension* at the curved surface, or *meniscus*, of the water at the top of the capillary tube. Surface tension is an upward-directed force that results from the tendency of the water surface to assume a shape of minimum area when in contact with another fluid (in this case, air) with which it does not mix. If the pores in the soil are small, as in clay and silt, capillary rise above the water table occurs just as in a single capillary tube. The capillary effect varies with soil type; a small-to-nonexistent rise occurs in gravel, whereas rises of several meters or more are possible in fine-grained soils with high silt and clay contents. The fluid pressure, then, will be positive (greater than atmospheric) below the water table in the phreatic zone, equal to atmospheric at the water table, and negative (less than atmospheric) in the capillary fringe and in the unsaturated zone.

### *Ground-Water Flow Systems and Flow Nets*

Figure 11-5a shows a cross section through an area of hilly topography. The position of the water table, high beneath the hills and near the surface in the valley, gives a good indication of the distribution of head. Ground water will flow from the areas where head is highest, called *recharge areas*, to areas where

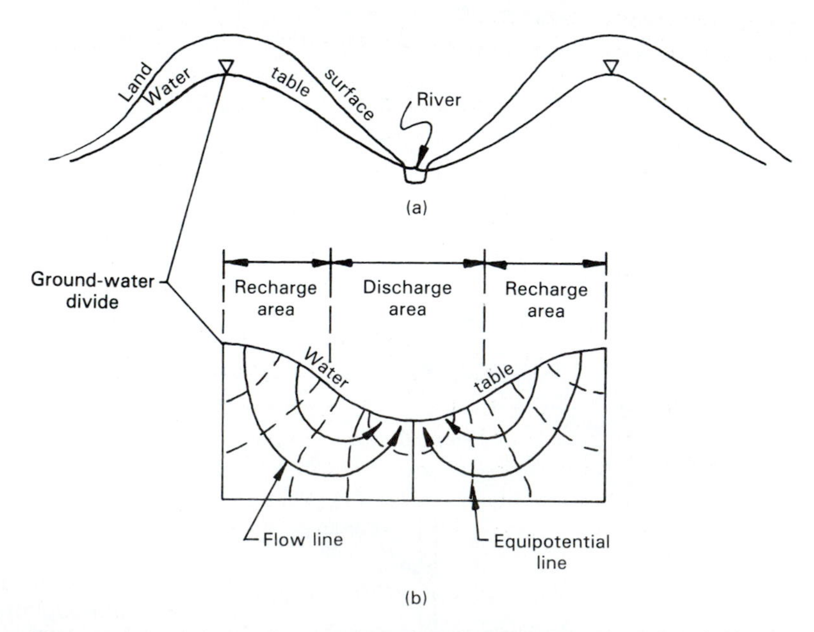


**Figure 11-5** (a) A cross section showing the relationship of the water table to topography. (b) A flow net showing the position of flow lines and equipotential lines in the flow system.

head is lowest, called *discharge* areas. The discharge area in Figure 11-5a is a river channel. Ground water emerges as springs and seeps along the sides and bottoms of the channel, thereby sustaining the flow of the stream during dry periods. The component of stream flow derived from ground water is the *base flow*. Rivers that flow all year because of ground-water discharge are *perennial streams*, while streams that flow only during periods of rainfall that generate surface runoff are *ephemeral streams*.

Aside from recognizing that ground water will somehow flow from the recharge areas to the discharge areas in Figure 11-5a, we would find it useful to know exactly what paths the ground water will follow in this flow system. If we make some simplifying assumptions, the problem can be solved mathematically and graphically. A solution requires specified boundary conditions for the region of flow. For this flow system we can specify *no-flow* boundaries, that is, boundaries normal to which there is no flow, at the bottom and sides of the flow region in Figure 11-5b. The left and right no-flow boundaries are considered to be imaginary because if the hills are symmetrical, a *ground-water divide* will develop near the crest of the hill and serve as a plane of symmetry. Ground water will tend to flow away from this vertical plane as it moves toward discharge points in valleys on opposite sides of the hill. The lower no-flow boundary could represent an impermeable bed of rock at this depth in the flow system.

With the boundary conditions that we have established, we can determine the head at any point by developing and solving equations that describe flow in this system, using numerical methods that require computers to calculate the distribution of head, or by using graphical techniques. Once we know the distribution of head, we can draw contour lines called *equipotential lines* on the diagram as in Figure 11-5b. Equipotential lines connect points of equal head. The configuration of the equipotential lines will allow us to draw one other group of lines, *flow lines*. These lines, which are drawn perpendicular to equipotentials, indicate the paths that ground water will follow in this flow system, the objective of this exercise. Thus the resulting *flow net* indicates that under these conditions, ground water travels along long, curving paths from a recharge point to a discharge point. A flow net drawn for an area of glacial topography in western Michigan is shown in Figure 11-6. A flow net such as this is only an approximation of ground-water flow because of the complexity of the subsurface geology. Most flow systems involve materials that are *heterogeneous* to some extent; that is, materials that vary in hydraulic conductivity. If these variations are relatively simple and continuous, as, for example, in a sequence of beds of known hydraulic conductivity extending through the entire region of flow, the flow net can represent their effects upon ground-water movement. If the hydraulic conductivity variations are discontinuous, or if the subsurface geology is not well known, however, the flow net is only a generalized depiction of ground-water flow.

When ground-water flow in an area is predominantly horizontal, a flow net can also be drawn on a map or plan view. The contours of head used in this case are simply contours of the water-table elevation (Figure 11-7). The flow lines, which show the direction of ground-water movement, are very useful in predicting the movement of pollutants that may be introduced into the flow system by a landfill or other waste-disposal site.

### Recharge and Discharge Processes

The input to the dynamic, circulating ground-water flow systems below the earth's surface is recharge in the form of rainfall and snowmelt that percolates

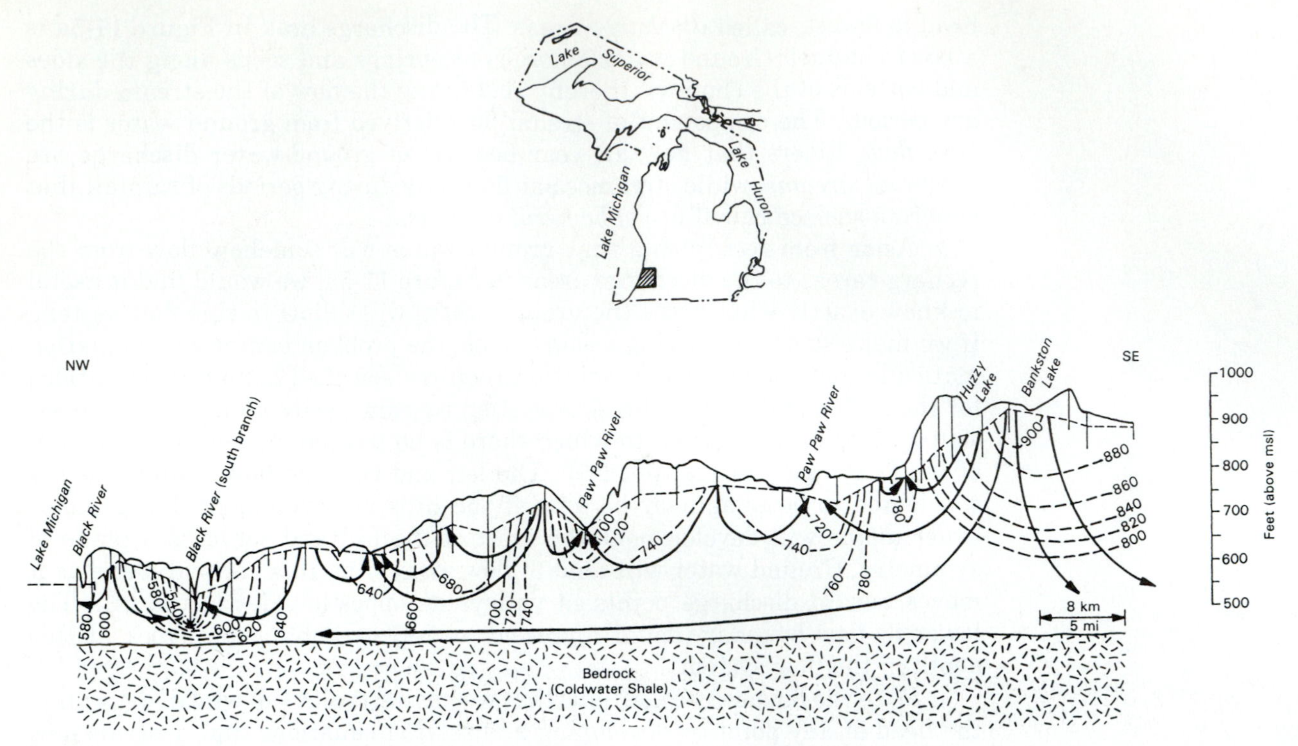


**Figure 11-6** Conceptual ground-water flow net through a series of ridges and valleys formed by glacial advances in western Michigan. (From Straw et al., 1993.)

downward from the surface to the water table. Without recharge, the water table would steadily drop, wells would go dry, and the great ground-water reservoirs of the earth would be depleted.

When rainfall hits the ground surface or when snow melts, several possible pathways are available for water movement (Figure 11-8). If the water sinks

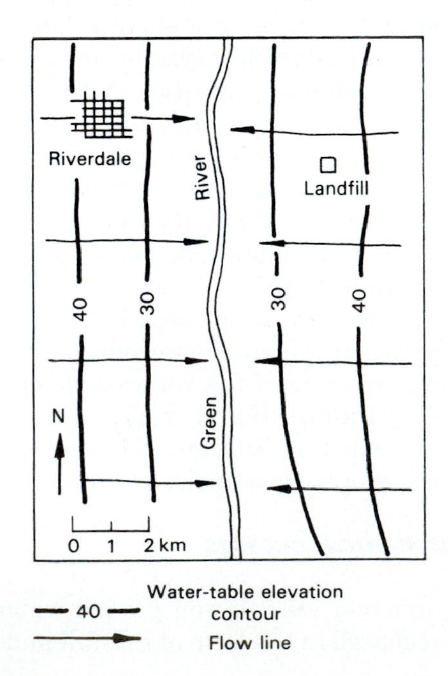


**Figure 11-7** A flow net drawn on a map of an area. The water-table contours indicate flow from the east and west sides of the area toward a discharge area in the river valley.

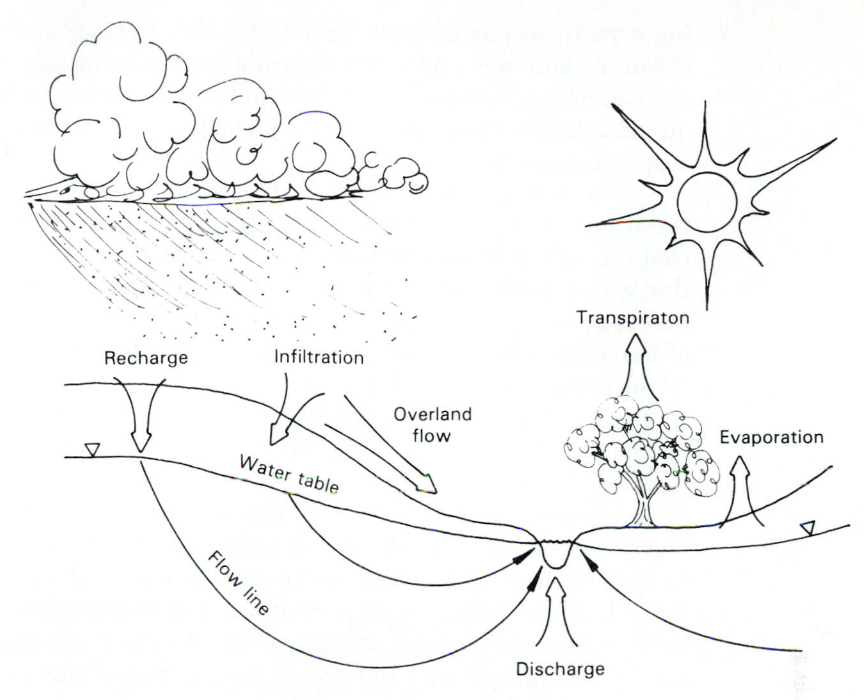


**Figure 11-8** The interaction between atmospheric water and ground water.

into the soil, *infiltration*, has occurred. Soils differ greatly in the rate at which water is absorbed, which is called the *infiltration capacity*. Coarse-grained soils usually have higher infiltration capacities than fine-grained soil. The infiltration capacity is also influenced by such other factors as slope, type of vegetation, and the existing soil-moisture condition. As infiltration continues, the infiltration capacity decreases in a manner shown in Figure 11-9. The limiting value of the infiltration capacity is the hydraulic conductivity of the soil, which is reached when the soil becomes saturated. After the infiltration capacity drops below the rainfall rate, water ponds on the surface and soon after begins to flow across the surface to lower elevations if the land is sloping. This process, *overland flow*, is more common in dry regions with fine-grained soils that have low infiltration capacities.

The infiltration of water into the ground does not necessarily mean that ground-water recharge will result, because there are two processes that can recycle water back to the atmosphere. Direct *evaporation* of the water from the soil is one mechanism. The other, *transpiration* (Figure 11-9), is the utilization of soil water and the release of water vapor by plants. Some plants, includ-

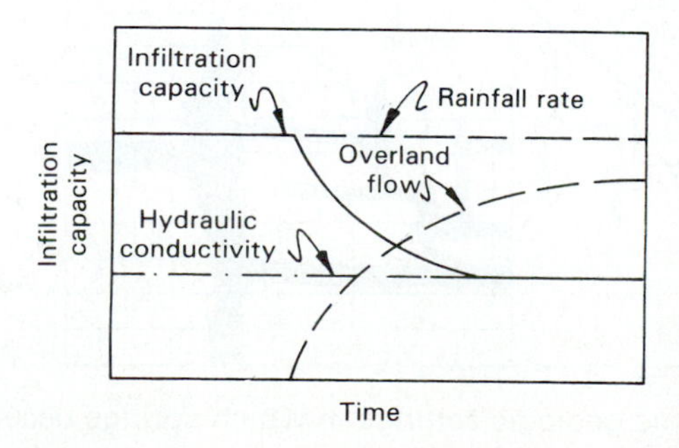


**Figure 11-9** The decrease in infiltration capacity that occurs after a zone of saturation develops near the soil surface. The minimum value of infiltration capacity is hydraulic conductivity. (From R. A. Freeze and J. A. Cherry, *Groundwater*, copyright © 1979 by Prentice-Hall, Inc., Englewood Cliffs, N.J.)

ing certain species of trees, have the ability to take up large amounts of water through their roots and then return a high percentage of it to the atmosphere through their leaves. Evaporation and transpiration are so difficult to separate quantitatively that they are frequently discussed together under the term *evapotranspiration*.

The amount of water available for ground-water recharge is, therefore, the fraction not lost by overland flow or evapotranspiration. Even part of the remaining water may be held in the unsaturated zone and not actually reach the water table. Some water must move downward to the water table for recharge to occur. In many areas recharge may take place only for short periods of the year. The rise in the water table during the spring and early summer, when precipitation is high and evapotranspiration low, is the most common indication of ground-water recharge. In some arid and semiarid areas this may be the only period of recharge in a typical year, aside from occasional heavy thunderstorms.

Ground-water discharge is the opposite of recharge; in discharge areas water moves from the saturated zone into the unsaturated zone and perhaps to land surface. If the near-surface sediments are homogeneous and the water table is high, discharge areas will occur in topographic lows where the water table intersects the land surface, as in lakes and streams. The elevation of the stream or lake actually represents the position of the water table at that point. A surficial body of water is not required, however; sometimes discharge areas can be recognized by indications of a high water table and abundant evapotranspiration. Evidence for this type of discharge area includes persistent swampy conditions during all or most of the year, plants like willow and cottonwood that extend their roots below the water table to obtain moisture (*phreatophytes*),

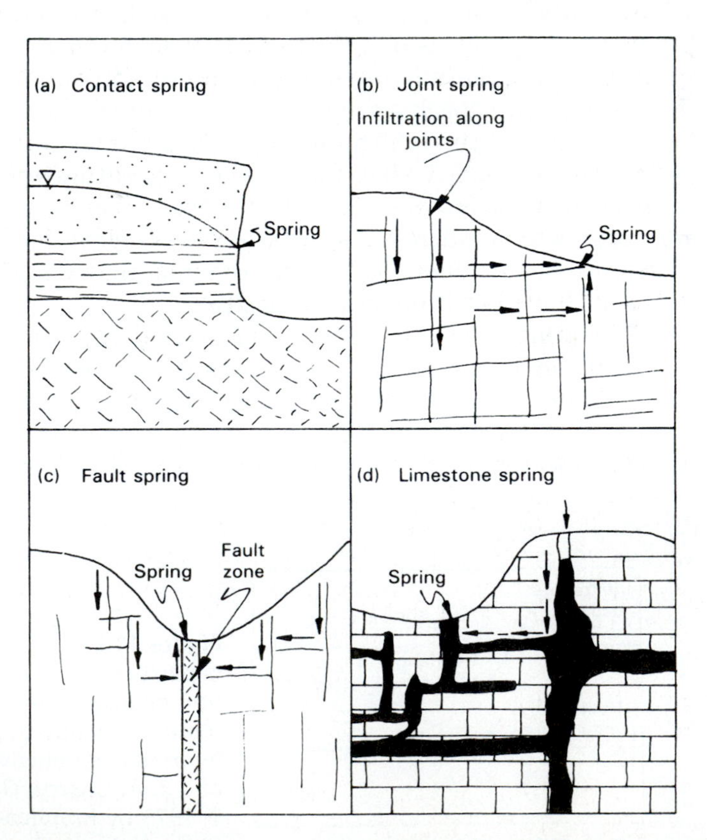


**Figure 11-10** Some geologic settings in which springs occur.

**Figure 11-11** Comal Springs, near San Antonio, Texas, discharges over 8000 L/s of water from a limestone aquifer system.

and saline or alkaline soil conditions. Saline and alkaline soils indicate evaporation of ground water from the soil, with the dissolved salts carried by ground water left behind in the soil. In some areas, particularly where the water table is deep, ground-water recharge, rather than discharge, occurs from topographic low points. In these situations most of the recharge is derived from ephemeral stream channels or lakes.

*Springs*, which are localized discharge points, represent a more obvious type of discharge. Springs can vary greatly in flow rate, ranging from barely a trickle to more than 10 $m^3$/s. Some general types of springs are shown in Figure 11-10. In stratigraphic sequences with alternating permeable and nonpermeable beds, *contact springs* can develop. Joints, faults, and fractures commonly provide conduits for the upward movement of ground water to the surface. The high temperature and highly mineralized chemical composition of some springs attest to the depths to which some of these conduits extend. Some of the most spectacular springs are located in areas of soluble bedrock like limestone. Ground water chemically reacts with these rocks and, over thousands of years, can dissolve large amounts of rock, forming subsurface caves and passages. Tremendous amounts of water may issue from springs in limestone terrains (Figure 11-11).

## GROUND-WATER RESOURCES

The most obvious reason for studying ground water is its value as a resource. In some parts of the United States, particularly in rural areas, ground water constitutes the only water supply. Surface water, the alternative source, when available, must be extensively treated before drinking and is quite often limited in quantity. Increases in demand for water in the United States will have to be met predominantly by expanding utilization of ground water.

### *Aquifers*

An *aquifer* is a saturated body of rock or soil that transmits economically significant quantities of ground water. The most important property of an aquifer is its hydraulic conductivity, for it is the rate at which water can move to the screen of a pumping well to replace the water pumped out that determines the yield of the well. This rate of movement, under natural or pumping conditions, is controlled by the hydraulic conductivity. Geologic materials that constitute productive aquifers include sand and gravel, highly fractured rocks, and soluble rocks containing large subsurface openings for ground-water movement and storage.

Materials that do not transmit economically significant quantities of ground water are known as *aquitards*. Aquitards are usually composed of dense, nonfractured rock units or sedimentary beds of silt and clay. Sedimentary rock terrains often consist of alternating aquifers and aquitards. Exploration for ground water in these areas involves location of the permeable rock horizons. Some materials are referred to as *aquicludes*, units that transmit no ground water. It is unlikey, however, that true aquicludes exist in nature, because all materials seem to convey some ground water, however small the amount.

Aquifer type is determined by the presence or absence of an overlying aquitard. Aquifers that lack overlying aquitards and have the water table as their upper boundary are *unconfined aquifers*. These often occur in surficial deposits of sand and gravel. Because the water table is the upper boundary, contour lines of water-table elevation drawn on a map indicate the direction of flow of ground water in an unconfined aquifer. As shown in Figure 11-12, the slope of the water table is a close approximation of the hydraulic gradient, providing the flow in the aquifer is nearly horizontal. Similarly, the elevation of the water table is closely equivalent to the hydraulic head in an unconfined aquifer.

*Confined aquifers* are bounded above and below by aquitards. As a result of this confinement, there are important differences between the flow of ground water in confined aquifers and the flow in unconfined aquifers. The water level, or head, in a well in a confined aquifer may be above or below the water table; in fact, there is no specified relationship between the head in a confined aquifer

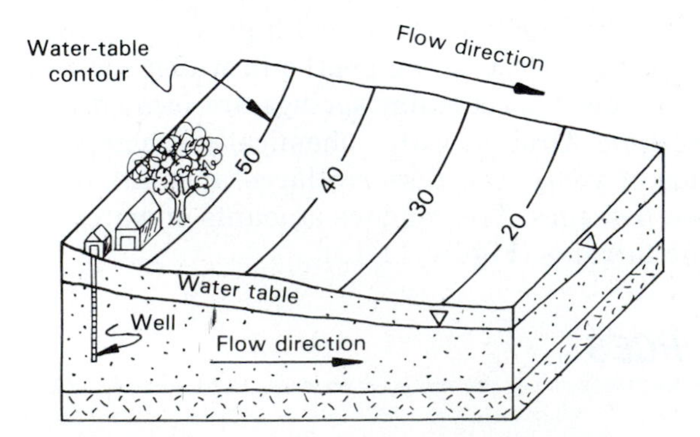


**Figure 11-12** An unconfined aquifer showing water-table contours, which are elevations above sea level of the water table, projected onto the surface.

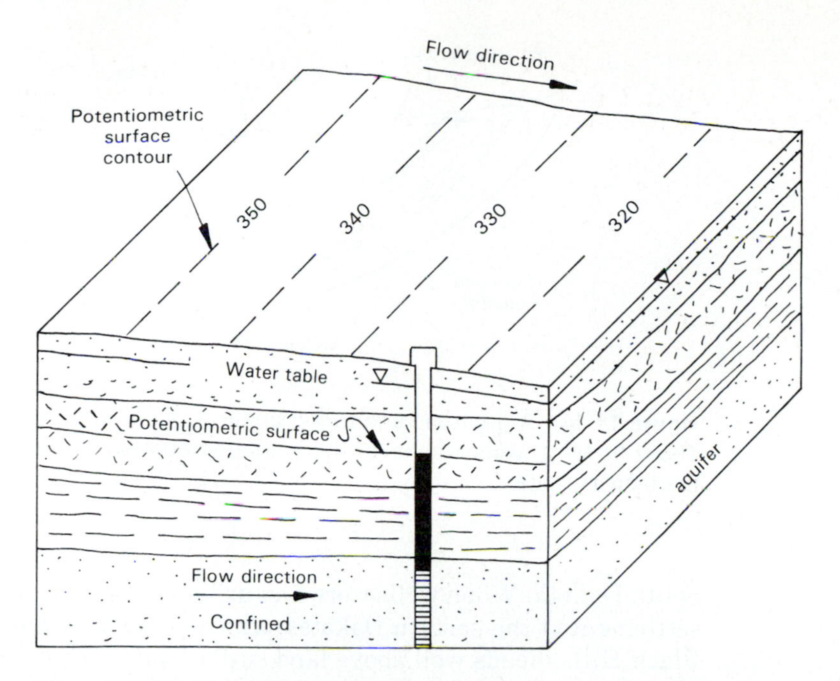


**Figure 11-13** A confined aquifer, in which the water level in a well rises to the potentiometric surface. Contours of the potentiometric surface are shown on the land surface.

and the water table. This distinction between the two aquifer types is shown in Figure 11-13.

If the heads in a confined aquifer are contoured on a map, as previously shown for the water table in unconfined aquifers, the contours will define an imaginary surface called the *potentiometric surface* (Figure 11-13). The potentiometric-surface map is similar to the water-table contour map in that it indicates the distribution of head and therefore the direction of water movement in a confined aquifer. The major difference between the potentiometric surface and the water table is that the potentiometric surface is not an actual surface in the ground analogous to the water table; it is a hypothetical level defined by the elevations of water levels in wells in confined aquifers.

Confined aquifers are also known as *artesian aquifers*. The fluid pressure in some artesian aquifers may be so high that the head, or potentiometric surface, is above the land surface. A well completed in this type of aquifer will flow without the aid of a pump; these wells are known as *flowing artesian wells*. Flowing wells are not unlike the plumbing system in a house in that the water will flow out under its own pressure as long as the faucet is left open.

Geologic and topographic conditions combine to produce confined aquifers. In Figure 11-14, a classic example of a confined aquifer is illustrated. The aquifer in this case is exposed at the land surface at the left end of the cross section, where recharge occurs. To the right, the aquifer dips downward beneath a thick aquitard. The potentiometric surface shown above the aquifer indicates that just to the right of the recharge area, the aquifer develops confined conditions. The surface topography of the area determines where flowing artesian conditions will occur; as land surface slopes to the right more rapidly than the potentiometric surface, flowing well conditions are common. Figure 11-14 is actually a simplified version of the regional ground-water flow system in the Great Plains of the United States. The confined aquifer system, the Dakota Sandstone of Cretaceous age, crops out in the Black Hills region of western

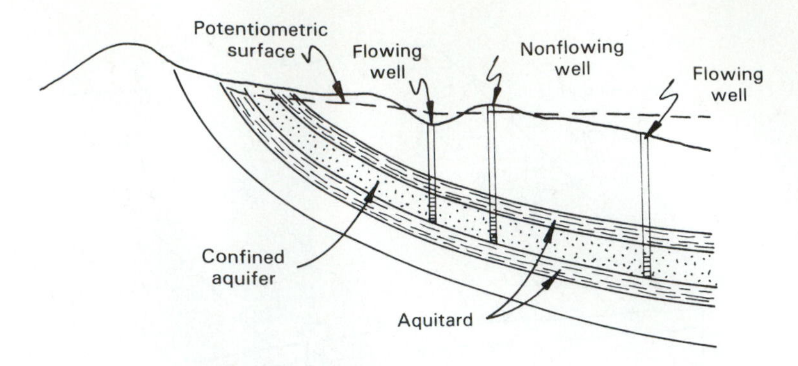


**Figure 11-14** A geologic and topographic configuration that produces flowing wells whenever the potentiometric surface is above the land surface.

South Dakota. When wells were first drilled into the Dakota Sandstone during settlement of the eastern Dakotas, which is much lower in elevation than the Black Hills, heads well above land surface were encountered. Heads declined over the years as many flowing wells discharged water from the Dakota Sandstone aquifer.

## EXAMPLE 11-2

The first well completed in the Dakota aquifer in South Dakota was drilled at Aberdeen by W. E. Swan in 1881. The well was 1100 feet in depth and had a closed-in pressure of 180 psi. (This is the pressure that developed at land surface when the well was closed and not allowed to flow.) How high above land surface was the potentiometric surface?

***Solution.*** Fluid pressure is equal to the unit weight of the fluid times the height of the fluid column above the point of interest.

$$p = \gamma_{water} \times h$$

Since the unit weight of water in English units is usually reported as 62.4 lb/ft$^3$, pressure in psi can be converted to lb/ft$^2$.

$$180 \text{ lb/in.}^2 \times 144 \text{ in.}^2/\text{ft}^2 = 25{,}920 \text{ lb/ft}^2$$

The true unit weight of Dakota aquifer ground water would be slightly more because it was somewhat mineralized. Head above land surface in feet can now be determined.

$$h = \frac{p}{\gamma} = \frac{25{,}920 \text{ lb/ft}^2}{62.4 \text{ lb/ft}^3} = 415.4 \text{ ft}$$

Thus the potentiometric surface, or the level to which water would rise in a well (if the well were extended up into the air), was significantly above land surface in parts of eastern South Dakota.

A final aquifer type is the *perched aquifer*. Beds of material with low hydraulic conductivity in the unsaturated zone lead to the establishment of perched conditions (Figure 11-15). Meandering stream deposits (discussed later in this chapter) commonly contain perched aquifers. The downward flow of water through the unsaturated zone toward the regional water table is retarded by the clay unit. A localized water table and a perched aquifer then develop

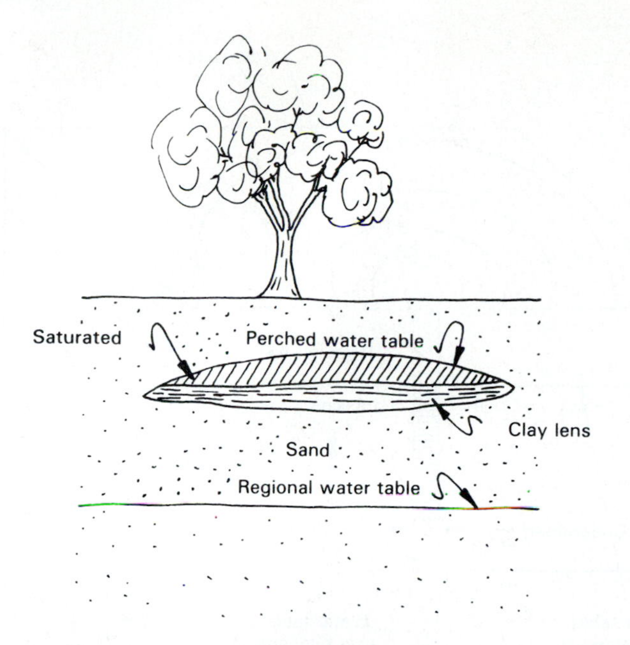


**Figure 11-15** Development of a perched aquifer above a low-permeability unit that occurs within the unsaturated zone of a unit of higher permeability.

above the clay. Perched aquifers cannot usually sustain high well yields because of their limited areal extent.

### *Production of Water from Aquifers*

Unlike piezometers, wells are designed to maximize the yield of aquifers. Larger diameters and greater lengths of screen are therefore utilized. If a well penetrating an unconfined aquifer is limited to that aquifer, and if vertical components of flow in the aquifer are not significant, the *static*, or unpumped water level in a well may be a close approximation to the water table. Similarly, if a well in a confined aquifer is isolated from ground water in overlying and underlying rock units, the water level in the well may be a good indication of the potentiometric surface in the aquifer. But if the well is pumped, changes in the hydraulic regime of the aquifer occur.

When an unconfined aquifer is pumped, the response to this stress is a decline of the water table in the vicinity of the well. These effects are usually assumed to be equal in all directions, so the area of influence of pumping is a cone-shaped region with greatest decline in the water table adjacent to the well (Figure 11-16). The affected region changes from a saturated to an unsaturated state during the drop in the water table. Because of its shape it is known as a *cone of depression*. As pumping continues, the cone expands radially outward from the well. The actual shape and dimensions of the cone depend on the hydraulic conductivity, thickness, and storage properties of the aquifer. If the pumping rate is not excessive, the cone of depression will eventually reach an equilibrium position. If pumping is excessive, the water table will continue to drop, causing increased pumping costs. The situation in which ground-water withdrawals exceed recharge is referred to as *ground-water mining*. The implications of this situation are discussed in Case Study 11-1.

In a confined aquifer the effects of pumping are somewhat different. Here the cone of depression develops in the potentiometric surface without any

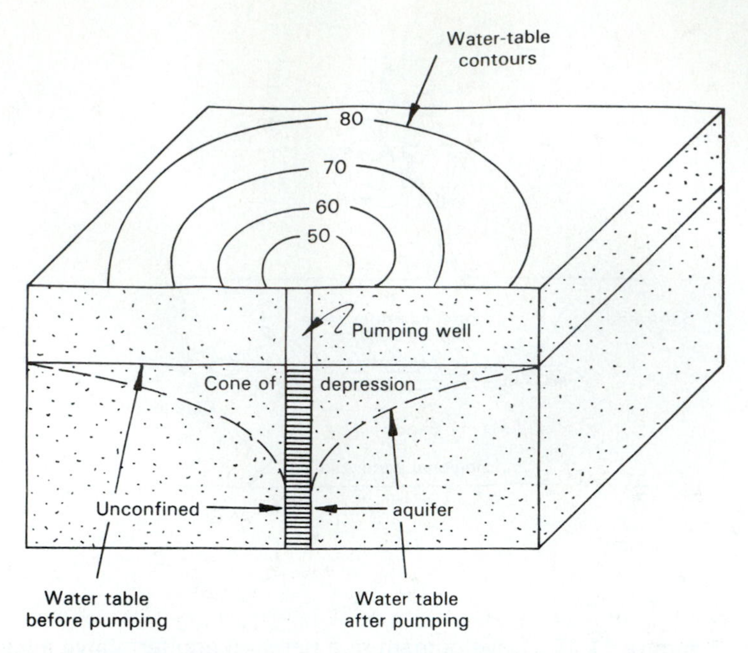


**Figure 11-16** Development of a cone of depression in an unconfined aquifer by lowering of the water table.

dewatering of the aquifer (Figure 11-17). To explain the response of the aquifer to the lowering of the potentiometric surface we must invoke the *Principle of Effective Stress*, a concept that has many applications in geology. Referring to Figure 11-18, consider the distribution of stress in a confined aquifer. The total weight of rock, soil, and water above the aquifer exerted per unit area on the upper surface of the aquifer is designated as the *total stress* ($\sigma_T$). This stress

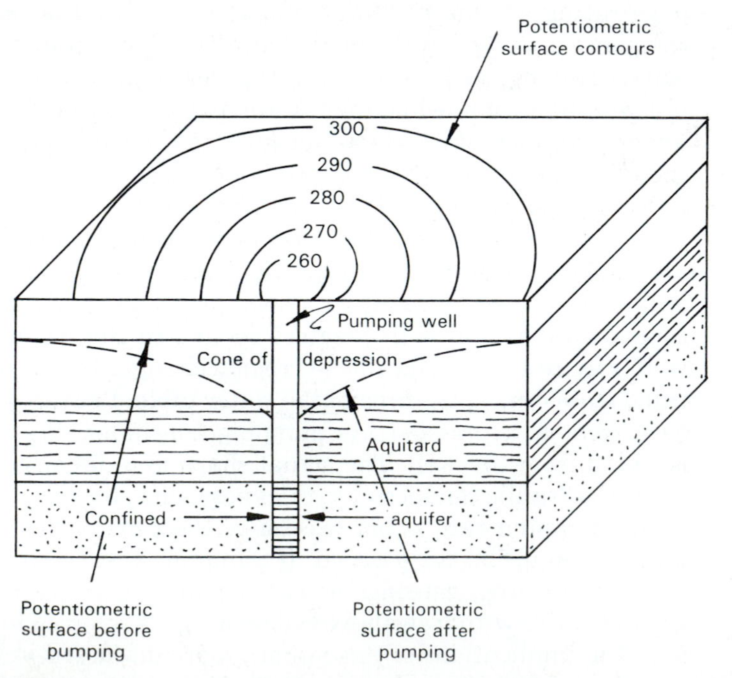


**Figure 11-17** The cone of depression produced in the potentiometric surface by pumping a confined aquifer.

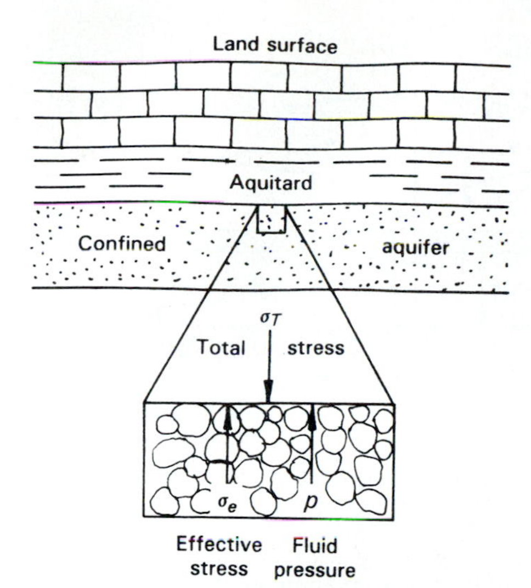


**Figure 11-18** Distribution of stress in a confined aquifer. Pumping lowers the fluid pressure and increases the effective stress.

is opposed in the aquifer by two stresses, the fluid pressure ($p$), which we have previously described, and a stress transmitted through the solid grains of the aquifer, the *effective stress* ($\sigma_e$). The equilibrium relationship defined by this group of stresses can be defined as the effective stress equation,

$$\sigma_T = \sigma_e + p \quad \text{(Eq. 11-5)}$$

Despite its simplicity, this equation is of utmost importance in many areas of geology and engineering. When a confined aquifer is pumped and a cone of depression develops in the potentiometric surface, the fluid pressure is reduced. Since the total stress exerted on the aquifer is unchanged, the effective stress equation requires that the effective stress increases. When the grain-to-grain pressure in a material increases, the material tends to compact due to closer packing of the particles. This reduction in porosity is not possible unless water within the pores is expelled. Since the head has been lowered in the vicinity of the well, the pore water moves radially toward the well and the aquifer compacts vertically. Therefore, even though large amounts of water are removed from a confined aquifer, it remains saturated. An additional mechanism that accounts for water produced in the well is the expansion of water that occurs when the fluid pressure is decreased. As in unconfined aquifers, the cone of depression in the potentiometric surface in a confined aquifer expands with pumping until it approaches an equilibrium condition.

The size of the cone of depression can be estimated using a form of Darcy's Law if some assumptions are made. For a confined aquifer of thickness B (Figure 11-19), let us assume that the piezometric surface is horizontal and that the aquifer extends to infinity in all directions. Let us further assume that the screen of the well installed in the aquifer penetrates the entire thickness of the aquifer. This assumption means that ground-water flow to the well will be horizontal and radial from all directions in the aquifer. Darcy's Law can now be written for flow passing through the surface area of a cylinder of radius $r$ and depth $B$ (Figure 11-20):

$$Q = KA\frac{dh}{dr} = K2\pi rB\frac{dh}{dr} \quad \text{(Eq. 11-6)}$$

This equation is similar to the form of Darcy's Law that we used previously,

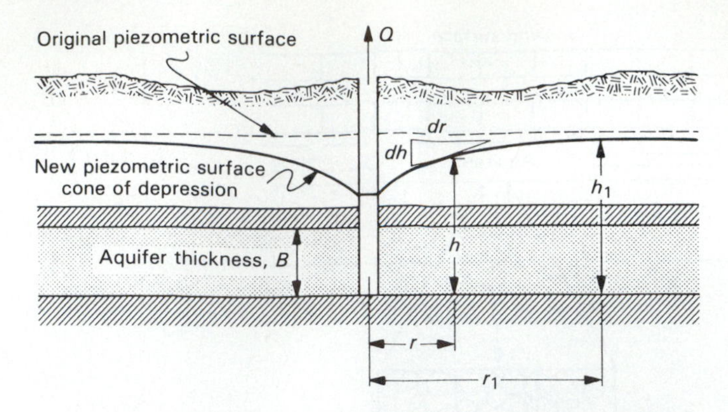


**Figure 11-19** Diagram showing parameters used in Equation 11-6.

except the area used is the surface area of a cylinder and the hydraulic gradient is written in radial form. Equation 11-6 can now be written as an integral:

$$\int_r^{r_1} Q \frac{dr}{r} = 2\pi B \int_h^{h_1} dh \qquad \text{(Eq. 11-7)}$$

If we then integrate between arbitrary values $r$ and $r_1$, which have heads of $h$ and $h_1$, we obtain

$$Q \ln \frac{r_1}{r} = 2\pi KB(h_1 - h) \qquad \text{(Eq. 11-8)}$$

or

$$Q = \frac{2\pi KB(h_1 - h)}{\ln (r_1/r)} \qquad \text{(Eq. 11-9)}$$

A similar equation can be derived for an unconfined aquifer. The only difference is that the height of the cylinder is a variable because the saturated thickness of the aquifer decreases as the cone of depression is formed. This equation can be stated as

$$Q = \frac{\pi K(h_1^2 - h^2)}{\ln (r_1/r)} \qquad \text{(Eq. 11-10)}$$

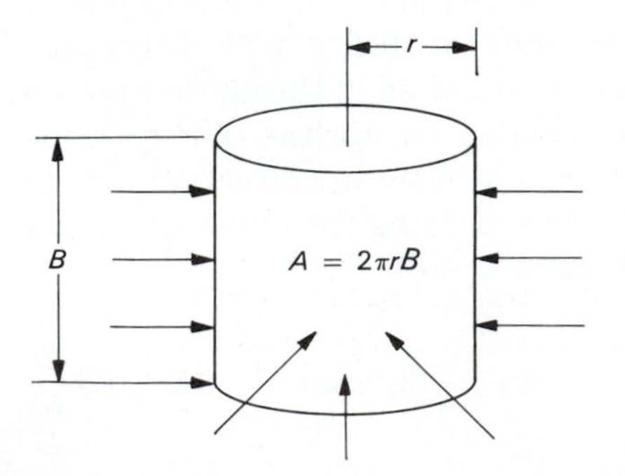


**Figure 11-20** Cylinder illustrating radial flow of ground water to well.

**EXAMPLE 11-3**

A 3 m thick confined aquifer is pumped at a rate of 1 L/s until the cone of depression in the potentiometric surface has reached equilibrium. The original potentiometric surface was at an elevation of 50 m. An observation well located 30 m from the pumping well has a drawdown of 1 m. If the hydraulic conductivity of the aquifer is $10^{-4}$ m/s and the diameter of the pumping well is 0.1 m, what is the drawdown in the potentiometric surface at the well?

***Solution.*** Using the radius of the well, $r_w$, as the value of $r$, use Equation 11-9:

$$Q = \frac{2\pi KB(h_1 - h_w)}{\ln(r_1/r_w)}$$

$$0.001 \text{ m}^3/\text{s} = \frac{2(3.14)(10^{-4}\text{ m/s})(3\text{m})(49\text{m} - h_w)}{\ln(30\text{m}/0.05\text{m})}$$

Solving for $h_w$ gives

$$h_w = 45.7 \text{ m}$$

and the drawdown is equal to

$$50 \text{ m} - 45.7 \text{ m} = 4.3 \text{ m}$$

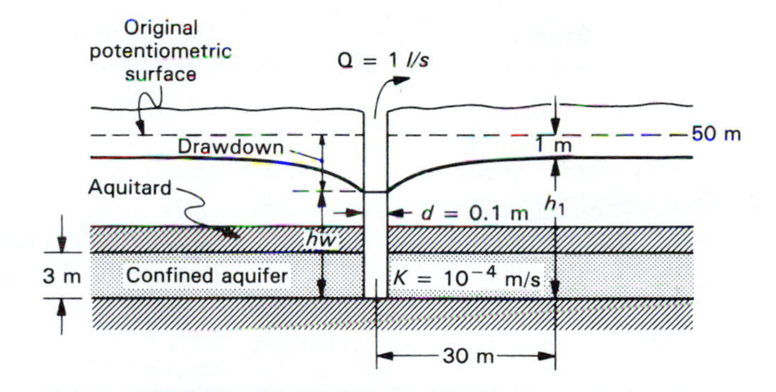


**CASE STUDY 11-1**

**Ground-Water Mining and the High Plains Aquifer**

Problems arise during production of ground water when an equilibrium condition in the cone of depression is not established. Under heavy pumpage, particularly when multiple wells penetrate the same aquifer, steady declines in the water table or potentiometric surface are noted. Drawdowns are accentuated when cones of depression from nearby wells overlap. A good example of this unfortunate circumstance is the High Plains Aquifer, a shallow, mostly unconfined aquifer that occurs throughout much of the Great Plains region (Figure 11-21). The High Plains aquifer has been extensively pumped to support irrigated farming in this region since the agricultural disaster caused by the 1930s drought. The development of irrigation has progressed to such an extent that annual ground-water withdrawals now exceed the annual recharge to the aquifer by 2 to 200 times. The result is a steady decline in water

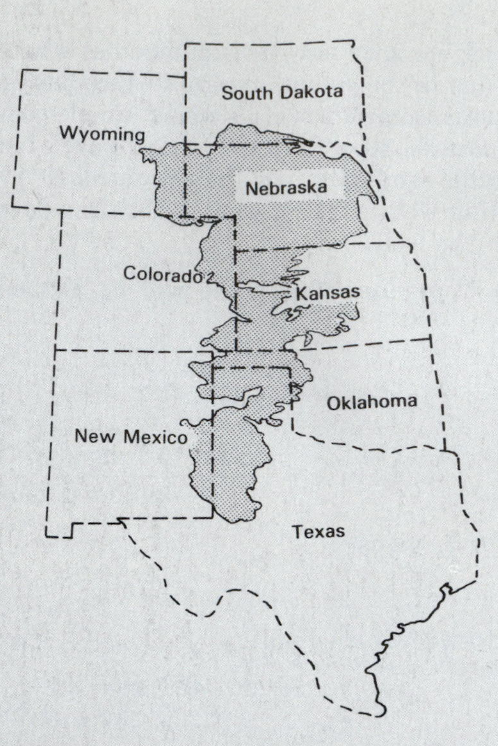


**Figure 11-21** Location of the High Plains Aquifer. (From U.S. Geological Survey Professional Paper 1400-B.)

levels. These declines exceed 30 m in some areas. In effect, the water is being "mined" like any other nonrenewable resource. On a short-term basis the effect of water-table declines is to increase costs because of the greater pumping lift required. On a long-term basis, however, parts of the High Plains aquifer will be in serious trouble within the next 20 years. The point may be reached where irrigated farming will no longer be possible in some areas. This will have severe consequences for the affected areas.

### *Geologic Setting of Aquifers*

Aquifers occur in vastly different geological settings, making exploration for ground water challenging. Among the more common settings for aquifers are the sedimentary deposits of river systems. Such *alluvial aquifers* lie within valleys of different types or within rock units that were originally deposited in river valleys. Many characteristics of the aquifer can be explained by classification of the river system that deposited the aquifer sediments as *braided* or *meandering* (Figure 11-22). Braided segments flow in numerous interconnected channel segments within a broad, shallow channel. The channel sediments deposited by these rivers are predominantly coarse grained, resulting in productive aquifers if sufficient thicknesses are present. Meandering streams differ from braided streams in that they deposit large quantities of *overbank* or *flood-plain sediment* on flat surfaces (flood plains) adjacent to the sinuous channels. Unlike channel sediment, overbank sediment does not have good aquifer potential because it is finer grained. Thus an aquifer in alluvial sediments of a meandering river system may consist of narrow bodies of channel sediment surrounded by aquitards composed of less permeable overbank sediment; how-

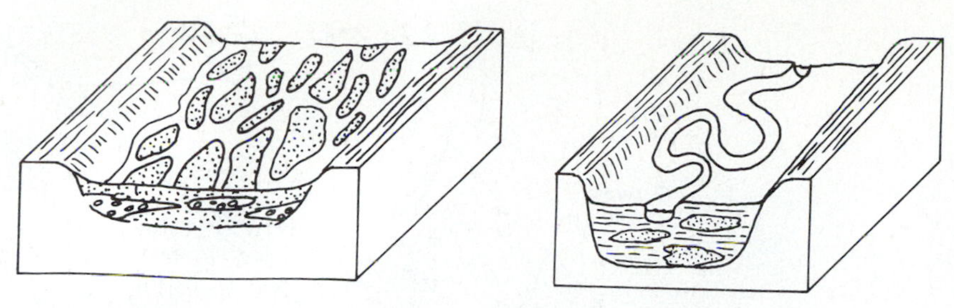

**Figure 11-22** Aquifers produced by braided (left) and meandering (right) rivers.

ever, braided-river deposits may underlie the meandering-river sediment in some places.

In regions where structural deformation has produced alternating mountain ranges and valleys by movement along faults, alluvium of *tectonic-valley aquifers* is present. The alluvium is deposited by streams eroding the rising mountain blocks. A generalized tectonic-valley setting, similar to those in the Basin-and-Range Province of the western United States, is illustrated in Figure 11-23. In arid regions, tectonic-valley aquifers may provide an excellent supply of ground water, although development of the aquifers must be carefully managed because of the low rate of recharge.

Although predominantly of alluvial origin, *glacial aquifers* also include shoreline and deltaic deposits of glacial lakes. Rivers carrying meltwater from glaciers are generally braided, high-discharge streams that flow on top of and within glaciers as well as away from glacier margins. Sorted alluvial sediment deposited by streams flowing away from glacial margins, called *outwash*, may be laid down over broad, gently sloping plains known as *outwash plains,* or they may be confined to valleys in deposits called *valley trains* (Figure 11-24).

Some glacial aquifers consist of valleys filled with alluvium buried by glacial sediment (till) deposited during a later glacial advance. These *buried valleys*, as shown in Figure 11-25, form some of the most productive aquifers in glaciated regions.

Other types of aquifers existing in unconsolidated sediment include *coastal-plain* aquifers (Figure 11-26), which are commonly composed of perme-

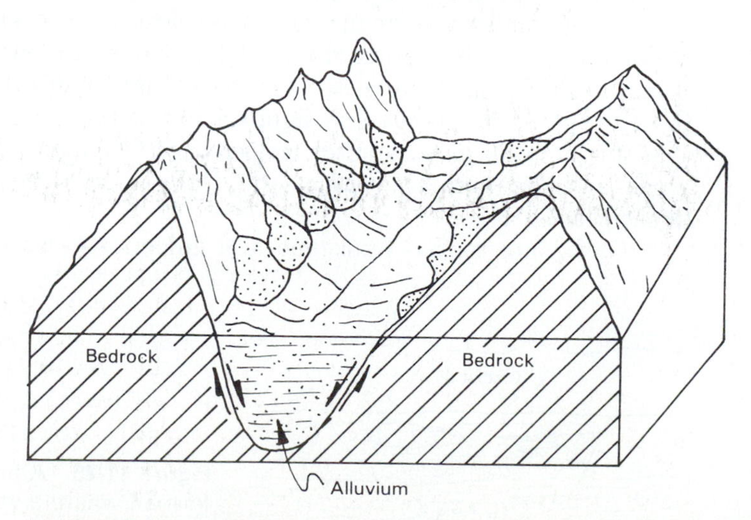


**Figure 11-23** A tectonic-valley aquifer formed by deposition of alluvium in a down-faulted basin bounded by uplifted mountain blocks.

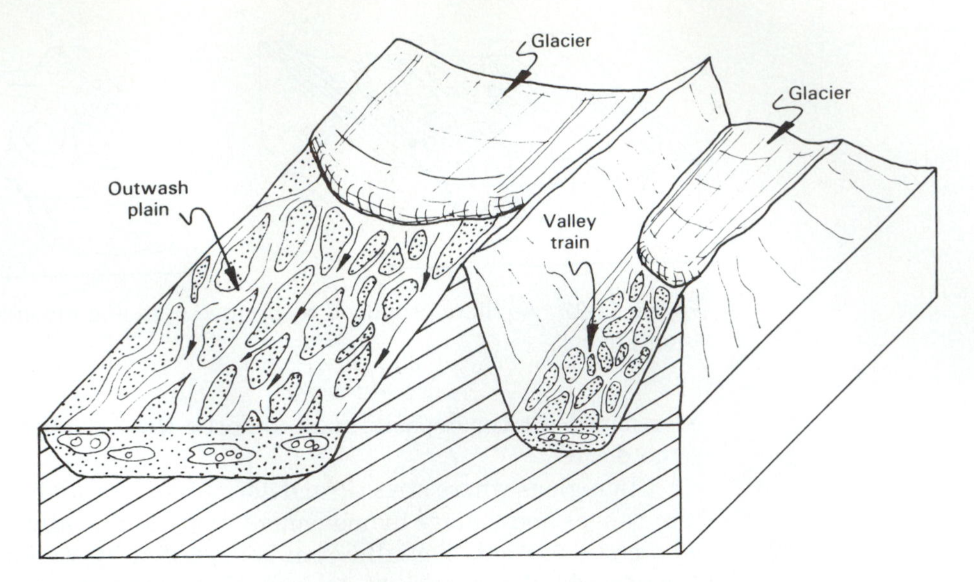


**Figure 11-24** Deposits of glacial-meltwater streams that form productive aquifers.

able beach sediment deposited when sea level was higher at some time in the geologic past. Aquifers in modern coastal areas are characterized by a zone of interface between fresh ground water and saline ground water derived from the ocean. The depth to the interface can be estimated by an approach called the *Ghyben-Herzberg relation*, which states that the depth below sea level of the interface at any point inland from the shore is equal to 40 times the elevation of the water table above sea level at that point. Thus if the water table is 2 m above sea level, the saltwater-freshwater interface would lie 80 m below sea level. When the aquifers are developed by pumping, the lowering of the water table causes the upward and landward encroachment of the saline ground water. In the above example, if pumping from wells in a coastal community lowered the water table by 1 m, the saltwater-freshwater interface would rise to an elevation of −40 m (40 m below sea level). Increasing salinity of drinking water can therefore be a serious problem for coastal cities that depend upon coastal-plain aquifers for their water supply.

Aquifers also develop in many types of rock. Sedimentary rock formations are among the most important rock aquifers in many parts of the world. Sandstone and conglomerate rock units have high hydraulic conductivities and therefore constitute productive aquifers unless the rock is highly cemented, so that its potential as an aquifer is decreased. We have already mentioned the great artesian aquifer system, the Dakota aquifer, which underlies the Great Plains.

Carbonate rocks—limestone and dolomite—also contain high-yielding

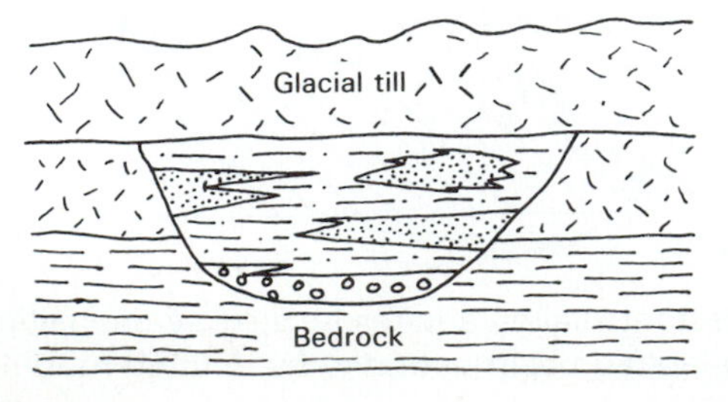


**Figure 11-25** A buried-valley aquifer formed where a valley filled with alluvium is covered by glacial sediment (till) from a later glacial advance.

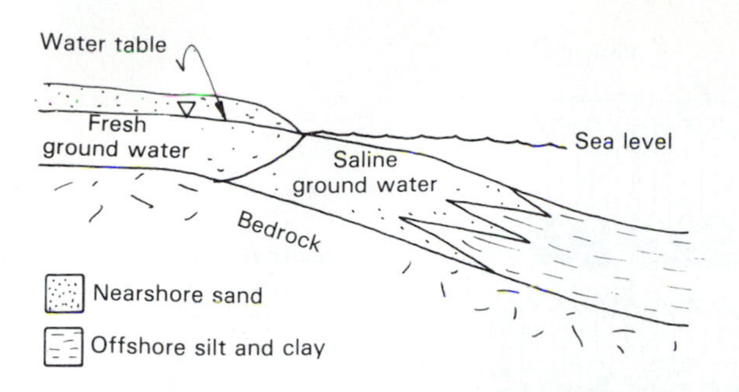


**Figure 11-26** A coastal-plain aquifer showing the boundary between fresh and saline ground water.

aquifers in many places. The suitability of these rocks as aquifers is determined by the amount of dissolution that has taken place during ground-water circulation. Cavernous limestones can yield phenomenal amounts of water if they occur below the water table.

Many rocks either have a low initial hydraulic conductivity because of their fine grain size or have a greatly reduced porosity because of cementing. These rocks may also function as aquifers, however, if they are fractured. Even shales can yield high quantities of ground water if the network of joints, faults, and other fractures is dense, closely spaced, and interconnected. Coals also sometimes constitute aquifers through the development of such *fracture permeability*. Fractures are also important in carbonate rocks, because it is along these cracks that water circulation and, therefore, dissolution of the soluble calcium carbonate proceeds. A fractured rock aquifer is shown in Figure 11-27. One important aspect of ground-water production from fractured rocks is the problem of intersecting the fractures with a well screen. Well yields vary greatly in fractured rocks, depending on the number, size, and orientation of fractures in contact with the screen.

Igneous and metamorphic rocks, with several exceptions, are dense, interlocking aggregates of crystals with low initial porosity and permeability. Fracturing is the only mechanism that allows significant production from wells. Among the exceptions to this generalization are lava flows with vesicular texture; these rocks may be very porous and highly permeable. Because vesicular zones often occur near the tops of the lava flows (Figure 11-28), ground-water exploration is commonly focused on contacts between flows. A sequence of lava flows may therefore contain aquifers at the flow contacts, separated by aquitards formed by the main body of the lava flows.

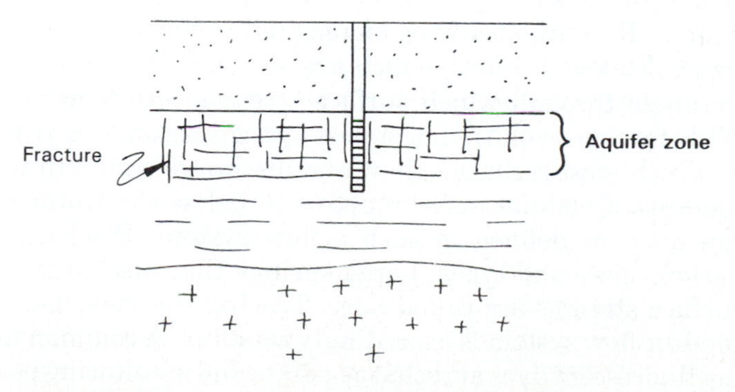


**Figure 11-27** A fractured-rock aquifer yielding water to wells from a fractured zone in a shale rock unit.

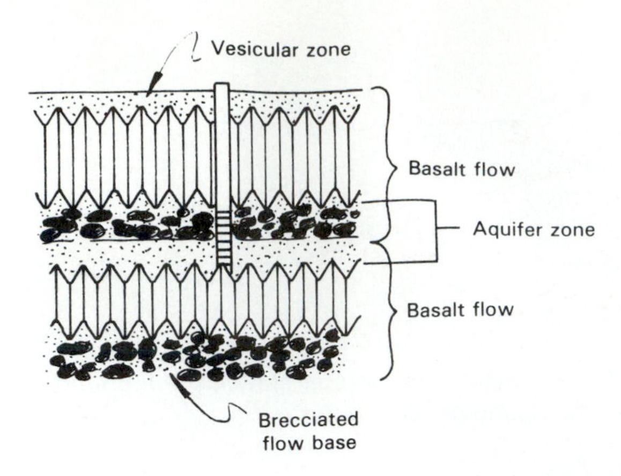


**Figure 11-28** An aquifer present at the contact between two basalt flows.

## GROUND WATER IN KARST REGIONS

The movement of ground water in carbonate-rock terrains differs from almost every other type of geological setting. The unique hydrogeological properties of these regions owe their origin to the solution of limestone and dolomite by surface and ground water. Following the exposure of carbonate rocks at or near the surface, the landscape undergoes a gradual transition in which the surface runoff and drainage of water is replaced by subsurface drainage. The unique topography that characterizes such areas is known as *karst*. A common form of karst topography associated with the surface exposure of nearly flat-lying beds of limestone is shown in Figure 11-29. The land surface consists of a mosaic of shallow circular depressions called *sinkholes*, which make up a sinkhole plain. Except during short periods following heavy precipitation, there may be no surface runoff on a sinkhole plain. Variations of the topography shown in Figure 11-29 occur when limestone is overlain by beds of nonsoluble rock types.

Ground-water movement in karst aquifers has distinct aspects of recharge, flow, and discharge. In early stages of karst development, subsurface hydrology is similar to other aquifers. Ground-water movement is slow and follows minute cracks and joints within the rock. This condition is known as a *diffuse flow system* (Figure 11-30). Over time, flow paths grow unequally by solution so that flow becomes concentrated in subsurface conduits to form a *mixed flow system* (Figure 11-30), which retains some aspects of the diffuse system. Recharge becomes increasingly concentrated in point sources such as sinkholes and *swallets*, which are solution or collapse openings along stream channels through which surface streams lose flow to subsurface drainages. With further evolution, the flow system becomes a true *conduit flow system*, in which most recharge, flow, and discharge occurs in well-defined subsurface conduits. Conduits may lie above or below the water table, if a water table can even be defined in such a flow system. Discharge occurs in springs of various sizes and types. Large springs that discharge subsurface drainage to surface streams are called *rises*. Tracing the movement of ground water in a conduit flow system is exceedingly complex. A common method involves injecting fluorescent dyes at recharge points and monitoring possible discharge points for the emergence of the dye.

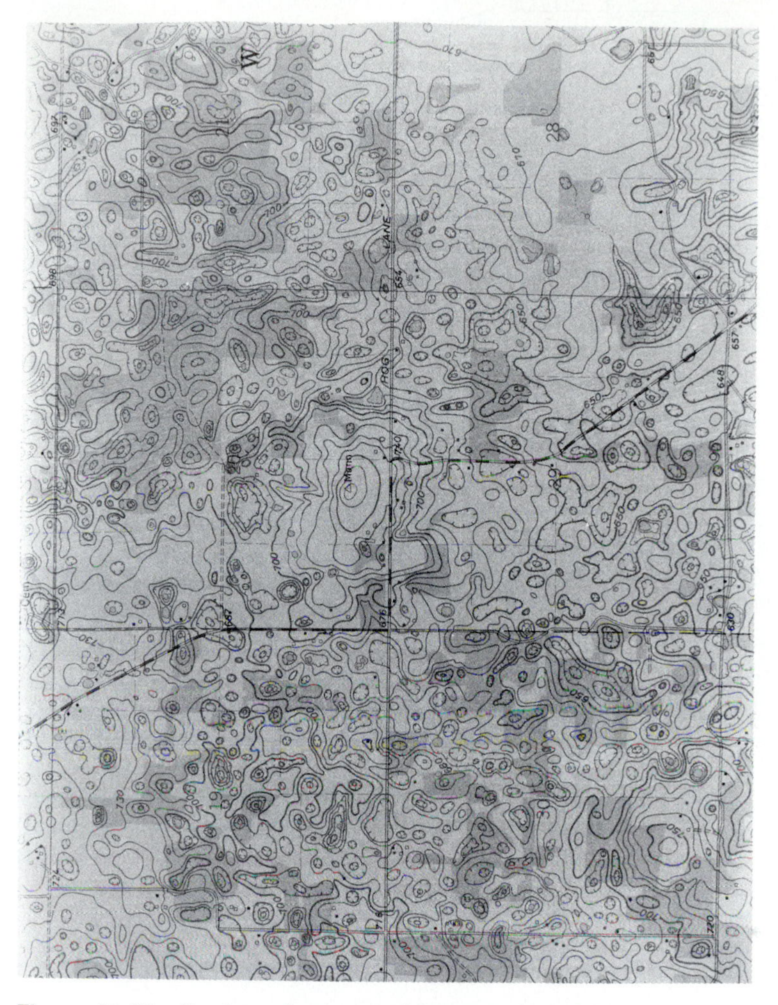

**Figure 11-29** Portion of topographic map of the Mitchell Plain (portion of the Corydon West 7.5 min Quadrangle), southern Indiana. Each depression is a sinkhole on the karst plain.

An idealized cross section of a conduit flow system in the Mammoth Cave region of Kentucky is shown in Figure 11-31. Ground water flows from recharge areas in the southeast to a discharge point in the northwest (Turnhole spring), even though the surface topography is higher in the northwest! Notice that conduit flow is above the potentiometric surface toward the recharge end of the flow system and below the potentiometric surface throughout much of the downgradient area to the northwest. The strong control exerted by bedding on conduit location and orientation in these gently dipping rocks is apparent from the diagram.

Caves are simply large abandoned conduits formed by ground-water flow. Abandoned conduits are sometimes enlarged by roof collapse to form larger subsurface openings called *rooms*. Studies of major cave systems have shown that their evolution is quite complex and requires hundreds of thousands or even millions of years to achieve their current stage of development. Distinct levels of cave conduits can sometimes be related to episodes of landscape erosion, glacial cycles, or other geologic events.

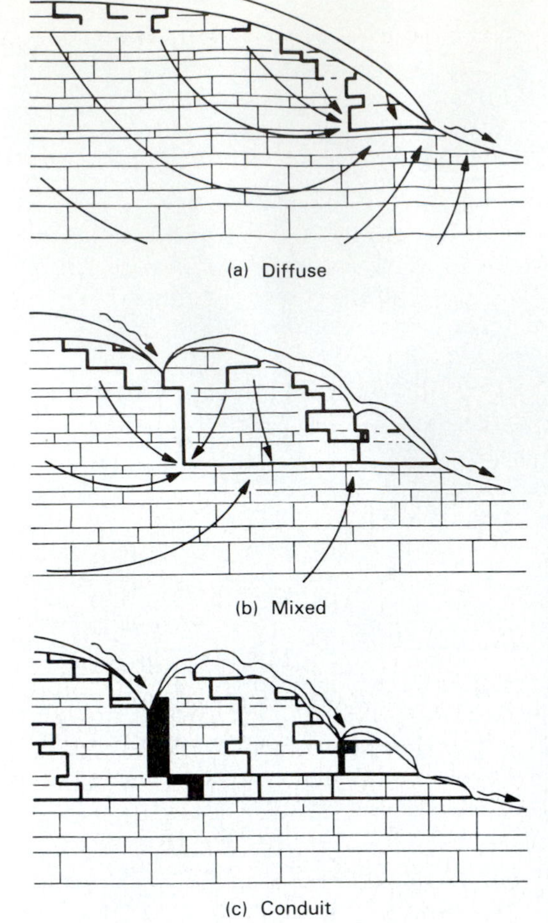


**Figure 11-30** Cross sections showing diffuse, mixed, and conduit ground-water flow systems. Heavy black lines are cave passages. Flow systems evolve over time from diffuse to conduit. (From J. F. Quinlan and R. O. Ewers. Ground water flow in limestone terranes: strategy, rationale, and procedure for reliable, efficient monitoring of ground water quality in karst areas, in *Ground Water Flow in the Mammoth Cave Area, Kentucky, with Emphasis on Principles, Contaminant Dispersal, Instrumentation for Monitoring Water Quality, and Other Methods of Study*, National Water Well Association.)

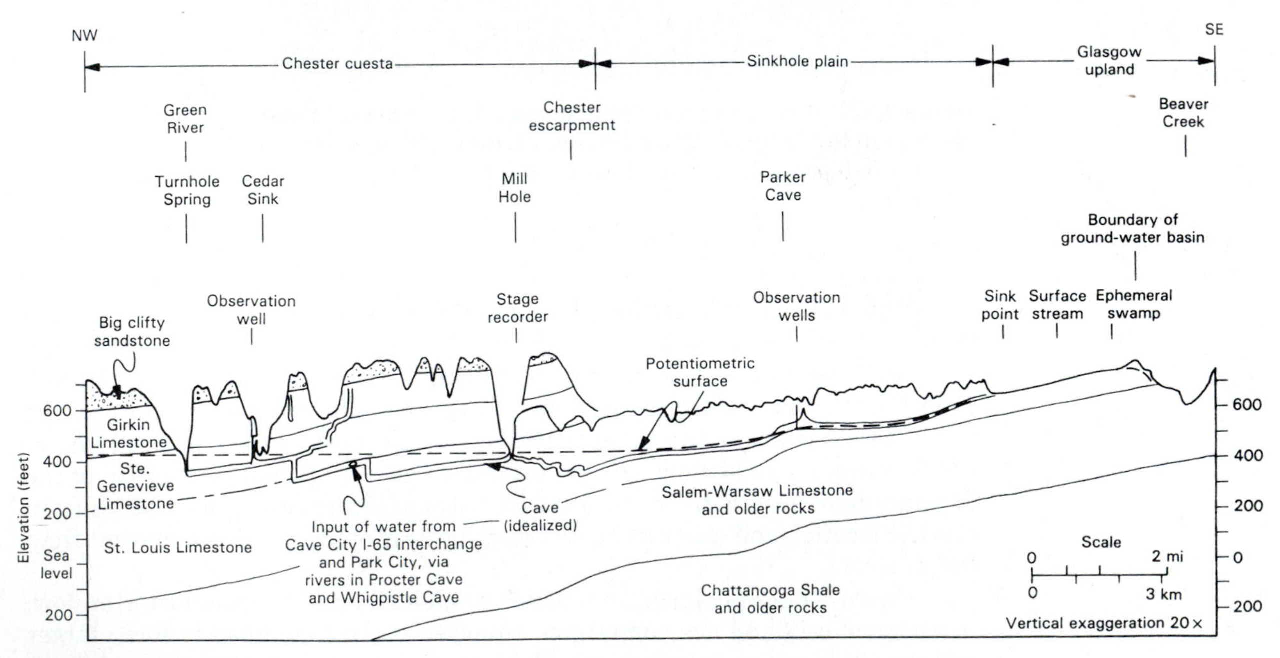


**Figure 11-31** Cross section showing ground-water flow in conduit flow system in Mammoth Cave region. (From J. F. Quinlan and R. O. Ewers. Ground water flow in limestone terranes: strategy, rationale, and procedure for reliable, efficient monitoring of ground water quality in karst areas, in *Ground Water Flow in the Mammoth Cave Area, Kentucky, with Emphasis on Principles, Contaminant Dispersal, Instrumentation for Monitoring Water Quality, and Other Methods of Study*, National Water Well Association.)

## CASE STUDY 11-2

# Ground-Water Supply from a Karst Aquifer

The Edwards aquifer in Texas (Figure 11-32) is certainly one of the most important aquifers in the United States. In the San Antonio region (Figure 11-32), the aquifer supplies nearly all the municipal, domestic, and agricultural water needs. Within this area, San Antonio alone, the tenth largest city in the United States, has a population of approximately 1 million. Because there are no other sources of water in the area, the U.S. Environmental Protection Agency (EPA) has designated the Edwards aquifer as a Sole Source Aquifer.

Three major hydrological divisions of the aquifer are recognized: the catchment area, the recharge area, and the artesian area (Figure 11-33). In the catchment area, surface streams carry water southward to the recharge area. The recharge zone coincides with a fault zone, the Balcones fault zone, and a steep topographic escarpment (Figure 11-34). Karst topography has developed in the recharge zone, and streams that collect water from the catchment area lose water to the ground-water flow system as they flow down the escarpment (Figure 11-35). Dams have been built on streams in the recharge area to enhance recharge to the aquifer. Some of these dams are designed to trap floodwater only, for recharge purposes, and do not contain a year-round reservoir. Natural discharge from the aquifer includes several large springs located near the base of the Balcones escarpment. The largest of these springs, Comal springs (Figure 11-11), is located near the town of New Braunfels (Figure 11-34). With an average discharge of 8000 L/s, Comal springs is the largest spring in Texas.

In the recharge area, the aquifer is under unconfined conditions. South of the recharge area, in the Gulf coastal plain, the aquifer dips below younger formations and becomes confined to form the artesian area of the aquifer (Figure 11-33). Major withdrawals from wells are made in this area for the City of San Antonio and smaller communities. Porosity and perme-

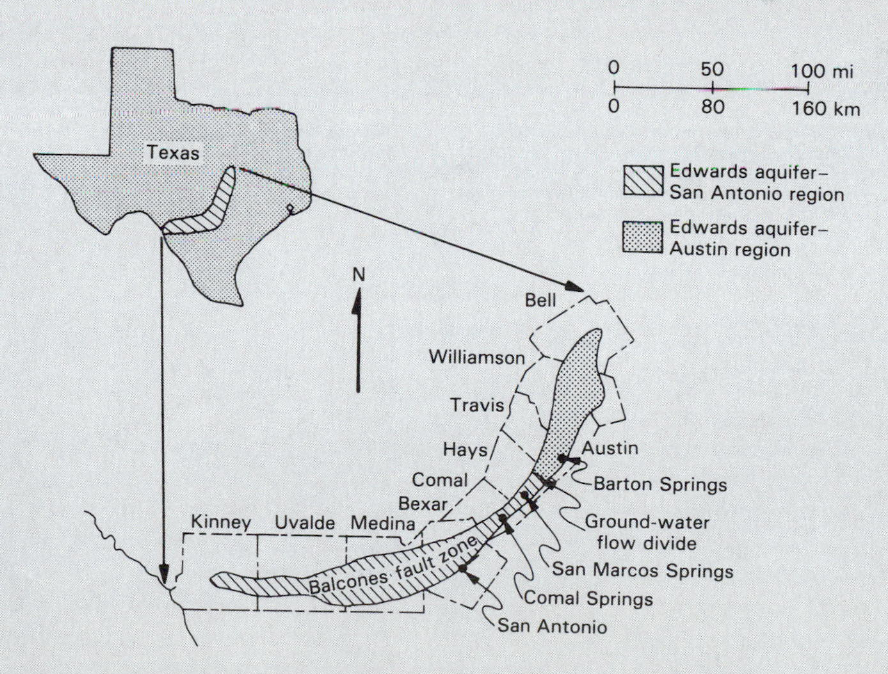


**Figure 11-32** Location of Edwards aquifer. (From R. K. Senger and C. W. Kreitler, Bureau of Economic Geology, University of Texas at Austin, Report of Investigations 141.)

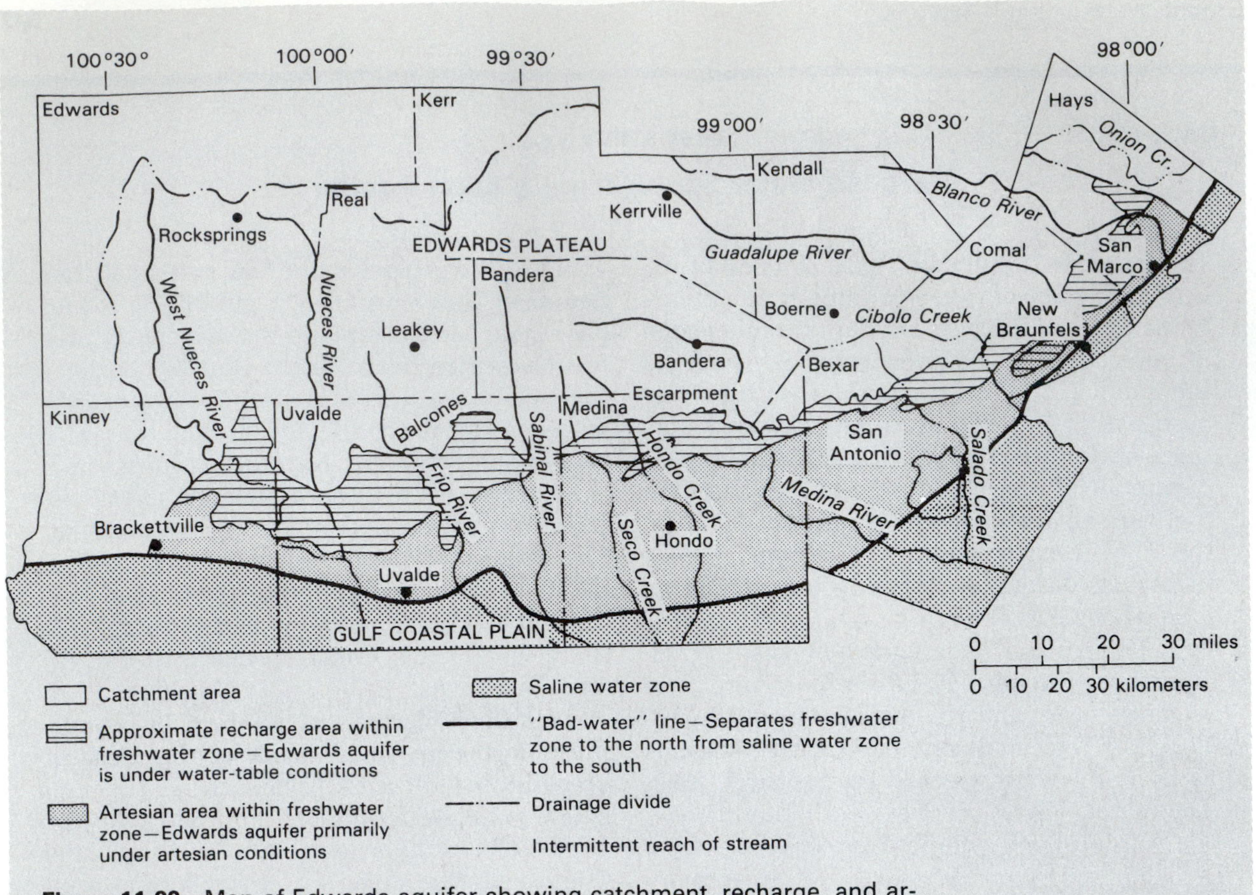


**Figure 11-33** Map of Edwards aquifer showing catchment, recharge, and artesian areas. (From C. R. Burchett et al., *The Edwards Aquifer, Extremely Productive, But . . .*, U.S. Geological Survey.)

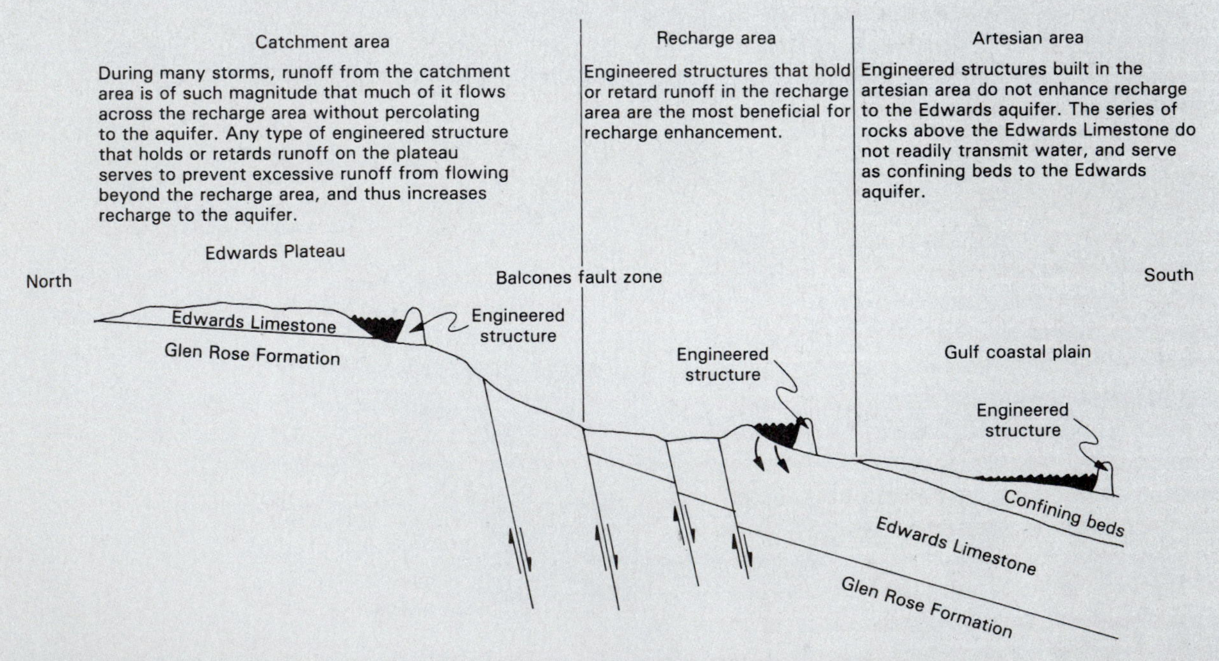


**Figure 11-34** Cross section of Edwards aquifer, showing components of flow system. (From C. R. Burchett et al., *The Edwards Aquifer, Extremely Productive, But . . .*, U.S. Geological Survey.)

**Figure 11-35** Sinkhole in recharge area of Edwards aquifer.

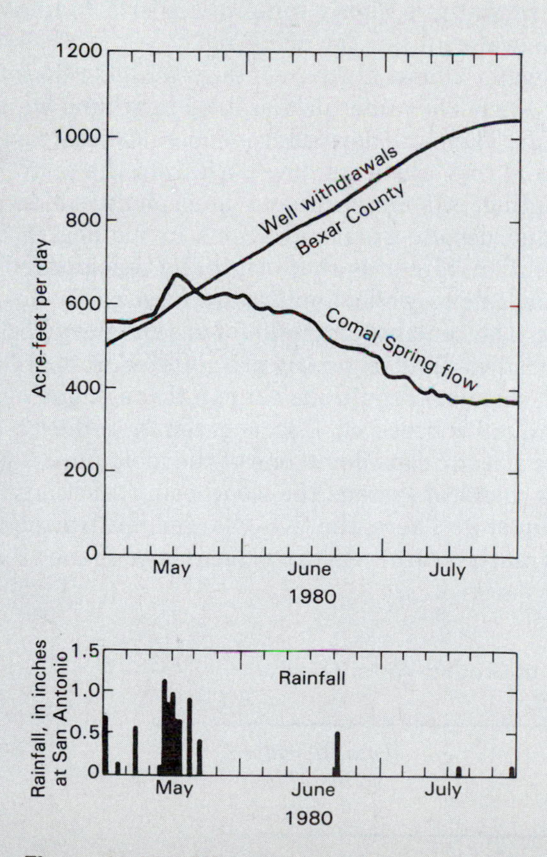


**Figure 11-36** Effects of pumping from Edwards aquifer on discharge at Comal Springs. (From C. R. Burchett et al., *The Edwards Aquifer, Extremely Productive, But . . .* , U.S. Geological Survey.)

ability in the confined area are very high due to the effects of the solution of the limestone. The southern boundary of the aquifer is a zone known as the "Bad Water Line" (Figure 11-33). This line represents a change in water quality rather than a termination of the aquifer rocks. Ground water south of the Bad Water Line is too mineralized for drinking purposes. One concern of water managers is that overpumping of the aquifer will cause mineralized water from the bad water zone to migrate up-dip into supply wells for the City of San Antonio. This would cause serious problems for the metropolitan area.

Although the Edwards aquifer is a prolific source of ground water, droughts in the region can cause significant depletion of the water supply. One consequence of drought conditions is that flow to Comal springs declines (Figure 11-36) as a result of decreased recharge and increased pumpage from the aquifer. Flow of the spring actually ceased for a short time in 1956. Maintenance of flow at Comal springs is a high priority because the park containing the springs is a valued scenic and recreational area.

## GROUND-WATER QUALITY

Quality, or chemical composition of ground water is as important as quantity in terms of ground-water supply. The most productive aquifer in the world will be useless if the water is highly mineralized or contains chemical compounds that make the water unsafe for drinking.

Ground water almost always contains a higher dissolved mineral content than surface water. The minerals dissolved in ground water, measured as *total dissolved solids* (TDS), are derived from chemical reactions between the water and the soil and rock along its flow path from the point of infiltration to the point of sampling. All minerals and amorphous solids are soluble to some extent in water, despite great variations in the degree of solubility and the rates of dissolution. Minerals that originally precipitated from water, such as halite, gypsum, calcite, and dolomite, are most easily dissolved by circulating ground water. The cementing agents of sedimentary rock, including calcium carbonate and silica, that originally precipitated from ground water can return to solution as chemical conditions change through geologic time. A classification of water quality based on TDS is given in Table 11-1.

We have already considered one of the most important components of the ground-water chemical system, the weathering reactions that characterize the soil zone (Chapter 9). There, the $CO_2$-charged infiltrating water dissolves some minerals and reacts with others to produce clays. In the process, silica, bicarbon-

**TABLE 11-1**

**Classification of Ground-Water Quality Based on TDS**

| *Water type* | *Total dissolved solids (TDS) mg/L* |
|---|---|
| Fresh | 0–1000 |
| Brackish | 1000–10,000 |
| Saline | 10,000–100,000 |
| Brine | >100,000 |

**TABLE 11-2**

**Classification of Dissolved Inorganic Constituents in Ground Water**

| | |
|---|---|
| *Major constituents (>5 mg/L)* | |
| Bicarbonate | Silica |
| Calcium | Sodium |
| Chloride | Sulfate |
| Magnesium | |
| *Minor constituents (0.01–10.0 mg/L)* | |
| Boron | Nitrate |
| Carbonate | Potassium |
| Fluoride | Strontium |
| Iron | |
| *Trace constituents (<0.1 mg/L)* | |
| Aluminum | Nickel |
| Antimony | Niobium |
| Arsenic | Phosphate |
| Barium | Platinum |
| Beryllium | Radium |
| Bismuth | Rubidium |
| Bromide | Ruthenium |
| Cadmium | Scandium |
| Cobalt | Selenium |
| Copper | Silver |
| Gallium | Thallium |
| Germanium | Thorium |
| Gold | Tin |
| Indium | Titanium |
| Iodide | Tungsten |
| Lanthanum | Uranium |
| Lead | Vanadium |
| Lithium | Ytterbium |
| Manganese | Zinc |
| Molybdenum | Zirconium |

SOURCE: From R. A. Freeze and J. A. Cherry, *Groundwater*, copyright © 1979 by Prentice-Hall, Inc., Englewood Cliffs, N.J.

ate, sulfate, and some cations may reach the water table during recharge. In the aquifer, additional chemical processes may modify the composition of the initial recharge.

Ultimately, the chemical compounds and concentrations in ground water at any point in the flow system are determined by the type of solid materials it has encountered along the way. A water sample taken from an aquifer in limestone and dolomite will contain high concentrations of calcium, magnesium, and bicarbonate. The calcium and magnesium ions impart a condition called *hardness* to the water. Hard water, although not harmful to drink (and perhaps even beneficial to health), produces several undesirable effects, such as scale on plumbing fixtures and boilers, and a lack of soap suds. Waters in contact with gypsum will be rich in sulfate. Sedimentary basins containing highly soluble rocks will yield ground water with very high concentrations of sodium and chloride. Iron is another parameter of concern for reasons other than health. Iron concentrations above 0.3 mg/L will produce stains and solid precipitates on plumbing and clothing. The relative proportions of inorganic elements, ions, and compounds are listed in Table 11-2. In terms of concentration, various combinations of a small number of major constituents dominate ground-water quality.

### *Concentration Units*

The most commonly encountered concentrations are *mass concentrations*. Laboratory analyses of ground water usually report concentrations in mg/L, which means the mass of the solute in milligrams per liter of solution. For dilute ground waters that fall within the fresh category in Table 11-1, milligrams per liter are numerically equal to *parts per million* (ppm) because a liter of water has a mass of $10^6$ milligrams. As ground water becomes progressively more mineralized, this approximate equality becomes less accurate. *Molar concentrations* are expressed as moles per liter or millimoles per liter, in which a mole is the gram molecular weight of a solute in a liter of solution. *Normal concentrations*, such as equivalents per liter (EPL), are derived by multiplying the molarity by the valence of the solute species. All three systems of concentration units are used in some situations.

Conversion of units is often necessary. To convert mg/L to molar concentrations, the following formula is used:

$$\frac{\text{moles}}{\text{L}} = \frac{\text{mg/L (ppm)}}{1000 \times \text{formula weight}} \qquad \text{(Eq. 11-11)}$$

It follows that

$$\frac{\text{mg}}{\text{L}}\text{(ppm)} = \frac{\text{moles}}{\text{L}} \times 1000 \times \text{formula weight} \qquad \text{(Eq. 11-12)}$$

**EXAMPLE 11-4**

A water sample from an aquifer contains 60 mg/L calcium. What are the molar and normal concentrations?

***Solution.*** Since calcium has an atomic weight of 40 g/mole, then

$$\frac{\text{moles}}{\text{L}} = \frac{60 \text{ mg/L}}{1000 \times 40} = 0.0015 \frac{\text{moles}}{\text{L}}$$

In solution, dissolved calcium is dominated by the calcium ion with a valence of +2. Therefore,

$$\frac{\text{equivalents}}{\text{L}} = 0.015 \frac{\text{moles}}{\text{L}} \times 2 = 0.003 \text{ EPL}$$

### *Graphical Displays of Ground-Water Quality*

Ground-water concentrations can be plotted on several types of diagrams in order to create a visual image of water quality. This is particularly useful when comparing multiple water-quality analyses or looking for trends in an aquifer. The *stiff graph* (Figure 11-37) is constructed by plotting certain concentrations on horizontal axes. The units used, equivalents per million, are numerically equal to equivalents per liter. Each geometric figure represents one water analysis. With practice, water-quality changes are apparent at a glance. The five Stiff diagrams on Figure 11-37, for example, represent samples taken along a ground-water flow path in which part of the aquifer has been contaminated.

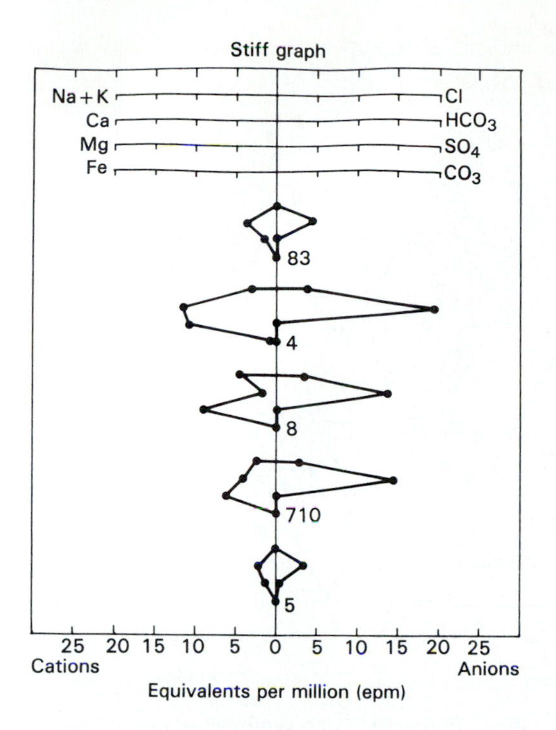


**Figure 11-37** Stiff diagrams. Sequence from top to bottom shows changes in water quality caused by contamination from a waste-disposal site.

The figures at the top and bottom have the same shape and represent the natural water quality. The three figures between show the constituents affected by the contamination, as well as the changes in concentration.

Another type of graph is the *Piper diagram* (Figure 11-38), which contains two triangles and a diamond. Cations are plotted as a point on the left triangle and anions on the right. The values are calculated as percentages of cations or anions in equivalents per liter. For example, each apex of the cation triangle represents 100% calcium, magnesium, or sodium plus potassium. Waters that contain mixtures of the three plot somewhere in the center of the diagram. If a water plots near one of the apices, the water can then be classified as calcium, magnesium, or sodium plus potassium type. The same procedure is followed in plotting and classifing the anions. Each water analysis is then represented by one point on the cation triangle and one point on the anion triangle. These points can be projected onto the diamond along lines parallel to the sloping lines shown (Figure 11-38). A point on the diamond is plotted where the two lines intersect. Many ground-water analyses can be plotted on a Piper diagram. Clusters or linear trends will sometimes provide information on chemical processes operating in the aquifer.

### *Chemical Evolution*

Systematic changes in ground-water quality are sometimes observed in ground-water flow systems. The changes, which can be called *chemical evolution* of ground water, are most prevalent in flow systems that penetrate deep into thick sedimentary rock sequences. Gradual increase in TDS is the most common trend along deep flow paths. Along with higher TDS, dominant ions change with distance in the direction of flow. In the recharge area, bicarbonate ($HCO_3^-$)

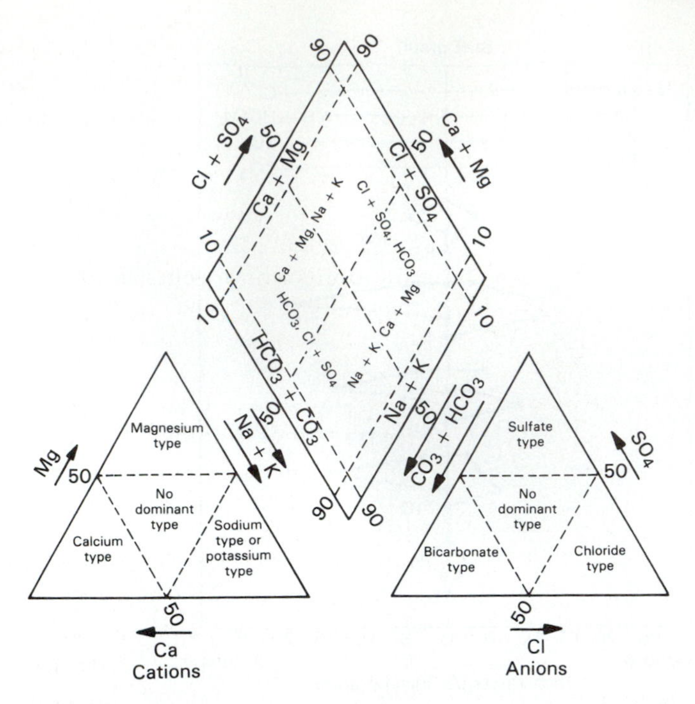


**Figure 11-38** Piper diagram. Each water analysis includes one point in the cation triangle and one point in the anion triangle. These points are projected into the diamond to form one point where lines intersect.

is usually the dominant anion. As the ground water moves deeper along the flow path, it is more likely to encounter evaporite-bearing rocks such as gypsum and halite. An idealized evolutionary sequence would proceed from bicarbonate to sulfate to chloride as dominant anions. Evolutionary trends in cations are not as well developed because of ion exchange and other chemical processes. Figure 11-39 shows the possible relationships between water quality and flow

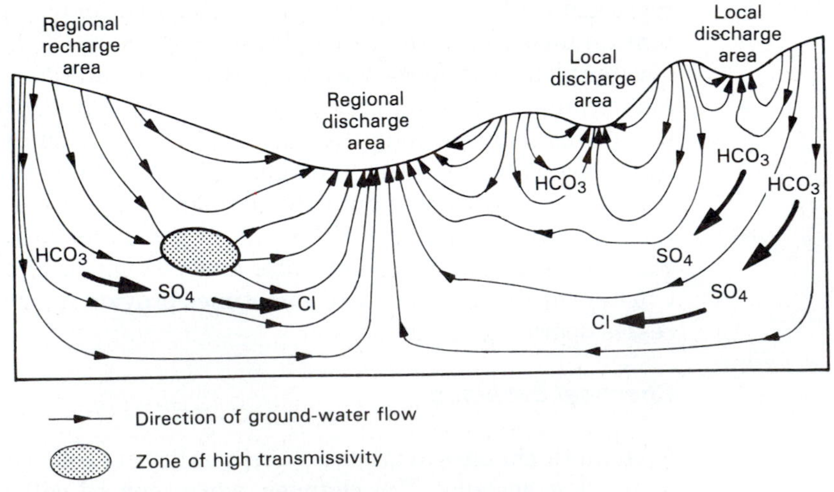


**Figure 11-39** Chemical evolution of ground water in sedimentary basins. Left side has uniformly sloping water table and predominance of regional flow. Right side has hummocky water table and local, intermediate, and regional flow systems. (Modified from J. Tóth, *Proceedings, First Canadian/American Conference on Hydrogeology.*)

systems. The left side of the diagram has a uniformly sloping surface that yields bicarbonate-type waters in the upper left corner of the cross section; sulfate-type waters appear in the center; and chloride-type waters are beneath the discharge area. This schematic is based on the assumption that the appropriate rock types are present to supply the anions shown. On the right is a cross section with a hilly, or hummocky, land surface profile. This situation tends to produce an abundance of *local flow systems*, which involve recharge at individual topographic highs and discharge at adjacent topographic lows. The local flow systems dominate the shallow region of the cross section and commonly contain only bicarbonate-type ground water. Flow is relatively rapid in the local flow systems and any salts that may have been present are flushed out of the system. The longer *intermediate* and *regional* flow systems differ in water quality from the local systems. Regional flow systems begin at the highest recharge area (at the upper right-hand corner of the cross section) and follow deep paths to the lowest discharge point at the center of the diagram. Intermediate systems are recharged at intermediate topographic highs but descend lower than the local flow systems. Water quality is bicarbonate type in the intermediate and regional systems, but evolves to sulfate and chloride types with distance along the flow path. Flow velocities are extremely slow in these deep systems and salts are not flushed out. TDS can reach the brine level (Table 11-1) in deep basins, and residence times on the order of millions of years are possible. Some of the water in these deep formations may even have originated as seawater trapped within the sediments during deposition.

## GROUND WATER AND CONSTRUCTION

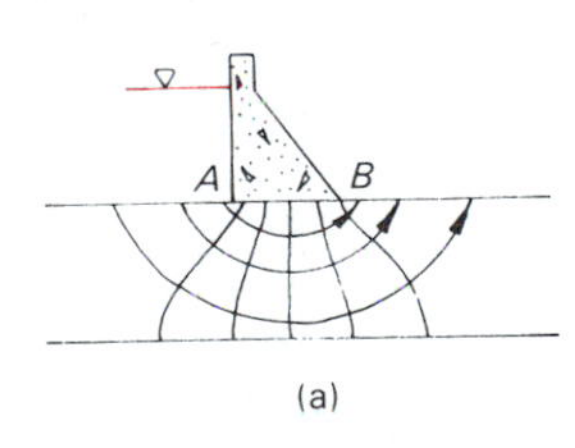


(a)

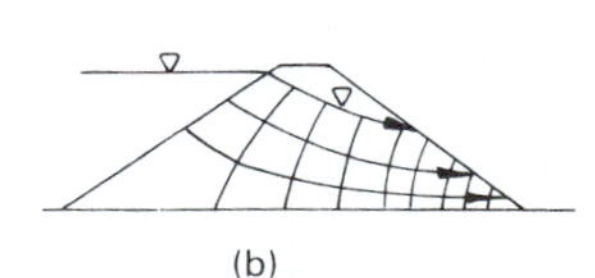

(b)

**Figure 11-40** Flow nets showing ground-water flow systems beneath impermeable masonry dam (a) and through permeable earth dam (b). (From R. A. Freeze and J. A. Cherry, *Groundwater*, copyright © 1979 by Prentice-Hall, Inc., Englewood Cliffs, N.J.)

In contrast to hydrogeologists who devote their careers to finding and developing ground-water resources, engineers involved in construction of dams, tunnels, buildings, and highways spend much of their time trying to control or dispose of ground water.

Dams illustrate the potential problems of ground-water movement, or seepage, quite well. Dams can be designed to prevent internal seepage, as in the case of masonry dams, or, as in the case of earth dams, to allow controlled seepage through the embankment at a safe rate. In the former case, excessive seepage beneath or around the dam is the major concern. Because the impoundment of the reservoir behind the dam increases the head within the ground-water flow system, extremely high fluid pressures decrease the stability of the dam by generating uplift pressures on the base of the structure or by causing internal erosion, or piping, of material near the downstream toe of the dam, thus tending to undermine it. Most dam failures have resulted from inadequate control of ground-water seepage through and beneath the structure. The St. Francis Dam (Case Study 5-1) and the Teton Dam (Case Study 11-3) are examples.

Ground-water seepage beneath and through dams is studied by constructing flow nets of the type shown in Figure 11-40. From the flow net, seepage pressure and gradient can be calculated so that the dam design can be modified if necessary. Seepage control beneath dams is sometimes accomplished by injection of cement *grout* under high pressure into closely spaced boreholes before the dam is constructed (Case Study 7-1). After the grout penetrates cracks and joints in the rock, it sets up, providing a barrier to ground-water seepage. Various design measures are also incorporated in permeable earth dams for seepage control.

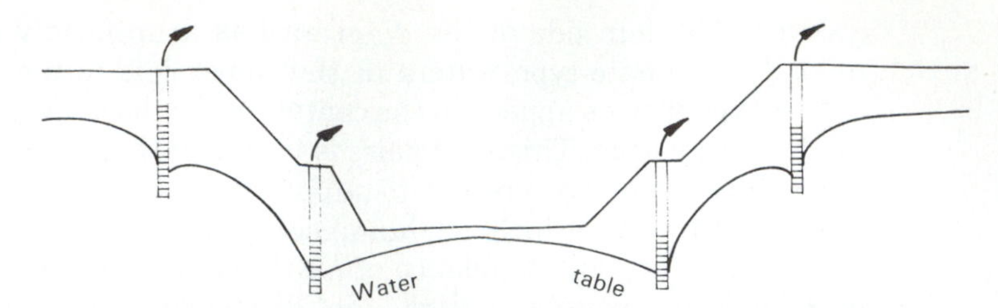


**Figure 11-41** Dewatering by well points to lower the water table below the base of an excavation and prevent excessive inflow of water into the excavation.

Ground-water inflow into tunnels and excavations often presents difficult engineering problems. In tunneling through rock, workers often suddenly encounter water in joints, fractures, or fault zones. Many lives are lost because of tunnel flooding when water under high pressure unexpectedly bursts into the tunnel. Seepage into excavations often leads to slumping and sliding of the walls of the cut. In Chapter 12 we will consider the importance of ground water in landslide phenomena of all types. If seepage into an excavation is determined to be undesirably rapid, *dewatering* techniques can be applied. One method, as shown in Figure 11-41, is to install a system of shallow pumping wells, or *well points*, around the excavation to lower the water table until the project can be completed.

**CASE STUDY 11-3**

## Failure of the Teton Dam

The failure of the Teton Dam in 1976 has proved to be one of the most costly dam failures in history. Constructed on the Teton River in southern Idaho, the dam was designed to provide water for irrigation and also to function as a flood control facility. The reservoir impounded by the dam was 17 miles long and had a capacity of 200,000 acre-feet. Construction of dam was completed in 1975, and filling of the reservoir began in the fall of that year. On June 5, 1976, before reservoir filling was complete, the dam failed with very little warning, sending a flood wave down the Teton River. The first indications of problems were found on June 3, when two small seeps were detected 1000 and 1500 feet downstream from the dam. On June 4th, another small seep was found 150 feet downstream of the right abutment, but as the day ended, no serious leaks were evident. The next morning, the first workers to arrive at the dam found small leaks at the toe of the right abutment and about halfway up the downstream face. As the morning progressed, the situation quickly became critical. About 10:30 A.M., the higher leak suddenly began to flow at a much greater rate following the collapse of a portion of the surface of the dam into a void below (Figure 11-42). Bulldozers were sent over the dam to push rock into the hole in an attempt to stop the leak. Instead, the flow grew stronger and at 11:30, the dozers slid into the hole and were washed downstream. Luckily, the operators were able to escape. The now growing hole progressed quickly up the face of the dam toward the crest, and at 11:55, the crest was breached and total failure began (Figure 11-43). In only 5 hours, the reservoir was completely emptied, releasing a discharge to the river below equal to the Mississippi River in flood.

The consequences of the dam failure included the loss of 11 lives and the displacement of 25,000 people from their homes due to the

**Figure 11-42** Increase in flow from leak near right abutment about 10:30 A.M. Dozers were lost in pit about this time. (From U.S. Dept. of Interior, *Failure of Teton Dam.*)

effects of the flooding. The monetary costs of the disaster were staggering. The $86 million cost of the dam, which was not rebuilt, was only the beginning. Damage claims paid by the U.S. Bureau of Reclamation were in the neighborhood of $400 million.

How could a dam designed and built with methods used many times in the past fail before it was filled for the first time? To answer this question, we must examine the geology of the site and the design of the dam.

The dam was constructed in an area of

**Figure 11-43** Breach of the crest of the dam about 11:55 A.M. Great increase in discharge rate. Complete failure came minutes after. (From U.S. Dept. of Interior, *Failure of Teton Dam.*)

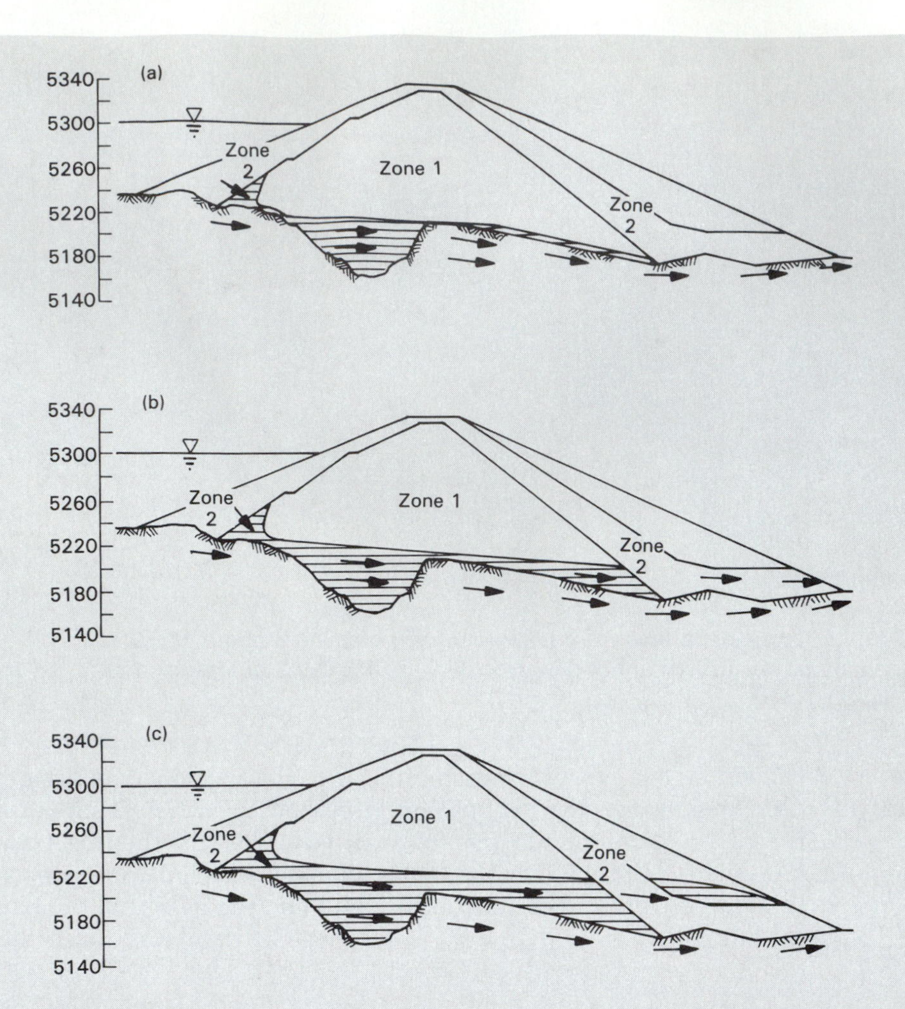


**Figure 11-44** Progressive failure of dam by enlarging area of seepage. Cross-ruled area and arrows show location of seepage and piping. Sinkhole formation and collapse of surface shown in (e). Black area represents material collapsed from above. Zone 1: silt core of dam; Zone 2: coaser alluvial cover. (From U.S. Dept. of Interior, *Failure of Teton Dam.*)

Late Cenozoic rhyolite welded ash-flow tuffs. These volcanic pyroclastic materials, which were thoroughly studied prior to design, were known to be highly fractured and jointed. Extensive grouting was carried out prior to construction, particularly in a cutoff trench excavated into the rhyolite tuff beneath the dam (Figure 11-44). The dam was designed as a zoned earth-fill embankment, with a core (Zone 1) of compacted silt derived from wind-blown deposits nearby. The selection of this nonplastic material may have contributed to the failure. Coarser alluvium (Zone 2) was emplaced over the core to protect it from surface erosion.

Although the exact cause of failure may never be known, the sequence of events illustrated in Figure 11-44 is a plausible explanation for the failure. The hypothesis includes the supposition that the Zone 1 material in the core of the dam was subject to a certain amount of cracking during settlement of the embankment. A more plastic core material would not have been susceptible to cracking. The result of the cracking would be to create preferential pathways for water movement. Figure 11-44a shows seepage through the fractured rock beneath the dam as well as through the Zone 1 soil in the cutoff trench. Because of the high

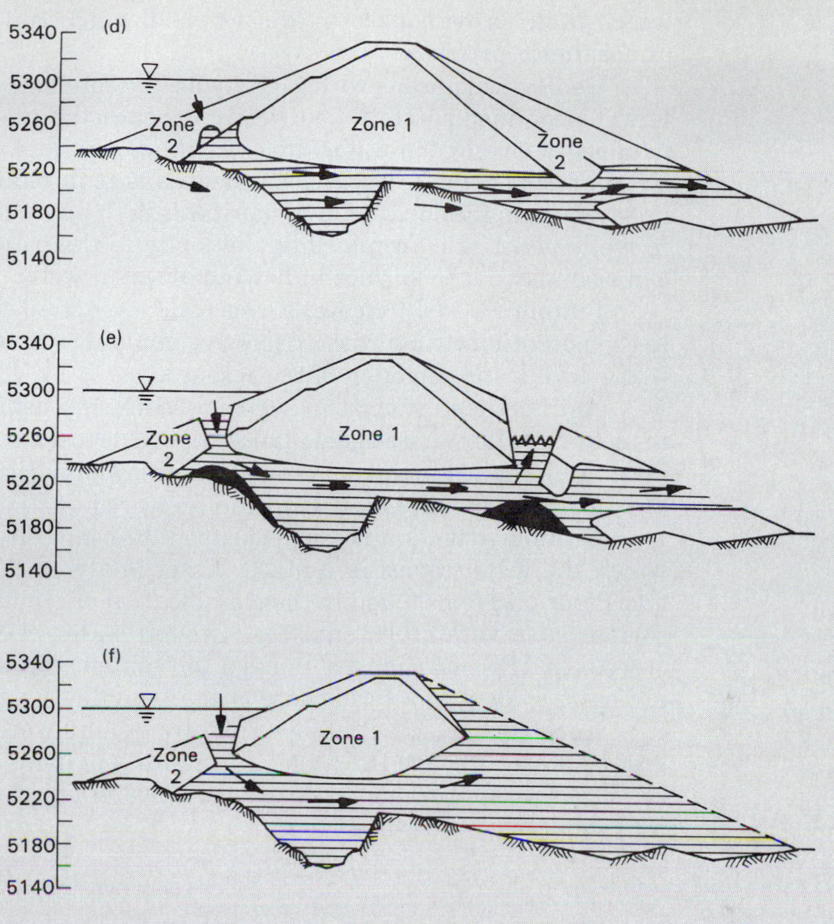


**Figure 11-44** (cont.)

hydraulic gradients caused by the reservoir behind the dam, Zone 1 soil at the contact with bedrock downstream from the cutoff trench began to be eroded by piping. The danger of a situation such as this is that removal of soil concentrates more flow in the zone of piping, leading to still more ground-water flow. This situation is illustrated in Figure 11-44c and 11-44d, where the zone of piping eventually breaks through Zone 2 soil, leading to the formation of sinkholes and collapse on both the upstream and downstream faces of the dam. After this point, failure is inevitable, as more and more of the core is washed away. Among the lessons learned from the Teton Dam failure is that even more thorough geologic investigations of dam sites are necessary prior to and during design and construction of these critical structures.

## SUMMARY AND CONCLUSIONS

Ground water is a vital resource; development and management of these resources are the keys to any major expansion of our water supplies in the future. Ground-water flow from recharge areas to discharge areas is governed by the parameters of Darcy's Law: aquifer cross sectional area, hydraulic gradient, and hydraulic conductivity. Recharge involves the downward movement of water through the unsaturated zone to the water table. The water table is recognized as the level at which fluid pressure is equal to atmospheric pressure.

Geologic materials with high hydraulic conductivity are known as aquifers. Unconfined aquifers lie directly beneath the unsaturated zone and are bounded above by the water table. Confined aquifers, on the other hand, can occur at considerable depths. The conditions required for development of confined aquifers include confining aquitards both above and below. Water levels in wells penetrating confined aquifers rise to the potentiometric surface and can rise above land surface in flowing artesian wells.

Pumping of wells creates a cone of depression in either the water table or in the potentiometric surface. Excessive pumping can lead to steadily dropping water levels and depletion of the resource.

Aquifers are common in surficial sediments, which are especially thick in tectonic valleys and coastal plains. Glacial deposits contain important unconfined aquifers in alluvial outwash sediments and buried-valley deposits. Aquifers in rocks often are located within fractured and jointed zones.

Ground-water quality depends upon the composition of rocks or soils with which the water comes in contact. Water analyses can be displayed on Stiff and Piper diagrams to aid in the classification of ground water and in making comparisons with other analyses. Ground water chemically evolves by increases in TDS and changes in major ion content. These evolutionary changes are most evident in intermediate or regional flow systems.

Construction activities often require ground-water control. Uncontrolled ground water seepage threatens the stability of dams and leads to hazardous conditions in tunnels and excavations. Dewatering methods may be necessary for completion of the project.

## PROBLEMS

1. For what reasons would you need to use Darcy's Law in the field rather than in the laboratory?
2. Flow occurs through a sand-filled pipe with a cross-sectional area of 60 $cm^2$. If $K$ is $10^{-3}$ cm/s and the gradient is 0.01, what is the discharge from the pipe?
3. Water seeps vertically downward from the bottom of a lake through a uniform material to an aquifer below. The discharge is 0.005 $ft^3$/s through a 1 $ft^2$ area of the lake bottom. The lake is 50 ft deep and the elevations shown on the diagram are given in feet above sea level. What is the hydraulic conductivity of the material between the lake and the aquifer? Assume there is no horizontal flow in the aquifer. (Hint: Hydraulic head at lake bottom is equal to the elevation of the lake.)

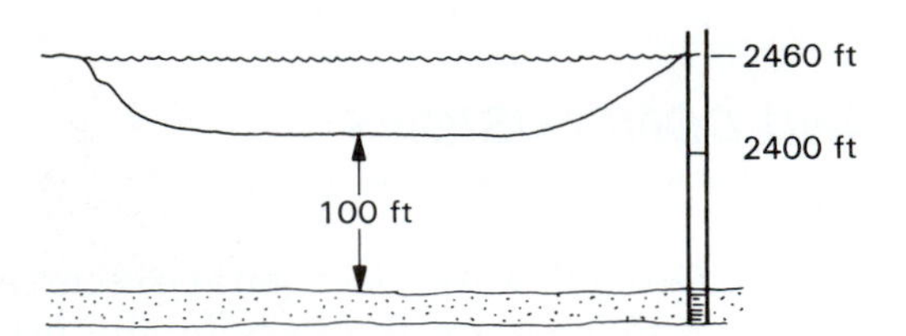


4. The bottoms of piezometers A and B are on the same flow line and 1000 m apart. The bottom of piezometer A is at 130 m and its water level is at 160 m. The

bottom of piezometer B is at 100 m and the water level is 150 m. What is the hydraulic gradient?

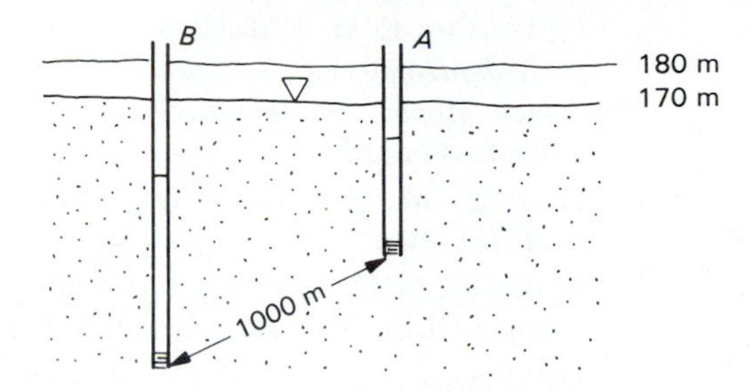


5. Why isn't the water table defined as the upper limit of the saturated zone?
6. Is it correct to say that ground water flows parallel to the water table? Why or why not?
7. What is an aquifer? Why aren't all saturated rocks and soils considered to be aquifers?
8. How do confined and unconfined aquifers differ in their response to pumping?
9. A well drilled into an artesian aquifer is 30 m deep. The fluid pressure in the aquifer is 392 kPa. Will the well be a flowing well?
10. The water table of a saturated, unconfined aquifer coincides with land surface. The aquifer is 52 ft thick and the material has a unit wet weight of 135 lb/ft$^3$. What is the effective stress at the base of the aquifer?
11. What factors control the amount of ground-water recharge?
12. What rock units or sediments constitute the most productive aquifers in your area? How does ground water impact the economy of the region?
13. How do ground-water flow systems in karst aquifers differ from those in other aquifers?
14. What would be the chemical characteristics of ground water from a limestone aquifer?
15. Water analyses are given below for samples from two wells in different areas. Calculate the molarity and normality of each constituent, and plot them on Stiff and Piper diagrams.

| *Ion (mg/L)* | *Well A* | *Well B* |
|---|---|---|
| $HCO_3^-$ | 311 | 159 |
| $SO_4^{2-}$ | 17.8 | 1902 |
| $NO_3^-$ | 0.34 | 7 |
| $K^+$ | 0.65 | 41 |
| $Na^+$ | 2.26 | 574 |
| $Mg^+$ | 22.4 | 171 |
| Fe (total) | 0.95 | — |

16. Why is ground water sometimes a problem at construction sites? How can the problems be solved?

## REFERENCES AND SUGGESTIONS FOR FURTHER READING

BURCHETT, C. R., P. L. RETTMAN, and C. W. BONING. 1986. *The Edwards Aquifer, Extremely Productive, But . . . .* U.S. Geological Survey in cooperation with the Edwards Underground Water District.

FREEZE, R. A., and J. A. CHERRY. 1979. *Groundwater*. Englewood Cliffs, N.J.: Prentice-Hall, Inc.

GUTENTAG, E. D., F. J. HEIMES, N. C. KROTHE, R. R. LUCKEY, and J. B. WEEKS. 1984. *Geohydrology of the High Plains Aquifer in Parts of Colorado, Kansas, Nebraska, New Mexico, Oklahoma, South Dakota, Texas, and Wyoming*. U.S. Geological Survey Professional Paper 1420-B.

HUBERT, M. K. 1940. The theory of ground water motion. *Journal of Geology*, 48:785–944.

MASTERS, G. M. 1991. *Introduction to Environmental Engineering and Science*. Englewood Cliffs, N.J.: Prentice-Hall, Inc.

QUNILAN, J. F., and R. O. EWERS. 1986. *Ground Water Flow in the Mammoth Cave Area, Kentucky, with Emphasis on Principles, Contaminant Dispersal, Instrumentation for Monitoring Water Quality, and Other Methods of Study*, Field Trip Guide Book. Dublin, Ohio: National Water Well Association.

SENGER, R. K., and C. W. KREITLER. 1984. *Hydrogeology of the Edwards Aquifer, Austin Area, Central Texas*. Bureau of Economic Geology, University of Texas at Austin, Report of Investigations No. 141.

STRAW, W. T., R. N. PASSERO, and A. E. KEHEW. 1993. Conceptual hydrogeologic glacial facies models: Implications for aquifers and agrichemicals in southwest Michigan, in, *Environmental Impacts of Agricultural Activities: Hydrogeologic Investigations and Modeling*, Y. Eckstein and A. Zaporozec, eds. Water Environment Federation, Alexandria, Va. pp. 35–68.

TÓTH, J. 1984. The Role of Regional Gravity Flow in the Chemical and Thermal Evolution of Ground Water, in *First Canadian/American Conference on Hydrogeology, Practical Applications of Ground Water Geochemistry*, B. Hitchon and E. I. Wallick, eds. Dublin, Ohio: National Water Well Association.

U.S. DEPARTMENT OF INTERIOR TETON DAM FAILURE REVIEW GROUP. 1977. *Failure of Teton Dam*. Superintendent of Documents, U.S. Government Printing Office, Stock No. 024-003-00112-1.

WHITE, W. B., and E. L. WHITE. 1989. *Karst Hydrology*. New York: Van Nostrand Reinhold.

# 12

# Mass Movement and Slope Stability

*Mass movement* is the collective name for a variety of processes involving the downslope motion of soil and rock materials under the influence of gravity. Damages resulting from such movements have been estimated to exceed $1 billion per year in the United States. An even greater cost is the loss of life that commonly accompanies major slope movements. Prevention of damage from slope movements requires recognition and avoidance of potentially unstable slopes. Where construction takes place on slopes, a detailed geologic investigation coupled with a thorough engineering analysis of the stability of rocks and soils that underlie the slope must be the initial phase of the project.

## TYPES OF SLOPE MOVEMENTS

Mass movements are classified according to the type of movement and the type of slope material involved (Figure 12-1). Slope materials are divided into bedrock, soil composed of predominantly coarse particles (debris), and soil composed of predominantly fine clasts (earth). Six types of movement are utilized in the classification. Each slope movement, therefore, is given a two-part name that relates the type of movement and the type of material. Many slope movements cannot be assigned to a single process and thus must be included in the "complex" category, an indication that more than one type of movement has occurred.

Missing from the classification is any indication of the velocity of movement. Within each category, rates can range from imperceptibly slow to freight-train velocities (Figure 12-2). The prediction of the particular types of movement that can be expected in rocks and soils in different geologic settings is a very important aspect of the analysis of slope stability that should be carried out during site selection and design of construction projects.

<table>
<tr><td colspan="3" rowspan="3">Type of movement</td><td colspan="3">Type of material</td></tr>
<tr><td rowspan="2">Bedrock</td><td colspan="2">Engineering soils</td></tr>
<tr><td>Predominantly coarse</td><td>Predominantly fine</td></tr>
<tr><td colspan="3">Falls</td><td>Rock fall</td><td>Debris fall</td><td>Earth fall</td></tr>
<tr><td colspan="3">Topples</td><td>Rock topple</td><td>Debris topple</td><td>Earth topple</td></tr>
<tr><td rowspan="3">Slides</td><td>Rotational</td><td rowspan="2">Few units</td><td>Rock slump</td><td>Debris slump</td><td>Earth slump</td></tr>
<tr><td rowspan="2">Translational</td><td>Rock block slide</td><td>Debris block slide</td><td>Earth block slide</td></tr>
<tr><td>Many units</td><td>Rock slide</td><td>Debris slide</td><td>Earth slide</td></tr>
<tr><td colspan="3">Lateral spreads</td><td>Rock spread</td><td>Debris spread</td><td>Earth spread</td></tr>
<tr><td colspan="3">Flows</td><td>Rock flow<br>(deep creep)</td><td colspan="2">Debris flow | Earth flow<br>(soil creep)</td></tr>
<tr><td colspan="3">Complex</td><td colspan="3">Combination of two or more principal types of movement</td></tr>
</table>

**Figure 12-1** Classification of slope movements. (From D. J. Varnes, 1978, Slope movement types and processes, in *Landslides: Analysis and Control*, R. L. Schuster and R. J. Krizek, eds., TRB Special Report 176, Transportation Research Board, National Research Council, Washington, D.C.)

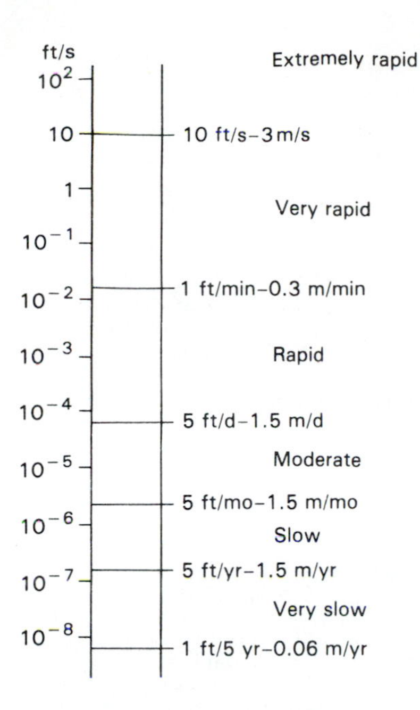


**Figure 12-2** Velocity scale for slope movements. (From D. J. Varnes, 1978, Slope movement types and processes, in *Landslides: Analysis and Control*, R. L. Schuster and R. J. Krizek, eds., TRB Special Report 176, Transportation Research Board, National Research Council, Washington, D.C.)

### *Falls and Topples*

When a mass of rock, debris, or soil separates from a steeply sloping surface and rapidly moves downslope by free fall, bounding, or rolling, the movement is termed a *fall*. These phenomena range from massive bodies of rock on mountain peaks set in motion by earthquakes (Figure 12-3) to small blocks of soil that fall down a river bank when lateral erosion by the stream undercuts the bank

**Figure 12-3** Damage to town of Friuli, Italy, from rockfall caused by earthquake in 1976. (Photo courtesy of Mario Panizza.)

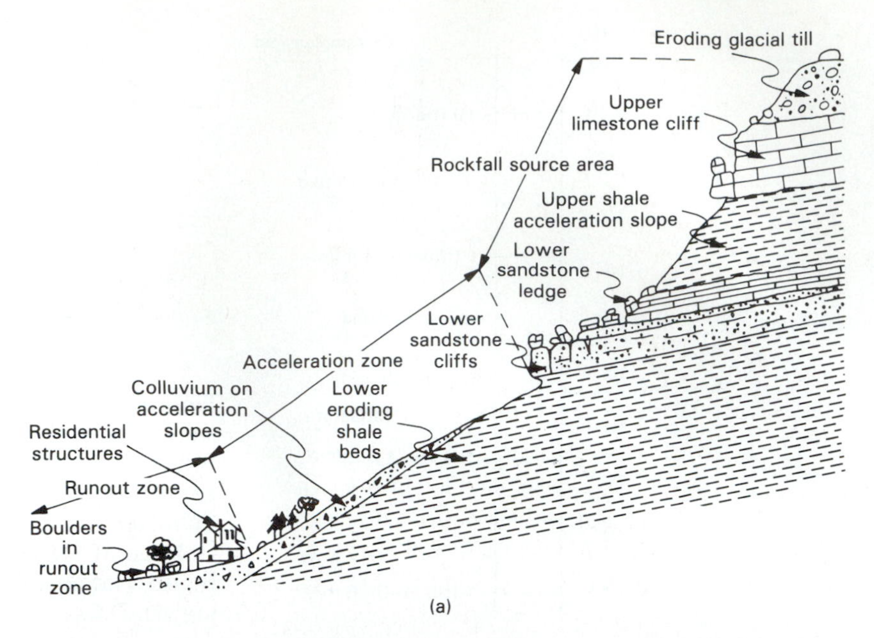


(a)

(b)

**Figure 12-4** Diagram (a) and photo (b) of slope in Vail Colorado that produces rockfalls. (From B. K. Stover, 1988, Booth Creek Rockfall Hazard Area, in *Field Trip Guidebook*, Geological Society of America.)

to the point when the overhanging section collapses. Hazards are greatest in narrow mountain valleys with steep rock walls. Residental construction in areas of rockfall hazard has accompanied population growth in the Rocky Mountains, primarily associated with recreational development. The resort town of Vail, Colorado, provides an excellent example. Development had encroached into hazardous areas prior to adoption of hazard mapping and zoning. Figure 12-4a shows the slope geology of a rockfall area near a ski resort. Nearly vertical cliffs form at the outcrop of resistant formations near the top of the 300-m high slope. Rockfalls are generated by weathering processes, including freeze and thaw. Rock slabs that break off from the upper limestone cliff gain momentum by rolling down the upper shale acceleration zone and are launched

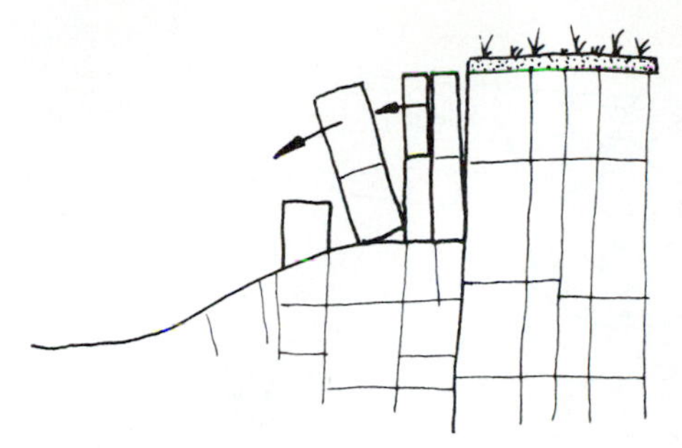

**Figure 12-5** Topples occurring in jointed bedrock. (Modified from D. J. Varnes, 1978, Slope movement types and processes, in *Landslides: Analysis and Control*, R. L. Schuster and R. J. Krizek, eds., TRB Special Report 176, Transportation Research Board, National Research Council, Washington, D.C.)

into the air over the vertical sandstone ledge below. Rocks from both source areas accelerate on the steep slope below the sandstone ledge and come to rest in the runout zone, where houses are located (Figure 12-4b).

A *topple* is a rotational movement that occurs as a block of material pivots forward about a fixed point near the base of the block (Figure 12-5). Topples develop in rock or cohesive-soil slopes divided into blocks by vertical fractures or joints oriented parallel to the slope face. Horizontal discontinuities, such as bedding, may affect the process if differential erosion allows a more resistant column of overlying rock to be undermined (Figure 12-6). The thin slabs of material are pushed outward by lateral forces exerted by adjacent material or by water freezing and expanding in the cracks. If the base of the slab is fixed, the lateral forces cause an overturning moment on the slab, and a topple may occur. The gradual pivoting motion prior to final failure may take a long period of time.

### Slides

One of the most common types of movement is *sliding*—that is, shearing displacement between two masses of material along a surface or within a thin zone of failure. The basic difference between slides and flows is that slides

**Figure 12-6** Column of rock beginning to pivot downslope. A topple will eventually occur. Petrified Forest National Park, Arizona.

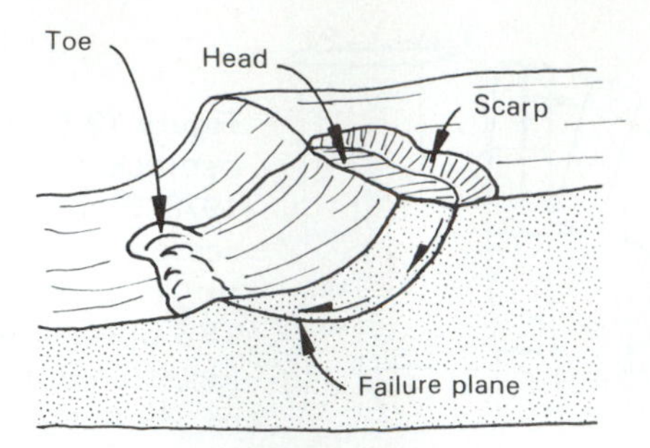


**Figure 12-7** Slumps involve rotational sliding along a curved failure surface.

move as a unit with little or no deformation within the sliding mass, whereas *flows* are thoroughly deformed internally during movement.

Slide types are further divided by the nature of the failure surface. Failure surfaces are often curved or circular. These slides, or *slumps*, are very common in soils or rocks of low shear strength (Figure 12-7). The rotational movement of the slump mass or block may result in a significant tilt of the upper surface of the moving block, or *head*, backward toward the exposed upper part of the failure plane, which is called the *scarp* (Figure 12-7). Trees, telephone poles, and other objects on top of the sliding blocks are also tilted by the rotational movement (Figure 12-8). Slumps can attain truly immense proportions. Recent mapping of the sea floor around the island of Hawaii, for example, has shown that huge slumps extend from the coasts out onto the sea floor for more than 100 km. Major fault scarps on land, some with more than 700 m displacement, are actually the head scarps of the gigantic submarine slumps (Figure 12-9).

If failure surfaces are planar rather than curved, the resulting movement is a *translational slide*. *Rock block slides*, consisting of the motion of a mass of rock that remains in a small number of blocks, can be distinguished from *rock slides* (Figure 12-10), which are composed of multiple rock fragments of various sizes. The actual configuration of the slide mass usually is determined by joints and planes of foliation or bedding. The most hazardous situation

**Figure 12-8** Telephone poles tilted by the rotational movement of a slump located in northern Idaho.

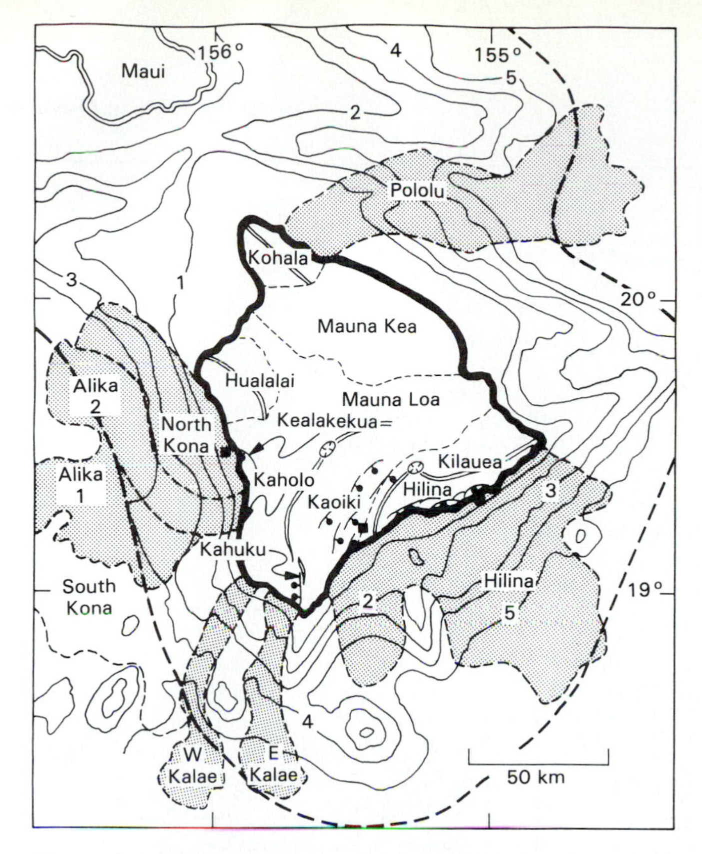


**Figure 12-9** Island of Hawaii showing huge submarine slumps (stippled pattern). Contours are water depth in kilometers. Faults on land (shown by fine lines with ball on downdropped side) are scarps of slumps. (From J. G. Moore and R. K. Mark, 1992, Morphology of the island of Hawaii, *GSA Today*, 2:257–262.)

**Figure 12-10** The scar of a rock slide near Lake Tahoe that occurred in granitic igneous rocks.

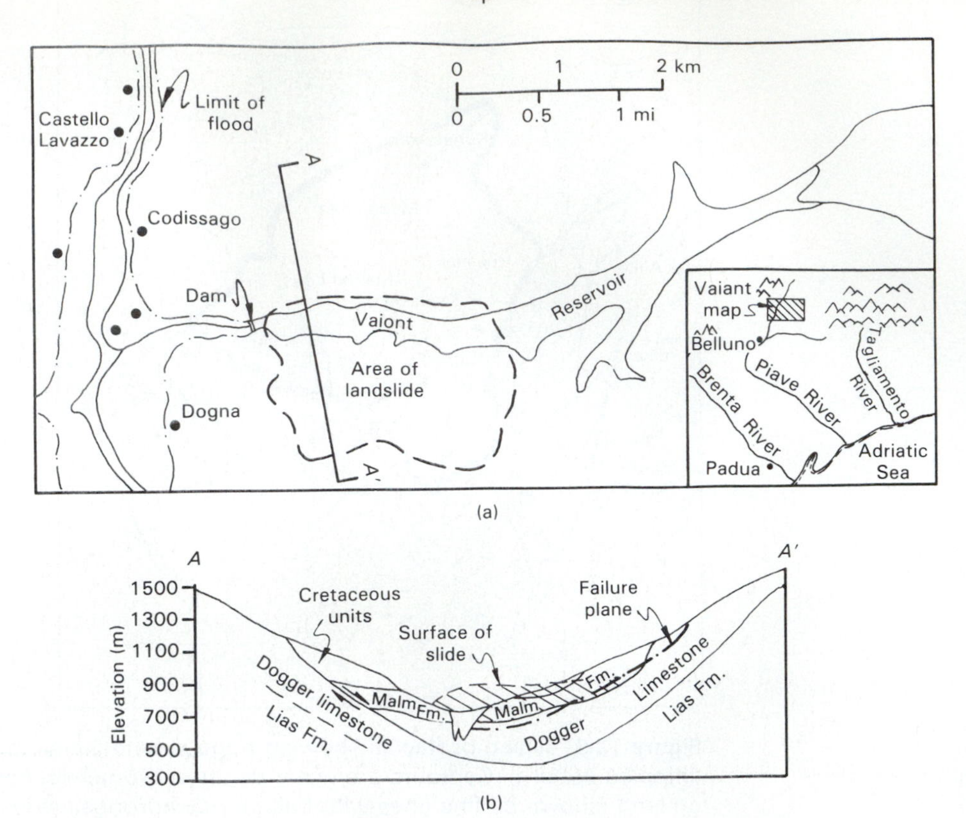


**Figure 12-11** (a) Map of Vaiont Reservoir area showing extent of the landslide of October 1963. Black dots show the locations of villages in the flooded area. (b) Cross section of the valley showing the structure and stratigraphy of the slide area. (Modified from G. A. Kiersch, 1964, Vaiont Reservoir disaster, *Civil Engineering*, 34:32–39.)

occurs when predominant planes of weakness dip in the same direction as the slope. A disastrous slide that at least began with translational movement localized along planes of weakness dipping toward the center of a valley occurred in the Italian Alps in 1963. During the slide 240 million cubic meters of rock and soil traveled rapidly downslope into a reservoir impounded by Vaiont Dam, the world's highest thin-arch dam (267 m) (Figure 12-11). The slide, which filled the reservoir for a distance of 2 km, generated a huge wave of water that overtopped the dam and flooded the valley below. Nearly 3000 people were killed by a wall of water as much as 70 m high. The destruction would have been even greater had the dam not remained intact during the tremendous stress placed upon it. A major factor in the Vaiont slide was the presence of planes of weakness in the valley wall rocks dipping toward the valley floor (Figure 12-11b). These planes included bedding planes in the sedimentary rock units at the site, as well as fractures produced by unloading of the rocks in the slopes—fractures similar to those caused by exfoliation. Other conditions contributing to the Vaiont disaster were increased pore pressures in the slopes resulting from seepage from the reservoir and heavy rains that preceded the final failure of the slope.

Debris slides are common on slopes where bedrock is overlain by *colluvium*, a mixture of residual and transported soil slowly moving downslope under

**Figure 12-12** Shallow debris slide near Monterey, California, resulting from the severe winter storms of 1982–1983. (G. F. Wieczorek; photo courtesy of U.S. Geological Survey.)

the influence of gravity (Figure 12-12). The failure plane in these situations is either the interface between less-weathered bedrock and the debris above or a plane within the debris. The common association of excess pore-water pressure and slope movement is recognized in this type of setting. The effect of increased pore-water pressure is to produce a buoyant force acting upward on the slide mass, thus decreasing the natural stability of the slope. In addition, heavy rainfall can increase the weight of the potential slide mass by saturating the upper part of the debris and creating a perched water table within the slope debris. The role of water will be examined in more detail later in this discussion.

A type of translational slide that caused major damage in Anchorage, Alaska, during the 1964 earthquake is illustrated in Figure 12-13. Translational movement was initiated above a plane in the weak Bootlegger Cove clay. Here the triggering mechanism was the violent ground shaking by a major earthquake, although the subsurface materials created a potentially unstable situation.

**Figure 12-13** A complex translational slide in the Turnagain Heights area of Anchorage, Alaska, resulting from the 1964 earthquake. Notice that some of the houses remained nearly intact during the predominantly lateral movement. (W. R. Hansen; photo courtesy of U.S. Geological Survey.)

### *Lateral Spreads*

The slow-to-rapid lateral extensional movements of rock or soil masses are known as *lateral spreads*. Liquefaction and flowage of a weak soil layer within a slope is the cause of most lateral spreads in the debris and earth categories of the slope-movement classification. The stronger material above the failure is rafted along without intense deformation, although it may be broken into blocks that can subside or rotate as the spread progresses (Figure 12-14). The most susceptible materials are sensitive clays (Chapter 10), like the Bootlegger Cove clay in Anchorage. These clays exhibit the tendency to instantaneously lose shear strength upon rearrangement of the clay particles (remolding).

The danger of quick clays (clays with the highest sensitivity values) has been proven by many spreading failures in such places as the St. Lawrence River valley and the fjords of Scandinavia, in addition to the Anchorage region. A hypothesis for the mechanism of failure of the clays in these areas is based on their common geologic histories. When glaciers advanced to coastal regions, the crust was depressed under the great weight of the continental ice sheets. Offshore marine clays deposited during these times were later raised above sea level by the slow, but continuous rebound of the crust after the glaciers retreated. Because of their origin, the clays were originally deposited with a flocculated structure (Figure 12-15). After the clay deposits became exposed to nonmarine conditions, the salty pore water was gradually displaced by fresher water derived from rainfall and snowmelt. This freshwater then began

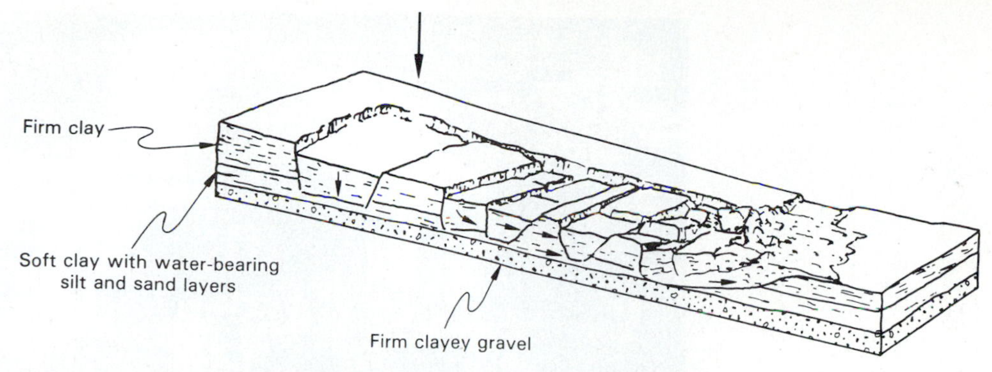


**Figure 12-14** A lateral spread involves the rapid outward movement of blocks of firm material because of the liquefaction and flowage in weaker material below. (From D. J. Varnes, 1978, Slope movement types and processes, in *Landslides: Analysis and Control*, R. L. Schuster and R. J. Krizek, eds., TRB Special Report 176, Transportation Research Board, National Research Council, Washington, D.C.)

to leach away the salt ions that provided the bonds between particles, or to cause geochemical reactions that weakened the particle bonds. By this process a very unstable situation is created. In its flocculated state the clay has significant shear strength. However, after leaching or geochemical changes, the clay fabric can be instantaneously changed from a flocculated state to a dispersed state. The dispersed clay has only a fraction of the shear strength it previously possessed and will behave as a liquid, with the ability to flow on extremely low slope angles. Thus lateral spreads are generated.

Earthquakes often provide the shock that induces the sudden liquefaction of quick clays. Other possible causes are construction blasting and the vibrations caused by the movement of heavy equipment. Once a lateral spread is

**Figure 12-15** Scanning electron microscope image of flocculated fabric of Bootlegger Cove Clay, which liquefied in the Anchorage earthquake. (From R. G. Updike et al., 1988, U.S. Geological Survey Bulletin 1817.)

**Figure 12-16** Aerial view of the Nicolet, Québec, lateral spread of 1955. Buildings and trees were rafted along in an upright position. (Henry Laliberté; photo courtesy of Governement du Québec, Ministère de l'Environnement.)

initiated, buildings, and everything else on the land surface, are rafted along above the flowing clay with the more coherent surficial material. A good example of this type of movement is the lateral spread that occurred in Nicolet, Québec, in 1955 (Figure 12-16). The costs of this event were three lives lost and several millions of dollars in property damage.

The lateral spreads just described involve the rapid flowage of earth materials. They are assigned to a separate category in the classification because of their predominantly lateral motion and because of the thick blocks of surficial material that are passively transported along above the flowing clay like boxes on a conveyor belt. There is really a complete transition from earth slides like the Anchorage slope failure, where remolding of sensitive clay occurred in very thin zones within more competent clay; to lateral spreads, where the zone of flowage is thicker; to certain types of rapid earth flows, where the entire mass of sediment is flowing. It is this latter type of movement to which we now turn our attention.

### *Flows*

The processes of flows are complex physical phenomena. Flow of various types of rock and soil may exhibit viscous or plastic behavior, as well as variations and combinations of both. The overriding criterion, however, is that flow must involve continuous internal deformation of the moving material.

When the flow travels at velocities at the slow end of the velocity scale, the process is called *creep*. Creep may occur in rock or surficial debris. In rock,

**Figure 12-17** Creep in weathered bedrock, Maryland. The initially vertical beds at the top of the exposure have been tilted by slow downslope movement. (G. W. Stose; photo courtesy of U.S. Geological Survey.)

creeplike flow can be a very slow, steady process, persisting over long periods of time. Alternatively, slow creep may accelerate to the point where a dramatic failure results. Instruments can be installed in vertical bore holes to measure downslope movement; it can also be observed by the displacement of trees or fences. Creep had been monitored in the slopes above the Vaiont reservoir for 6 months preceding the main slide. The rapid acceleration of the movement alerted officials to the impending disaster. Unfortunately, desperate attempts to draw down the reservoir were too late to avert the slide.

A common type of creep is often observed within weathered bedrock and soil on steep slopes (Figure 12-17). This process displays a pronounced seasonal variation characterized by expansion of soil materials perpendicular to the slope (Figure 12-18). The expansion, primarily due to freezing or swelling caused by increases in water content, is only the first phase of the movement. When either thawing or shrinkage takes place, the movement is vertically downward under the controlling influence of gravity. Thus during each cycle of expansion and contraction the soil undergoes a small downslope component of movement.

The more rapid examples of flow phenomena are variously termed *debris*, *earth*, or *mud* flows. These processes involve the rapid to very rapid flow of materials down steep slopes (Figure 12-19). The similarity in process among the flow of debris, lava, and glacial ice is reflected in the production of similar landforms by all three flow phenomena. Like glacial ice and lava, debris and

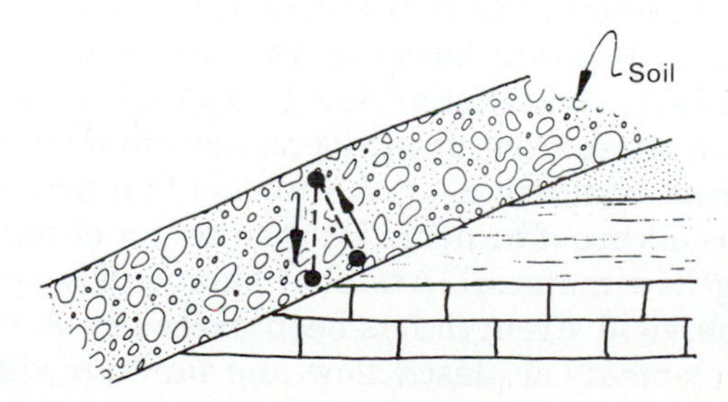


**Figure 12-18** Creep is often a seasonal process in which freezing or swelling causes movement of particles perpendicular to the slope. Upon thawing or shrinkage, vertical movement results in a net downslope component of motion.

**Figure 12-19** An earth flow that occurred in 1953 in Okanogan County, Washington. The flow lasted 7 minutes and destroyed a house. The source of the flow is the hill in the background. (F. O. Jones; photo courtesy of U.S. Geological Survey.)

earth flows follow channels, spread out to form lobate *toes* when not confined to channels (Figures 12-19), and develop conspicuous *lateral ridges* along the sides of the channels (Figure 12-20).

Many observations of debris flows have confirmed that the flows have a high water content, and thus are highly fluid, yet can transport boulders and other large objects many times the size of anything that even the fastest rivers can move. While in transit, debris flows treat boulders and other surficial objects with remarkable care. For example, large boulders that apparently have been transported intact within a debris flow have broken apart along fractures by weathering processes within a relatively short time period after deposition by the flow. These observations require careful analysis to determine the mechanical nature of debris-flow processes.

The visual similarity of rivers and rapid debris flows seems to suggest a similar flow mechanism. Thinking in these terms, we could consider debris flows simply to be rivers of high-density, highly viscous fluid in which boulders are supported by *turbulent eddies* (random velocity fluctuations) in the flow. Laboratory tests have shown, however, that dense debris flows are *laminar* rather than turbulent, and laminar flow lacks high-velocity turbulent eddies (Chapter 13). As an alternative, plastic behavior could be called upon to explain the concentration of boulders near the tops of the flow masses. Plastic flow could prevent the sinking of boulders into the center of the flow by the strength of the supporting flow material. An ideal plastic, however, would accelerate infinitely once the yield strength has been exceeded. A reasonable composite model combines elements of plastic flow and laminar viscous flow. This type

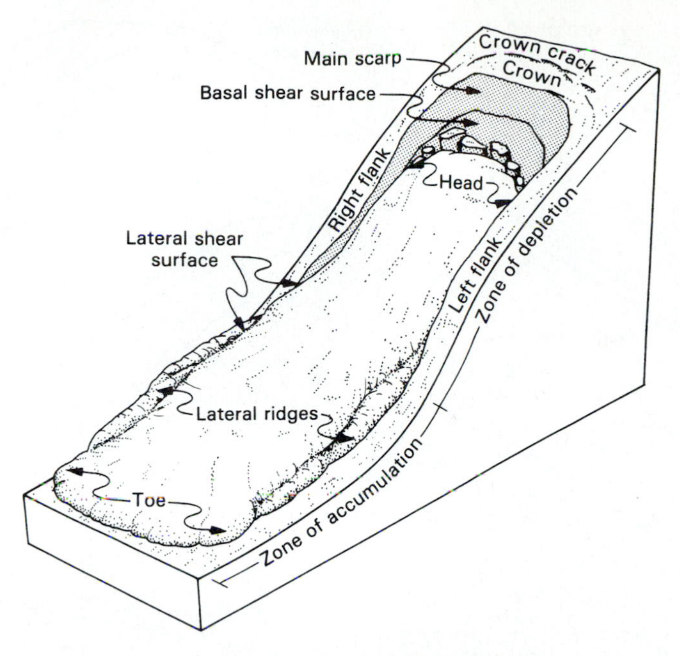


**Figure 12-20** Morphology and features of an earth flow. (From D. K. Keefer and A. M. Johnson, 1983, *Earth Flows: Morphology, Mobilization, and Movement*, U.S. Geological Survey Professional Paper 1264.)

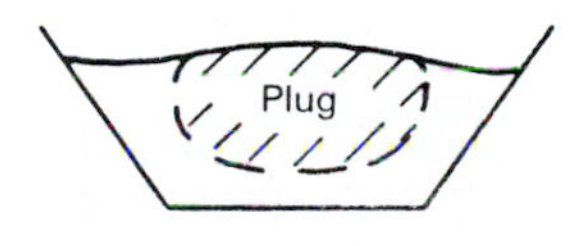


**Figure 12-21** Cross section of channelized debris flow showing the development of a nondeforming plug at the center of the flow.

of behavior is the type exhibited by a Bingham substance (Chapter 6). It seems to best fit the observations of debris flows and debris-flow deposits with respect to their ability to transport boulders in a laminar flow regime.

A further implication of the flow of a Bingham material is the existence of a rigid "plug" near the center of the channel and at the top of the flow (Figure 12-21). This plug is rafted along as a nondeforming unit in the debris flow because the shear stress within the plug is less than the strength of the material.

### *Complex Slope Movements*

Slope failures in nature that involve a single type of movement are probably the exception rather than the rule. A very common occurrence is the initiation of a slope movement as a slide, followed by conversion to a flow during the course of the event. A slump-earth flow, for example is illustrated in Figure 12-22. Topographic features resulting from the slump include a *main scarp,* separating the *crown* from the *head*, as well as one or more *minor scarps.* Beyond the main body in the downslope direction the lobate *toe* indicates that the dominant type of movement in this section is flowage. *Transverse ridges* in bands parallel to the toe cross the earthflow.

Some of the greatest natural disasters associated with slope failures have resulted from complex slope movements that began as rock falls or rock slides and then transformed into rapid debris flows, or avalanches, along their downslope paths. Particularly deadly examples of these events originated from the glacier-mantled Peruvian peak Nevados Huascaran in 1962 and 1970 (Figure 12-23). The first avalanche traveled 14 km in 5 minutes and buried the village of Ranrahirca, killing 3500 residents in the process. An even larger avalanche swept down the mountain in 1970, overtopping a ridge too high for the 1962

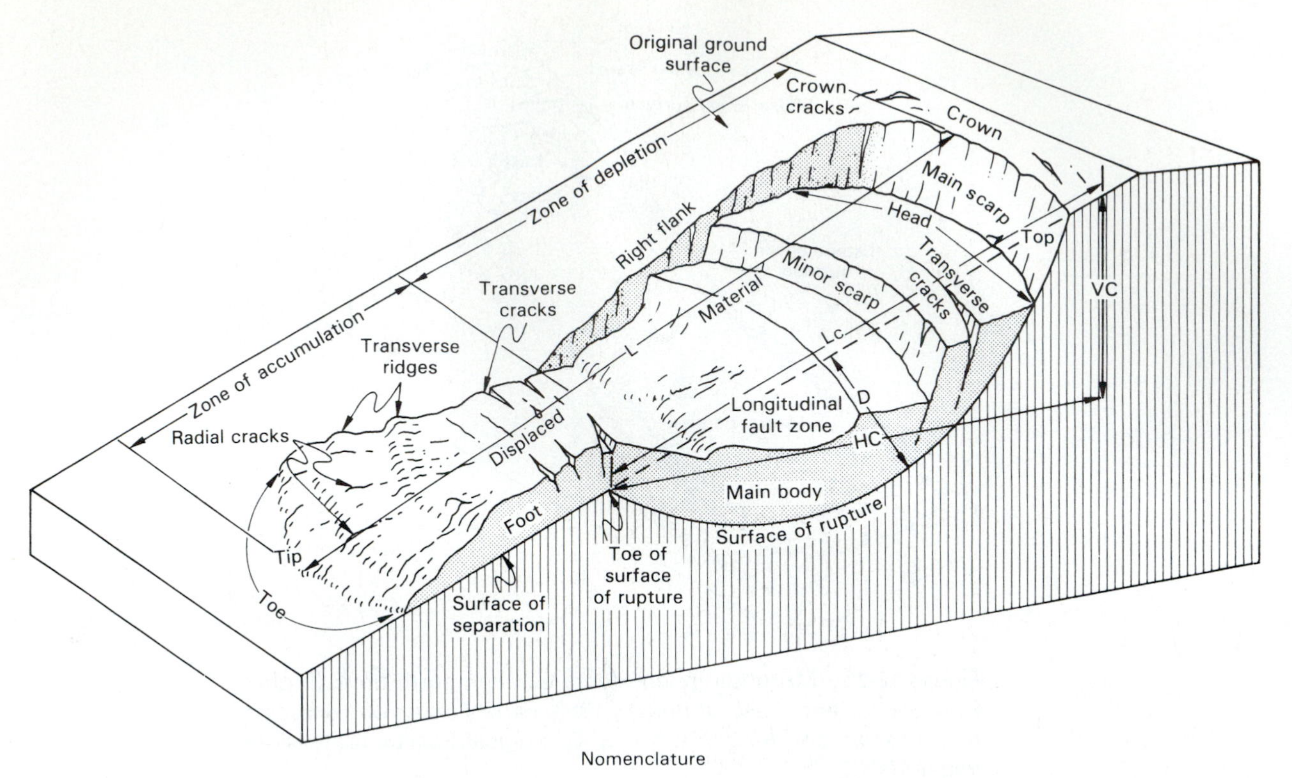


Nomenclature

Main scarp—A steep surface on the undisturbed ground around the periphery of the slide, caused by the movement of slide material away from undisturbed ground. The projection of the scarp surface under the displaced material becomes the surface of rupture.

Minor scarp—A steep surface on the displaced material produced by differential movements within the sliding mass.

Head—The upper parts of the slide material along the contact between the displaced material and the main scarp.

Top—The highest point of contact between the displaced material and the main scarp.

Toe of surface of rupture—The intersection (sometimes buried) between the lower part of the surface of rupture and the original ground surface.

Toe—The margin of displaced material most distant from the main scarp.

Tip—The point on the toe most distant from the top of the slide.

Foot—That portion of the displaced material that lies downslope from the toe of the surface of rupture.

Main body—That part of the displaced material that overlies the surface of rupture between the main scarp and toe of the surface rupture.

Flank—The side of the landslide.

Crown—The material that is still in place, practically undisplaced and adjacent to the to the highest parts of the main scarp.

Original ground surface—The slope that existed before the movement which is being considered took place. If this is the surface of an older landslide, that fact should be stated.

Left and right—Compass directions are preferable in describing a slide, but if right and left are used they refer to the slide as viewed from the crown.

Surface of separation—The surface separating displaced material from stable material but not known to have been a surface on which failure occurred.

Displaced material—The material that has moved away from its original position on the slope. It may be in a deformed or undeformed state.

Zone of depletion—The area within which the displaced material lies below the original ground surface.

Zone of accumulation—The area within which the displaced material lies above the original ground surface.

**Figure 12-22** Diagram of a slump-earth flow. (From D. J. Varnes, 1978, Slope movement types and processes, in *Landslides: Analysis and Control*, R. L. Schuster and R. J. Krizek, eds., TRB Special Report 176, Transportation Research Board, National Research Council.)

avalanche, and devastated the town of Yungay. More than 18,000 lives were lost on this tragic day. Unfortunately, rock avalanches are also common in the United States and Canada in the Rocky and Cordilleran Mountain ranges. The Madison Canyon "slide" of 1959 (Figure 12-24), which was triggered by an earthquake, created a lake as it flowed across a valley and blocked a river. A lake quickly developed in the valley behind the debris dam, raising fears that the rapid rise of water level would soon cause a catastrophic failure of the natural dam. Quick work by the U.S. Army Corps of Engineers, which excavated an emergency spillway through the debris, prevented the dam from failing and causing massive floods downstream.

**Figure 12-23** Aerial view of Nevados Huascaran area, Peru. Rock avalanches in 1962 and 1970 buried the villages of Yungay and Ranrahirca, killing thousands of people. (Photo courtesy of U.S. Geological Survey.)

**Figure 12-24** The Madison Canyon "slide." Actually a rock avalanche, this devastating slope movement blocked the Madison River to form a lake and buried a campground at the base of the valley on the side opposite the avalanche.

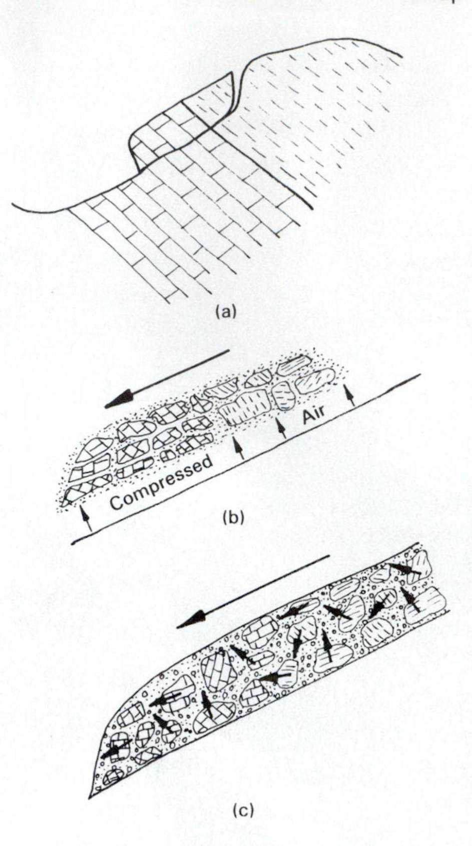


**Figure 12-25** Possible flow mechanisms of rock avalanches: (a) A block of rock on a steep mountain slope is suddenly launched into motion by an earthquake or other triggering factor. (b) Travel of the debris by sliding on a layer of compressed air. Initial stratigraphic order within the moving mass is preserved. (c) Movement of the avalanche by grain flow. After suspension of blocks in the air, motion and stratigraphic order are maintained by inertial collisions between trailing blocks and those fragments ahead of them. Acoustic fluidization is similar to (c) except that rocks are closer together.

The mechanics of movement of rock avalanches have been debated by geologists in recent years. One theory, which was widely accepted in the 1960s and 1970s, attributed the mass movements to a sliding mechanism (Shreve, 1968). According to this hypothesis, the mass moves as a relatively intact block that slides above a thin layer of compressed air. The compressed air would sufficiently lower the coefficient of friction to allow the great runout distances observed in these movements (Figure 12-25b). A sliding mechanism was invoked to explain the observation that the stratigraphic distribution of rock types in the mass of material at the base of the slope was identical with the distribution of rock types in the source area of the avalanche high upon the mountain. In other words, the mass appeared to have moved without extensive internal disruption. This stratigraphic order within the moving mass could be explained if the mass moved as an intact block (Figure 12-25b). Stratigraphic order would not be preserved in a viscous flow because the material near the top of the flow moves faster than material near the base. Thorough mixing of rock types would therefore be expected in viscous flow.

An alternative hypothesis was later put forth (even though it was first suggested in the 1880s by Swiss geologist Albert Heim) to explain the problem of stratigraphic order. In this type of flow, called *grain flow*, solid grains dispersed throughout the avalanche achieve the ability to flow without being carried along by a flowing viscous fluid (Hsu, 1975). Although both water and compressed air may be involved in the flow, their presence is not necessary for the flow to occur. This is best illustrated by the identification of giant rock avalanches on the Moon and Mars, bodies without an earthlike atmosphere or water. As grains collide in a grain flow, they transfer momentum to grains ahead of them (Figure 12-25c) and therefore do not pass their counterparts in the direction of flow. Thus stratigraphic order of rock types will be preserved in a grain flow. This model seems to fit evidence observed in rock avalanches better than the compressed air hypothesis because of the topographic form of the deposits. Rock-avalanche deposits have the physical appearance of flow, including lobate toes and transverse flow ridges.

Still another hypothesis suggests a process called *acoustic fluidization* for the mobility of rock avalanches (Melosh, 1987). The process is similar to grain flow, except that the particles are in closer contact and are fluidized by the propagation of elastic (acoustic) waves rather than by collisions. The elastic waves, which are similar to seismic waves, are generated as the mass of debris initially moves over an irregular surface. As the waves propagate through the flowing mass, they transmit pressure in the same way as particle collisions.

Controversy over the mechanics of movement of rock avalanches is likely to continue because none of the above hypotheses has yet been proven.

**CASE STUDY 12-1**

## The Frank Slide, Southwestern Alberta

One of the best known North American examples of a rock avalanche occurred in 1903 in the Canadian Rockies near the mining town of Frank, Alberta. The avalanche started suddenly when a 30 million $m^3$ mass of rock broke loose from Turtle Mountain and flowed across the valley below (Figure 12-26). In all, an area of 3 $km^2$ was buried in rock debris to a depth of 14 m. The entire event took only 100 seconds. Part of the town of Frank was destroyed and 70 people were killed. The entrance to a coal mine near the base of the slope was buried, but miners trapped below the debris were able to dig their way out.

It is likely that the structure of Turtle Mountain played an important role in the launching of the avalanche. The crest of the mountain forms the core of a large anticline (Figure 12-27). Thus on the eastern side, beds of limestone were steeply dipping in the same direction

**Figure 12-26** Photo of Frank "slide". Whitish area at base of slope is the area of deposition of rubble from the rock avalanche. (Photo courtesy of C. B. Beaty.)

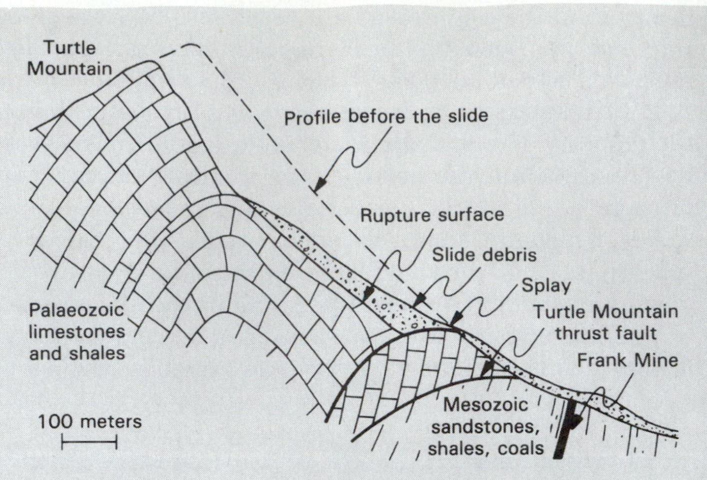


**Figure 12-27** Cross section of Turtle Mountain, Alberta, showing conditions before and after the rock avalanche. (From D. M. Cruden and C. B. Beaty, 1987, The Frank Slide, southwestern Alberta, *Geological Society of America Centennial Field Guide—Rocky Mountain Section.*)

as the slope of the mountain. Fractures near the region of maximum curvature of the anticline could also have been important in facilitating weathering processes that may have weakened the rock mass. The triggering event is not known but freeze and thaw may have been involved. The avalanches occurred on a very cold spring morning following several weeks of warm, wet weather. These conditions would have caused freezing of meltwater from the heavy snowpack on the mountain.

All the mechanisms discussed in the preceding section have been used to explain the mechanics of movement of the Frank slide. The appearance of the debris (Figure 12-26) is strongly suggestive of some type of flow. The lobate form of the rubble has been described as resembling a sack of flour spilled onto the ground.

Further research into the cause and mechanics of the Frank slide and other rock avalanches may yet yield an accurate understanding of these awe-inspiring natural processes. As mountainous areas become more and more developed the ability to predict the locations and the effects of rock avalanches will achieve great importance.

## CAUSES OF SLOPE MOVEMENTS

Determining the cause of a slope movement may be more difficult than it appears. Rarely is there a single cause; in most cases, although there may be an obvious triggering mechanism, the interaction of many variables controls the initiation of a slope movement.

A basic approach to evaluating slope stability in terms of simple sliding mechanisms is to identify the forces tending to cause movement, or *driving forces*, and those that tend to resist failure, or *resisting forces*. The ratio of the resisting forces to the driving forces is a quantitative indication of stability known as the *safety factor*. Thus

$$\text{safety factor (S.F.)} = \frac{\Sigma \text{ resisting forces}}{\Sigma \text{ driving forces}} \qquad \text{(Eq. 12-1)}$$

The greater the safety factor, the less likely the slope is to fail. The minimum value of the safety factor is 1 because below this value the driving forces are greater than the resisting forces, and the slide will be in progress. When the safety factor is exactly 1, the slope is in a state of incipient failure and any increase in driving forces or decrease in resisting forces will serve to trigger the movement.

### *Geologic Setting of Slopes*

Before considering factors that effect changes in the safety factor of a slope, it is worthwhile to identify some of the conditions that render a slope susceptible to mass movement. The slope angle, for example, plays an important role in establishing the initial ratio of resisting forces to driving forces. This can be illustrated by analyzing the forces acting on a block of rock resting upon a slope (Figure 12-28a). The weight of rock mass, $W$, acting vertically downward can be resolved into a force acting parallel to the slope in the downslope direction, $W \sin \theta$, and a force acting normal to the slope, $W \cos \theta$. Opposing the block's tendency to slide down the slope is the frictional force $F$, which is dependent upon the normal force and the coefficient of friction ($\mu$) between the rock and the slope, so that $F = \mu(W \cos \theta)$. As the slope angle increases (Figure 12-28b), the force acting downslope increases and the normal force decreases. When the force acting downslope, or driving force, equals the frictional resisting force, motion of the block is incipient.

The lithology of the materials composing the slope is probably the most important component of the slope-stability problem. Geologic materials of low strength or materials that tend to weather to materials of low strength present the greatest hazard. Included in this group are clays, shales, certain volcanic tuffs, and rocks containing soft platy minerals such as mica. Clays, and other rocks and soils that contain clays, are particularly notorious for their association with slope failures. Clays have an initially low strength and tend to become even weaker with increases in water content. The ability of clays to absorb water and swell can lead to a significant loss in strength by the materials on a slope.

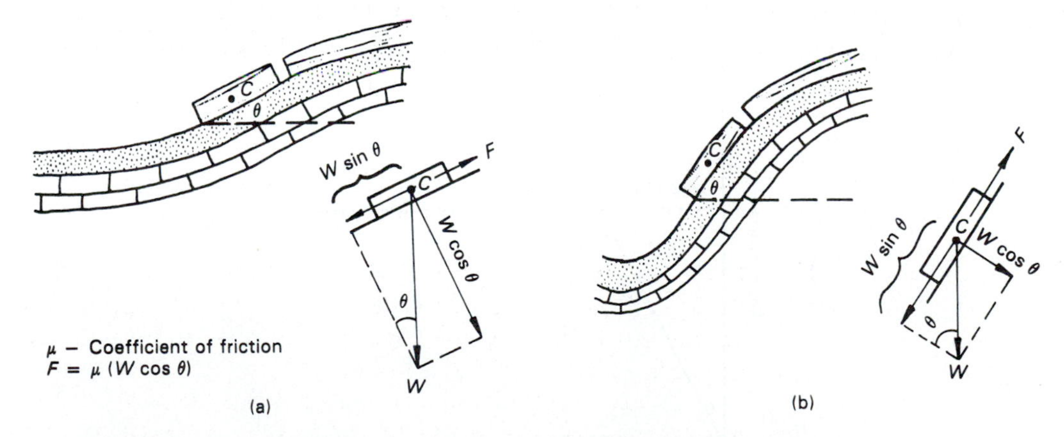


**Figure 12-28** Forces acting on a block of rock tending to slide downslope. In (a) the weight of the block is resolved into a downslope force ($W \sin \theta$) and a normal force ($W \cos \theta$). Downslope movement is opposed by the frictional force, $F$, which is equal to the coefficient of friction times the normal force. The same block is more likely to slide on a steeper slope (b) because the downslope force is greater and the normal force is less than in (a).

A special type of slope failure occurs in clays that are classified as *overconsolidated*. Overconsolidated clays have been subjected to large overburden loads during their geologic history. Under these loads clays become denser by compaction and expulsion of pore water. When overburden pressures are eased, either by erosion of overlying materials or by excavation, the compacted clays tend to expand and, in the process, develop a network of fractures, or *fissures*. The clay is then classified as overconsolidated because it currently supports a lower weight of overburden than it has at some time in its geologic past. The effect of these fissures is to lower the strength of the clay along the discontinuities. When small samples are taken for lab strength tests, they are frequently trimmed from intact blocks of clay between the fissures. The strength values obtained from these tests, described as the *peak strength*, give a misleading estimate of the strength of the clay mass, because mass movements on the slope are initiated by shear failure along the fissures. The strength mobilized along these discontinuities, the *residual strength*, is significantly lower than the strength measured from intact samples (Figure 12-29). There are numerous examples of excavated slopes such as road and railway cuts that have failed because the slope angle was designed using the peak instead of the residual strength.

In rock slopes a situation similar to that with overconsolidated clays exists, because it is the discontinuities in the rock that control the stability of the rock mass. Discontinuities along which strength is reduced are common in almost every type of rock. In sedimentary rocks, bedding planes constitute significant planes of weakness within the rock mass. In terms of slope stability, the orientation of the planes is critical. Bedding that dips in the same direction as the slope and at about the same amount is the most dangerous situation (Case Studies 6-1 and 12-1). Road cuts and other excavations can decrease the stability of rock slopes by increasing the stress or decreasing strength along bedding planes in several different ways.

Planes of weakness other than bedding that are present in all rock types include joints and faults. Careful preliminary studies of slopes are necessary to determine the spacing, width, and orientation of these fractures.

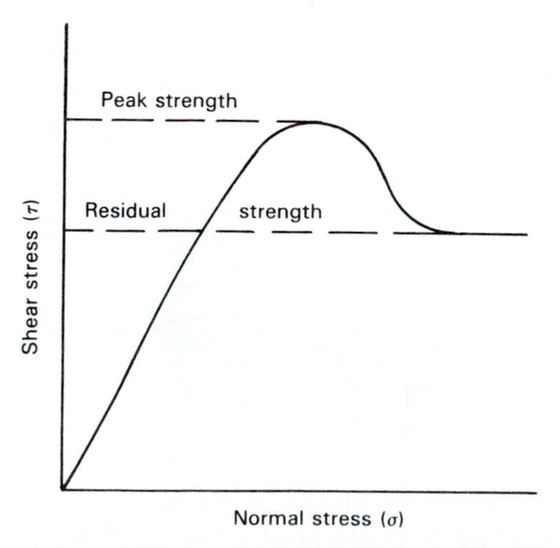


**Figure 12-29** In overconsolidated clays, the residual strength—the resistance to failure along existing fissures in the clay—is less than the peak strength. The peak strength is the resistance to failure obtained by tests on nonfissured, massive samples of clay.

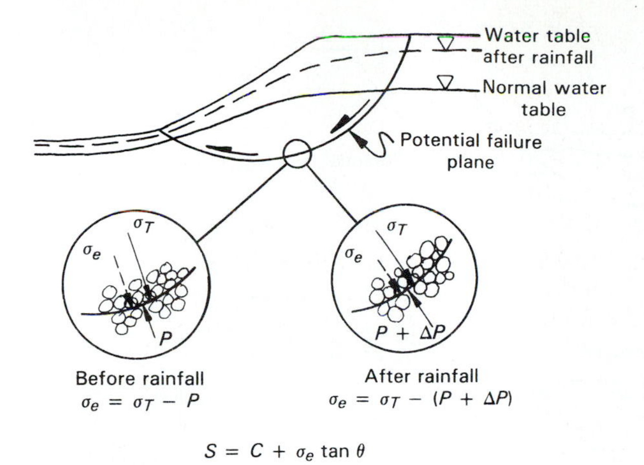


**Figure 12-30** Effect of pore water on slope movements. Rainfall causes a rise in the water table and increases pore pressure ($P + \Delta P$) throughout the subsurface. The stresses on a potential failure plane show the magnitude of effective stress ($\sigma_e$). Strength ($S$) is proportional to effective stress. After rainfall (right), effective stress decreases because of the increase in pore pressure. Strength of the soil along the potential failure plane is therefore lower.

### *Influence of Water*

Water is just as important as type and attitude of slope materials as a major controlling factor in the occurrence of mass movements. Its importance can be seen in the fact that most rapid mass movements occur during and after periods of heavy rainfall.

Water can interact with slope materials in several different ways. An obvious result of the addition of water to a slope is an increase in the weight of a potential sliding block. With reference to Figure 12-28, the effect of the addition of water is to increase both the downslope force $W \sin \theta$ and the normal force $W \cos \theta$. Although these changes will not drastically alter the safety factor of the slope, the water has a totally different effect on the normal force than was suggested previously. To understand this relationship, we must recall the effective stress equation presented in Chapter 11. In Figure 12-30, the stresses are shown on a potential failure plane within the slope material. Because the fluid pressure acts upward against the soil and rock load above the failure plane, the effective stress equation can be rearranged as

$$\sigma_e = \sigma_T - P \qquad \text{(Eq. 12-2)}$$

to show the amount of load borne by the soil particles. The name *effective stress* is appropriate because it determines the frictional resistance to sliding that could be mobilized by the soil particles along a potential failure plane.

Figure 12-30 illustrates the application of the effective stress equation to sliding-type slope movements. The fluid pressure is determined by the thickness of the saturated zone above the failure plane. So when the fluid pressure *increases*, as it would when the water table rises in response to heavy rainfall, the effective stress equation indicates that the effective stress will *decrease*. The rise in water table has therefore decreased the *effective* normal stress ($\sigma_e$) resisting sliding movement along the failure plane.

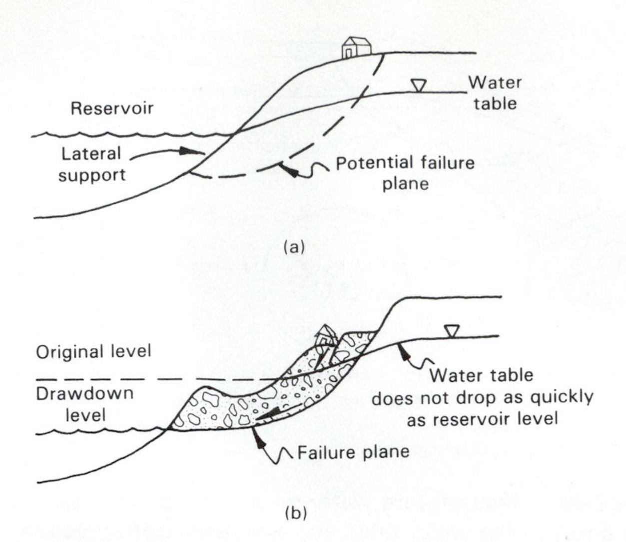


**Figure 12-31** Reservoir bank failure due to rapid drawdown of reservoir level. When the reservoir is drawn down (b), lateral support provided by the water is lost. Pore pressure in the slope remains high because the water table adjusts to its new level more slowly than the reservoir. These conditions combine to cause slope failure.

When effective stress is inserted into the Mohr-Coulomb equation for shear strength, the equation becomes

$$S = C + \sigma_e \tan \phi \qquad \text{(Eq. 12-3)}$$

It is obvious from Equations 12-2 and 12-3 that the shear strength of a material will decrease when the fluid pressure increases.

Another way to think of the effect of fluid pressure on sliding is to compare the process with the buoyant force exerted on an object submerged in water. Just as the submerged weight of any object is less than its weight in air, the effective weight of soil particles submerged by a water table rise will be less than their unsaturated weight. Thus, the resisting forces on the failure plane are decreased and the safety factor is reduced because the driving forces are unchanged or even increased by the weight of the water added to the soil.

A problem encountered with reservoirs provides an interesting application of the relationship of pore pressure and mass movements (Figure 12-31). The water table in the slopes adjacent to a reservoir rises and falls to adjust to the water level in the impoundment. When the water table rises, the decrease in resisting forces is offset by the lateral support exerted by water in the reservoir on the reservoir banks (Figure 12-31a). If the water level in the reservoir is rapidly lowered, perhaps in anticipation of a flood that must be contained or partially contained in the reservoir, the lateral support provided by the water is removed. The water table in the adjacent slopes is much slower to respond, and the high pore pressures now tend to cause slumping of bank material into the reservoir (Figure 12-31b).

### *Processes That Reduce the Safety Factor*

A natural or constructed slope exists under a certain ratio of resisting and driving forces (the safety factor) at any particular time. Many processes, however, acting over long or short periods of time, may decrease the slope's safety

**TABLE 12-1**

**Processes Causing Changes in the Safety Factor**

*Processes that cause increased shear stress*

1. Removal of lateral support
   a. Erosion by rivers
   b. Previous slope movements such as slumps that create new slopes
   c. Human modifications of slopes such as cuts, pits, canals, open-pit mines
2. Addition of weight to the slope (surcharge)
   a. Accumulation of rain and snow
   b. Increase in vegetation
   c. Construction of fill
   d. Stockpiling of ore, tailings (mine wastes), and other wastes
   e. Weight of buildings and other structures
   f. Weight of water from leaking pipelines, sewers, canals, and reservoirs
3. Earthquakes
4. Regional tilting
5. Removal of underlying support
   a. Undercutting by rivers and waves
   b. Construction of underground mines and tunnels
   c. Swelling of clays

*Processes that reduce shear strength*

1. Physical and chemical weathering processes
   a. Softening of fissured clays
   b. Physical disintegration of granular rocks such as granite or sandstone by frost action or thermal expansion
   c. Swelling of clays accompanied by loss of cohesion
   d. Drying of clays resulting in cracks that allow rapid infiltration of water
   e. Dissolution of cement
2. Increases in fluid pressure within soil
3. Miscellaneous
   a. Weakening due to progressive creep
   b. Actions of tree roots and burrowing animals

factor. Recognition and identification of these processes are necessary for the prediction of slope instability. In Table 12-1 processes are grouped into those that increase the shear stress on a failure plane and those that reduce shear strength within the materials on a slope.

One of the most common ways to increase the shear stress on a potential failure plane is to remove material from the base of the slope. As shown in Figure 12-32, construction of the road cut has removed material from an area where the failure plane is nearly horizontal or even curving upward. Material in this part of the potential slide mass acts against movement by resisting the tendency for rotational sliding. When the material is removed, the driving forces on the remainder of the slide mass are increased. In addition, the decrease in area of the failure plane after the excavation results in less shear strength developed along the plane. In this way, the resisting forces have been decreased. The combination of the increased shear stress and the decreased shear strength significantly lowers the safety factor of the slope.

The effect of *surcharge* (overloading) applied upslope from the line of action of the center of gravity of the sliding mass is to increase the rotational tendency of the mass. Surcharge loads introduced by humans, including mine wastes and construction fill, frequently initiate slope failures (Figure 12-33).

Processes that decrease the resistance to failure, with the exception of increases in fluid pressure, are usually long-term phenomena. These changes gradually modify the stability of the slope for many years prior to movement. The safety factor is then lowered to the point where a sudden triggering mecha-

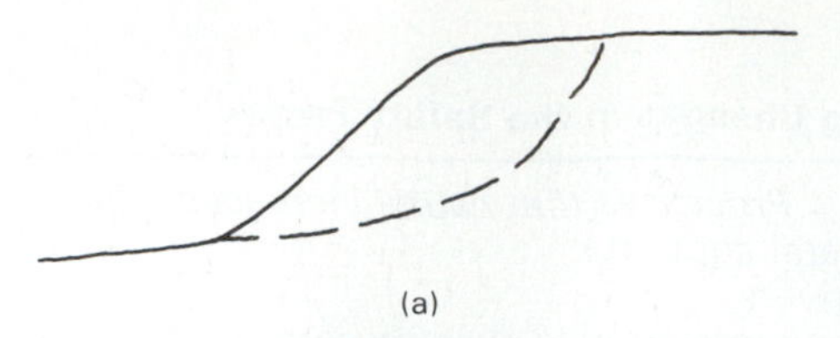


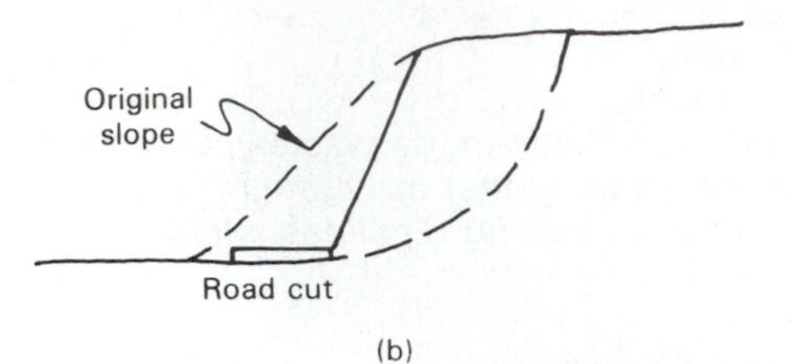


**Figure 12-32** Stability of the slope shown in (a) is decreased by removal of material from the base of the slope and by steepening the slope (b).

nism, an earthquake (Figure 12-34) or unusually wet period, for example, rapidly equalizes the driving forces and resisting forces, causing failure.

The large number of factors affecting both the driving and resisting forces supports the statement made earlier concerning the complexity of mass movements. Investigations of mass movements made after the fact as well as those aimed at predicting stability relationships, must not overlook any of the long- or short-term geologic and human influences on slopes.

**Figure 12-33** Slumping of surface mine spoils into a newly excavated mine pit was caused by piling too much spoil material at the edge of the pit.

## SLOPE STABILITY ANALYSIS AND DESIGN

The most important step preceding the construction of any engineering project that will alter or interact in any way with a slope is to determine the stability of the slope, both in its natural state as well as in its altered state after completion of the project. Embankments, earth dams, and other constructed

**Figure 12-34** A huge debris avalanche covering part of Sherman glacier, Alaska. The avalanche was triggered by the 1964 earthquake. (Austin Post; photo courtesy of U.S. Geological Survey.)

slopes must likewise be designed with slope stability in mind. We must begin our discussion of stability analysis, therefore, with some comments about the existing state of stability of natural slopes in and near the area of excavation.

### *Recognition of Unstable Slopes*

Areas of potential slope instability are most easily defined if evidence exists for previous slope movements. These areas can be delineated and evaluated by preliminary geologic studies. Several techniques can be used. First, all existing geologic reports of an area should be examined. Some geologic maps are specifically made for the purpose of identifying hazardous areas (Figure 12-35). Other maps show existing slope movements or the distribution of rock and surficial material types that are known to be associated with slope movements.

If more detailed studies of the project area are needed, these should be initiated by analysis of topographic maps and aerial photographs. Air photos are invaluable for the analysis and interpretation of landforms related to mass movements. Recognizable features include scarps, slump shoulders, and disrupted drainage patterns. Hummocky topography, a random pattern of hills

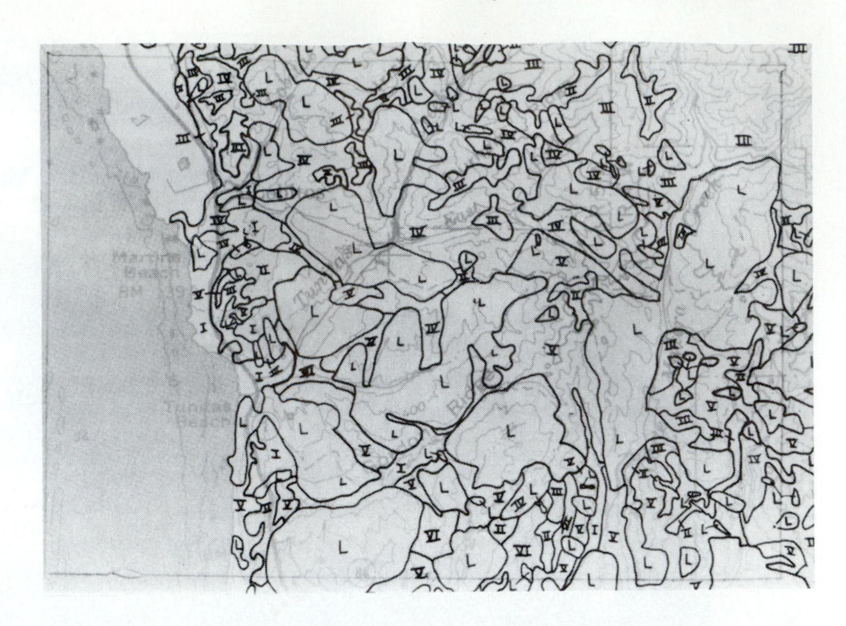

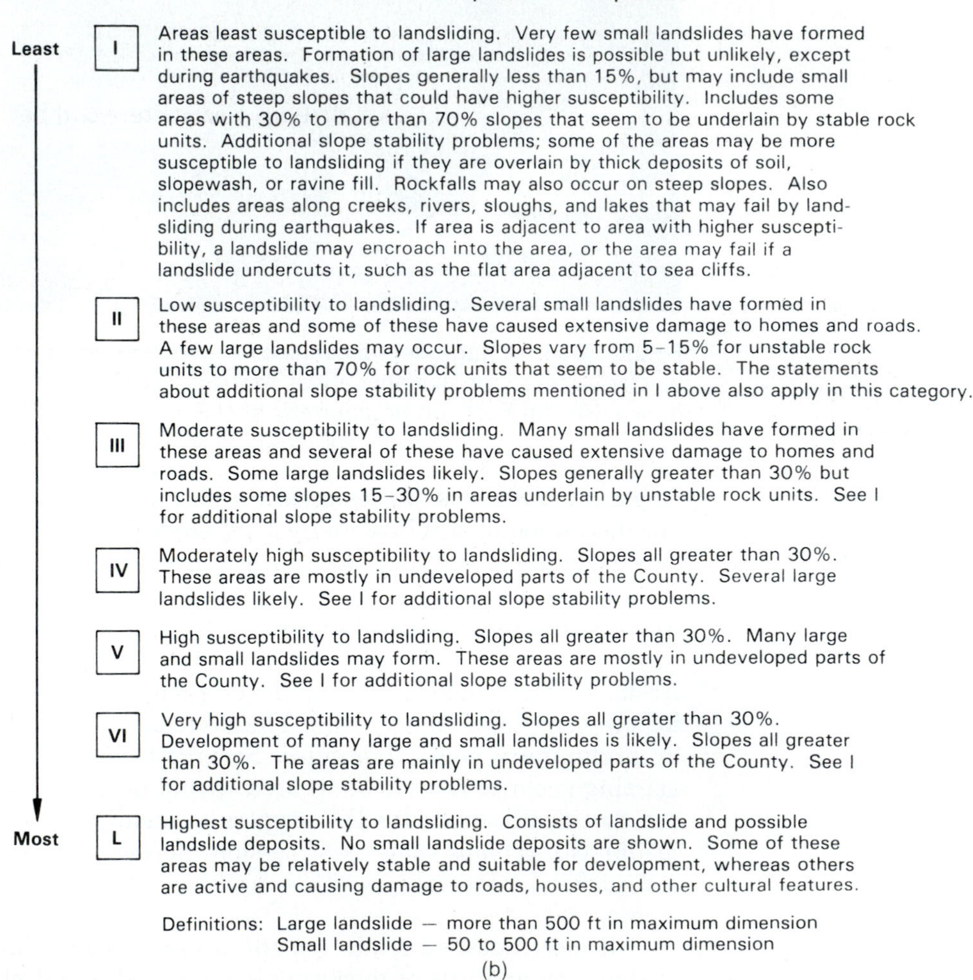


**Figure 12-35** A portion of a landslide-susceptibility map for San Mateo County, California. (From E. E. Brabb, E. H. Pampeyan, and M. G. Bonilla, 1972, *Landslide Susceptibility in San Mateo County, California*. U.S. Geological Survey Miscellaneous Field Studies Map MF-360.)

and depressions containing ponds or wetlands, is a good indicator of recent slump or flow activity. Deposits of flows occur as lobate or tongue-shaped landforms in association with such other features as lateral levees and transverse flow ridges. Other relevant slope conditions, including fracture and joint patterns, the locations of springs and seeps, and the oversteepening of slopes by rivers and waves, can be noted on air photos.

Field studies complete the site investigation. Detailed geologic maps of the area can be made by examining rock and soil exposures and plotting data on topographic base maps. Subsurface information is frequently needed for evaluation and design. Some types of data can be obtained using surface geophysical techniques like seismic refraction and earth resistivity. In seismic refraction, seismic energy is generated near land surface using hammers or explosions. Seismic waves are refracted from harder layers of rock or soil at depth back to the surface where they are recorded and analyzed to calculate the depth of the subsurface layer. Earth resistivity is a technique based on the variable resistance of subsurface materials to an electrical current passed through the ground. The most accurate method available for obtaining subsurface data (and also the most expensive) is test drilling. There are many methods and types of equipment that can be used for various subsurface conditions. Samples taken during drilling are useful not only for determining the distribution of soil and rock units but also for conducting lab tests to characterize the properties of the materials. For slope-stability analysis, the shear strength of the soil, including its lateral and vertical variations, is a necessary type of data. After test holes have been drilled, piezometers can be installed for monitoring ground-water conditions.

### *Stability Analysis*

The methods used for analyzing stability are based on finding the safety factor for a particular slope movement. For soil slopes composed of uniform, cohesive clays, rotational slump is considered to be the dominant failure mechanism. Successive circular failure surfaces are evaluated in order to determine the potential failure surface with the lowest safety factor. If this failure surface, the *critical circle*, has a safety factor near 1.0, the slope is considered to be unstable and must be redesigned or modified. Several trial surfaces and their computed safety factors are shown in Figure 12-36.

One of the most basic methods of calculating the safety factor for circular failure surfaces is shown in Figure 12-37. The safety factor for the indicated failure surface is expressed as the ratio of the resisting moments to the driving moments about the center of rotation, point $O$. Each cross section of the slope, such as the one illustrated, can be analyzed separately by treating the failure surface as a line rather than as a plane. The resisting moment is the product of the soil shear strength, $S$, which resists downslope movement of the mass, times the length, $L$, of the failure circle, times the moment arm, $R$, about the center of rotation. The driving moment is expressed as the weight, $W$, of the slump mass acting through its center of gravity times its moment arm, $X$, about the center of rotation. The safety factor for this particular slump would then be

$$\text{S.F.} = \frac{\Sigma \text{ resisting moments}}{\Sigma \text{ driving moments}} = \frac{SLR}{WX} \qquad \text{(Eq. 12-4)}$$

Higher or lower values of safety factors will be obtained by considering different slope angles and different trial surfaces. More sophisticated methods of analysis

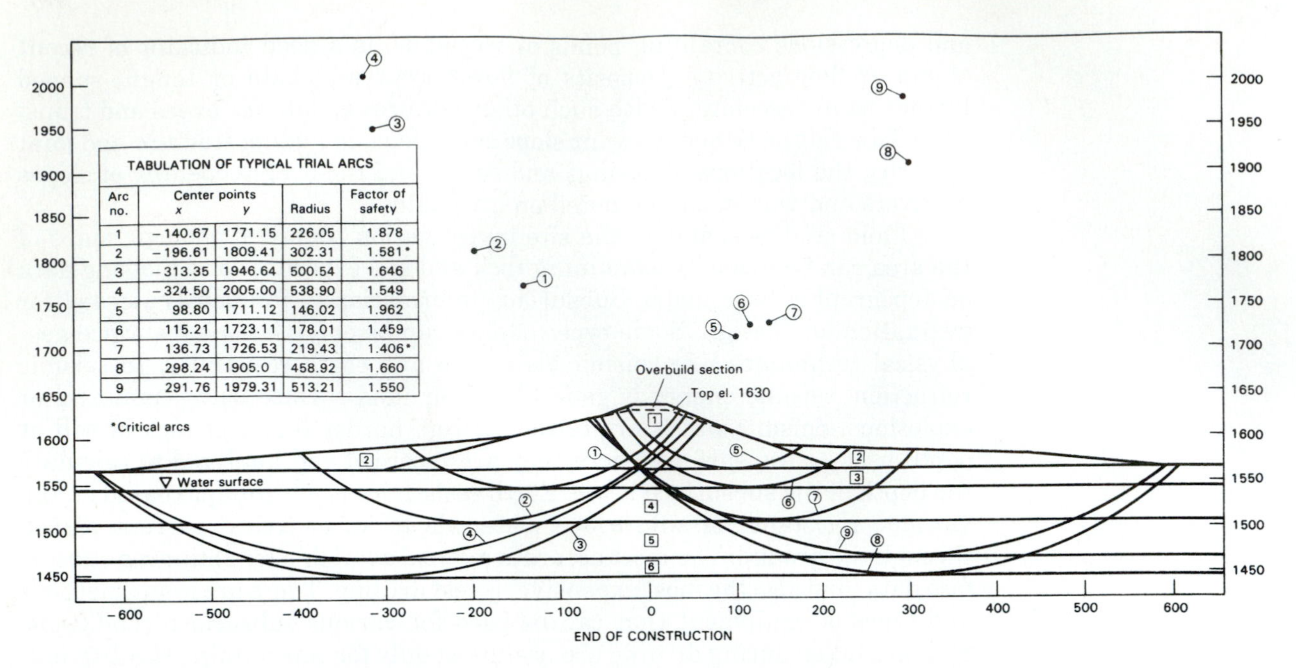

TABULATION OF TYPICAL TRIAL ARCS

| Arc no. | Center points $x$ | Center points $y$ | Radius | Factor of safety |
|---|---|---|---|---|
| 1 | −140.67 | 1771.15 | 226.05 | 1.878 |
| 2 | −196.61 | 1809.41 | 302.31 | 1.581* |
| 3 | −313.35 | 1946.64 | 500.54 | 1.646 |
| 4 | −324.50 | 2005.00 | 538.90 | 1.549 |
| 5 | 98.80 | 1711.12 | 146.02 | 1.962 |
| 6 | 115.21 | 1723.11 | 178.01 | 1.459 |
| 7 | 136.73 | 1726.53 | 219.43 | 1.406* |
| 8 | 298.24 | 1905.0 | 458.92 | 1.660 |
| 9 | 291.76 | 1979.31 | 513.21 | 1.550 |

*Critical arcs



**Figure 12-36** Slope-stability analysis of a proposed earth dam. The factor of safety is calculated for each trial arc. Centers of the arcs are plotted above the dam. Trial arcs 2 and 7 are the critical arcs. (From Flood Control Burlington Dam, Design Memo No. 2, U.S. Army Corps of Engineers, 1978.)

are used in most slope-stability studies, and slope-stability computer programs make the rapid evaluation of many trial surfaces possible.

The analysis of rock slopes first requires a detailed examination of the discontinuities in the rock mass because failures will be controlled by the spacing and geometry of these planes of weakness. The difficulty of determining an exact failure mechanism in these slopes, because of the complexity of the discontinuities, makes slope-stability analyses of rock slopes considerably less accurate than many soil-slope analyses.

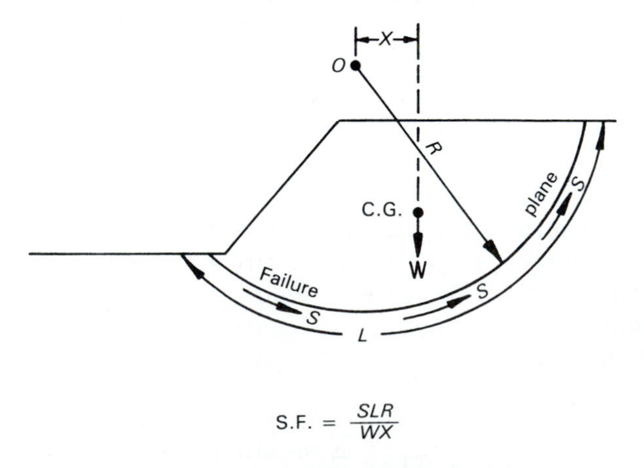



$$\text{S.F.} = \frac{SLR}{WX}$$

**Figure 12-37** The Swedish circle method gives the safety factor of a slope for a particular failure plane as the ratio of the resisting moment ($SLR$) to the driving moment ($WX$).

**EXAMPLE 12-1**

A cut made during the construction of a highway was made with the geometry shown below. Soon after construction, the slope failed by rotational slump. The soil in the slope is a uniform saturated clay with unit weight (density) of 1.75 Mg/m$^3$. Test drilling showed that the failure surface approximates a circular arc, $AB$, with a radius of 20 m. The angle formed by radii connecting the ends of the failure surface is 80 degrees. Using a reconstruction of the original profile and the failure circle, the center of gravity ($c$) of the slump area was located as shown. The area was found to be 102.3 m$^2$ and the moment arm, $X$, about the center of the circle was 4.38 m. Estimate the shear strength of the soil at the time of failure.

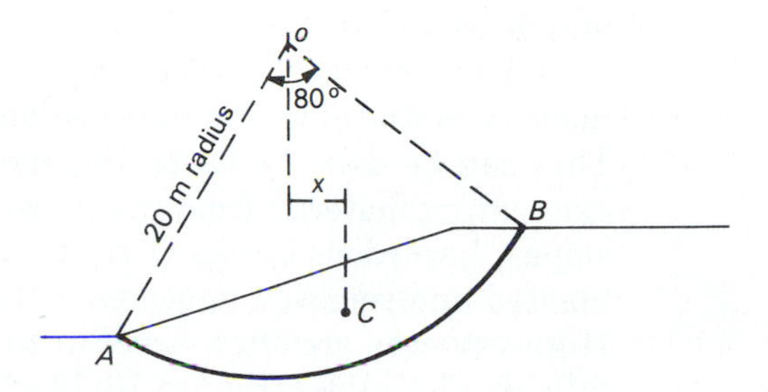


***Solution.*** To apply Equation 12-4, we must determine the weight of soil involved in the slump and the length of the failure arc, $L$. The mass of a volume of soil with the known cross-sectional area and a depth of 1 m measured perpendicular to the page is:

$$1.75 \text{ Mg/m}^3 \times 102.3 \text{ m}^2 \times 1\text{m} = 179 \text{ Mg}$$

The weight of this volume of soil is

$$179 \text{ Mg} \times 9.81 \text{ m/s}^2 = 1756 \text{ kN}$$

The length of the failure arc can be determined using the properties of a circle. Since angle $AOB$ is 80 degrees, the fraction of a semicircle is 80/180 = 0.44. If the perimeter of a semicircle is $\pi R$, the length of arc $AB$ is 0.44 $\pi R$, or

$$0.44 \times 3.14 \times 20 \text{ m} = 27.63 \text{ m}$$

At the time of failure, the safety factor must have declined to a value of 1.0. Therefore, from Equation 12-4,

$$1.0 = \frac{SLR}{WX}$$

and

$$S = 1.0\frac{WX}{LR} = \frac{1.0 \times 1756 \text{ kN} \times 4.38 \text{ m}}{27.63 \text{ m} \times 20 \text{ m} \times 1 \text{ m}} = 13.9 \text{ kN/m}^2$$

The 1 m in the denominator is the width of the unit depth of the cross section. This shear-strength value is equivalent to the cohesion measured in an undrained test (Chapter 10). It can only be used in the field for estimates of slope stability while undrained conditions are approximated in the slope.

### *Preventative and Remedial Measures*

Once a slope fails, measures usually must be taken to correct the problems and stabilize the slope. These measures are very expensive and, in addition, often follow costly damages to highways, buildings, and other projects. It is much more desirable to prevent such losses by taking steps to stabilize slopes before failure occurs.

Potentially hazardous slopes commonly can be identified in preliminary geologic investigations. The simplest method to avoid slope failures, but not always the most practical, is to avoid unstable areas for construction of engineering projects. If a project must be built in a potentially unstable area, however, there are a number of ways to increase the stability of slopes. The methods are the opposite of processes that tend to increase the driving forces and decrease the resisting forces of the slope.

The most common method of reducing the driving forces is to reduce the mass of material acting to cause downslope movement above a failure plane. This can be done by flattening the slope (decreasing the slope angle) or by excavating material from the upper part of the slope (Figure 12-38). For rock slopes, hazardous blocks of rock bounded by joints or bedding planes can be blasted and removed to decrease the possibility of rock slides at a later date. Highway cuts are often *benched* to decrease the potential damage from rock falls or rock slides (Figure 12-38b).

The resisting forces of a slope can be increased in several ways. Since one of the most important processes that reduces resisting forces is the increase in pore-water pressure, dewatering, or the drainage of a slope, constitutes one of the most effective mechanisms for increasing resisting forces. Drainage includes both controlling the surface-water movement across a slope as well as decreasing the internal water within a slope. Surface water and shallow subsurface water can be directed from the slope by drainage ditches and interceptor drains (Figure 12-39). Deeper drainage devices can also be installed. Horizontal drains (Figure 12-39) bring water to the slope face, where it can be safely removed from the slope. An alternative method of deep internal drainage is the use of wells that are continually or intermittently pumped.

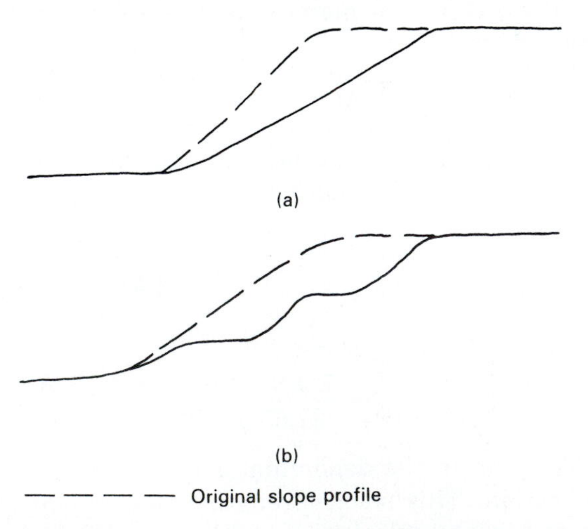


**Figure 12-38** Techniques for increasing slope stability include (a) decreasing the slope angle and (b) benching the slope.

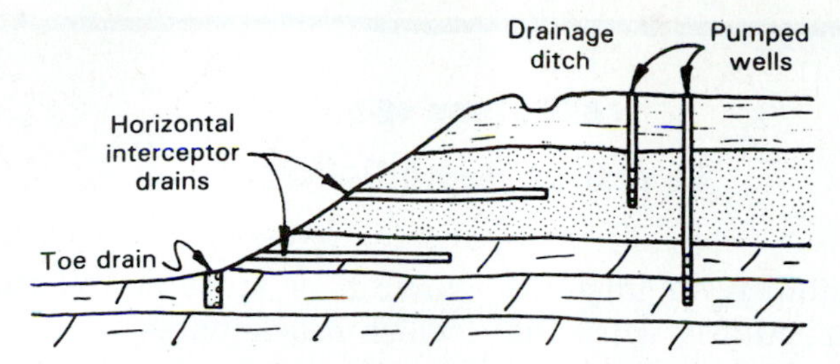


**Figure 12-39** Various methods of drainage control and dewatering used to increase stability.

In addition to the beneficial effect of drainage, resisting forces can also be increased by the construction of various types of walls or fills at the base of the slope (Figure 12-40). Among several types of retaining walls used, walls composed of piles are particularly useful for resisting failure along planes located below the base of the slope (Figure 12-40b). Closely spaced piles made of timber, concrete, or other materials anchor the unstable material above the failure surface to more stable beds of rock or soil below.

Where more space is available, *buttress* or *counterweight* fills can be emplaced at the base of the slope (Figure 12-40a). The fills should be composed of rock or soils with good strength and drainage characteristics. The use of counterweight fills is sometimes called "loading the toe." Increases or decreases of mass can therefore be used to stabilize slopes. Removal of slope material from the upper part of the slope and addition of suitable fill to the lower part of the slope produces a more desirable safety factor.

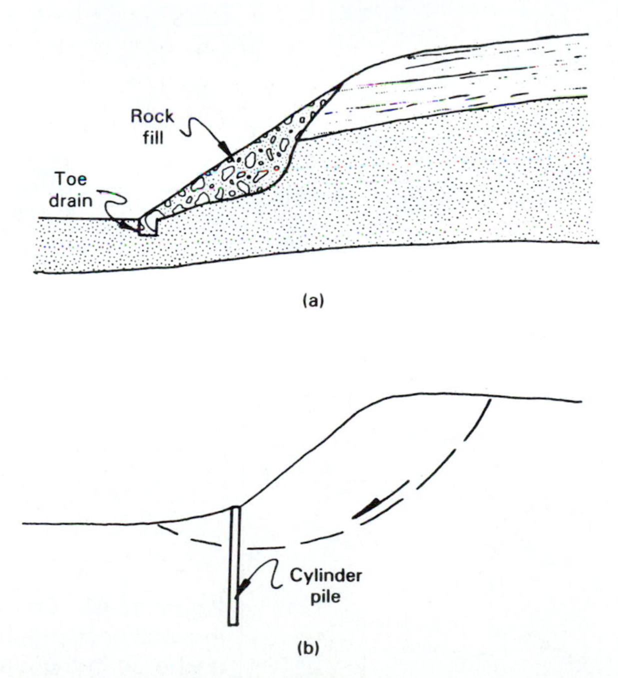


**Figure 12-40** Retaining walls are often constructed at slope bases to prevent failure: (a) rock buttress, (b) cylinder-pile wall.

## CASE STUDY 12-2
## Landslide Remediation

The costs necessary to stabilize an unstable slope can be very high, particularly when the slope movement affects a highly developed area. Cincinnati, Ohio, is a city with serious slope-stability problems, which result from the presence of weak surficial materials along the steep sides of the Ohio River valley. The bedrock beneath the city includes shales that weather to form an unstable mantle of colluvium up to 15 m thick. *Colluvium* is a slope deposit created by the gradual weathering and slow downslope movement of debris derived from bedrock. The strength of the clay-rich colluvium is much lower than the shale from which it originates.

During preliminary construction of interchanges for an interstate highway, excavations in colluvium were made at the base of a hill known as Mount Adams. Slowly moving slides soon developed in the colluvium upslope from the excavation, damaging numerous buildings, streets, and utilities. In an attempt to stabilize the hillside, a 300-m-long retaining wall was constructed at the base of the slope (Figure 12-41).

The retaining wall utilizes a unique de-

**Figure 12-41** Cylinder-pile retaining wall under construction at the base of Mount Adams in Cincinnati, Ohio.

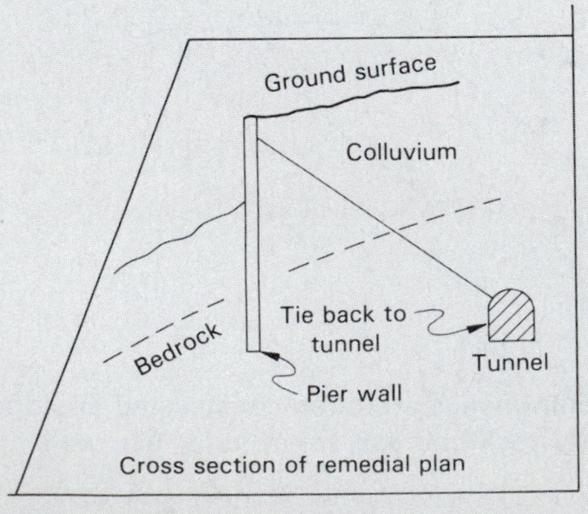


**Figure 12-42** Cross section of the retaining wall showing the supporting tie backs anchored in the tunnel excavated in bedrock. (From Fleming et al., 1981, *Engineering Geology of the Cincinnati Area*, American Geological Institute.)

sign involving cylinder piles and tiebacks to a tunnel constructed deep beneath the hill (Figure 12-42). Most of the cylinder piles in the wall are 2.1 m in diameter and are located on 2.4-m centers. The piles, which are constructed of reinforced concrete, extend to depths of more than 20 m, well into bedrock beneath the colluvium. Additional support is provided by high-strength wire tie backs that are anchored in a tunnel that runs parallel to the retaining wall. The $22 million total cost of the wall, which was completed in 1982, was shared between the federal government and the city. The retaining wall project was at that time the most expensive landslide-stabilization project in the United States.

## SUMMARY AND CONCLUSIONS

Mass movements include falls, topples, slides, lateral spreads, and flows. Combinations of more than one process are common. Rotational slides (slumps) have a circular failure surface and occur in slopes underlain by materials with low shear strength. Translational slides in rock or debris have a planar failure surface that often occurs at the contact of bedrock with overlying unconsolidated material. Processes involving liquefaction and flow of a zone within a slope composed of quick clay, accompanied by transport of more competent material above, like packages on a conveyor belt, are called lateral spreads. Flow—slope movements with internal deformation—are mechanically complex. Earth and debris flows approximate the behavior of a Bingham substance, a combination of viscous and plastic flow that is laminar and can transport large boulders within a plug of constant velocity at the center of the channel. Catastrophic rock avalanches may create a fluid composed of rock fragments, a phenomenon known as grain flow.

Determining the causes of specific slope movements involves the evaluation of a large number of interacting variables. Slopes that are intrinsically susceptible to mass movements contain materials with low shear strength. Clays are commonly involved; quick and overconsolidated clays present special problems. Failures of rock slopes are controlled by planes of weakness within the rock.

Slope stability can be described as a ratio of resisting forces to driving forces. Removal of lateral support, addition of weight, and earthquakes tend to increase the driving forces, while increases in pore pressure and various weathering processes decrease the resisting forces. If these processes can be identified by preliminary geologic investigations, various design and construction procedures can be used to counteract potential declines in the safety factor or actually to increase the safety factor of the slope.

A better understanding of slope practices can help prevent loss of life and property in addition to minimizing the long-term costs of engineering projects.

### PROBLEMS

1. On what parameters is the classification of mass movements based?
2. Describe the surface features of a slump.
3. What conditions are conducive to the development of debris slides?

4. How can lateral spreads move long distances without excessive deformation of the blocks involved in the movement?
5. What is the basic mechanical difference between slides and flows?
6. A detached 3000 $m^3$ block of rock rests on a rock slope of 30 degrees. The porosity of the rock mass is 10% and the specific gravity of the rock minerals is 2.65. The coefficient of friction is 0.15. Calculate the driving and resisting forces acting on the block. Is the block stable?
7. Why can the movement of a debris flow be explained by the deformation of a Bingham substance?
8. Identify and explain the major hypotheses put forth for the movement of catastrophic rock-slide and debris avalanches?
9. What is meant by grain flow?
10. How does the safety factor measure the stability of a slope?
11. Explain why slope movements are commonly preceded by a decrease in effective stress?
12. The total stress on a potential failure plane within a soil mass on a slope is 1650 lb/$ft^2$ and the fluid pressure is 312 lb/$ft^2$. The cohesion of the soil is 1000 lb/$ft^2$ and the angle of internal friction is 28°. If the water table rises by 2 ft, what is the change in strength along the failure plane?
13. What are overconsolidated clays?
14. A potential failure surface within a slope is shown below. The failure surface is a circular arc with a radius of 30 ft and the angle formed by radii connecting the ends of the failure surface is 110 degrees. The weight of the potential slump block is 43,000 lb per foot of depth perpendicular to the cross-section, and it acts through a center of gravity that is offset from the center of the failure circle by 10 ft. If the shear strength of the soil is 400 lb/$ft^2$, what is the safety factor for this potential failure surface?

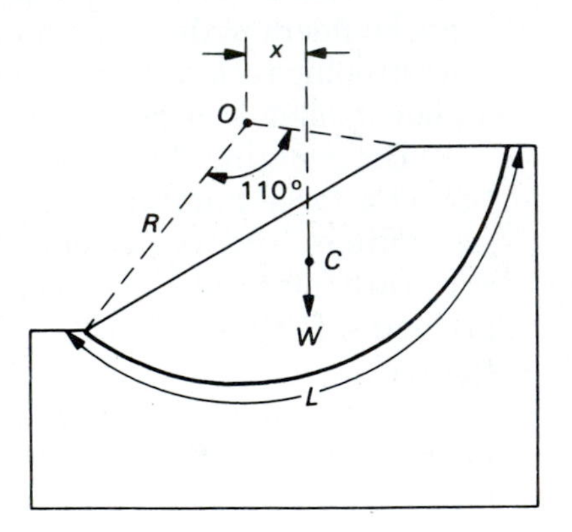


15. What are the specific objectives of remedial actions after a slope failure has taken place?

## REFERENCES AND SUGGESTIONS FOR FURTHER READING

CRUDEN, D. M., and C. B. BEATY. 1987. The Frank Slide, southwestern Alberta. *Geological Society of America Centennial Field Guide—Rocky Mountain Section.* Boulder, Colo.: Geological Society of America, pp. 15–18.

FLEMING, R. W., A. M. JOHNSON, and J. E. HOUGH. 1981. *Engineering Geology of the Cincinnati Area: GSA Cincinnati '81 Field Trip Guidebooks*, pt. 3, T. C. Roberts, ed. American Geological Institute.

HADLEY, J. B. 1964. *Landslides and Related Phenomena Accompanying the Hebgen Lake Earthquake of August 17, 1959*. U.S. Geological Survey Professional Paper 435, pp. 107–138.

HANSEN, W. R. 1966. *Effects of the Earthquake of March 27, 1964 at Anchorage, Alaska*. U.S. Geological Survey Professional Paper 542-A.

HSU, K. J. 1975. Catastrophic debris streams (sturztroms) generated by rockfalls. *Geological Society of America Bulletin*, 86:129–140.

JOHNSON, A. M. 1970. *Physical Processes in Geology*. San Francisco: Freeman, Cooper.

KEEFER, D. K., and A. M. JOHNSON. 1983. *Earth Flows: Morphology, Mobilization, and Movement*. U.S. Geological Survey Professional Paper 1264.

KIERSCH, G. A. 1964. Vaiont Reservoir disaster. *Civil Engineering*, 34:32–39.

MELOSH, H. J. 1987. The mechanics of large rock avalanches. *Geological Society of America Reviews in Engineering Geology*, v. VII, pp. 41–49.

MOORE, J. G., and R. K. MARK. 1992. Morphology of the island of Hawaii. *GSA Today*, 2:260–262.

PLAFKER, G., and G. E. ERIKSEN. 1978. Nevados Huascaran avalanches, Peru, in *Rockslides and Avalanches: 1. Natural Phenomena*, B. Voight, ed. Amsterdam: Elsevier, pp. 277–314.

SCHUSTER, R. L. 1978. Introduction, in *Landslides, Analysis, and Control*, R. L. Schuster and R. J. Krizek, eds. Transportation Research Board Special Report 176. National Research Council, National Academy of Sciences.

SCHUSTER, R. L., and R. J. KRIZEK, eds. 1978. *Landslides, Analysis, and Control*. Transportation Research Board Special Report 176. National Research Council, National Academy of Sciences.

SHREVE, R. L. 1968. *The Blackhawk Landslide*. Geological Society of America Special Paper 108.

STOVER, B. K. 1988. Booth Creek Rockfall Hazard Area, in *Field Trip Guidebook*, Geological Society of America, Colorado School of Mines Professional Contribution, 12:395–401.

UPDIKE, R. G., H. W. OLSEN, H. R. SCHMOLL, Y. K. KHARAKA, and K. H. STOKOE, II. 1988. *Geologic and Geotechnical Conditions Adjacent to the Turnagain Heights Landslide, Anchorage, Alaska*. U.S. Geological Survey Professional Paper 1817.

U.S. ARMY CORPS OF ENGINEERS, ST. PAUL DISTRICT. 1978. Flood Control Burlington Dam, Souris River, North Dakota. Design Memorandum no. 2, Phase 2: Project Design, Appendix B—Geology and Soils.

# 13

# Rivers

Throughout history, rivers have played a crucial role in human development. These threads of civilization have served as transportation routes, borders for political land divisions, and sources of water for drinking, industry, farming, and waste disposal since the earliest days of human existence. But along with their benefits, the practical problems associated with living near rivers—flooding, bridge construction, pollution, and so forth—have been with us equally as long. One might suppose that our intimate association with rivers would have led to the ability to harness the benefits of rivers, while avoiding the hazards and drawbacks of river life. On the contrary, flood damages in the United States have increased during the past few decades despite the construction of flood-control projects on an unprecedented scale in the twentieth century. The devastating floods along the Mississippi River and its tributaries in 1993 are a prime example. The lesson of these events is clear: Our ability to control geologic processes is limited. In the future we must use our growing understanding of rivers to find alternatives to traditional solutions.

## *RIVER BASIN HYDROLOGY AND MORPHOLOGY*

### *The Drainage Basin*

A river by itself is only one component of the hydrologic cycle. The potential energy possessed by water that falls as rain and snow on the continents has the capacity to accomplish a great deal of work during its return flow to the ocean. We have already considered the fraction of this water that moves through the subsurface; now we must examine the water that travels on the surface of the earth. The work done by surface water is truly immense. Over millions of years, continents are gradually worn down by rivers along with other weathering and erosional processes and are transported to the sea as small particles and dissolved constituents.

A useful approach to the study of river, or *fluvial*, systems is based on the concept of a *drainage basin* (Figure 13-1). A drainage basin is an area containing an integrated network of stream segments that join together to form successively larger streams until one channel carries the entire surface flow out of the basin at its outlet. The outlet of the Mississippi Basin is, therefore, the point where the river discharges into the Gulf of Mexico. Within the Mississippi Basin, however, many smaller basins can be defined. The Missouri Basin, for example, consists of the areas drained by the Missouri River and all its tributaries upstream from its confluence with the Mississippi River. The boundaries of river basins are *drainage divides* (Figure 13-1), because they separate adjacent basins. Drainage divides, which are usually ridges or other topographically high areas, also separate tributary streams within a drainage basin. Thus a river basin can be thought of as a well-defined area bounded on all sides by divides and within which all streams and rivers contribute to the flow of the main, or *trunk*, stream that leaves the basin.

### *Stream Patterns*

Numerous analyses of the spatial arrangement of streams in drainage basins have shown that stream networks conform to a small number of *stream patterns* (Figure 13-2). Specific stream patterns develop in response to the initial topography of an area and the distribution of rock types of varying erosional resistance. *Dendritic* patterns are characteristic of moderate slopes and soil and

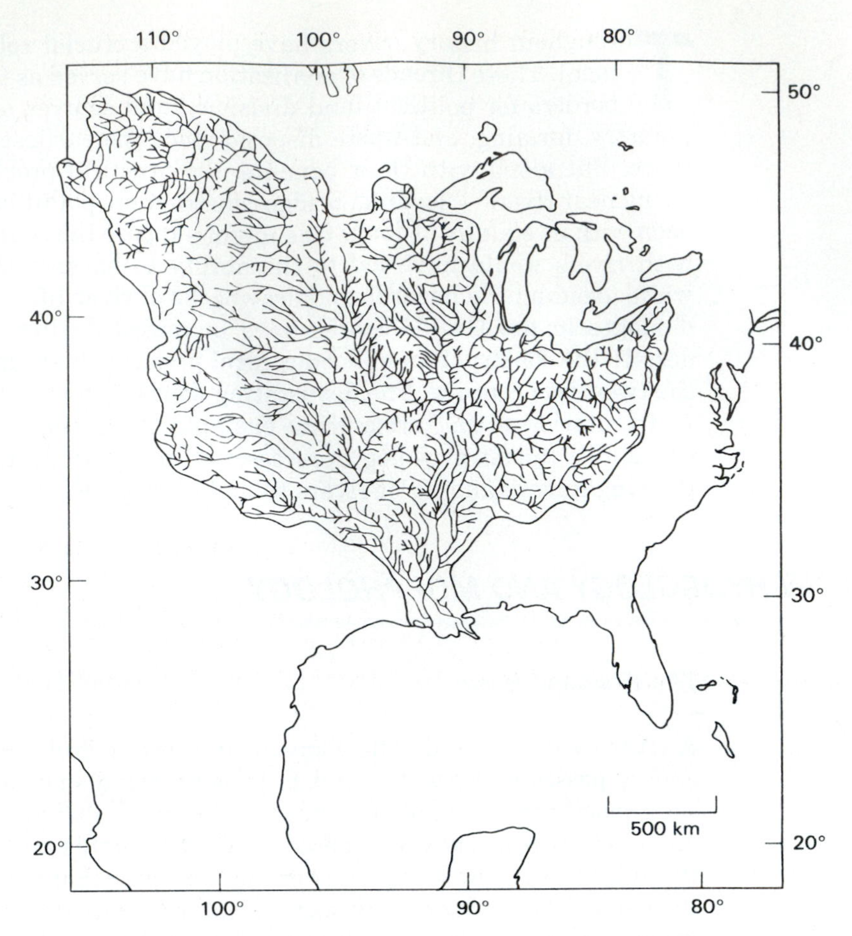


**Figure 13-1** The Mississippi River drainage basin, which covers most of the central United States. (From R. K. Matthews, *Dynamic Stratigraphy*, 2d ed., copyright © 1984 by Prentice-Hall, Inc., Englewood Cliffs, N.J.)

rocks of fairly uniform resistance to stream erosion. These same rock types can develop a *parallel* pattern on steeper slopes. A different pattern results when the rock units are arranged in linear bands of alternating weak and strong lithologies. Rocks with a low resistance to erosion are preferentially exploited by streams, so that, in a *trellis* pattern, larger streams occupy zones of nonresistant rocks, while small, short stream segments drain the resistant,

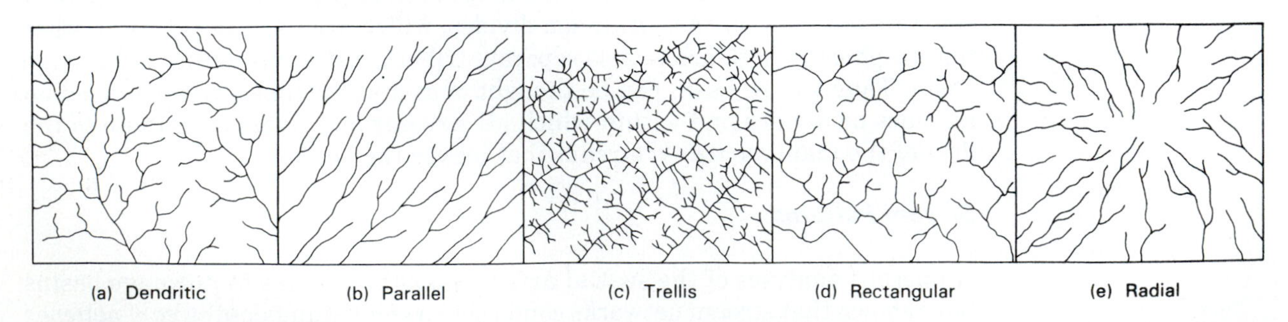


**Figure 13-2** Common drainage patterns. (From A. D. Howard, 1967, *American Association of Petroleum Geologists Bulletin*, 51:2246–2259. Used by permission of the American Association of Petroleum Geologists.)

topographically higher beds. Folded sedimentary rock sequences commonly display trellis stream patterns. *Rectangular* patterns are characteristic of rocks of uniform resistance cut by perpendicular (orthogonal) joint sets. Streams preferentially occupy joint traces because of the lowered resistance caused by fracturing and weathering along the joints. The final major stream pattern, *radial*, develops on isolated topographic uplands; volcanoes, domes, and buttes may display radial drainage.

### *Stream Order and Drainage Density*

The stream pattern is only one aspect of the stream networks that occupy drainage basins. Early researchers of drainage-basin relationships devised a system of ordering stream segments within a drainage basin (Figure 13-3). The system consists of progressively higher order numbers, starting from the small streams near the drainage divides to the larger streams at the center of the basin. *First-order* streams are defined as those stream segments with no tributaries. When two first-order stream segments join, a *second-order* stream is formed. The resulting stream ordering system is useful because of the consistent proportional relationships between stream order and other basin characteristics. For example, Figure 13-4 indicates that stream order is proportional to the logarithm of the number of stream segments, the length of stream segments, and the drainage basin area. These relationships, showing that the number of stream segments decreases and the length of stream segments increases with increasing stream order, are common enough to be considered empirical laws of drainage networks. In a similar fashion, the area of a particular drainage basin is directly proportional to stream order.

Another important relationship in drainage basins is the *drainage density*. This parameter is defined as the total length of all stream segments divided by the area of the basin; it is therefore a measure of the number of stream channels per unit area of a drainage basin. Drainage density is extremely important because it influences the conveyance of water through the stream network during floods and lesser runoff events. Drainage density is controlled

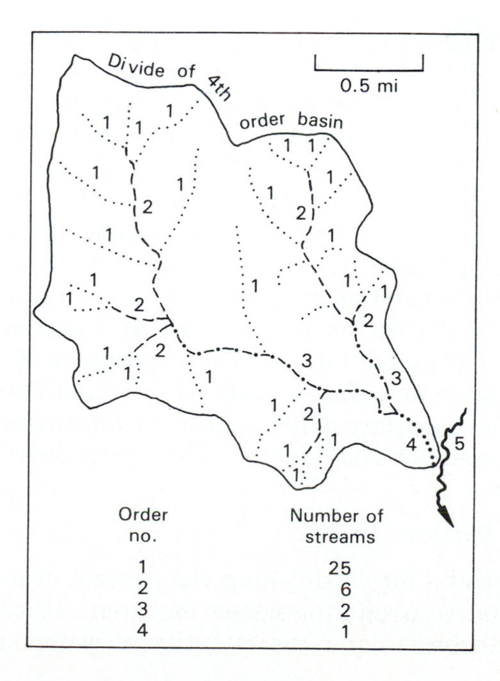


**Figure 13-3** Strahler's method of determining stream order. (From A. N. Strahler, 1957, *Transactions, American Geophysical Union*. Used by permission of the American Geophysical Union.)

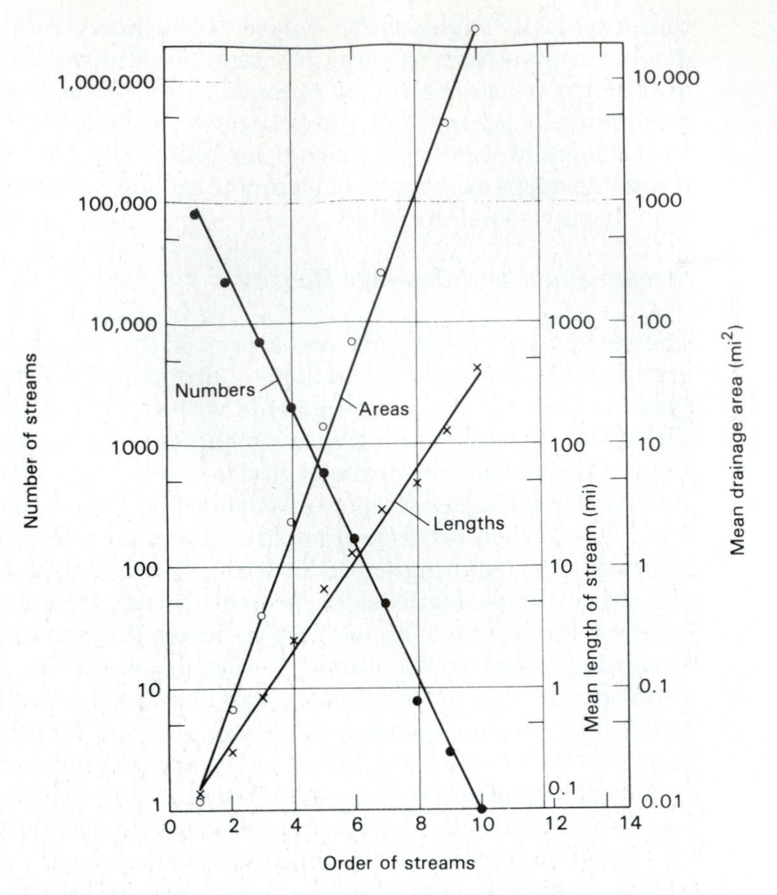


**Figure 13-4** Relationship of stream order to other parameters in the Susquehanna River drainage basin. With increasing stream order, the number of streams decreases, whereas the drainage-basin area and stream length increase. (From L. M. Brush, Jr., 1961, U.S. Geological Survey Professional Paper 282-F.)

by the interaction between climate and geology. The geologic factors include the permeability of the rocks and soils at the basin surface. High-permeability materials usually have high infiltration capacities. The more rain or snow melt that infiltrates, the less is available for surface runoff. A low-density, or *coarse-texture*, drainage basin is the result (Figure 13-5). Low infiltration capacity, on the other hand, causes more surface runoff. More drainage channels are eroded and a high-density, or *fine-texture,* drainage basin develops. Climate influences drainage density by controlling the amount and type of vegetation present. The heavy vegetation cover of humid climatic regions increases infiltration relative to the barren slopes of an arid or semiarid region, where overland flow is much more intense. The importance of drainage density lies in the percentage of rain that travels as overland flow versus that which infiltrates into the subsurface. A fine-texture drainage basin can transmit more water through the stream channels at a faster rate. Severe and frequent floods can be the result.

### The Hydrologic Budget

One important reason for establishing the concept of a drainage basin is to develop a quantitative accounting system for water in the basin. The *hydrologic budget* of a watershed relates the quantity of water supplied to the basin

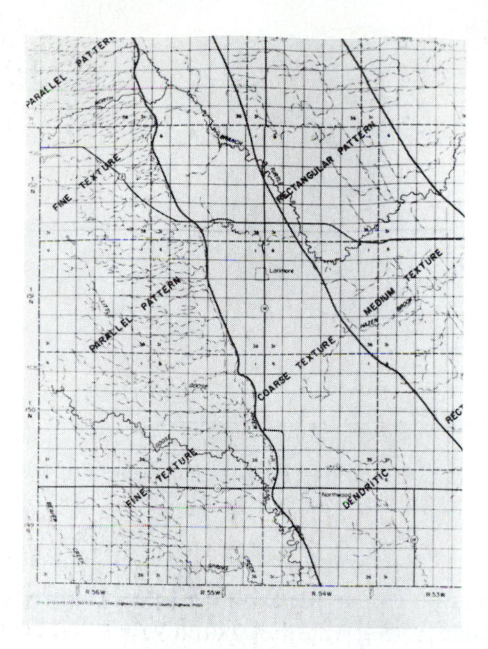


**Figure 13-5** Stream drainage in a portion of Grand Forks County, North Dakota. The area of coarse texture is underlain by sand with high permeability. The area of fine texture is a slope underlain by material with low permeability. (From D. E. Hansen and J. Kume, 1970, North Dakota Geological Survey Bulletin 53, Part 1.)

through precipitation to the quantity of water leaving the basin. Assuming that there are no ground-water inflows or outflows from a basin and that the equality refers to a long-term average, the hydrologic budget equation can be expressed as

$$P = Q + E \quad \text{(Eq. 13-1)}$$

where $P$ represents precipitation, $Q$ is the runoff, and $E$ is the evapotranspiration. The *runoff* term in the equation includes both the surface runoff from the basin and the ground-water component of runoff, which is the base flow of the streams in the basin (Chapter 11). Since precipitation and runoff can be measured fairly accurately, evapotranspiration is usually calculated as the difference between precipitation and runoff.

One of the major objectives of engineering hydrology is the prediction of the amount of flow in a river that will result from a particular precipitation event. This information, which is needed for the design of bridges, dams, and flood-control structures, can be obtained only through an understanding of the processes operating in the drainage basin. The geology, topography, climate, and vegetation interact to produce a specific network of streams in a drainage basin. These factors also control the movement of water from a particular storm through the basin.

### *Development of Drainage Networks*

Similar statistical relationships between variables in drainage basins throughout the world prove that rivers erode the valleys in which they flow over long periods of time. One of the major processes in the development of drainage networks is *headward erosion.* All streams exhibit the tendency to extend their channels in a headward, or upstream, direction. If streams are followed upstream to the point where first-order tributary channels just begin, the process of headward erosion can best be observed. The channel frequently ends on a hillslope with a concave profile (Figure 13-6). This part of the channel collects converging overland flow from three sides of the slopes above during

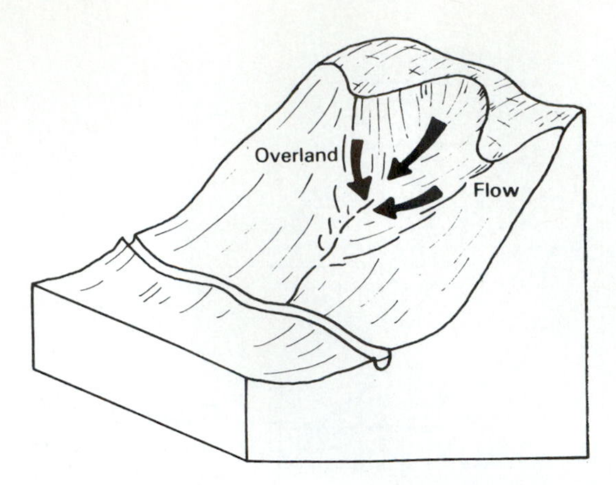


**Figure 13-6** Focusing of overland flow at the heads of first-order tributaries. Headward erosion is the result.

heavy rainfall events. Consequently, erosion is intense at this point and the channel is extended upslope. As streams erode headward on both sides of a drainage divide, the drainage divide is gradually lowered in elevation.

Headward erosion is related to the process of *stream capture*. Streams erode headward at different rates because of the erosional resistance of the rocks or sediments they are flowing over or because one stream has a steeper slope than another and therefore deepens its channel at a greater rate. As a stream is actively eroding its channel headward, it may intersect another stream (Figure 13-7). Because of its greater slope and energy, the first stream may divert, or capture, the second stream and all its tributaries above the point of intersection. The river that has lost its upstream sections is said to have been *beheaded*. It is left with a much smaller discharge and drainage basin than it had prior to the capture. Drainage basins can by stream capture become larger or smaller through geologic time.

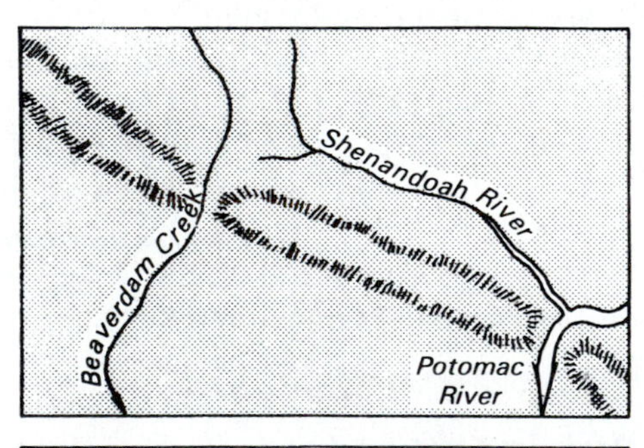


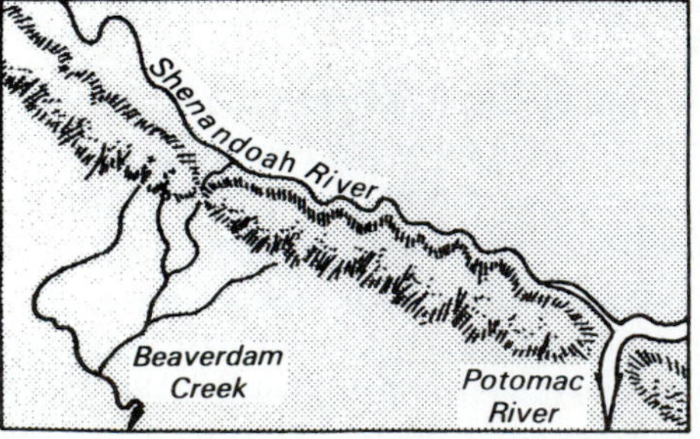


**Figure 13-7** An example of stream capture in Virginia. Headward erosion by a tributary of the Shenandoah River captured the upper part of Beaverdam Creek. (From T. L. McKnight, *Physical Geography: A Landscape Appreciation*, copyright © 1984 by Prentice-Hall, Inc., Englewood Cliffs, N.J.)

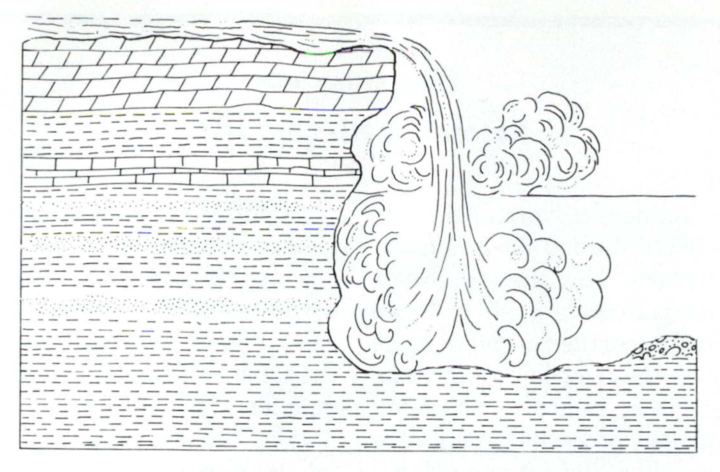

**Figure 13-8** A waterfall formed by erosion of a weak formation underlying a more resistant formation. (From S. Judson, M. E. Kauffman, and L. D. Leet, *Physical Geology*, 7th ed., copyright © 1987 by Prentice-Hall, Inc., Englewood Cliffs, N.J.)

If a stream is actively downcutting and deepening its valley, it may encounter contacts between rock types of very different erosional resistance. These conditions can lead to the formation of a waterfall, particularly if a more resistant rock is underlain by a less resistant rock. As shown in Figure 13-8, the underlying nonresistant rocks are rapidly eroded, thereby removing underlying support for the resistant rocks. Eventually, the resistant rocks collapse to restore a more stable slope profile. This process is continuous and, with time, the waterfall migrates upstream, while maintaining its form. Niagara Falls is a classic example of the process (Figure 13-9).

**Figure 13-9** Niagara Falls, a waterfall formed in the manner suggested by Figure 13-8. Horseshoe Falls is in the foreground. Smaller falls in background on the right side of the gorge is the American Falls. (J. R. Balsley; photo courtesy of U.S. Geological Survey.)

## CASE STUDY 13-1
# Retreat of Niagara Falls

As one of the scenic wonders of North America, Niagara Falls has been studied by geologists for more than 150 years. The Niagara River (Figure 13-10), which is unusually short for its discharge, begins at the northeast end of Lake Erie, where it carries the overflow from Lake Erie and the upper Great Lakes to Lake Ontario. Thereafter, the flow proceeds through the St. Lawrence River valley and into the North Atlantic Ocean. From its beginning at Lake Erie, the Niagara descends 23 m to Niagara Falls, where it drops vertically 51 m. From the base of the falls, it then descends another 23 m in a deep gorge to Lake Ontario.

Water diversions for hydroelectric power generation have decreased the flow of the river over the falls to about 50% of its natural mean discharge of 5721 $m^3/s$ during daylight hours. After tourist viewing hours are over, flow over the falls is decreased even further to about one quarter of its natural amount. The falls are divided into the American Falls and the much larger Horseshoe Falls on the Canadian side (Figure 13-9). More than 90% of the flow descends over the Horseshoe Falls.

The existence of Niagara Falls is due to Niagara Escarpment, a steep topographic escarpment which separates the Lake Ontario

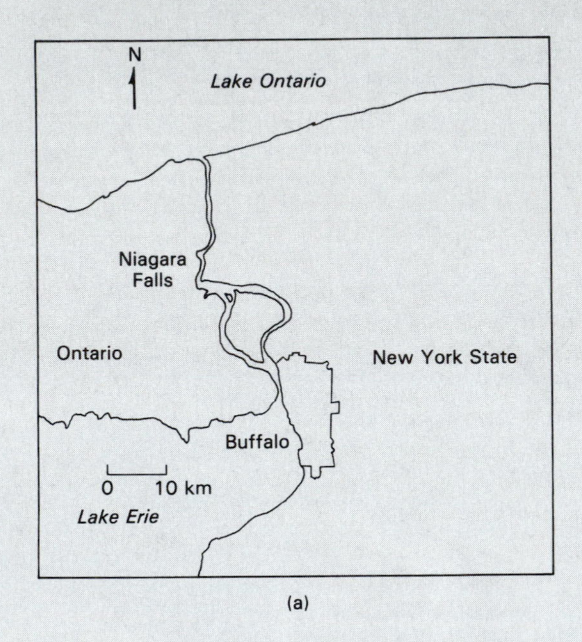


(a)

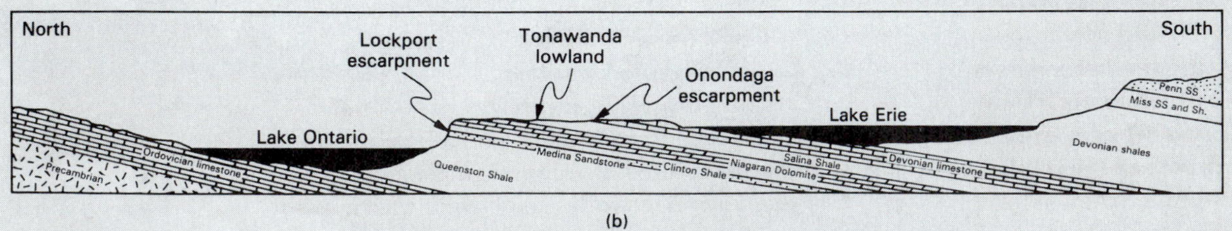


(b)

**Figure 13-10** (a) Location of Niagara Falls. (b) Cross section through western end of Lake Ontario and eastern end of Lake Erie basin. (Modified from P. E. Calkin and P. J. Barnett, 1990, Glacial geology of the eastern Lake Erie basin, in *Quaternary Environs of Lakes Erie and Ontario*, D. I. McKenzie, ed., Escart Press, Waterloo, Ont.)

and Lake Erie basins (Figure 13-10). The escarpment is capped by the resistant Lockport Formation of Silurian age. This dense dolomite is directly underlain by less resistant shales of the Rochester Formation, which is part of the unit labeled as the Clinton Shale on Figure 13-10.

The cutting of the Niagara Gorge began during the last retreat of the Pleistocene glaciers about 12,000 years ago. At that time meltwater from the ice sheet produced greater discharge through the Niagara River and the lower gorge was rapidly cut by headward retreat of Niagara Falls from the Niagara escarpment at a rate of approximately 1.6 m/yr. Meltwater flow decreased about 10,000 years ago and the rate of retreat of the falls was consequently lower. The upper gorge, which is about 3700 m long was cut at an approximate rate of 1 m/yr. Since the seventeenth century, retreat of the falls has been measured by repeated surveys. Successive positions of the Horseshoe Falls are shown in Figure 13-11. The actual rate of recession may be very irregular. Several relict crest lines are shown on Figure 13-11 from the time that the position of the Horseshoe Falls was in the vicinity of the American Falls. A horizontal arch spanning the gorge was probably the most stable form of cross section. During the time that an arch existed, the rate of retreat was relatively low. Separation of the falls about 600 years ago was initiated by the formation of a notch (Figure 13-11), which greatly accelerated the rate of retreat. As a result, Horseshoe Falls separated from the American Falls and has since retreated at a much greater rate due to its higher proportion of the total discharge.

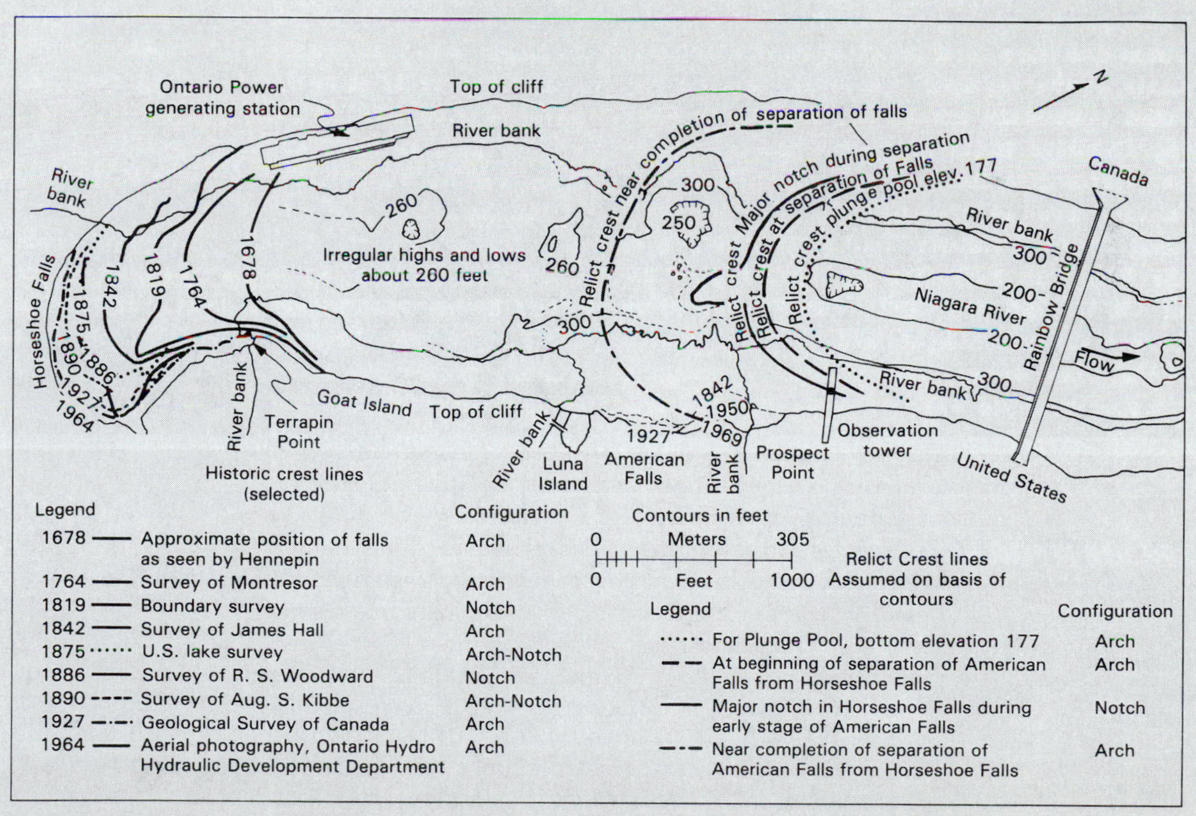


**Figure 13-11** Historic and relict crest lines of Niagara Falls. (From P. E. Calkin and P. J. Barnett, 1990, Glacial geology of the eastern Lake Erie basin, in *Quaternary Environs of Lakes Erie and Ontario*, D. I. McKenzie, ed., Escart Press, Waterloo, Ont. After S. S. Philbrick.)

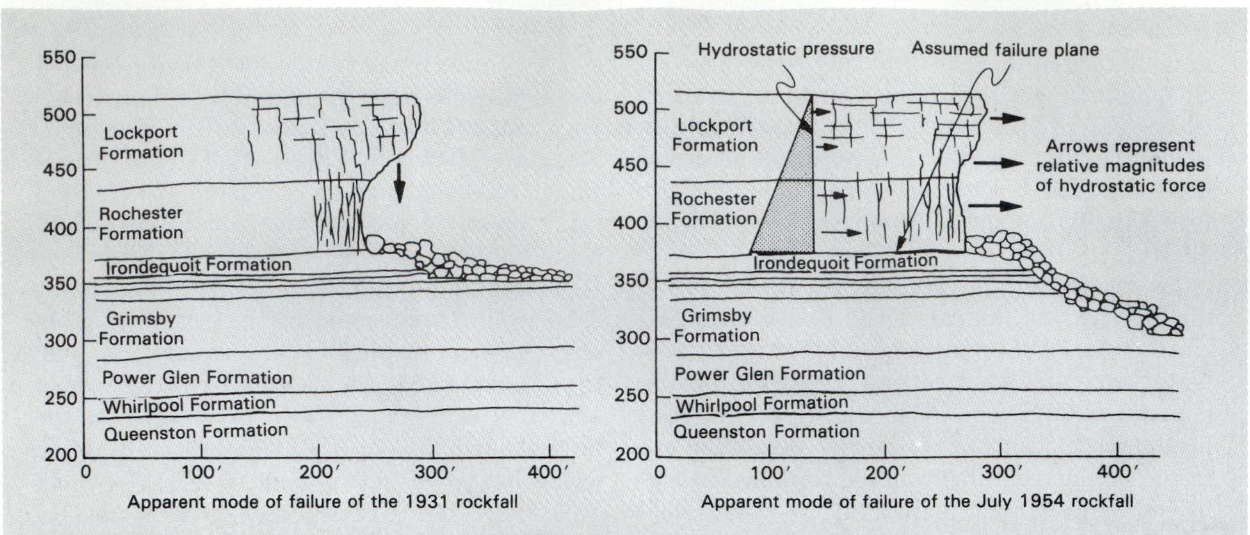


**Figure 13-12** Failure mechanisms, American Falls. (From American Falls International Board, 1974.)

The specific processes that contribute to retreat of the falls include weakening due to wetting and drying of the Rochester shale, freeze and thaw, and penetration of water into joints in the Lockport Dolomite, causing elevated hydrostatic pressures at depth. The more rapid erosion of the Rochester Shale produces an overhang of Lockport Dolomite. The actual failures occur as large rockfalls when the overhang finally collapses. Two large rockfalls at the American Falls are shown in Figure 13-12. On the left, the 1931 rockfall occurred by gradual undermining of the dolomite by erosion of the shale beneath. When the overhanging block of dolomite was unable to support itself, it collapsed. The 1954 rockfall (Figure 13-12, right), however, involved horizontal shearing within the shale. Hydrostatic pressure (hydraulic head), which increases with depth in joints and fractures as shown in the diagram, may have contributed to this failure.

Huge dolomite blocks from rockfalls at the American Falls accumulate as talus at the base of the American Falls because the relatively low discharge is insufficient to rapidly erode and transport them. No such accumulations are found at the base of the powerful Horseshoe Falls, however, because rockfall debris is quickly pulverized and carried away by the larger current flows.

## STREAM HYDRAULICS

### Flow Types

We are all familiar with the variety of flow conditions that exist in river channels. Although the rapid rush of water down a boulder-filled mountain stream channel appears very different from the sluggish, almost imperceptible, movement of a winding coastal-plain river, the flow conditions in all rivers are the balance between the gravitational driving forces and the resistance of the stream channel and the flowing water itself. Two basic flow conditions can be defined by considering the flow paths of individual water particles. At low velocities, the parallel flow paths of adjacent particles constitute *laminar* flow. Figure 13-13 illustrates laminar flow using *streamlines,* which are imaginary

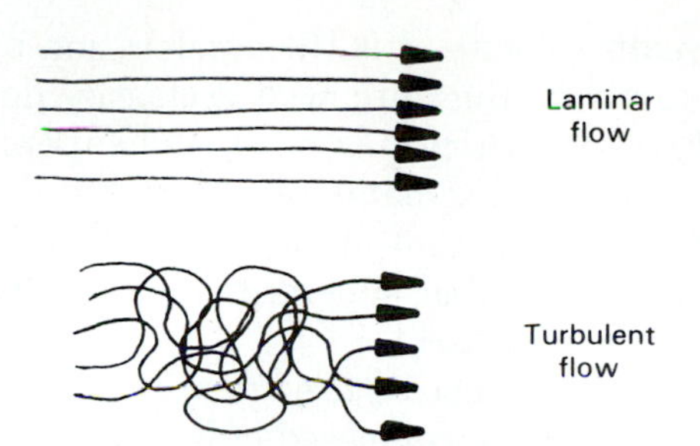


**Figure 13-13** Laminar and turbulent flow.

lines parallel to the velocity vectors of representative water particles. Resistance to flow under laminar conditions is generated by viscous interaction between particles within the flow.

As velocity increases, the parallel laminar flow paths can no longer be maintained. Instead, random velocity fluctuations in all directions arise within the flow. These perturbations of flow are called *turbulence* and the resulting flow is *turbulent* flow. Turbulence generates additional resistance by these secondary motions, initiating a component of viscosity called *eddy viscosity*.

A dimensionless number called the *Reynolds number* ($Re$) is used to distinguish laminar from turbulent flow. Its definition is

$$Re = \frac{\rho_w v R}{\mu} \quad \text{(Eq. 13-2)}$$

where $\rho_w$ is the density of water, $v$ is velocity, $\mu$ is dynamic viscosity, and $R$ is the hydraulic radius of the channel. Hydraulic radius, $R$, is a geometric property of the river channel defined as the cross-sectional area of the channel, $A$, divided by the wetted perimeter, $P$:

$$R = \frac{A}{P} \quad \text{(Eq. 13-3)}$$

In many natural channels, cross-sectional area can be approximated by the product of width and depth (Figure 13-14), and the wetted perimeter as $P = W + 2D$. The boundary between laminar and turbulent flow occurs within the Reynolds number range of 500–750. In nature, laminar flow usually occurs in ground-water flow systems, while even the slowest of rivers fall into the range of turbulent flow.

A second dimensionless number used to characterize the flow of rivers is the *Froude number* ($Fr$):

$$F_r = \frac{v}{\sqrt{gD}} \quad \text{(Eq. 13-4)}$$

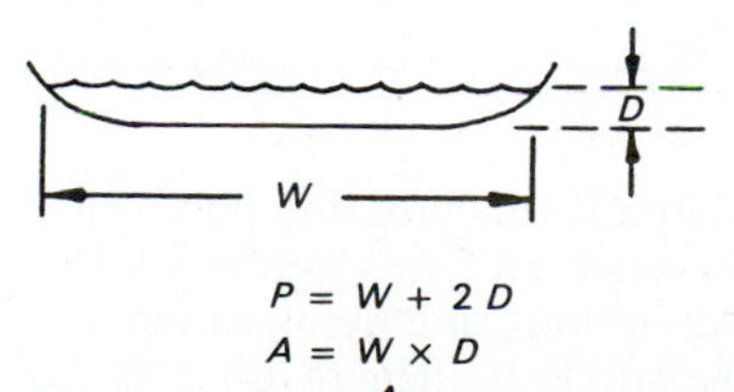


**Figure 13-14** Hydraulic geometry of a natural stream channel.

in which $v$ is stream velocity, $g$ is the acceleration of gravity, and $D$ is depth of flow. Froude number values are used to classify flow into categories defined as subcritical ($Fr < 1$), critical ($Fr \sim 1$), and supercritical or rapid ($Fr > 1$). Since velocities of natural streams rarely exceed several meters per second, examination of Equation 13-4 indicates that only shallow, high-velocity streams can achieve critical or supercritical flow. Gravelly mountain streams occasionally meet these criteria.

Froude numbers can also be used to predict the type of *bedform* that will develop on a stream bottom composed of loose, cohesionless sediment such as sand. Bedforms include wavelike accumulations of bed material that form in response to the frictional drag exerted by the flowing water on the particles lying on the stream bed. The continued flow of water over the bedforms erodes, deposits, and transports sand, so that the entire train of sand waves migrates across the bed, usually in a downstream direction, while maintaining its wave-like form. Figure 13-15 illustrates that, with gradually increasing flow veloci-

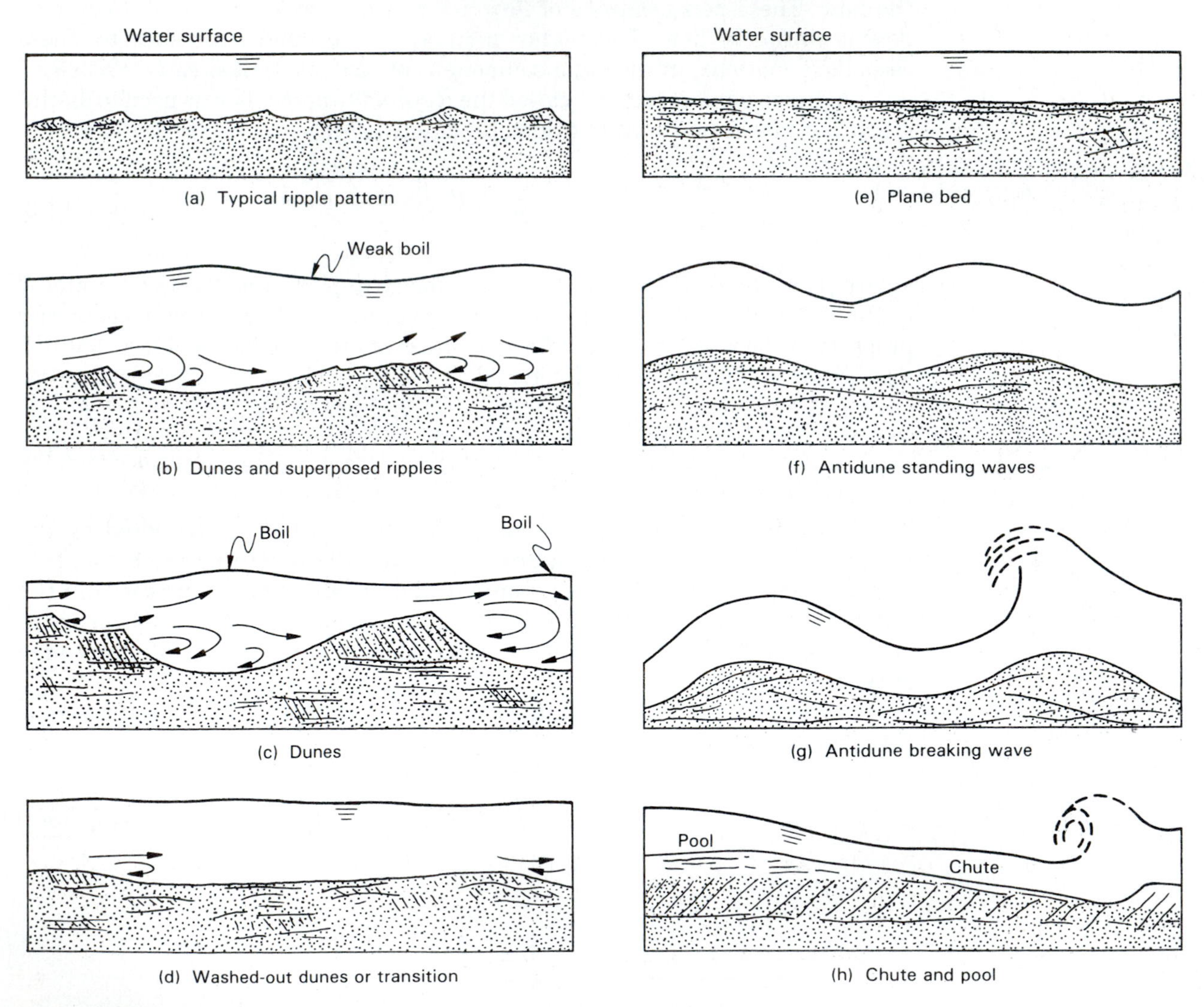


**Figure 13-15** Sequence of bedforms associated with increasing flow velocities and Froude numbers. Froude number is less than 1 for diagrams (a) through (c), approximately 1 for diagrams (d) and (e), and greater than 1 for diagrams (f) through (h). (From D. B. Simons and E. V. Richardson, 1966, U.S. Geological Survey Professional Paper 422-J.)

ties and Froude numbers, bedforms progress through a sequence of *ripples*, *dunes*, plane-bed conditions, and *antidunes*. Ripples are small bedforms less than 60 cm in length from trough to trough, whereas dunes and antidunes are larger bedforms that usually range in size from 60 cm to several meters in length. Antidunes migrate in an upstream direction because particles are eroded from the downstream side of one antidune and deposited on the upstream side of the adjacent bedform.

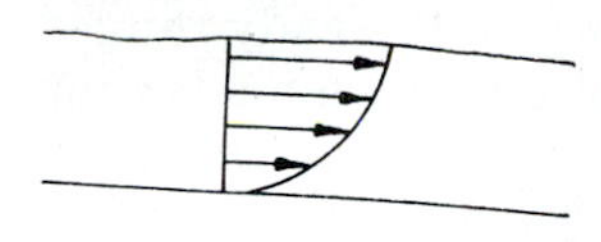

Longitudinal cross section

(a)

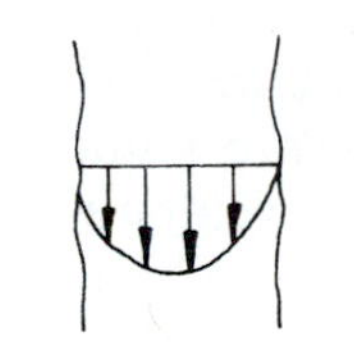

Plan view

(b)

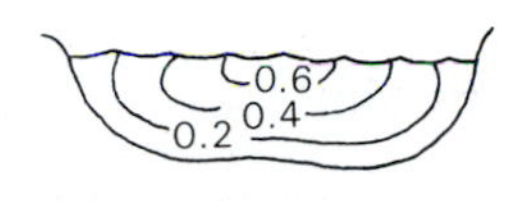


Transverse cross section

(c)

**Figure 13-16** Velocity distribution in a river channel.

### *Discharge and Velocity*

The amount of water moving past a certain point in a channel in a given amount of time is a very important characteristic of flow. This quantity, known as *discharge*, is measured perpendicular to the direction of flow. The usual units of measurement for discharge are cubic meters per second. Flow is considered to be either *steady* or *unsteady*, depending on whether discharge is constant or variable with time at a certain point. *Uniform* flow exists when the water depth is constant along a stream course and the slope of the water surface is equal and parallel to the slope of the bed. Even though steady, uniform flow is never truly present in a natural river, this assumption is often made for short periods of time so that uniform flow equations can be used.

The velocity of a stream at any point in the channel is determined by the slope of the channel and the distance from the point in the flow to the sides or bed of the channel. The influence of the bed on velocity is shown in Figure 13-16. There is a very thin layer of water adjacent to the bed in which the water velocity is zero. The variation in velocity with vertical distance above the bed is shown in Figure 13-16a. The channel sides exert a similar influence on velocity (Figure 13-16b). These two effects are combined in Figure 13-16c, which shows contours of relative velocity (lines connecting points with an equal fraction of the maximum velocity). It is clear that the maximum velocity occurs at the water surface in the middle of the channel. The mean velocity of the stream ($v$) occurs somewhere below the water surface in the middle of the channel. It is sometimes assumed that the velocity at six-tenths of the depth as measured downward from the water surface in the middle of the stream is the mean velocity of the cross section. If this velocity can be measured or estimated, it is possible to determine the discharge of the river at a point using the formula

$$Q = Av \qquad \text{(Eq. 13-5)}$$

where $Q$ is the discharge in cubic meters per second, $A$ is the cross-sectional area in square meters, and $v$ is the mean velocity in meters per second.

Other commonly used flow formulas can be derived by equating the driving forces of stream flow (gravity) with the resisting forces (channel boundaries) and solving for the resulting velocity. For example, the Chezy formula,

$$v = C\sqrt{RS} \qquad \text{(Eq. 13-6)}$$

expresses the velocity ($v$) in terms of the hydraulic radius ($R$), the slope ($S$), and a constant ($C$). The Manning equation (for SI units),

$$v = \frac{1}{n}R^{2/3}S^{1/2} \qquad \text{(Eq. 13-7)}$$

is a similar approach, although it includes a parameter $n$, which describes the roughness of the stream bed. A value for *Manning's n*, which typically ranges from 0.01 to 0.07, can be chosen from published descriptions of channels that consider variation in grain size of channel-bottom material, vegetation along the stream banks, and other factors. Both the Manning and Chezy formulas are used for natural streams even though they assume uniform flow.

**EXAMPLE 13-1**

During a flood, the water depth at a stream station was 2 m. The slope of the water-surface profile was 0.005 and the Manning's $n$ value was 0.05. If the channel cross section can be approximated as a rectangle with a width of 20 m, what was the discharge?

***Solution.*** From Equation 13-3,

$$R = \frac{A}{P} = \frac{(20\text{ m} \times 2\text{ m})}{(20\text{ m} + 4\text{ m})} = 1.7\text{ m}$$

Mean velocity can be obtained from the Manning equation:

$$v = \frac{1}{n}R^{2/3}S^{1/2} = \frac{1}{(0.05)}(1.7)^{2/3}(0.005)^{1/2} = 2\text{ m/s}$$

Using Equation 13-5, discharge is equal to

$$Q = Av = (40\text{ m}^2)(2\text{ m/s}) = 80\text{ m}^3\text{/s}$$

## *STREAM SEDIMENT*

Part of the work that streams accomplish is the transportation of sediment produced by weathering and erosional processes. Over millions of years, entire mountain ranges are leveled by these processes. In a particular reach of a stream, sediment can simply be transported by the stream. Alternatively, the stream, by the forces it exerts on its channel boundaries, can *entrain* particles from the channel sides and bottom into the flow, or it can deposit sediment that is being carried. In this way rivers can deepen or fill their channels over time.

### *Entrainment*

Erosion of a stream's sides or bottom can be accomplished in several ways. If the boundary material is composed of soluble rock such as limestone, solution by the stream is possible. In addition, the sediment carried by the stream can do erosional work. Particularly in the case of large particles that slide and roll across the stream bottom, *abrasion* can occur. Laboratory experiments indicate that streams with a high sediment load are more erosive under some conditions than sediment-deficient streams. When large particles are dropped to the stream bottom as a result of velocity fluctuations, the impact can be great enough to break off projecting fragments of the stream-bed material.

Although streams may actively erode their channels by the preceding mechanisms, the process of *entrainment* usually refers to the ability of the river to pick up loose particles of material from the bed by the direct hydraulic action of flowing water. As the stream flows in its channel, the moving water exerts shearing forces upon the stationary bed. If these forces are larger than

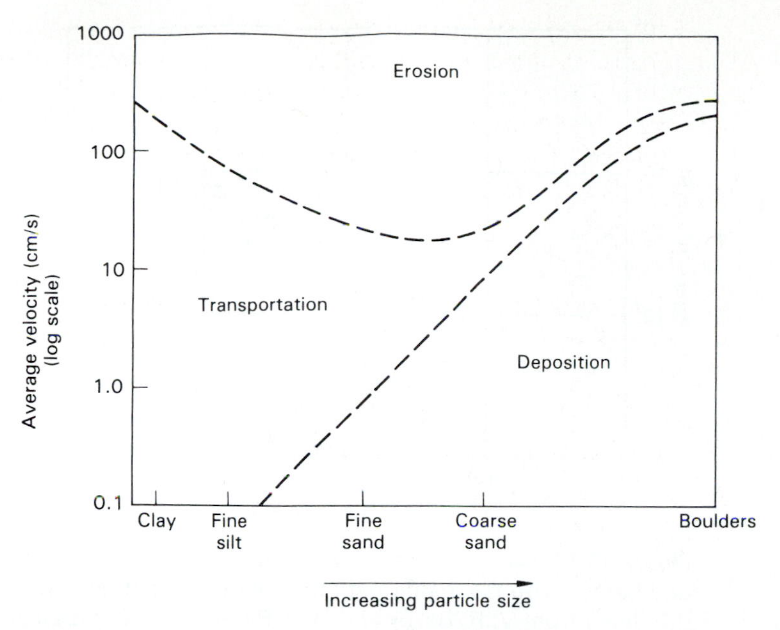


**Figure 13-17** The Hjulstom diagram, relating stream velocity to particle size for conditions of erosion, transportation, and deposition.

the gravitational forces tending to hold a particular particle on the stream bottom, the particle will be picked up off the bed and entrained into the flow.

Flume experiments, in which river processes are simulated by the flow of water in long tanks, demonstrate that as the velocity increases, progressively larger particles are entrained from the bed. The Hjulstrom diagram, shown in Figure 13-17, relates water velocity and particle size. The dashed lines divide the areas into fields of erosion, transportation, and sedimentation. One important point shown by the diagram is that higher velocity is required to entrain a particle of a given size than is needed to transport the particle once it has entered the flow. Another aspect of the curves that initially may not seem logical is that the velocity needed for erosion increases as particle size decreases near the left side of the diagram. This effect can be explained by the properties of clay particles that fall in that size range. The cohesion between clay particles leads to the entrainment of clay-particle aggregates into the flow rather than individual clay particles.

The relationship between velocity and particle size is often difficult to work with because of the variation of velocity in the channel. For a particle of a certain size, it is the flow velocity just above the bed, or the *critical bed velocity*, that must be large enough to move the particle. The relationship between the critical bed velocity and the mean velocity of the stream is not always clear. For these reasons the relationship between particle size and *critical shear stress* is often used instead of velocity. The critical shear stress, $\tau_c$, or the shear stress just large enough to move a particle of a certain size, is expressed as

$$\tau_c = \rho_w g R S \qquad \text{(Eq. 13-8)}$$

where $\rho_w$ is the water density, $g$ is gravity, $R$ is hydraulic radius, and $S$ is the energy slope of the stream. The energy slope can be approximated by the mean slope of the stream bed and the hydraulic radius can be approximated by the stream depth for wide, relatively shallow streams. This approach presents

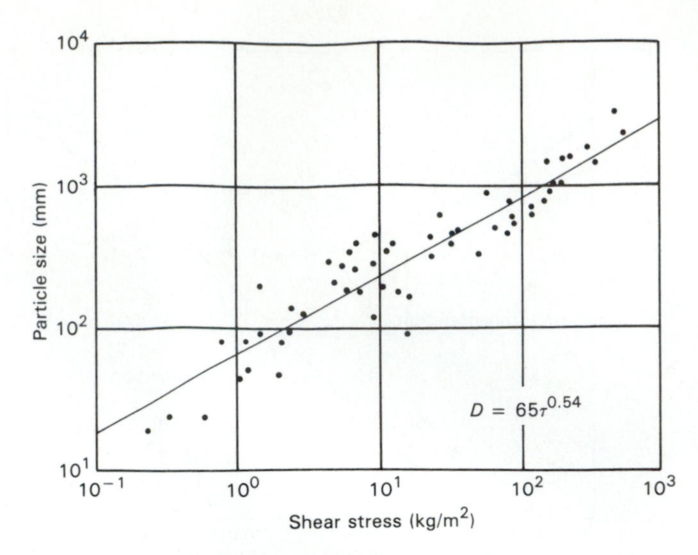


**Figure 13-18** An empirical relationship between shear stress ($\tau$) and particle size ($D$), using data from natural streams and canals. (Modified from V. R. Baker and D. F. Ritter, 1975, Geological Society of America Bulletin, 86:975–978.)

fewer problems in application because hydraulic radius and slope can be measured more accurately than critical bed velocity. An empirical relationship between critical shear stress and particle size is shown in Figure 13-18 for data from natural streams and canals.

### *Transportation*

The total amount of sediment a stream carries is known as its *load*. Sediment load can be divided into three types based on the process by which it is transported.

*Dissolved load* is carried totally in solution by a stream. Much of the dissolved load of a stream is derived from the base flow—that is, the ground-water component of the total stream flow. During its relatively long residence time in the ground-water flow system, the base-flow component of stream flow chemically reacts with the solid materials it encounters through the various types of weathering processes. From these reactions it derives the dissolved load that it carries to the surface stream in a ground-water discharge area. The surface runoff component of a stream is generally lower in dissolved solids content because of its short travel time to the stream channel.

Of the nondissolved load of a river, the smaller particles are carried as *suspended load*. This form of transport implies that particles are prevented from settling to the bottom of the stream by the movement of the water. The rate at which particles tend to fall through a static liquid column is known as the *terminal velocity*. This is a constant value that represents a balance between the opposing forces of gravity and viscous resistance by the fluid. The equation relating these forces is known as *Stokes' Law*, from which the terminal velocity can be stated as (in SI units)

$$v_t = \frac{1}{18} D^2 \frac{(\rho_s - \rho_w)g}{\mu} \qquad \text{(Eq. 13-9)}$$

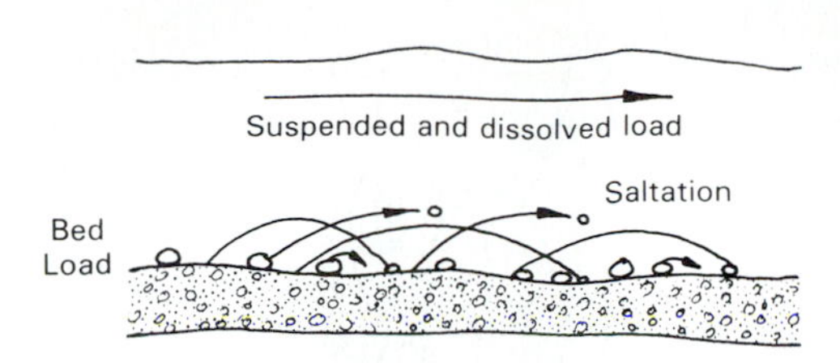


**Figure 13-19** Transportation of stream sediment includes rolling and sliding of bed load, saltation of small bed-load particles, and suspended and dissolved load transport.

where $D$ is particle diameter, $\rho_s$ and $\rho_w$ are the densities of the particle and fluid, $g$ is the acceleration of gravity, and $\mu$ is the dynamic viscosity of the fluid. Stokes' Law indicates that the terminal velocity of a particle is proportional to the square of the particle diameter.

For the particle to remain in suspension, an upward velocity greater than the terminal velocity must be imparted to the particle. Upward velocities are provided by turbulent eddies and other velocity fluctuations distributed throughout the stream. These turbulent effects are proportional to the mean velocity of the stream, so that the higher the stream velocity, the larger the particle size that can be carried in suspension.

The final type of load is the fraction of the total load that generally remains in contact with the bed during transportation. *Bed load* includes the particles that are too large to be transported by suspension. These particles move by rolling or sliding along the bed or by a process known as *saltation*, in which turbulent eddies can temporarily lift a particle off the bed. When the eddy dissipates, the particle can no longer be kept in suspension, and is dropped to the bed (Figure 13-19). The particle thus moves downstream in a series of small jumps or bounces.

Two general terms characterize the load transported by a stream. *Competence* refers to the largest particle that a stream can carry. It is a function of the velocity and shear stress exerted by the stream. *Capacity* is the total amount of sediment that a stream can carry by all mechanisms. Thus mountain streams have high competence but low capacity, and large rivers like the lower Mississippi may have high capacity but low competence.

## DEPOSITIONAL PROCESSES

When the velocity of a stream decreases, particles begin to drop from suspension or cease movement along the stream bed. Deposition of this type is usually a very temporary situation. Sediment is merely stored in the channel for short periods of time until the velocity increases again. Periodic fluctuations in stream velocity are common.

More permanent deposition within and adjacent to a stream channel is also possible. When a river floods, for example, a river overflows its channel and inundates the flat *flood plain* that adjoins the channel. Velocity drops significantly once water leaves the channel. Flood plains are characterized by parallel layers of fine-grained sediment deposited by successive floods.

Climatic conditions or the existence of an excessive sediment supply can promote a gradual accumulation of sediment in a river valley. The nature of the deposits corresponds to the particular type of channel pattern developed

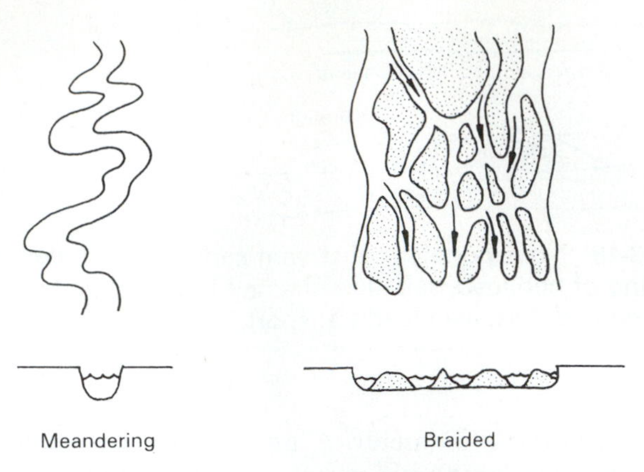


**Figure 13-20** Map and cross-sectional views of meandering and braided streams.

by the river. Channel patterns, which can be classified as *straight*, *meandering*, and *braided*, will be discussed in the next section.

A different type of depositional control is provided by decreases in channel gradient (slope). Gradient and velocity decrease where a stream valley extends from a mountain range out onto a plain and when a river enters a body of standing water like the ocean or a lake. Continuous sediment deposition occurs at these locations.

### *Meandering Streams*

The factors that control the type of channel pattern include the gradient of the channel, the type of material the channel is incising, and the type and amount of sediment that the river carries. Straight channels are not very common, although meandering streams can have straight segments. Meandering streams, the most common type, tend to develop in gently sloping areas with low sediment supply and a high percentage of cohesive sediment. The channel pattern also correlates with the cross-sectional shape of the channel

**Figure 13-21** Vertical air photo of a meandering stream with high sinuosity in eastern North Dakota.

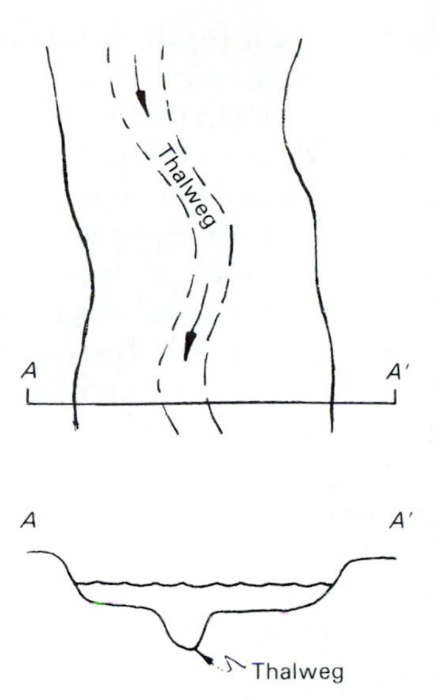


**Figure 13-22** Meandering thalweg in a nearly straight channel.

(Figure 13-20), with meandering streams characterized by narrow, deep channels.

The intensity of meandering is quantified by the use of the parameter called *sinuosity*, which is measured as the ratio of channel length to valley length. Sinuosity is apparently related to the type of material that composes the channel banks, the type of sediment load transported, and the gradient. The highly sinuous river shown in Figure 13-21 has a low gradient, is cut into sediment with a high percentage of silt and clay, and transports mostly fine-grained sediment as suspended load rather than bed load. Streams with the opposite characteristics have a much lower sinuosity.

The tendency of a stream to meander appears to be a basic property of fluid flow, although the factors that cause meandering are not totally understood. Even in a straight channel (Figure 13-22), the *thalweg*, the deepest part of the channel, establishes a meandering pattern. Straight excavated channels like canals or drainage ditches often erode their banks in order to develop a natural meandering pattern.

The flow of water in a meandering channel is shown in Figure 13-23. The position of the thalweg impinges on the outside banks of the meanders just downstream from the point of maximum curvature. Because the thalweg follows the deepest part of the channel, it also represents the point of highest

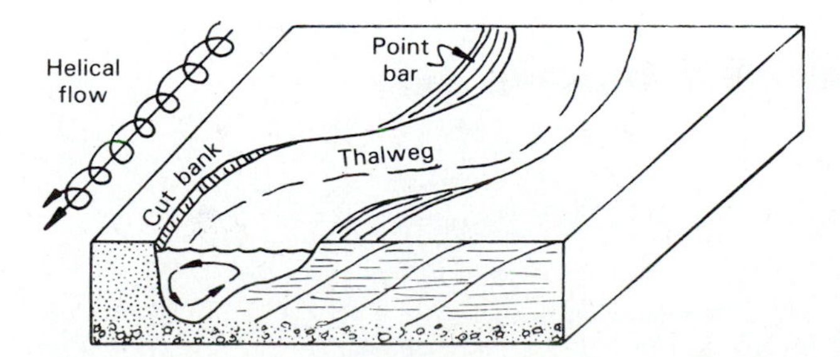


**Figure 13-23** In a meandering channel, erosion occurs on the outer sides, or cut banks, of meander bends. Point-bar deposition is present on the inner bends. A secondary flow pattern superimposed on the downstream motion gives the overall flow a helical pattern.

velocity in the channel. As a result of this higher velocity, the meandering channel tends to be eroded on the outside bank of the meander to form a *cut bank*. A secondary pattern of circulation is also imposed upon the flow in the vicinity of meanders. In the cross section of the lower part of Figure 13-23, the secondary flow is shown to move across the channel toward the cut bank at the water surface and then back along the stream bottom. Because the water is also moving downstream, the resulting overall motion resembles a corkscrew and is termed *helical flow*. On the inner bank of the meander, where the depth and velocity are less, deposition of the coarser fraction of stream sediment occurs. The resulting sediment accumulation is called a *point bar*.

This pattern of erosion and deposition of a meandering stream causes lateral migration of the channel meanders across the flood plain. The characteristic set of landforms that occur on the flood plains of meandering rivers (Figure 13-24) records the sequence of events of recent geologic time. Many of the topographic features are related to the lateral migration of meanders. The continuous process of cut-bank erosion and point-bar deposition produces a series of parallel concentric ridges and swales called *meander scrolls* (Figure 13-24). The ridges represent old point bars and therefore are composed of coarser sediment. The intervening swales are underlain by organic-rich, fine-grained sediment. Lateral migration of meanders cannot occur indefinitely because rivers tend to maintain a relatively constant value of sinuosity. The pattern of erosion and deposition in a meander leads to growth of the meander followed by its abandonment. After abandonment, or meander *cutoff*, the old channel loop, now bypassed by the river, becomes an *oxbow lake* (Figures 13-25 and 13-26). Eventually, the oxbow lake fills with highly organic silty and clayey sediments. These old channel fills, or *clay plugs*, provide foundation problems in construction because of the high compressibility of the sediment.

The flood plain is so-named for a good reason, for stream discharge periodically exceeds the channel capacity and flood water flows out of the channel and onto the adjacent surfaces. Typically, streams attain bankfull flow every 1 to 2 years on the average. The flood plain is the river system's natural storage area for the water and sediment available when the discharge exceeds bankfull flow.

Because flooding is intimately involved in the development of a flood plain, flood-plain sediment can be divided into two types. The dominantly coarser sediment in meander scrolls related to the lateral migration of meanders is classified as a *lateral accretion* deposit, and the sediment deposited by floods is known as a *vertical accretion* deposit. Vertical accretion deposits are

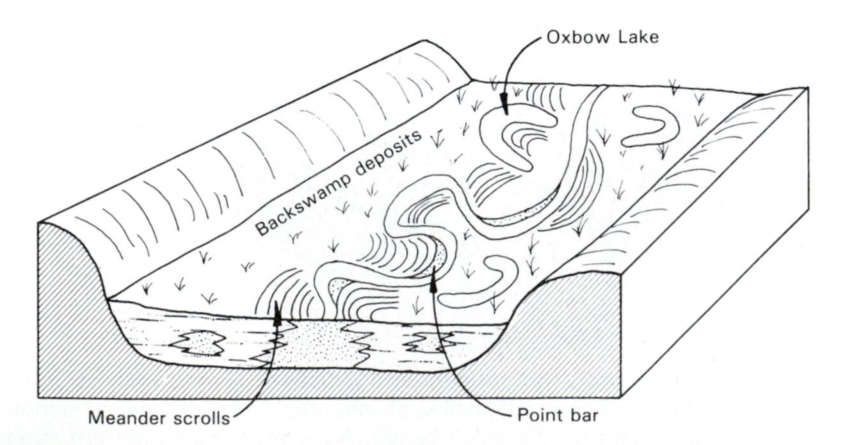


**Figure 13-24** Landforms associated with a meandering river.

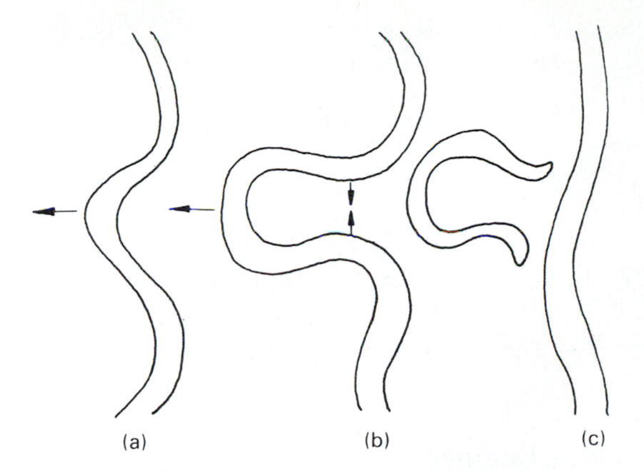


**Figure 13-25** Meander migration leading to abandonment and formation of an oxbow lake. Arrows show direction of cut-bank migration.

fine grained because of the velocity decrease that accompanies flow out of the channel onto the broad flood plain. The coarsest available sediment is deposited adjacent to the channel margins to form *natural levees*, as shown in Figure 13-27. Grain size decreases with distance from the channel. The areas behind the natural levees are often called *backswamps* because of their organic-rich, fine-grained sediment, high water table, and swampy vegetation.

The complex distribution of sediments in the alluvial valley of a meandering stream presents many engineering problems because flood plains often are

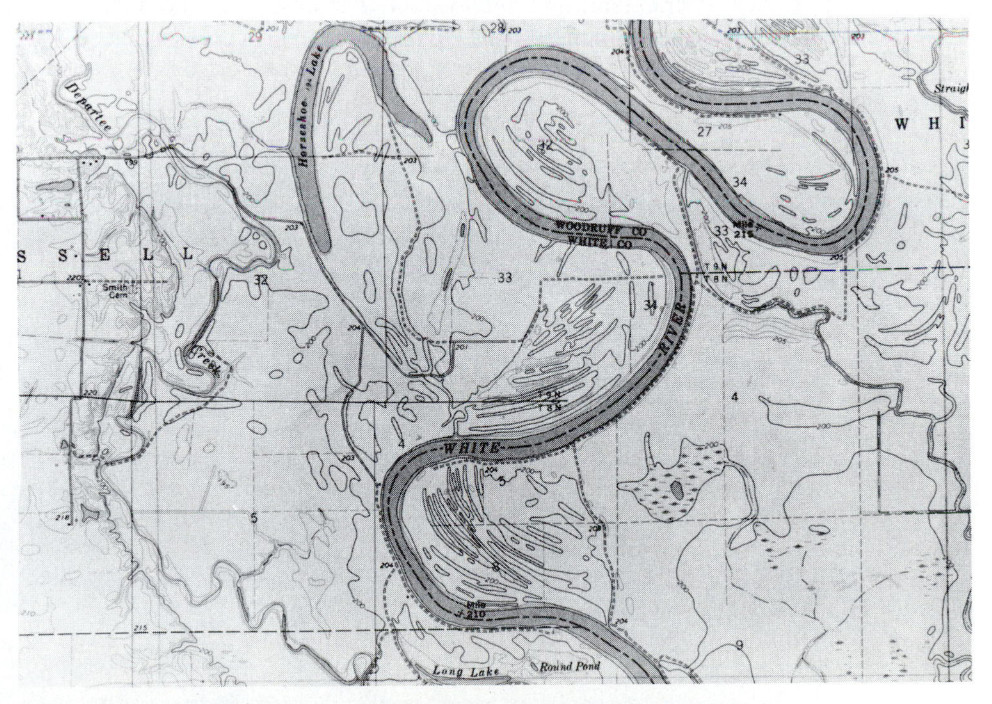


**Figure 13-26** Topographic map of part of the White River flood plain in Arkansas, showing meander scrolls and an oxbow lake. The river will soon form another meander cutoff in the upper right part of the map. (Augusta SW Quadrangle, Arkansas, U.S. Geological Survey, 7.5-minute series.)

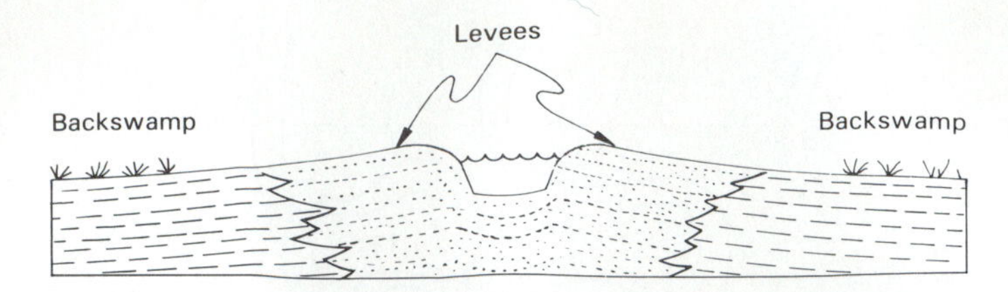


**Figure 13-27** Natural levee deposits on a flood plain. (From S. Judson, M. E. Kauffman, and L. D. Leet, *Physical Geology*, 7th ed., copyright © 1987 by Prentice-Hall, Inc., Englewood Cliffs, N.J.)

highly developed. Engineering problems arise both in the control of the periodic floods that inundate these areas and in the construction attempted on flood plains. High water tables frequently pose troublesome conditions. In addition, the rapid lateral and vertical variation in sediment type requires a great deal of site investigation and soils testing for foundation design. The engineering properties associated with specific fluvial landforms and deposits are listed in Table 13-1.

### *Braided Streams*

Streams that flow in broad, shallow channels of low sinuosity and consist of multiple subchannels separated by islands or bars are called *braided streams* (Figures 13-20 and 13-28). The channel pattern within a braided stream—where subchannels continually divide and recombine around obstacles in the main channel—can be referred to as *anastomosing*. The anastomosing channel network of a braided stream forms in response to a different set of controlling variables from those associated with meandering streams.

**Figure 13-28** A braided stream in Alaska. (T. L. Péwé; photo courtesy of U.S. Geological Survey.)

**TABLE 13-1**

**Engineering Properties of Alluvial Deposits in the Lower Mississippi Valley**

| *Environment* | *Soil texture (USC classification)* | *Natural water content (percent)* | *Liquid limit* | *Plasticity index* | *Cohesion (lb/ft$^2$)* | *Angle of internal friction (degrees)* |
|---|---|---|---|---|---|---|
| Natural levee | Clays (CL) | 25–35 | 30–45 | 15–25 | 360–1200 | 0 |
| | Silts (ML) | 15–35 | Nonplastic (NP)–35 | NP–5 | 180–700 | 10–35 |
| Point bar (ridges) | Silts (ML) and Silty Sands (SM) | 25–45 | 30–45 | 10–25 | 0–850 | 25–35 |
| Abandoned Channel | Clays (CL and CH) | 30–95 | 30–100 | 10–65 | 300–1200 | 0 |
| Backswamp | Clays (CH) | 25–70 | 40–115 | 25–100 | 400–2500 | 0 |
| Swamp | Organic Clay (OH) | 110–265 | 135–200 | 100–165 | Very low | |
| Marsh | Peat (Pt) | 160–465 | 250–500 | 150–400 | Very low | |
| Lacustrine | Clay (CH) | 45–165 | 85–115 | 65–95 | 75–150 | 0 |
| Beach | Sand (SP) | Saturated | NP | NP | 0 | 30 |

SOURCE: From C. R. Kolb and W. G. Shockley, 1957. Used by permission of the American Society of Civil Engineers.

Conditions that favor the formation of braided streams include an abundant source of coarse-grained sediment, a high gradient, and cohesionless, nonresistant sediment in the channel banks. The cross-sectional profile that develops under these conditions is broad and shallow (Figure 13-29). In a broad shallow stream, higher velocities are closer to the stream bed, and transport of coarse particles is facilitated.

The formation of braids involves the deposition of bars of sediment within the channel. A bar is a large-scale bedform. Once initiated, bars grow and migrate downstream because sediment is transported over the bar and then deposited on the downstream end, where the depth increases. As the bar grows, it occupies more of the channel cross-sectional area that was previously available for water. As a result, the stream must erode its banks laterally to enlarge the channel to maintain a sufficient capacity for the discharge (Figure 13-29). As the discharge drops from a period of very high flow when much of the bed load was in transportation, previously deposited bars are dissected by erosion and the sediment is redistributed through the system. The channel pattern in a braided stream constantly changes with fluctuations in discharge.

### *Alluvial Fans*

If the gradient of a stream suddenly decreases, Equations 13-6 and 13-7 indicate that the velocity will decrease if other channel parameters remain the same. A river responds to a decrease in velocity by depositing the coarser fraction of its sediment load. A good example of a gradient decrease is the point where a stream emerges from a steep mountain front onto the floor of the adjacent basin. The stream gradient and velocity are suddenly decreased and sediment is deposited to form a hemiconical landform called an *alluvial fan* (Figure 13-30).

The velocity decrease is also partly caused by the increase in width of the channel on the fan surface. The width can increase to the point where flow is no longer channelized but instead moves as a thin layer across a broad surface. This type of flow is known as *sheet flow*. If the fan surface is permeable, water rapidly infiltrates into the fan, and overland flow ceases. Other geomor-

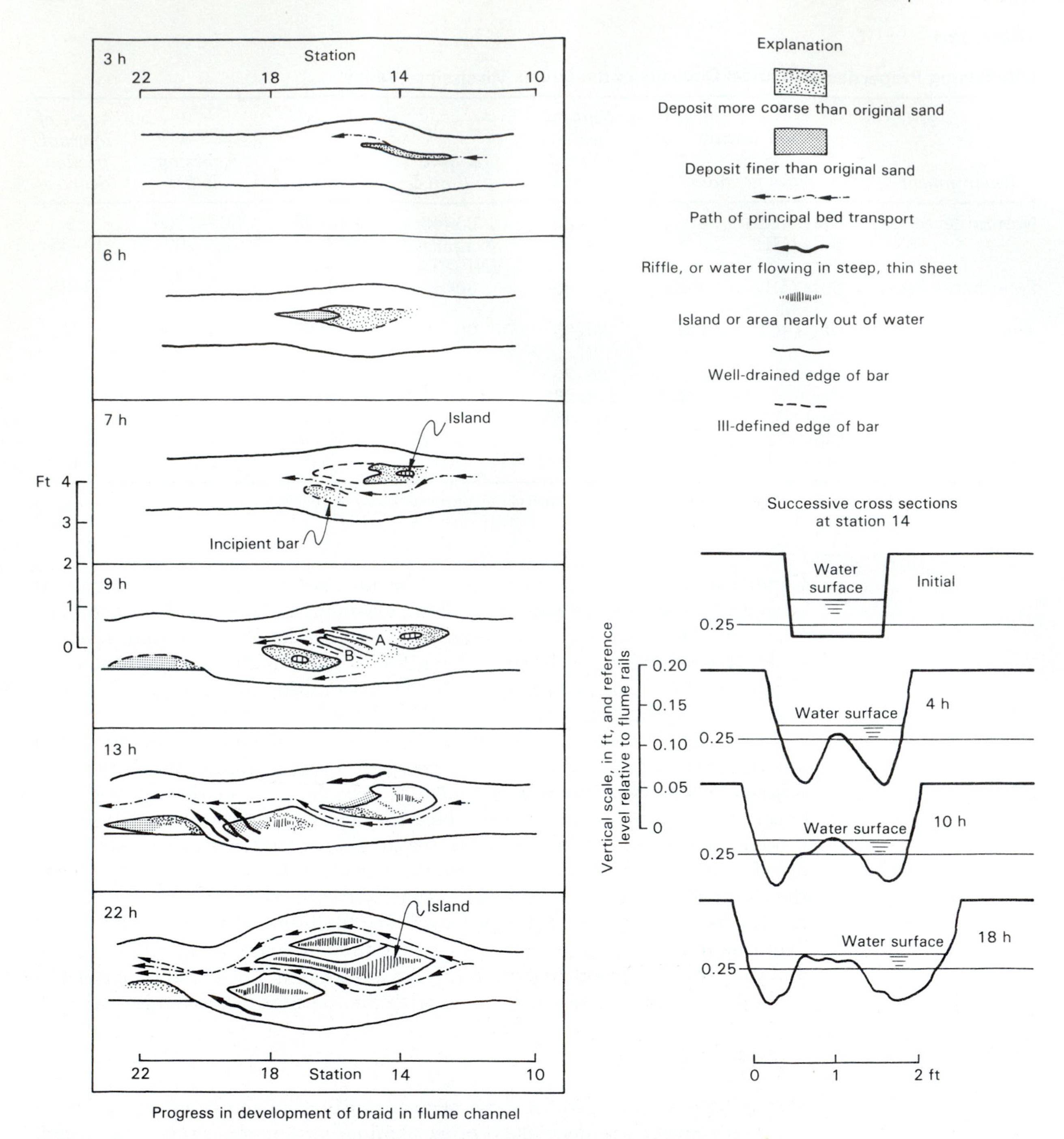


**Figure 13-29** Development and modification of braid bars in a flume channel. Note changes in cross section during flow period. (From L. B. Leopold and M. G. Wolman, 1957, U.S. Geological Survey Professional Paper 282-B.)

phologic processes, such as debris flows, are also recognized as being important on many fans.

Alluvial fans are best developed where recent faulting has produced steep mountain fronts with adjacent basins (Figure 13-31). Arid climatic conditions, where flow in drainage networks occurs mainly during infrequent rainfall events of large magnitude, also facilitate the deposition of alluvial fans. The

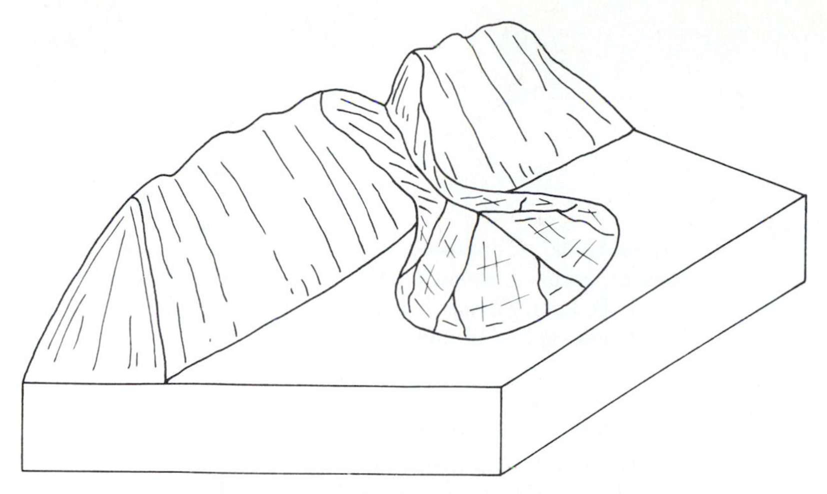

**Figure 13-30** Alluvial fans form at places where stream gradients decrease and channel widths increase.

hemiconical shape of some alluvial fans makes them easily recognizable on topographic maps.

### *Deltas*

Where a river enters a lake or the sea, a *delta*—the subaqueous counterpart of an alluvial fan—is constructed. Deposition occurs in a delta because the river current slows and eventually dissipates as it moves progressively offshore. The gradual reduction in velocity is reflected in the distribution of sediment types in a delta, with deposition of coarse bed-load materials in the vicinity of the mouth of a river. As the river deposits its sediment load, the delta extends or *progrades* in a seaward direction.

**Figure 13-31** Alluvial fans in Death Valley National Monument, California. (H. E. Malde; photo courtesy of U.S. Geological Survey.)

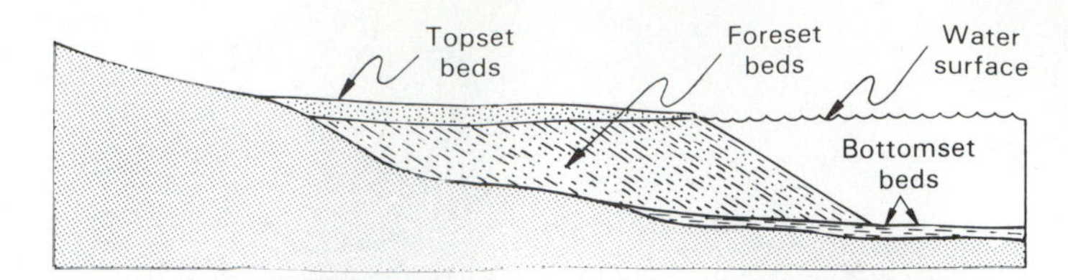


**Figure 13-32** Idealized sedimentation in a delta, including topset, foreset, and bottomset beds. (From S. Judson, M. E. Kauffman, and L. D. Leet, *Physical Geology*, 7th ed., copyright © 1987 by Prentice-Hall, Inc., Englewood Cliffs, N.J.)

Deltaic environments are highly variable. Marine deltas are influenced by the fluvial input of water and sediment, the wave energy imparted to the coastline, and tidal currents. The relative dominance of one of these three factors determines the form and depositional processes of the delta. A basic type of delta sedimentation recognized by G. K. Gilbert more than 100 years ago is shown in Figure 13-32. This type of delta is limited to lakes. As the delta is gradually constructed, the river extends across accumulating delta sediments to a slope called the *foreset slope.* Coarse sediments are deposited on the foreset slope in long, dipping *foreset beds* (Figures 13-32 and 13-33). Progradation (offshore migration) of the delta results from the continual accumulation of foreset beds. Beyond the foreset slope, slow-moving currents carry the finer fractions of the river's suspended load across the lake floor, where they are eventually deposited in nearly horizontal beds called *bottomset beds.* As the river extends its channel across the prograding delta to the foreset slope, it deposits horizontal beds called *topset beds.*

As a river approaches a delta, the single main channel divides repeatedly into a network of *distributary* channels, which transport the sediments across the delta (Figure 13-34). Distributary channels are maintained by natural levees that project slightly above the delta surface. Sediment is transported

**Figure 13-33** Deltaic foreset beds. (C. S. Denny; photo courtesy of U.S. Geological Survey.)

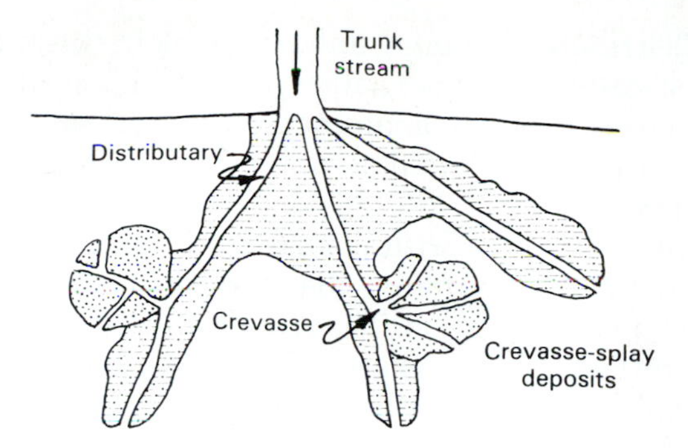


**Figure 13-34** In a delta, the trunk stream branches into distributaries. When distributaries are breached, crevasse-splay sediments are deposited.

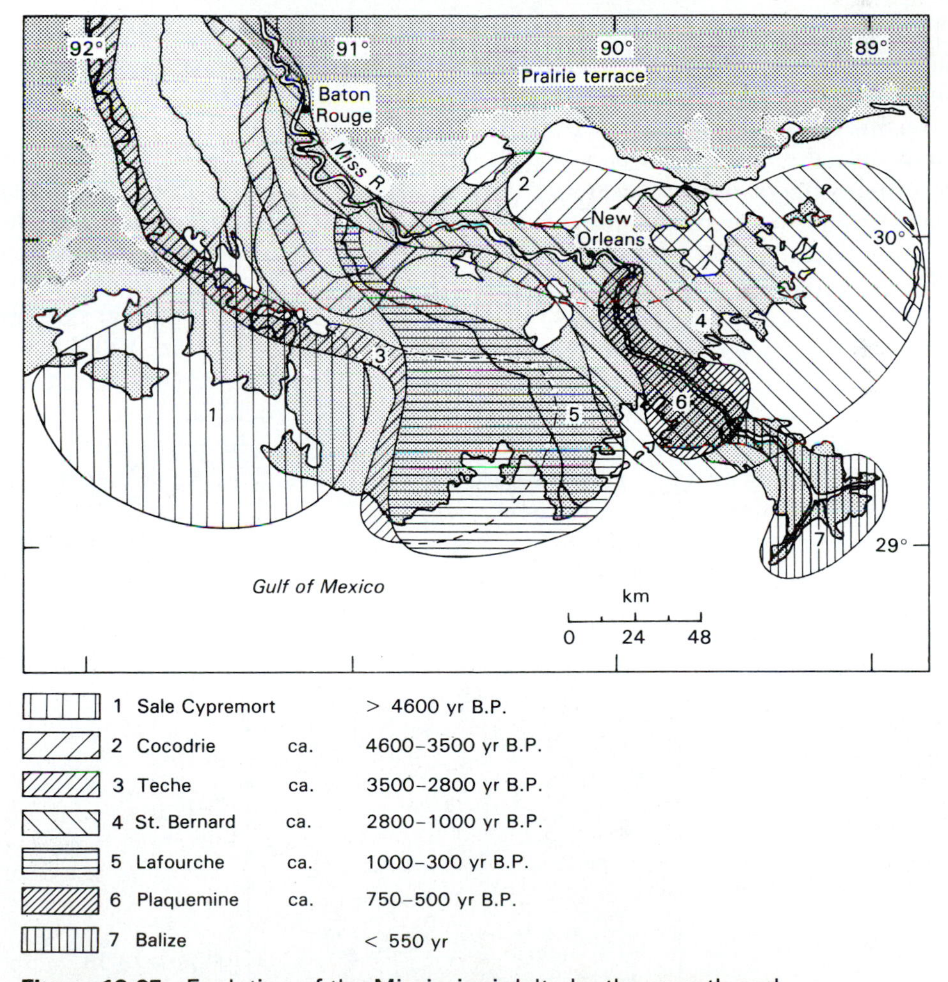


**Figure 13-35** Evolution of the Mississippi delta by the growth and abandonment of delta lobes. (From J. P. Morgan, 1970, Deltas—a résumé, *Journal of Geologic Education*, 18, National Association of Geology Teachers.)

through the distributary channel unless a break in the natural levee is present. These breaks, called *crevasses*, allow the diversion of water and sediment through the natural levee to a lateral position between distributary channels. The minidelta that is deposited as water flows through the crevasse is called a *crevasse splay*.

Sediment is deposited within a particular distributary network for a certain period of time. Eventually, that network becomes raised in elevation because of sediment deposition. The river will then abandon the active part of the delta in preference to an adjacent part of the delta with a steeper gradient. The Mississippi River delta is a good example of this process. Over the past 5000 years, active sedimentation has successively deposited seven major lobes (Figure 13-35).

## CASE STUDY 13-2

## Future Shift of the Mississippi River?

The history of the Mississippi River delta indicates that the position of the river at any particular time is a transient situation. As a river and its distributary system build up a delta lobe by sediment deposition on natural levees, crevasse splays, and in backswamps, the delta lobe gradually becomes higher in elevation than adjacent, inactive parts of the delta. The river finally reacts by changing its course to begin deposition of a new lobe of the delta.

For some years, it has been evident that the Mississippi River is beginning to shift its course to one of its distributaries, the Atchafalaya River (Chen 1983). If this event occurred, the Atchafalaya (Figure 13-36) would become the main channel of the Mississippi and the current channel would become a distributary with a greatly reduced flow. The consequences of such a shift would be disastrous to the Louisiana cities of New Orleans and Baton Rouge because of the loss of shipping trade and also fresh water. The greatly reduced flow of the main channel of the Mississippi would not provide the cities with a sufficient supply of fresh water, and, in fact, salt water from the Gulf of Mexico would move up the river channel. Saltwater encroachment would occur because the Mississippi channel bottom is below sea level for more than 100 km upriver from Baton Rouge. Dramatic changes would also occur in the Atchafalaya Basin. The town of Morgan City, already a flood-prone area, would be flooded out of existence.

In anticipation of a channel shift, the U.S. Army Corps of Engineers constructed the Old River Control Structure (ORCS) in the 1950s,

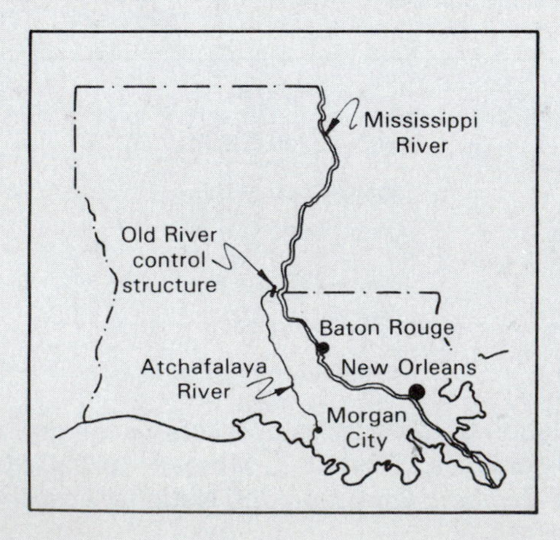


**Figure 13-36** Map of Louisiana showing the Mississippi River and the Atchafalaya River, the valley to which the Mississippi may alter its course.

a series of locks, dams, and levees at the head of the Atchafalaya to maintain flow down the main channel. Damage to the ORCS by severe floods in 1973 necessitated construction of the $219 million Auxiliary Control Structure to preserve the ability of the ORCS to block the channel shift into the Atchafalaya.

Thus the stage is set for a continuing struggle between the Mississippi River and the U.S. Army Corps of Engineers. As the present course of the Mississippi becomes more and more unstable with respect to a course to a different part of the delta, more and more extensive (and expensive) control measures will be required. Only time will reveal the outcome of this situation.

## EQUILIBRIUM IN RIVER SYSTEMS

### The Graded Stream

River systems provide the primary means by which continents elevated to great heights by tectonic processes are gradually worn down to low-relief plains projecting only slightly above sea level. Throughout this process, rivers transport sediment from the topographically high areas of continents to the ocean. Along the way, stream channels progressively grow in size to accommodate the larger discharge supplied by the addition of tributary streams carrying the flow from drainage basins of lower stream order. Many studies of river systems have indicated that sediment grain size decreases in a downstream direction. The sediment transported by a stream ranges from large boulders in mountainous areas to silt and clay at the river's mouth. The slope, or gradient, of a stream also decreases progressively in the downstream direction. The change in gradient of a stream from headwaters to mouth is known as the *longitudinal profile*. As shown in Figure 13-37, longitudinal profiles have a characteristic concave shape. *Base level* is the elevation that the downstream segment of the curve asymptotically approaches. Sea level is the ultimate base level for all streams, although temporary base levels may be established by lakes or reservoirs at intermediate positions along a stream's course.

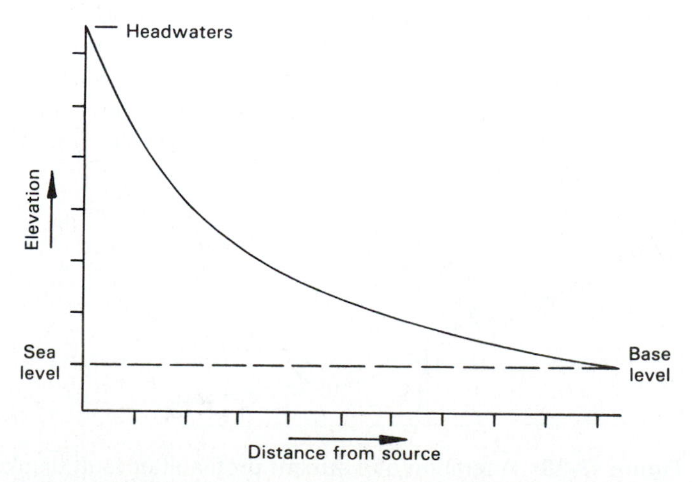


**Figure 13-37** Idealized longitudinal profile of a stream. The gradient decreases as the stream approaches base level.

It has been generally believed that the concave longitudinal profiles of most streams represent a state of equilibrium in the river system. Under this assumption the longitudinal profile of a river adjusts to provide the velocity necessary to transport the sediment supplied to the channel. An equilibrium, or *graded*, stream tends to maintain its profile for a period of time during which there are no drastic changes in the prevailing climatic or tectonic conditions affecting the basin. Although the idea of equilibrium in stream systems has merit, it is now known that the interactions between the numerous variables controlling the discharge, sediment load, and profile of rivers are complex. For example, the velocity of a river actually increases in the downstream direction despite the decrease in slope evident in the longitudinal profile.

### Causes and Effects of Disequilibrium

The concept of equilibrium must be considered in terms of the period of time over which it is proposed. The graded stream represents equilibrium during some intermediate-length period of time (tens to hundreds of years), during which the slope and channel characteristics are adjusted to the discharge and sediment load supplied to the channel segment. There are many factors that tend to disturb the fragile balance that may exist.

Tectonic activity can be a major cause of disequilibrium in stream systems. An example of tectonic initiation of disequilibrium is shown in Figure 13-38. A fault has down-dropped part of a stream valley in relation to the upstream reach. The response of the stream to this disturbance is to reestablish an equilibrium condition. By erosion and downcutting upstream from the fault, and by deposition downstream from the fault, the stream can eventually develop a new equilibrium profile. Gradient is not the only variable that can be adjusted, however. Channel characteristics such as width and depth may also be altered.

Other tectonic influences on river equilibrium involve changes in base level. If a coastal region is uplifted so that the elevation of sea level relative to that particular coastline drops, streams must steepen their gradients in the vicinity of the coast in order to adjust to the changing conditions.

Climatic change is of vital importance to stream equilibrium. The climate in a drainage basin, along with the geology, controls the discharge and sediment load supplied to a river system. Increases or decreases in precipitation within a drainage basin, aside from short-term variations, will cause changes in vegetation and will also cause the streams to adjust their gradient and channel characteristics. The extreme manifestation of climatic change—an advance of a glacier into the drainage basin—can lead to drainage-system disruptions that persist for thousands of years after the glacier has retreated. The Missouri River and many of its tributaries, for example, flowed northward to Hudson

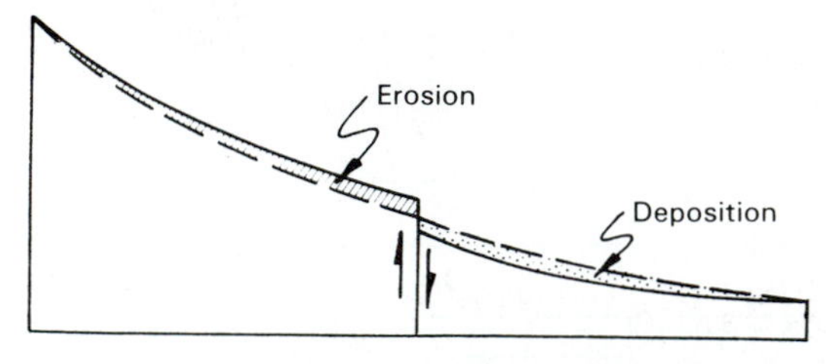


**Figure 13-38** Alteration of a stream profile after fault displacement. The stream reestablishes a smooth, concave profile by erosion above the fault and deposition downstream.

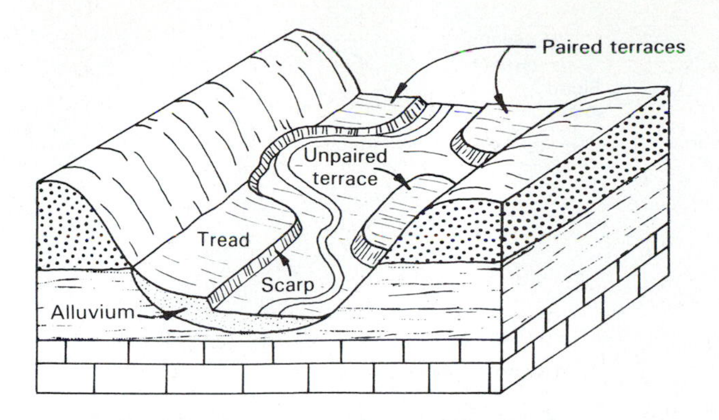


**Figure 13-39** Morphology of river terraces.

Bay before the Pleistocene Epoch. The southward advance of glaciers blocked and diverted the course of the Missouri into the Mississippi River drainage basin.

Stream Terraces. One of the best types of evidence available for inferring past changes in river system equilibrium is the existence of *stream terraces*, which are remnants of former flood plains that stand above modern flood plains in stream valleys. Terraces attest to a former state of equilibrium that was abandoned during a period of erosional instability. Figure 13-39 illustrates the general morphology of terraces. Terraces may be *paired* or *unpaired*, depending upon whether corresponding surfaces at the same elevation exist on opposite sides of the valley. Morphologically, the terrace consists of a flat surface, the *tread*, bordered by a slope called the *scarp*. Multiple terraces are common in many valleys (Figure 13-40).

**Figure 13-40** Multiple river terraces in the Madison River valley, Montana.

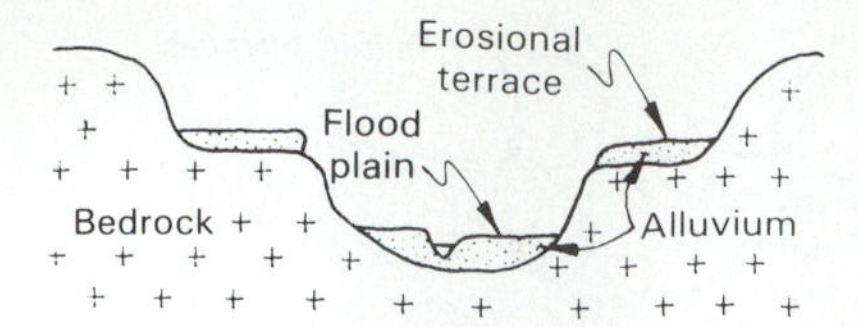


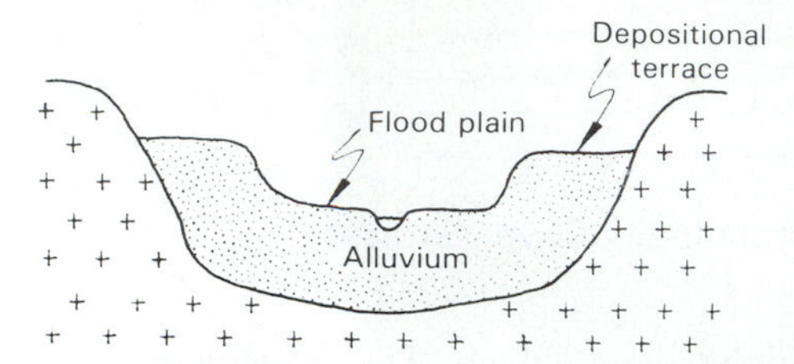


**Figure 13-41** Differences between erosional and depositional terraces. Depositional terraces are underlain by thick deposits of alluvium.

In terms of genetic processes, terraces consist of two types. During the formation of an *erosional terrace* (Figure 13-41), stream downcutting is dominant. Possible explanations for downcutting could include tectonic or climatic factors. The origin of *depositional terraces* includes a time interval in which the stream fills its valley with sediments. This period of *aggradation* can be caused by decreases in discharge, increases in sediment load, or changes in base level. A frequent cause of aggradation in northern latitudes was the abundant sediment load supplied to the stream system by Pleistocene glaciers. Following the period of aggradation, a stream must then erode its channel to form a depositional terrace. The sudden decrease in sediment load accompanying the retreat of a glacier or other climatic factors can lead to the incision of a previously deposited valley fill.

Human Interaction with Equilibrium. The changes in equilibrium we have considered thus far have not included those that are primarily induced by human activity. Among these, the most important examples are changes in river systems brought about by land-use changes in the drainage basin. The increased sediment yield caused by the conversion of natural areas to agricultural production and urban zones was discussed in Chapter 9 and illustrated in Figure 9-21. The consequences of increased sediment yield in a drainage basin can be severe. Sediment is temporarily stored in river channels, causing a decrease in channel capacity, which in turn results in more frequent and damaging flooding. The effect of the artificial drainage network in a city (storm sewers) is to supply more water to river channels in a shorter time, as compared with rivers in nonurbanized areas. The magnitude of flooding from a given rainstorm is also increased in an urbanized area. The river's response to greater sediment load, as well as greater discharge, is to enlarge its channel to accommodate the increased flow. Serious bank erosion may result, with flood-plain landowners sustaining property losses.

Another class of human adjustments to equilibrium includes such engineering works as dams and levees that seek to control the river for such purposes as flood protection, water supply, power supply, and recreation. Reservoirs interrupt the longitudinal profile of streams by establishing temporary base levels. Figure 13-42 shows some of the effects. The sediment load of the stream is dropped as it flows into the reservoir. The decrease in gradient causes aggradation upstream from the impoundment as well, as the stream attempts to reconstruct an equilibrium profile utilizing the reservoir as base level. The outflow from the reservoir, having deposited its sediment load, is likely to

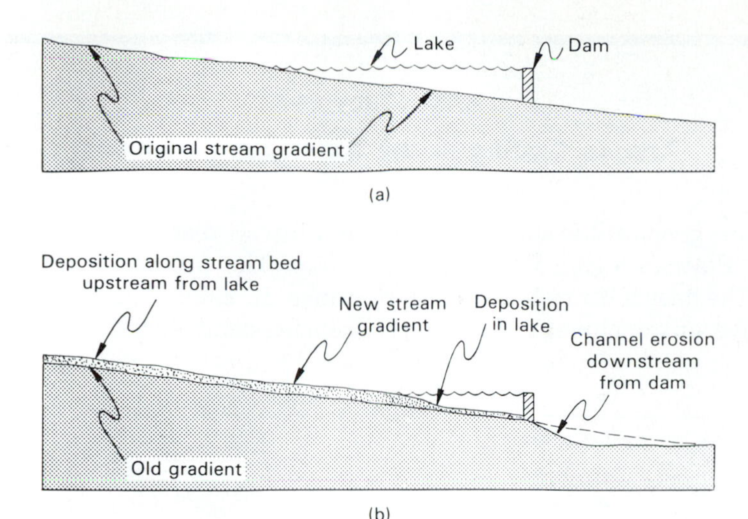


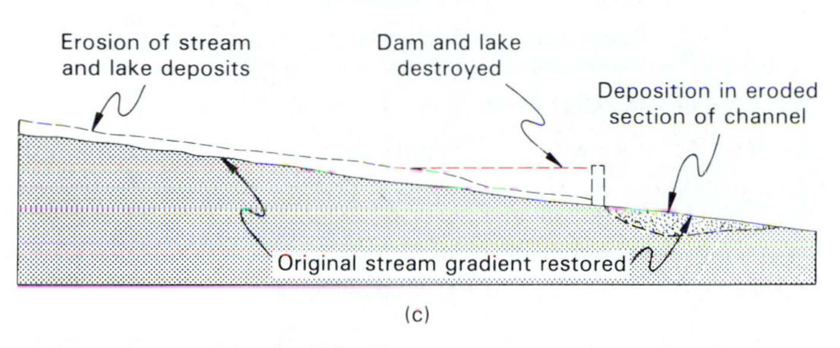


**Figure 13-42** Construction or removal of reservoirs causes streams to adjust their gradients in order to approach a new condition of equilibrium. (From S. Judson, M. E. Kauffman, and L. D. Leet, *Physical Geology*, 7th ed., copyright © 1987 by Prentice-Hall, Inc., Englewood Cliffs, N.J.)

erode as it flows down the remainder of the river channel and thus dissipate energy available for sediment transport. Dams along the Missouri and other rivers have required extensive downstream bank-stabilization projects to limit loss of flood-plain area by bank erosion.

Flood-control projects often involve alterations to river channels. These may include levees and other structures to contain the flow within the channel so that flooding will be minimized. Alternatively, the channel may be enlarged, or its gradient steepened, in order to convey flood water at a greater discharge or higher velocity through a particular reach. Unfortunately, these changes often are answered by undesirable and costly adjustments by the river system. Levees may decrease flooding along one reach of the channel, but because natural flood-plain storage of water is decreased, they may aggravate the downstream flood problem. The increase in channel size or gradient for attempted control of the river is called *channelization*. This disturbance of the river's equilibrium is propagated both upstream and downstream from the channelized reach. An interesting example documented by Daniels (1960) is the Willow River in Iowa. After artificial straightening of the channel for flood-control purposes, the gradient was increased to the point where the river began to incise its channel. Eventually, the destabilized channel was twice as wide and deep as its original size. Changes of this type are transmitted upstream in a river system. Tributary down cutting following channelization can lead to the destruction of farm land throughout the basin.

## CASE STUDY 13-3
## Arroyo Cutting in the American Southwest

The most significant geomorphic change in the semiarid Colorado Plateau region of the American Southwest in the past century has been the rapid and extensive entrenchment of streams in alluvial reaches to form deep valleys, or *arroyos*. The process was well documented because it caused great hardship among the ranchers and farmers that had just recently set-

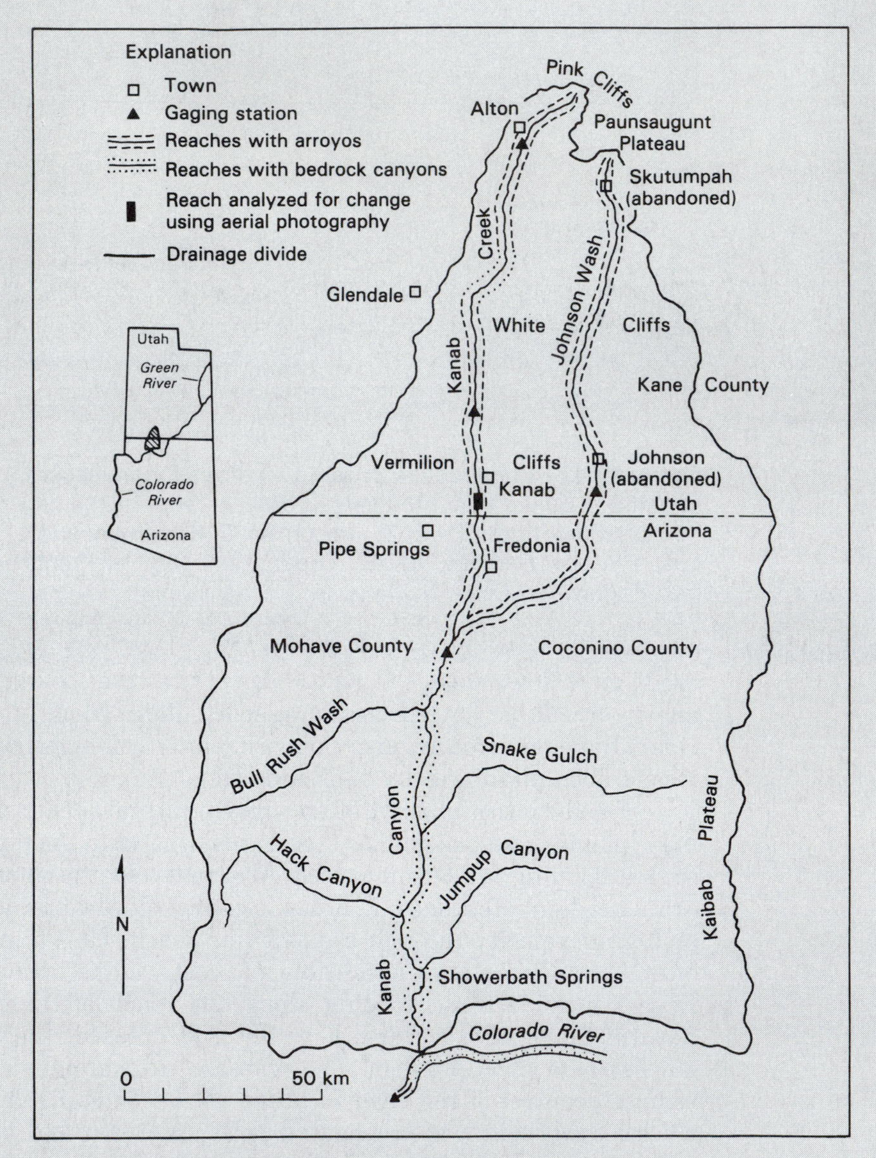


**Figure 13-43** Map of the Kanab Creek drainage basin. (From R. H. Webb, S. S. Smith, and V. A. S. McCord, 1992, Arroyo cutting at Kanab Creek, Kanab, Utah, in *Paleoflood Hydrology of the Southern Colorado Plateau: Field Trip Guidebook*, J. Martinez-Goytre and W. M. Phillips, eds., Arizona Laboratory for Paleohydrological and Hydroclimatological Analysis, University of Arizona, Tucson, Ariz.)

tled the area in the mid-nineteenth century. A prime example of arroyo cutting is the history of Kanab Arroyo (Webb et al., 1992), a tributary of the Colorado River located in southern Utah and northern Arizona (Figure 13-43).

Kanab Creek, and its tributary Johnson Wash, alternate between reaches that flow in bedrock canyons and reaches that flow across thick alluvial deposits (Figure 13-43). The small town of Kanab, Utah, which was settled in the 1860s, lies along an alluvial reach of Kanab Creek. At the time of settlement, the stream flowed in shallow channels that the settlers could literally step across. The topography along the stream consisted of broad meadows with a rich cover of natural grasses. The water table was only about 1 meter below land surface and watering holes for stock could be created easily by digging a shallow hole. In places, drainage ditches had to be dug in order to drain the land to cut grasses for hay. The conditions were ideal for grazing, and cattle and sheep were quickly introduced to the area.

Beginning in 1882, a series of unusually large floods caused drastic changes in Kanab Creek. By 1885 the channel, which was several meters deep only a few years before, had been downcut to a depth 18 m and a width of 21 m. A series of dams was built along the valley to supply domestic and irrigation needs for water, but these all failed in subsequent floods. By 1900, the arroyo had reached its current dimensions, 25 m deep and 80–120 m wide (Figure 13-44). A similar sequence of events has occurred on other streams in the Southwest. Erosion was so severe in places that towns were abandoned. Bedrock reaches of the streams, however, were only slightly altered, in contrast to the extreme changes in the alluvial reaches.

The question that now must be answered is, what was the cause of this remarkable disequilibrium in Kanab Creek and other streams? A hypothesis that has been put forth several times is that overgrazing and other agricultural practices caused a decrease in vegetation and an increase in soil erosion. This explanation is attractive because the entrenchment began soon after settlement and agriculture were introduced. Other evidence, however, leads to a different conclusion. Analysis of climatic data from stations in the Southwest with long-time records suggests that precipitation in the 1880s and subsequent decades was unusually high. This may have been related to an anomalous El Niño event, a condition in the Pacific Ocean that alters weather patterns over broad areas. The causes for arroyo cutting could therefore have been climatic changes on a global or hemispheric scale. Additional supporting evidence comes from studies of the alluvial sediments in the walls of the arroyos. These studies, using radiocarbon dating of organic matter, have documented cycles of arroyo cutting and filling events dating back several thou-

**Figure 13-44** Appearance of Kanab Arroyo in 1992. Arroyo is about 25 m deep and 80–120 m wide.

sand years, long before the impacts of grazing occurred.

The final verdict on the cause of arroyo cutting is still not known. The case for natural climatic cycles is becoming stronger, but it is also possible that the effects were accentuated or localized by agriculture, dam building, and other human interferences with the complex fluvial system.

## FLOODING

Flooding is the most serious hazard posed by river processes (Figure 13-45). In 1972, 238 people were killed by flooding in Rapid City, South Dakota. During the same year, flooding in the eastern United States caused by precipitation from Hurricane Agnes claimed 113 lives and resulted in more than $3 billion in property damage. More recently, residents of the upper Mississippi valley will never forget the disastrous flood of spring and summer 1993, which was the largest recorded flood for many stations on the river. The explanation for the unprecedented flooding is that an unusual weather pattern became established over the region in the spring and stubbornly remained for the entire summer. The rains were generated in the frontal region between a cool air mass over the Great Plains and a warm humid air mass to the south. A large high-pressure system over the southeastern United States blocked the movement of the storms out of the upper Mississippi basin. The flooding began as spring rains in April and May—about 40 percent greater than average—fell onto soils that were saturated from the effects of heavy rains in autumn of the year before. Average rainfall totals in June and July were twice and two to three times the average, respectively (Figure 13-46). Individual areas experienced astonishing amounts of rainfall. For example, Skidmore, Missouri, reported 25.35 inches of rain in July and Leon, Iowa, received 20.68 inches for that month. These areas normally record about 4 inches in July and only about 35 inches for the entire year.

The result of the flooding was the inundation of farm land and urban areas on a vast scale. Levees built to protect these areas against moderate floods were obviously unable to cope with a flood of this magnitude. Although

**Figure 13-45** Flooding periodically damages structures built on unprotected flood plains.

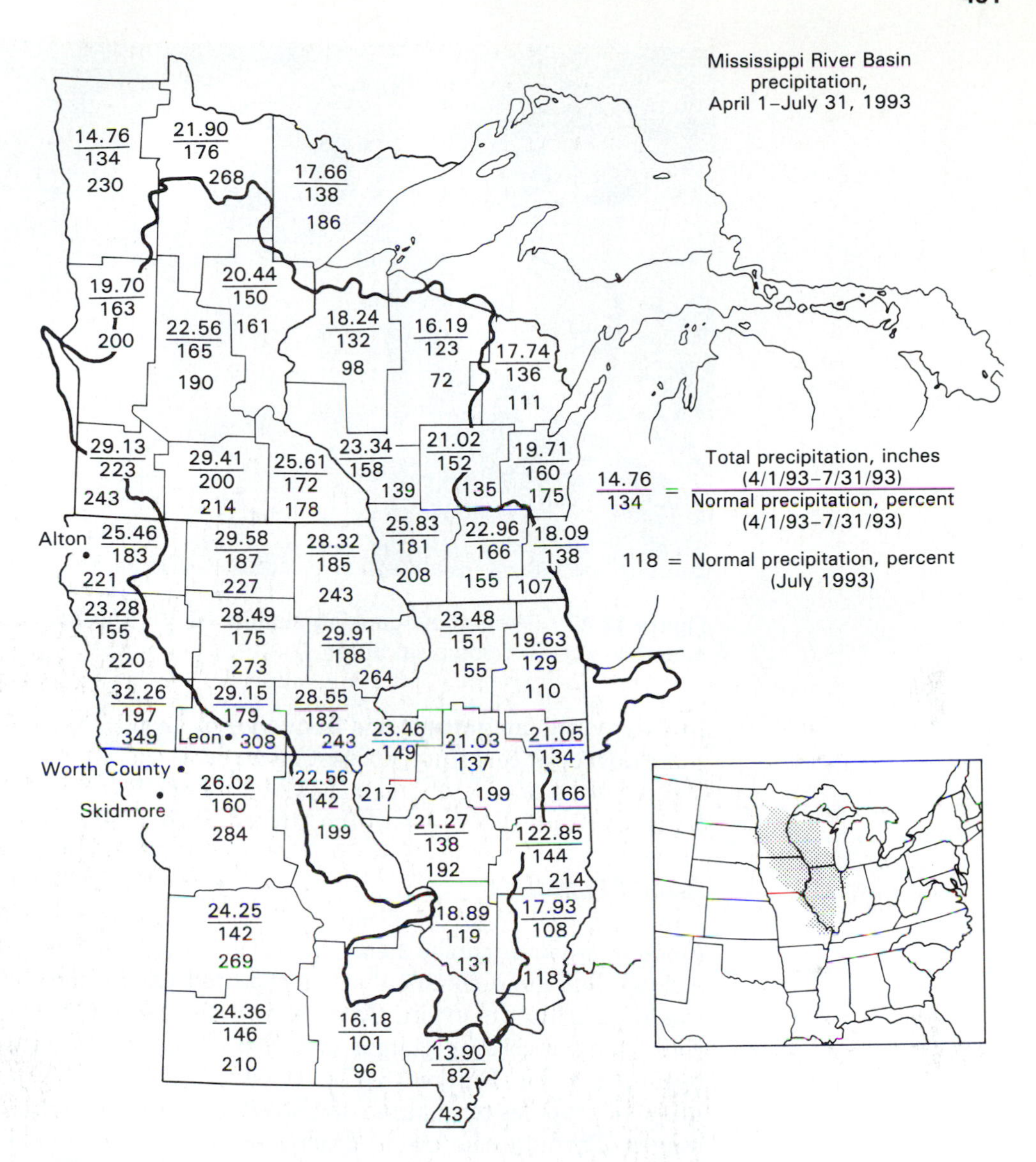


**Figure 13-46** Upper Mississippi River basin precipitation, April 1 to July 31, 1993. (From *Currents*, vol. 9, no. 1, fall 1993. Champaign, Ill.: Illinois State Water Survey.)

the U.S. Army Corps of Engineers was criticized both for levees that failed as well as for aggravating flood potential by its system of river management projects throughout the basin, the criticism was unwarranted in this case. A flood of this size would have caused problems for any concievable protective measures that could have been built. Nature's lesson is clear in this regard: We can protect ourselves against moderate events, but when extreme events occur (which will invariably happen over time), there is little that we can do other than get out of the way.

Flooding typifies many hazardous geologic processes in that the obstacles to mitigating the threat to human life and property are largely psychological rather than of a hydrologic or engineering nature. The public perception of floods is that they are "acts of God" that are impossible to predict or explain. People are reluctant to relocate their homes and businesses out of flood-prone areas because of events that they believe may never happen in their lifetimes.

**Figure 13-47** Streamflow gauging station. (B. W. Thomsen; photo courtesy of U.S. Geological Survey.)

In reality, the magnitude and frequency of floods can usually be predicted with some degree of certainty, at least on a statistical basis. In addition, flood-prone areas can be accurately identified and mapped. In fact, flood-plain maps exist for most developed areas in the United States.

### Flood Magnitude

*Flood magnitude* can be measured as the elevation to which a river rises during a flood, but more commonly it is reported as the maximum discharge of the stream during the event. Changes in river height, or *stage*, are measured by continuous-recording gauges installed along streams throughout most developed countries (Figure 13-47). At each gauging station, measurements are made of cross-sectional area and stream velocity through the range of possible stages. Through the use of Equation 13-5, a *rating curve* can be constructed for each station relating river stage to discharge. With the rating curve and the stage measurement made during a flood, a stream hydrograph can be made to show the relationship between discharge and time (Figure 13-48). The stage height hydrograph of a large flood in Grand Forks, North Dakota, in 1966 is shown in Figure 13-49). The crest of the 1966 flood represents a discharge of 55,000 cubic feet per second.

Variables that control the magnitude of a flood can be divided into two types: transient factors that depend on the meteorological conditions, and permanent factors that deal with the hydrology and geology of the drainage basin. There are a large number of meteorological conditions that must be considered. If the flood is caused by a rainfall event, the factors include the size, direction, and rate of movement of the storm; the intensity, duration, and location of the rain; and the moisture conditions prevailing in the soil before the storm. If the flood is caused by a snow-melt event, transient factors include amount of snow, rate of melting, and the soil-moisture conditions. Permanent controls within a basin include type and distribution of vegetation, topography, type of soil and rock at the surface, and type of drainage network present in the basin. From the number of factors involved, it is easy to understand why it is so difficult to predict the magnitude of a particular flood so that safety precautions

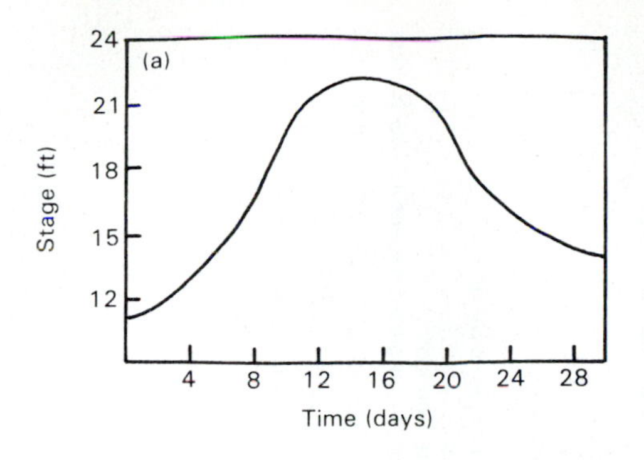


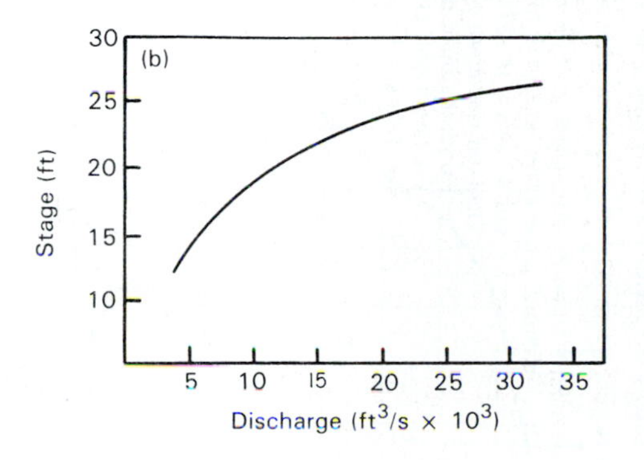


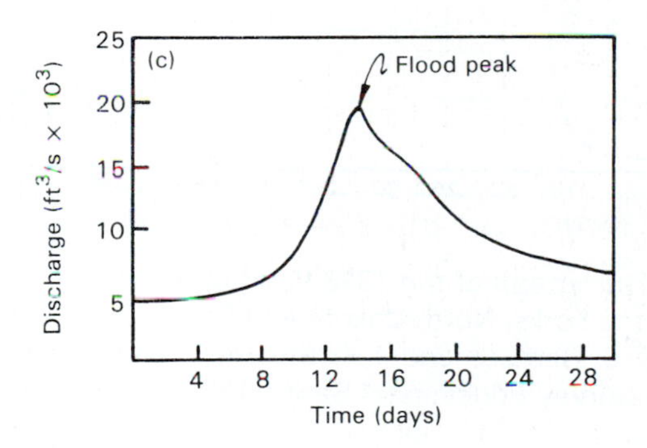


**Figure 13-48** (a) Hydrograph of river stage versus time during a flood. (b) A rating curve for a gauging station. (c) A hydrograph showing discharge versus time constructed from (a) and (b).

can be taken. The most basic method of predicting the peak discharge from a particular rainfall event is by use of equations such as the *Rational Equation*, which can be stated as

$$Q = CIA \qquad \text{(Eq. 13-10)}$$

where $Q$ is peak discharge in cubic feet per second, $C$ is a runoff coefficient whose values are given in Table 13-2, $I$ is the rainfall intensity in inches per hour, and $A$ is basin area in acres.

### *Flood Frequency*

Perhaps as important as predicting the magnitude of a flood generated by a particular rainfall event is the problem of predicting the frequency of large floods. Because it is impossible to predict when a flood of a certain magnitude

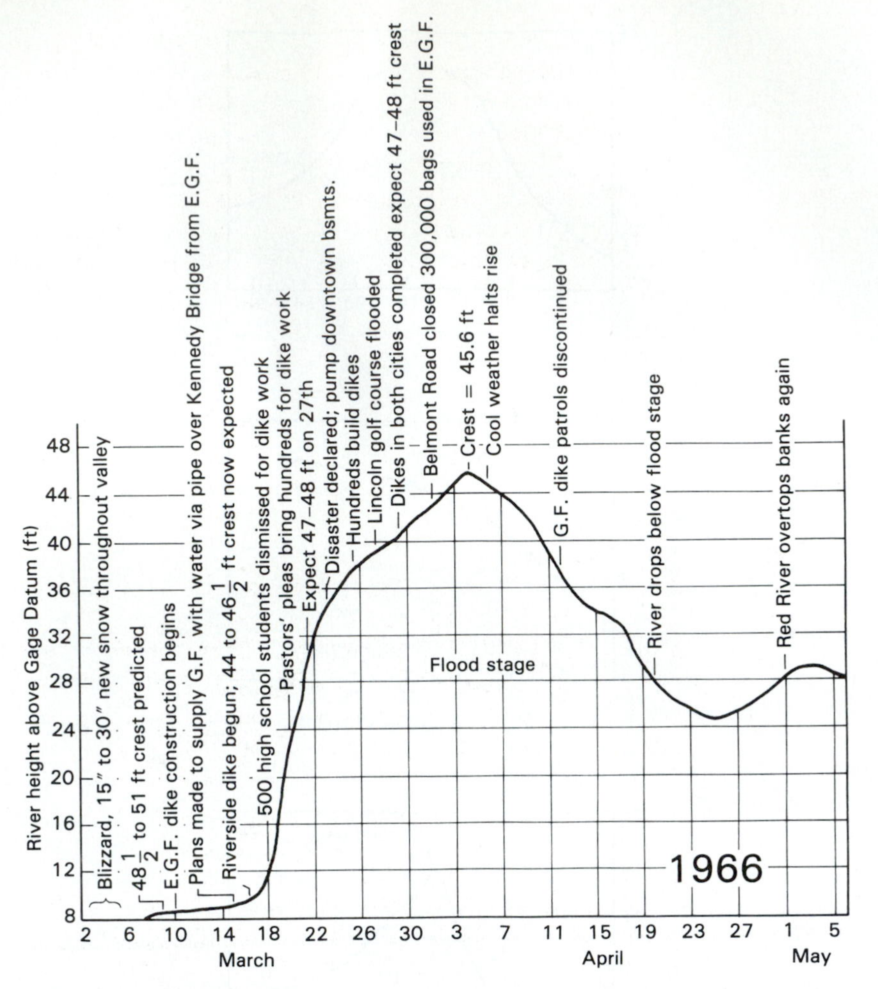


**Figure 13-49** Hydrograph of the 1966 flood of the Red River of the North at Grand Forks, North Dakota and East Grand Forks, Minnesota. (From S. S. Harrison and J. P. Bluemle, 1980, North Dakota Geological Survey Educational Series 12.)

**TABLE 13-2**

**Values of *C* (runoff coefficient) for the Rational Equation**

| *C* | *Land cover or land use* |
|---|---|
| 0.1–0.3 | Forest |
| 0.1–0.5 | Grasslands |
| 0.3–0.6 | Row crops, plowed ground |
| 0.5–0.8 | Bare ground, smooth |
| 0.5–0.9 | Suburban lands |
| 0.7–0.95 | Urban lands |
| 0.9–0.99 | Water body, full |
| 0.0 | Water body, empty |

SOURCE: From C. C. Mathewson, *Engineering Geology*, copyright © 1981 by Charles E. Merrill Publishing Co., Columbus, Ohio.

will occur, the only alternative is to attempt to determine the probability of a particular flood within a given year. A general relationship that seems to hold true is that the magnitude of natural processes is inversely proportional to the probability of their occurrence. Thus the larger the flood under consideration, the less likely it is to happen in a particular amount of time.

Another way of defining frequency is in terms of *recurrence interval* (R.I.), which is the average period of time between floods of a certain magnitude. It is related to probability through the following relationship:

$$\frac{1}{p} = \text{R.I.} \qquad \text{(Eq. 13-11)}$$

Thus if a flood has a probability ($p$) of 0.01 of occurring in a single year, its recurrence interval is 100 years. It cannot be overemphasized that the recurrence interval refers to a statistical average over a long period of time. A town that experiences a 100-year flood has the same risk of a similar flood in every following year.

Flood-frequency curves are established from gauging-station records to show the relationship between magnitude and frequency. A basic way of constructing these diagrams is to plot magnitude versus recurrence interval for the maximum discharge for each year of the historical record (Figure 13-50). Recurrence interval (R.I.) is established by ranking each yearly discharge from highest to lowest and using the relationship

$$\text{R.I.} = \frac{N+1}{M} \qquad \text{(Eq. 13-12)}$$

where $N$ is the number of years of record and $M$ is the magnitude rank. This method can be successful only after many years of record are available. After the points are plotted on a flood-frequency diagram, a curve is fitted to the points so that the record can be extrapolated to higher recurrence intervals. Standard equations are available for fitting the plotted points. The greater the

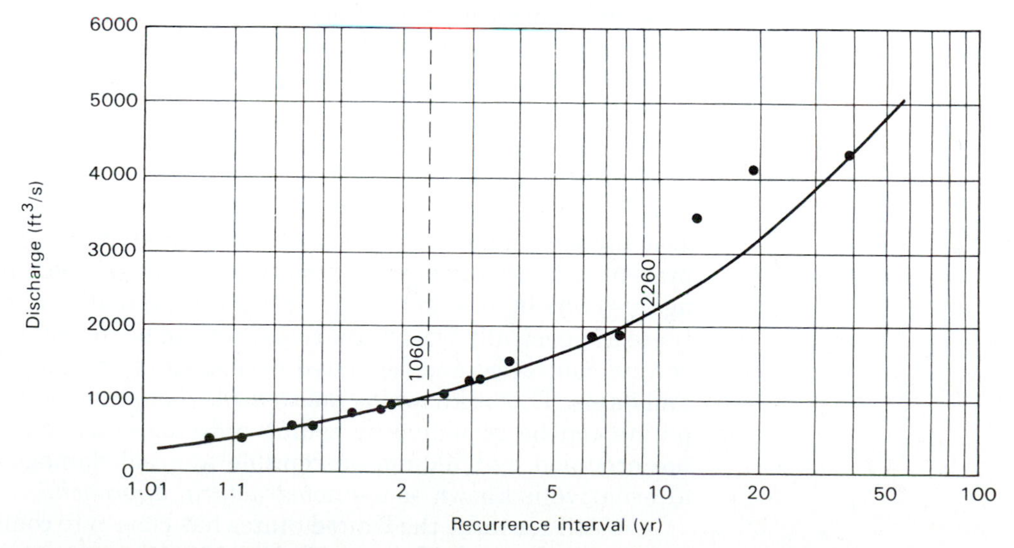


**Figure 13-50** A flood-frequency diagram for Big Piney Run near Salisbury, Pennsylvania. (From T. Dalrymple, 1960, U.S. Geological Survey Water Supply Paper 1543-A.)

extrapolation, however, the more uncertain are the predicted flood magnitudes and recurrence intervals.

**EXAMPLE 13-2**

The table below shows the five largest annual flood discharges for Seneca Creek at Dawson, Maryland, for the period 1928–1958. Calculate the recurrence interval for the 1953 flood.

| *Year* | *Discharge* ($m^3/s$) | *Rank Order* |
|---|---|---|
| 1956 | 427.7 | 1 |
| 1933 | 263.3 | 2 |
| 1953 | 207.6 | 3 |
| 1928 | 107.6 | 4 |
| 1958 | 103.1 | 5 |

***Solution.*** The record contains 31 years of annual flood peaks. For the 1953 flood,

$$\text{R.I.} = \frac{N+1}{M} = \frac{31+1}{3} = 10.7 \text{ years}$$

### *Effect of Land-Use Changes*

The flood-frequency analysis presented involves the assumption that flood magnitudes are random samples from a population of possible events and that the underlying factors that cause floods do not change. As Case Study 13-3 illustrated, however, we know that yearly mean temperature and precipitation amounts can change through time, making floods more or less frequent and severe for a certain period of time. Land-use changes that have been mentioned also influence flood magnitude and frequency. Most important among possible land-use changes is urbanization. The cumulative effects of urbanization, including less infiltration and more rapid runoff, can alter a flood hydrograph in the manner shown in Figure 13-51. The result is flooding of greater magnitude and shorter lag time. Such floods are termed *flashy*. The flood-frequency analysis made from natural conditions in the stream basin before urbanization may not accurately predict the magnitude-frequency relationships after urbanization.

### *Flood Control*

The unpredictability of floods and the great damages that they can cause present a major dilemma to society. How can we best protect our population against a nebulous and poorly understood threat? In answer to this question there are basically two alternatives that can be considered. First, river systems can be controlled and regulated by means of channel improvements, levees, and dams. This is known as the *structural approach*. Second, land use on flood plains can be regulated by zoning ordinances so that flood-prone areas are not occupied by buildings susceptible to great damage during flooding. This alternative is known as the *nonstructural approach*.

Until recently, the United States has chosen to combat floods through the structural approach. Costa and Baker (1981) report that the federal government had spent more than $14 billion on flood control between 1936 and 1981. The advantage of the structural approach is that flood-control projects can offer

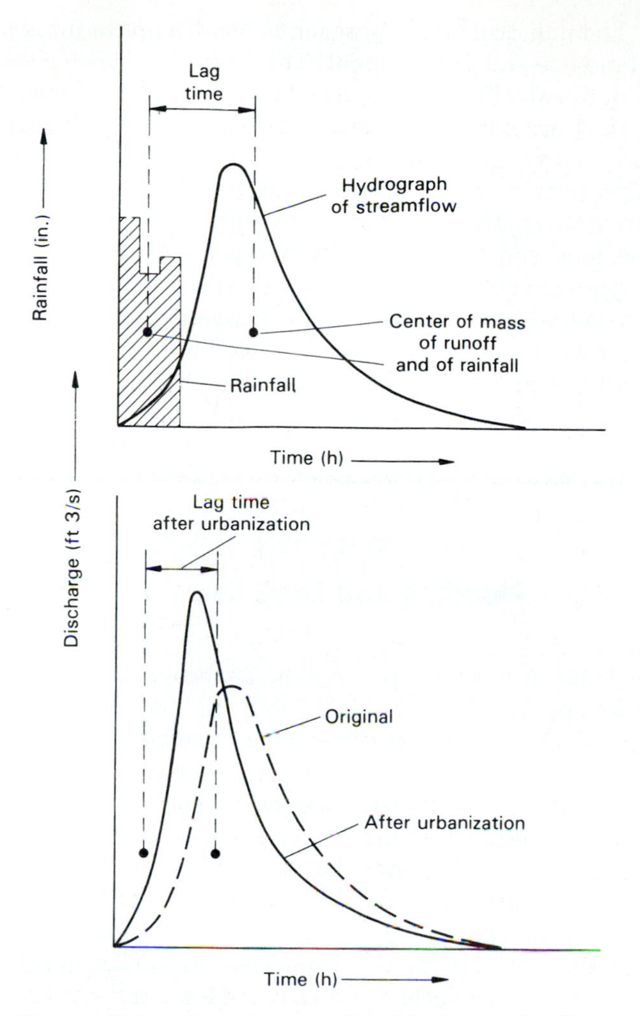


**Figure 13-51** Changes in a flood hydrograph after urbanization of a drainage basin. (After L. B. Leopold, 1968, U.S. Geological Survey Circular 554.)

protection against floods of a certain magnitude (the design flow) if they are properly designed and constructed. In addition, they are politically attractive to legislators because they create jobs in a flood-prone area and are perceived by the public as a positive step toward solving the problem. Unfortunately, there are a number of disadvantages to flood-control structures. Most important, a flood greater than the magnitude for which the structure is designed can cause greater damage than would have occurred without the structure. If a dam fails, for example, the resulting flood can assume catastrophic proportions. Most dam failures have been caused by inadequate site investigations or design relative to the geologic conditions at the site. The St. Francis and Teton Dam failures are prime examples.

We have already discussed the potential effect of altering river channels by levees, channelization, or other projects. These projects change the natural balance of the stream and cause it to adjust other variables in the system.

A final comment about structural measures is that they may actually encourage development of flood-prone areas. If people believe that there is no danger from floods, they will readily develop flood plains. This only increases the potential for loss and damage in the event of a rare flood of large magnitude.

The nonstructural approach is an attempt to integrate geologic processes into land use and development. The method of flood protection in this approach is to delineate flood-prone areas and restrict development in those areas to uses that are compatible with periodic flooding. In urban areas, these uses include parks, golf courses, and other natural or open areas. In addition to flood protection, these zones enhance the aesthetic value of a river in a crowded urban setting. The costs for land acquisition for these uses may even be less in the long run than the costs of repeated flood damage. The limitations to this approach are that it requires the relocation of homes and businesses. This is considered by some to be an infringement of their right to live wherever they please. In highly developed areas, it may be possible to apply this approach only after a major flood has occurred.

## CASE STUDY 13-4
## Flooding and Land Use

Two examples that illustrate differing approaches to flood control are provided by the cities of Winnipeg, Manitoba, and Rapid City, South Dakota.

The city of Winnipeg lies at the confluence of the Assiniboine river and the Red River of the North (Figure 13-52), both of which flood regularly. Topographically, the area is extremely flat because it is the floor of a glacial lake basin. Large areas of the city are inundated by floods because of the shallow valleys of the rivers and the low relief of the areas adjacent to the rivers. Floods in Winnipeg generally occur only during the spring snow melt. Snow depths and the rate of melting are critical factors in determining whether or not a flood will be forthcoming. After the devastating flood of 1950, the city government decided to construct a floodway to divert flood waters from the Red River of the North around the city. An inlet control structure in the channel of the Red River consists of a gate that can be raised to block the flow of the river. Water then rises behind the structure and flows into the floodway. The artificial channel, constructed in 1966, has an average depth of 9 m, a maximum width of 18 m, and can carry a maximum discharge of 1700 $m^3/s$ (Figure 13-53). The floodway, along with other structural and nonstructural flood-protection measures, has been successful in preventing flood damage in the city.

Rapid City, South Dakota, lies along

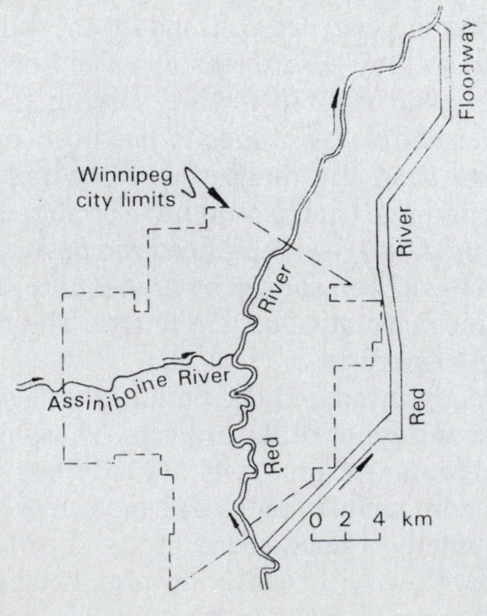


**Figure 13-52** Map of the Winnipeg, Manitoba, area showing the location of the Red River floodway.

**Figure 13-53** The Red River floodway.

Rapid Creek at the base of the Black Hills. On the night of June 9, 1972, heavy thunderstorms over the Black Hills dumped as much as 38 cm of rain in less than 6 h. The result was a 100-year flood that killed 238 people and cost $128 million in property damage (Figure 13-54). In response to this event, the city adopted a nonstructural plan for flood control (Rahn, 1984). The flood plain today consists of parks, natural areas, golf courses, and other recreational-area developments. A few large commercial buildings were allowed to remain. The main objective was to remove private dwellings from the flood-prone area. As the mayor of Rapid City remarked, "Anyone who sleeps on a flood plain is crazy." Unfortunately, outside of the city limits, areas along Rapid Creek that were devastated by the 1972 flood were rebuilt with the same type of structures in the same hazardous locations. New construction also has occurred in the flood plain. The "act of God" mentality is difficult to combat.

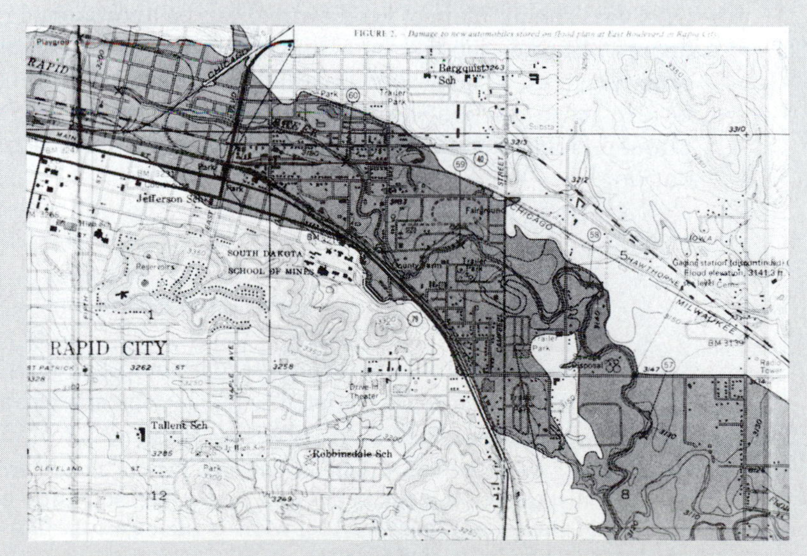


**Figure 13-54** Map of part of Rapid City, South Dakota, showing extent of flooding (dark area) June 9–10, 1972. (From O. J. Larimer, 1973, U.S. Geological Survey Hydrologic Investigations Atlas 511.)

## SUMMARY AND CONCLUSIONS

Individual rivers are components of an integrated stream network in a drainage basin. A stream network develops a particular spatial stream pattern that is controlled by the geology, soils, topography, climate, and vegetation in a drainage basin. The concept of stream order has led to quantitative analysis of the variables characterizing stream networks. A hydrologic budget can be established for a drainage basin as a whole.

The flow of water in rivers can be studied using hydraulic parameters such as Reynolds and Froude numbers. The Chezy and Manning equations relate the velocity of the stream to the gradient, channel shape, and boundary roughness. The velocity and shear stress, in turn, control the entrainment and transportation of stream sediment. Sediment is transported as bed load, suspended load, and dissolved load until the velocity drops and deposition removes the suspended and bed load. Deposition and other stream characteristics depend on whether the river is meandering or braided. These stream types result in different sediment distributions and landforms. Alluvial fans and deltas develop at points of major change in gradient in stream systems, where much of the stream load is deposited in a valley or in an ocean or lake.

All the geologic and climatic variables that control the discharge and sediment load in a stream can be considered as a system that can achieve a state of equilibrium when considered over a period of years. If natural or artificial changes in equilibrium are induced, the system attempts to reestablish an equilibrium condition. In this process many of the variables may be affected. Stream terraces constitute evidence that natural changes in equilibrium have occurred. Urbanization, dams, and other human alterations to the system may produce far-reaching changes to the channel and to the drainage basin.

Flooding is a natural, periodic process in river systems. Flood magnitude depends on a large number of geologic and climatic factors. Flood-frequency graphs show the inverse relationship between flood magnitude and recurrence interval. The problem of flood control can be approached in two ways. A certain degree of protection can be achieved by the construction of dams, levees, and other projects. This method is called the structural approach. A better alternative in most cases is the nonstructural approach, whereby land use on flood plains is restricted to those uses that are most compatible with periodic inundation.

### PROBLEMS

1. Why do hydrologists use the drainage basin in the study of river systems?
2. How is drainage density related to the geology of a drainage basin?
3. How does stream capture modify river systems?
4. If Niagara Falls maintains an average rate of headward retreat of 1 m/yr, and the distance to the outlet of Lake Erie is 29 km, when will the falls reach Lake Erie? Based on Figure 13-10b, what is the eventual fate of Lake Erie?
5. How do the Reynold's and Froude numbers characterize flow in rivers?
6. If the depth of a stream is 2 m, what must the velocity be to attain critical flow?
7. What factors determine the velocity of flow in a stream?
8. A particular channel has a semicircular cross-section with a radius of 5 m. If the

slope is $10^{-3}$ and Manning's roughness coefficient is 0.04, what is the stream discharge?

9. Describe the two parameters that are used to predict entrainment of particles as flow conditions change.
10. A channel cross-section has a hydraulic radius of 1.6 m and a slope of 0.002. Calculate the critical shear stress exerted by the stream and the largest particle size that could be transported, based on the relationship shown in Figure 13-18. (Figure 13-18 uses units of $kg_{force}/m^2$. One $kg_f$ = 9.8 N.)
11. Describe the ways in which sediment is transported in streams.
12. Compare and contrast braided and meandering stream systems.
13. What are the similarities and differences between deltas and alluvial fans?
14. How can a stream be considered to be in a state of equilibrium when discharge, velocity, and other parameters are constantly changing?
15. How do human activities affect equilibrium?
16. Two nearby drainage basins have areas of 5 acres. One basin is urbanized and the other basin is forested. Calculate the stream discharge from each basin for a 1 in./h rainfall.
17. A steam gauge has 89 years of record. What is the recurrence interval of the fourth largest flood recorded?
18. What is the magnitude of the 50-year flood on Big Piney Run, Pennsylvania? What is the magnitude of the 100-year flood? What are the problems involved in making the latter estimate?

## REFERENCES AND SUGGESTIONS FOR FURTHER READING


AMERICAN FALLS INTERNATIONAL BOARD. 1974. *Preservation and Enhancement of the American Falls at Niagara.* Final Report to Niagara International Joint Commission (Appendix C—Geology and Rock Mechanics, and Appendix D—Hydraulics.)

BAKER, V. R., and D. F. RITTER. 1975. Competence of rivers to transport coarse bedload material. *Geological Society of America Bulletin,* 86:975–978.

BLOOM, A. L. 1978. *Geomorphology.* Englewood Cliffs, N.J.: Prentice-Hall, Inc.

BRUSH, L. M., Jr. 1961. *Drainage Basins, Channels, and Flow Characteristics of Selected Streams in Central Pennsylvania.* U.S. Geological Survey Professional Paper 282-F.

CALKIN, P. E., and P. J. BARNETT. 1990. Glacial geology of the eastern Lake Erie basin, in *Quaternary Environs of Lakes Erie and Ontario,* D. I. McKenzie, ed. Waterloo, Ont.: Escart Press.

CALKIN, P. E., and T. A. WILKINSON. 1982. Glacial and engineering geology aspects of the Niagara Falls and Gorge, in *Geology of the Northern Appalachian Basin, Western New York,* E. J. Buehler and P. E. Calkin, eds. Field Trips Guidebook for the New York State Geological Association, 54th Annual Meeting, Amherst, N.Y., pp. 247–245.

CHEN, A. 1983. Dammed if they do and dammed if they don't. *Science News,* 123:204–206.

COSTA, J. E., and V. R. BAKER. 1981. *Surficial Geology: Building with the Earth.* New York: John Wiley.

DALRYMPLE, T. 1960. *Flood-frequency Analysis.* U.S. Geological Survey Water Supply Paper 1543-A.

DANIELS, R. B. 1960. Entrenchment of the Willow drainage ditch, Harrison County, Iowa. *American Journal of Science,* 225:161–176.

HANSEN, D. E., and J. Kume. 1979. *Geology and Ground Water Resources of Grand Forks County, North Dakota. Part I: Geology.* North Dakota Geological Survey Bulletin 53, and North Dakota State Water Commission County Ground Water Studies 13.

HARRISON, S. S., and J. P. BLUEMLE. 1980. *Flooding in the Grand Forks–East Grand Forks Area.* North Dakota Geological Survey, Educational Series 12.

HOWARD, A. D. 1967. Drainage analysis in geologic interpretation: A summation. *American Association of Petroleum Geologists Bulletin*, 51:2246–2259.

JUDSON, S., M. E. KAUFFMAN, and L. D. LEET. *Physical Geology*, 7th ed. Englewood Cliffs, N.J.: Prentice-Hall, Inc.

KOLB, C. R., and W. G. SHOCKLEY. 1957. Mississippi Valley geology: Its engineering significance. *Journal of Soil Mechanics and Foundations*, American Society of Civil Engineers, 83 (No. SM3):1–14.

LARIMER, O. J. 1973. *Flood of June 9–10, 1972, at Rapid City, South Dakota.* U.S. Geological Survey Hydrologic Investigations Atlas 511.

LEOPOLD, L. B. 1968. *Hydrology for Urban Land Planning: A Guidebook on the Hydrologic Effects of Urban Land Use.* U.S. Geological Survey Circular 554.

LEOPOLD, L. B., and M. G. WOLMAN. 1957. *River Channel Patterns: Braided, Meandering, and Straight.* U.S. Geological Survey Professional Paper 282-B.

MATTHEWS, R. K. 1984. *Dynamic Stratigraphy*, 2d ed. Englewood Cliffs, N.J.: Prentice-Hall, Inc.

MATHEWSON, C. C. 1981. *Engineering Geology.* Columbus, Ohio: Charles E. Merrill.

MCKNIGHT, T. L. 1984. *Physical Geography: A Landscape Appreciation.* Englewood Cliffs, N.J.: Prentice-Hall, Inc.

MORGAN, J. P. 1970. Deltas: A résumé. *Journal of Geologic Education*, 18:107–117.

RAHN, P. H. 1984. Flood-plain management program in Rapid City, South Dakota. *Geological Society of America Bulletin*, 95:838–843.

RENDER, F. W. 1970. Geohydrology of the metropolitan Winnipeg area as related to groundwater supply and construction. *Canadian Geotechnical Journal*, 7:244–274.

RITTER, D. F. 1986. *Process Geomorphology*, 2d ed. Dubuque, Iowa: Wm. C. Brown.

SIMONS, D. B., and E. V. RICHARDSON. 1966. *Resistance to Flow in Alluvial Channels.* U.S. Geological Survey Professional Paper 422-J.

STRAHLER, A. N. 1957. Quantitative analysis of watershed geomorphology. *Transactions, American Geophysical Union*, 38:913–920.

WEBB, R. H., S. S. SMITH, and V. A. S. MCCORD. 1992. Arroyo cutting at Kanab Creek, Kanab, Utah, in *Paleoflood Hydrology of the Southern Colorado Plateau: Field Trip Guidebook*, J. Martinez-Goytre and W. M. Phillips, eds. Tucson, Ariz.: Arizona Laboratory for Paleohydrological and Hydroclimatological Analysis, University of Arizona, pp. 17–39.

# 14

# Oceans and Coasts

The importance of the oceans can be appreciated when we realize that more than 70% of the earth's surface is covered by water. The significance of the oceans, however, goes well beyond their areal extent. Life on the earth began in the oceans. Today, microscopic organisms in the sea form the base of the oceanic food chain, which supports the population of fish and other higher organisms in the oceans. This huge reservoir of food is desperately needed by the exponentially growing human population.

Circulation of the oceans exerts an important influence on the earth's climate. Our ability to sustain agriculture on land can be directly related to interactions between oceanic and atmospheric circulation.

In recent decades we have begun to tap the vast storehouse of mineral and energy resources in the oceans. Throughout geologic time the remains of tiny marine organisms have accumulated to form the earth's supplies of petroleum. Exploration for petroleum is now focusing on offshore targets. The problems involved in the construction of offshore drilling platforms and other structures make an understanding of the oceans a matter of practical importance.

Coastlines, where the oceans and continents meet, are among the most dynamic areas of the earth. The combination of active geologic processes and dense development of coastal areas leads to a wide spectrum of engineering and environmental problems. The recent occurrence of large oil spills from tanker accidents has highlighted the need to understand the geologic, chemical, and biological interactions in coastal waters.

## THE OCEAN BASINS

### Oceanic Circulation

Large-scale motions of seawater are partly driven by prevailing winds over the sea surface. Surface currents, the Gulf Stream, for example, travel long distances and transport huge quantities of water. Individual currents are actually parts of huge loops called *gyres* (Figure 14-1). These currents are important mechanisms for transferring heat, in the form of warm water, from the equatorial regions toward the poles.

The oceans also undergo vertical circulation. When surface currents diverge, or flow away from a continent, deep water moves upward to replace the surface water. This movement, known as *upwelling* (Figure 14-2), is of great economic importance because deep water is rich in nutrients. Organisms, including fish, proliferate in areas of upwelling. The opposite of upwelling is sinking. Vertical movement is also driven by density differences. Cold polar water sinks and moves very slowly toward the equator. North Atlantic deep water is forced to flow over the even denser Antarctic bottom water (Figure 14-2).

### Topography

The modern study of the oceans began with the voyage of the research vessel H.M.S. *Challenger* in 1872. Prior to this voyage, the ocean basins were assumed to be broad, featureless plains. Instead, the *Challenger* expeditions brought back evidence of huge submarine mountain ranges, deep trenches, and other topographic features. Detailed study of the ocean basins accelerated during

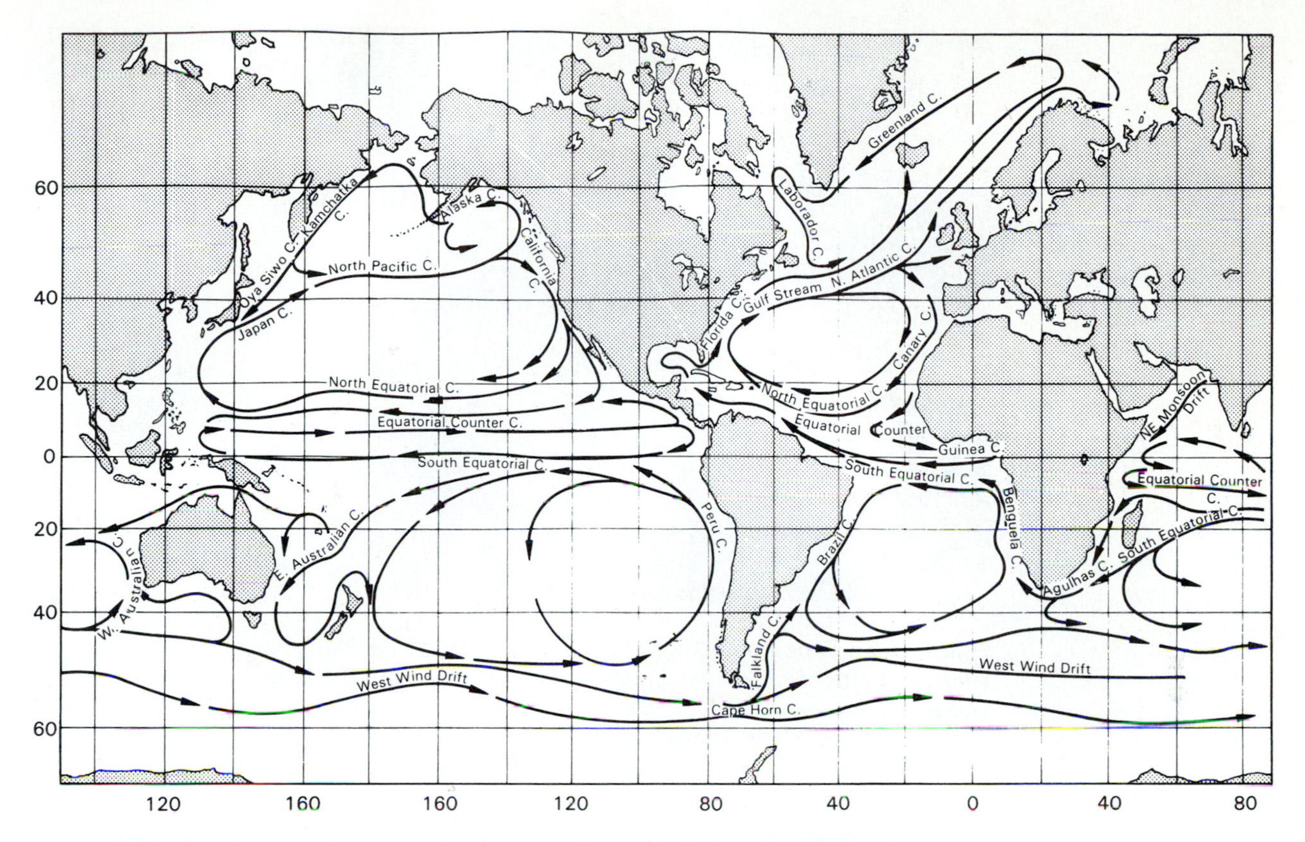


**Figure 14-1** Current movement in the oceans. (From S. Judson, M. E. Kauffman, and L. D. Leet, *Physical Geology* 7th ed., copyright © 1987 by Prentice-Hall, Inc., Englewood Cliffs, N.J.)

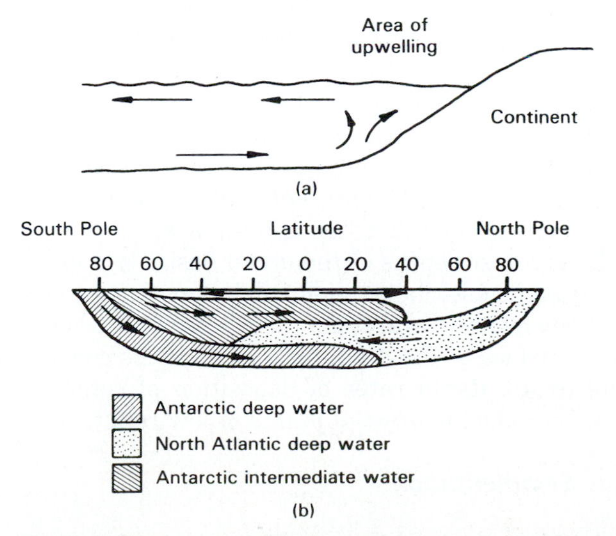


**Figure 14-2** (a) Upwelling caused by surface flow away from a continent. (b) Vertical distribution of selected water masses.

**Figure 14-3** Topographic features of the ocean basins. Notice position and size of midoceanic ridges. (Reproduced from the *World Ocean Floor* panorama by Bruce C. Heezen and Marie Tharp, copyright © 1977 by Marie Tharp.)

and after World War II. This work has provided much of the evidence for plate tectonics.

The topography of ocean basins is dominated by midoceanic ridges (Figure 14-3). These ridges rival the size and extent of continental mountain ranges. At the crest of the midoceanic ridges lie the faulted rift valleys, where new oceanic crust is formed in the sea-floor spreading process.

The ocean bottom slopes downward away from the rugged midoceanic ridges into a region of more subdued relief containing *abyssal hills*. The relatively high elevation of the midoceanic ridges is the result of volcanism and thermal expansion of the lithosphere; as newly formed crustal material moves away from the ridge crest it cools and subsides. Isolated volcanic mountains, or *seamounts*, rise above the floor of the oceans in many places. Some seamounts have flat tops rather than the typical conical shape of a volcano. Drilling of these flat-topped seamounts, termed *guyots*, has produced evidence to explain their emergence in the geologic past. The flat-topped form is attributed to wave erosion near sea level. The present depth of guyots is an indication of the great amount of subsidence that has taken place.

The greatest depths of the ocean basins are located at the *trenches*, long narrow troughs associated with subduction zones. Oceanic crust is pushed or dragged downward toward the mantle beneath these linear troughs.

The surface of the ocean bottom changes considerably in the vicinity of the continents. Rapid rates of deposition of continental sediment bury the landscape, producing *abyssal plains* of low relief.

### *Oceanic Sedimentation*

Sediments deposited on the ocean floors beyond the influence of the continents, called *pelagic sediments*, consist of inorganic and organic types. The inorganic pelagic sediments are composed mostly of clay-size ($<0.004$ mm) particles of

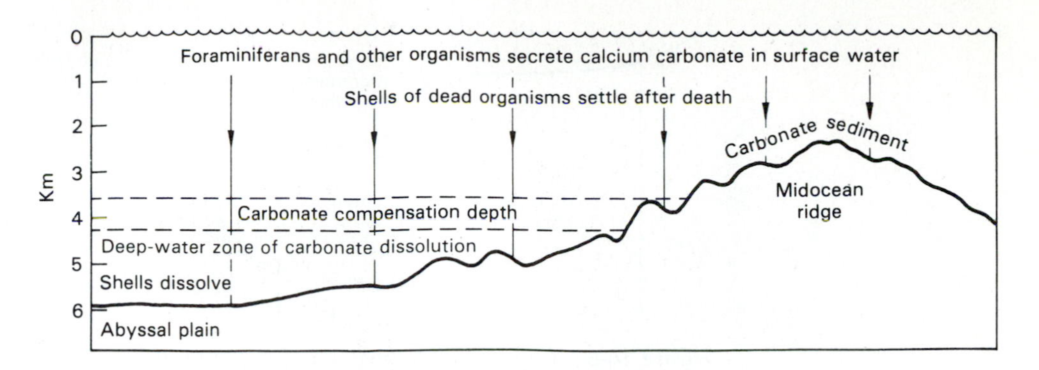


**Figure 14-4** Carbonate compensation depth in the oceans. (From F. Press and R. Siever, *Earth*, 3d ed., copyright © 1982 by W. H. Freeman and Co., San Francisco.)

quartz, feldspar, clay, and other common rock-forming minerals. These particles are originally derived from the continents but can be transported great distances by ocean currents or winds because of their extremely small size. Large areas of the ocean floors are covered by red or brown muds composed of the inorganic pelagic sediments. Sedimentation rates are very low, often less than 1 cm per 1000 years.

Organic sediments are also deposited on the ocean floor. These include the shells of microscopic organisms living near the surface of the ocean, as well as fecal pellets and other organic remains. The shells secreted by organisms are composed of calcium carbonate or silica. The shells of organisms that secrete calcium carbonate, such as the one-celled foraminifera, are the most abundant in the oceans, although these shells are never found in sediments deposited below a water depth of approximately 4 km. The reason for this is that deeper waters tend to be undersaturated with respect to calcium carbonate, and therefore these deep waters steadily dissolve shells that fall to the bottom. The depth at which seawater becomes undersaturated is called the *carbonate compensation depth* (Figure 14-4). Carbonate-rich *foraminiferal oozes* are common where the ocean bottom lies above this critical level. *Radiolarian* or *diatom oozes* are deposited where organisms that secrete siliceous shells are abundant. These sediments are not affected by the carbonate compensation depth.

## CONTINENTAL MARGINS

### Topography

The margins of the continents have several topographic divisions (Figure 14-5). Extending seaward from the coastline are gently sloping platforms called *continental shelves*. Continental shelves vary in width from less than 2 km to more than 1000 km. Although arbitrary depth limits have been used to define the extent of the shelves, the best method is to define the limit of the continental shelf as the point where the slope suddenly increases. A water-depth range of 200 to 650 m will include a large percentage of the shelves.

Continental shelves are correctly considered to be parts of the continents submerged by the present elevation of sea level. A variety of sediment types covers the shelves, although the sediments were reworked and redistributed by waves and currents during the postglacial sea-level rise. Valleys called

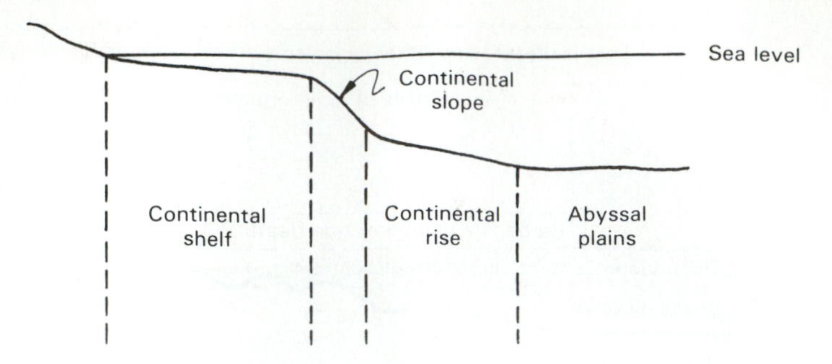


**Figure 14-5** Topographic divisions of the continental margins.

*submarine canyons* cross the continental shelves. Most of these valleys are the offshore continuation of river valleys on land. Many submarine canyons were cut across the continental shelves during Pleistocene time when rivers flowed across the exposed shelves.

The economic value of the continental shelves is very great because of the fishing industry that they sustain and the petroleum reserves that they overlie. Construction of offshore drilling platforms on the continental shelves is an engineering problem that requires detailed knowledge of the distribution and engineering properties of shelf sediments. The platforms must be designed to withstand the huge waves that are generated as storms pass over the continental shelves.

The continental shelves terminate at the upper margins of the *continental slopes*. The most steeply sloping segments of the continental slopes average 70 m per kilometer in comparison to the mean value of about 2 m per kilometer for the slope angle of the shelves. Submarine canyons dissect the continental slopes, although these valleys could not have been eroded by rivers because the sea level has never dropped low enough to expose them.

At the base of the continental slopes, the slope angle again decreases to a more gentle decline along the surface of the *continental rises*. The continental rises gradually merge seaward with the flat abyssal plains of the ocean floors.

### *Turbidity Currents*

The continental slopes and rises are blanketed by thick accumulations of continental sediment that was transported to the edge of the continental shelves by rivers and wave action during the recent geologic past. On the continental slopes these unconsolidated sediments are somewhat unstable at the greater slope angle found on that part of the continental margin. Slumps are common, sometimes triggered by earthquakes.

When slope movements occur on the continental slope, a very important type of current can be generated (Figure 14-6). The current consists of a dense cloud of suspended sediment that accelerates and flows rapidly downslope because it is more dense than the surrounding seawater. The name given to this type of flow is *turbidity current.* When turbidity currents flow out onto the continental rises, the velocity decreases because of the decrease in slope. Particles begin to drop out of suspension in order of decreasing size as the current slows, so that graded bedding develops in the layer of sediment deposited by the turbidity current.

Dramatic evidence for the existence of turbidity currents was provided when data were analyzed from the 1929 earthquake of the Grand Banks of Newfoundland. The data included the times of breakage of submarine transat-

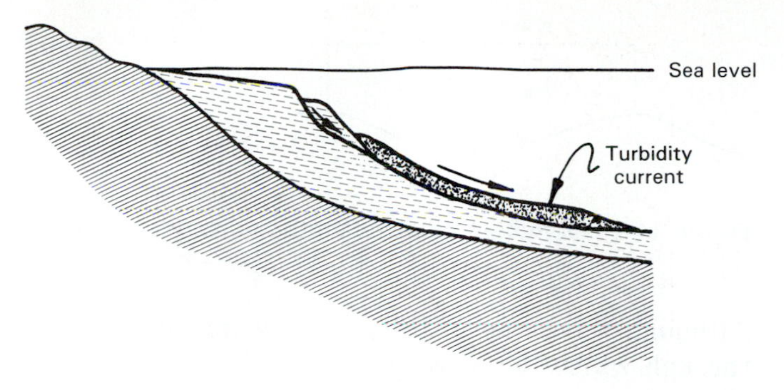


**Figure 14-6** Generation of a turbidity current by slumping in the sediments of the continental slope.

lantic telephone cables following the earthquake. Some of the cables were broken hours after the event, so the earthquake was ruled out a direct cause. Cables were broken sequentially with increasing distance from a point on the continental slope. The only reasonable explanation for this pattern of cable breakage is that the earthquake generated a powerful turbidity current that flowed down the continental slope at a velocity of 40 to 55 km/h, systematically breaking each cable it crossed along its path.

Turbidity currents are important because they transport huge quantities of continental sediment onto the adjacent abyssal plains to form *deep-sea fans.* The submarine canyons incised into the continental slopes are most likely eroded by these dense, sediment-laden flows. Ancient sequences of marine sedimentary rocks exposed on land have now been interpreted to be the result of turbidity-current deposition (Chapter 4). These sequences of turbidites consist of alternating layers of sandstone or siltstone and shale (Figure 4-24). The coarse-grained beds display graded bedding and are interpreted to be deposits of turbidity currents. The interbedded shales represent normal pelagic sedimentation between turbidity currents. The wide distribution of these sedimentary sequences attests to the importance of turbidity currents in marine sedimentation along continental margins.

## SHORELINES

### Waves

Among the processes that shape our coastlines, those involving waves are probably the most important. The impingement of waves upon a shoreline governs coastal erosion, the formation of beaches, and the distribution of sediment in the shoreline environment. Coastal engineering and management therefore must be based on an understanding of wave processes.

Generation and Movement. The basic parameters used to describe wave motion are illustrated in Figure 14-7. *Wavelength* is the distance between adjacent crests or troughs. The amount of time between passage of successive crests relative to a fixed point is the *wave period.* The *wave velocity*, $C$, then, is simply the wavelength, $L$, divided by the period, $T$:

$$C = \frac{L}{T} \quad \text{(Eq. 14-1)}$$

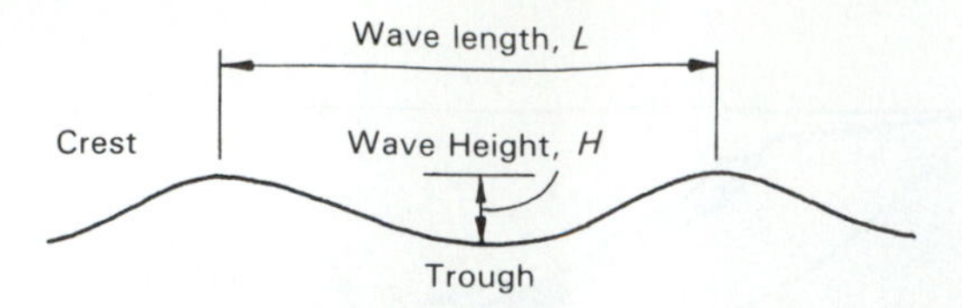


**Figure 14-7** Definition of wavelength and wave height.

Although waves move through the water at a particular velocity, the water through which the waves propagate does not travel forward in the direction of wave movement. Instead, particles of water beneath a passing wave describe circular orbits with decreasing diameters as depth increases below the surface (Figure 14-8). This pattern of motion can be observed in the behavior of a cork bobbing in the water as waves pass beneath. When we speak of wave velocity, then, we are referring to the velocity of the moving waveform rather than the water. Additional parameters needed to characterize wave movement are *wave height*—the vertical distance from trough to crest—and water depth. The relationship between water depth and wavelength is an important aspect of wave motion.

Waves are produced by the transfer of energy from wind to the water surface. The amount of energy transferred, which is reflected in the size and period of the waves generated, is dependent on three parameters illustrated in Figure 14-9. The storm duration and wind speed are functions of the size and intensity of the storm. The *fetch*, the distance over which the wind blows, is a function of the storm and the body of water. Fetch is a limiting factor for the generation of waves in small lakes and ponds.

The generation of waves by storms at sea produces a complex group, or spectrum, of waves, with large variation in period. Under the influence of the generating storm, these *sea waves* are highly irregular. Individual wave crests are random summations of smaller waves of different periods.

The amount of energy transferred to the water from the wind is a function of the wave period and the square of the wave height. Long-period waves, those with periods of 20 seconds or more, are generated only when wind speed, storm duration, and fetch are all at their maximum values. Wave heights of 30 m or more have been observed in the open ocean. As waves move out of the area of generation, wave groups of similar periods sort themselves out and the wave motion becomes more regular. These waves, which travel great distances from their area of generation, are termed *swell*. Although the equations that describe wave motion are fairly complex, simplifications can be made under certain

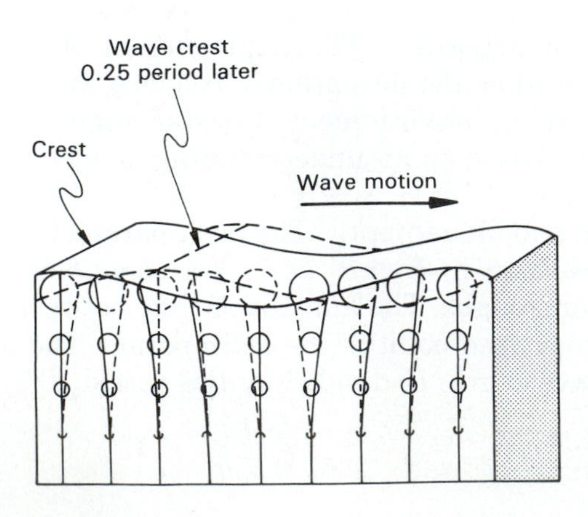


**Figure 14-8** The circular motion of water particles in waves in deep water. (After U.S. Hydrographic Office Publication 604, 1951.)

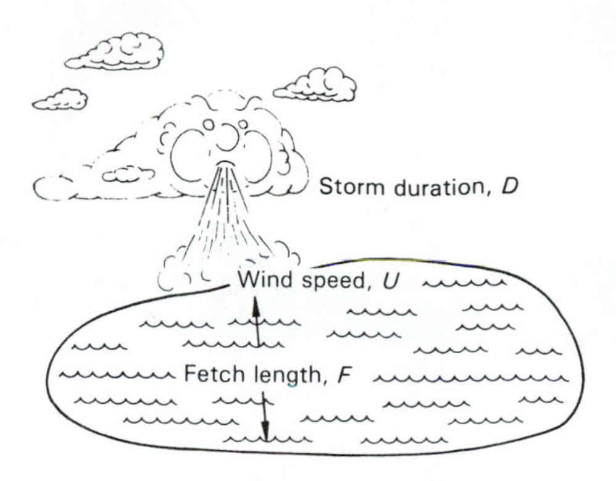


**Figure 14-9** Factors that determine the size and period of wind waves. (From P. D. Komar, *Beach Processes and Sedimentation*, copyright © 1976 by Prentice-Hall, Inc., Englewood Cliffs, N.J.)

conditions. For example, in deep water—in which the depth is greater than one-half the wavelength—both wavelength and velocity can be expressed as functions of period. Thus

$$L = \frac{gT^2}{2\pi} \quad \text{(Eq. 14-2)}$$

where $g$ is the acceleration of gravity. By substituting Equation 14-2 into 14-1, wave velocity can be expressed as a function of period:

$$C = \frac{gT}{2\pi} \quad \text{(Eq. 14-3)}$$

After conversion to swell, wave groups of similar period travel forward until they encounter a coastline. Along this path the waves lose energy by several processes, including lateral spreading of the waves from the area affected by the storm to a large area of the open ocean. The potential coastal erosion or damage to structures depends upon the characteristics of the waves originally generated by the storm and the changes to the wave groups that occur along their path to the coast. Charts are available for prediction of wave heights and periods from known storm characteristics.

Breaking Waves. As waves approach shore, they undergo a gradual transition from sinusoidal profiles with low, rounded crests to waves with sharp peaks and higher wave heights (Figure 14-10). This transitional process begins when the water depth decreases to about one-half the wavelength. As the waveform begins to interact with the bottom, the circular orbits of water

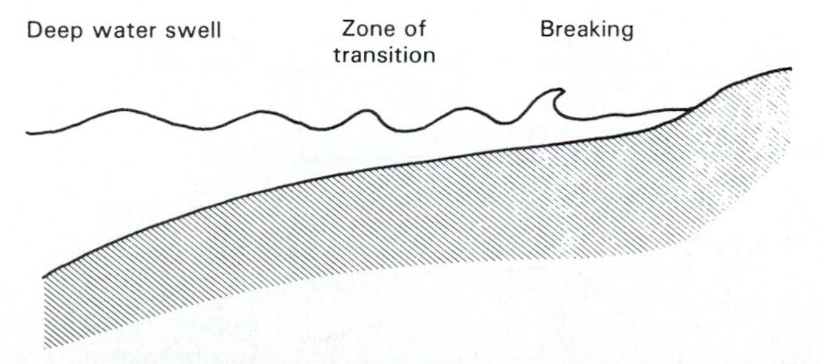


**Figure 14-10** Changes in waves as they approach shore. (From P. D. Komar, *Beach Processes and Sedimentation*, copyright © 1976 by Prentice-Hall, Inc., Englewood Cliffs, N.J.)

**Figure 14-11** Waves breaking on shore. (R. Dolan; photo courtesy of U.S. Geological Survey.)

particles (Figure 14-8) become compressed to elliptical shapes. The wavelength and the velocity both decrease, while the period remains the same. As the wavelength decreases, the wave crest must build in order to maintain a constant mass of water within a single wavelength. The ratio of the wave height to wavelength, called *wave steepness*, gradually increases until a critical value of stability is exceeded. This occurs when the steepness reaches a value of about $\frac{1}{7}$. At this point the wave breaks (Figure 14-11) because it has become too steep to maintain its shape. In addition, at the point of breakage the horizontal orbital velocity at the crest exceeds the velocity of the waveform, and the crest collapses ahead of the wave.

Most breakers can be classified as either *spilling* breakers or *plunging* breakers (Figure 14-12). Spilling breakers, in which the crest collapses into bubbles and foam that rush down the shoreward face of the wave, occur over gently sloping bottoms. The plunging breaker creates a spectacular visual effect as the crest curls outward and downward in front of the wave, forming a tunnel for a few seconds prior to total collapse of the wave.

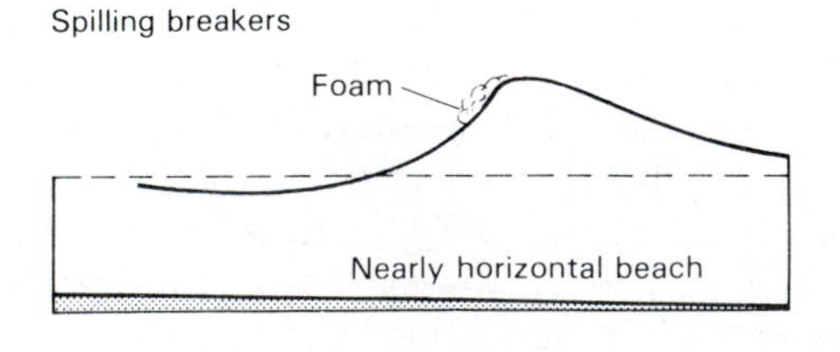


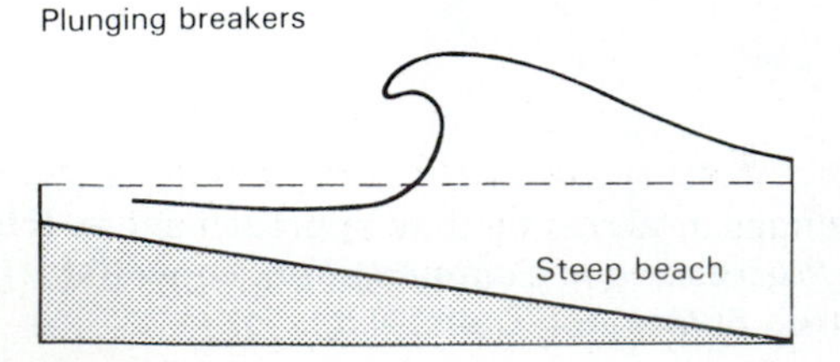


**Figure 14-12** Form of spilling and plunging breakers. (From P. D. Komar, *Beach Processes and Sedimentation*, copyright © 1976 by Prentice-Hall, Inc., Englewood Cliffs, N.J.)

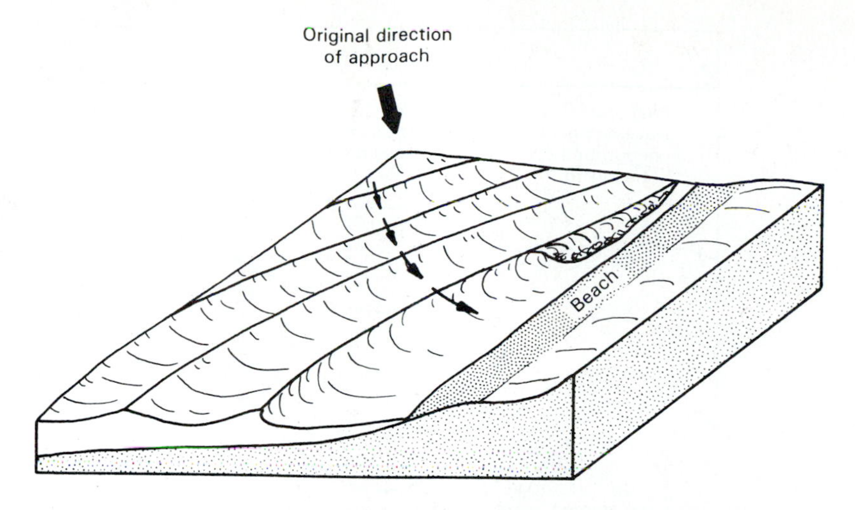


**Figure 14-13** Refraction of wave crests approaching shore at an angle to become parallel with beach.

Wave Refraction. The decrease in velocity experienced by waves as they enter shallow water results in a change in direction of the wave called *wave refraction*. Along a straight shoreline in which the offshore bottom contours parallel the shore, wave refraction is evident when waves approach the coast at an angle (Figure 14-13). The portion of a wave crest that reaches shallow water first begins to decrease in velocity compared with the portion of the same wave that has not yet reached shallow water. Consequently, the seaward part of the wave travels faster and the wave crest appears to bend, so that it becomes nearly parallel to the beach by the time it breaks.

Wave refraction has an important influence on erosional and depositional processes affecting irregular coastlines (Figure 14-14). In a coastline composed of headlands, or points jutting out into the ocean between adjacent bays, the bottom contours often parallel the shoreline. In other words, waves approaching the shore first begin to interact with the bottom directly offshore from the headland because water is shallower there relative to the bay on either side. As shown in Figure 14-14, wave refraction tends to focus the wave rays (lines drawn perpendicular to wave crests) toward the headland. The spacing of the wave rays, which are equally spaced and parallel in open water, is an indication of the amount of wave energy expended against a particular part of the coastline. Close spacing in the vicinity of headlands indicates that wave energy is concentrated in those areas. The rate of shore erosion, therefore, is much greater on headlands than in adjacent bays. The low-energy environment of bays, in fact, often results in net deposition of sediment there. This explains why headlands are commonly rock and lack good beaches, whereas nearby bays contain broad expanses of sandy beaches. If the shoreline materials are nonresistant to erosion, these processes tend to straighten out an initially irregular coast by erosion of headlands and deposition in bays.

### *Tides*

In coastal areas, tides are an important factor in many processes. This is particularly true where the conditions and forces that influence tides combine to produce a large difference in elevation, or *tidal range*, between high and low tides. The Bay of Fundy between New Brunswick and Nova Scotia, for

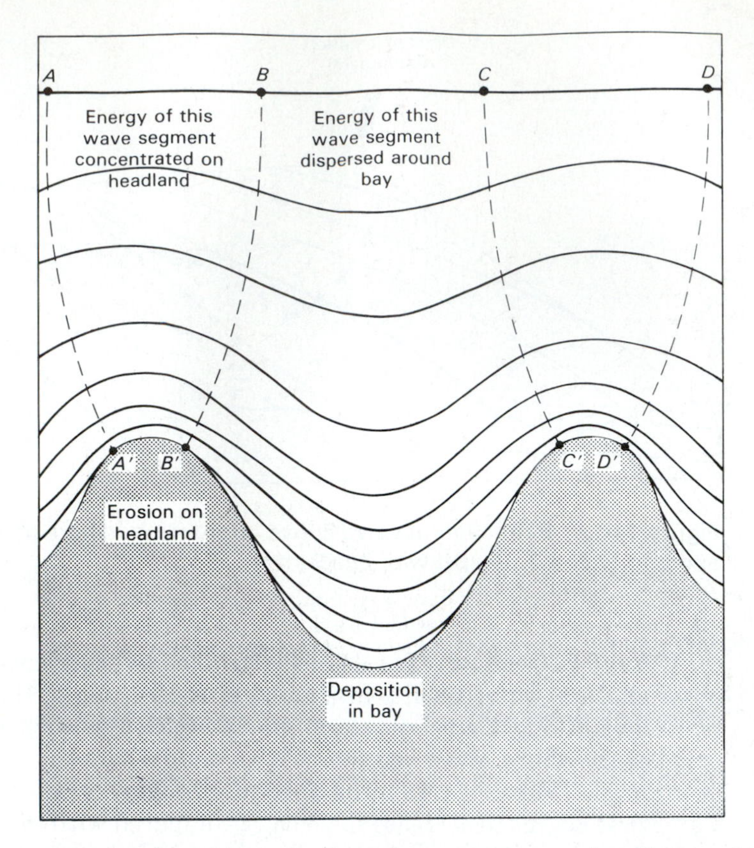


**Figure 14-14** Concentration of wave energy on headlands of an irregular shoreline. The energy of wave segments *AB* and *CD* is distributed over a short distance at the headland, whereas the energy of segment *BC* is dispersed over a long shoreline distance in the bay. (From S. Judson, M. E. Kauffman, and L. D. Leet, *Physical Geology*, 7th ed., copyright © 1987 by Prentice-Hall, Inc., Englewood Cliffs, N.J.)

example, has a maximum tidal range of 16 m, the highest in the world (Figure 14-15).

Although many factors influence tides, the dominant control is the gravitational attraction between the earth and the moon. In terms of Newton's law of gravitation, the net attractive force can be expressed as

$$F = \frac{M_m M_e}{R^2} G \qquad \text{(Eq. 14-4)}$$

The terms $M_m$ and $M_e$ represent the masses of the moon and the earth, respectively, and $R$ is the distance between the center of the earth and the moon. $G$ is the universal gravitation constant.

Although Equation 14-4 represents the net attractive force between the earth and the moon, the attractive force between the moon and a unit of mass on the earth's surface is slightly different from the net force because the distance $R$ between the earth's surface and the moon is not the same as the distance between the centers of the two bodies. It is the difference between the total force of attraction and the local attractive force that is responsible for tides.

If the earth were covered with a uniform layer of water over its surface, tidal forces would produce the effect shown in Figure 14-16. Two tidal bulges

**Figure 14-15** Bay of Fundy at low (a) and high (b) tide. (Photo courtesy of Tourism New Brunswick, Canada.)

would be created. The bulge facing the the moon would be generated because the water on the moon's side of the earth is attracted more than the earth as a whole. The bulge on the opposite side of the earth is produced in a similar way: The moon's attraction for water on the opposite side of the earth is less than its attraction for the main body of the earth.

As the earth rotates, the tidal bulges travel over the surface. A fixed point on the earth would therefore experience two high and two low tides in a little more than 24 h because the moon revolves around the earth in the same direction as the earth's rotation. It therefore takes an additional 50.47 min for the fixed point on the earth to "catch up" to its original position relative to the moon because the moon also has moved along its path around the earth during the day.

A number of factors complicate this simplified explanation of tide-generating forces. One of these factors is the influence of the sun. Despite its huge mass, the sun's great distance from the earth lessens its effect upon the tides.

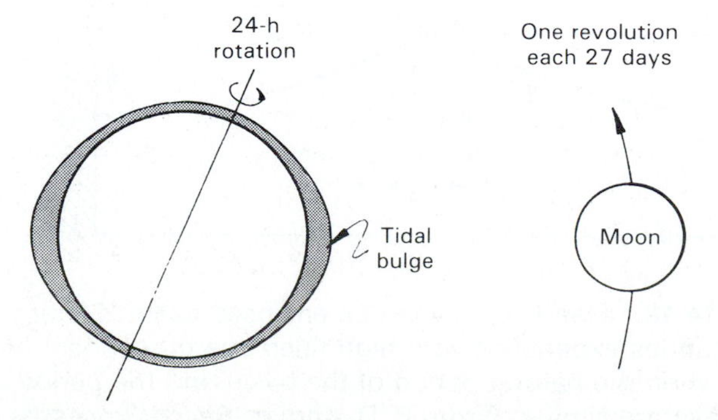


**Figure 14-16** Tidal bulge produced on opposite sides of the earth by the gravitational attraction of the moon.

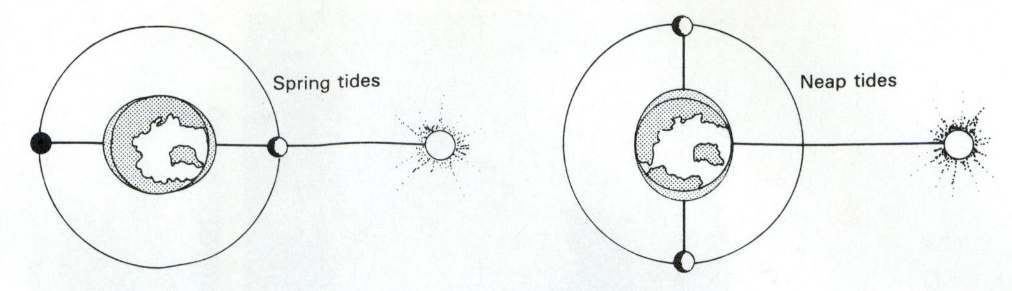


**Figure 14-17** Origin of spring and neap tides. (From E. A. Hay and A. L. McAlester, *Physical Geology: Principles and Perspectives*, 2d ed., copyright © 1984 by Prentice-Hall, Inc., Englewood Cliffs, N.J.)

The influence it does have, however, must be considered in terms of the positions of the three bodies. When the earth, the moon, and the sun are aligned (Figure 14-17), the size of the tidal bulge on the earth is maximized. These periodic high tides are called *spring tides*, although they have nothing to do with the seasons. When the sun and the moon are aligned at right angles with respect to the earth (Figure 14-17), the lower tides produced are called *neap tides*.

The factors discussed above still do not explain the huge tidal ranges of the Bay of Fundy and other areas. For the reasons that cause these tides we must consider the distribution of continents and ocean basins and the topography of bays and estuaries along the coastline.

Because the earth is not covered by a uniform layer of water, tidal movements must occur in the confined basins that contain the world's oceans. An ocean basin can be likened to a rectangular container of water that is suddenly disturbed by being lifted at one end (Figure 14-18), forming a wave of water that moves toward the opposite side of the tank. When it reaches the side, it is reflected back toward its original side. The process continues and a standing wave called a *seiche* is generated as the wave moves back and forth across the tank. The period of the wave is determined by the length and depth of the tank. In other words, every container, including ocean basins, bays, and estuaries, has a characteristic natural period. As the earth rotates, tides move across ocean basins as long-period waves. If the period of a *tidal wave* (not to be confused with the misnamed "tidal waves" that cause great destruction along coasts) is similar to the natural period of the basin in which it moves, resonance occurs and the wave is amplified. A process of this type takes place in bays such as

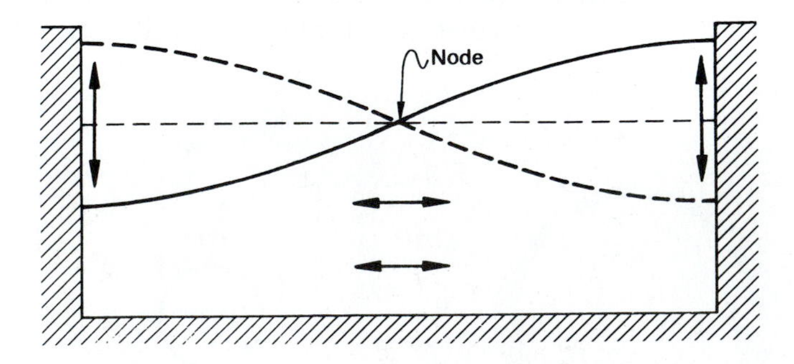


**Figure 14-18** Standing wave in an enclosed basin. Certain bays and estuaries experience very high tides due to resonant amplification when the natural period of the basin and the period of the tidal wave are similar. (From P. D. Komar, *Beach Processes and Sedimentation*, copyright © 1976 by Prentice-Hall, Inc., Englewood Cliffs, N.J.)

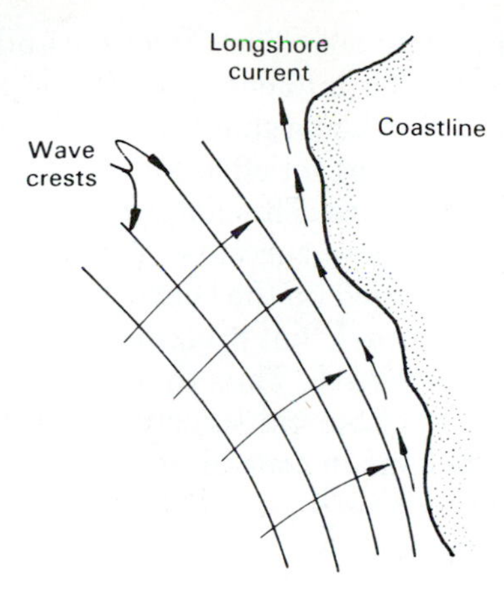


**Figure 14-19** Longshore current associated with waves approaching coast at an angle.

the Bay of Fundy; the natural period of the bay is close to the tidal period, so a resonant standing wave moves back and forth across the bay. This accounts for the very large tides observed there.

### *Longshore Currents*

Although wave refraction tends to change the direction of wave movement so that waves approach the shore more directly, an oblique approach angle is often observed (Figure 14-19). This movement results in a net flow of water, or *longshore current*, parallel to the shore. Longshore currents are apparent to swimmers in the surf zone who may suddenly realize that they have drifted along the shore for some distance from the point on the beach at which they swam out into the water. Rather than continuing indefinitely along the shore, longshore currents are often broken into cells (Figure 14-20) that are bounded by rapidly moving currents flowing directly away from the beach in a seaward direction. These *rip currents* are dangerous because of their ability to carry swimmers out to sea. Many swimmers make the mistake of swimming directly against the rip current toward shore. Instead, a person caught in a rip current should swim laterally, because rip currents are narrow zones that can be evaded by swimming along the shore for a short distance. Bottom topography can control the position of rip currents; for example, they often occur near the

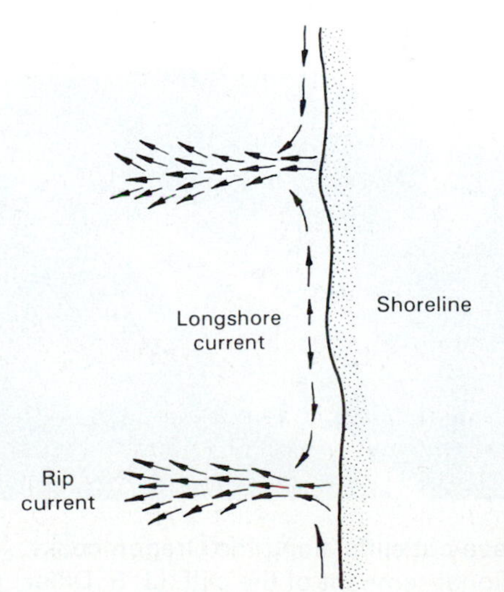


**Figure 14-20** A longshore current cell bounded by rip currents.

heads of submarine canyons. The direction and characteristics of incoming waves, however, can also influence the development of longshore cells. Rip currents have been observed on shores with a very regular bottom topography.

One of the most important effects of longshore currents is the transportation of sand along the shore. This *longshore drift* is the movement of sand in the shore zone by longshore currents. Transport actually occurs in a zig-zag fashion. Sediment is carried up the beach at the oblique angle of the approaching wave by the wave *swash,* but it is carried back to the water at right angles to the beach in the *backwash.* Thus there is a component of motion along the shore. The interaction of coastal structures with longshore drift is one of the most significant problems in coastal engineering.

### Coastal Morphology

Coastlines are characterized by continuous change. The changes that occur present a great challenge to coastal engineers and managers because coastal processes can be rapid and powerful. In addition to the drastic changes that can be caused by a single damaging storm, long-term processes are also at work. These include worldwide changes in sea level or regional changes in the elevation of the land relative to sea level caused by tectonic forces. A study of the geomorphology of a coast can frequently indicate the types of processes that are currently influencing the coast.

Erosional Features. Coastlines rapidly eroding under the relentless attack of wave erosion provide examples of steep *wave-cut cliffs* (Figure 14-21). The cliffs maintain steep slope angles as they retreat because they are undermined at their bases by wave erosion that removes support for the overlying material. A rockfall, slump, or other type of mass movement can occur when the safety factor of the slope decreases to a value of 1.0 (Figure 14-22). The debris that accumulates at the base of the slope is further eroded and transported away from shore by wave action. The flat surface at the base of wave-cut cliffs beveled by wave erosion is called a *wave-cut platform* (Figure 14-23).

**Figure 14-21** Wave-cut cliffs along the Oregon coast. A stack remains as an erosional remnant of the cliff. (J. S. Diller; photo courtesy of U.S. Geological Survey.)

**Figure 14-22** A mass movement (arrow) caused by coastal erosion. (J. T. McGill; photo courtesy of U.S. Geological Survey.)

As the coastline retreats, resistant rock units may project out of the water as *stacks* or *arches* (Figure 14-21). Eventually these remnants are also worn down and removed by wave erosion.

Depositional Features. In addition to erosional processes, shorelines undergo deposition from waves, wind, and longshore currents. The landforms that result from these processes are dominant along some coastlines. Some of these features are illustrated in Figure 14-24. A detailed discussion of the beach, probably the most important depositional landform, will be reserved for the following section.

The transportation of sand by longshore currents is responsible for the deposition of a *spit*, a thin strip of beach that extends from a point or headland across the mouth of a bay or other coastal indentation. Frequently, the ends of spits curve inward toward the bay, forming a *hooked spit*.

The presence of an offshore island close to the shore creates a low-energy zone shoreward of the island that is somewhat protected from wave processes.

**Figure 14-23** Wave cut platform in Pleistocene carbonate reef rock, Puerto Rico. (C. A. Kaye; photo courtesy of U.S. Geological Survey.)

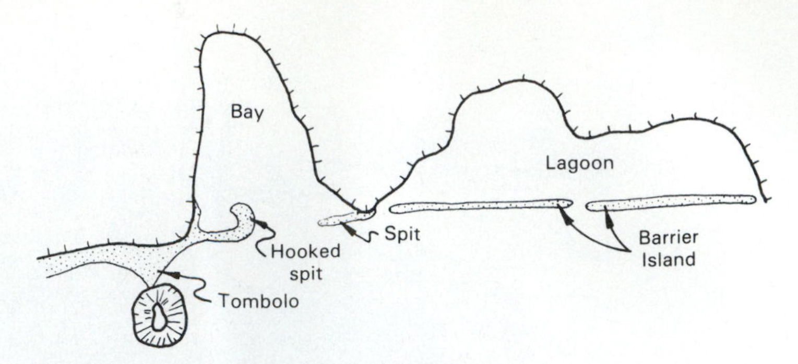


**Figure 14-24** Depositional landforms along coasts.

A spit commonly forms in such a protected location, and when the spit connects the beach and the offshore island, it is given the name *tombolo* (Figure 14-24).

Most depositional coastlines are characterized by long, narrow islands that lie just offshore and trend parallel to the coast. Such *barrier islands* are common off the East and Gulf coasts of the United States (Figures 14-24 and 14-25). Periodic breaks in the barrier islands, or *tidal inlets*, allow tidal flow between the ocean and the narrow *lagoons* that lie between the barrier islands and the mainland.

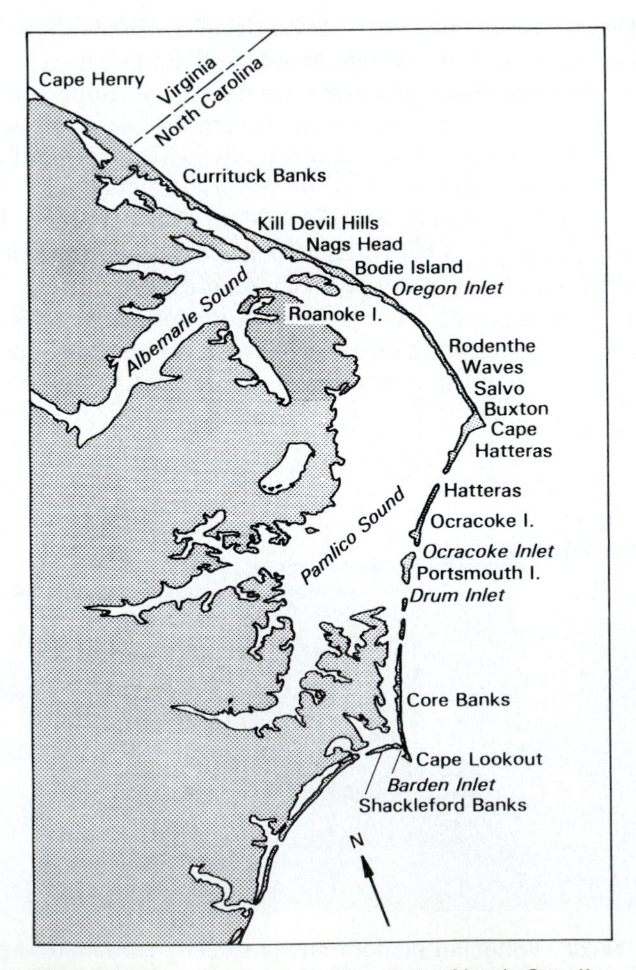


**Figure 14-25** Barrier islands off the North Carolina mainland.

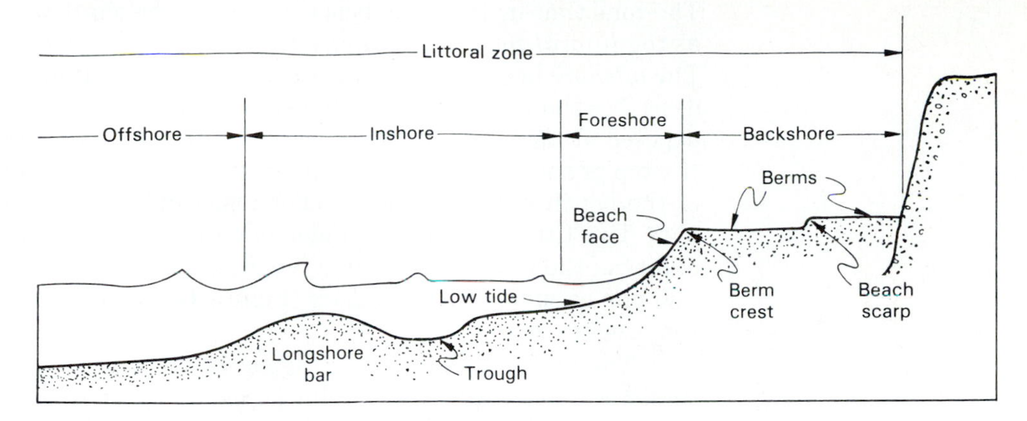


**Figure 14-26** Topographic divisions of the beach profile. (From P. D. Komar, *Beach Processes and Sedimentation*, copyright © 1976 by Prentice-Hall, Inc., Englewood Cliffs, N.J.)

The origin of barrier islands has been a matter of considerable geologic debate. One of the most likely recent hypotheses relates the formation of barrier islands to the recent rise in sea level accompanying the melting of Pleistocene glaciers. Linear ridges may have been built along coastal plains by wave or wind processes and isolated as barrier islands when the low-lying areas shoreward of the ridges were flooded to form the lagoons. Once formed, barrier islands can migrate in response to changes in sea level and other factors.

### *Beaches*

The *beach* is the actual point of interaction between the sea and land. As such, its form adjusts constantly to the energy expended upon it by tides, wind, and waves. Because the term beach describes only the part of the shoreline that is exposed above the mean low-tide position, we must expand our zone of interest to consider all the significant processes in the nearshore environment.

**Figure 14-27** Longshore bars (arrows) running parallel to the Lake Michigan shoreline.

The zone that includes the beach as well as the shallow offshore area is known as the *littoral zone.* Its subdivisions and landforms are shown in Figure 14-26. The *offshore* begins just at the breaker zone and extends seaward. The *inshore* describes the zone between the breaker zone and the lowest point on the beach exposed at low tide. The *foreshore* includes the area from the low-tide point to the top of the *berm crest,* and the term *backshore* is applied to the remainder of the beach shoreward to a major topographic break such as a dune or sea cliff. The littoral profile includes one or more flat *berms* on the shoreward extent of the beach, the sloping *beach face* that is under constant wave action, and one or more *longshore bars* (Figure 14-27) and *troughs* in the vicinity of the breaker zone.

The profile illustrated in Figure 14-26 is a very generalized representation. The beach, or littoral, profile is constantly changing in response to changing wave and tide conditions. The beach must therefore be thought of as a dynamic system in which changes in wave inputs initiate responses that tend to establish a new condition of equilibrium under those particular wave conditions. The adjustments that take place in order to reach equilibrium are changes in the beach profile. The size of the particles making up the beach is also an important factor in this process.

An example of this system at work is the change in profile that occurs from the predominantly swell-wave condition that exists in summer to the influence of storm waves that are more common in winter. Figure 14-28 shows the profiles that develop under each of these conditions. The broad berm that characterizes the summer profile is a depositional landform that is gradually built up by the onshore sand transport by swell waves. A berm is a very desirable attribute for a beach, not only because it provides wide expanses for sunbathers, but because it protects the sea cliff or dunes from wave erosion. The powerful storm waves of winter attack and erode the berm that was constructed during the summer. The sand that was incorporated into the berm is transported offshore and deposited as part of one or more longshore bars. Longshore bars are long, narrow submerged ridges of sand that develop in the vicinity of large breakers (Figure 14-27). Without the protection of the berm, storm waves are free to attack the sea cliff. Many beachfront property owners have learned about this dynamic system as they watched destructive storm waves pounding against the foundations of their houses. In the summer the process is again reversed: Gentle swell waves transport sand from the longshore bars up onto the beach to form a new berm.

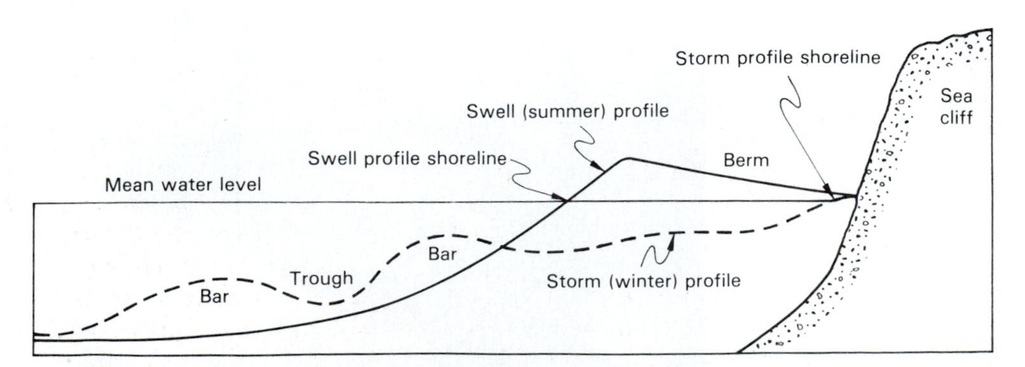


**Figure 14-28** The summer and winter beach profiles—the response of a beach to different energy conditions. (From P. D. Komar, *Beach Processes and Sedimentation*, copyright © 1976 by Prentice-Hall, Inc., Englewood Cliffs, N.J.)

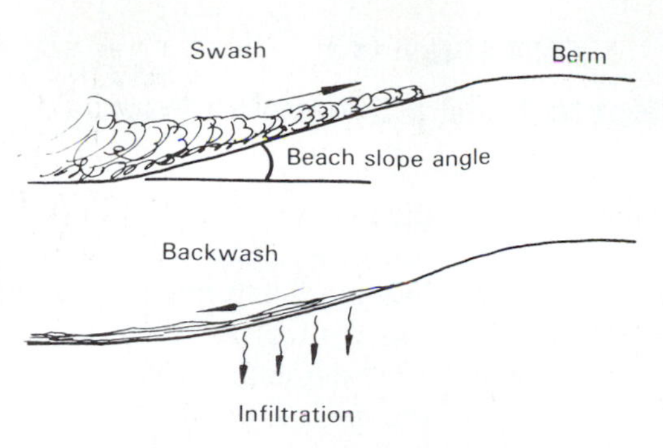


**Figure 14-29** The stronger swash transports sediment up the beach face. Backwash is weakened by infiltration of water into the beach. In order for an equilibrium profile to develop, equal amounts of sediment must be transported up and down the beach face. The coarser the particles, the steeper the beach slope angle must be.

Another manifestation of equilibrium between form and process is the angle of slope of the beach face. This angle is related to the size of the incoming waves and the particle size and angularity of the beach material. Even though strong waves transport sand from a beach in the offshore direction, they also have the ability to transport large particles toward the beach. Thus an exposed beach that is frequently subjected to strong waves is commonly composed of coarse particles.

The movement of a wave onto the beach face is illustrated in Figure 14-29. Swash transports particles upslope until deposition when the velocity decreases. Before the return flow down the beach face in the backwash can occur, some of the water infiltrates into the beach face. More water infiltrates into the beach as the particle size increases. Because of this loss of water, the backwash tends to be weaker than the swash. In order for an equilibrium profile to develop, the same amount of sand must be transported down the beach slope as up. Backwash must be assisted by an additional force, namely, the downslope component of gravity. The larger the beach particles are, the steeper the slope must be, because of infiltration losses. Beaches composed of large particles, such as the beach illustrated in Figure 14-30, must therefore develop steeper slopes in order to establish an equilibrium state.

**Figure 14-30** A beach in Iceland composed of large boulders because of the high wave energy.

## CASE STUDY 14-1

## Where Not to Build a Swimming Pool

Location of buildings in areas of active geologic processes usually leads to rapid destruction. The elaborate pool in Figure 14-31a extends too far out onto a beach along the Virginia coastline. The movement of waves onto the beach adjacent to the structure is obvious in the photograph. Construction of anything on a beach indicates a total lack of understanding of coastal processes. The tranquil waves of summertime become much more powerful and destructive in the winter. A storm of moderate intensity in 1979 demonstrated the foolishness of constructing such a project (Figure 14-31b).

(a)

(b)

**Figure 14-31** (a) A structure that extends out onto a beach is vulnerable to storm damage. (b) The same house after a storm of moderate intensity. (C. Alston; photos courtesy of Virginia Sea Grant at VIMS.)

### *Types of Coasts*

Classification of coasts is useful because the particular type of landforms present on a coast gives an indication of geologic processes that are currently active or have been significant during the development of the coast.

*Depositional coasts* (Figure 14-24) are characterized by deltas, barrier islands, and other depositional features. Gently sloping coastal plains underlain by nonresistant materials favor the development of depositional coasts. The east and Gulf coasts of the United States are good examples of depositional coasts.

*Glacial coasts* are produced by glacial erosion of valleys that meet the coastline. Glacial valleys tend to have very steep walls and a U-shaped cross-sectional profile. Scouring of these valleys took place at a time when sea level was much lower than at present, so the valleys extend well below present sea

**Figure 14-32** The steep valley walls of the Norwegian fjords were produced by glacial erosion. (Photo courtesy of Norwegian Tourist Board.)

**Figure 14-33** Wave-cut terraces on the west side of San Clemente island, California. (John S. Shelton.)

level. The postglacial rise in sea level drowned these glacial valleys, producing the long, narrow bays with steep valley sides called *fjords*. The coast of Norway is famous for the picturesque scenery of its fjords (Figure 14-32).

Some coasts have been raised or uplifted relative to sea level. These *emergent coasts* can be produced in several ways. Tectonic uplift associated with faulting is an important mechanism. Parts of the California coast have a series of raised wave-cut platforms (Figure 14-33), indicating that the coast has been subjected to a succession of tectonic episodes. Each platform represents a temporary period of stability.

Glaciated regions may also have emergent coastlines. In this instance, the mechanism producing emergence is *isostatic rebound*. The huge weight of a glacier several kilometers thick causes the rigid crust to sink slightly deeper into the plastic mantle of the earth. Isostatic rebound is the readjustment of the crust after the glacier's retreat. The fact that glaciated regions are still rising relative to sea level 10,000 years after the last ice retreat demonstrates that isostatic rebound is a very slow process. A series of beaches or other shoreline features above sea level is an indication of an emergent coastline produced by postglacial rebound.

A final type of coastline is the *biogenic coast*, characterized by the presence of carbonate reef structures in the shallow offshore zone along the coast. Carbonate reefs are composed of the shells of living organisms called corals that grow in colonies, where individual organisms are connected to form a three-dimensional reef framework.

Coral reefs are classified according to their position relative to the mainland. Reefs that actually connect with the shoreline are called *fringing reefs*. If the reef is separated from the coast by a lagoon, the name *barrier reef* is used. A circular reef without the presence of land is known as an *atoll*. More than 100 years ago, Charles Darwin provided a perceptive hypothesis for the origin of atolls that has been confirmed by deep-sea drilling only in the decades since World War II. Noting that most islands in the tropics of the Pacific Ocean were of volcanic origin, Darwin proposed that fringing reefs originally developed around these volcanoes (Figure 14-34). Gradual subsidence of the

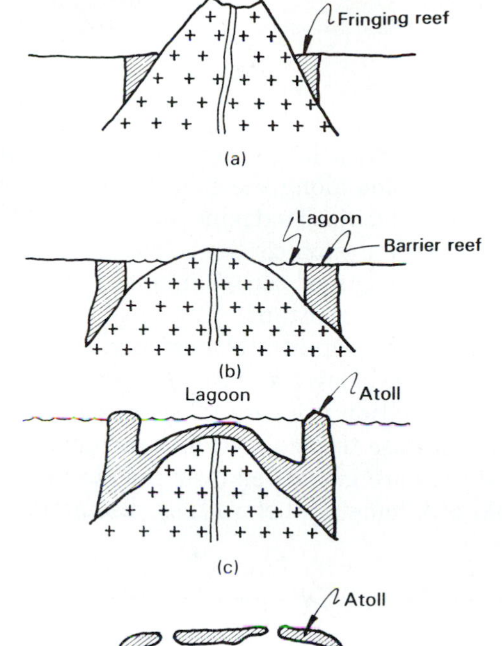


**Figure 14-34** Evolution of atolls. (a) Fringing reefs form around an active volcano. (b) Reefs grow vertically upward to form barrier reefs as volcano subsides and is eroded. (c) Further subsidence leaves only reefs above sea level. (d) Map view of atoll.

ocean crust caused the volcanoes to sink until only a small island remained surrounded by barrier reefs. Upward growth of the reef structure is necessary to counteract the subsidence of the crust, because corals can live only in warm, shallow water. With continued subsidence, the eroded remnants of the volcano disappeared altogether, leaving only a circular atoll to mark its original position.

### *Sea-Level Changes*

Far from being a static surface, the elevation of sea level has undergone major fluctuations in geologic time. One of the major causes for *eustatic* sea level changes, those that are synchronous throughout the world, is a change in the mass of the polar ice caps. When polar ice caps expand and advance into the midlatitudes, as they did repeatedly during the Pleistocene Epoch, sea level drops accordingly. During the last glaciation, sea level dropped an estimated 100 m. While sea level was lowered, a greater area of the continental shelves was exposed than at present. Rivers eroded valleys across these broad coastal plains underlain by unconsolidated sediment. Later, as glaciers retreated and released huge volumes of water to the oceans, the rising sea level inundated these newly formed valleys, creating estuaries. An irregular coastline in which estuaries extend up previous river valleys is sometimes referred to as a *drowned* coastline. Chesapeake Bay along the eastern coast of the United States is an example of an estuary that was formed by the drowning of a river valley during the postglacial sea-level rise.

## COASTAL MANAGEMENT

### Coastal Hazards

Throughout human history, a large percentage of the population of coastal nations has chosen to reside along the coasts. Unfortunately, various coastal processes have produced events of extreme magnitude that have inflicted great disasters upon coastal population centers. Today, the steady migration of the U.S. population to the coasts of the "Sunbelt" makes an understanding of coastal hazards even more important.

*Tropical cyclones* are potentially destructive storm systems in which winds move around centers of low pressure in a circular fashion. In the Northern Hemisphere, wind circulation is counterclockwise around low-pressure areas. Regional names for intense tropical cyclones include hurricane, typhoon, and monsoon. These storms are generated over the ocean in the tropics and then migrate thousands of kilometers before they dissipate. In North and Central

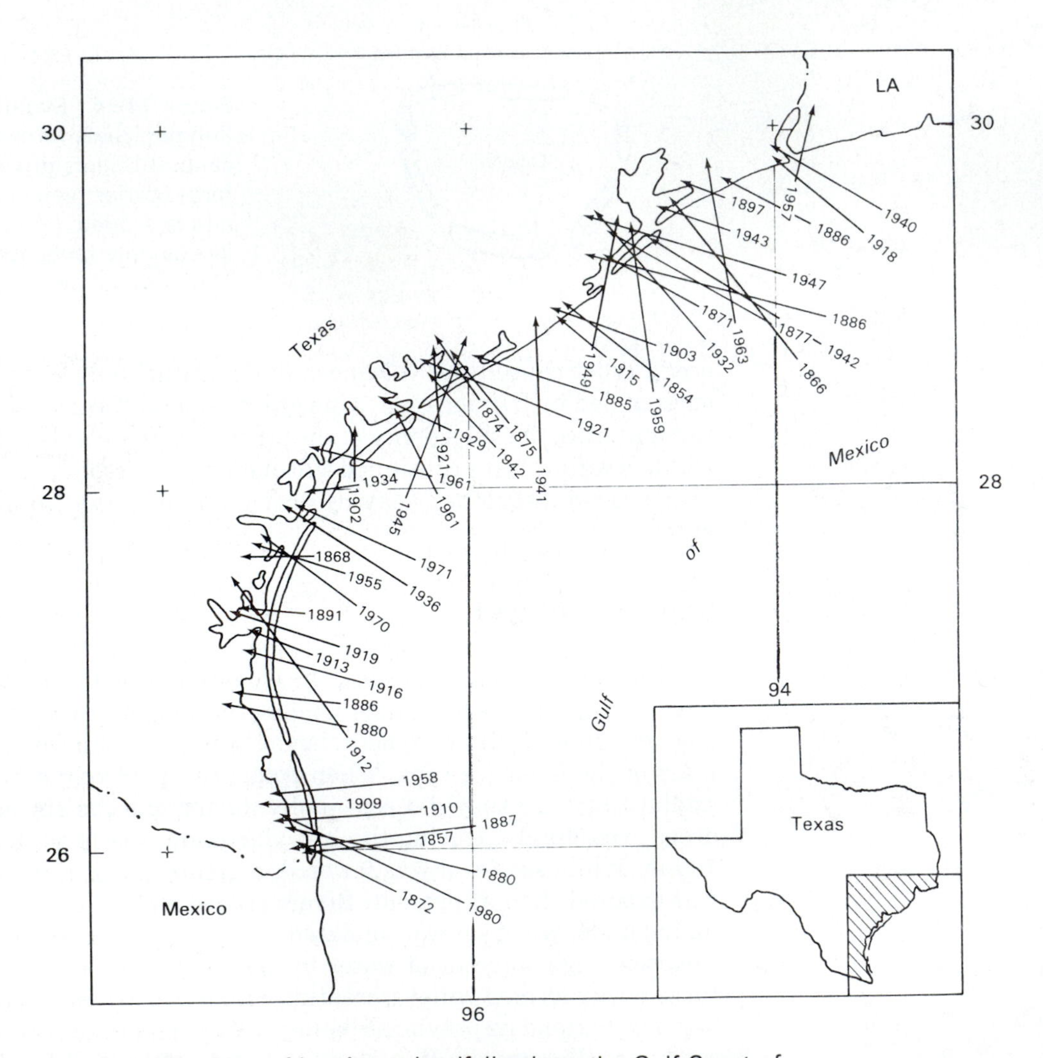


**Figure 14-35** Map of hurricane landfalls along the Gulf Coast of Texas between 1850 and 1980. (From J. P. Giardino, P. E. Isett, and E. B. Fish, 1984, *Environmental Geology and Water Sciences*, 6. Used by permission of Springer-Verlag, Inc., New York.)

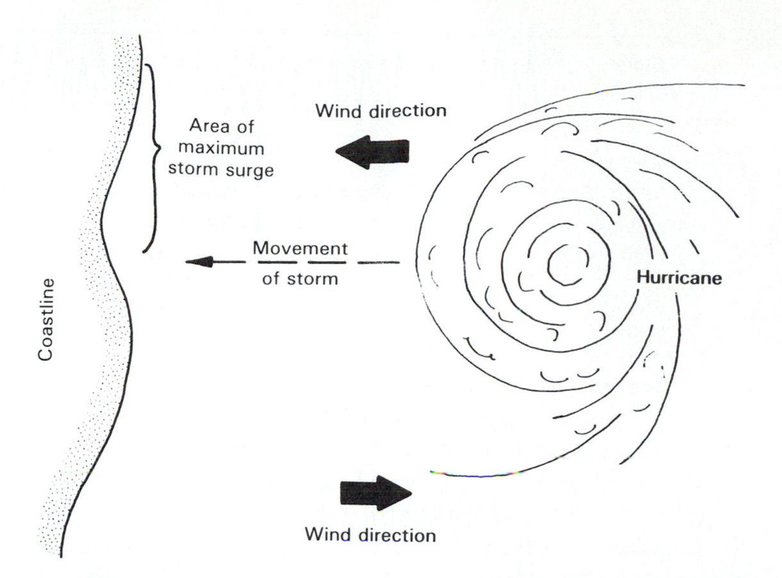


**Figure 14-36** A hurricane approaching shore in the Northern Hemisphere will produce maximum storm surge to the right side of the storm center as viewed in the direction of movement of the storm.

America, the Gulf Coast region is the prime land target of hurricanes. The map of part of this coastal area (Figure 14-35) indicates that the twentieth-century distribution of storms has been quite uniform.

The arrival of a hurricane at a particular coastline is preceded by a 1 to 1.5 m rise in water level called a *forerunner.* The forerunner is caused by swell waves produced by the storm in its generating area, as much as several thousand kilometers distant from the coast at the time of arrival of the abnormally high forerunner swells.

Most damage from a tropical cyclone is produced by the *storm surge*, a rapid rise in water level generated by the winds of the storm as it approaches shore. Because of the counterclockwise circulation of winds, the storm surge reaches its maximum to the right (north, in the northern hemisphere) of the storm center, where the winds are blowing in the direction of the storm's movement (Figure 14-36). The actual rise in water level in a storm surge is also a function of the tides at the time the storm strikes the shore (Figure 14-37). Thus the magnitude of storm damage on the shore depends on the time of day and month at which the maximum storm surge arrives. A rising trend in sea level will gradually lead to higher storm surges under the same storm and tidal conditions. Effects of the storm surge include flooding of low-lying coastal areas and intense wave erosion of beaches, dunes, sea cliffs, and any structures that happen to be in the way (Figure 14-38). Additional effects of hurricanes include wind damage to structures and coastal flooding from the intense rainfall that occurs as the storm moves onshore.

Barrier islands are particularly susceptible to damage from large storms. In the absence of human modification, barrier islands are commonly narrow, low-lying ridges. During large storms, water can flow over the island to the lagoon, thus dissipating the energy of the storm waves. The erosion that accompanies these *washovers* is often intense enough to cut a new inlet through the island. Severe damage is inflicted on roads, bridges, buildings, and utilities in washover zones. An example of storm effects on barrier islands is shown in Figure 14-39.

Perhaps the most dreaded coastal hazard is the arrival of a wave or group

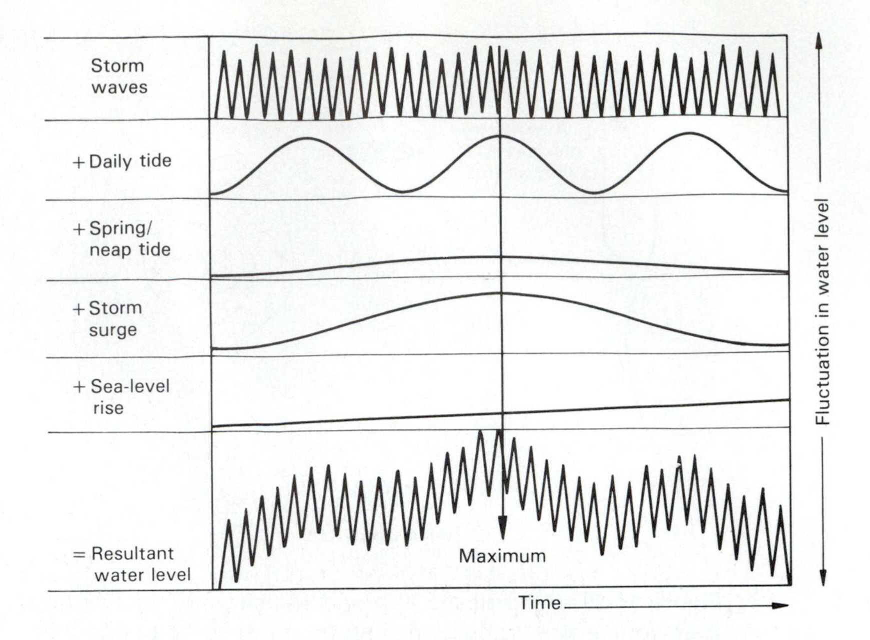


**Figure 14-37** Dependence of water-level rise in storm surge on tides and other factors. (From R. Dolan and H. Lins, U.S. Geological Survey Professional Paper 1177-B.)

of waves that can rise above the normal water level by 10 m or more. These waves are sometimes incorrectly referred to as tidal waves; however, they have nothing to do with astronomical tides. The actual cause of the waves is the rapid displacement of water by submarine earthquakes, volcanic eruptions, or submarine landslides. Because of their common association with earthquakes, they are called seismic sea waves or, from the Japanese term, *tsunamis*. Tsunamis move through the deep water of the ocean basins at velocities of several hundred meters per second. At these rates they can cross major oceans in a matter of hours. In deep water, tsunamis are barely detectable because their amplitudes reach only several meters. They are also characterized by wavelengths ranging from 150 to 250 km and periods of 10 to 60 min. When tsunamis encounter shallow water, however, they undergo the same transformations as

**Figure 14-38** Severe beach erosion along the Virginia coastline by a storm. (Scott Hardaway; photo courtesy of Virginia Sea Grant at VIMS.)

**Figure 14-39** Washover damage near Cape Hatteras, North Carolina. (R. Dolan; photo courtesy of U.S. Geological Survey.)

**Figure 14-40** Damage at Seward, Alaska, from the tsunami generated by the 1964 Alaska earthquake. (NOAA/EDIS.)

wind waves. The wave height increases to the point where the run-up is able to inundate nearshore areas and cause great damage (Figure 14-40). A tsunami generated during the eruption of the volcano Krakatoa in the Pacific in 1883 killed 36,000 people in Java and Sumatra. Catastrophes with similar losses of life at other times occurred in Japan and Portugal. In the Pacific Ocean a large destructive tsunami has an average return period of about 10 years. Smaller tsunamis happen several times per year in the Pacific.

## CASE STUDY 14-2

## "The Storm(s) of the Century"

Early in the morning of September 22, 1989, Hurricane Hugo, a category 4 storm, made landfall near Charleston, South Carolina. The devastation it produced represents the most costly natural disaster in the history of the United States. Three years later, a storm of similar magnitude, Hurricane Andrew, struck the coast of Florida (Figure 14-41). Despite a similar degree of destruction, the effects of Hugo and Andrew were quite different. For an explanation for the differences in these storms, we must examine the storms themselves, and the geology of the coastlines.

The winds of both storms exceeded 131 mph; gusts may have been significantly higher. Hugo's maximum storm surge was about 20 ft, while Andrew's surge was much less—in the range of 8 to 10 ft. Herein lies the major differ-

**Figure 14-41** Radar image of Hurricane Andrew approaching the coast of Florida. (Image courtesy of Harold Wanless.)

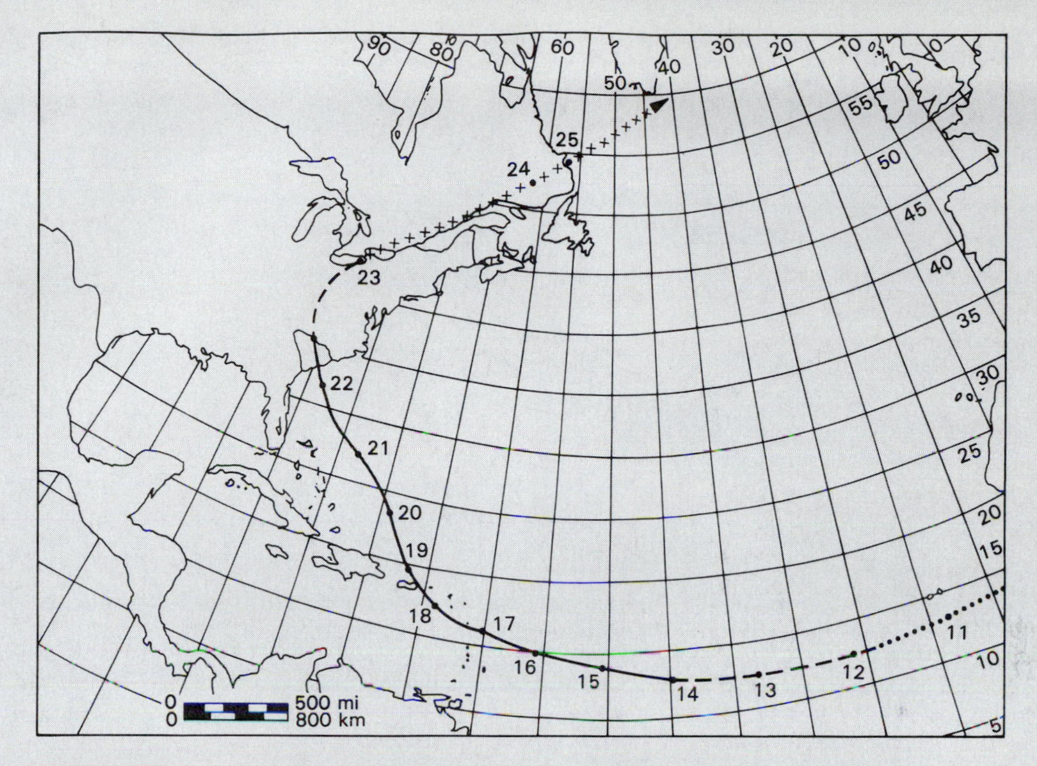


**Figure 14-42** Storm track of Hurricane Hugo, September 1989. Line type indicates storm severity. Dots = depression; dashes = tropical storm; solid line = hurricane; ++ = extratropical cyclone. (From R. H. Willoughby and P. G. Nystrom, Jr., 1990, *South Carolina Geology*, 33.)

ences of the storms. Hugo caused major damage to beaches by erosion and redeposition of sediment. Structures that happened to be within the zone of impact of the storm surge were damaged or destroyed. Andrew, on the other hand, caused most of its damage by winds to structures inland as well as on the beach. The geologic effects of Andrew were therefore much less than Hugo's.

There are several differences in the storm tracks of the two hurricanes that controlled their effects. Andrew crossed the Bahama platform prior to hitting the Florida coast. Some of the wave energy was dissipated over this shallow carbonate shelf. In addition, the Florida coast itself is very flat at the point of the hurricane's landfall, a condition less conducive to the formation of large waves. Hugo's greater damage can be attributed to the longer fetch between Puerto Rico and South Carolina and the steeper South Carolina coastline.

Hugo's storm track is shown in Figure 14-42. Winds were distributed around the eye, which came ashore at Charleston, as shown in Figure 14-43. As a result, the maximum storm surge was north of the eye. Several barrier islands along the coast were heavily impacted. Hugo's most significant effect was beach erosion. Beach profiles were flattened and lowered by erosion of sand (Figure 14-44). The erosion extended landward to natural dunes, which were truncated and steepened. Any structures that happened to be seaward of the natural dunes were weakened by removal of sand from around the pile foundations. Structures that were located landward of forested ridges and higher in elevation sustained much less damage. Buildings near the shore were commonly protected by riprap consisting of large granite blocks. In some places, the riprap afforded some protection, but in others, 4 to 6 foot blocks were moved shoreward, and the buildings were de-

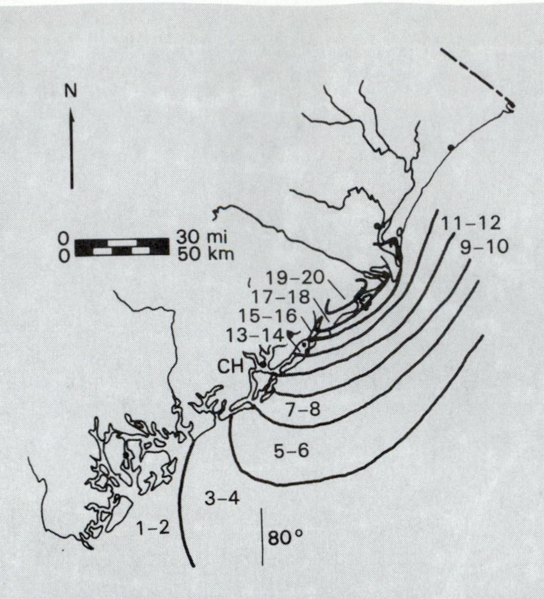


**Figure 14-43** Approximate storm surge contours (ft) for Hurricane Hugo. (From R. H. Willoughby and P. G. Nystrom, Jr., 1990, *South Carolina Geology*, 33.)

**Figure 14-44** Flat storm beach produced by erosion of sand and damage to buildings at Surfside Beach, near Charleston, S.C. by Hurricane Hugo. (From R. H. Willoughby and P. G. Nystrom, Jr., 1990, *South Carolina Geology*, 33.)

**Figure 14-45** Destruction of road and damage to building foundations by Hurricane Hugo at northern Folly Island, S.C. (From R. W. Willoughby and P. G. Nystrom, Jr., 1990, *South Carolina Geology*, 33.)

stroyed. Damage was also sustained to roads and other structures (Figure 14-45).

Sand eroded from the beaches was deposited both offshore in offshore bars and landward to form washover fans and storm ridges. Normal waves and currents began to redistribute the sand as soon as the storm passed. A new inlet was cut in a narrow section of one barrier island as the storm surge, temporarily impounded landward of the island, escaped back to the sea. The inlet was subsequently filled by the U.S. Army Corps of Engineers.

The damage by Hugo was predictable. Until development is tailored to the natural configuration of the shoreline, and prevented on beaches, hurricanes will continue to take a terrible toll of life and property.

### *Coastal Engineering Structures*

Coastal engineering works date back to biblical times. Recently, attempts have been made to control and manage coastal processes on a larger scale. Some of these projects have failed because of a lack of understanding of coastal processes. Coastal systems tend at all times to establish a condition of equilibrium between form and process. Whenever natural processes are interfered with, adjustments throughout the system are a result. These adjustments often take the form of undesirable erosion and deposition of sediment.

Coastal engineering structures are designed to enhance or maintain navigation, to prevent erosion, or to promote deposition. *Jetties* are walls built perpendicular to the shoreline in pairs at the mouths of inlets (Figure 14-46). Their purpose is to prevent blockage of the inlet by longshore drift. In doing so, they may prevent the movement of sand by longshore currents past the jetty. This will cause deposition of sand along the up-drift side of the jetty and erosion of sand from the down-drift side.

Inlets serve as the location for the inflow and outflow of tidal currents. Therefore, the size and channel geometry imposed by jetties must be carefully designed to provide the proper velocity of tidal currents. If the channel opening is too large, deposition of sand and shoaling of the inlet channel occurs, and dredging is required to maintain the necessary channel depth. If the opening is too small, the high current velocity causes scouring of the channel and potential undermining and failure of the jetties.

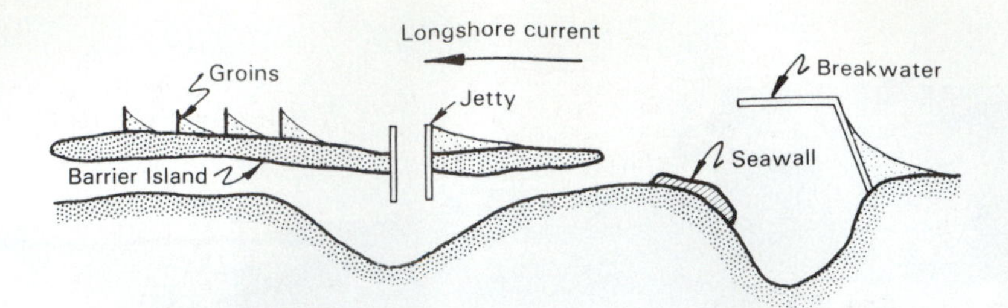


**Figure 14-46** Types of coastal engineering structures.

When adequate natural harbors are lacking, *breakwaters* can be constructed (Figure 14-46). Breakwaters provide protection from waves, but they also interact with the longshore drift in the same manner as jetties. A classic example of this interaction is the Santa Barbara, California, breakwater. As illustrated in Figure 14-47, enlargement of the beach on the up-drift side of the breakwater occurred until longshore currents were able to transport sand along the breakwater and deposit it directly into the harbor. Dredging was then required to maintain the harbor. At the same time that the deposition was occurring in the harbor, erosion of the beach was in progress on the down-drift side of the harbor. Eventually, shoreline dwellings were destroyed by wave erosion because the sand source for the beaches had been eliminated.

Detached breakwaters have also been constructed (Figure 14-48) in an attempt to provide protection from waves while at the same time allowing the passage of longshore drift. These structures usually fail to achieve their purpose because the protected area behind the breakwater becomes a low-energy zone for the longshore drift, and sand deposition ensues at that site.

*Seawalls* are structures built parallel to the shore to prevent erosion and slumping of sea cliffs or bluffs. Construction materials range from concrete to

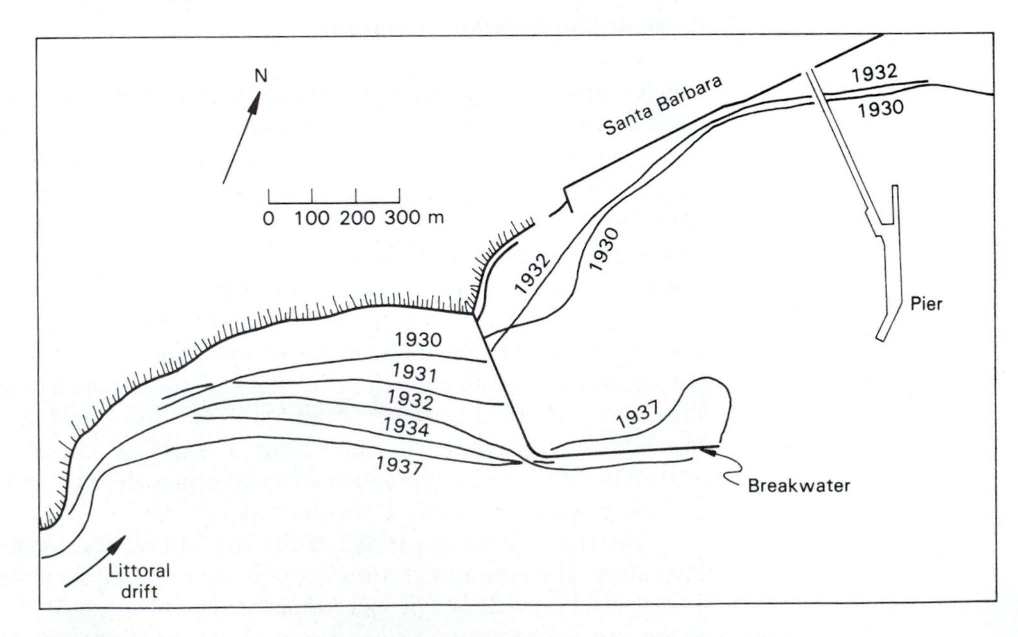


**Figure 14-47** Patterns of erosion and deposition produced by the Santa Barbara breakwater. (After J. W. Johnson, 1957, *Journal of Waterways and Harbors Division*, 83, American Society of Civil Engineers.)

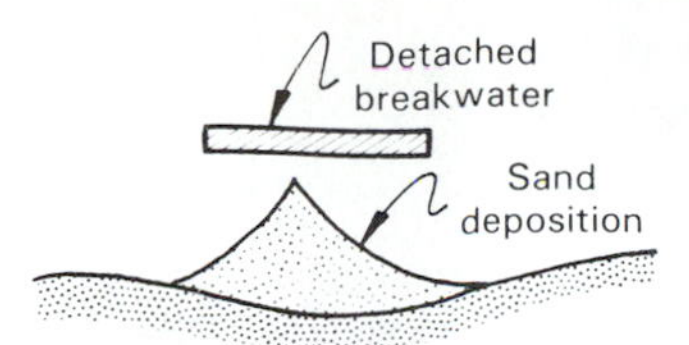


**Figure 14-48** Construction of detached breakwaters usually results in sand deposition between the structure and the shoreline.

timber to riprap (Figure 14-49). Seawalls provide short-term protection against moderate events, but erosion continues on both sides of the protected zone.

*Groins* (Figure 14-50) are walls built perpendicular to the beach for the purpose of trapping sand from longshore drift. Although a beach can be considerably widened by the construction of a groin, down-drift erosion is also induced. This may necessitate a series of groins along a particular segment of the coastline.

### *Implications for Coastal Development*

With an understanding of the natural processes occurring along a coastline, development that minimizes the potential for damage from hazardous processes can be implemented. Individual processes and the area that they may affect must be identified. When this is accomplished, zoning regulations can be utilized to prevent development of particularly hazardous areas. In some cases, development of marginal areas may be allowed if specific protective design criteria are met. In assessment of risks, however, the possibility of extreme events must not be neglected. The magnitude and frequency of these occurrences is a major uncertainty in any risk analysis.

In evaluation of coastline processes, historical records are of invaluable assistance. Maps and charts show changes in shoreline configuration during the period of record. Air photos are particularly useful in this respect.

The lessons of past coastal management projects must not be lost when considering future plans. In the following case study some of the problems of large-scale coastal management schemes are illustrated.

**Figure 14-49** Tiered seawall designed to prevent erosion along the shore of Lake Michigan.

**Figure 14-50** Aerial view of groin field along the Maryland coast. Arrows show individual groins in foreground. (R. Dolan; photo courtesy of U.S. Geological Survey.)

## CASE HISTORY 14-3
## Human Interference with Coastal Processes

The coast of North Carolina is a depositional coast with a nearly continual system of barrier islands (Figure 14-25). An illustrative comparison of altered and unaltered sections of these islands has been provided by Dolan, Godfrey, and Odum (1973).

A typical profile of a barrier island in its natural state is shown in Figure 14-51. This profile, consisting of a broad berm and low irregular dunes, is adjusted to the impact of large storms. Wave energy is efficiently dissipated by frequent overwash over the low natural dunes. Salt-tolerant vegetation species well adapted to periodic overwash inhabit the dunes and overwash terraces behind the dunes.

Although the barrier islands are well adjusted to extreme events that characterize their environment, frequent overwash is not compatible with human occupation of the islands. To rectify this situation, nearly 1000 km of sand fence was constructed behind the beach berm in the 1930s. The purpose of the fences was to trap sand, thereby creating a system of high, artificial barrier dunes to protect the islands from overwash. Non–salt-tolerant vegetation was introduced behind the dunes and fertilized in order to stabilize the dunes.

These measures accomplished the short-term objective of making the islands safe for development. Unfortunately, the disturbance of the natural equilibrium state initiated serious long-term changes in the beach profile. Figure 14-51 shows the altered profile 30 years after construction of the artificial dunes. The major change has been the destruction of the original beach berm. The artificial barrier to wave overwash concentrated wave energy upon the beach. Beach widths decreased from an average

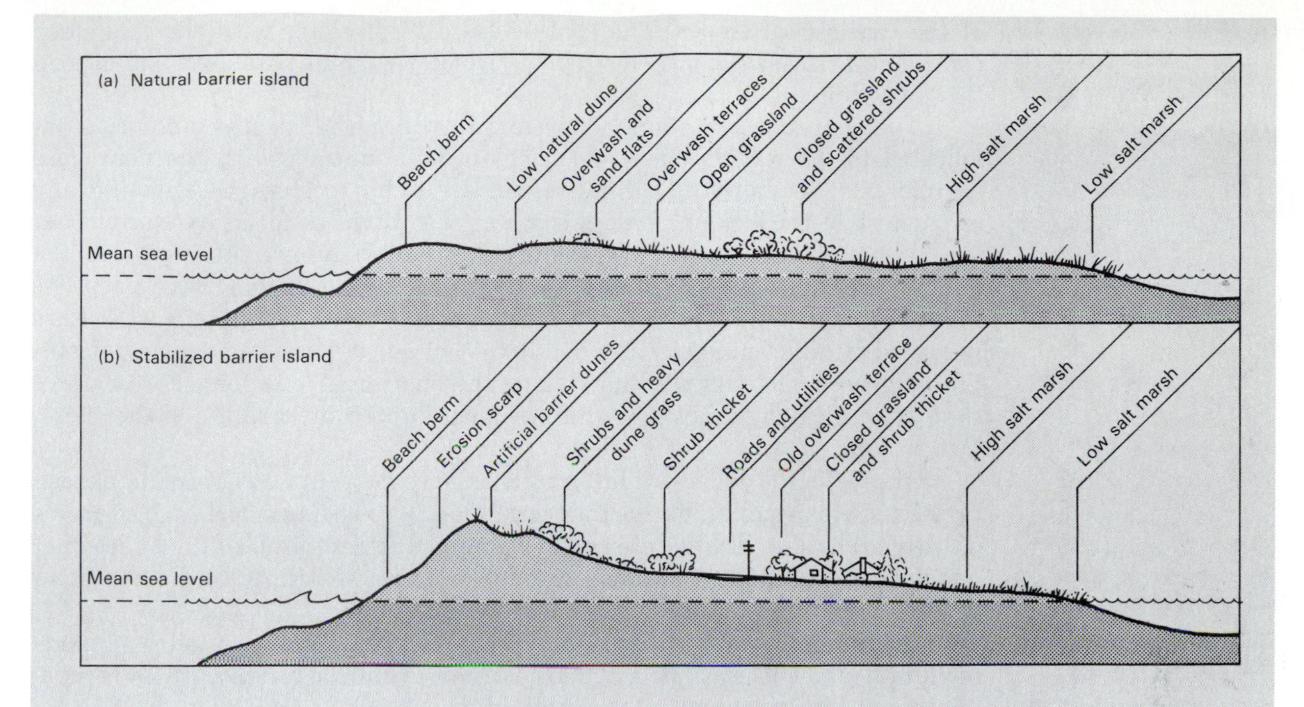


**Figure 14-51** Cross-sectional profiles of barrier islands in their (a) natural state and (b) after construction of barrier dunes. (From R. Dolan and H. Lins, U.S. Geological Survey Professional Paper 1177-B.)

of 150 m to less than 30 m. In some areas wave attack has removed the beaches altogether and has begun to destroy the artificial dune system. This process will continue in the future. When the artificial dune barrier is severed, overwash will again occur, only now roads and buildings located behind the dunes will be affected. The vegetation will die when flooded by saltwater.

The attempted stabilization of the North Carolina barrier islands is an excellent example of the effect of interference with coastal processes. The structures that were built behind the dunes, which were intended to provide protection, are now increasingly threatened by large storms. The cost of maintaining the developed areas will include the cost of building even more extensive and expensive protective structures.

## SUMMARY AND CONCLUSIONS

The oceans support a great quantity and diversity of living organisms. Life on land is affected by the ocean's influence on climate. The circulation pattern of the ocean is composed of large gyres driven by atmospheric circulations that assist in heat transfer between the poles and equator.

Ocean-basin floors have a varied topographic configuration. Midoceanic ridges slope downward through regions of abyssal hills to sediment-covered abyssal plains. Pelagic sediments include inorganic muds as well as organic oozes composed of either calcium carbonate or silica. Calcium carbonate oozes are absent below the carbonate compensation depth.

The continental margins are covered with thick sedimentary sequences. The continental shelves slope gently seaward to a break in slope at the tops of the continental slopes. The continental slopes, along with the less steep continental rises, are characterized by turbidity-current transport and deposition of sediment.

The shoreline is a dynamic system in which the coastal landforms are adjusted to the waves, tides, and currents that interact with the nearshore sediments. The shoreline profile responds rapidly to changes affecting any component of the system. Waves begin as sea waves induced by storms over the ocean and travel toward the shoreline as more regular swell waves. As the circular orbits of wave motion begin to interact with the bottom, wave steepness increases until the waves become unstable and break. Although waves are refracted by differences in water depth, the oblique approach of waves initiates the longshore currents that flow along the shoreline. The longshore drift of sand accompanying longshore currents is an important variable in the shoreline system.

The action of waves, tides, and longshore currents controls the depositional and erosional landforms that occur along a coastline. The beach responds quickly to changes in wave energy, developing different profiles under summer and winter conditions. Coastlines can be classified according to the dominant processes that have shaped their morphology.

The dynamic nature of coastal geologic processes makes coastal management difficult. The possibility of such extreme events as hurricanes and tsunamis is not always taken into consideration in coastal development. Coastal engineering structures are utilized in an attempt to control certain coastal processes, but the long-term effects of these structures often cause damage to the beach or coastline because there is often a lack of understanding of the way in which natural processes interact within the coastal system.

## PROBLEMS

1. Give a general overview of the topography of the ocean basins.
2. What is the cause and mode of travel of turbidity currents?
3. Why do waves break?
4. What effects could wave refraction have upon shoreline development?
5. Why is there such a large tidal range from place to place and over time at a particular place?
6. What are the engineering implications of longshore currents?
7. Describe the typical beach profile and explain how it changes during the year.
8. What are the hazards associated with tropical cyclones?
9. If you were asked to design an artifical harbor, what types of geologic, hydrologic, and oceanographic data would you need to gather before design and construction begin?

## REFERENCES AND SUGGESTIONS FOR FURTHER READING

DOLAN, R., and H. LINS. 1985. *The Outer Banks of North Carolina.* U.S. Geological Survey Professional Paper 1177-B.

DOLAN, R., P. J. GODFREY, and W. E. ODUM. 1973. Man's impact on the barrier islands of North Carolina. *American Scientist,* 61:152–162.

GIARDINO, J. R., P. E. ISETT, and E. B. FISH. 1984. Impact of Hurricane Allen on the morphology of Padre Island, Texas. *Environmental Geology and Water Sciences*, 6:39–43.

HAY, E. A., and A. L. MCALESTER. 1984. *Physical Geology: Principles and Perspectives*, 2d ed. Englewood Cliffs, N.J.: Prentice-Hall, Inc.

JOHNSON, J. W. 1975. The littoral drift problem at shoreline harbors. *Journal of Waterways and Harbors Division*, 83:1–37. American Society of Civil Engineers.

JUDSON, S., M. E. KAUFFMAN, and L. D. LEET. 1987. *Physical Geology*, 7th ed. Englewood Cliffs, N.J.: Prentice-Hall, Inc.

KOMAR, P. D. 1976. *Beach Processes and Sedimentation*. Englewood Cliffs, N.J.: Prentice-Hall, Inc.

PRESS, F., and R. SIEVER. 1982. *Earth*, 3d ed. San Francisco: W. H. Freeman.

WILLOUGHBY, R. H., and P. G. NYSTROM, JR. 1990. Some effects of Hurricane Hugo on shoreline landforms, South Carolina coast, Columbia, September 21–22, 1989. *South Carolina Geology*, 33:39–58.

# 15

# Glacial Processes and Permafrost

Those of us who live in the northern states of the United States or in other high-latitude regions would be hard pressed to select a process more important in shaping our present landscape than glaciation. Although glaciers currently cover only about 10% of the earth's land surface, more than 30% of the surface was buried by ice during the repeated advances of continental glaciers during the Pleistocene Epoch (Figure 15-1). The importance of glaciers, however, goes well beyond the areas that actually were glaciated. The modern drainage system in many nonglaciated parts of the United States owes its origin to glaciation. For example, the evolution of the Mississippi Valley is intimately related to the advances and retreats of continental glaciers. During retreat, floods of meltwater were funneled through the Mississippi Valley into the Gulf of Mexico. In addition, the modern Mississippi would be a much smaller stream without the contribution of water from the Missouri River, which flowed north to Hudson Bay before glaciation. The present valley of the Missouri was established by diversion of meltwater streams along the margin of the ice sheet.

The great coastal cities of the world are dependent upon the current position of sea level for their economic existence. Any sustained climatic change involving advances or retreats of existing glaciers will either leave these ports high and dry or partially submerged. Currently, a rising trend in sea level is causing concern in many parts of the world. Some scientists believe that an increase in carbon dioxide and other gases in the atmosphere, caused by the burning of fossil fuels, will produce warmer conditions and even higher sea levels.

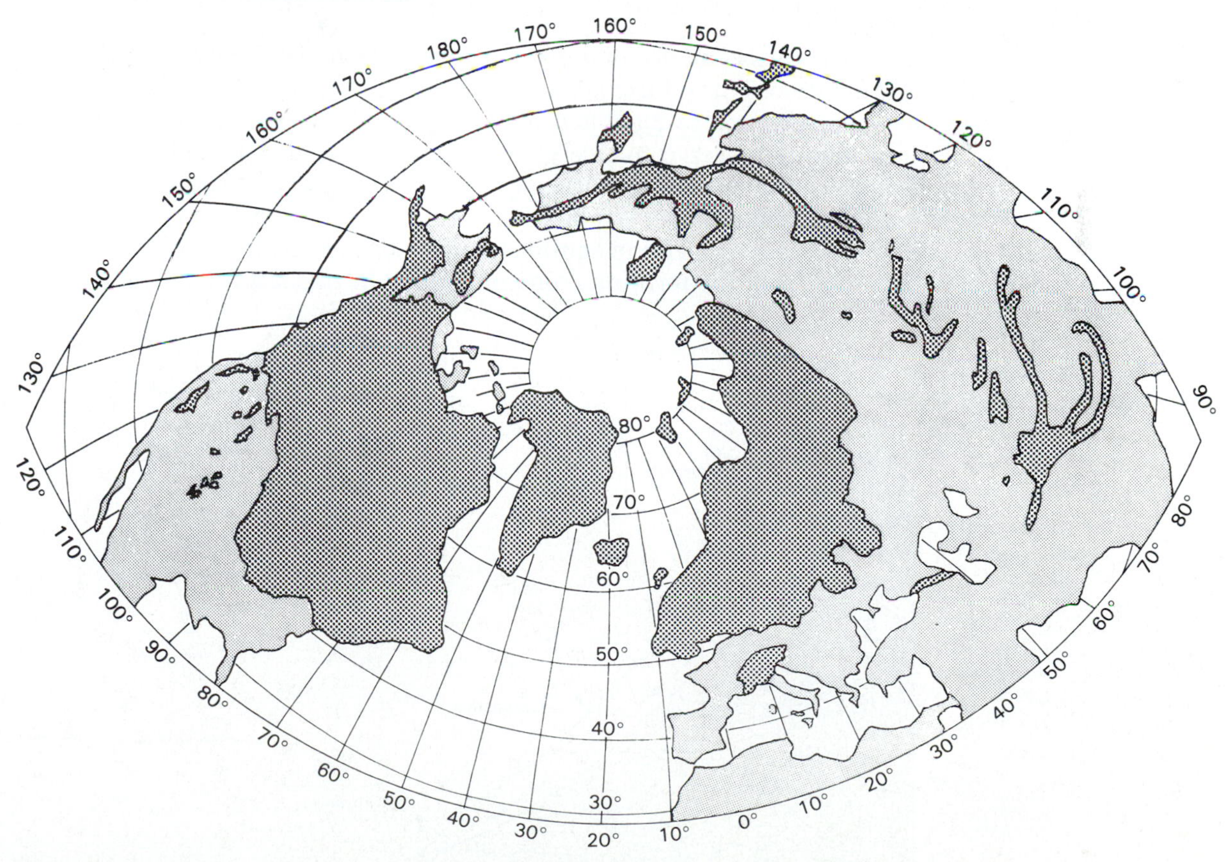


**Figure 15-1** Extent of Pleistocene glaciations (darker areas) in the Northern Hemisphere. (After E. Antevs, 1929, *Geological Society of America Bulletin*, 40.)

## GLACIERS

Snowflakes fall to the earth in a variety of delicate, complex crystalline forms. On the ground, several processes begin to transform the new-fallen snow. Melting and sublimation (the direct conversion of ice to water vapor) occur. Compaction by new layers of snow reduces the volume and increases the density of the underlying material. Gradually, snow is converted to *firn*, a denser granular substance. Above an elevation called the *snow line*, some of each winter's snow remains after the summer melting season. If the net gain in snowfall is significant over a period of years, a growing body of firn and snow may become a *glacier*. A glacier is composed of an interlocking network of ice crystals formed by additional compression and recrystallization of firn. In addition, a glacier is sufficiently massive to flow downslope under its own weight or to have previously moved in this manner.

### Mass Balance

As a glacier flows downslope from the area in which there is a net accumulation of mass, it experiences progressively warmer temperatures, which cause a net loss of mass. Thus a glacier can be divided into two zones: a *zone of accumulation* and a *zone of ablation* (Figure 15-2). Ablation is a term that includes both melting and sublimation in the zone of net loss of mass. The boundary between the zones of accumulation and ablation is called the *equilibrium line*.

The mass balance of a glacier is the relationship between accumulation and ablation. This relationship determines the position of the glacier's *terminus* or *snout* (Figure 15-2). When total accumulation exceeds total ablation, the terminus advances; conversely, the terminus retreats when ablation is greater than accumulation. A condition of equilibrium, or exact balance between accumulation and ablation, is indicated by a stable terminus. It must be emphasized, however, that a glacier in equilibrium may be in continual motion even though the position of the terminus remains stable. In this situation, the loss in mass

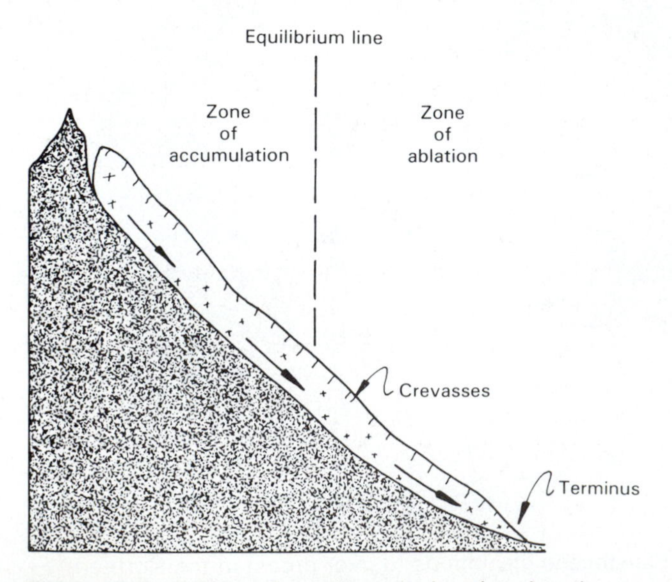


**Figure 15-2** Profile of a valley glacier showing the components of its mass balance.

from the ablation area is exactly compensated for by the downslope flowage from the zone of accumulation.

### *Movement*

The flow of glaciers is a complex phenomenon involving a combination of several physical processes. One component of movement is derived from the visco-plastic internal deformation of ice, or the tendency of ice to flow downslope under its own weight when a sufficient thickness is achieved. Glacial flow involves elements of both viscous and plastic behavior (Chapter 6). A critical (yield) stress must be exceeded for a plastic material to deform. The yield stress for ice is about 100 kPa (1 bar). When glacier ice is subjected to this magnitude of stress, which can be imposed by the downslope component of the weight of the glacier, the ice will flow.

The plastic flow of ice also contains an element of viscous behavior in which the flow velocity is proportional to the amount of shear stress. An equation called *Glen's Flow Law* approximates the visco-plastic flow of ice. This equation can be stated as

$$\dot{\gamma} = A\tau^n \qquad \text{(Eq. 15-1)}$$

where $\dot{\gamma}$ is the strain rate or velocity of the ice, $\tau$ is the shear stress, and $A$ and $n$ are empirical constants. The coefficient $A$ is temperature dependent, and the value of $n$ ranges between 1.9 and 4.5. Like the flow of water in streams, the flow of ice is opposed by friction between the bed and the glacial ice. Therefore, velocity is greatest at the ice surface and decreases toward the bed (Figure 15-3). An important difference between glacier and stream flow, however, is that the flow of glacial ice is laminar rather than turbulent. The increase in velocity toward the surface of the ice seems to conflict with Glen's Flow Law, which indicates that strain rate should be higher near the base of the glacier because the weight of the overlying ice, and therefore the shear stress, is higher there. This incongruity can be explained if the glacier is thought of as a vertical sequence of parallel layers, each with an individual strain rate. The lowest layer in the stack would indeed have the highest strain

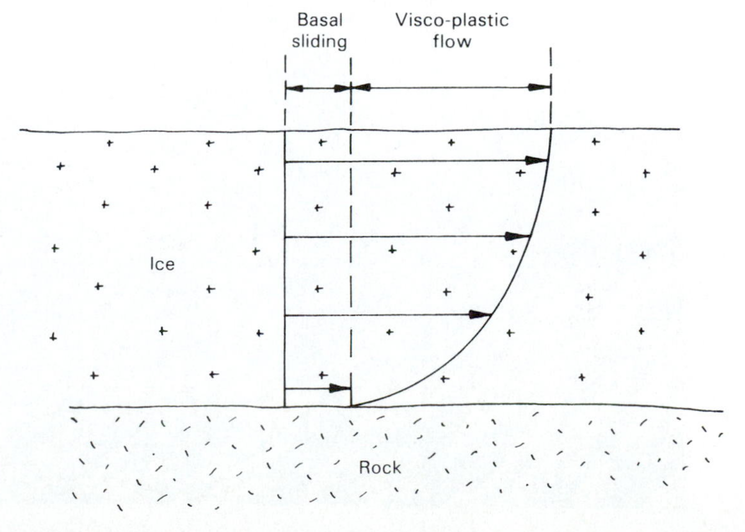


**Figure 15-3** Velocity profile of a glacier showing the two components of motion.

rate because of the weight of the overlying layers of ice. Total velocity, however, would be the summation of the individual strain rates for all layers of ice in the stack. Therefore, although the lowest layer would have the highest individual velocity, each overlying layer would move at a rate determined by the sum of its own individual strain rate plus the cumulative strain rates of all layers beneath. Thus velocity does increase from the base to the surface even though strain rate decreases in that direction.

One complication of the flow process is caused by the brittle rupture of the upper portion of the glacier. Because of the lower pressure near the surface of the glacier, a network of vertical cracks called *crevasses* is commonly present (Figures 15-2 and 15-4), particularly where the glacier passes over a steeply sloping bedrock surface.

A second major component of glacier movement, which may take place along with visco-plastic flow, is *basal sliding*. The tendency of a glacier to slide as a block above its bed is dependent upon the thermal condition at the base of the ice. Because the thermal conductivity of ice is low, a glacier several kilometers thick acts as an insulating blanket to trap geothermal heat that is constantly flowing upward toward land surface. This heat may be sufficient to melt a small amount of ice at the base of the glacier. The melting point of ice is lowered as pressure increases, so ice will melt at a lower temperature at the base of a glacier than at the surface. Thus glaciers may be divided into two major types: *cold-based*, which contain no water at their beds, and *wet-based*, which do contain basal water. Cold-based glaciers are at a temperature below the pressure melting point of ice throughout their vertical profile and are therefore frozen to their beds. For this reason, cold-based glaciers move by visco-plastic flow only. Wet-based glaciers, on the other hand, are in effect decoupled from their beds by the presence of water between the ice and the bed. These glaciers may therefore slide upon their beds. Frictional heat produced by sliding provides an additional basal heat source. Typical estimates of basal sliding rates range between 10 and 25% of the total velocity of the glacier. Glaciers whose movement is intermediate between wet-based and cold-based

**Figure 15-4** Crevasses on the surface of Columbia glacier, Alaska. (M. R. Meier; photo courtesy of U.S. Geological Survey.)

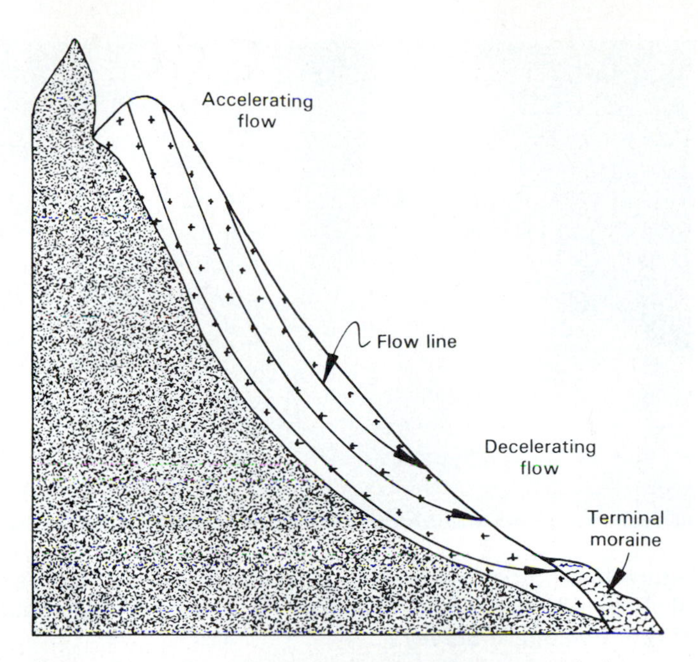


**Figure 15-5** Flow lines converge in the accumulation area (accelerating flow) and diverge in the ablation zone (decelerating flow). Where flow lines curve upward near terminus, sediment is brought to glacier surface and concentrated by ablation.

are common. These include glaciers that are frozen to their beds in some areas and unfrozen in others. A common situation is to find a glacier unfrozen beneath its interior and frozen near its snout.

A recent hypothesis provides another mechanism of glacier movement. This idea attributes a significant component of total ice movement to deformation of weak, subglacial sediment. Thus shearing within weak, saturated sediment beds, rather than glacial sliding upon a bed, would be the dominant process in the advance of some modern as well as ancient glaciers. This hypothesis is controversial among glaciologists and glacial geologists.

An important aspect of glacial flow is the difference between accelerating and decelerating flow. Accelerating flow is common above the equilibrium line, where ice is thickening, and decelerating flow occurs below the equilibrium line, where the ice is either thinning or advancing upslope (Figure 15-5). A consequence of changes in flow velocity is that flow lines—imaginary lines oriented in the direction of movement at a particular point in the glacier—are not always parallel. Flow lines converge in areas of accelerating flow and diverge in areas of decelerating flow. At glacier margins, where the ice is thin and flow is accelerating, flow lines are directed upward toward the ice surface. This may cause thrusting along shear planes within the ice of active ice over stagnant or more slowly moving ice at the margin. Decelerating flow also provides a mechanism for bringing rock and sediment eroded from the bed upward into the ice and for concentrating it at the glacier's margin.

### *Erosion*

The amount of erosion accomplished by a glacier flowing over an area of land can vary from insignificant to deep scouring of the landscape. Variables that control glacial erosion are complex: They include the thermal condition of the glacier bed, the topography and lithology of the bed material, and many others.

**Figure 15-6** Striations on bedrock produced by glacial erosion, Isle Royale National Park, Michigan. (N. K. Huber; photo courtesy of U.S. Geological Survey.)

The processes of erosion, however, are relatively few in number.

The sliding movement of a wet-based glacier over a bedrock surface is unmistakably preserved by the polished and grooved bed evident after glacial retreat. Elongated grooves and other types of marks on a glaciated rock surface are an indication of glacial *abrasion*. The erosional marks vary from striations (Figure 15-6) on a single outcrop to grooves a meter or more in width and depth that can be traced for kilometers on the land surface. Abrasion is accomplished by rock fragments that are dragged along the bed rather than scoured by the ice itself. This crushing and grinding action reduces the particle size of transported rock fragments. The marks produced by erosion are an excellent indicator of ice-movement direction. Abrasion is not effective in cold-based glaciers because the ice is not sliding over its bed.

Incorporation of large blocks of bedrock into the ice is possible by a process known as *plucking*. Plucking is most prevalent in wet-based glaciers and occurs as the ice moves over bedrock obstacles that protrude above the surface of the bed (Figure 15-7). The process is controlled by the pressure distribution around the obstacle. On the up-glacier side, high pressures develop as the ice contacts

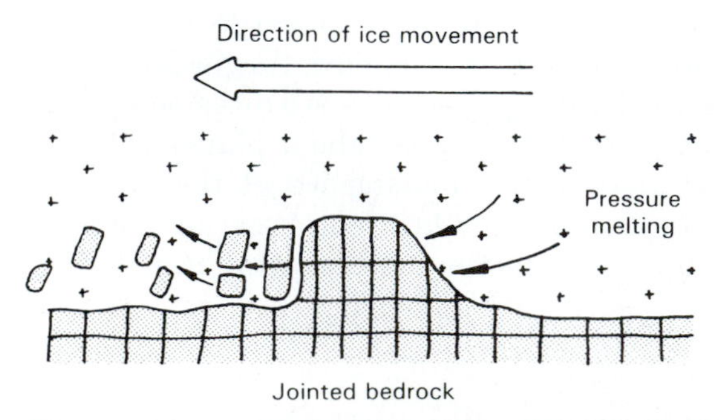


**Figure 15-7** Plucking takes place when ice melts on the upstream side of bed obstacles, and then refreezes within joints and fractures on the downstream side. Large blocks can then be incorporated into the flowing ice.

**Figure 15-8** Roche Moutonnée, a glacial landform produced by erosion and plucking. Glacial flow from left to right. (Jasper National Park, Alberta.)

the protrusion. If the pressure-melting temperature of the ice is reached, ice melts and flows around the obstacle to the region of lower pressure on the opposite side. Here the water refreezes around and within the joints and cracks of the rock. This weakens the rock mass, and larger fragments can be broken from the down-glacier rock faces and transported within the ice. Rock outcrops subjected to plucking during glaciation frequently have a characteristic shape. In profile, this shape consists of an abraded, gently sloping up-glacier—*stoss*—side, which terminates at a near-vertical, or *lee*, face formed by plucking on the down-glacier side. These asymmetric landforms are known as *roches moutonnées* (Figure 15-8).

In the marginal area of a glacier, large-scale erosion of rock or sediment may occur by *ice-thrusting*, particularly when this part of the glacier is frozen to its bed (Figure 15-9). Shearing of the rock or sediment takes place below the bed, mobilizing slabs of material up to tens of meters thick and kilometers in length. These blocks then are thrust upward into the ice along the shear planes that develop within decelerating flow zones. The ice-thrust blocks are transported and deposited as isolated hills or ridges marking the former ice

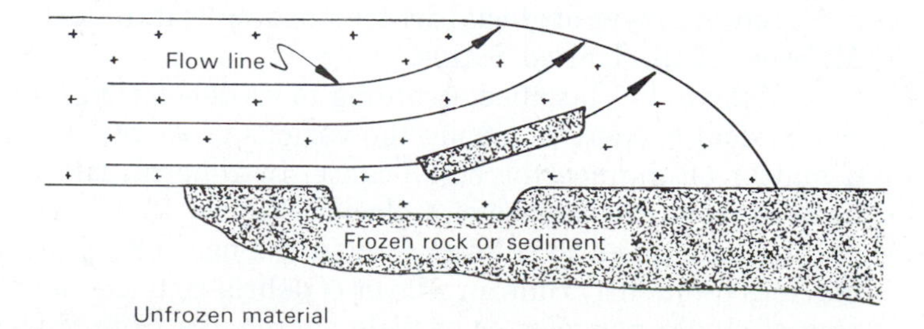


**Figure 15-9** Ice thrusting takes place near the glacier terminus when the sediment or rock is frozen. Large slabs or blocks are carried up into the ice along shear planes oriented parallel to the diverging flow lines.

**Figure 15-10** Folds and faults in glacial sediment produced by thrusting, near Moose Jaw, Saskatchewan. Contacts of folded bed of clayey glacial sediment shown with dashed lines.

margins. The stratigraphy within these deposits is extremely complex (Case Study 15-1), because the ice-thrust masses of rock or sediment are highly deformed by folding and faulting during erosion and deposition (Figure 15-10).

### *Deposition*

*Drift* is an umbrella term for all deposits associated with glaciers. These materials include sediment deposited directly by the ice itself as well as sediment deposited on, within, beneath, or in front of the ice by streams or in lakes. Glacial drift must be subdivided, however, to account for the various depositional processes and characteristics of the resulting sediment (Table 15-1).

Poorly sorted, nonstratified sediment deposited directly in contact with glacial ice is called *till* (Figure 15-11). The lack of sorting and stratification in till, which contains particles ranging in size from clay to boulders, usually makes it easy to differentiate till from other sedimentary deposits. Till is the most common type of glacial drift; for example, it underlies most of the northern Midwest of the United States.

Till can be classified according to its depositional process; the differences in physical properties among the various types of till make their recognition a matter of engineering significance. In general, till can be deposited at the base or from the surface of a glacier (Figure 15-12). *Basal*, or *lodgement*, till is the name given to till deposited at the base of a glacier. Wet-based, sliding glaciers frequently contain a layer of debris-rich ice just above the bed. Deposition of single particles or particle aggregates from this zone can occur if the frictional resistance to movement between the particles and the bed is greater than the shear stress exerted by the glacier. Melting of ice in the debris-rich zone is another important component in the lodgement process. The resulting

**TABLE 15-1**

**Types and Characteristics of Glacial Drift**

Till: Nonsorted, nonstratified sediment deposited by glacial ice at the base, from the interior, or from the top of a glacier.

Basal (lodgement) till: Dense, compact till deposited at the base of a wet-based glacier.

Ablation till: Till deposited as debris accumulates on top of a glacier and is gradually lowered as the ice melts from beneath. Tends to be less dense than basal till and more variable in grain size and sorting.

Ice-contact stratified drift: Drift sorted and stratified by meltwater in contact with glacial ice. Bedding disrupted by subsequent melting.

Outwash: Coarse-grained sediment deposited by meltwater in front of a glacier in outwash plains or valley trains.

Lacustrine sediment: Deposits of glacial lakes, mostly fine grained and often varved. Coarse-grained deposits occur in fans and deltas associated with the lakes.

Eolian sediment: Wind-blown sediment derived from other forms of drift. Very well sorted; grain size ranges from sand to silt (loess).

Glacial-marine sediment: Mostly fine-grained, clay-rich sediment deposited by glaciers that advance to the coastline. Fine and coarse material dropped from calving ice and floating ice bergs.

**Figure 15-11** Exposure of glacial till, Pierce County, Washington. Notice poor sorting and lack of stratification. (D. R. Crandell; photo courtesy of U.S. Geological Survey.)

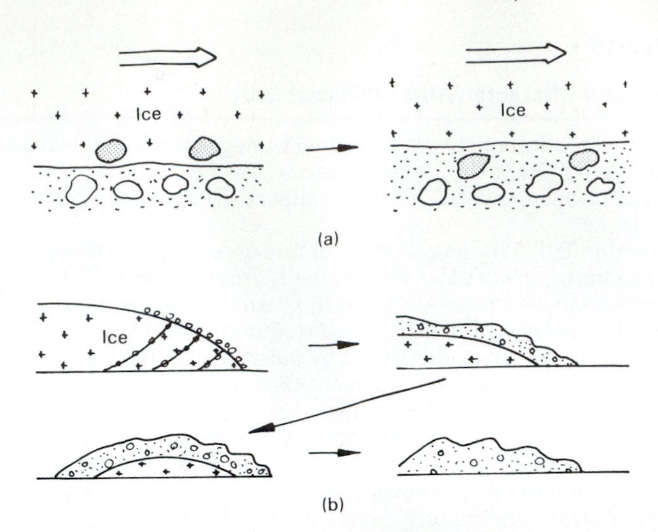


**Figure 15-12** Deposition of (a) lodgement till and (b) ablation till. In (a), particles transported near the base of the sliding ice are separated from the ice and added to the bed material below, because of frictional resistance to transport and pressure melting. Ablation till (b) originates from sediment accumulated on the glacier surface. It becomes thicker and is deposited as the ice melts from below.

lodgement till is a very densely compacted material because of the great weight of the overlying glacier.

*Ablation* till originates as debris is brought to the glacier surface by thrusting or flow along the upward-trending flow lines in the ablation area of the glacier (Figure 15-5). Debris is concentrated on the surface as the ice surface melts slowly downward. In the melting process, the debris is continually reworked by meltwater and mass movement (Figure 15-12). Debris flows are common as the saturated, clay-rich sediment will flow down even gentle slopes. Deposition is not complete until all the underlying ice is melted. This depositional process results in a loosely compacted till that may be better sorted than basal till. These two types of till therefore have major differences in particle-size distribution, density, strength, and compressibility.

During the summer, abundant meltwater is produced on the surface of a glacier, as it was in great quantities during the retreat of the continental ice sheets in past glaciations. This water moves toward the ice margin either on top of, within, or beneath the glacier. Tunnels within or beneath the glacier commonly convey meltwater to the terminus. Lakes on top of the ice may also be present from time to time. The sediment deposited by these streams or in ponded water is known as *ice-contact stratified drift.* This material is highly variable in grain size, sorting, thickness, and areal extent. In surface exposures, ice-contact stratified drift often can be recognized by folding, faulting, and other deformational structures produced during the melting of the glacier after deposition of the sediment. Isolated bodies, lenses, or beds of ice-contact stratified drift may be buried or interbedded with till in a glacial sequence. Location of these deposits is relevant to excavation, foundation design, and the successful siting of water wells.

Once the meltwater from the glacier reaches the snout, it may either be ponded in *proglacial lakes* or flow along or away from the ice in meltwater streams. The sediment deposited in these environments is known as glacial-lacustrine sediment, and *outwash,* respectively. When continental ice sheets retreated, large volumes of water were trapped in proglacial lakes. The largest

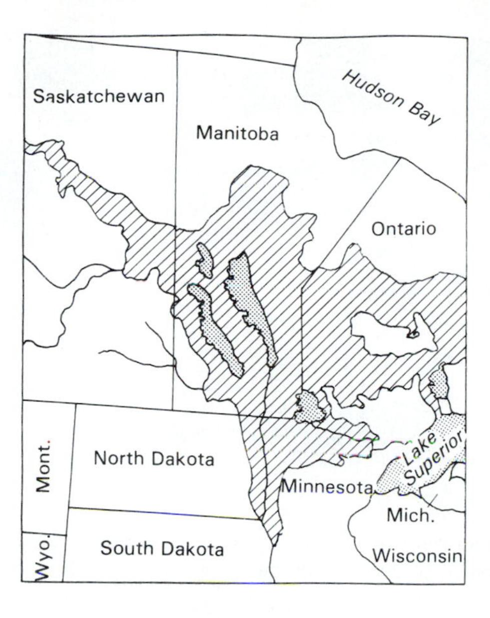


**Figure 15-13** Map of glacial Lake Agassiz at its maximum extent. The dark areas are modern remnants of the lake. (From J. P. Bluemle, 1991, *The Face of North Dakota,* North Dakota Geological Survey Educational Series 21.)

of these in North America was glacial Lake Agassiz (Figure 15-13). Glacial-lacustrine sediment is mostly fine-grained, bedded silt and clay. Low strength and high compressibility are characteristic of these sediments. Glacial-lacustrine sediment is sometimes interpreted to be *varved*. Varves are thin, alternating laminations of clay and silt or sand (Figure 15-14). The dark-colored clay beds are thought to be deposited during the winter, when the lakes are ice covered; and the coarser, light-colored silt or sand beds are deposited during the summer, when glacial meltwater carries more sediment into the lake. Each varve pair, or couplet, therefore implies one year's sedimentation in the lake.

Glacial outwash (Figure 15-15) is alluvial sediment deposited by braided meltwater streams. Outwash is coarse grained and may be quite thick in the valleys that carried meltwater. Outwash is a good source of aggregate material for road construction, concrete, and other purposes. Outwash deposits are also highly permeable and may contain excellent aquifers.

**Figure 15-14** Varved glacial lake deposits (light- and dark-colored banded unit), Sanpoil Valley, Washington. The rock at center (arrow) was dropped from an iceberg floating on the lake.

**Figure 15-15** Cross-bedded, coarse-grained outwash exposed in a gravel pit, north-central North Dakota.

*Eolian* sediment and *glacial-marine* sediment are less common but may be extremely important in some areas. Eolian sediment is derived from outwash and other types of glacial drift and then transported and deposited by wind. It is very well sorted and varies in grain size from sand to silt. Deposits of eolian silt are called *loess*. Loess has the unusual property (for nonlithified sediment) of forming stable vertical cuts. The stability is provided by bonds between clay particles composed of calcium carbonate, clay, or moisture. If these soils become saturated or subjected to excessive disturbance, the previously stable slopes may fail, or compaction may result (Chapter 10). Glacial-marine sediment is deposited where glaciers terminate at the ocean. Large blocks of ice calve off from the glacier and float away. As they melt, debris sinks to the sea floor and is deposited as glacial-marine sediment. This material is clay rich and stratified, but it also contains large particles that were incorporated into the ice.

## TYPES OF GLACIERS

The major types of glaciers are shown in Table 15-2. *Ice sheets* and *ice caps* are large expanses of ice that have a dome-shaped profile (Figure 15-16) with a much greater thickness near the center than at the edge. Ice builds up to such great thicknesses near the center of the dome, that it completely buries the topography, including whole mountain ranges. Expansion and contraction of the dome is therefore not controlled by topography, but instead by the mass

**TABLE 15-2**

**Types of glaciers**

| |
|---|
| Ice sheets and ice caps. |
| Icefields. |
| Valley glaciers and outlet glaciers. |
| Cirque glaciers. |
| Ice shelves. |

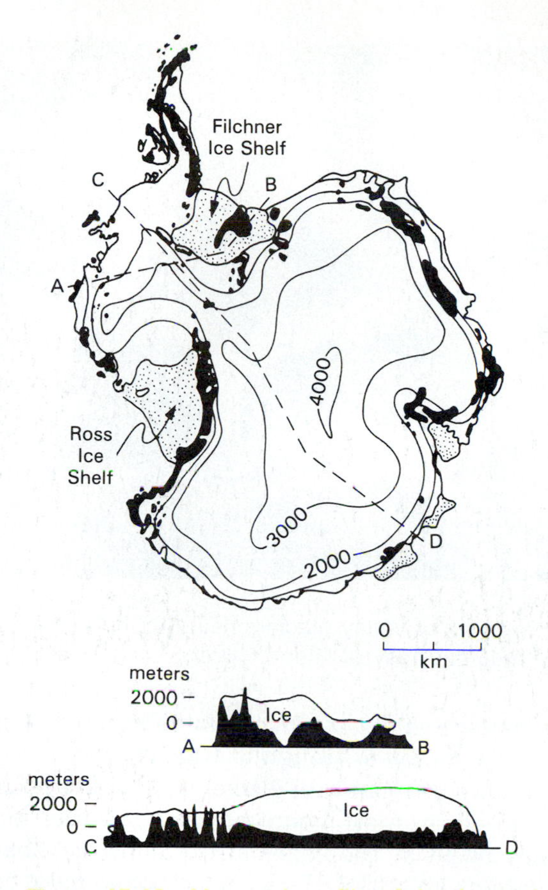


**Figure 15-16** Map and profile of the Antarctic ice sheet.

balance of the ice sheet. The difference between an ice sheet and an ice cap is simply based on size. Ice sheets must be greater than 50,000 $km^2$ in size. Only two ice sheets, the Antarctic and Greenland ice sheets, remain on the earth at the present time. During the Pleistocene Epoch, other ice sheets included the Laurentide, which was centered on Hudson Bay and covered most of Canada and the northern United States east of the Rocky Mountains.

*Icefields* are also very large, wide glaciers, but lack the dome-shaped profile of ice sheets and ice caps. Icefields may have a gently undulating surface profile and mountain peaks may project above the glacier (Figure 15-17).

**Figure 15-17** View from top of Wapta icefield, Alberta.

**Figure 15-18** Network of valley glaciers near Mount McKinley (Denali), Alaska. (Photo courtesy of Glenn Oliver.)

*Valley glaciers* and *outlet glaciers* are long narrow glaciers whose shape is controlled by the valley or trough in which they flow (Figure 15-18). Valley glaciers, also known as mountain glaciers and alpine glaciers, flow downslope in narrow valleys from high mountain peaks. Outlet glaciers are similar in shape, but they begin at the edges of ice fields, ice sheets, or ice caps. The valley that contains an outlet glacier may extend back beneath the ice cap or ice sheet as a subglacial trough, producing a narrow zone of more rapid flow in the ice sheet called an *ice stream*.

Common only in Antarctica, *ice shelves* are thin slabs of ice that extend beyond the coastline in bays and float over the denser salt water, while maintaining a connection with glaciers on shore (Figure 15-16).

### *Valley Glaciers*

Along with vanished ice sheets from past glaciations, modern valley glaciers are much more limited in size and extent than they were during the Pleistocene Epoch. Today, extensive valley glaciers develop only in high latitudes in areas of substantial precipitation. These glaciers form networks analogous to river systems, with tributary glaciers joining larger glaciers that occupy major valleys.

A region that has contained valley glaciers can be recognized by characteristic landforms of both erosional and depositional origin. The most distinctive erosional landform is the shape of the glaciated valley itself. Glacial erosion modifies the cross-sectional profile of a valley from the V-shape common to valleys produced by stream erosion to a U-shape, characteristic of glaciated valleys (Figure 15-19). The U-shaped profile of glaciated valleys results in valley sides that are steeper than their counterparts in V-shaped stream valleys. Glaciated valleys are said to be oversteepened because the steep valley sides are prone to rockfall, rockslide, and other forms of mass wasting. These conditions may be hazardous for highway construction or other development in glaciated valleys.

In a glaciated mountainous area, valley glaciers extend, or once extended downslope in all directions from the central region of highest elevation. In

**Figure 15-19** The U-shape of this valley in Yosemite National Park indicates that it was once occupied by a valley glacier. (Photo courtesy of Dexter Perkins III.)

addition to U-shaped valleys, other erosional landforms are characteristic to these areas. The bowl-shaped depression at the base of a mountain peak where glaciers form and begin to flow down valley is called a *cirque* (Figure 15-20). If a glacier is confined by temperature or lack of precipitation to the cirque, it is known as a *cirque glacier* (Table 15-2). Due to erosion by the cirque glacier and the action of physical weathering processes such as freeze and thaw, cirques are gradually enlarged by headward erosion and deepened. Small lakes, or *tarns*, are common in the scoured floors of cirques that formerly contained glacial ice. The narrow ridge that separates two glaciated valleys often is quite steep and sharp. It is called an *arête*, a French word meaning "sharp edge." When a peak contains cirques on several sides, headward erosion in the cirques begins to modify the topography of the peak itself. A gap produced in a ridge by the headward erosion of glaciers in two cirques from opposite sides of the ridge is a *col*. Where more than two glaciers erode headward at the base of a mountain peak, an isolated, sharply pointed pinnacle known as a *horn* is formed (Figure 15-20). The Matterhorn in the Alps of Switzerland is a horn.

The rock and sediment eroded from the cirque and valley walls by plucking and abrasion is transported down valley toward the glacier terminus and is deposited as drift in several types of landforms. A ridge of drift deposited by the glacier at its point of maximum advance is called a *terminal moraine* (Figure 15-21). Terminal moraines are usually composed of till and ice-contact stratified drift and may be deposited by subglacial or ablation processes. Frequently, meltwater emerges from the terminus in tunnels or from the glacier surface and deposits outwash in front of the glacier. These deposits are formed by braided fluvial systems because of the extremely high sediment load, and they may occupy the entire width of the valley bottom (Figure 11-24). In this case the outwash deposit is known as a *valley train*. Valley-train sediments are composed of coarse sand and gravel and decrease rapidly in grain size with distance from the glacier (Figure 13-28).

As a valley glacier retreats, the terminal moraine may impound a proglacial lake (Figure 15-22). In addition, during a period of net retreat, the glacier terminus may stabilize or even readvance short distances. The result may be a series of concentric *recessional* moraines in a glaciated valley.

(a)

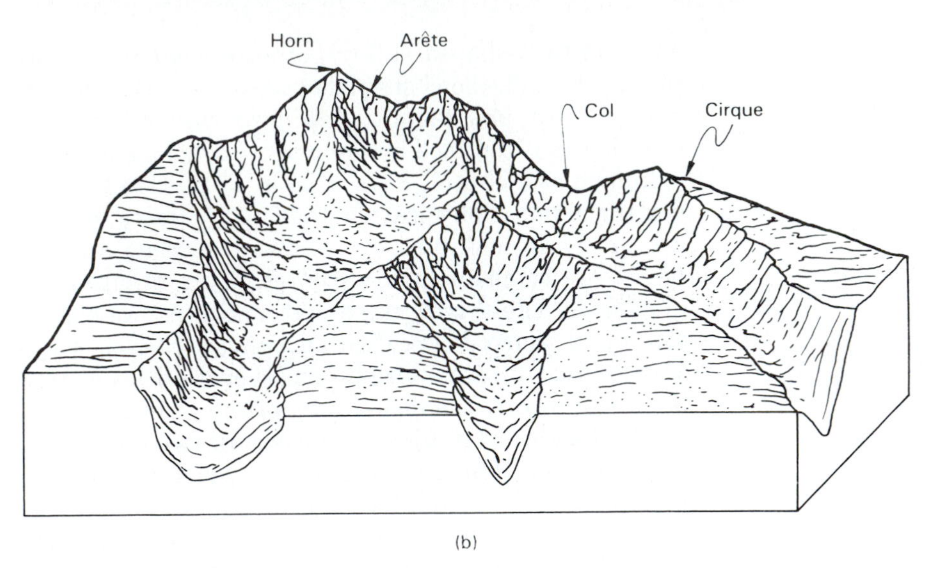


(b)

**Figure 15-20** (a) Cirques and arêtes in the Sierra Nevada Mountains, California. (F. E. Matthes; photo courtesy of U.S. Geological Survey.) (b) Diagram of valley-glacier landforms. (From W. M. Davis, 1911, reproduced by permission from the *Annals of the Association of American Geographers*, 1:57.)

A narrow ridge of drift is also deposited along the lateral edge of valley glaciers and of outlet glaciers against the valley wall. The elevation of these *lateral moraines* is an indication of the maximum elevation of a particular glacier in the valley (Figure 15-23). In a series of active, coalescing valley glaciers, the lateral moraines from glaciers in tributary valleys can be traced as longitudinal debris bands in the glacier in the main valley. These *medial moraines* (Figure 15-24) may or may not be preserved as ridges extending down the center of the valley after glacial retreat.

**Figure 15-21** A terminal moraine (arrow) near Ennis, Montana deposited by a valley glacier emerging from the U-shaped valley in the background. Faint channel markings can be seen on the gently sloping outwash plain in front of the moraine.

### *Ice Sheets*

The distribution of deposits of Pleistocene glaciers in the Northern Hemisphere (Figure 15-1) indicates that huge ice sheets, also known as continental glaciers, were not limited to high latitudes. These immense glaciers flowed southward across the Canadian border into the United States (Figure 15-25). The largest ice sheet, the Laurentide, penetrated southward to the approximate positions of the Ohio and Missouri Rivers. It was as much as 4 km thick at its center in the Hudson Bay region.

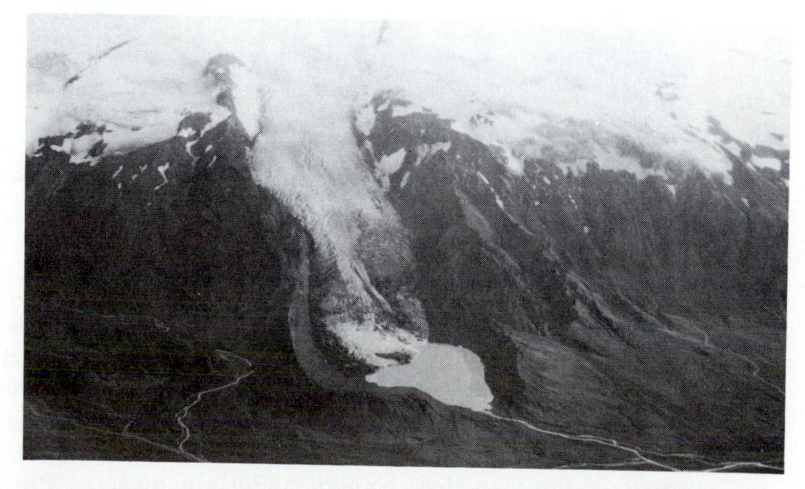

**Figure 15-22** A small proglacial lake in southern Iceland impounded between a retreating outlet glacier and its terminal moraine.

**Figure 15-23** Lateral moraine (arrow) at edge of Athabasca glacier, Alberta.

**Figure 15-24** Medial moraines (*M*) in Ruth Glacier, Alaska. (Photo courtesy of Glenn Oliver.)

The processes of erosion and deposition associated with continental ice sheets are similar to those of valley glaciers, although they vary tremendously throughout the extent of the glaciated area. Landforms and deposits present in any particular area are the result of the characteristics of the topography, bedrock material, and the glacier-bed thermal regime. For example, broad areas of the Canadian Shield, underlain by hard Precambrian igneous and metamorphic rocks, were subjected to relatively shallow erosion. Other regions, such as the basins of the Great Lakes, were deeply scoured into the bedrock,

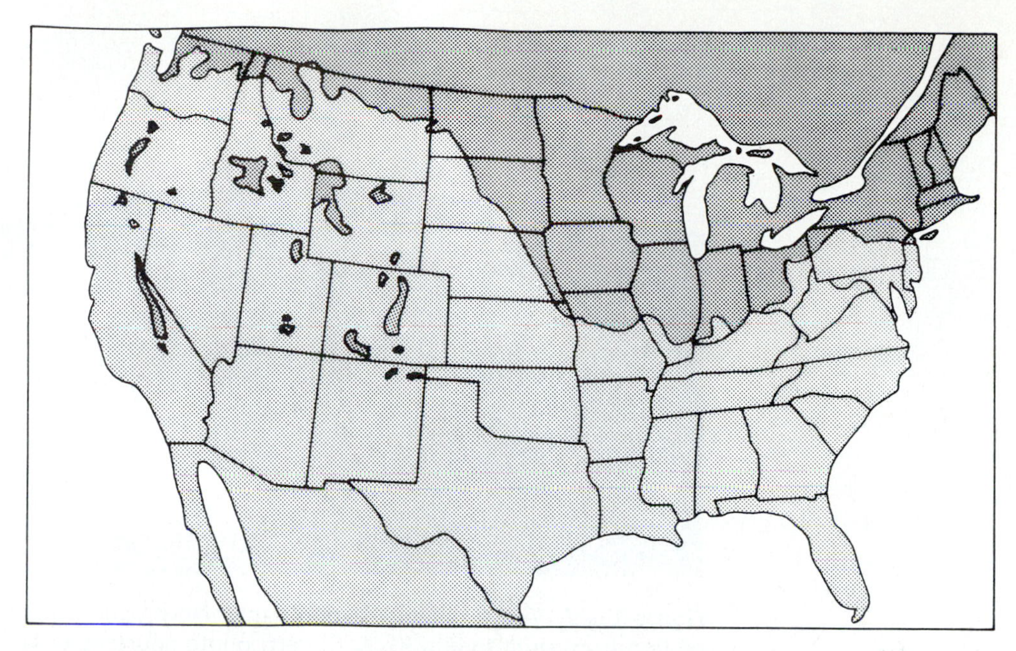

**Figure 15-25** Extent of Pleistocene glaciation in the United States (darker area). (From R. F. Flint, 1945, *Glacial Map of North America*, Geological Society of America Special Paper 60.)

and deposits of glacial drift tens of meters to more than a hundred meters thick blanket some parts of the plains of the United States and Canada.

Deposits of both lodgement till and ablation till are common. Thick accumulations of ablation till are common in terminal moraines and in other areas of decelerating flow, where large amounts of sediment were sheared upward into the ice and concentrated at the glacier surface. The topography of ablation-till deposits is called *hummocky* (Figure 15-26) because it consists of alternating mounds (hummocks) and depressions, filled with lakes or swamps. Local relief can be as much as tens of meters in areas of thick ablation till. These deposits represent thick covers of sediment emplaced on top of stagnant glacial ice that slowly melted from beneath.

**Figure 15-26** High-relief hummocky topography in southern Saskatchewan resulting from deposition of thick ablation till.

**Figure 15-27** A drumlin in Wayne County, New York. Ice flow direction from right to left. (G. K. Gilbert; photo courtesy of U.S. Geological Survey.)

Glacial meltwater deposits are concentrated in *outwash plains* (Figure 11-24), where braided fluvial systems transport sediment away from ice margins. Broad plains of outwash sediment often contain abundant resources of shallow groundwater. Unfortunately, these unconfined aquifers are highly susceptible to contamination.

Several types of continental glacial landforms deserve special mention. *Drumlins* are streamlined hills produced by wet-based, sliding glaciers (Figure 15-27). These hills, which occur in groups or fields (Figure 15-28), are the subject of more research and debate than perhaps any other glacial landform. Characteristics include an asymmetrical "streamlined" shape elongated in the direction of glacier flow and a composition of till sometimes deposited over a core of bedrock. The controversy surrounding the origin of drumlins may have arisen because there probably are multiple origins for these landforms. Drumlins may be formed by erosion, deposition, or a combination of both.

The origin of several glacial landforms is related to deposition of meltwater flowing within or beneath the ice. An *esker* (Figure 15-29) is a sinuous ridge composed of poorly sorted sand and gravel. Eskers are formed by meltwater streams flowing in tunnels within or beneath a glacier. Because the stream channel is confined by ice walls and roof, the deposits remain as a raised mound after the ice melts. The sinuous, meandering shape of eskers is similar to subaerial stream channels. *Kames* are isolated, commonly conical hills of sand and gravel (Figure 15-30). One way in which kames are formed involves the flow of meltwater into a cavity in the ice similar to a sinkhole in karst topography. Gradually, the hole is filled by the stream deposits.

## ENGINEERING IN GLACIATED REGIONS

Glaciated regions present several types of problems relevant to construction and other types of land use. In general, these include problems associated with differences in engineering properties among various types of deposits, lateral

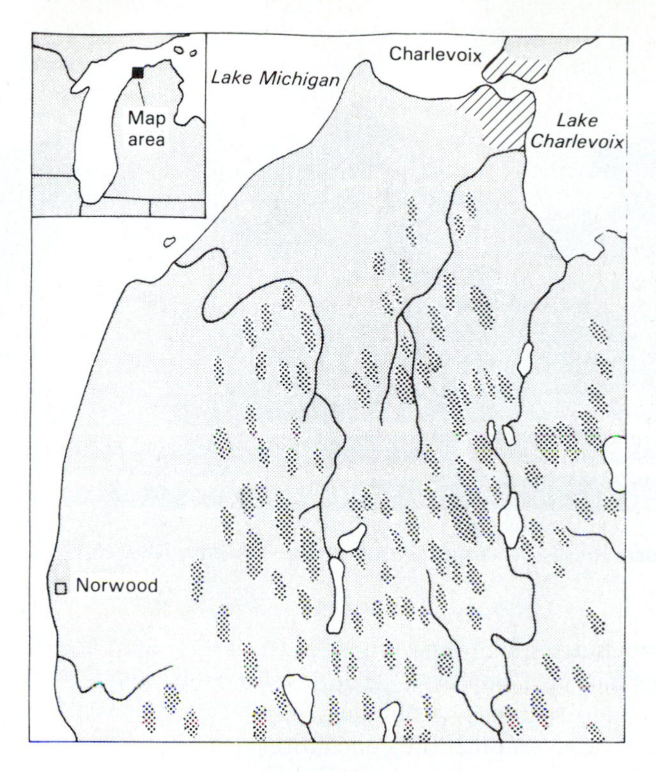


**Figure 15-28** Map of drumlins in northwestern Michigan. Ice moved toward the south-southeast. (From F. Leverett and F. B. Taylor, 1915, U.S. Geological Survey Monograph 53.)

**Figure 15-29** The Dahlen esker (arrows) located in eastern North Dakota. (Photo courtesy of J. R. Reid.)

and vertical variability in subsurface stratigraphy, and variations in depth to bedrock.

Of major importance are differences in engineering properties among till, glacial-lacustrine, glacial-marine, and outwash deposits. Glacial-lacustrine deposits are clayey, compressible materials with low shear strength. The role of

**Figure 15-30** A kame located in Erie County, New York.

these materials in the collapse of the Transcona, Manitoba, grain elevator was described in Chapter 10. The low bearing capacity of glacial-lake sediments may require piles or other deep foundation types for heavy structures. Piles are usually driven until they encounter a dense till or bedrock. Other engineering problems associated with glacial-lacustrine deposits include settlement, shallow water tables, shrink-swell behavior, and instability of natural or constructed slopes.

Glacial-marine deposits are similar in texture to glacial-lacustrine sediments. The special slope-stability problems associated with these soils were discussed in Chapters 10 and 12. The Nicolet, Québec slope failure (Figure 12-16) is an excellent example of the construction hazards that must be considered.

The geotechnical properties and behavior of till are dependent upon its mode of deposition. Lodgement tills become dense and overconsolidated under the load of the overlying glacier. As a result, they tend to have favorable strength and compressibility in comparison with ablation tills. On the negative side, lodgement tills may be so dense that excavation may require ripping or blasting.

An important characteristic of lodgement, and to some extent ablation tills, is that they are commonly observed to be jointed or fissured (Figure 15-31). In the case of lodgement tills, fissures may have been caused by unloading and rebound during and after the retreat of the glacier. Shrinkage associated with drying after deposition is another possible cause of fissure formation. Fissures have several important effects upon the engineering properties of the soil mass. First, fissures weaken the material. This can be shown by comparing the results of strength tests on samples of two sizes from the same construction site (Figure 15-32). The strength values measured from the larger samples are lower because of the greater number of fissures included in the sample. Fissures also influence the permeability of glacial deposits. Laboratory permeability tests conducted on small intact samples consistently yield lower permeability values than field tests because of the absence of fissures in the small lab samples. These discontinuities also provide preferred paths of movement for water through the deposit. A consequence of this condition is that tills may not be so favorable for waste-disposal sites as they may appear based on lab permeability tests on small samples.

**Figure 15-31** Exposure of a joint in till, southwestern North Dakota. The surfaces of the joint are covered by manganese oxide.

The presence of isolated large boulders can sometimes prove troublesome in construction projects involving till. Difficulties associated with excavation, test drilling, and pile driving have been experienced. Boulders may be concentrated in *boulder pavements* between till units in sequences containing multiple tills.

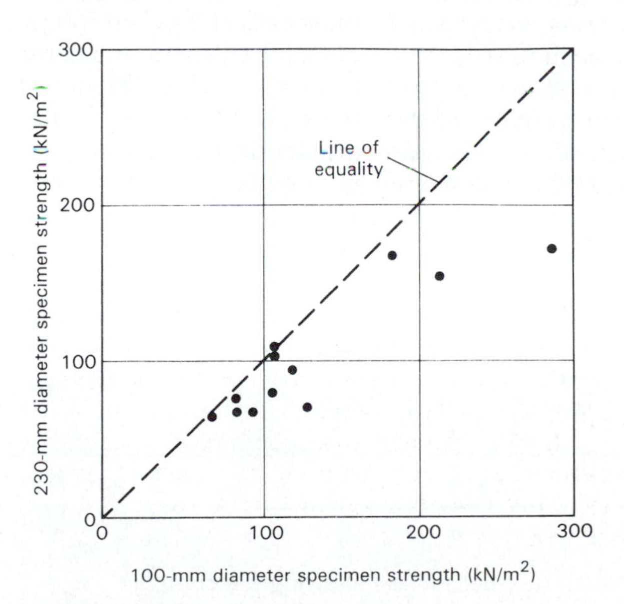


**Figure 15-32** Effect of fissures on strength of till. Samples of two different sizes from identical materials were tested. Points plot below the line of equality because the smaller samples are stronger than the larger samples, which have more fissures. (From W. F. Anderson, Foundation engineering in glaciated terrain, in *Glacial Geology: An Introduction for Scientists and Engineers*, N. Eyles, ed., copyright © 1983 by Pergamon Press, Ltd., Oxford.)

Outwash deposits generally are composed of cohesionless materials of variable sorting and high permeability. They frequently constitute excellent unconfined aquifers and favorable soils for irrigation if they occur at the surface. Outwash deposits are of great value as construction materials unless they contain excessive amounts of poor-quality material, such as shale. Therefore, states and provinces expend considerable effort in exploration for and evaluation of these deposits.

Sand and gravel deposits have moderate to high strength and low compressibility. Therefore, they provide high bearing capacities for foundations. Problems arise with high water inflow in excavations below the water table. Dewatering techniques (Chapter 11) may then be required. Glaciofluvial materials make very poor waste-disposal settings because of their high permeability and their potential utilization as sources of ground-water supplies.

## CASE STUDY 15-1

## Subsurface Complexity in Glacial Terrain

Complex glacial stratigraphy beneath deceptively simple glacial landforms can cause frustrating and expensive foundation problems. Boston's historic Beacon Hill was traditionally interpreted as a drumlin. Under this assumption, soils composed mainly of till with favorable conditions for foundations and excavations were predicted. The site investigations for several large modern buildings, however, revealed a much more complicated subsurface stratigraphy (Kaye, 1976). One of these structures was a garage constructed near the south end of the cross section shown in Figure 15-33. Till was expected to be present for the entire 13-m depth of the foundation excavation. Instead, subsurface folds containing gravel were encountered. Ground-water inflow from these gravels into the excavation required the installation of an expensive dewatering system and caused months of construction delay. The construction of a large office building on the north side of the hill was also affected by inadequate subsurface information. The building was designed to rest upon piles driven into dense till beneath a surficial bed of clay. When test piles were driven, the surficial clay was found to extend to a much greater depth than anticipated because of folding and faulting of the section. The piles were

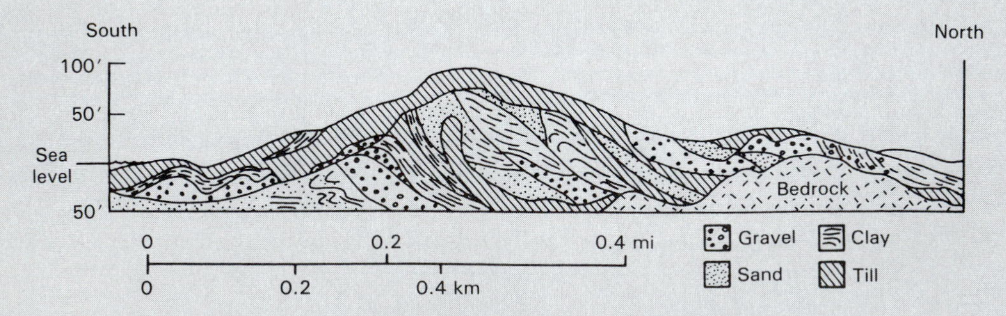


**Figure 15-33** Cross section of Beacon Hill showing the complex structure and stratigraphy. Ice advanced from the north and transported frozen slabs of bed sediment, which were emplaced by ice thrusting at the glacier terminus near Beacon Hill. (From C. A. Kaye, 1976, Beacon hill end moraine, Boston: A new explanation of an important urban feature, in Geological Society of America Special Paper 174, D. A. Coates, ed.)

driven 15 m beyond the predicted depth of the till unit with very little resistance to penetration.

Data gathered from these and other projects indicated that Beacon Hill is not a drumlin composed mostly of till. The hill is stratigraphically and structurally complex. Beneath a surficial layer of till, north-dipping slabs of sand, gravel, clay, and till are stacked against each other like books standing tilted to one side on a bookshelf. The depositional units are separated by thrust faults, and beds are intensely folded in some parts of the hill.

A reasonable explanation for the origin of Beacon Hill is that it consists of a series of individual thrust sheets emplaced by a glacier frozen to its bed near its terminus. Under this hypothesis, slices of frozen bed sediment were sheared into the ice and transported as intact slabs to the ice margin. There, the blocks were thrust upward against other slabs of subglacial bed material. Therefore, the landform is a type of terminal moraine rather than a drumlin. Construction experience in this area exemplifies the necessity for detailed subsurface site investigations in glaciated terrain.

## THE PLEISTOCENE EPOCH

The timing and causes of glaciations in the Pleistocene Epoch have been subject to much debate since the idea of a former great Ice Age was proposed by Swiss naturalist Louis Agassiz in the mid-nineteenth century. The beginning of the Pleistocene Epoch, the time at which late Cenozoic glaciations began, has been pushed back in time by evidence of older and older glaciations; it now stands between 1.5 and 3.0 million years, depending upon which particular chronology one accepts.

The most compelling evidence concerning the timing of glacial advances and retreats has been provided by studies of deep-sea sediment cores done in the past several decades. Deep-ocean floors are the only places on the earth that preserve a continuous record through the Pleistocene Epoch. The type of evidence used to decipher Pleistocene climates is the oxygen isotope content of microfossils in the sediments. This method is based on the assumption that the $^{18}O/^{16}O$ ratio present in the shells secreted by organisms is in equilibrium with the isotope ratio for seawater at the time of secretion. It is also known that the oxygen isotope ratio is temperature dependent. Thus ocean-bottom sediments contain a continuous record of temperature changes recorded in the oxygen isotope ratios of microfossils. The changes in mean surface seawater temperature between glacial and nonglacial periods are usually within several degrees Celsius. Fluctuations in oxygen isotope ratios thus indicate fluctuations in Pleistocene seawater temperature. These studies suggest that the Pleistocene Epoch was a period of time that fluctuated repeatedly between glacial and interglacial climates (Figure 15-34). Glacial periods tended to last 80,000

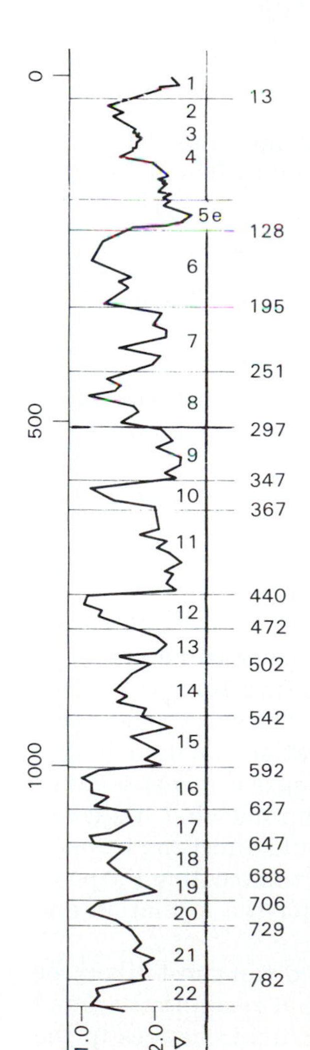


**Figure 15-34** Oxygen isotope variations during the last 800,000 years as measured in deep-sea sediments. Age in thousands of years on right, depth of core in cm on left. Periods or stages numbered in center. Curve moves to right of graph during interglacial periods and to left during glacial periods. (From N. Eyles, W. R. Dearman, and T. D. Douglas, Glacial land systems in Britain and North America, in *Glacial Geology: An Introduction for Engineers and Scientists*, N. Eyles, ed., copyright © 1983 by Pergamon Press, Ltd., Oxford. Data from N. J. Shackleton and N. D. Opdyke, 1973, *Quaternary Research*, 3.)

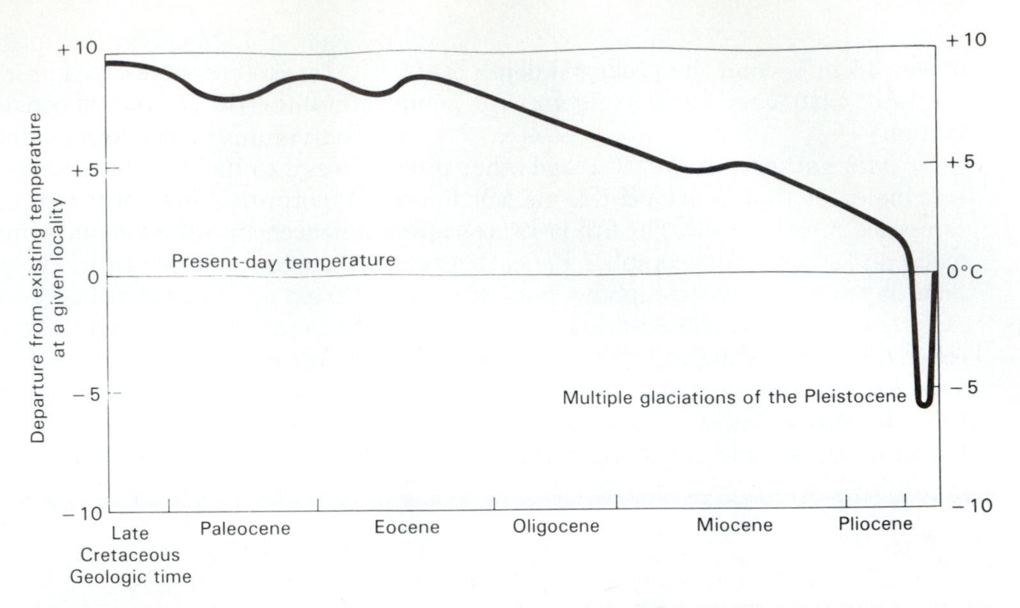


**Figure 15-35** Temperature trends in Cenozoic time. The gradual decline prior to the Pleistocene may be due to plate interactions and movements. (From S. Judson, M. E. Kauffman, and L.D. Leet, *Physical Geology*, 7th ed., copyright © 1987 by Prentice-Hall, Inc., Englewood Cliffs, N.J.)

to 90,000 years, interspersed with warmer interglacial intervals 10,000 to 15,000 years in length. The Holocene Epoch, the last 10,000 years of geologic time in which we now live, may be simply another interglacial period to be followed in several thousand years by another glaciation.

The number of glaciations revealed by oxygen isotope studies is much larger than the traditional model developed by studies of glacial deposits on land. Figure 15-34 indicates that at least eight major glacial periods occurred during the last 700,000 years. In contrast, studies dating back to the nineteenth century have recognized four glacial periods. In the United States, these *stages* have been named the Nebraskan, Kansan, Illinoisan, and Wisconsinan, listed in chronological order from oldest to youngest. The names are derived from the areas in which the deposits were first discovered or are most abundant. It is now realized that the concept of four glaciations is much too simple. The deposits of additional glaciations could be present and inaccurately dated or could have been removed by erosion.

The debate as to why the earth's climate suddenly began to fluctuate widely in late Cenozoic time is not fully resolved. Two major trends must be explained: a slow, gradual decrease in the earth's temperature during Cenozoic time and the climatic oscillations during the Pleistocene. The gradual decrease in temperature (Figure 15-35) may be related to the elevation and distribution of continents. Plate movements have resulted in a grouping of continents closer to the polar regions. In addition, these land masses are somewhat higher in elevation than at times in the past. These conditions would affect the transfer of heat by oceanic currents from the equator to the poles, thus cooling the polar regions, and also positioning the continents favorably for the formation and expansion of midlatitude ice sheets.

The fluctuations in climate during the Pleistocene can most likely be explained by periodic changes in the earth's rotation and revolution around the sun. These planetary movements affect climate by slight changes in the tilt of the earth's axis or its distance from the sun. A very close match has been obtained by comparing the periods of several types of orbital variations with the climatic changes indicated by deep-sea sediments. Thus it is reasonable to explain the glaciations of the Pleistocene Epoch by periodic changes in the earth's movement superimposed on a gradually cooling climate caused by plate movements. Volcanic eruptions, $CO_2$ fluctuations in the atmosphere, changes in oceanic circulation, and various other mechanisms have been incorporated into hypotheses for glaciation, either alone or along with other processes.

## PERMAFROST AND RELATED CONDITIONS

The development of petroleum resources in arctic regions and the transport of oil southward through pipelines have required new concepts of construction in cold regions. The likelihood of continued resource development in these regions requires a better understanding of this unique environment.

Construction problems in arctic regions result from the presence of permanently frozen ground, or *permafrost.* The area of permafrost on the earth is extensive. Approximately 20% of the earth's land surface is underlain by permafrost, including 50% of Canada and the former U.S.S.R. and 85% of Alaska.

### Extent and Subsurface Conditions

Permafrost areas can be classified as either continuous or discontinuous (Figure 15-36). Discontinuous permafrost can be identified by adjacent tracts of frozen and unfrozen ground. Continuous permafrost extends to great depths at high latitudes and gradually thins to the south (Figure 15-37).

Immediately below the land surface is the *suprapermafrost zone,* which is several meters thick and may contain pockets of material called *taliks* that never freeze, as well as ground that thaws during the summer and refreezes in the winter (Figure 15-38). This upper part of the suprapermafrost layer is called the *active zone.* At the base of the suprapermafrost layer is the *permafrost table.* This is the upper boundary of the zone that remains frozen year-round, although taliks also may be present below the permafrost table.

### Freeze and Thaw

Processes that produce distinctive landforms in climates subject to ground freezing are called *periglacial* processes. These processes, which are dominated by freeze and thaw activity, occur in any cold climate whether or not a glacier is nearby. Therefore, areas of high elevation in the midlatitudes experience periglacial activity. During Pleistocene time, periglacial climates extended southward as far and even beyond the advance of the ice sheets. Periglacial, like glacial, phenomena can reach equatorial regions if the elevation is sufficiently high. Although the effects of periglacial phenomena during the Pleistocene are sometimes relevant to engineering projects in the midlatitudes, we will focus the bulk of our discussion on the regions of current permafrost. In addition to the direct effects of freeze and thaw, special types of mass movement occur under periglacial conditions. They are discussed in the following section.

The effects of ground freezing are extremely variable. Lithology of the soil is an important factor. Under appropriate conditions, isolated *ice lenses* form in the soil and begin to grow by drawing moisture by capillary movement toward the expanding ice mass. Soils of predominantly silt-size particle compo-

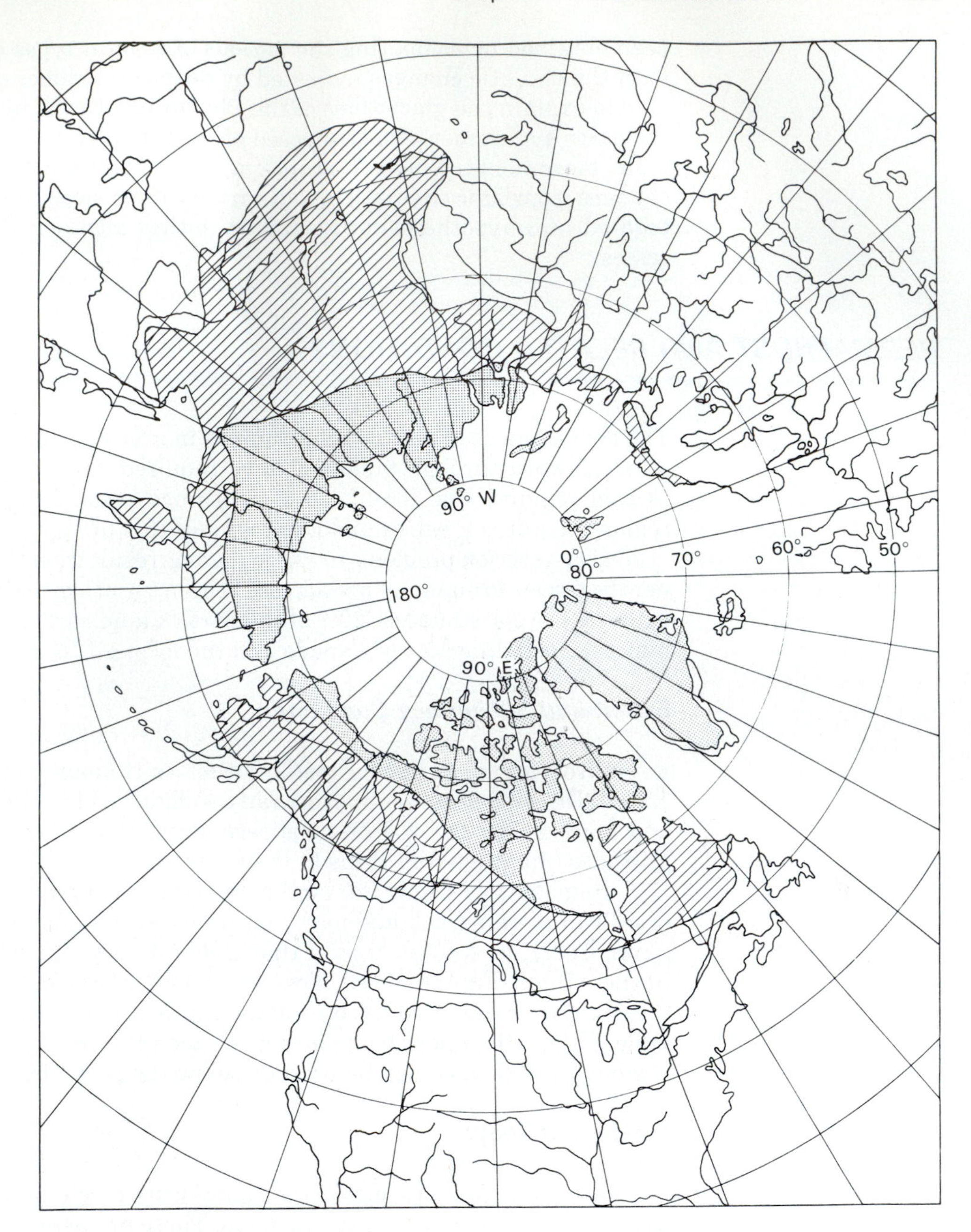


**Figure 15-36** Extent of permafrost in the Northern Hemisphere. Stippled area represents continuous permafrost; cross-hatching indicates discontinuous permafrost. (From O.J. Ferrians et al., 1969, U.S. Geological Survey Professional Paper 678.)

sition are most susceptible to ice-lens formation. Clay-rich soils retard the movement of soil water toward the ice lens, and coarse-grained soils contain pores that are too large for capillary movement. *Frost heave*, or uplift of the land surface, is most damaging in silty, frost-susceptible soils (Figure 15-39). Frost heave is not limited to permafrost regions, but is a problem in any area where temperatures dip below freezing for prolonged periods in the winter.

As the ground freezes in the cold season, vertical cracks may form by thermal contraction of the soil. These cracks, which extend below the bottom

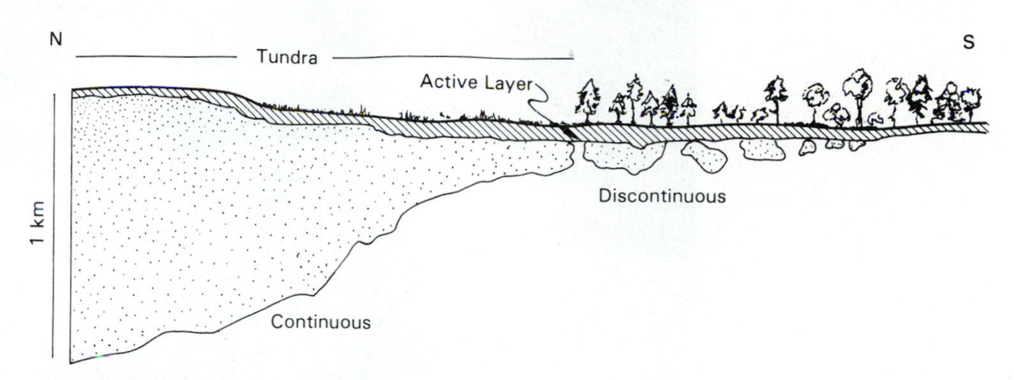


**Figure 15-37** Distribution of permafrost along a northwest transect in northwest Canada. Numbers above the diagram represent degrees of latitude. (From N. Eyles and M. A. Paul, Landforms and sediments resulting from former periglacial climates, in *Glacial Geology: An Introduction for Engineers and Scientists*, N. Eyles, ed., copyright © 1983 by Pergamon Press, Ltd., Oxford.)

of the active zone, commonly form a polygonal pattern on the land surface. In the summer, meltwater from the active zone percolates downward into the crack and freezes. Over a period of years, large ice wedges can develop by repeated frost cracking in the same location (Figure 15-40). The polygons at the surface are known as *ice-wedge polygons* (Figure 15-41). They are a good indication of permafrost conditions. Other characteristic permafrost landforms include *pingos* and *patterned ground*. Pingos are large circular or conical mounds, tens of meters high, that form by the vertical growth and heaving of an ice core. Patterned ground develops by the sorting of surficial particles into coarse and fine fractions (Figure 15-42). The resulting shapes include polygons, circles, and stripes.

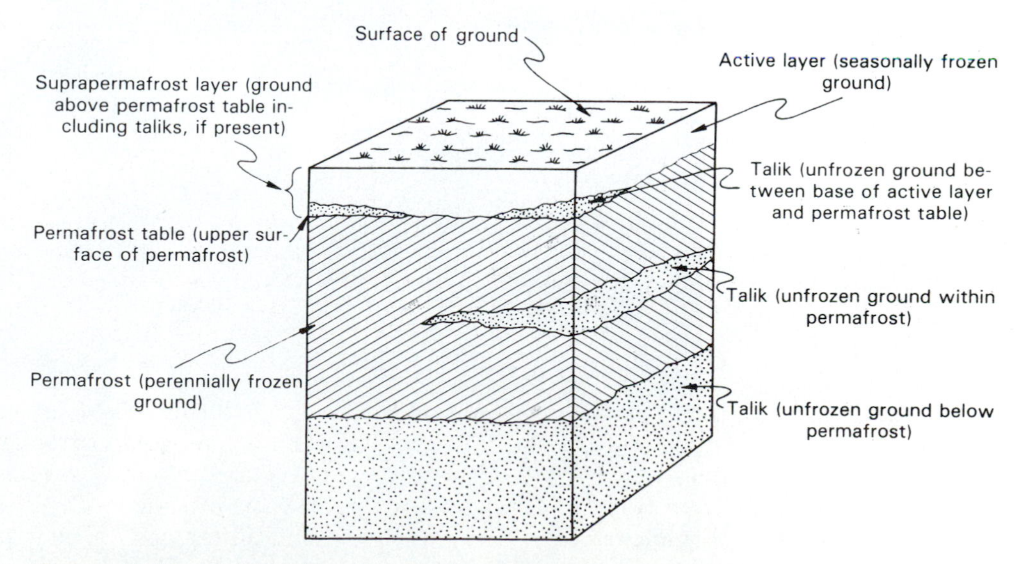


**Figure 15-38** Subdivision of permafrost below land surface. (From O. J. Ferrians et al., 1969, U.S. Geological Survey Professional Paper 678.)

**Figure 15-39** Cracking of a cement slab by frost heave, Grand Forks, North Dakota. Notice bulging of the slab in the vicinity of the crack.

### *Mass Movement*

A characteristic form of mass movement in permafrost terrain is known as *gelifluction*. During the summer, meltwater in the active zone is prevented from draining downward by the impermeable frozen soil below. Therefore, the soil within the active zone becomes saturated and flows as lobate shallow masses on very gentle slopes. The velocities of the gelifluction lobes are variable, from imperceptibly slow to tens of centimeters per day. Internally, gelifluction masses contain poorly sorted debris with a wide range of particle sizes.

**Figure 15-40** Ice wedge in soft sediment, Yukon Territory. (T. L. Péwé; photo courtesy of U.S. Geological Survey.)

**Figure 15-41** Aerial view of ice wedge polygons near Barrow, Alaska. (T. L. Péwé; photo courtesy of U.S. Geological Survey.)

Alpine regions contain moving masses of debris called *rock glaciers* (Figure 15-43), which originate at high elevations in cirques or on slopes leading to mountain peaks. They consist of angular blocks of rubble that are produced by frost action on rock outcrops and accumulate on slopes by rockfall, snow avalanches, or other processes. Rock glaciers resemble true glaciers in form and movement. In addition to rock debris, they also contain ice, either as a core beneath the debris or as an interstitial matrix between rock fragments.

**Figure 15-42** Patterned ground near Keflavik, Iceland, formed by sorting of surficial material. Coarse particles outline the polygons.

**Figure 15-43** An active rock glacier moving downslope, Copper River region, Alaska. (F. H. Moffit; photo courtesy of U.S. Geological Survey.)

**Figure 15-44** Disruption of railroad resulting from permafrost thaw after removal of vegetation; near Valdez, Alaska. (O. J. Ferrians, Jr.; photo courtesy of U.S. Geological Survey.)

### *Engineering Problems in Permafrost Regions*

Engineering problems in permafrost terrains result from unwanted thawing of frozen ground and frost heaving during seasonal freezing of the active zone. Problems are most severe in low-lying areas underlain by fine-grained soils. Upon thawing, these soils become water-logged and lose bearing capacity because the frozen soil beneath the active zone prevents drainage. A lesson learned from many mistakes is that permafrost exists in a fragile state of equilibrium. Vegetation insulates the frozen ground beneath, and if it is removed, thawing and depression of the permafrost table will result. Roads and railroads are constructed by building an insulating pad of non–frost-susceptible soil on the ground after vegetation has been removed. Failure to design roads correctly leads to expensive failures (Figure 15-44).

Buildings are often constructed on insulated piles driven below the permafrost table. The bottom of the floor is elevated about a meter above land surface to allow for the circulation of air and dissipation of heat.

## CASE STUDY 15-2
## The Trans-Alaska Pipeline

Discovery of a huge oil field near Prudhoe Bay on the north coast of Alaska in 1968 posed a unique challenge to the United States. Here was a golden opportunity to reduce the increasing dependence on foreign oil, a factor that became even more apparent during the energy crisis of 1973. However, Prudhoe Bay is accessible to oil tankers for only a short period each year. The alternative was to build a 1300 km pipeline across the state of Alaska (Figure 15-45). Williams (1986) has summarized the history of this and other pipeline projects.

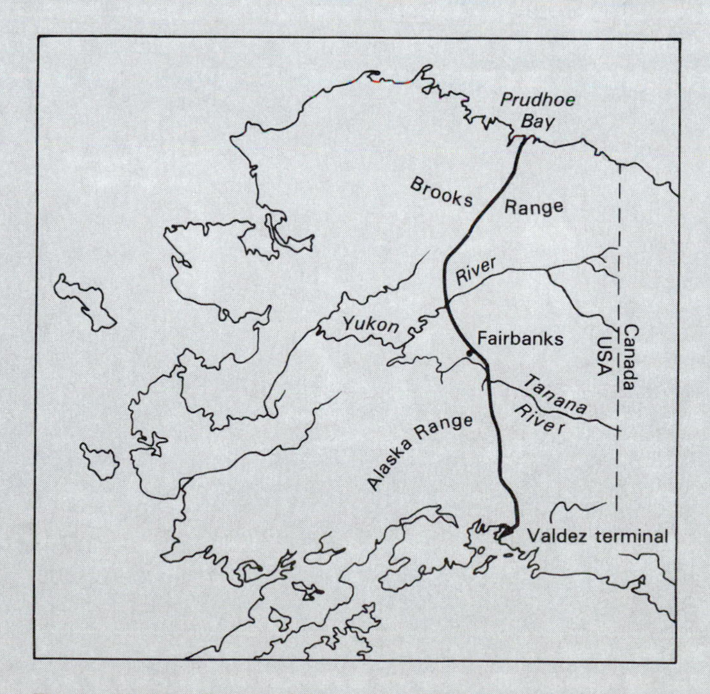


**Figure 15-45** Route of the trans-Alaska pipeline.

The trans-Alaska pipeline route was mostly wilderness and crossed two mountain ranges, several large rivers, and vast areas of permafrost. Construction problems in an uninhabited area with an inhospitable climate, not to mention the special construction techniques that would be required for the permafrost, were quickly appreciated. The environmental movement sensed an ecological disaster in the destruction of wilderness and the disruption of wildlife such as caribou that freely migrate across the route. When it was finally built at a cost of $7 billion, more than eight times the original estimate, the pipeline had become the largest private engineering project in history.

The major dilemma in design of the pipeline was the necessity to conduct oil at a temperature of 65°C through a 48-in. diameter steel pipe without thawing the permafrost. The engineers realized that destruction of the permafrost would cause land subsidence and the possibility of rupture of the pipeline. The potential of rupture from ground shaking due to earthquakes was also taken into consideration. After extensive test drilling along the route to assess the soil and permafrost conditions, the decision was made to construct about one-half of the line (610 km) above ground to prevent contact between the pipe and the frozen soil. Of the remainder, 640 km were buried in areas where soil conditions were favorable and frost heave or subsidence was not anticipated to be a problem. Eleven km of the pipeline in sensitive environmental areas were buried using special techniques including insulation and refrigeration around the outside of the insulated pipe to prevent permafrost thaw. Bridges were used to conduct the pipeline across the larger rivers and the pipeline was buried beneath others.

The design of the above ground pipeline supports is shown in Figure 15-46. Figure 15-47 is a photo of the pipeline. The pipe is supported by two vertical support members (VSMs) and a horizontal support beam. The VSMs are steel piles that are driven into the ground to depths ranging from 8 to 20 m. A critical component of the VSMs are sealed, 2-in. diameter heat pipes containing anhydrous ammonia refrigerant. In the winter the refrigerant evaporates from the lower part of the heat pipe and condenses at the top because the ground is warmer than the air. The heat is then released to the air by use of heat exchangers at the top of the VSMs. The evaporation of the refrigerant below the soil surface cools the permafrost, similar to the evaporation of sweat cooling the skin, to the point where the soil remains frozen during the warmer summer conditions. Thermal expansion and contraction of the pipe is accommodated at bends in the line at 250- to 600-m intervals. Connection of the pipe to the support beam allows up to 4 m of lateral movement of the line; vertical movement along the VSMs is also possible. This de-

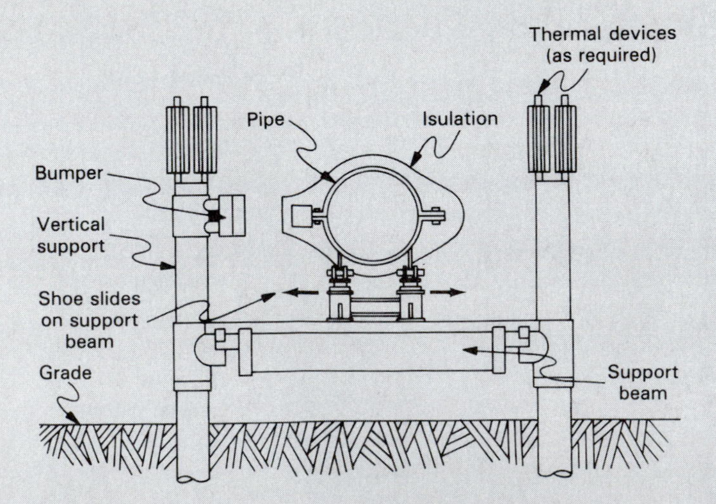


**Figure 15-46** Design of above ground pipeline support. (From P. J. Williams, 1986, *Pipeline & Permafrost: Science in a Cold Climate*, Ottawa: Carleton University Press.)

**Figure 15-47** Photo of above ground section of pipeline near Paxson, Alaska. (Photo courtesy of Glenn Oliver.)

sign protects against both ground thaw and disruption as well as earthquake ground motion. The equipment road running along the pipeline was constructed on a thick gravel mat to insulate the ground from temperature changes associated with the removal of vegetation.

To date, the pipeline has functioned as designed and disruption of the environment has been minimal for a project of this scope. Wildlife migration concerns were adequately addressed by the refrigerated buried sections and other measures. Oil production from Prudhoe Bay is now in decline. New reserves in Alaska have not been discovered. Major oil and gas reserves are present, however, in arctic Canada. Since the completion of the trans-Alaska pipeline, one other long Canadian pipeline in permafrost was constructed. Others in both the United States and Canada have been proposed, but not built. The success of the trans-Alaska pipeline was based on a knowledge of the geology of frozen ground. Future projects will continue to rely on geology to solve the unique problems of construction in cold climates.

## SUMMARY AND CONCLUSIONS

An understanding of glacial processes and landforms is important because of the large area covered by glaciers during the Pleistocene Epoch. The expansion and contraction of a glacier is determined by its mass balance—the relationship between accumulation and ablation. The position of the terminus is controlled by this relationship even though the ice may be flowing downslope continuously. Flow occurs by visco-plastic deformation and basal sliding. Both types of movement occur in wet-based glaciers, whereas only visco-plastic flow is characteristic of cold-based glaciers.

Glacial erosion includes abrasion and plucking by sliding glaciers. Glaciers frozen to their beds can incorporate huge slabs of bed material into the ice by thrusting. Eroded rock and sediment are transported and deposited at a later time. Deposition may take place beneath the ice to form lodgement till or above the ice to form ablation till. Other types of glacial drift are deposited by glacial meltwater streams as outwash or in proglacial lakes as lacustrine sediment.

Valley glaciers produce a characteristic set of erosional landforms in mountainous terrain. Chief among these is the U-shaped valley with over-steepened valley walls. Eroded sediment is deposited in terminal, lateral, and medial moraines. Modern glaciers cannot compare with the huge continental ice sheets that buried the landscape repeatedly during Pleistocene time. The landforms and deposits of these glaciers are the substrate upon which all human activities take place in northern latitudes.

Each type of glacial deposit has unique engineering characteristics. Glacial outwash deposits may contain valuable resources of aggregate material and ground water. Glacial lacustrine sediment and tills present varied construction conditions depending upon their method of deposition, thickness, stratigraphy, and many other factors. Lodgement and ablation tills differ with respect to density, strength, and compressibility.

The factors that led to repeated glaciations extending into the midlatitudes during Pleistocene time are not fully known; however, it is thought likely that periodic perturbations in the earth's motion can explain the climatic cycles of cooling and warming. Long-term climatic changes were probably caused by lateral movement and elevation of the continents by plate interactions.

Permafrost causes severe engineering problems in cold regions, which cover more than 20% of the earth's surface. Most of these problems are related to instability of the active layer of the soil. Differential settlement and mass movement are common during thaw periods, and frost heave takes place during freezing of the active layer. These conditions require special design and construction procedures for almost all engineering projects.

## PROBLEMS

1. Summarize the mechanics of glacier movement.
2. What effect does the condition of the base of the glacier have upon movement?
3. Contrast the different types of glacial erosion.
4. How is till deposited?
5. What types of sediment other than till are associated with glaciers?
6. What topographic features indicate whether or not a mountainous area was glaciated during Pleistocene time?
7. What construction problems can be expected in glaciated terrain?
8. What is the best explanation for Pleistocene glaciations?
9. What landforms suggest the presence of permafrost?
10. Discuss the problems of construction in areas of frozen ground.

## REFERENCES AND SUGGESTIONS FOR FURTHER READING

ANTEVS, E. 1929. *Maps of the Pleistocene Glaciations.* Geological Society of America Bulletin, 40:635.

BLUEMLE, J. P. 1991. *The Face of North Dakota.* North Dakota Geological Survey Educational Series 21.

DAVIS, W. M. 1911. The Colorado Front Range. *Annals of the Association of American Geographers*, v. 1.

EASTERBROOK, D. J. 1993. *Surface Processes and Landforms*. New York: Macmillan.

EYLES, N. 1983. *Glacial Geology: An Introduction of Scientists and Engineers*. Elmsford, N.Y.: Pergamon Press.

FERRIANS, O. J., R. KACHADOORIAN, and G. W. GREENE. 1969. *Permafrost and Related Engineering Problems in Alaska*. U.S. Geological Survey Special Paper 678.

FLINT, R. F. 1945. *Glacial Map of North America*. Geological Society of America Special Paper 60.

JUDSON, S., M. E. KAUFFMAN, and L. D. LEET. 1987. *Physical Geology*, 7th ed. Englewood Cliffs, N.J.: Prentice-Hall, Inc.

KAYE, C. A. 1976. Beacon Hill end moraine, Boston: New explanation of an important urban feature, in *Urban Geomorphology*, D. A. Coates, ed. Geological Society of America Special Paper 174, pp. 1–20.

LEVERETT, F., and F. B. TAYLOR. 1915. *The Pleistocene of Indiana and Michigan and the History of the Great Lakes*. U.S. Geological Survey Monograph 53.

SUGDEN, D. E., and B. S. JOHN. 1976. *Glaciers and Landscape*. New York: John Wiley.

WILLIAMS, P. J. 1986. *Pipelines and Permafrost: Science in a Cold Climate*. Ottawa: Carleton University Press.

# 16

# Subsurface Contamination and Remediation

Contamination of soil and water is a problem common to all human societies. Surface-water contamination probably has been recognized throughout the history of civilization. In industrialized countries, treatment of surface water for drinking became common in the late nineteenth century and health problems linked to impure drinking water are now rare. Underdeveloped countries, however, are still faced with a lack of safe drinking water.

During the past several decades a new problem has come to light: contamination of the subsurface. This problem is much more serious in developed countries because of their history of industrialization and the wide range of chemicals that have been introduced, either by accident or by design, to the underground environment. Much of the damage was caused by ignorance rather than intentionally. We simply did not comprehend the degree to which contaminants could migrate beneath the land surface or the difficulty we would encounter in tracing and removing them after discovery.

The response to subsurface contamination in the United States, as in many other countries, has been a massive effort to define the extent of contamination and to remediate the subsurface—starting with the most contaminated sites. This environmental campaign has been driven by governmental regulations dealing with waste handling and disposal, as well as with many other potentially contaminating activities. One side effect of this endeavor has been a boom in the employment market for engineers and environmental scientists that specialize in detection, monitoring, and remediation of subsurface contamination. A new industry has been created and the performance of this work seems sure to continue well into the twenty-first century.

## SOURCES OF SUBSURFACE CONTAMINATION

The range of activities that cause subsurface contamination is much larger than most engineers and scientists would have guessed even a few years ago. A representative listing in Table 16-1 is divided into categories based on whether the source of the contamination originates on the land surface, within the unsaturated zone, or below the water table. Several of these activities are illustrated in Figure 16-1.

### *Subsurface Contamination from Land Surface*

Contamination derived from land-surface activities includes a variety of common practices and occurrences. Infiltration of contaminated surface water may take place when shallow water-supply wells draw water from an alluvial aquifer adjacent to a stream (Figure 16-2). The cone of depression imposed by pumping the well or well field creates a gradient on the water table directed toward the well. In fact, wells are deliberately installed near streams and rivers in order to induce recharge from the stream and to provide high yields with low drawdowns. Unfortunately, contamination of the well field can occur if the stream is polluted.

Land disposal or stockpiling of wastes or materials is a common practice. Liquid or solid wastes from sewage treatment plants, food processing companies, and other sources are applied to agricultural lands, golf courses, and other areas to serve as fertilizer and irrigation. The objective is to allow chemical and biological processes in the soil, along with plant uptake, to break down the waste products into harmless substances. Although the concept may be

**TABLE 16-1**

**Sources of Ground-Water Contamination**

Ground-Water Quality Problems that Originate on the Land Surface
1. Infiltration of contaminated surface water
2. Land disposal of solid and liquid waste materials
3. Stockpiles, tailings, and spoil
4. Dumps
5. Salt spreading on roads
6. Animal feedlots
7. Fertilizers and pesticides
8. Accidental spills
9. Particulate matter from airborne sources
10. Composting of leaves and other yard wastes

Ground-Water Quality Problems that Originate Above the Water Table
1. Septic tanks, cesspools, and privies
2. Surface impoundments
3. Landfills
4. Waste disposal in excavations
5. Leakage from underground storage tanks
6. Leakage from underground pipelines
7. Artificial recharge
8. Sumps and dry wells
9. Graveyards

Ground-Water Quality Problems that Originate Below the Water Table
1. Waste disposal in wet excavations
2. Agricultural drainage wells and canals
3. Well disposal of wastes
4. Underground storage
5. Secondary recovery
6. Mines
7. Exploratory wells and test holes
8. Abandoned wells
9. Water supply wells
10 Ground-water development

SOURCE: Modified from U.S. EPA, 1990, *Ground Water, Volume 1: Ground Water and Contamination.* EPA/625/6-90/016a.

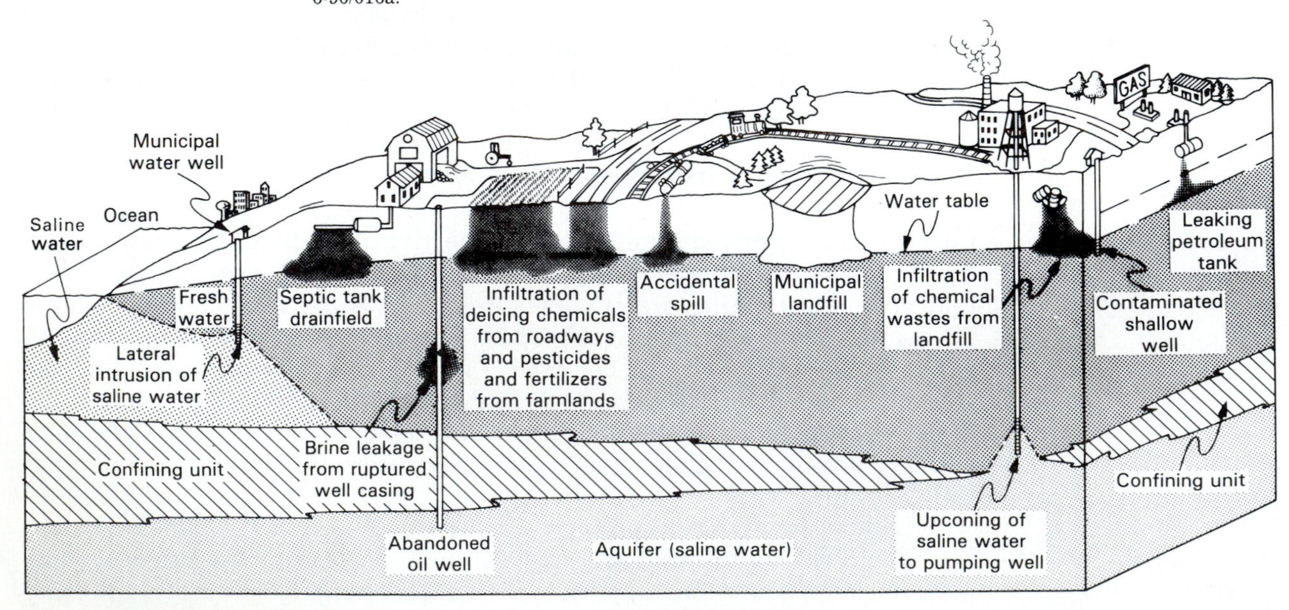


**Figure 16-1** Major sources and processes causing subsurface contamination. (From C. W. Fetter, *Contaminant Hydrogeology*, copyright © 1993 by Macmillan Publishing Company. Reprinted with permission.)

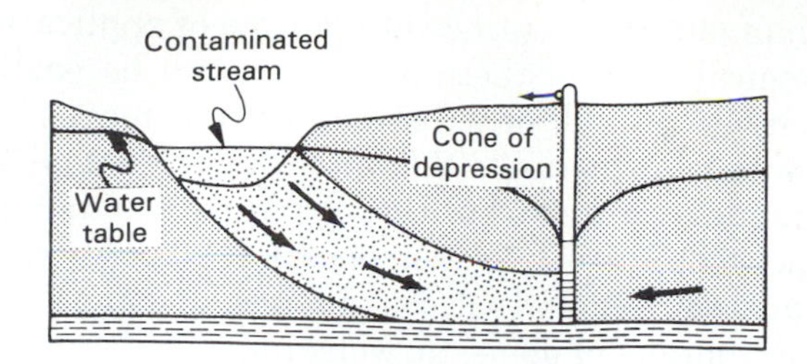


**Figure 16-2** Contamination of a well by induced infiltration from a stream.

generally successful, if any of the wastes are water soluble and mobile, they are typically carried into the subsurface. A problem may arise if the fields are underlain by shallow aquifers.

Stockpiles include substances such as road salt, which is highly soluble in water, and which may enter the subsurface easily. Dumps could contain a wide variety of potential contaminants. One recent regulatory trend is to exclude leaves and yard wastes from landfills in order to save landfill space. An alternative for these materials is large-scale surface composting. Because these leaves are collected from yards and streets, they may contain unwanted substances, such as animal wastes and petroleum contaminants.

Uncontrolled dumping is prohibited in most industrialized countries today, but old dumps may still constitute a threat of subsurface contamination. The spreading of salt on highways is a widespread practice in northern climates. In addition to causing deterioration of bridges, automobiles, and the roadway itself, the salt quickly leaches below the land surface. Figure 16-3 shows the steady rise in chloride, an ion derived from salt, in three city wells in a municipal well field in Kalamazoo, Michigan. The increases correlate with increases in the amount of salt applied to the city streets during the time period.

Fertilizers and pesticides yield several different types of contaminants. Nitrogen fertilizers are applied to stimulate plant growth, but often in greater

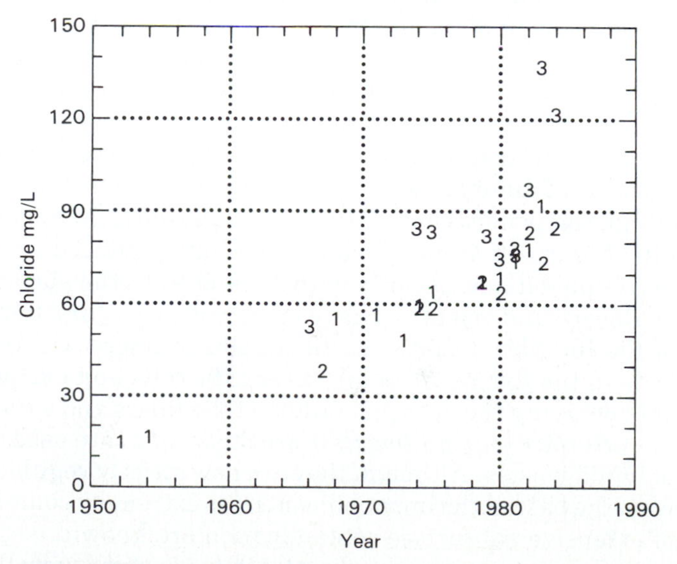


**Figure 16-3** Chloride concentrations in three wells (1, 2, and 3) in municipal well field in Kalamazoo, Michigan. (Data from Kalamazoo Dept. of Public Utilities.)

quantities than plants can utilize at the time of application. Nitrate, the most common chemical form of these fertilizers, can be easily leached below the plant root zone by rainfall or irrigation. Once it moves below the root zone, it usually continues downward to the water table. Most pesticides and herbicides are not as mobile as nitrate, but they are toxic at much lower concentrations. The subsurface mobility of these compounds is strongly related to the characteristics of the unsaturated and saturated zones. In some karst aquifers, for example, a high percentage of domestic wells in agricultural areas have detectable concentrations of pesticides and herbicides. Fertilizers and pesticides, along with road salt and land application of wastes, are known as *nonpoint sources*, because they are applied to large surface areas. The regulatory and remedial strategies for nonpoint-source contaminants are therefore different than point sources of wastes, such as landfills.

Accidental spills, which occur at manufacturing plants, and on roads and railroads, are disturbingly common. Containment and cleanup are relatively effective if initiated soon after the spill, but if the spill is discovered long afterward, the extent of migration of the pollutant can be great.

Particulate matter from airborne sources is thought to be minor compared with some pollutants from other sources of contamination, but it can be very serious in localized areas. Examples include the accumulation of heavy metals around smelters and exhaust fumes in urban areas that are associated with auto emissions.

### Subsurface Contamination Sources in the Unsaturated Zone

Contamination originating in the unsaturated zone includes some of the most common types. Shallow burial of wastes has been the most common method of disposal of most solid waste for many years. Experience has shown that the leaching of these wastes from the unsaturated zone to the saturated zone has been very common. In the United States, solid wastes are now separated for disposal into generally inert materials, municipal and household wastes, and hazardous wastes. Disposal in landfills of the latter type, in what are known as *secured landfills*, is subject to the most stringent regulations and requirements, including detailed site investigations, engineered leachate barriers, and regular monitoring of the unsaturated and saturated zones (Figure 16-4). *Sanitary landfills*, the type still used for municipal waste, can be designed in several ways (Figure 16-5). All designs include the construction of cells of compacted refuse with layers of soil added to minimize contaminant movement and to limit the infiltration of water (Figure 16-6). The current trend is toward greater isolation of these wastes by the use of impermeable liners and covers. Additional design components may include *leachate collection systems* in the secured landfills (Figure 16-4), in which any leachate produced in the landfills is drained to a sump on one corner and then pumped to the surface for treatment. Some engineers and scientists have criticized these "dry tombs," however, because of the likelihood that even thick, plastic liners will tear or rupture at some time in the future. Thus, by sealing the cells and excluding water, we may be only delaying the eventual failure of the liners and subsequent contamination.

*Surface impoundments* (Case Study 16-1) are used for storage and disposal of liquid wastes. Although they are now strictly regulated, in the past seepage from the base of the impoundments was extremely common and many examples of extensive subsurface contamination are known.

Septic systems, including leach fields and dry wells, are used for disposal of household liquid waste in unsewered rural and suburban locations. Waste solids are retained in a septic tank (Figure 16-1) and liquids are discharged

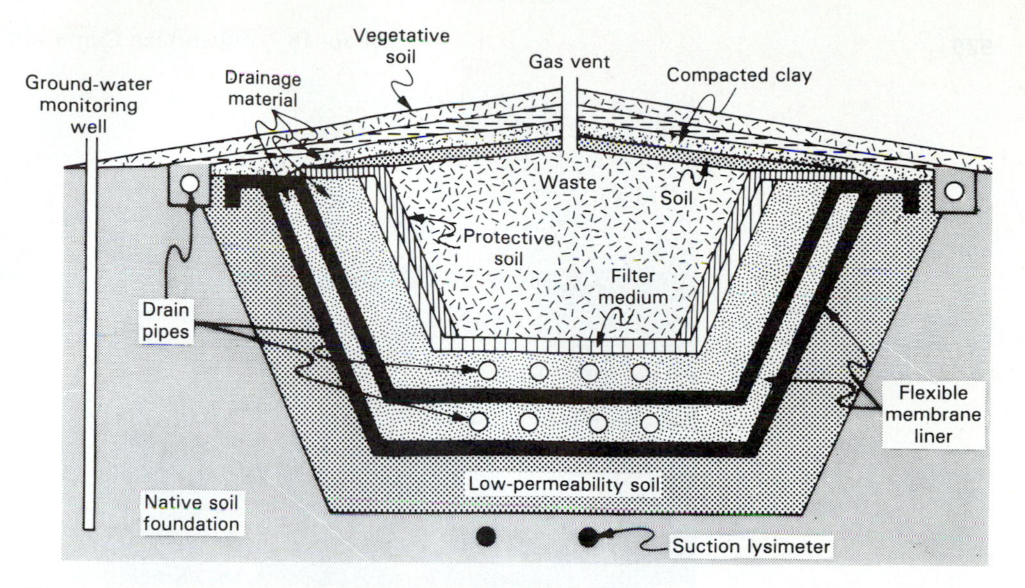


**Figure 16-4** Design components of a secured landfill.

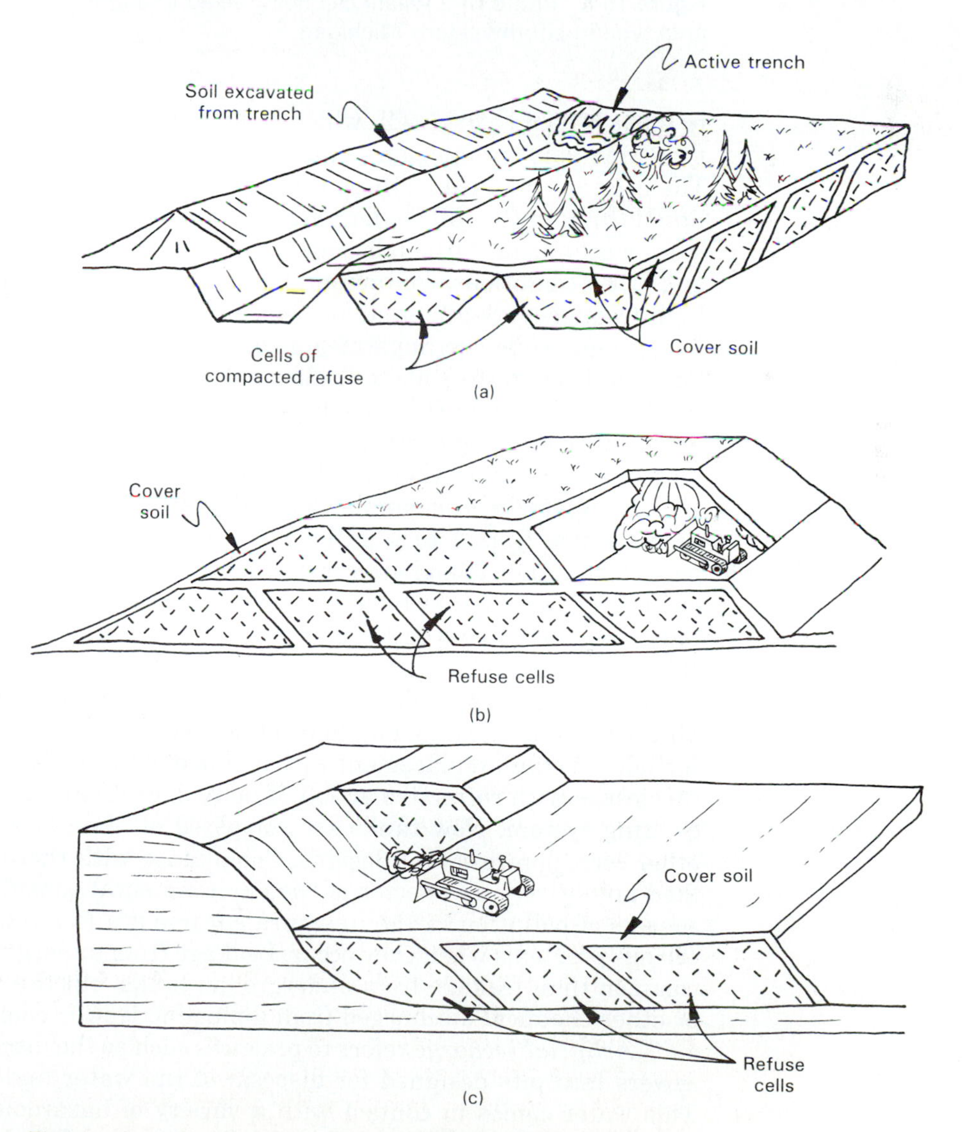


**Figure 16-5** Types of sanitary landfills include (a) the trench method, (b) the area method, and (c) the slope method.

**Figure 16-6** Photo of a waste cell being filled in a landfill of the area type in southwestern Michigan.

to a leach field or dry well, which is simply a large buried container of some type with holes in the sides, to promote infiltration of the liquid into the soil. The original premise of these systems was that the movement of the waste liquid through the soil would renovate the water quality and serve as a treatment mechanism. Although some treatment does occur, depending on soil type and other conditions, contaminants are known to reach ground water with a high frequency. Septic systems that are installed in unsuitable sites, where for example soils are impermeable or bedrock is close to the surface, tend to back up through the soils to the ground surface, creating an unsanitary situation. These systems are said to have "failed." In reality, the successful systems, with higher leaching capacity, are as much or more of a long-term environmental threat because of their impacts on ground water.

Estimates of the number of underground storage tanks in the United States range from 2 to 4 million. The vast majority of these tanks contain petroleum products and are constructed of steel. It is possible that as many as one-third of these tanks have caused contamination either through leaks in the tank or connecting piping or by spills and overfills occurring during refueling. Numerous tank excavations have shown that corrosion and leakage are widespread among older tanks. Federal regulations imposed in 1984 required that all tanks in the country had to meet new standards by 1998. These standards included (1) tank replacement with a tank of improved performance standards, (2) closure with corrective action, if needed, or (3) an approved upgrade to the existing system. New tanks are composed of fiberglass-reinforced plastic or other corrosion-resistant designs. Compliance with these regulations has created an entirely new sector of the environmental consulting industry. Other sources of pollution in the unsaturated zone are less common but may cause serious problems when they occur. Leakage from underground pipelines is less common than leakage from underground tanks, but the amount of petroleum or other chemical discharged from a pipeline leak is commonly very large.

*Artificial recharge* refers to practices such as the discharge of urban storm sewers into pits designed for disposal of the water and/or aquifer recharge. This water comes in contact with a variety of hazardous and nonhazardous compounds during runoff from roads, driveways, and parking lots. Even graveyards are a potential source of contamination. Of concern here are biological and chemical contaminants. For example, in the late 1800s and early 1900s, arsenic in very high concentrations was used as an embalming fluid.

### *Subsurface Contamination Sources in the Saturated Zone*

Many practices involving the saturated zone can lead to direct contamination of ground water without passage through the unsaturated zone. A common activity of the past, for example, was to use abandoned excavations associated with mines and gravel production for waste disposal. If the abandoned excavation is deeper than the water table, the emplaced waste materials will become saturated upon recovery of the water table. If the excavation is not reclaimed or backfilled, it will partially fill with water. Any waste materials will therefore be in direct contact with the ground-water flow system.

Agricultural drainage wells and canals may involve such practices as draining contaminated shallow ground water to deeper aquifers or inducing the upward flow of highly mineralized water as the water table is lowered by drainage. Canals excavated along coastal areas may lead to the inflow and infiltration of salt water during tidal fluctuations.

Disposal of hazardous chemicals in wells was based on injection-well technology developed in the oil industry. There, it has been traditionally used as a means of disposing brines produced with the oil, thereby overcoming the problems encountered with the disposal of brines in surface impoundments. The chemical industry adopted the method as a cheaper alternative than waste-water treatment or surface disposal. Disposal, or *injection, wells* are commonly hundreds to thousands of meters deep (Figure 16-7). Favorable conditions occur where a permeable zone containing nonpotable water is overlain by one or more aquitards to prevent upward migration of the waste water into usable aquifers. Problems occur due to failure of the well casing or the sealing material around the casing, which may allow contaminants to move out of the target horizon. Other pathways for contaminants include fractures produced in the protecting aquitard caused by high injection pressures or by abandoned, unplugged, or poorly plugged wells that may have penetrated the aquitard. Abandoned test holes, which are commonly drilled in both petroleum and mineral exploration, can also facilitate migration of contaminants in the subsurface.

Mines, even without the addition of wastes, can be environmental problems. The introduction of oxygen to anoxic zones in the subsurface can cause undesirable reactions with minerals leading to the production of toxic leachates. Sulfide minerals, which occur both in coal-bearing rocks and in metallic ores, are commonly the culprits. Oxidation of sulfides will produce sulfuric acid, and the leachates formed in this manner are corrosive and highly mineralized. Migration out of the mines leads to degradation of both ground water and surface water. One potential effect of a decrease in pH is the mobilization of toxic metals, which are more soluble in acid solutions than in neutral or basic waters.

Water supply wells themselves can sometimes be a source of contamination. Problems are most commonly linked to poor well construction. If the *annulus* (the part of the well bore outside the casing) is not properly sealed, contaminants from surface or subsurface sources have a direct vertical pathway to the aquifer. Flooding is a good example of how contaminated surface water could be brought into contact with the top of a well. The main water treatment plant for Des Moines, Iowa, was shut down during the summer floods of 1993 when floodwaters inundated the municipal wells. Ground-water development can lead to problems when pumping lowers the water table or potentiometric surface to such a degree that water from an undesirable source begins to

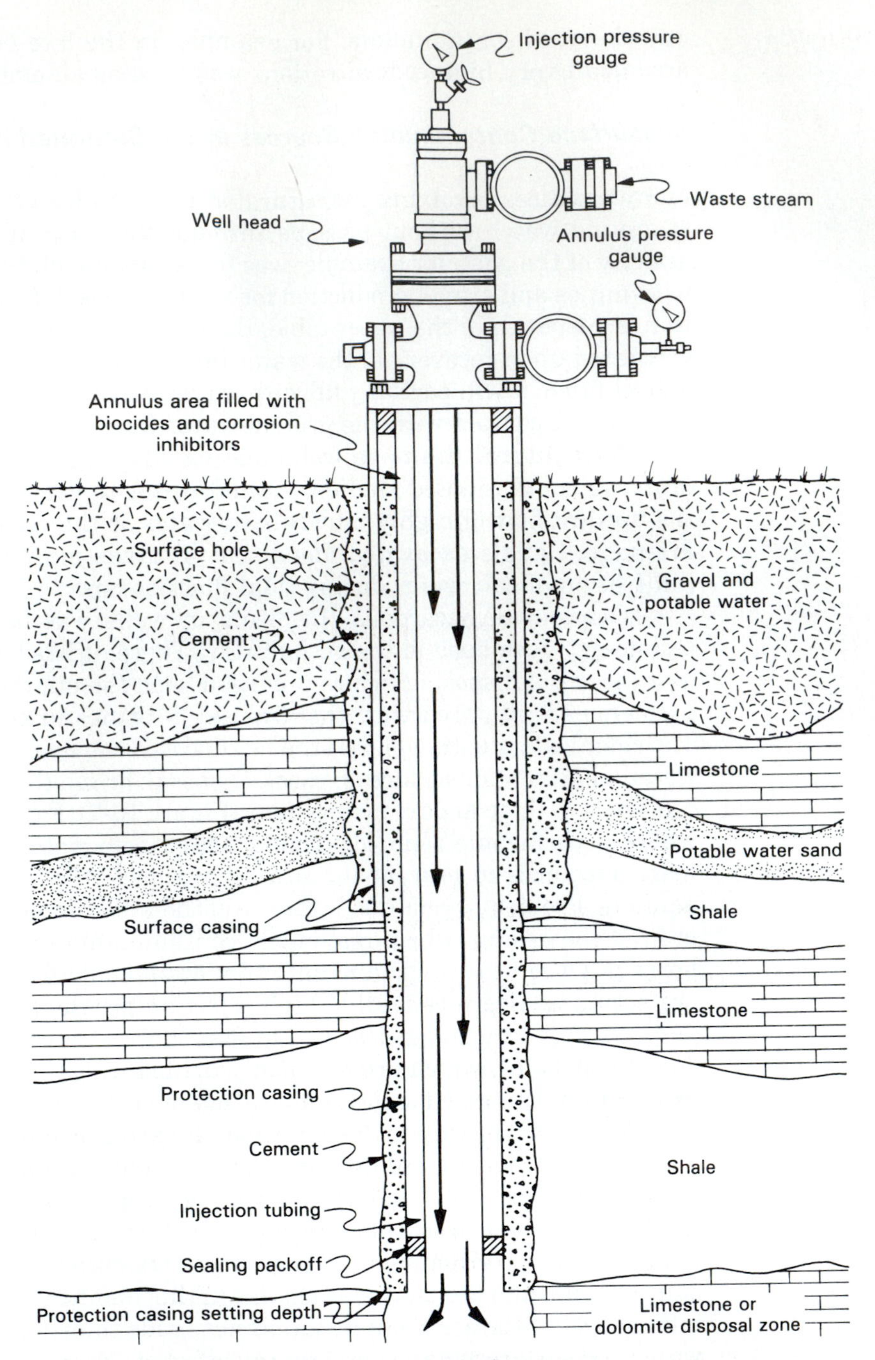


**Figure 16-7** Design features of a hazardous waste-disposal well. (From C. A. Wentz, *Hazardous Waste Management*, copyright © 1989 by McGraw-Hill, New York.)

migrate toward the well. The poor quality water may be drawn through an aquitard from a different aquifer, from an offshore source by subsurface migration of sea water, or from other possible locations.

The number and diversity of activities and practices that lead to groundwater contamination is a serious concern for environmental regulators and water managers. Through additional research into these processes, we will be able to minimize their effects in the future.

## ENVIRONMENTAL REGULATIONS

Recent discoveries and the study of subsurface contamination have made it clear that many human activities must be subject to strict regulations so as to prevent future contamination, as well as to redress the effects of past practices. Several of the most important environmental laws dealing with subsurface contamination in the United States are summarized subsequently. Regulations based on these laws are administered by the U.S. Environmental Protection Agency (EPA).

### *Clean Water Act (CWA)*

The *Clean Water Act* (CWA) of 1977 was designed to improve treatment of waste-water discharge to the environment. Although most waste water is discharged to surface water bodies, we have already seen how contaminated surface water can impact ground water. Waste-water treatment is described in terms of three possible phases: *primary*, *secondary*, and *tertiary*. Primary treatment involves physical processes such as screening and settling. Floating and suspended solids are removed in several steps. Primary treatment results in the accumulation of *sludge*, which requires further treatment and disposal. Prior to the Clean Water Act, which requires secondary treatment of waste water, primary treatment was the only type utilized in many communities.

Remaining in waste water after primary treatment are many undesirable substances, including both organics and inorganics. The organic content of waste water is characterized by the *Biochemical Oxygen Demand* (BOD), which is a measure of the amount of oxygen required by aerobic microorganisms to break down the organic compounds to less harmful substances such as carbon dioxide. The purpose of secondary treatment is to remove at least 85% of the BOD. At this level of treatment, normal municipal waste water can be discharged to a surface-water body. Although there are several technologies used for secondary treatment, the most basic and inexpensive is the *waste stabilization lagoon*, which simply is a surface impoundment in which the waste water is contained until BOD removal is complete. Treatment is based on the exchange of carbon dioxide and oxygen between bacteria and algae (Figure 16-8). The bacteria (both aerobic and anaerobic species are present in some lagoons) break down the BOD to carbon dioxide, which promotes the growth of algae. The algae give off oxygen in return, to sustain the population of bacteria. Waste stabilization lagoons are effective if they operate as designed. A problem arises, however, if the bottom of the impoundment is permeable enough to allow significant seepage downward toward the water table. Examples of ground-water contamination from this situation have been described in the literature.

A variety of chemical methods can be used in tertiary treatment. If necessary, very high quality water can be produced.

### *Safe Drinking Water Act (SDWA)*

Public drinking water supply systems were regulated by the *Safe Drinking Water Act* of 1974 and its amendments passed in 1986. One reason that this law is important is that it established drinking water standards for public supplies. These standards are based on the potential health risks of various water-borne substances. For cancer risk, for example, the standards are based

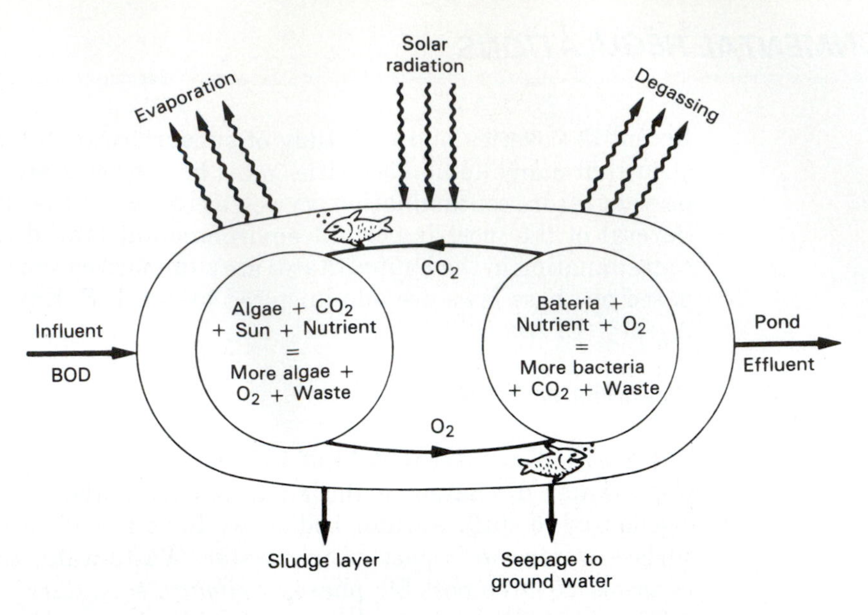


**Figure 16-8** Schematic of physical, chemical, and biological processes operating in a waste stabilization lagoon.

on rodent exposure studies that are extrapolated to human beings. The methods assume a lifetime exposure to the water. The risk level used by the EPA is $10^{-6}$. This means that the concentration of a particular carcinogen must be less than a level that would cause one additional cancer death in 1 million people. The methods used to arrive at these levels are extremely conservative and have generated controversy in the scientific community.

Three types of standards are reported: maximum contaminant-level goals (MCLGs), maximum contaminant levels (MCLs), and secondary maximum contaminant levels (SMCLs). The maximum contaminant-level goals are nonenforceable standards set at the most conservative levels with respect to adverse health effects. All carcinogens, for example, have MCLGs of zero. The maximum contaminant levels are the concentrations that must be met for water supply systems, and the secondary maximum contaminant levels are nonenforceable standards set for aesthetic reasons. SMCLs include constituents such as iron and hardness, which may cause taste, staining, or scale problems, but do not pose a known health risk.

Regulated contaminants are divided into four categories: organic chemicals, inorganic chemicals, microbiological contaminants, and radionuclides. The organic category, which contains by far the largest number of compounds, is further subdivided into *synthetic organic chemicals*, *trihalomethanes*, and *volatile organic chemicals*. Synthetic organics include pesticides and herbicides along with certain industrial chemicals. Trihalomethanes comprise a group of compounds that are formed during the disinfection of drinking water by the reaction between chlorine and organic compounds. It is somewhat disheartening to learn that carcinogenic compounds are being produced even in the treatment of drinking water. Volatile organics are compounds that readily vaporize at room temperature.

The number of chemicals regulated under the SDWA has increased drastically in the past few years. A number of municipal water systems have detected contamination that was not previously known, simply by testing for these compounds. An abbreviated list of drinking-water standards is given in Table 16-2.

**TABLE 16-2**

**Selected Drinking-Water Standard**

| *Chemical* | *MCLG* (*μg/L*) | *MCL* (*μg/L*) | *SMCL* (*μg/L*) |
|---|---|---|---|
| *Synthetic organic chemicals* | | | |
| Alachlor | 0 | 2 | |
| Aldicarb | 1 | 3 | |
| Atrazine | 3 | 3 | |
| Benzene | 0 | 5 | |
| Benzo[*a*]pyrene | 0 | 0.2 | |
| Butylbenzyl phthalate | 100 | 100 | |
| Carbofuran | 40 | 40 | |
| Carbontetrachloride | 0 | 5 | |
| Chlorodane | 0 | 2 | |
| Chrysene | 0 | 0.2 | |
| Dibromochloropropane (DBCP) | 0 | 0.2 | |
| *o*-Dichlorobenzene | 600 | 600 | 10 |
| *p*-Dichlorobenzene | 75 | 75 | 5 |
| 1,2-Dichloroethane | 0 | 5 | |
| 1,1-Dichloroethylene | 7 | 7 | |
| *cis*-1,2-Dichloroethylene | 70 | 70 | |
| *trans*-1,2-Dichloroethylene | 100 | 100 | |
| 1,2-Dichloropropane | 0 | 5 | |
| 2,4-Dichlorophenoxyacetic acid (2,4-D) | 70 | 70 | |
| Di(ethylhexyl)phthalate | 0 | 6 | |
| Ethylbenzene | 700 | 700 | 30 |
| Ethylene dibromide (EDB) | 0 | 0.05 | |
| Hexachlorobenzene | 0 | 1 | |
| Lindane | 0.2 | 0.2 | |
| Methoxychlor | 40 | 40 | |
| Methylene chloride | 0 | 5 | |
| PCBs as decachlorobiphenol | 0 | 0.5 | |
| Styrene | 100 | 100 | 10 |
| 2,3,7,8-TCDD (dioxin) | 0 | $5 \times 110^{-8}$ | |
| Tetrachloroethylene | 0 | 5 | |
| 1,2,4-Trichlorobenzene | 70 | 70 | |
| 1,1,2-Trichloroethane | 3 | 5 | |
| Trichloroethylene (TCE) | 0 | 5 | |
| 1,1,1-Trichloroethane | 200 | 200 | |
| Toluene | 1000 | 1000 | 40 |
| Toxaphene | 0 | 3 | |
| 2-(2,4,5-Trichlorophenoxy)-propionic acid (2,4,5-TP, or Silvex) | 50 | 50 | |
| Vinyl chloride | 0 | 2 | |
| Xylenes (total) | 10,000 | 10,000 | 20 |
| *Inorganic chemicals* | | | |
| Aluminum | | | 50–200 |
| Antimony | 6 | 6 | |
| Arsenic | 50 | 50 | |
| Asbestos (fibers per liter) | $7 \times 10^6$ | $7 \times 10^6$ | |
| Barium | 2000 | 2000 | |
| Beryllium | 4 | 4 | |
| Cadmium | 5 | 5 | |
| Chloride | | | 250,000 |
| Chromium | 100 | 100 | |
| Copper | 1,300 | 1,300 | |
| Cyanide | 200 | 200 | |
| Fluoride | 4,000 | 4,000 | 2,000 |
| Iron | | | 300 |
| Lead | 0 | 15 | |
| Manganese | | | 50–200 |
| Mercury | 2 | 2 | |
| Nickel | 100 | 100 | |

**TABLE 16-2**

*(Continued)*

| *Chemical* | *MCLG* (μg/L) | *MCL* (μg/L) | *SMCL* (μg/L) |
|---|---|---|---|
| *Inorganic chemicals (continued)* | | | |
| Nitrate (as N) | 10,000 | 10,000 | |
| Nitrite (as N) | 1,000 | 1,000 | |
| pH | | | 6.5–8.5 (pH units) |
| Selenium | 50 | 50 | |
| Silver | | | 100 |
| Sulfate | | | 250,000 |
| Total dissolved solids | | | 500,000 |
| Zinc | | | 5,000 |
| *Microbiological parameters* | | | |
| *Giardia lamblia* | 0 organisms | | |
| *Legionella* | 0 organisms | | |
| Heterotrophic bacteria | 0 organisms | | |
| Viruses | 0 organisms | | |
| *Radionuclides* | | | |
| Radium 226 | 0 | 20 pCi/L | |
| Radium 228 | 0 | 20 pCi/L | |
| Radon 222 | 0 | 300 pCi/L | |
| Uranium | 0 | 20 μg/L (30 pCi/L) | |

SOURCE: U.S. EPA.

### *Resource Conservation and Recovery Act (RCRA)*

The Resource Conservation and Recovery Act was enacted in 1976 to regulate the generation, transportation, storage, and disposal of hazardous wastes. The act created a system for tracking hazardous wastes from "cradle-to-grave," which is based on a manifest produced by the waste generator at the time it leaves the facility. This manifest contains information about the amount and type of waste, as well as identifying all companies involved in the transportation and ultimate disposal of the waste (Figure 16-9). With respect to disposal of hazardous waste, RCRA deals only with sites that are currently active. Amendments to RCRA in 1984 expanded its scope to include, among other components, the regulation of underground storage tanks (USTs).

Under RCRA, the classification of a waste as hazardous is based on specific criteria. A test known as the Toxicity Characteristic Leaching Procedure (TCLP) includes a list of 39 organic and inorganic compounds. If a threshold value of any of these is exceeded, the waste is classified as hazardous and is therefore subject to RCRA.

Among the many provisions of RCRA, the regulations of most relevance to subsurface contamination are those that apply to the final disposal of the waste. Each site is regulated through an elaborate permit system that regulates the design of the facility as well as the ground water and other monitoring systems. Design components similar to those shown in Figure 16-4 are required. The ground-water monitoring system consists of monitor wells located both upgradient and downgradient of the facility (Figure 16-10). Thus if any constituents contained in the waste are detected in ground water downgradient of the site, they can be compared with the upgradient concentrations to determine if a significant increase has occurred. Statistical procedures are specified to quantify whether or not the increase is "significant."

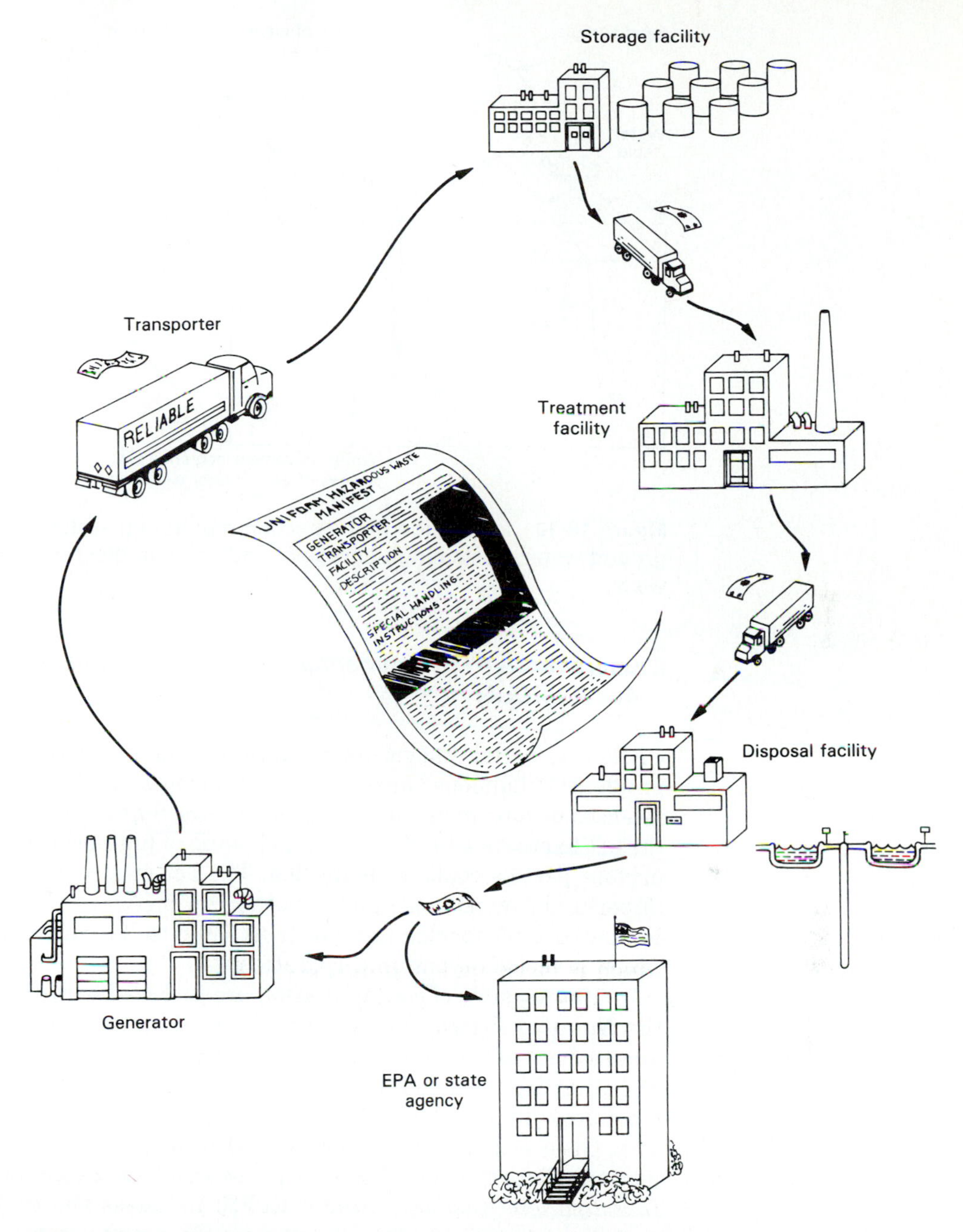


**Figure 16-9** RCRA manifest used for "cradle-to-grave" tracking of hazardous wastes. (From Masters, *Introduction to Environment Engineering and Science*, copyright © 1991 by Prentice-Hall, Inc., Englewood Cliffs, N.J.)

Three phases of RCRA ground-water monitoring exist. *Detection monitoring* is the initial phase, which is carried out until a significant increase is documented. If this should happen, the owner/operator must begin a *compliance phase* of monitoring, which involves a more extensive hydrogeological investigation in order to determine the extent and severity of contamination. Finally, a *corrective action* may be required to remediate the contamination. When a disposal facility is closed, RCRA regulations do not end. The facility must submit a closure plan which includes continued monitoring for a period of time that usually lasts for 30 years.

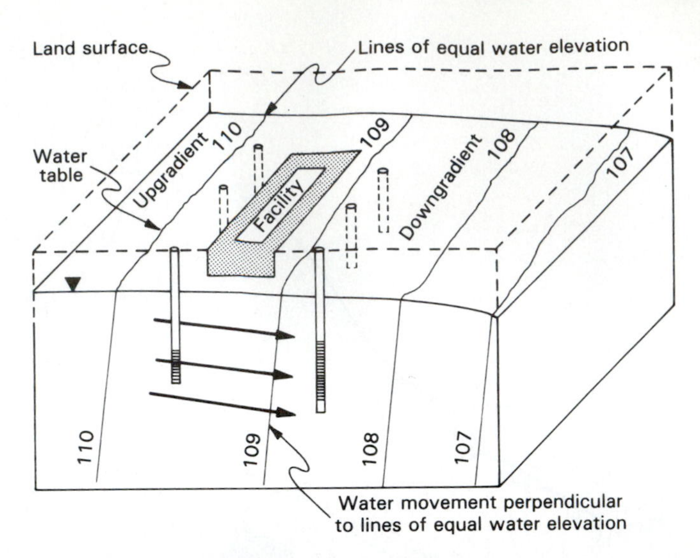


**Figure 16-10** Installation of a detection monitoring system for ground water, containing upgradient and downgradient monitor wells.

### *Comprehensive Environmental Response, Compensation, and Liabilities Act (CERCLA)*

CERCLA, better known as *Superfund*, was enacted in 1980 to handle the problem of abandoned hazardous waste disposal facilities along with emergency releases of hazardous materials. The term *Superfund* refers to a provision in the act to create a fund to initiate cleanup of high priority sites until the liable private parties could be identified. The act was reauthorized in 1986 by the Superfund Amendments and Reauthorization Act (SARA), which expanded the funds available for cleanup. The trust fund itself is derived from environmental taxes of various types on corporations.

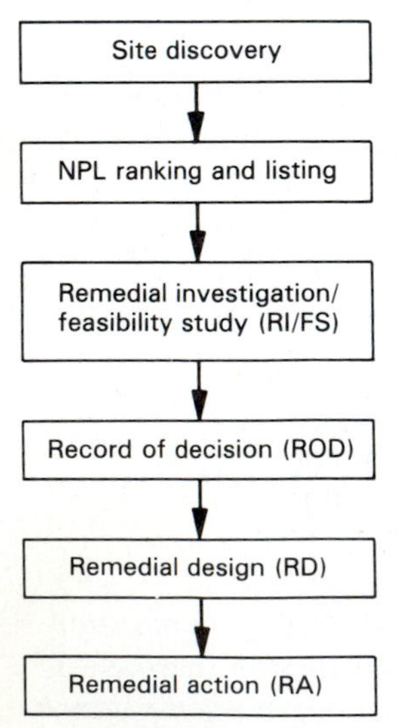


**Figure 16-11** Steps in the Superfund remedial process.

Known sites of contamination are included on a rank-ordered list called the *National Priority List* (NPL). The ranking is based on the characteristics and toxicity of the contaminants, the risk factors created for nearby residents, and other factors. Funds allocated for investigation and cleanup are then prioritized according to the NPL.

Once a site is listed on the NPL, a complex series of steps is set in motion (Figure 16-11). The first phase involves the completion of a *Remedial Investigation/Feasibility Study* (RI/FS) to assess the contamination problem and evaluate remediation alternatives. These are extensive site investigations including installation and sampling of many monitor wells. Additional steps include the *Record of Decision*, which reports the remedial alternative selected, the *Remedial Design*, and the *Remedial Action*, which is the actual cleanup procedure. This process can take up to five years or more before the Remedial Action is even started, and decades for the completion of the cleanup, if a large volume of aquifer is contaminated and ground water must be pumped and treated.

As the technical procedure is put into action, EPA is also pursuing a legal strategy to identify the *Potentially Responsible Parties* (PRPs) and forcing them to pay for the costs of cleanup. PRPs can include past or present owners of the site, as well as generators and transporters of the waste. If the PRPs (sometimes there are many of them) refuse to participate in the cleanup, EPA can legally collect more than three times the costs of the project. These amounts are not trivial as it is not unusual for remedial costs to reach tens of millions of dollars. The legal complexities of these operations have raised criticisms that the Superfund program spends an inordinate amount on legal costs as compared with the amount actually spent on cleanup.

## SUBSURFACE FATE AND TRANSPORT OF CONTAMINANTS

The movement of contaminants through the subsurface environment, which is known as *mass transport*, takes place by many complex and somewhat poorly understood processes. Because of the complexities of these phenomena, our discussion will serve only as a brief introduction to this important topic.

The most basic distinction to make with regard to mass transport mechanisms is the difference between aqueous and nonaqueous phases. *Aqueous-phase* transport occurs when the contaminant is dissolved in water, whereas *nonaqueous-phase* transport refers to the movement of a contaminant in a phase that is *immiscible* with water. Nonaqueous-phase liquids, or *NAPLs* (rhymes with "apples"), include petroleum products and other organic fluids that are spilled or leaked into the subsurface. The presence of NAPLs introduces additional complexity in description and prediction of transport of subsurface contaminants. Most of the difficulties come into play with mass transport through the unsaturated zone. Transport of NAPLs through the saturated zone is less common because it can only occur when the NAPL is denser than water. In this case, it is referred to as a *DNAPL*, a dense nonaqueous-phase liquid. Petroleum products, among the most common nonaqueous-phase contaminants, are less dense than water and are therefore prevented from moving through the saturated zone as a distinct phase. These petroleum-derived fluids are known as *LNAPLs*, or light nonaqueous-phase liquids.

### Aqueous-Phase Mass Transport

From a physical standpoint, there is little difference between the transport of an aqueous solution containing dissolved contaminants and the movement of pure water in the subsurface. Exceptions to this generalization include instances where there are large quantities of highly soluble compounds. When evaporite minerals dissolve in water, for example, the aqueous solution can become significantly more dense than pure water because of the high solubility of these minerals. Density would then become an important factor in transport, in addition to the normal hydraulic gradients that we discussed in Chapter 11. Most contaminants, however, are much less soluble, and we can assume that the solutes do not have a significant effect on the movement of the aqueous fluid.

All inorganic and organic substances will dissolve in water to some extent. The range in solubility, however, is extremely large. The solution of minerals was discussed in several previous chapters. Minerals in the unsaturated zone will dissolve in water until a state of chemical equilibrium is reached, at which time solution of the mineral will cease. Inorganic contaminants are no different. Leachate plumes in ground-water beneath landfills in which building materials such as drywall (which contains gypsum) were disposed of, commonly contain high concentrations of calcium and sulfate. The solubility of heavy metals and other elements is highly dependent upon the geochemical conditions of the leaching solution, including the pH and the *oxidation-reduction potential* (*redox potential*). Iron, for example is much more soluble when the pH and redox

**TABLE 16-3**

**Properties of Common Ground-Water Subsurface Contaminants**

| | *Property*[a] | | | |
|---|---|---|---|---|
| *Compound* | *Density, g/mL* | *Solubility, mg/L* | *Henry's*[b] *constant, atm* | $Log_{10}K_{ow}$ |
| Trichloroethene | 1.4 | 1 100 | 550 | 2.29 |
| Tetrachloroethene | 1.63 | 200 | 1 100 | 2.88 |
| Chloroform | 1.49 | 8 200 | 170 | 1.95 |
| Benzene | 0.876 | 1 808 | 240 | 2.01 |
| Toluene | 0.876 | 535 | 308 | 2.69 |
| 1,2-Dichlorobenzene | 1.305 | 145 | 90 | 3.38 |
| Phenol | 1.07 | 93 000[c] | 0.04 | 1.49 |
| 1,1-Dichloroethene | 1.013 | 250 | 1 400 | 0.73 |
| 1,1,1-Trichloroethane | 1.435 | 480 | 860 | 2.49 |
| Vinyl chloride | gas | 1 100 | 35 500 | 0.60 |
| Methyl ethyl ketone | 0.805 | 260 000 | 1.5 | 0.26 |
| Acetone | 0.79 | 1 000 000 | 1.0 | −0.24 |
| Ethylene dibromide | 2.18 | 3 400 | 26 | 1.80 |

SOURCE: Bouwer et al., 1988.

[a] Values for temperature of 20 to 25°C.

[b] Values should be considered representative of literature data: wide ranges in values are common.

[c] Depends on pH.

potential are low. In the unsaturated zone, the pH is commonly close to neutral and the redox potential is high due to the presence of oxygen from the atmosphere. Therefore, iron concentrations are usually low in well-aerated unsaturated-zone solutions.

The subsurface behavior of organic compounds depends upon several physical properties. Like the inorganics, the solubility of organic compounds varies over a wide range (Table 16-3). Solubility is an indication of the tendency of an organic compound to form an immiscible phase. Differences in solubility are due in part to the nature of chemical bonds in the compound. Compounds with a predominance of nonpolar, covalent bonds are less soluble in water, which is a strongly polar substance. These compounds are more likely to form a nonaqueous phase.

Another diagnostic property is called the *octanol-water partition coefficient.* This quantity, $K_{ow}$, is expressed as

$$K_{ow} = \frac{C_{oc}}{C_w} \qquad \text{(Eq. 16-1)}$$

where $C_{oc}$ is the equilibrium concentration of the compound in question dissolved in octanol, and $C_w$ is the equilibrium concentration in water. The higher the value of $K_{ow}$, which is usually expressed as a log value, the more likely it is to form a nonaqueous phase. Compounds with high log $K_{ow}$ values are known as *hydrophobic* compounds.

The *Henry's constant* is useful in predicting the tendency of an organic compound dissolved in water to volatilize into the gas phase. This is particularly important in the unsaturated zone, where organic vapors can collect and migrate separately from their transport in the aqueous phase. Table 16-3 indicates, for example, that benzene is much more volatile than phenol. The differences in volatility between organic compounds have an important bearing upon the remedial method used.

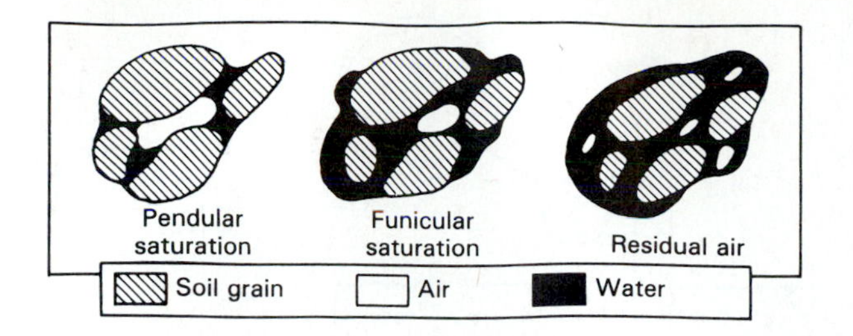


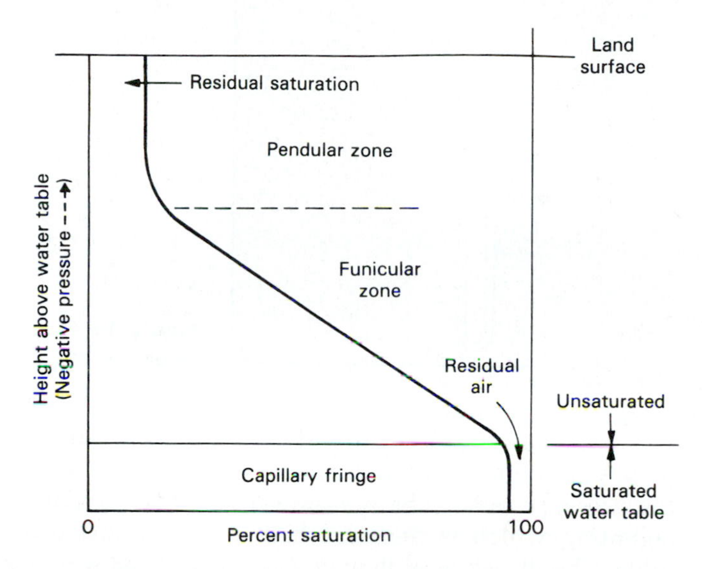


**Figure 16-12** Soil water retention curve and classification of soil moisture zones. (From Abdul, 1988.)

Unsaturated Zone Transport. Because the majority of contaminants originate above the water table, we must first discuss the movement of water (along with dissolved contaminants) through the unsaturated zone. In order to do this, we must expand upon the comments made about the unsaturated zone in Chapter 11. In that discussion, it was mentioned that the fluid pressure in unsaturated zone water is less than atmospheric. These negative pressures, which hold the water against gravity by suction or tension, are known as the *matric potential*. The distribution of matric potential is shown in Figure 16-12, in which matric potential is plotted against the percent saturation. At the water table, where the fluid pressure is equal to atmospheric (matric potential = 0), the pores are saturated with water. Matric potential decreases above the water table, but within the capillary fringe, the pores remain saturated. Above the capillary fringe, the amount of water in pores decreases as the amount of air increases. This is known as the *funicular zone*. Water dominates pore volumes within this region (Figure 16-12). At some point the amount of water within the pores achieves a constant residual value as the matric potential decreases. In this region, called the *pendular zone*, pore water is limited to pore necks and may form only a very thin continuous layer over the particle surfaces. The bulk of the pore space is occupied by air.

The matric potential of the soil is measured by a device called a *tensiometer* (Figure 16-13). The tensiometer consists of a tube with a porous cup on the end that is inserted into the soil. The tube is totally filled with water and connected to a vacuum gauge at the top. Water is drawn out through the porous cup into the soil, creating a vacuum in the tube, which is measured by the gauge.

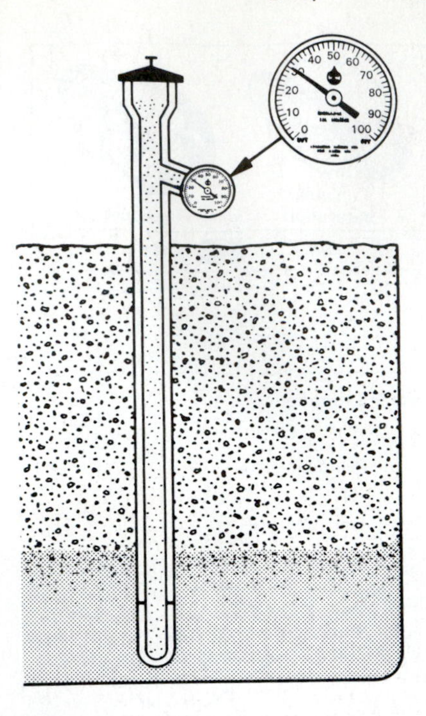


**Figure 16-13** Schematic diagram of a tensiometer.

Like flow in the saturated zone, flow in the unsaturated zone is controlled by the gradient of hydraulic head, which is the sum of the elevation head and the pressure head. The pressure head in the unsaturated zone is the matric potential, which is negative rather than positive as it is below the water table. The direction of flow in the unsaturated zone can be determined by the installation of two adjacent tensiometers at different depths.

### EXAMPLE 16-1

Tensiometer A, which is inserted to a depth of 50 cm below land surface, has a measured matric potential of $-100$ cm. Tensiometer B, which is inserted at a depth of 70 cm, has a matric potential of $-50$ cm. Is water moving upward or downward in the unsaturated zone?

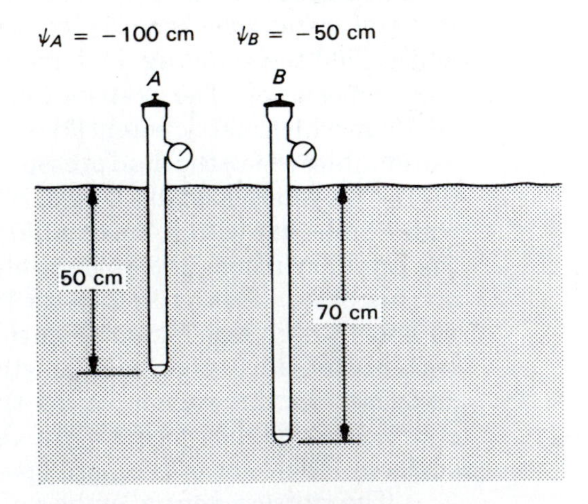


***Solution.*** Total head or potential is the sum of elevation head and matric potential. If we use the ground surface as our datum and use depth for the elevation head, elevation head, $z$, is negative. The total head, $\theta = \psi + z$. For tensiometer A, $\theta = -100 + (-50) = -150$ cm. For tensiometer

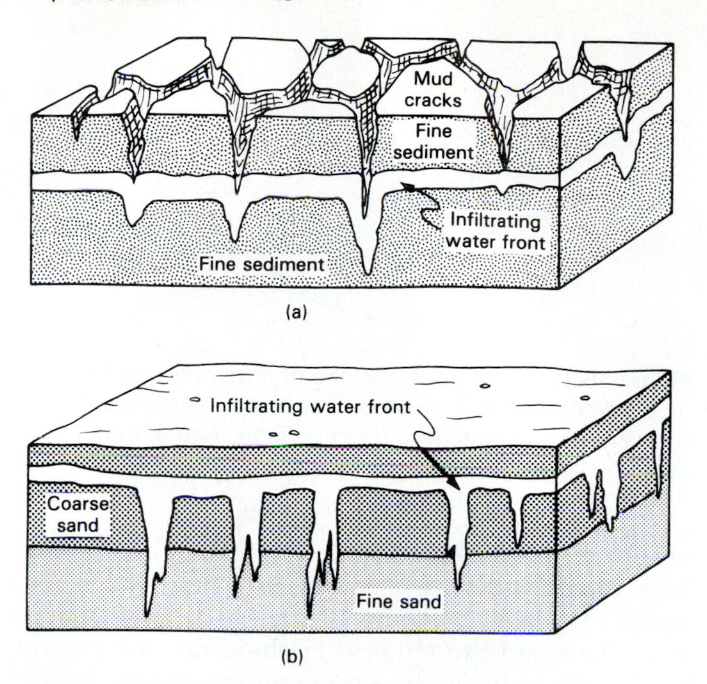


**Figure 16-14** Contrast between unsaturated flow through (a) uniform soils with isolated macropores and (b) fingering at a hydraulic conductivity boundary.

B, $\theta = -50 + (-70) = -120$ cm. Since the head at tensiometer B (−120 cm) is larger than the head at tensiometer A (−150 cm), the flow is in the direction of B to A, or upward.

When a substantial amount of rain falls to the surface, the infiltrating water flows into the pore spaces and increases the percent saturation. As a result the matric potential is increased (becomes less negative), and water moves downward through the soil. If the soil is homogeneous, the moisture will move downward as a slug, with a well defined *wetting front* (Figure 16-14). Any contaminants in solution would therefore move toward the water table at about the same rate. It is more likely, however, that the soil is not homogenous, but exhibits *preferential flow*. Preferential flow refers to the presence of more permeable pathways in the soil through which water and contaminants can move more rapidly than the wetting front. Preferential flow can follow *macropores*, which are large openings (such as root casts, animal burrows or shrinkage cracks), or it can occur through a process known as *fingering* (Figure 16-14), in which pathways of more subtle variation in permeability are exploited. Fingering is commonly observed to occur at the contact between a fine-grained soil and a coarse-grained soil beneath. The significance of preferential flow to mass transport is that it provides a very rapid journey for contaminants through the unsaturated zone to the water table.

Saturated Zone Transport. Once contaminants reach the water table, they travel in the direction of ground-water flow. If the contaminants originated as a point source, they move within the saturated zone as a coherent mass of degraded ground water known as a *contaminant,* or *leachate, plume* (Figure 16-15). The direction of movement may be horizontal, vertical, or inclined, depending on the gradient of the flow system. If the contaminant plume is the approximate density of the ground water, the major control on the direction

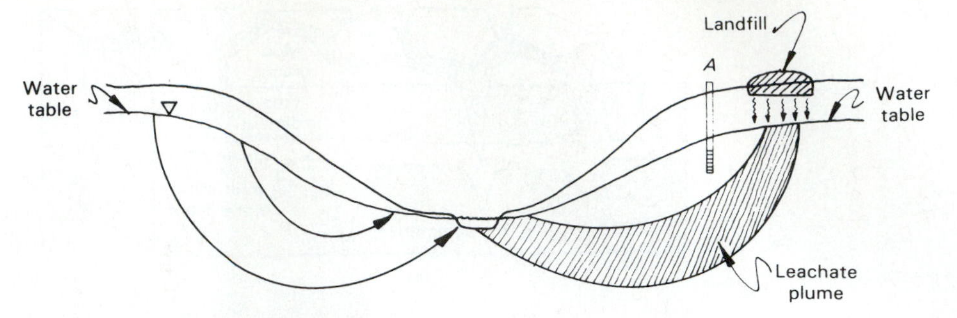


**Figure 16-15** Control of leachate-plume movement by the hydraulic gradient. A well at *A* would not be contaminated because of the vertical component of ground-water motion in the flow system.

of movement will be the hydraulic gradient; if the plume is significantly denser, however, it will have a downward component of motion separate from the hydraulic gradient (Figure 16-16).

There are two physical mechanisms that control the movement of dissolved contaminants in a contaminant plume. *Advection* is the transport of contaminants by the bulk movement of the flowing water, in response to the

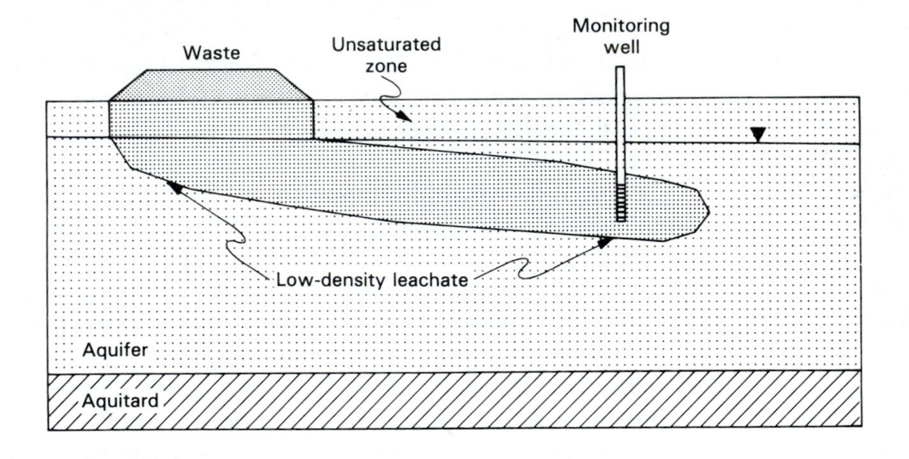


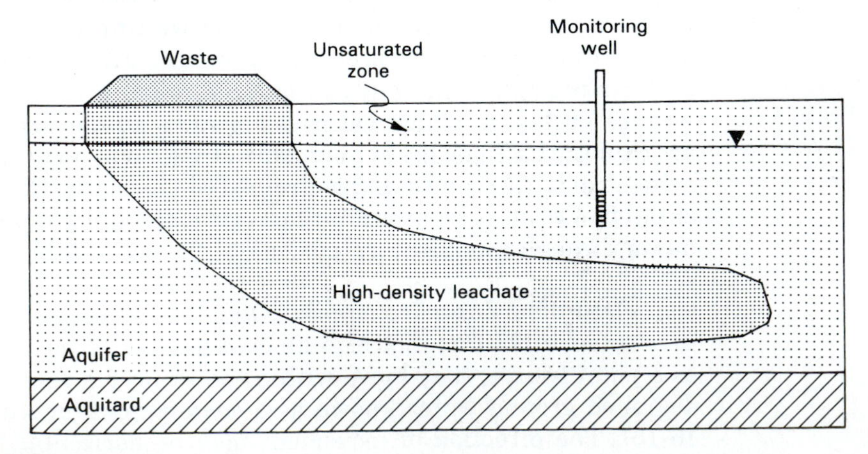


**Figure 16-16** Modification of contaminant flow by increased density of leachate.

hydraulic gradient. The rate of movement of contaminants can be estimated by determining the *average linear velocity* of the ground water. This velocity is described as *average* because of the nature of flow in porous media. Variation in size of the microscopic flow channels followed by individual particles of water leads to a difference in velocity, depending on which particular path a particle of water follows. Since it is not possible to determine the infinite variety of possible flow paths, an average velocity is used. When an average linear velocity is reported, it must be kept in mind that some ground-water, and therefore dissolved contaminants, will be traveling along certain flow paths at higher velocities. The average linear velocity can be calculated by the formula

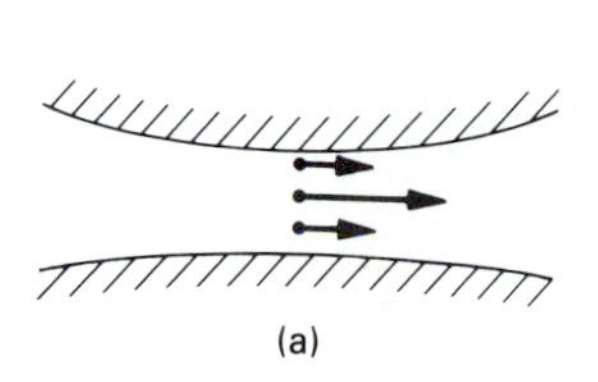


$$\overline{v} = \frac{K}{n}\frac{dh}{dl} \quad \text{(Eq. 16-2)}$$

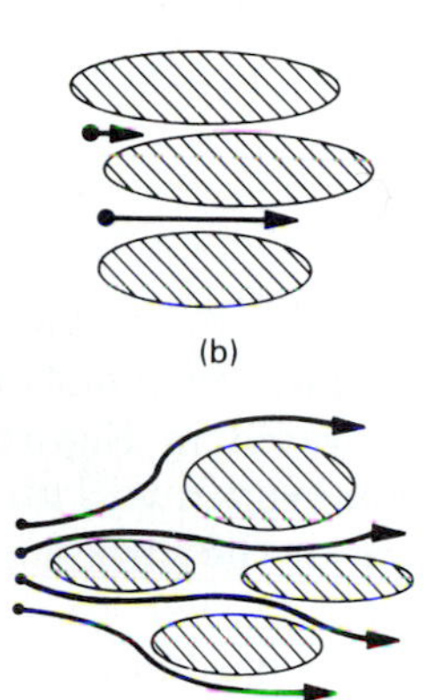


Figure 16-17 Mechanisms of mechanical dispersion operating at the scale of individual pores and grains. (From Gillham and Cherry, 1982.)

where $\overline{v}$ is average linear velocity, $K$ is hydraulic conductivity, $n$ is the porosity of the aquifer, and dh/dl is the hydraulic gradient.

*Hydrodynamic dispersion*, the second physical transport mechanism, results in the spreading of the zone of contamination along the flow path. On a microscopic scale, several effects tend to move contaminants away from the center of mass of the plume (Figure 16-17). These effects include (a) higher flow rates in the center of a pore than along the margins, (b) more rapid flow through larger pores, and (c) lateral expansion of the contaminant because water molecules must flow around aquifer grains. The overall effect of hydrodynamic dispersion is to spread the contaminant over a larger volume of aquifer, both in the direction of flow and laterally to the direction of flow. When the aquifer is heterogenous, that is, when it is composed of beds or laminae of different hydraulic conductivity, even more spreading of the plume occurs. Figure 16-18 illustrates an aquifer that has lenses of high hydraulic conductivity. The contaminant moves more rapidly through these lenses with hydrodynamic dispersion superimposed on this preferential flow. The plume is spread in the direction of flow by this process.

Despite the spreading effect of hydrodynamic dispersion, the influence is much less than the mixing caused by turbulent flow in a river. It is correct to think of leachate plumes moving in compact masses through aquifers, with relatively minor spreading caused by hydrodynamic dispersion and aquifer heterogeneity.

Because dispersion spreads a fixed mass of contaminant over a larger volume of aquifer, a reduction in contaminant concentration occurs. If the particular contaminant is very toxic at low concentrations, the dispersive tendency will have an undesirable result because more area is affected even though the concentration is decreased.

The ultimate path of contaminants in ground water can be predicted only by knowing the nature of the flow system. Water-table and potentiometric-

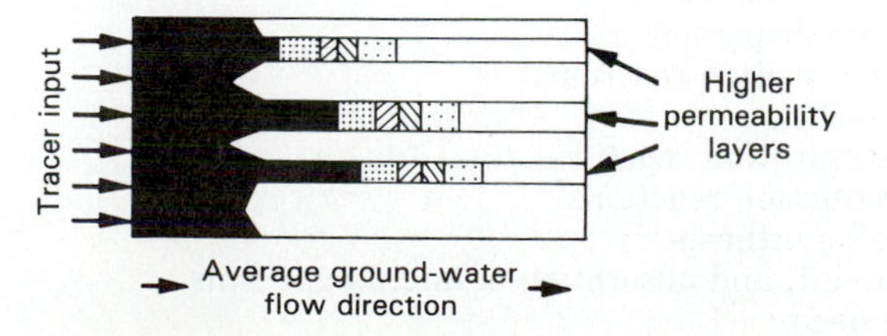


Figure 16-18 More rapid flow of contaminants in thin lenses of high hydraulic conductivity.

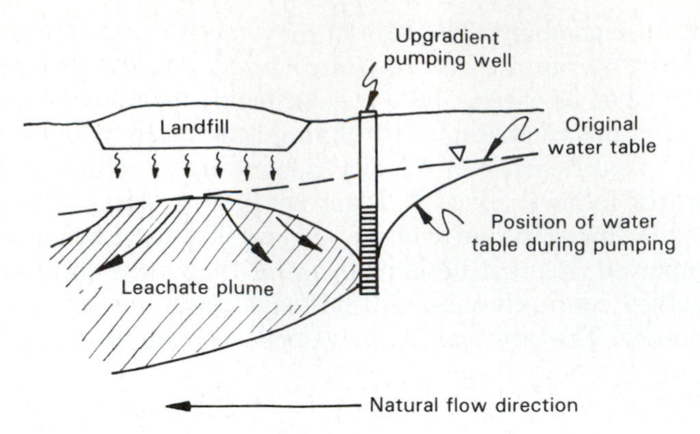


**Figure 16-19** Contamination of an upgradient pumping well due to the gradient reversal caused by pumping.

surface contour maps will give an indication of the horizontal component of movement. The importance of vertical components is illustrated in Figure 16-15, where a contaminant source is added to a flow system. Well A will not be contaminated because it is too shallow, even though it is in the direction of flow of ground water from the waste-disposal site.

Contamination of a well can occur even when the natural gradient suggests that ground-water flow will carry contaminants away from the well. A pumping well located upgradient of a disposal site, like the one shown in Figure 16-19, can become contaminated because the cone of depression it creates reverses the natural gradient and induces movement of the contaminated water toward the well.

Attenuation Mechanisms. As dissolved contaminants move through the unsaturated and saturated zones, a number of physical, chemical, and biological processes may alter the concentration of the solute. Chemical processes, including acid-base and oxidation-reduction reactions, can change the chemical species of the contaminant, and in doing so, alter its mobility.

Some important processes that decrease the concentration of contaminants in the subsurface, or *attenuation mechanisms*, are listed in Table 16-4. One of these, *adsorption*, is particularly important. Adsorption, the attraction of an ion to the surface of a solid particle in the unsaturated or saturated zones, is really a form of *ion exchange*. In order to maintain an overall neutral charge, the particles of clay and other substances that attract ions must release ions

**TABLE 16-4**

| Attenuation Mechanisms |
|---|
| Hydrodynamic dispersion |
| Adsorption-desorption reactions |
| Acid-base reactions |
| Solution-precipitation reactions |
| Oxidation-reduction reactions |
| Microbial cell synthesis |
| Filtering, die-off, and adsorption of microorganisms |
| Radioactive decay |
| Biodegradation |

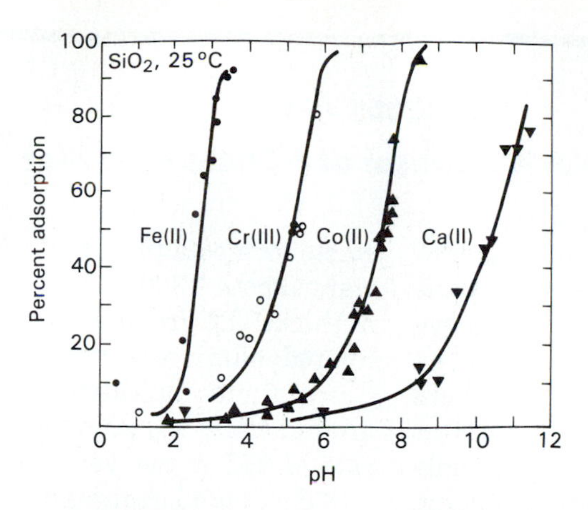


**Figure 16-20** Adsorption of metals to metal-oxide surfaces as a function of pH. (From R. O. James and T. W. Healy, 1972, *Journal of Colloid and Interface Science*, vol. 40.)

of the same charge. The most common reaction involves the exchange of cations. Contaminant cations are drawn to negatively charged particle surfaces in exchange for other cations released from the particle surface to the groundwater solution. Clays have the strongest ability to effect this exchange reaction, although other material, including iron oxide coatings or particles, may also participate in adsorption reactions. Heavy-metal cations can be maintained at low dissolved concentrations by adsorption, although the pH of the solution is very important. Figure 16-20 shows that adsorption of each metal cation is controlled by a relatively narrow range of pH values. As the pH rises above the critical value, the environment changes from a condition where the metal is very mobile to a state where a high percentage of the metal is adsorbed to metal oxides.

Adsorption of nonpolar organics is controlled by solid organic matter in the aquifer. Even though only minor amounts of organics may be present, nonpolar dissolved organic molecules have a strong affinity for these substances. The log $K_{ow}$ value for the compound is a good indication of its mobility in ground water. The higher the log $K_{ow}$, the less mobile the compound will be due to adsorption on solid organics.

*Biodegradation*, the breakdown of organic compounds by microorganisms in the contaminant zone, is also important in the attenuation of organic contaminants. Here again, the chemical environment is very important in the process. For example, *aromatic* compounds such as benzene, toluene, and xylene, which are common components of gasoline, dissolve in water but can be actively broken down to carbon dioxide under oxidizing conditions. These conditions are most common in the unsaturated zone. Biodegradation of the aromatics is much less rapid in the saturated zone. Other contaminants, like the chlorinated solvents trichloroethene (TCE) and tetrachloroethene (also known as perchloroethene, or PCE), biodegrade more rapidly under reducing conditions. Therefore, these compounds are resistant to biodegradation in the unsaturated zone, but do break down more readily below the water table.

Contaminants may also be removed from solution by chemical precipitation, radioactive decay, and the die-off of microorganisms. Hydrodynamic dispersion is also an attenuation mechanism because spreading of the contaminant reduces its concentration.

## CASE STUDY 16-1
## Industrial Waste Disposal of a Shallow Aquifer

Liquid chemical wastes were discharged to unlined surface impoundments at many locations in the past. The contaminants may move largely by aqueous-phase transport through the unsaturated zone to the saturated zone, where they form a contaminant plume. Perlmutter and Lieber (1970) presented an early example of this type of contamination on Long Island, New York. During and after World War II, an aircraft factory discharged aluminum-plating wastes into the disposal ponds shown in Figure 16-21. Because the area is underlain by a highly permeable unconfined aquifer with a high water table, the wastes rapidly infiltrated downward to the water table. From this point, the waste formed a plume of contaminated ground water about 1300 m long, 300 m wide, and 21 m thick. The gradual spreading resulting from hydrodynamic dispersion is evident on the map. Calculations indicated that the contaminated ground water moves rapidly at an approximate rate of 165 m per year. Some of the contaminated water discharges into Massapequa Creek in the southern part of the area.

Figure 16-22 shows a subsurface cross section of the plume. It is apparent that contamination is confined to the upper glacial aquifer and that movement is predominantly horizontal. The contours on the cross section show the distribution of hexavalent chromium in 1962. Hex-

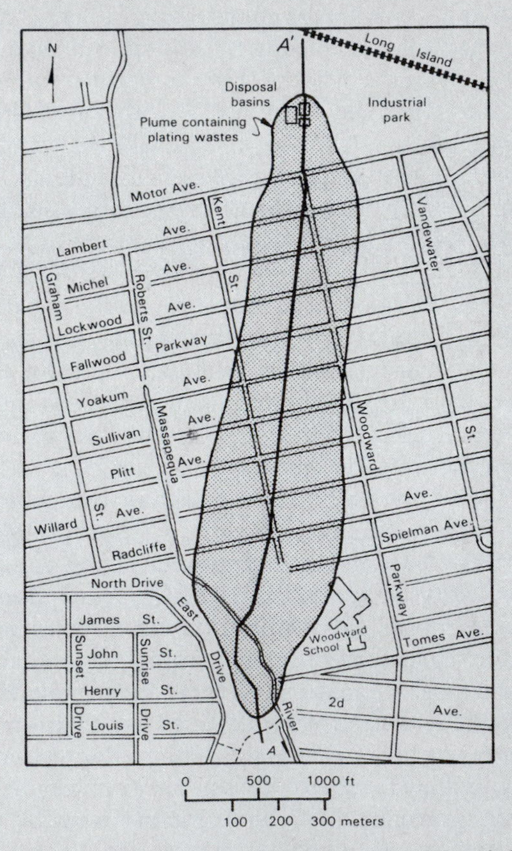


**Figure 16-21** Map of part of Long Island, New York, showing size and extent of plume of contaminated ground water originating from disposal pits at a metal-plating facility. (From N. M. Perlmutter and M. Lieber, 1970, U.S. Geological Survey Water Supply Paper 1879-G.)

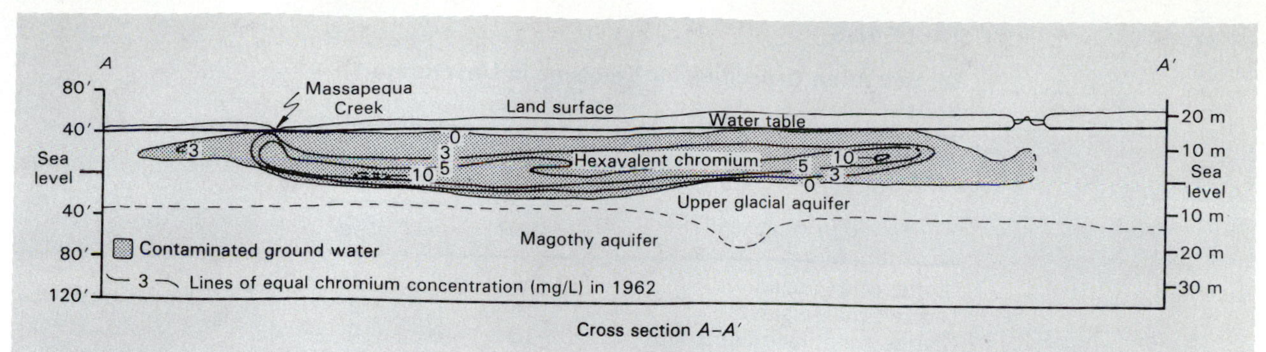


**Figure 16-22** Cross-section *A-A'* (location shown on Figure 16-21) through the center of the contaminant plume. Contours show the concentrations of hexavalent chromium in milligrams per liter. (From N. M. Perlmutter and M. Lieber, 1970, U.S. Geological Survey Water Supply Paper 1879-G.)

avalent chromium is a toxic metal with an MCL of only 0.05 mg/L. At the levels present in the aquifer, use of the water for drinking would pose a very serious health threat. In addition to chromium, the plume contained other toxic contaminants.

### *Nonaqueous-Phase Mass Transport*

Unsaturated Zone Transport. The movement of a NAPL through the unsaturated zone is a complex phenomenon because it involves three or more phases: the NAPL, water, and air. If part of the NAPL evaporates into the soil atmosphere, a fourth phase is present. Additionally, the NAPL, or certain compounds contained within it, can dissolve into the pore water contained in the unsaturated zone. The discovery that spills and leaks have contaminated the subsurface on a routine basis has stimulated research into the properties and processes of flow through the unsaturated zone.

The flow of a NAPL through an unsaturated porous medium follows the same basic principles as does the flow of water. Below the level of saturation of the pores by the fluid, a capillary suction is exerted upon the NAPL similar to that exerted upon water, except that the adhesion of most nonpolar organics to soil surfaces is less than the adhesion of water. NAPLs are also subject to a condition similar to water in the pendular zone (Figure 16-12). Here water is said to be at *residual saturation* because the percent saturation does not decrease with decreases in the matric potential (greater suction). As a NAPL moves through the subsurface, it also must exceed the level of residual saturation prior to further movement. This level of NAPL will remain in the soil after drainage of the mobile fraction. The residual saturation depends upon the soil type (Table 16-5, where oil residual capacity is a measure of residual saturation). Fine-grained soils have a much higher residual saturation than coarse-grained soils.

The NAPL in a two- or three-phase unsaturated system is referred to as the *nonwetting fluid*, because water usually covers all solid surfaces within the medium. Water is known as the *wetting fluid*. The movement of a NAPL through a water-wet porous medium differs from movement through an initially dry medium. As a petroleum product, the nonwetting fluid, is released

**TABLE 16-5**

**Oil Retention Capacities for Kerosene in Unsaturated Soils**

| *Soil Type* | *Oil Retention Capacity (R)* | |
|---|---|---|
| | *L/m³* | *g/yd³* |
| Stone, coarse sand | 5 | 1 |
| Gravel, coarse sand | 8 | 2 |
| Coarse sand, medium sand | 15 | 3 |
| Medium sand, fine sand | 25 | 5 |
| Fine sand, silt | 40 | 8 |

SOURCE: CONCAWE 1979.

into the unsaturated zone, it begins to move downward under the influence of gravitational forces. These are resisted by capillary forces during the development of residual saturation. The NAPL may move downward as a distinct interface, although preferential flow is common. In order to progress through the medium, the NAPL must displace the existing pore fluids. This is relatively easy in the pendular zone, where air-filled pores are the norm. The rate of movement is largely a function of the viscosity of the NAPL, which can vary over a large range, and the grain size of the soil. Fine-grained soils tend to have a higher residual saturation of water and it is therefore more difficult to drive a NAPL through them. When the NAPL encounters the funicular zone, its rate of movement slows, because pressure must be built up to overcome the resistance of the higher saturation percentage of water.

Several methods are available for prediction of the depth of penetration of NAPL into the unsaturated zone. For example, the equation

$$D = \frac{1000V_{hc}}{ARC} \qquad \text{(Eq. 16-3)}$$

gives the approximate depth of penetration, $D$, in terms of $V_{hc}$, the volume of discharged NAPL in barrels (1 barrel = 42 gal), the area of infiltration, $A$, the oil retention capacity or residual saturation, $R$ (Table 16-5), and a correction factor, $C$, based on the viscosity of the NAPL (0.5 for gasoline to 2.0 for light fuel oil).

As the downward moving NAPL finally reaches the top of the capillary fringe, a change in behavior occurs. If the source of the spill is large enough to provide a relatively continuous downward movement of product, and the product is an LNAPL (which would include gasoline, heating oil, kerosene, jet fuel, and aviation gas), the petroleum product begins to form a pool, or *pancake*, at the top of the capillary fringe (Figure 16-23). The layer is known as *free product* because it is not held under capillary pressure and is free to migrate along the top of the capillary fringe in the direction of slope of the water table. The top of the layer of free product forms an *oil table*, because product below is under positive pressure. With sufficient source, the pancake layer can actually depress the capillary fringe so that free product rests directly upon the water table. The layer of free product can remain in its subsurface position for a long period of time. Even after the source is terminated, a column of NAPL at a level of residual saturation will remain in the unsaturated zone above the free product.

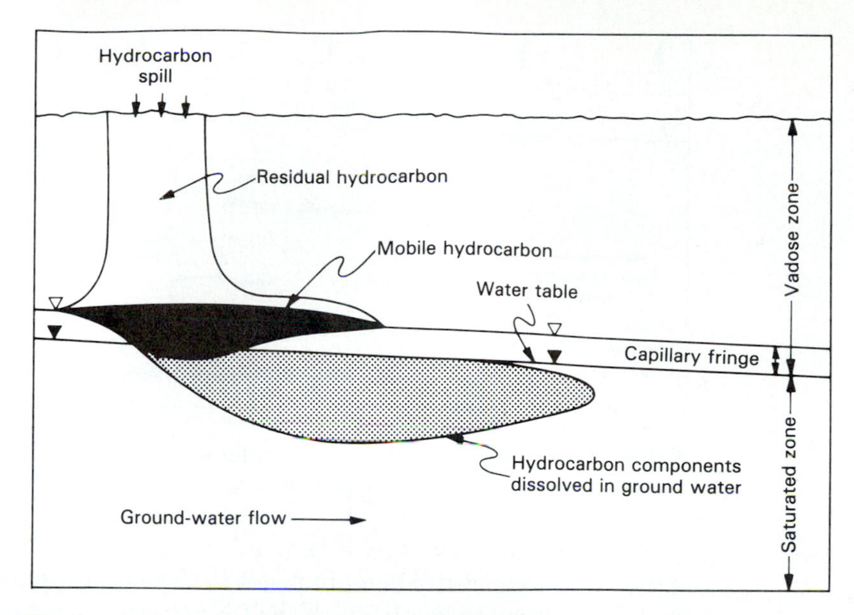


**Figure 16-23** Subsurface migration of LNAPL contaminants.

Several other transport processes occur once an LNAPL has reached the water table to form a free product layer. One serious problem is that the more soluble components of the hydrocarbon will dissolve and form a ground-water contaminant plume that moves downgradient (Figure 16-23), typically for a much longer distance than the extent of the free product. In gasoline, the dissolved contaminants include benzene, toluene, ethylbenzene, and xylene. Together these are referred to as *BTEX*. Additional processes include volatilization of hydrocarbon vapors from the free product into the unsaturated zone above. These gases can migrate within the unsaturated zone into basements and pose explosion hazards. The discovery of free product in the unsaturated zone and/or BTEX plumes in ground water has become all too common since the advent of the UST program initiated under RCRA. More likely than not, the corner gas station that has been in business for several decades has had or will have a petroleum product release of some type.

Saturated Zone Transport. Transport of NAPLs in the unsaturated zone is limited to DNAPLs, those products denser than water (Table 16-3). The most common compounds involved in DNAPL spills and leaks are the chlorinated hydrocarbons, synthetic compounds such as 1,1,1-trichloroethane, carbon tetrachloride, and tetrachloroethene. These compounds are very commonly used as solvents and degreasers in industry, among other uses. The dry cleaning industry is a very good example of an industry that uses compounds of this type.

Migration of DNAPs through the unsaturated zone is similar to the movement of LNAPLs. Once the residual saturation is exceeded, the product moves vertically downward to the water table. At that point however, its behavior differs from LNAPLs. Although a layer of DNAPL may initially form at the top of the saturated zone, under sufficient pressure, the product will penetrate the water table and sink because of its higher density to the first layer of low permeability material in the aquifer (Figure 16-24). The DNAPL will form a layer of free product on top of the low permeability layer, and it can then migrate in the direction of slope of the layer. This direction is not necessarily in the direction of ground-water flow and, in fact, can be in the opposite direction

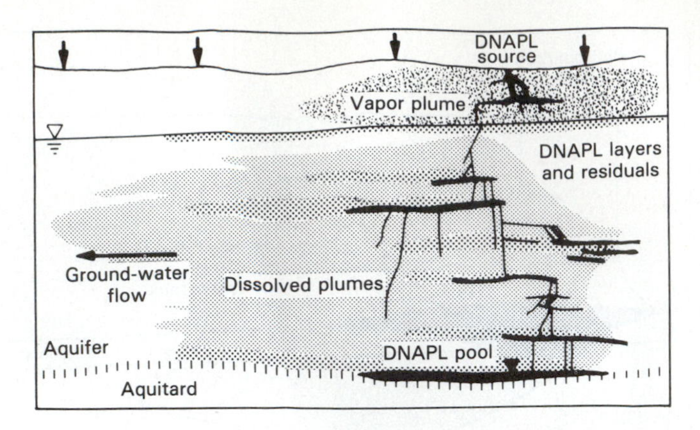


**Figure 16-24** Migration of DNAPL contaminants in the subsurface. (Diagram courtesy of Stan Feenstra.)

if the low-permeability layer happens to slope that way. Above the free product pool, the column through which the DNAPL moved retains product at residual saturation. Like LNAPLs, DNAPLs may be somewhat soluble and will therefore form contaminant plumes (Figure 16-24). Because the hydrocarbon is dissolved at low levels, these plumes move under the hydraulic gradient and may not display density control.

Location of DNAPL free product layers is extremely difficult, particularly if the surface source is unknown. Since the free product layer does not follow the hydraulic gradient, a detailed knowledge of the subsurface geology must be developed by test drilling and other methods to track the movement of the free product layer.

### CASE STUDY 16-2

## The Lawrence Livermore National Laboratory (LLNL) Superfund Site

The degree of complexity of Superfund projects can reach immense proportions, not only in terms of the hydrogeology of the site, but also in terms of the site's history and contamination. The Lawrence Livermore National Laboratory (LLNL) typifies the staggering task required in investigation and remediation of these sites (Thorpe et al., 1990).

The LLNL is a weapons research laboratory located southeast of Oakland, California (Figure 16-25). The site encompasses about 800 acres, including the facility itself and a buffer zone. The history of the site begins in World War II, when it was acquired by the U.S. Navy for use as an air station. Extensive repair and overhaul operations were conducted until the end of the war. Numerous organic solvents were used in these operations and, as was common at that time, handling and disposal was haphazard. Large amounts of solvents were undoubtably released at this time. In 1951, the property was transferred to the Atomic Energy Commission (AEC) for use as a thermonuclear weapons design facility. Later, the site was passed onto successors of the AEC, including the Energy Research and Development Agency and the Department of Energy. Weapons research continued under the management of the University of California, along with nonmilitary research. Hazardous materials of many types were handled at the site during this era.

Contamination was first detected in the early 1980s. An almost continuous series of hydrogeologic studies led to LLNL's inclusion on

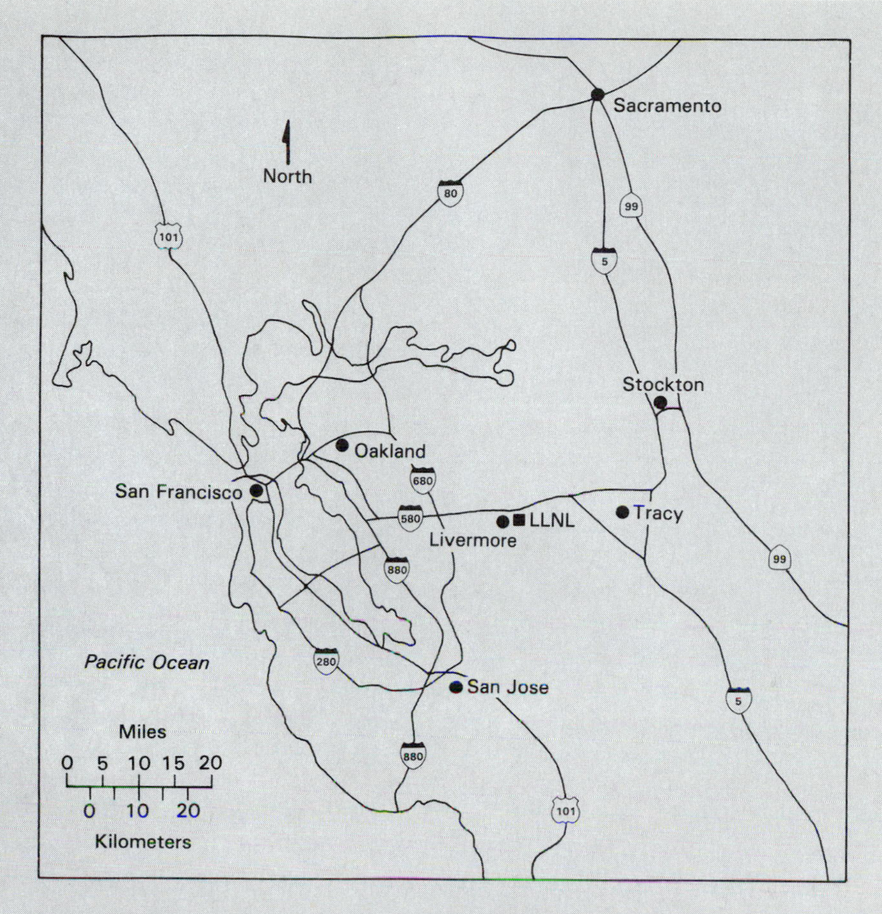


**Figure 16-25** Location of Lawrence Livermore National Laboratory.

the NPL under the Superfund program in 1987. Volatile organic compounds (VOCs) were identified as the most serious contaminants, although a gasoline leak of 17,000 gallons over a 20-year period was documented for one portion of the site. Out of 130 possible sites for VOC releases, 14 were considered to be the most significant and were designated for further study. Hundreds of borings and monitor wells were installed. The purpose of the borings was to define the hydrogeological setting of the site and to determine the subsurface distribution of contaminants.

The LLNL is built upon several coalescing alluvial fans. The uppermost subsurface sediments beneath the site are composed of a complex series of alluvial units, consisting of alternating beds of fine and coarse materials. The coarser sediments comprise channel fills, formed as stream channels repeatedly migrated across the growing alluvial fan. The subsurface stratigraphy creates a very complex pattern of migration for contaminants. Overall contamination appears to be limited to several large, diffuse and overlapping plumes. Total VOC concentrations are shown in Figure 16-26, and the ranges in concentration for the major contaminants are shown in Table 16-6.

A sense of the vertical complexity of the site is given by Figure 16-27, which shows the generalized distribution of lenses of high hydraulic conductivity along one cross section. High concentrations of VOCs occur within these zones. Figure 16-27 suggests the difficulty of characterizing the three-dimensional distribution of contaminants.

Remediation of the site will be based on a pump and treat approach (Hoffman, 1993) that will include source removal, reinjection of treated water, and dynamic management of the remedial system in order to minimize the time required for cleanup.

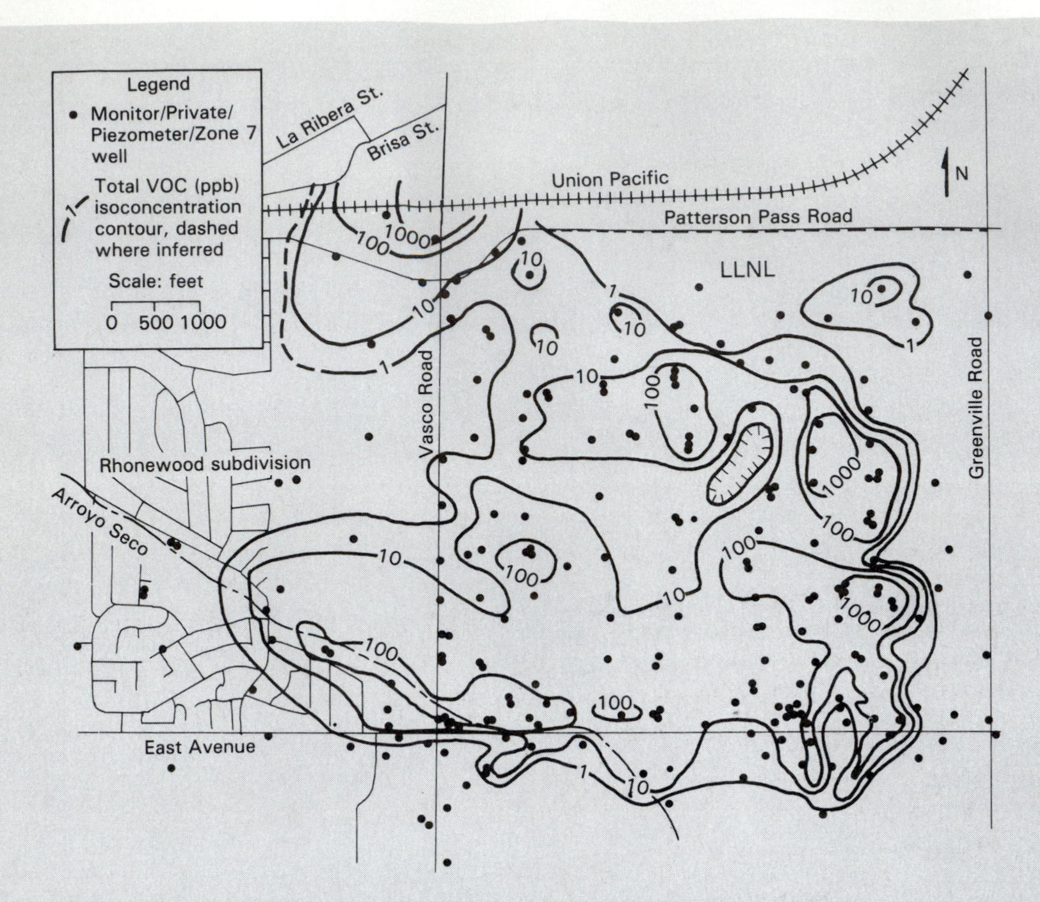


**Figure 16-26** Contour map showing concentrations of total VOCs in ground water at LLNL site, November 1991. (From Hoffman, 1993.)

**TABLE 16-6**

**Summary of California State Action Levels and Maximum VOC Concentrations in Ground Water from LLNL Monitor Wells, June 15, 1988–June 15, 1989**

| *VOC* | *EPA MCL (ppb)* | *DHS MCL (ppb)* | *DHS action level (ppb)* | *Maximum LLNL concentration (ppb)* |
|---|---|---|---|---|
| PCE | None | 5 | None | 1800 |
| TCE | 5 | 5 | None | 5200 |
| cis 1,2-DCE | None | None | 6 | — |
| trans 1,2-DCE | None | None | 10 | 22 |
| 1,1-DCE | 7 | 6 | None | 640 |
| 1,1,1-TCA | 200 | 200 | None | 20 |
| 1,2-DCA | None | 0.5 | None | 200 |
| 1,1-DCA | None | None | 5 | 59 |
| Carbon tetrachloride | 5 | 0.5 | None | 140 |
| Chloroform | 100 | None | None | 650 |
| Freon-113 | None | None | 1200 | 950 |

SOURCE: Thorpe et al., 1990.

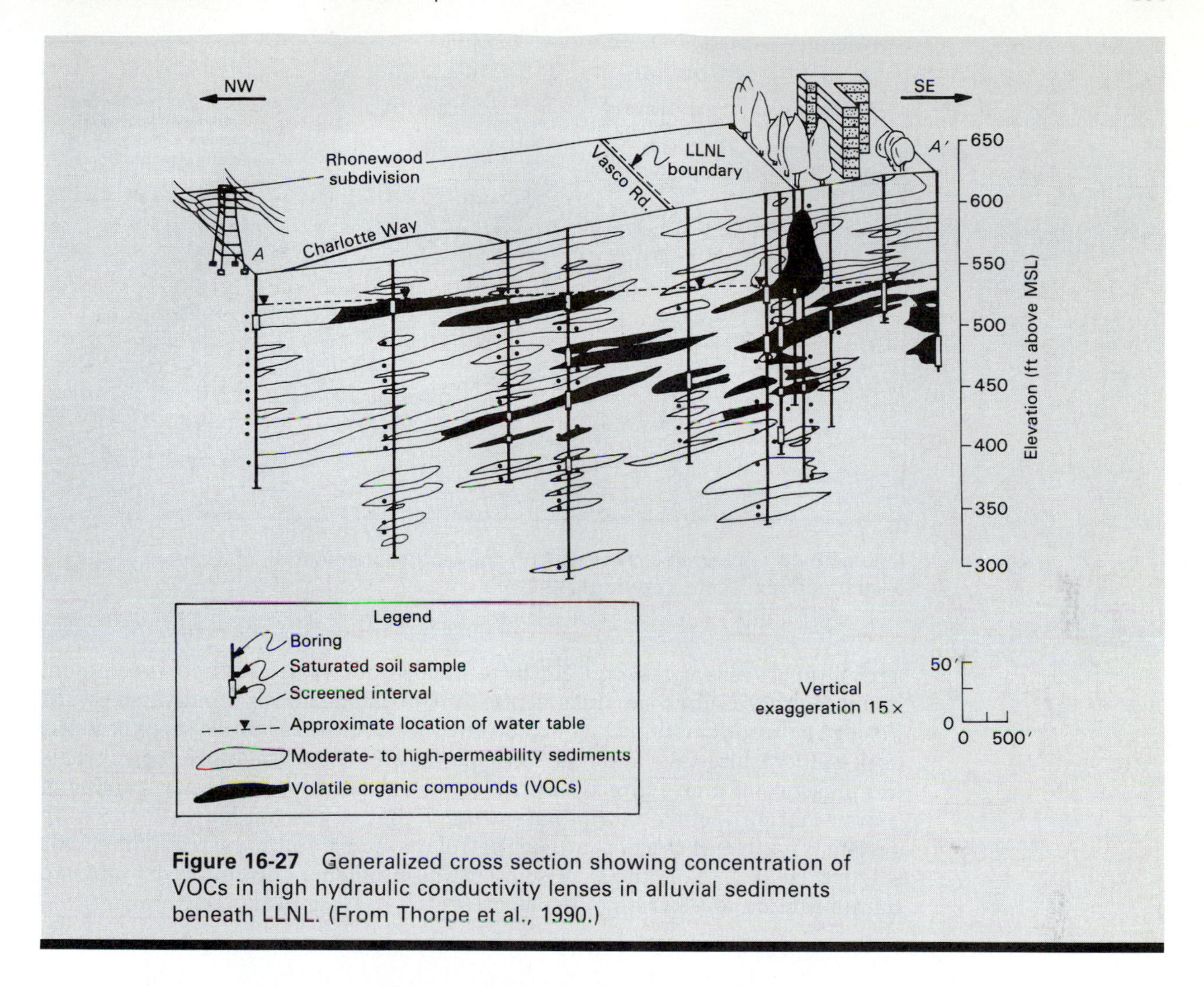


**Figure 16-27** Generalized cross section showing concentration of VOCs in high hydraulic conductivity lenses in alluvial sediments beneath LLNL. (From Thorpe et al., 1990.)

## *SUBSURFACE CONTAMINATION OF KARST AQUIFERS*

In the previous discussion of contaminant transport in the unsaturated and saturated zones, a basic assumption was made about the subsurface materials: The volume of material through which contaminants moved was considered to function as a granular porous medium, in which the flow of water and contaminants occurred through pores between closely packed solid particles. Preferential flow was mentioned as a condition in which larger interconnected pores are present. In some geologic media, however, most pore fluids move along linear discontinuities such as fractures, joints, or faults, and the intergranular flow is minor.

A well-developed karst aquifer with conduit flow represents the end member of this type of flow. Surface drainage moves rapidly through the unsaturated zone to a complex maze of subsurface conduits that may be partially or completely filled with water. Conduit flow is rapid, particularly during times of high precipitation. The consequences of this type of hydrological regime for contaminant movement are that surface contaminants move into the conduit flow system with very little opportunity for attenuation (Figure 16-28). In

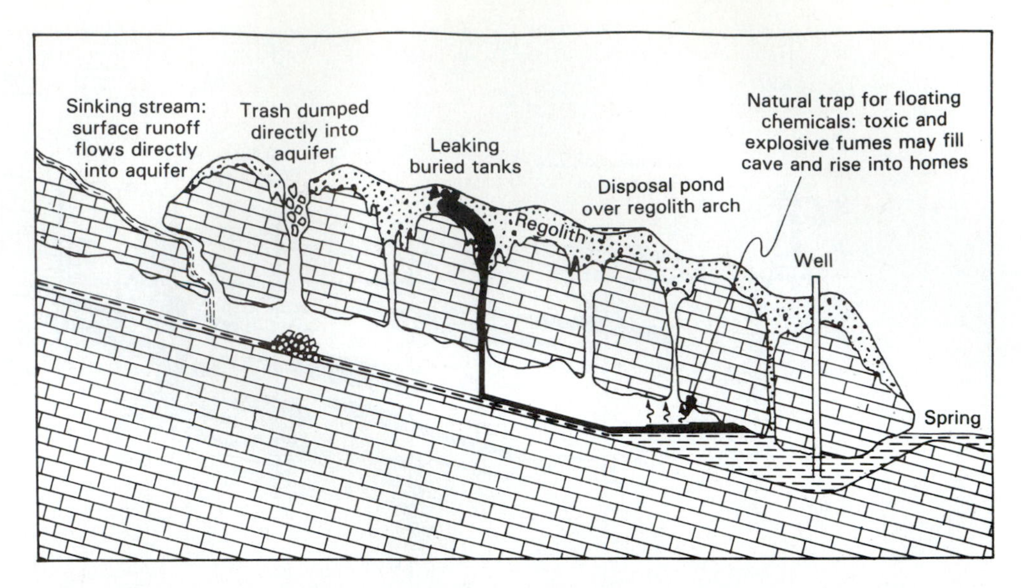


**Figure 16-28** Schematic cross section showing contamination of a karst aquifer. (From Crawford, 1986).

agricultural areas, contamination by pesticides and livestock waste is common. In unseweraged suburban areas, septic-tank drainage moves rapidly downward through solution cavities. Cave explorers have documented direct seeps of septic tank effluent into cave systems. Perhaps the worst circumstance, however, is the presence of a city directly over a karst aquifer system. Major sources of contamination include storm water runoff that is discharged into the cave system, and buried tanks that leak petroleum fuels and industrial chemicals (Figure 16-28). An example of some of the problems encountered by urban contamination of a karst aquifer is presented in Case Study 16-3.

### CASE STUDY 16-3

## Subsurface Migration of Toxic and Explosive Fumes in Bowling Green, Kentucky

Difficulties with subsurface contamination in Bowling Green, Kentucky, typify the problems that can arise with urban development over a karst aquifer (Crawford, 1986). The city partially overlies the Lost River ground-water basin, a karst aquifer system with well-developed conduit flow (Figure 16-29). Figure 16-28 schematically illustrates the movement of contaminants through the system. The focus of this case study is the problem of NAPL contaminants that float on the surface of underground streams in cave and conduit passages. Numerous sources for NAPL contaminants derive from the more than 500 buried tanks in the city.

As the contaminants float on the cave stream surfaces, toxic and explosive fumes evaporate into the cave atmosphere. In itself, this condition poses a nontrivial explosion threat. Storm drainage wells in parking lots and along streets provide an open connection to the cave systems. A discarded cigarette or other source could trigger an underground explosion that could progress along the cave system. Even more disturbing, perhaps, is the upward movement of fumes through solution cavities in the limestone to basements where they present a serious health hazard. Although the presence of hazardous fumes in houses in

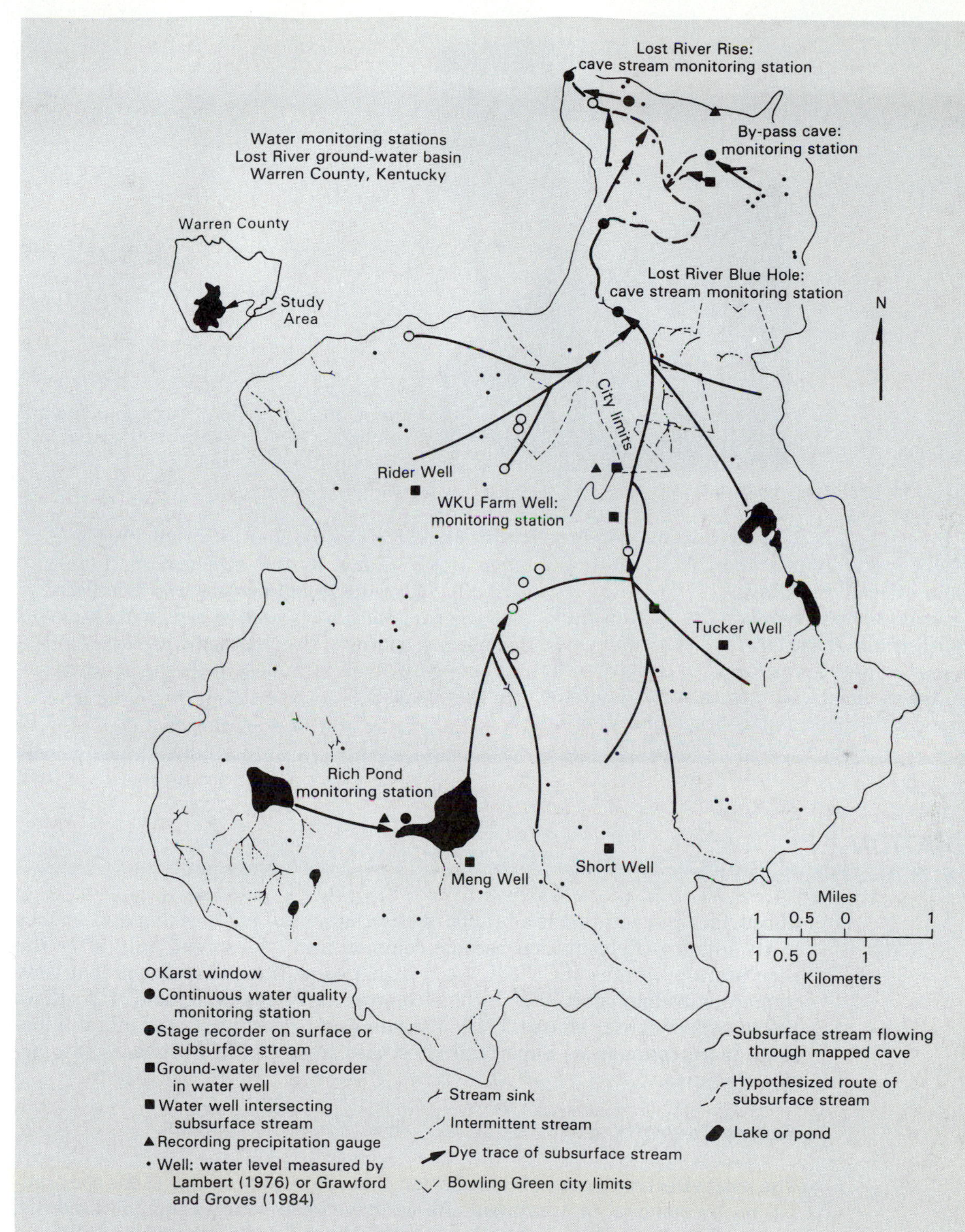


**Figure 16-29** Map of Lost River ground-water basin, which includes part of Bowling Green, Kentucky. (From Crawford, 1986.)

Bowling Green had occurred several times in the past, the problem became so serious in the mid-1980s that the U.S. EPA initiated an emergency response in 1985 under the Superfund program. Fumes were reported in more than 50 homes, along with several schools and commercial buildings. Two elementary schools were partially evacuated on several occasions. The

**Figure 16-30** Photo of hazardous fume ventilation pipe at elementary school in Bowling Green, Kentucky.

EPA investigation focused on the potential sources of contaminants and the mitigation of fumes in affected buildings.

Excavations were made around the buildings with the highest levels of toxic fumes. In each case, a crevice was found to lead from the base of the surficial soils downward through the limestone to the polluted cave system. Ventilation pipes were cemented into the crevices to vent toxic fumes to the atmosphere (Figure 16-30). The ventilation system was completed by the installation of exhaust fans with explosion-proof motors. This method was successful in reducing levels of fumes inside the buildings to safe levels.

## REMEDIATION

Subsurface remediation is a branch of environmental engineering and science in its infancy. Remediation became common and widespread only after the regulatory programs of CERCLA and RCRA came into effect. Since that time numerous technologies have become commercially available, and still others are in experimental phases. Large remediation projects will undoubtably become more common as Superfund sites pass from the RI/FS phases into active cleanups.

### Source Control

The most effective means of remediation is to control the source of the contamination by removal or treatment. As we have seen sources include landfills, impoundments, and buried tanks, among others. In the case of landfills and impoundments, the source is easy to locate. The remaining waste can be removed to a more secure site or treated. Liquid waste can be transported to a waste-water treatment plant. Evidence of a release from a buried tank is commonly detected when the tank is excavated for repair or replacement. Further source removal can be achieved by excavation of the soils beneath the tank that contain petroleum product in residual saturation. If left in the unsaturated zone, residual contaminants can slowly dissolve into infiltrating water that can continue to contaminate ground water for long periods of time. The common practice is to conduct a soils investigation including soil borings and lab examination of soil samples. The contaminated volume of soil can thus be delineated and subsequently excavated. Obviously, this is only cost effective for small soil volumes. For larger spills or leaks, onsite excavation, treatment, and replacement is becoming more common.

### Containment

Once the contaminant has reached the water table, remediation becomes more difficult. Because of the flow of ground water and the dispersion of the contaminant within the flow system, excavation of contaminated aquifer material is usually not feasible. One option for controlling the spread of the contaminant in the ground-water flow system is the *slurry wall* (Figure 16-31). A slurry wall is an impermeable barrier emplaced as viscous liquid (slurry) into a trench excavated around the contaminant source. This method is most practical when the aquifer is thin and close to the surface, so that the depth of excavation can be limited. The wall must also be tied in to an impermeable formation at the base of the aquifer so that contaminants may not escape beneath the wall. If the wall is constructed completely around the source, wells must be installed within the wall to prevent a rise in water table to the surface. The contaminated water pumped from the wells can then be treated and discharged.

### Pump and Treat

One of the most commonly used strategies for ground-water remediation is known as *pump and treat*. The method has several objectives. Pumping, from one or more wells, can be used to alter or reverse the hydraulic gradient in order to hydraulically control the plume. For example, an extraction well located near the source of contamination could reverse the hydraulic gradient due to its cone of depression, so that ground-water flow and contaminants move back toward the source rather than away from the source. Contaminants at an industrial site, if discovered soon enough after the release, for example, could be prevented from migration offsite, where they could potentially reach water-supply wells.

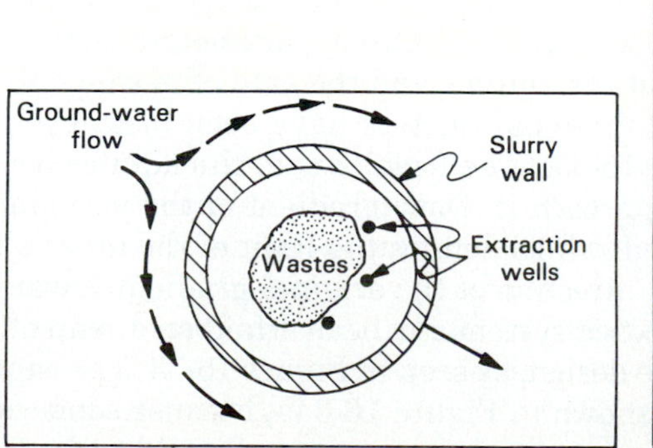


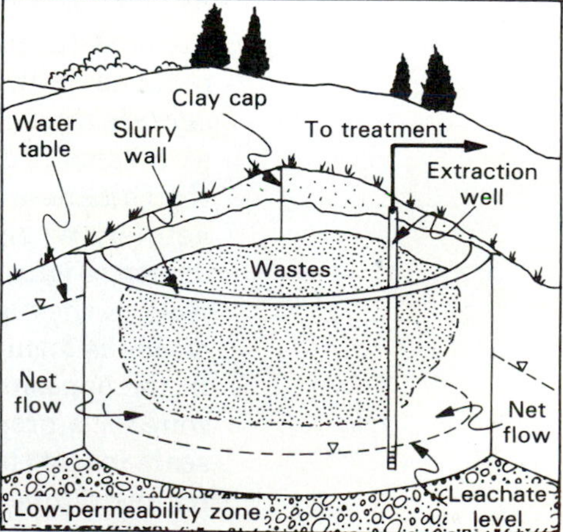


**Figure 16-31** Diagram of slurry-wall used to contain ground-water contamination plume.

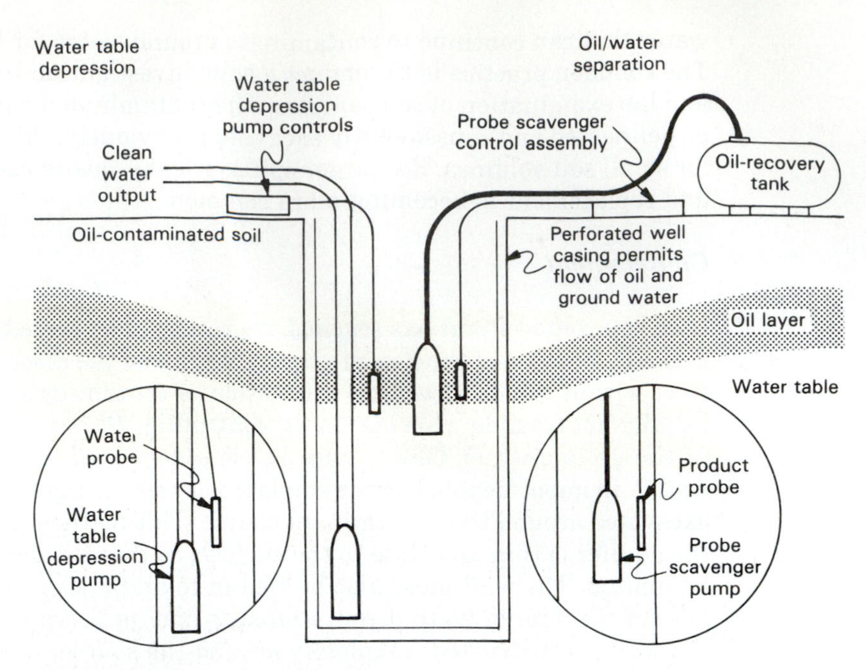


**Figure 16-32** Dual pump system used to remediate petroleum product contamination. Deeper pump creates cone of depression, causing flow of free product to well. Upper pump removes free product. (From U.S. EPA, 1989.)

If the contamination involves a layer of LNAPL free product at the top of the saturated zone, special considerations are necessary. One common design utilizes a well that is screened over a large interval including the water table and a dual pump system (Figure 16-32). A standard submersible pump is installed at the base of the well to pump water and create a cone of depression and flow of free product toward the well. In the zone of free product, a product pump controlled by an indicator probe is used to remove the LNAPL. Under certain conditions, reuse of the product is even possible.

Capture Zones. The choice of a pump and treat strategy for ground-water cleanup is followed by an analysis of the aquifer and plume in order to design a pumping well system that is capable of intercepting the entire plume, while at the same time minimizing the amount of clean aquifer water that is pumped to the treatment system. This process includes the delineation of a *capture zone*, a region of flow that is intercepted by one or more wells pumping at a certain discharge rate. Other parameters that must be determined are the thickness of the aquifer and the rate of movement of ground water in the aquifer. When these parameters have been measured, a capture zone can be drawn (Figure 16-33). The flow lines in the aquifer bend toward the pumping well as they approach it. Downgradient of the pumping well, a ground-water divide is formed, which defines the limit of the area in which water is flowing in the opposite direction of the original gradient toward the well. The capture zone for a proposed system can be overlaid on a map of the plume at the same scale to test the design, as seen in Figure 16-34. The capture zone is inadequate for the plume shown in Figure 16-34a, because some contaminant will escape around the margins of the capture zone. Possible solutions would be to increase the pumping rate of the well or add one or more additional wells spaced along the $y$ axis to enlarge the dimensions of the capture zone. Figure 16-34b illus-

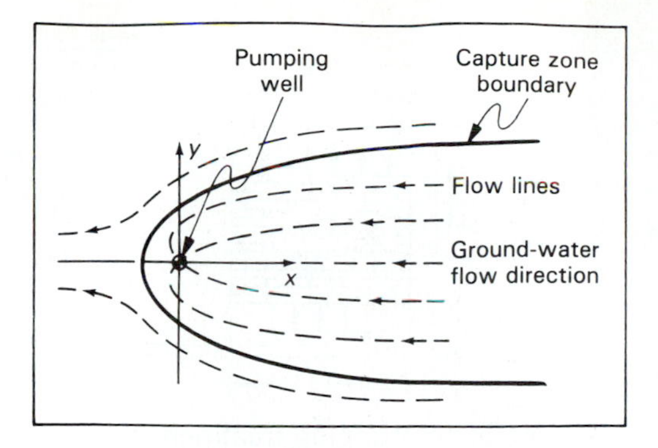


**Figure 16-33** Capture zone imposed by pumping one or more wells along *Y* axis. All water within boundaries flows to well.

trates a case in which the capture zone is too large for the plume. The system will be inefficient in this case because a large amount of uncontaminated water will be pumped along with the contaminated water. A lower pumping rate could be used to remedy this problem. A well-designed capture zone is shown in Figure 16-34c, because the lateral margins of the capture zone just surround the edges of the plume.

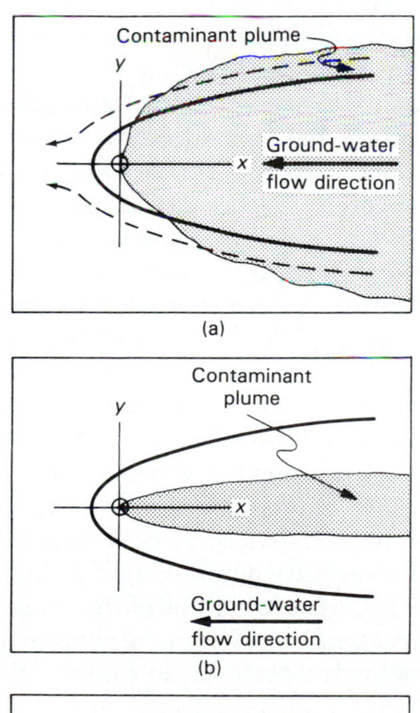


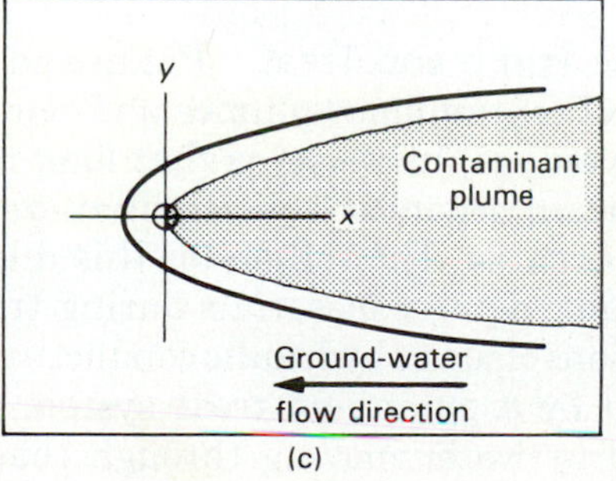


**Figure 16-34** Relationship of capture zone to contaminant plume. (a) Capture zone is inadequate to capture entire plume. (b) Capture zone is too large and unnecessary volume of clean water is pumped through treatment system. (c) Capture zone is correctly sized for contaminant plume.

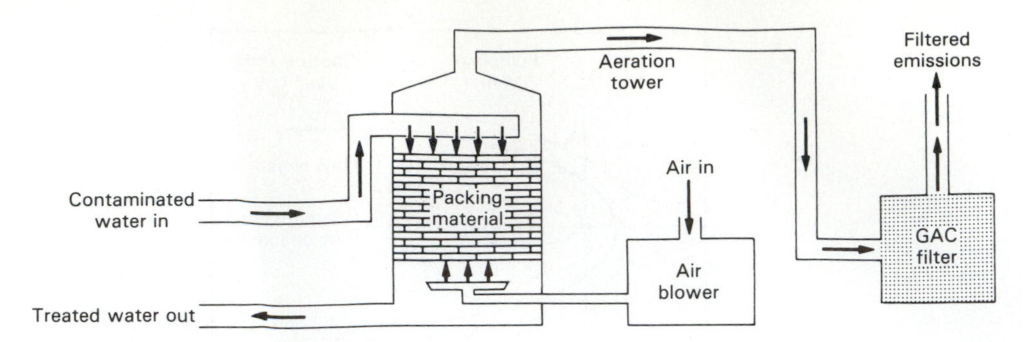


**Figure 16-35** Schematic of air stripping treatment process for VOCs coupled with GAC filter for air-quality control.

Treatment methods. Numerous technologies are available for treatment of contaminated ground water. Treatment systems are commonly constructed at the site, so that treated water can be discharged to surface drainages or reinjected into the aquifer. At times, untreated water can be discharged to a sanitary sewer and treated at the waste-water treatment plant.

The most frequently encountered hazardous contaminants in ground water are the volatile organics, which include the chlorinated solvent compounds as well as the BTEX compounds derived from petroleum products. The treatment method selected for a particular contaminant must be based on the chemical characteristics of the compound. *Carbon adsorption* is a method based on adsorption of the compound during flow of the contaminant through a canister containing a carbon medium, which is called *granular activated carbon* (GAC). GAC has a very high surface area and is an effective absorbent for nonpolar, hydrophobic organics. The log $K_{ow}$ value of the compound (Table 16-3) is an indication of this property. Carbon adsorption would be well suited for benzene, with a high log $K_{ow}$, but would not work for acetone, which has a very low log $K_{ow}$.

The volatile nature of VOCs is used in several methods of treatment including *air* and *steam stripping*. In a typical system, contaminated water infiltrates downward through an air-stripping tower (Figure 16-35), while air is blown up through the tower. The tower is filled with a permeable packing material that causes agitation and turbulence of the water. This promotes separation of the volatiles from the water and the air stream carries them out of the top of the tower where they can be released to the atmosphere if the concentrations are low, or captured by GAC or other methods, if the concentrations are high. The feasibility of air stripping can be predicted by the Henry's constant (Table 16-3). On the lower end of the range of Henry's constant values, steam can be used rather than air to strip the less volatile compounds. Specifications for these methods are shown in Figure 16-36.

Limitations to Pump and Treat. Pump and treat remedial systems are useful for control of contaminant plumes and removal of some contaminants. Their main disadvantage, however, is that long periods of time are required for removal of certain organics from aquifers to the levels required of safe drinking water standards. The reason for this dilemma is that contaminants adsorb onto favorable aquifer materials during transport and sometimes also penetrate into regions of lower hydraulic conductivity. When the high hydraulic gradients imposed by a pump and treat system take effect, the flow to the well is dominated by water moving through the zones of highest hydraulic

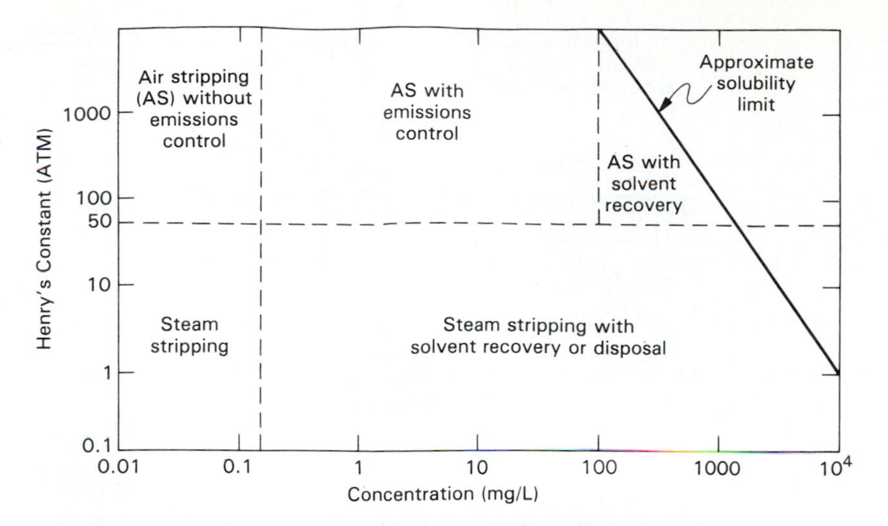


**Figure 16-36** Feasibility of air and steam stripping based on concentration and Henry's constant of contaminants. (From Bouwer et al., 1988.)

conductivity. Dissolved contaminants contained in the pores of these zones are quickly flushed out, leading to a large initial decrease in concentrations measured from the pumping well and monitoring wells. These concentrations tend to level out, however, at low levels and may remain at those levels indefinitely. These remaining contaminants represent the adsorbed molecules that are slowly released to the ground water moving toward the well. Estimates of pumping times of decades are not uncommon to achieve drinking water standards for some contaminated ground waters. The net result is that large amounts of money are required to remove minute quantities of contaminants. Alternative remedial methods that could achieve cleanup goals more rapidly are being actively sought.

### Bioremediation

Among the many promising techniques for aquifer remediation, *bioremediation* encompasses a wide variety of processes that may be used on a routine basis in the future. The basic premise of bioremediation is to allow microorganisms to breakdown organic contaminants to harmless compounds such as carbon dioxide. As simple as this concept appears, there are many problems to overcome to ensure the success of this method. Bioremediation can be used in a pump and treat process, in which the water is treated above ground, or it can be used in the subsurface, which is known as *in situ bioremediation.* Microorganisms native to the environment may be used or appropriate microorganisms can be introduced into the environment to do the work.

Bioremediation usually involves an oxidation reaction, in which an organic substrate donates electrons during oxidation to electron acceptors such as oxygen. Other nutrients are also required for the proliferation of the microorganisms. Petroleum hydrocarbons serve as excellent substrates and natural bioremediation occurs to some extent at nearly all contaminant sites. A lack of electron acceptors is the usual problem that prevents the rapid and complete breakdown of the compounds. Attempts to stimulate the organisms by providing electron acceptors and nutrients are called *enhanced bioremediation.* A schematic of this process is shown in Figure 16-37. Water pumped from the

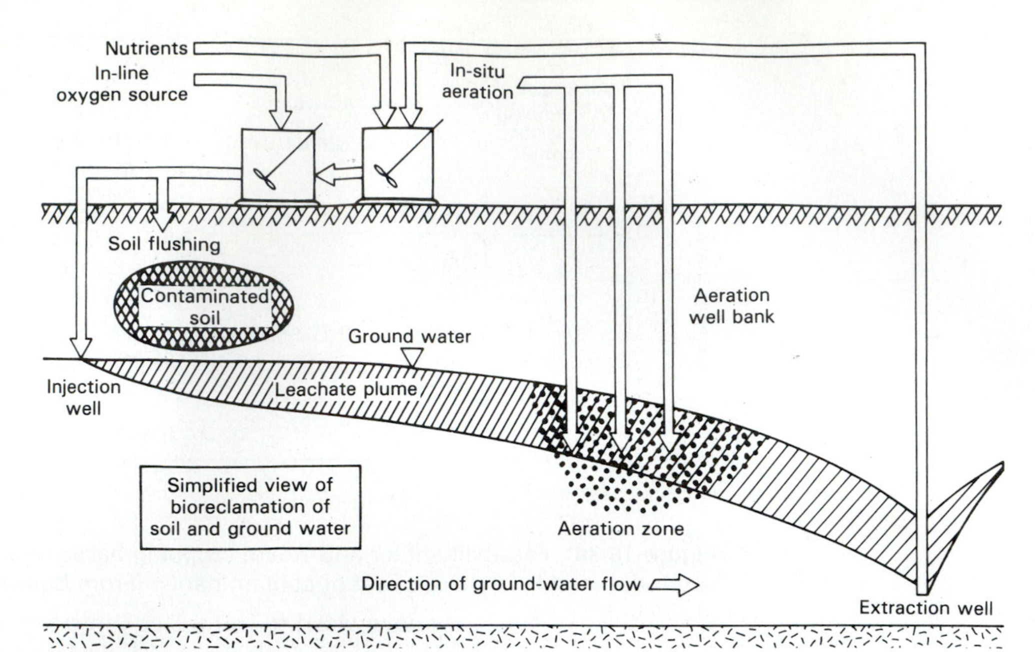


**Figure 16-37** Schematic diagram of enhanced bioremediation system. Ground-water pumped from extraction well is recycled through contaminant source after addition of oxygen and nutrients. Oxygen also pumped directly to contaminant plume. (From EPA, 1989.)

contaminant plume is reinjected into the aquifer after addition of oxygen and nutrients. Oxygen is also injected directly into the subsurface in order to accelerate the process.

Contaminant plumes that contain only low levels of toxic chlorinated compounds pose a somewhat different problem. The concentrations of organics are so low that they cannot support the growth of an active population of microorganisms. One solution in this case is to add both the electron donor and the electron acceptor to the plume. Methane is a commonly used electron donor. The objective of this technique is to stimulate a population of microorganisms large enough to break down the trace-level organics along with the added electron donor. This process is called *co-metabolism.*

There are many technical problems in bioremediation, which include dispensing the appropriate microorganisms, as well as all the other compounds and conditions necessary for the process to be successful at the proper location in the subsurface. Active interdisciplinary research involving microbiologists, organic chemists, environmental engineers, and hydrogeologists will almost certainly yield significant advances in this field in the next several decades.

## SUMMARY AND CONCLUSIONS

Subsurface contamination is the environmental frontier. Until the 1960s, almost all efforts were focused on surface water contamination. During the 1960s and 1970s, however, the increasing rate of accidental detections of ground-water contamination gradually raised an awareness that subsurface contamination was an extremely serious problem. The sources of contamination were found to be numerous and diverse. Some point sources, such as landfills and industrial waste impoundments were obvious suspects for contaminations. Others, however, including nonpoint-source contamination by agricultural activities and septic systems, buried storage tanks, and well injections, were largely overlooked in terms of the magnitude of subsurface contamination that they could cause.

These revelations led to a series of unprecedented federal laws and regulations, including the Clean Water Act (CWA), the Safe Drinking Water Act (SDWA), the Comprehensive Environmental Response and Liability Act (CERCLA), or "Superfund," and the Resource Conservation and Recovery Act (RCRA). The net effect of these acts was to initiate a massive program to prevent future contamination and to characterize and remediate existing and past sites of contamination.

The subsurface transport of contaminants can be divided into aqueous-phase and nonaqueous-phase processes. The presence of most contaminant sources above the water table requires an understanding of movement of water through the unsaturated zone. Fluids are held in this zone by the matric potential of the soil, opposing gravitational forces that tend to cause downward movement. Both inorganic and organic compounds dissolve in unsaturated zone fluids and migrate downward to the water table. At that point, transport is controlled by the hydraulic gradient, and the dissolved contaminants move in the direction of ground-water flow.

Dissolved contaminants are attenuated in the unsaturated and saturated zones by physical, chemical, and biological mechanisms, including hydrodynamic dispersion, adsorption, chemical reaction, and biodegradation. Common adsorbents include clay particles and metal oxides for inorganic cations, and organic solids for dissolved organic contaminants.

Nonaqueous-phase transport is typically observed beneath leaking underground tanks and spills of various compounds. LNAPL compounds will migrate downward to the top of the capillary fringe, where they spread out laterally in a pool of free product, which then migrates in the direction of slope of the water table. The unsaturated zone retains these fluids at a level of residual saturation for long periods of time. Volatilization can disperse hazardous vapors throughout the unsaturated zone. DNAPL compounds sink to the base of the aquifer and can form free product pools there. Migration will occur by controls other than the hydraulic gradient.

Karst aquifers present a unique set of conditions relative to contamination. Migration is much more rapid through the large conduits in both the unsaturated and saturated zones. Hazardous fumes accumulate in caves and migrate up through the unsaturated zone to cause environmental and health problems.

Remediation technologies have led to varying degrees of success. Source control or removal is the most effective measure. Pump and treat strategies can control the contaminant plume, but are inordinately slow in removing low concentrations of toxic organics from contaminant plumes. Bioremediation is perhaps the most likely approach to yield future successes in the rapid and effective remediation of contaminant plumes.

## PROBLEMS

1. Give two specific examples of nonpoint sources of contamination.
2. What are the differences between sanitary and secured landfills?
3. Why are successfully operating septic systems a threat to ground-water quality? What contaminants would you expect to find in ground-water from these systems?

4. What conditions must be met in order to install hazardous waste injection wells? How can these wells fail?
5. What types of sites do RCRA and CERCLA regulate?
6. Outline the steps that are taken to clean up sites under CERCLA.
7. Arrange the following chemicals in order of (a) decreasing tendency to volatilize from the water table, and (b) increasing tendency to sorb onto solid organic materials in an aquifer: (1)Benzene, (2) Phenol, (3) Vinyl chloride.
8. A tensiometer installed to a depth of 60 cm reads −200 cm. An adjacent tensiometer is 150 cm deep and reads −100 cm. Which way is the soil moisture moving?
9. Discuss ways in which solutions can move downward in the unsaturated zone more rapidly than the average rate of movement.
10. A nonreactive contaminant is released from a landfill into a sandy aquifer with $K = 10^{-4}$ m/s, $n = 0.35$, and a hydraulic gradient of 0.001. Neglecting dispersion, how many years will it take for the contaminant to reach a well located 2000 m downgradient along a flow line extending from the point of release on the water table?
11. A monitor well 50 m from a waste lagoon first shows the presence of a nonreactive contaminant from the lagoon 10 months after waste is released into the lagoon. If the lagoon intersects the water table, the porosity of the aquifer is 0.3, and the hydraulic gradient is $10^{-3.5}$, what is the hydraulic conductivity of the aquifer? Assume dispersion is too small to be significant.
12. Discuss ways in which aqueous-phase contaminants can be attenuated during transport.
13. Explain how the movement of NAPLs differs from aqueous-phase transport both above and below the water table.
14. What special problems of contaminant movement are noted in karst terrains?
15. Discuss the advantages and disadvantages of pump and treat remediation systems.

## REFERENCES AND SUGGESTIONS FOR FURTHER READING

ABDUL, S. A. 1988. Migration of petroleum products through sandy hydrogeologic systems. *Ground Water Monitoring Review*, 8, no. 4:73–81.

BOUWER, E., J. MERCER, M. KAVANAUGH, and F. DIGIANO. 1988. Coping with groundwater contamination. *Journal of the Water Pollution Control Federation*, 60:1415–1423.

CONCAWE. 1979. Protection of Groundwater from Oil Pollution. The Hague, Netherlands, NTIS PB82-174608.

CRAWFORD, N. C. 1986. *Karst Hydrologic Problems Associated with Urban Development: Ground Water Contamination, Hazardous Fumes, Sinkhole Flooding, and Sinkhole Collapse in the Bowling Green Area, Kentucky*, Field Trip Guidebook. Dublin, Ohio: National Water Well Association.

DEVINNY, J. S., L. G. EVERETT, J. C. S. LU, and R. L. STOLLAR. 1990. *Subsurface Migration of Hazardous Wastes*. New York: Van Nostrand Reinhold.

FEENSTRA, S., and J. A. CHERRY. 1988. Subsurface contamination by dense non-aqueous phase liquid (DNAPL) chemicals, in *Proceedings, International Groundwater Symposium on Hydrogeology of Cold and Temperate Climates and Hydrogeology of Mineralized Zones*, C. L. Lin, ed. International Association of Hydrogeologists, Canadian National Chapter, pp. 62–69.

FETTER, C. W. 1993. *Contaminant Hydrogeology*. New York: Macmillan.

GILLHAM, R. W., and J. A. CHERRY. 1982. Contaminant movement in saturated geologic deposits, in *Recent Trends in Hydrogeology*, T. N. Narasimham, ed. Geological Society of America Special Paper 189:31–62.

HOFFMAN, F. 1993. Ground-Water Remediation Using "Smart Pump and Treat," *Ground Water*, 31:98–106.

MASTERS, G. M. 1991. *Introduction to Environmental Engineering and Science*. Englewood Cliffs, N.J.: Prentice-Hall, Inc.

PERLMUTTER, N. M, and M. LIEBER. 1970. Disposal of Plating Wastes and Sewage Contamination in Ground Water and Surface Water, South Farmingdale-Massapequa Area, Nassau County, New York. U.S. Geological Survey Water Supply Paper 1879-G.

TESTA, S. M., and D. L. WINEGARDNER. 1991. *Restoration of Petroleum-Contaminated Aquifers*. Chelsea, Mich.: Lewis.

THORPE, R. K., W. F. ISHERWOOD, M. D. DRESEN, and C. P. WEBSTER-SCHOLTEN. 1990. *CERCLA Remedial Investigations Report for the LLNL Livermore Site*. Livermore, Calif.: Lawrence Livermore National Laboratory.

U.S. EPA. 1987. *Seminar on Transport and Fate of Contaminants in the Subsurface*. CERI-87-45.

U.S. EPA. 1989. *Seminar on Site Characterization for Subsurface Remediations*. CERI-89-224.

U.S. EPA. 1990. *Ground Water, Volume 1: Ground Water and Contamination*. EPA/625/6-90/016a.

WENTZ, C. A. 1989. *Hazardous Waste Management*. New York: McGraw-Hill.

APPENDIX

# Units and Conversion Factors

| *Quantity* | *SI Unit* | *U.S. Customary Unit* | | *Conversion Factors* |
|---|---|---|---|---|
| Length, $L$ | m | ft, in., yd, mi | 1 in. | = 2.54 cm |
| | | | 1 ft | = 0.3048 m |
| | | | 1 yd | = 0.9144 m |
| | | | 1 mi | = 1.6093 km |
| | | | 1 m | = 3.2808 ft |
| | | | 1 km | = 0.6214 mi |
| Area, $L^2$ | $m^2$, ha | $ft^2$, $in.^2$, $yd^2$, $mi^2$, acre | 1 $in.^2$ | = 6.452 $cm^2$ |
| | | | 1 $ft^2$ | = 0.0929 $m^2$ |
| | | | 1 acre | = 43,560 $ft^2$ |
| | | | | = 4046.85 $m^2$ |
| | | | | = 0.404685 ha |
| | | | 1 $mi^2$ | = 640 acre |
| | | | | = 2.604 $km^2$ |
| | | | | = 259 ha |
| | | | 1 $m^2$ | = 10.764 $ft^2$ |
| | | | 1 ha | = 2.471 acre |
| | | | | = 10,000 $m^2$ |
| Volume, $L^3$ | $m^3$, L | $ft^3$, $yd^3$, gal | 1 $ft^3$ | = 7.4805 U.S. gal |
| | | | | = 0.02832 $m^3$ |
| | | | | = 28.32 L |
| | | | 1 U.S. gal | = 0.134 $ft^3$ |
| | | | | = 0.003785 $m^3$ |
| | | | | = 3.785 L |
| | | | 1 $m^3$ | = 35.3147 $ft^3$ |
| | | | | = 264.172 gal |
| | | | | = 1000 L |
| | | | | = $10^6$ $cm^3$ |

| | | | | |
|---|---|---|---|---|
| Mass, $M$ | kg, g<br>tonne (metric) | slug, lbm<br>ton (short) | 1 slug | = 14.59 kg |
| | | | 1 kg | = 0.685 slug |
| | | | 1 lbm | = 0.453592 kg |
| | | | 1 kg | = 2.205 lbm |
| | | | 1 ton (short) | = 2000 lbm |
| | | | | = 907.2 kg |
| | | | | = 0.9072 ton (metric) |
| | | | 1 ton (metric) | = 1000 kg |
| | | | | = 2204 lbm |
| | | | | = 1.1023 ton (short) |
| Force, weight $\left(\frac{ML}{T^2}\right)$ | N, dyne | lb, ton | 1 N | = 0.2248 lb |
| | | | 1 kN | = 224.8 lb |
| | | | 1 lb | = 4.448 N |
| | | | 1 dyne | = $1 \times 10^{-5}$ N |
| | | | | = $2.25 \times 10^{-6}$ lb |
| | | | | = 1 g-cm/$s^2$ |
| Stress, pressure ($F/L^2$) | N/$m^2$, Pa, bar | lb/$ft^2$, atm<br>tons/$ft^2$<br>psi, tsf | 1 N/$m^2$ | = 0.0209 lb/$ft^2$ |
| | | | 1 Pa | = 1 N/$m^2$ |
| | | | 1 bar | = $10^5$ Pa |
| | | | 1 kN/$m^2$ | = 20.90 lb/$ft^2$ |
| | | | | = 0.145 psi |
| | | | 1 lb/$ft^2$ | = 47.88 N/$m^2$ |
| | | | 1 psi | = 6.895 kN/$m^2$ |
| | | | 1 atm | = 14.7 psi |
| | | | | = 1 bar |
| Density ($\rho$) ($M/L^3$) | kg/$m^3$, g/$cm^3$ | lbm/$ft^3$, slugs/$ft^3$ | 1 Mg/$m^3$ | = 62.4 lbm/$ft^3$ |
| | | | 1 lbm/$ft^3$ | = 0.0157 Mg/$m^3$ |
| | | | 1 g/$cm^3$ | = 62.4 lbm/$ft^3$ |
| | | | 1 kg/$m^3$ | = 0.062 lbm/$ft^3$ |
| Unit weight ($\gamma$) ($F/L^3$, $M/L^2T^2$) | kN/$m^3$<br>dynes/$cm^3$ | lb/$ft^3$ | 1 kN/$m^3$ | = 6.36 lb/$ft^3$ |
| | | | | = 0.0037 lb/$in.^3$ |
| | | | 1 lb/$ft^3$ | = 157 N/$m^3$ |
| Velocity ($L/T$) | m/s, cm/s | ft/s, ft/min<br>mi/hr (mph) | 1 m/s | = 3.28 ft/s |
| | | | 1 cm/s | = 1.97 ft/min |
| | | | 1 ft/s | = 0.305 m/s |
| | | | 1 mi/hr | = 1.61 km/hr |
| Acceleration ($L/T^2$) | m/$s^2$ | ft/$s^2$ | 1 m/$s^2$ | = 3.28 ft/$s^2$ |
| | | | 1 ft/$s^2$ | = 0.3048 m/$s^2$ |
| Flow rate (Discharge) ($L^3/T$) | $m^3$/s | $ft^3$/s, $ft^3$/min<br>U.S. gal/min | 1 $ft^3$/s | = 0.0283 $m^3$/s |
| | | | | = 448.8 gal/min |
| | | | 1 $ft^3$/min | = $4.72 \times 10^{-4}$ $m^3$/s |
| | | | | = 7.4805 gal/min |
| | | | 1 gal/min | = $6.31 \times 10^{-5}$ $m^3$/s |
| | | | 1 $m^3$/s | = 35.315 $ft^3$/s |
| | | | | = 2118.9 $ft^3$/min |
| Energy (F-L) | Joule (N-m) | ft-lb, cal<br>Btu, Kw-hr | 1 Btu | = 778 ft-lb |
| | | | | = 252 cal |
| | | | 1 Joule | = 1 N-m |
| | | | | = 0.73756 ft-lb |
| | | | 1 Kcal | = 4.185 kJ |

USEFUL CONSTANTS

$g = 9.8$ m/$s^2$
$= 980$ cm/$s^2$
$\rho$ (water) $= 1.0$ g/$cm^3$
$= 1000$ kg/$m^3$
$\gamma$ (water) $= 9.8 \times 10^3$ N/$m^3$
$= 62.4$ lb/$ft^3$

# Index

### B



### C

## D



## E



## F

## G



## H

## I



## J



## K



## L



## M

## N



## O



## P

## T



## U



## V

## W



## X



## Y

## UNIFIED SOIL CLASSIFICATION SYSTEM

| Major Divisions | | | Group Symbols[a] | Typical Names | Field Identification Procedures (excluding particles larger than 75 mm and basing fractions on estimated weights) | | |
|---|---|---|---|---|---|---|---|
| 1 | 2 | | 3 | 4 | 5 | | |
| Coarse-grained soils<br>More than half of material is *larger* than No. 200[b] (75 μm) sieve size.<br>The No. 200 sieve size is about the smallest particle visible to the naked eye. | Gravels<br>More than half of gravel fraction is larger than No. 4 sieve size (4.75 mm)<br>(for visual classification, 5 mm may be used as equivalent to the No. 4 sieve size) | Clean gravels (little or no fines) | GW | Well-graded gravels, gravel sand mixtures, little or no fines. | Wide range in grain sizes and substantial amounts of all intermediate particle sizes. | | |
| | | | GP | Poorly graded gravels, gravel-sand mixtures, little or no fines. | Predominantly one size or a range of sizes with some intermediate sizes missing. | | |
| | | Gravels with fines (appreciable amount of fines) | GM | Silty gravels, gravel-sand-silt mixtures. | Nonplastic fines or fines with low plasticity (for identification procedures see ML below). | | |
| | | | GC | Clayey gravels, gravel-sand-clay mixtures. | Plastic fines (for identification procedures see CL below). | | |
| | Sands<br>More than half of coarse fraction is smaller than No. 4 sieve size (4.75 mm) | Clean sands (little or no fines) | SW | Well-graded sands, gravelly sands, little or no fines. | Wide range in grain sizes and substantial amounts of all intermediate particle sizes. | | |
| | | | SP | Poorly graded sands, gravelly sands, little or no fines. | Predominantly one size or a range of sizes with some intermediate sizes missing. | | |
| | | Sands with fines (appreciable amount of fines) | SM | Silty sands, sand-silt mixtures. | Nonplastic fines or fines with low plasticity (for identification procedures see ML below). | | |
| | | | SC | Clayey sands, sand-clay mixtures. | Plastic fines (for identification procedures see CL below). | | |
| Fine-grained soils<br>More than half of material is *smaller* than No. 200 (75 μm) sieve size. | | | | | Identification Procedures on Fraction Smaller than No. 40 Sieve Size | | |
| | | | | | Dry strength (crushing characteristics) | Dilatancy (reaction to shaking) | Toughness (consistency near PL) |
| | Silts and clays<br>Liquid limit less than 50 | | ML | Inorganic silts and very fine sands, rock flour, silty or clayey fine sands or clayey silts with slight plasticity. | None to slight | Quick to slow | None |
| | | | CL | Inorganic clays of low to medium plasticity, gravelly clays, sandy clays, silty clays, lean clays. | Medium to high | None to very slow | Medium |
| | | | OL | Organic silts and organic silty clays of low plasticity. | Slight to medium | Slow | Slight |
| | Silts and clays<br>Liquid limit greater than 50 | | MH | Inorganic silts, micaceous or diatomaceous fine sandy or silty soils, elastic silts. | Slight to medium | Slow to none | Slight to medium |
| | | | CH | Inorganic clays of high plasticity, fat clays. | High to very high | None | High |
| | | | OH | Organic clays of medium to high plasticity, organic silts. | Medium to high | None to very slow | Slight to medium |
| Highly organic soils | | | Pt | Peat and other highly organic soils | Readily identified by color, odor, spongy feel, and frequently by fibrous texture. | | |

After U.S. Army Engineer Waterways Experiment Station (1960), "The Unified Soil Classification System," *Technical Memorandum* No. 3-357, Appendix A, Characteristics of Soil Groups Pertaining to Embankments and Foundations, 1953, and Appendix B, Characteristics of Soil Groups Pertaining to Roads and Airfields, 1957; and A. K. Howard (1977), "Laboratory Classification of Soils—Unified Soil Classification System," *Earth Sciences Training Manual* No. 4, U.S. Bureau of Reclamation, Denver, 56 pp.

[a]Boundary classifications: soils possessing characteristics of two groups are designated by combinations of group symbols. For example: GW-GC well-graded gravel sand mixture with clay binder.

[b]All sieve sizes on this chart are U.S. Standard.